Powers of 10: 10^x

x	Name	Symbol	x	Name	Symbol
−1	deci[a]	d	1	deca[a]	da
−2	centi[a]	c	2	hecto[a]	h
−3	milli	m	3	kilo	k
−6	micro	μ	6	mega	M
−9	nano	n	9	giga	G
−12	pico	p	12	tera	T
−15	femto	f	15	peta	P
−18	atto	a	18	exa	E

[a]To be avoided except for area and volume.

Table of Integrals

Indefinite	Definite

$$\int x^n \, dx = \frac{1}{n+1} x^{n+1} + c \, (n \neq -1) \qquad \int_0^\infty x^n e^{-ax} \, dx = \frac{n!}{a^{n+1}}$$

$$\int \frac{dx}{x} = \ln x + c \qquad \int_0^\infty e^{-ax^2} \, dx = \frac{1}{2}\sqrt{\frac{\pi}{a}}$$

$$\int \sin x \, dx = -\cos x + c \qquad \int_0^\infty x e^{-ax^2} \, dx = \frac{1}{2a}$$

$$\int \cos x \, dx = \sin x + c \qquad \int_0^\infty x^2 e^{-ax^2} \, dx = \frac{1}{4a}\sqrt{\frac{\pi}{a}}$$

$$\int e^{ax} \, dx = \frac{e^{ax}}{a} + c \qquad \int_0^\infty x^3 e^{-ax^2} \, dx = \frac{1}{2a^2}$$

$$\int x e^{ax} \, dx = \frac{e^{ax}}{a^2}(ax - 1) + c \qquad \int_0^\infty x^4 e^{-ax^2} \, dx = \frac{3}{8a^2}\sqrt{\frac{\pi}{a}}$$

$$\int \frac{dx}{(a + bx)} = \frac{1}{b} \ln(a + bx) + c \qquad \text{Also, see Table 11-3}$$

$$\int \frac{x \, dx}{(a + bx)} = \frac{1}{b^2}\{a + bx - a \ln(a + bx)\}$$

Physical Chemistry

Physical Chemistry

Joseph H. Noggle

University of Delaware

Little, Brown and Company
Boston   Toronto

Library of Congress Cataloging in Publication Data

Noggle, Joseph H., 1936-
  Physical chemistry.

  Includes bibliographical references.
  1. Chemistry, Physical and theoretical. I. Title.
QD453.2.N64   1984   541.3   84-20102
ISBN 0-316-61164-6

Library of Congress Catalog Card No. 84-20102

ISBN 0-316-61164-6

9 8 7 6 5 4 3 2 1

HAL

Published simultaneously in Canada
by Little, Brown & Company (Canada) Limited

Printed in the United States of America

Photo research by Susan Van Etten

CREDITS

Photos: Chapter 1, p. 5, J. D. van der Waals, The Center for History of Chemistry/The University
of Pennsylvania; Chapter 2, p. 72, James Prescott Joule, Aip Niels Bohr Library; Chapter 3,
p. 127, Rudolf Clausius, The Bettmann Archive; Chapter 4, p. 183, Josiah Willard Gibbs, The
Bettmann Archive; Chapter 5, p. 213, Ludwig Boltzmann, The Bettmann Archive; Chapter 6,
p. 263, Claude Louis Berthollet, The Bettmann Archive; Chapter 7, p. 316, Gilbert N. Lewis,
reproduced by permission University Archivist, The Bancroft Library, University of California,
Berkeley, 94720; Chapter 8, p. 375, Peter Debye, Fankuchen Collection/American Institute of
Physics; Chapter 9, p. 440, Albert Einstein, The Bettmann Archive; Chapter 10, p. 504, Henry
Eyring, Aip Niels Bohr Library/Physics Today Collection; Chapter 11, p. 577, Werner Heisenberg,
The Bettmann Archive; Erwin Schrödinger, Aip Niels Bohr Library/Photograph by Francis
Simon/Ullstein; Paul Adrien Maurice Dirac, Center for History of Physics/American Institute of
Physics; Niels Bohr, Aip Niels Bohr Library; Chapter 12, p. 661, Wolfgang Pauli, Aip Niels Bohr
Library/Goudsmit Collection; Chapter 13, p. 705, J. Robert Oppenheimer, The Bettmann
Archive; Chapter 14, p. 813, Robert S. Mulliken, courtesy, Robert S. Mulliken; Chapter 15
p. 861, Ilya Prigogine, Lon Cooper, Courtesy of University of Texas at Austin News &
Information Service.

Text: Figure 1.6. From J. O. Hirschfelder, C. F. Curtiss, and R. B. Bird, *Molecular Theory of
Gases and Liquids.* Copyright, 1954 by John Wiley & Sons, Inc. Reprinted by permission;
Figure 1.9. Reprinted with permission from Gouq-Jen Su, *Ind. Eng. Chem.*, *38*, 803 (1946).
Copyright 1946 American Chemical Society; Figure 1.10. From O. A. Hougen and K. M. Watson,
Chemical Process Principles, Part II. Copyright, 1947 by Olaf A. Hougen and Kenneth M.
Watson. Reprinted by permission of John M. Wiley & Sons, Inc; Figure 1.11. From E. A.
Guggenheim, *Thermodynamics*, 5th ed. © North-Holland Publishing Company, Amsterdam,

(continued on page 936)

To Carol

To the Instructor

The order of topics in this text is, for the most part, the traditional one found in most physical chemistry texts and, apparently, the one preferred by most instructors — namely, thermodynamics, kinetics, then quantum theory. However, as always, the instructor has the option of teaching the chapters in any order. Those who prefer another order may also find this book useful; the chapters on quantum theory and spectroscopy, 11 through 14, are relatively self-contained. One of the problems with teaching thermodynamics before quantum theory is the placement of statistical mechanics. On one hand, one would like to teach it with thermodynamics, where its utility is more immediately apparent to the student. On the other hand, the foundations of this topic cannot be understood without a firm background in quantum theory. *Physical Chemistry* attempts to resolve this dichotomy by placing statistical mechanics in *both* places. In Chapter 5, the major emphasis is on the calculation of thermodynamic functions. This not only provides a deeper insight into the molecular significance of macroscopic thermodynamics, but also gives an important motivation to the student for the coverage of quantum theory to come. Then, in Chapter 15, statistical mechanics is revisited with more rigorous derivations and additional applications. Those who do not like this approach can, with small loss, simply omit Chapter 5 and cover it later, together with Chapter 15; the subsequent applications (mostly in Chapter 6) can be avoided by a judicious choice of material and problems.

One of the things you will no doubt notice when reading the Table of Contents is the absence of the usual topical chapters such as "Solids," "Liquids," "Polymers," "Spectroscopy," "Surfaces," and the like. The reason for this absence is that *Physical Chemistry* is organized around theoretical principles, and such chapters would cut cross-grain. Closer inspection will reveal that many of these topics are there, but are distributed into various chapters. For example, surfaces are discussed in Chapters 4 and 10, and polymers are discussed in Chapters 3, 5, 7, 9, 10. Spectroscopy is integrated into the teaching of quantum theory. In the matter of choice of topical material and examples, physical chemistry is so broad that only an encyclopedia could cover all of the material that is interesting, relevant, and important. The choices are somewhat arbitrary, so, since I was the author, I made them. Generally, I tried to choose topics and applications that, in my judgment, would best fit in with the principal subject of the chapter and that would best provide an illustration of the theoretical principles the student could follow at that point. It is likely that you may disagree in some cases, or that your favorite topic has been omitted — on the other hand, I'm certain you will find some way to put it back in. This book is intended to be a teaching text, not an encyclopedia, and the students should not be left to think that any one book contains all they will need to know about physical chemistry; they should be encouraged to explore other resources. A major objective in this book is developing the students' vocabulary so that they can read more specialized books in the various areas of physical chemistry.

The coverage of quantum mechanics (Chapters 11 to 14) may, at first glance, appear to be excessively difficult and deep. However, closer inspection will reveal that many derivations are worked out in detail that in many other texts is left to the imagination of the student or to the travail of the instructor. The principal difficulty that students find with this approach is its linearity — each chapter must be understood well before the next one makes much sense. You can help them with this by encouraging them to review frequently; for example, it is a good idea to reread parts of Chapter 11 while studying Chapters 12 and 13. There has been a deliberate effort to limit the level of mathematics in these chapters. Operator algebra is used because it is relatively easy to teach and learn at this level. Differential equations and matrix algebra are avoided as much as possible.

With electronic calculators, not to mention computers, today's students have more computational power available than G. N. Lewis ever dreamed of — yet few of them know how to use this power effectively. For that reason, this text emphasizes numerical analysis to an unusual degree. Throughout, the student is encouraged, even required, to use techniques such as numerical integration, differentiation, root finding, and linear regression. These techniques are easily implemented with pocket calculators costing scarcely more than this book. Since they are used in a number of different chapters, they are discussed together in a general way in Appendix I. The use of numerical methods together with "real" data is, I believe, helpful in showing the student the relevance of mathematical formulas and abstract functions to experimental results. Particularly in thermodynamics, this approach permits the use of realistic models and avoids leaving the student under the impression that the world is an ideal gas.

Although the problems in this text take full advantage of the capabilities of the electronic calculator, they do not take into account the now ubiquitous microcomputer. This deficiency will be remedied by a separate volume, *The Microcomputer in Physical Chemistry,* which will be published in early 1985. A *Solutions Manual* with worked out solutions to most of the problems is also available.

Acknowledgments

Two people deserve credit for having encouraged me to write this book — my wife Carol and my friend Cecil. Far be it from me to choose between two redheads, so let them share the credit equally; however, it was Carol who had to bear the brunt of the execution, and I am most grateful for her patience and help. Also, major credit should go to the students of the University of Delaware who, through their insistent curiosity and probing questions, made me learn the subject better. Those who struggled through the early drafts deserve special mention. In the line of redheads, I would also like to thank Don Wetlaufer for his support and encouragement throughout the trying period in which this book was written.

A number of colleagues and students have critically read parts of the manuscript, made useful suggestions, and gratuitously corrected my errors. In particular, I would like to thank Cecil Dybowski, Bob Wood, Don Wetlaufer, Doug Ridge, and students Seth Digel, Eric Scharpf, John Townsend, Suzie Kretchmar, Ellen Yurek, and John Nahay. Many others contributed to a lesser extent, and I thank them together. Despite their efforts, I am certain that some errors remain, and can only hope that they are minor. The manuscript was typed efficiently and accurately by Rose O'Neill.

The referees also made numerous useful suggestions. They include Thomas J. Murphy, University of Maryland; David W. Pratt, University of Pittsburgh; Roland Roskos, University of Wisconsin, La Crosse; Don Secrest, University of Illinois at Urbana-Champaign; C. Daniel Cornwell, University of Wisconsin — Madison; Clifford E. Dykstra, University of Illinois at Urbana-Champaign; Sherril D. Christian, University of Oklahoma; Charles E. Reid, University of Florida; Charles S. Johnson, University of North Carolina at Chapel Hill; Richard L. Snow, Brigham Young University; Dewey Carpenter, Louisiana State University; Peter C. Jordan, Brandeis; Alan G. Marshall, Ohio State University.

Finally, when I began this book, Jim Moore told me to be sure to mention the parachor — I just did.

Physical chemistry is not a variety of chemistry; it is the chemistry of the future.

—Wilhelm Ostwald

...lecture courses that consisted of descriptive facts and recipes for separations bored me; my desire was to understand phenomena. I discovered physical chemistry on my own and was inspired.

—Joel H. Hildebrand

To the Student

Physical chemistry, like a table with four legs, is built upon four major theoretical principles; thermodynamics, kinetics (or, more generally, transport processes), quantum mechanics, and statistical mechanics. This is not all of physical chemistry, no more than a table is all legs. Physical chemistry is a widely diverse subject that cannot be summarized adequately in any brief definition; certainly there are important parts of physical chemistry that do not fit neatly into this quadrivium. But it is not a bad place to start, and certainly no education in the subject can fail to provide some foundation in each of these four areas. However, they do not, in this or any other text, receive equal time; usually thermodynamics and quantum theory, in that order, are covered more thoroughly than the other two subjects. This is because these two are, to some extent, prerequisite to kinetics and statistical mechanics. This emphasis does not reflect the relative importance of these areas within physical chemistry, let alone within chemistry as a whole, but only the exigencies of teaching an introductory course.

But a first course in physical chemistry is more than an introduction to four, or a hundred subjects. It is, in addition, a course on mathematical problem solving, with emphasis on chemical problems. This is a two-edged sword. Mathematics is the feature of this course that, for the majority of students, is the most difficult — and is in no small part responsible for the frightful reputation of physical chemistry courses in general. It is also the part of the course that is of greatest value — you may never need to integrate a heat capacity, measure a reaction rate, or analyze a spectrum, but the experience in quantitative problem solving will be of constant value.

But how does one learn to solve problems? There are some who feel that this is a subject in itself that can be taught in the abstract; this area is called *heuristics* (see, for example, G. Polya, "How to Solve It," 1957: Garden City, New York, Doubleday & Company, Inc.). Be that as it may, it is certain that problem solving is something that can also be learned by example and by practice. This book provides numerous worked-out examples to introduce you to the subject; these should be studied carefully as you encounter them in the text. The exercises in the text and the problems at the end of each chapter serve two purposes: to make you think about and review the material just covered and to provide you with practice in problem solving. You may have available worked-out solutions to some (or perhaps all) of these problems — but beware: You cannot learn to swim by watching, you must get into the water, and you cannot learn physical chemistry by reading other people's problem solutions. The effort in solving them yourself may be ten times greater, but the benefit will increase in proportion. It can be stated as a certainty that, as this text is written, if you read the words without doing the problems, you will receive only half of what is intended.

In connection with the question of mathematics, there are appendices at the end of the book covering a number of areas that cause problems for many students. You should, before beginning, familiarize yourself with what is there, and then use them as needed. When an appendix is referred to in the text, you should, unless you are confident of your background in that area, review the appropriate material before proceeding.

Good luck!

Contents

*Physical
Chemistry*

Let curious minds
Who would the air inspect,
On its elastic energy reflect.

–Sir Richard Blackmore (1712)

1

Properties of Matter

In large part, "properties of matter" is the subject of this entire book. This chapter broaches several important fundamentals. It also introduces a number of basic concepts, models, and techniques that will be used and, in some cases, developed further in subsequent chapters.

The properties of matter may be discussed on two levels, the macroscopic and the microscopic. At the *microscopic* level we examine the properties of atoms and molecules such as molecular size, shape, velocity, momentum, and intermolecular forces. At the *macroscopic* level we investigate properties of bulk matter, such as temperature, pressure, or viscosity, which may have no meaning whatever at the atomic-molecular level. Making the connection between microscopic and macroscopic is a major mission for physical chemistry. In this book we shall first be concerned mostly with macroscopic properties; then, beginning with Chapter 11, we look into the properties of atoms and molecules. Even when we are focusing on macroscopic properties, however, the microscopic picture will never be far in the background, since it can give us an insight into the "why" in the behavior of nature which would otherwise be unavailable.

1.1 Equations of State

The macroscopic properties of matter may be classified as either extensive or intensive. *Extensive* properties are proportional to the amount of material — for example: mass (W), volume (V), number of moles (n), heat capacity (C). In subsequent chapters we shall encounter many other extensive properties, such as energy and entropy.

Intensive properties depend on the nature of the material but not on the amount. Temperature and pressure are the most obvious examples, but there are many others, including viscosity, thermal conductivity, electrical conductivity, dielectric constant, magnetic susceptibility, and compressibility. In addition, any ratio of extensive properties is intensive; some important examples include density $(\rho = W/V)$, specific volume $(v = V/W)$, molar volume $(V_m = V/n)$, and specific heat (C/W).

These properties are interrelated; their mutual relationships can be expressed as a functional dependence in the mathematical sense, as, for example, $\rho(T)$, the "density as a function of temperature." (It would probably be useful to read the first section of Appendix I at this point.) In fact, for a pure, homogeneous material, only two intensive variables can be independent; the remaining variables must then be a function of these two. ["Pure" means that the entire material has a single chemical identity; by chemical standards, water is a pure substance and not a mixture of hydrogen and oxygen (let alone, a mixture of protons, neutrons, electrons, and so on). "Homogeneous" means that the entire material is uniform throughout with respect to all intensive properties; this implies that there is only one physical phase — solid, liquid, or vapor.]

This idea is expressed in the concept of the *state* of a material; we define the state of a pure, homogeneous material by giving the values of *any* two intensive properties (which then become the *independent variables*). The functional dependence of any other property on these variables is called an *equation of state*. A common (but by no means unique) choice for the independent variables is temperature and pressure (P, T). The viscosity (η), considered as a function of these variables, $\eta(P, T)$, is an equation of state. The common meaning of "equation of state" is an expression of the relationship among the variables P, V, and T, and we shall now consider this topic in detail. [For the preceding statements to be rigorous, we must exclude external fields — for example, electric, magnetic, or gravitational. Gravity cannot be easily excluded, but it varies only slightly, and, as we traverse the earth from Delaware to Mt. Everest, most properties of matter will not be altered significantly (assuming the same T, P). We also must, at times, be concerned with the state of subdivision of the material; the properties of water in a beaker may differ significantly from those of the same material dispersed as a fine aerosol because of the greatly different surface area; this question will be addressed in Chapter 4.]

The Ideal Gas Law

The simplest and most easily used of the equations of state for gases is the **ideal gas law:**

$$PV = nRT \tag{1.1a}$$

This equation resulted from the experimental work of Boyle, Charles, and Gay-Lussac and expresses the necessary functional relationship among the intensive variables (T, P) and the extensive variables (n, V). It is often useful to write it with only intensive variables such as the **molar volume:**

$$V_m = \frac{RT}{P} \tag{1.1b}$$

or the **density:**

$$\rho = \frac{PM}{RT} \tag{1.1c}$$

(M is the molecular weight of the gas.) The last two forms show clearly that only two intensive variables can be independent.

Early investigators were fortunate in their choices of gases and conditions for their experiments, for this "law" does not hold up under close examination. (There has even been some suggestion that Boyle may have fudged his results to make them come out. If so, this was not, in the context of the times, a dishonest act; early scientists were convinced that simple laws must be true and that any minor deviations must be the fault of the experiment. Modern students often get the same misconception.) Nonetheless, the concept of an *ideal gas,* a gas that obeys Eqs. (1.1) precisely, is useful on two levels.

On a practical level, the ideal gas law is simple, easy to use, and a reasonably good approximation to the actual behavior of a great number of gases at low to moderate pressures (roughly, $P < 10$ atm).

On another level, an ideal gas is a useful abstraction; it is a hypothetical material whose properties will be approached by real gases in the *limit* of low pressure or density. The precise statement of the ideal gas law is:

$$\lim_{P \to 0} (PV_m) = RT \tag{1.2}$$

In this form, the law is universally valid for all gases. [In fact, Eq. (1.2) is, with an appropriate choice of the constant R, the definition of the absolute temperature scale, T.]

Units

The use of Eqs. (1.1) should cause no problems for anyone who has passed general chemistry, but perhaps we should digress for a moment on the subject of *units*.

The fundamental units for the *system international* (*SI*) are: for volume, cubic meters (m^3); for pressure, pascals ($Pa = N\ m^{-2} = J\ m^{-3}$); for energy, joules ($J = kg\ m^2\ s^{-2}$); and for temperature, kelvins (K). With these units, the gas constant as used in Eq. (1.2) is $R = 8.3143\ J\ K^{-1}\ mol^{-1}$. Also, masses, including the molecular weight, must be in kilograms (kg); e.g., for H_2, $M = 0.0020158\ kg/mol$. The SI system has several awkward features insofar as it affects chemistry: (1) The unit of pressure is rather small; chemists tend to prefer the atmosphere (1 atm = 101,325 Pa). However the MPa (about 10 atm) is often used. (2) The unit of volume is rather large; chemists strongly prefer the cubic decimeter (also called the *liter*) with $10^3\ dm^3 = m^3$. The cubic centimeter is also useful with $dm^3 = 10^3\ cm^3$, $m^3 = 10^6\ cm^3$. In these units, the gas constant is $R = 0.08206\ dm^3\ atm\ K^{-1}\ mol^{-1} = 82.06\ cm^3\ atm\ K^{-1}\ mol^{-1}$. (3) The "gram mole," which requires molecular weights to be interpreted in grams, is in common use; however, the conversion to kilograms is no problem if you keep your wits about you. In the real world, one must be able to deal with a variety of units, consistent, logical, and otherwise; for that reason, we shall not be totally consistent but shall use the system of units that best fits the problem at hand and/or that for which the majority of data is available.

The "mole" is not exactly a unit; it has no dimensions, SI or otherwise. Rather it is a quantity or a number; saying that an energy is so many joules per mole is like giving the price of eggs per dozen. The molar volume of liquid water at 25°C, 1 atm, is, depending on your point of view, the volume occupied by one gram-molecular weight (18.01534 g) of water or the volume occupied by Avogadro's number ($L = 6.02217 \times 10^{23}$) of molecules. Think of Avogadro's number as a very large quantity, like a dozen, or a gross, or a thousand, but bigger. We shall indicate molar quantities by a subscript m (unless the symbol has no other defined meaning); therefore the qualifying phrase "per mole" is slightly redundant. If V_m is the volume of one mole of a material by definition, we can say (e.g., for H_2O at 25°C, 1 atm) $V_m = 18.07\ cm^3$ without ambiguity. Saying $V_m = 18.07\ cm^3/mol$ is not incorrect, but, like saying the price of a dozen eggs is a dollar per dozen, it is redundant. There is no harm in carrying this "unit" along in calculations, and it is sometimes useful for avoiding ambiguity. Similarly, since the symbol M denotes the mass of one mole of a substance, stating (for H_2O) $M = 18.01534\ g$ or $M = 0.01801534\ kg$ is perfectly

lacking in ambiguity. However, if you are using moles as a unit, the molecular weight must be written as kg/mol to be consistent. Now, let's get back to equations of state.

The van der Waals Equation

The search for more realistic and accurate equations of state has been an active one for over a century. One of the earliest improvements was that proposed by J. D. van der Waals:

$$P = \frac{nRT}{V - nb} - \frac{n^2 a}{V^2}$$

$$= \frac{RT}{V_m - b} - \frac{a}{V_m^2} \tag{1.3}$$

In this equation, a and b are empirical constants that are characteristic of the particular gas (see Table 1.1); R, by way of contrast, is a universal constant. The reasoning that led van der Waals to his equation was, roughly, as follows.

The ideal gas law is obeyed by real gases at very low concentrations, where the molecules never interact because they never meet. At finite concentrations, molecules will have attractive forces which will reduce the pressure; in effect, the molecules in the center of the container make the molecules near the wall "pull their punches" by attracting them toward themselves. This effect gives rise to the a term of

Table 1.1   Gas constants

Gas	van der Waals $a \left(\dfrac{dm^6 \ atm}{mol^2} \right)$	$b \left(\dfrac{dm^3}{mol} \right)$	Critical constants T_c (K)	P_c (atm)	$V_c \left(\dfrac{dm^3}{mol} \right)$	Dieterici constants $a \left(\dfrac{dm^6 \ atm}{mol^2} \right)$	$b \left(\dfrac{dm^3}{mol} \right)$
He	0.034	0.0237	5.3	2.26	0.0577	0.0453	0.0260
Ne	0.211	0.0171	44.5	25.9	0.0416	0.2787	0.0191
Ar	1.345	0.03219	151	48.0	0.0752	1.732	0.0349
Kr	2.318	0.03978	210.6	54.2	0.092	2.983	0.0432
H_2	0.244	0.0266	33.2	12.8	0.0650	0.3139	0.0288
N_2	1.39	0.0391	126.0	33.5	0.0900	1.727	0.0418
O_2	1.36	0.0318	154.3	49.7	0.0744	1.746	0.0345
Cl_2	6.49	0.0562	417	76.1	0.123	8.329	0.0609
NO	1.34	0.0279	183	65	0.058	1.878	0.0313
CO	1.49	0.0399	134	35	0.090	1.870	0.0425
CO_2	3.59	0.0427	304.2	73.0	0.0956	4.621	0.0463
SO_2	6.71	0.0564	430	77.7	0.123	8.674	0.0615
HCl	3.667	0.04081	325	81.6	0.0862	4.718	0.0442
H_2O	5.46	0.0305	647.1	217.7	0.0450	7.011	0.0330
NH_3	4.17	0.0371	405.5	111.5	0.0723	5.375	0.0404
CH_4	2.25	0.0428	190.6	45.8	0.0988	2.891	0.0462
C_2H_6	5.489	0.06380	305.2	49	0.136	—	—
C_2H_4	4.471	0.05714	282.9	50.9	0.1275	—	—
C_2H_2	4.390	0.05136	309	62	0.0601	—	—
C_6H_6	18.00	0.1154	561.6	47.7	0.257	—	—
$CHCl_3$	15.17	0.1022	536.6	54	0.24	—	—

Eq. (1.3). (The detailed reasoning which suggested that this term is proportional to V^{-2} and independent of temperature is complex and not totally convincing; we omit it here, as we shall shortly have a better way to discuss these forces.)

Another difference between real and ideal gases is that real gas molecules collide with each other. Van der Waals argued that the "volume" in the ideal gas law should be, not the volume of the container, but the volume available to a molecule for kinetic movement — that is, the portion of the container not occupied by other molecules. The centers of spheres having a diameter σ cannot approach closer than σ without colliding; we can picture this (Fig. 1.1) as an imaginary sphere of diameter 2σ (radius σ) around each sphere into which the center of another sphere cannot penetrate. Thus, the excluded volume per *pair* of spheres is $4\pi\sigma^3/3$; for L molecules ($L/2$ pairs), the excluded volume is:

$$b = 2\pi L \frac{\sigma^3}{3} \tag{1.4}$$

This, of course, is the b correction of van der Waals; even when molecules are not reasonably spherical, Eq. (1.4) can be used to calculate a "van der Waals diameter" (σ) which gives a crude, but reasonably accurate, estimate of the molecular size.

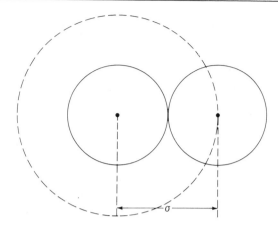

Figure 1.1 Excluded volume. A sphere of diameter σ creates an excluded volume, which is a sphere of *radius* σ, into which the center of another sphere of the same size may not penetrate.

Other Equations of State

Students often seem to have the impression that the van der Waals equation is "truth" while the ideal gas equation is its first approximation. In fact, Eq. (1.3) is only one of a number of semiempirical, two-constant equations of state. To put the matter in perspective, here are two more — the **Berthelot equation:**

$$P = \frac{nRT}{V - nb} - \frac{n^2 a}{TV^2} \tag{1.5}$$

and the **Dieterici equation:**

$$P = \frac{nRT}{V - nb} e^{-a/RTV_m} \tag{1.6}$$

In these equations, a and b are empirical constants with the same qualitative significance as the van der Waals a and b; however, the numerical values for a given gas would be different in each case. The van der Waals equation is, by far, the most popular of these three, and the van der Waals constants are tabulated for a great number of gases; but it is not, in general, any better than the other two. For some purposes one equation may be best; for another, a different equation; none of these three is perfect for all applications. The Dieterici equation in particular has been, perhaps unjustly, neglected — probably because it was difficult to use with a slide rule. Later in this chapter we shall see that it is, for some purposes, an attractive alternative to the van der Waals equation.

The search for an equation of state which is accurate for all purposes and conditions has been a failure in the sense that such equations tend to be so complicated, and

Table 1.2   Beattie-Bridgeman constants

(for volume in dm^3, pressure in atm, $R = 0.08206$ dm^3 atm K^{-1} mol^{-1})

Gas	A_0	a	B_0	b	$c \times 10^{-4}$
He	0.0216	0.05984	0.01400	0	0.004
Ne	0.2125	0.2196	0.02060	0	0.101
Ar	1.2907	0.02328	0.03931	0	5.99
Kr	2.4230	0.02865	0.05261	0	14.89
H_2	0.1975	−0.00506	0.02096	−0.04359	0.050
N_2	1.3445	0.02617	0.05046	−0.00691	4.20
O_2	1.4911	0.02562	0.04624	0.004208	4.80
CO_2	5.0065	0.07132	0.10476	0.07235	66.00
CH_4	2.2769	0.01855	0.05587	−0.01587	12.83
C_2H_6	5.8800	0.05861	0.09400	0.01915	90.0
C_2H_4	6.1520	0.04964	0.12156	0.03597	22.68

contain so many constants, that their use is very cumbersome. An example of a five-constant equation of state is the **Beattie-Bridgeman equation:**

$$PV_m = RT\left(1 - \frac{c}{T^3 V_m}\right)\left(1 + \frac{B_0}{V_m} - \frac{bB_0}{V_m^2}\right) - \frac{A_0}{V_m}\left(1 - \frac{a}{V_m}\right) \tag{1.7}$$

Constants for this equation are given in Table 1.2 for several gases.

Example: Calculate the pressure exerted by 15 g of H_2 in a volume of 5 dm^3 at 300 K using the van der Waals equation (constants, Table 1.1):

$$V_m = \frac{V}{n} = \frac{5(2.016)}{15} = 0.672 \text{ dm}^3$$

$$P = \frac{RT}{V_m - b} - \frac{a}{V_m^2}$$

$$= \frac{(0.08206 \text{ dm}^3 \text{ atm/K})(300 \text{ K})}{0.672 \text{ dm}^3 - 0.0266 \text{ dm}^3} - \frac{0.244 \text{ dm}^6 \text{ atm}}{(0.672 \text{ dm}^3)^2}$$

$$= 37.6 \text{ atm} \qquad \blacksquare$$

Example: Calculate the molar volume of H_2 gas at 40.0 atm, 300 K. This is a trickier problem; the van der Waals equation cannot be solved explicitly for volume. Equation (1.3) can be arranged into a cubic equation:

$$V_m^3 - \left(\frac{RT}{P} + b\right)V_m^2 + \left(\frac{a}{P}\right)V_m - \left(\frac{ab}{P}\right) = 0 \tag{1.8}$$

but this is rather difficult to solve. Usually the best method is to arrange it as:

$$V_m = \frac{RT}{P + \dfrac{a}{V_m^2}} + b \tag{1.9}$$

which can be solved by successive approximations, provided the volume term on the right-hand side (a/V_m^2) is small compared to P. The first approximation can be obtained from the ideal gas law:

$$V_m^{(1)} = \frac{0.08206(300)}{40.0} = 0.615 \text{ dm}^3/\text{mol}$$

This is then used to calculate the second approximation:

$$V_m^{(2)} = \frac{0.08206(300)}{40.0 + \dfrac{0.244}{(0.615)^2}} + 0.0266 = 0.6323 \text{ dm}^3/\text{mol}$$

The reader should demonstrate that the next two approximations are:

$$V_m^{(3)} = 0.6328 \text{ dm}^3, \qquad V_m^{(4)} = 0.6328 \text{ dm}^3$$

Clearly the answer has converged to four significant figures. If it is not obvious that this is the correct answer, calculate the pressure (as in the previous example) using $T = 300$ K, $V_m = 0.6328$ dm^3. The reader should work through the units of this example as an exercise. ∎

1.2 The Virial Series

Another approach to representing the PVT behavior of real gases starts with the definition of the **compressibility factor:**

$$z = \frac{PV_m}{RT} \tag{1.10}$$

This quantity is clearly equal to one for an ideal gas and, for real gases, will approach one as the pressure and concentration approach zero. In such a case we might represent z as a power series in the molar concentration $n/V = 1/V_m$:

$$z = 1 + B\left(\frac{n}{V}\right) + C\left(\frac{n}{V}\right)^2 + D\left(\frac{n}{V}\right)^3 + E\left(\frac{n}{V}\right)^4 + \cdots \tag{1.11}$$

The coefficients of this expansion are functions of T only and are called the *virial coefficients*.

The significance of the virial series is primarily theoretical. The *second virial coefficient, $B(T)$*, can be calculated from the forces between molecules interacting two at a time; we shall do this in Section 1.7 for some simple potentials. The *third virial coefficient, $C(T)$*, can be calculated by considering the interactions of molecules three at a time; this is a much more difficult calculation, and relatively less is known theoretically about this and the higher virial coefficients.

As a practical method for representing gas behavior, the virial equation has its limitations; in particular, it may fail to converge at high densities and/or at low temperatures. Nonetheless we shall find it is a very useful way to discuss moderate deviations of gases from ideality.

The compressibility factor can also be written as a power series in the pressure:

$$z = 1 + BP + CP^2 + DP^3 + \cdots \tag{1.12}$$

This is a less fundamental form of the virial series than Eq. (1.11) but more useful, since it treats pressure as the independent variable and can be solved explicitly for volume. We shall more often write it in the form:

$$PV_m = RT + \beta P + \gamma P^2 + \delta P^3 + \cdots \tag{1.13}$$

It should be obvious that these sets of virial coefficients are related as:

$$\beta = BRT, \qquad \gamma = CRT, \qquad \delta = DRT, \qquad \ldots \tag{1.14}$$

The relationship between the coefficients of Eq. (1.11) and those of (1.13) is a great deal less obvious. It can be shown by substitution of Eq. (1.13) into (1.11) that:

$$\beta = B, \qquad \gamma = \frac{C - B^2}{RT} \tag{1.15}$$

Since the second virial coefficients β and B are the same, we shall use the latter symbol all the time.

Orders of Approximation

In subsequent chapters we shall often use the ideal gas law; it is convenient, easy to use, and reasonably accurate in many circumstances. However, when using an approximate relationship, it is important to be aware of the size of the possible errors and the possibility of a total failure of the approximate equations. The ideal gas law provides a useful first-order approximation to gas behavior, but we can judge its validity only by calculating the second-order correction to the ideal gas prediction. The second-order approximation can conveniently be obtained using, for example, the van der Waals equation, but, as we have seen, this equation has the disadvantage that it cannot be solved explicitly for volume. Another convenient way to calculate the second-order approximation is to use the virial coefficients, particularly the second virial coefficient.

Suppose, in a particular calculation, an accuracy of 1% is required. If the correction calculated with a second virial coefficient is only 1%, you can be assured that the accuracy is better than that. If the correction is 10%, the error in the (corrected) calculation is probably no more than a few percent. Furthermore, you need only 10% accuracy in the virial coefficient to obtain such accuracy, since the error in the calculation due to this is 10% of 10%, or 1%. However, if the correction were much greater than 10%, the error could be large and unpredictable, and an alternative method should be sought.

Example: The molar volume of H_2 at 773.15 K, 100.0 atm, is 0.65245 dm³/mol; the second virial coefficient at this temperature is 0.017974 dm³/mol. If we estimated this volume using the ideal gas law, the result would be $V_m = 0.63442$ dm³/mol (error 3%). Suppose, using some method (such as those to be discussed below), we estimated the second virial coefficient as $B = 0.0216 \ \mathrm{dm^3/mol}$ (i.e., with an error of 20%). The second approximation to the molar volume would be, from Eq. (1.13):

$$V_m = \frac{RT}{P} + B$$

$$= 0.63442 + 0.0216 = 0.65602 \ \mathrm{dm^3/mol}$$

The second approximation is in error by only 0.5%, despite the 20% error in B. ∎

Estimation of Virial Coefficients

Since we shall be using Eq. (1.13) frequently in subsequent chapters (usually with only the B term), we shall need reliable methods for approximating the second virial coefficient. For gases for which the Beattie-Bridgeman constants have been determined, the following formulas are very accurate and relatively easy to use:

$$B(T) = B_0 - \frac{A_0}{RT} - \frac{c}{T^3}$$

$$C = \frac{A_0 a}{RT} - B_0 b - \frac{B_0 c}{T^3}, \qquad D = \frac{B_0 bc}{T^3} \qquad \textbf{(1.16)}$$

Since Beattie-Bridgeman constants are available only for a limited number of gases, the Berthelot second virial coefficient [not related to Eq. (1.5)] can be used to give reasonably accurate values from two easily measured parameters, the critical temperature (T_c) and critical pressure (P_c) of the gas:

$$B(T) = \frac{9RT_c}{128P_c}\left(1 - \frac{6T_c^2}{T^2}\right) \qquad \textbf{(1.17)}$$

(The critical constants of a gas, what they mean, and how they are measured will be discussed in the next section; some values are listed in Table 1.1.)

The virial coefficients can also be related to the van der Waals constants. The details of this derivation will be given next, not because the result is notably accurate (it is not), but because it will give us some insight into the meaning of the second virial coefficient.

First, the van der Waals equation (1.3) is arranged so that $z = PV_m/RT$ is on the left-hand side; with a little algebra you should be able to show:

$$z = \frac{V_m}{V_m - b} - \frac{a}{RTV_m}$$

The first term on the right-hand side can be also written as:

$$\frac{1}{1 - \dfrac{b}{V_m}}$$

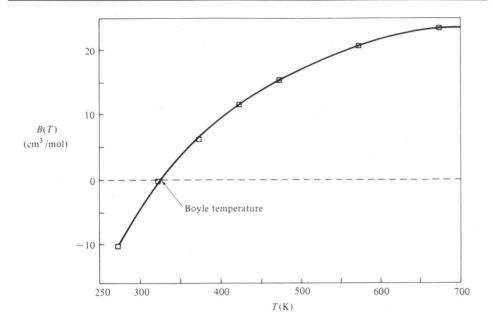

Figure 1.2   The second virial coefficient of nitrogen as a function of temperature. [Data from Holburn and Otto, *Z. Phys.*, 33, 5 (1925)]

If b/V_m is small compared to 1, this can be expanded in a power series (Appendix I):

$$\frac{1}{1 - \dfrac{b}{V_m}} = 1 + \frac{b}{V_m} + \left(\frac{b}{V_m}\right)^2 + \left(\frac{b}{V_m}\right)^3 + \cdots$$

This result is then used in the equation for z, and, if terms are collected by powers of $V_m^{-1} = n/V$, we get:

$$z = 1 + \left(b - \frac{a}{RT}\right)\left(\frac{n}{V}\right) + b^2\left(\frac{n}{V}\right)^2 + b^3\left(\frac{n}{V}\right)^3 + \cdots$$

A theorem in algebra states that two polynomials in the same variable (n/V in this case) are equal if, and only if, the coefficients of each power of the variable are equal. Comparison of this result to Eq. (1.11) shows:

$$B(T) = b - a/RT, \qquad C = b^2, \qquad D = b^3, \qquad \ldots \tag{1.18}$$

The van der Waals form of the second virial coefficient is qualitatively correct (Fig. 1.2); at high temperatures $B(T)$ is positive and roughly constant and at low temperatures it is negative. It also shows that $B(T)$ is affected by attractive forces (the a term) which make it negative and repulsive forces (the b term) which dominate and make it positive at high temperatures. Equation (1.18) for $B(T)$ is not, however, very

Table 1.3 Virial coefficients of gases calculated by various methods

	Observed*	van der Waals	Berthelot	Beattie-Bridgeman	Lennard-Jones 6-12 potential
H_2, 273.15 K:					
$B(cm^3)$	12.03	15.7	13.7	12.1	12.0
$C(cm^6)$	358	707	—	870	301
N_2, 223.15 K:					
$B(cm^3)$	-28.79	-36.8	-19.8	-26.7	-21.9
$C(cm^6)$	3570	1500	—	2080	1600
N_2, 473.15 K:					
$B(cm^3)$	14.763	3.3	12.5	15.4	16.5
$C(cm^6)$	1295	1500	—	1230	1300

*Data from S. Maron and J. Lando, *Fundamentals of Physical Chemistry* (New York: Macmillan, 1974). The third virial coefficients listed in this source are, in our notation, the γ of Eq. (1.13); the constants (C) above were calculated using Eq. (1.15), $C = RT\gamma + B^2$.

accurate, so we shall prefer to use either Eq. (1.16) or (1.17) when possible. Table 1.3 shows some sample calculations of $B(T)$ and C by various methods; the reader should check some of them for practice. (The Lennard-Jones calculation, the last column of Table 1.3, will be discussed in Section 1.7.)

The Boyle Temperature

For all gases, there is some temperature at which $B(T) = 0$. This is called the Boyle temperature (T_B), because at that temperature Boyle's law will be obeyed up to moderately large pressures. The reason is that the repulsive and attractive forces cancel each other at the Boyle temperature, giving nearly ideal behavior.

Exercise: Show that the Boyle temperature can be estimated from the van der Waals constants [Eq. (1.18)]:

$$T_B = \frac{a}{bR}$$

or from the critical temperature [Eq. (1.17)]:

$$T_B = \sqrt{6}\, T_c \qquad\qquad \blacksquare$$

Example: Estimate the Boyle temperature of N_2; the actual value is 324 K.
From van der Waals constants:

$$T_B = 1.39 \text{ dm}^6 \text{ atm}/(0.0391 \text{ dm}^3)(0.08206 \text{ dm}^3 \text{ atm/K}) = 433 \text{ K}$$

From the Berthelot second virial coefficient:

$$T_B = \sqrt{6}\,(126) = 309 \text{ K} \qquad\qquad \blacksquare$$

Example: Estimation of the Boyle temperature for the Beattie-Bridgeman second virial coefficient [Eq. (1.16)] requires solving a cubic equation; this is most easily done using the Newton-Raphson iteration discussed in Appendix I. Doing a little algebra first will simplify the procedure:

$$B(T) = B_0 - \frac{A_0}{RT} - \frac{c}{T^3} = 0$$

Multiply this equation by T^3 and divide by B_0:

$$f(T) = T^3 - \left(\frac{A_0}{RB_0}\right)T^2 - \frac{c}{B_0} = 0$$

For N_2, the constants of Table 1.2 give:

$$f(T) = T^3 - 324.7T^2 - 8.32 \times 10^5 = 0$$

$$f' = 3T^2 - 2(324.7)T$$

An easy way to get a first approximation is from the Berthelot virial coefficient, $T_B = \sqrt{6}T_c = 309$ K. For this temperature:

$$f(309) = -2.33 \times 10^6, \qquad f'(309) = 8.578 \times 10^4$$

The next approximation is, therefore:

$$T = 309 - (-2.33 \times 10^6/8.578 \times 10^4) = 336.2$$

The reader should demonstrate that the next approximation is 332.3; with another round of approximation, the final answer (to three significant figures) is:

$$T_B = 332 \text{ K} \qquad \blacksquare$$

1.3   Critical Behavior of Fluids

Consider the following experiment (see Fig. 1.3). A quantity of a substance is placed in a sealed glass tube; under the proper conditions there will be two phases present, a dense phase called the liquid and a less dense phase, the gas. A meniscus shows the separation of the two phases. If the tube contains a relatively small amount of material [Fig. 1.3(c)], when it is heated the liquid will evaporate and the meniscus will go down and ultimately disappear. If the tube is nearly full [Fig. 1.3(a)], when it is heated the liquid will expand and the meniscus will rise (think of the behavior of a thermometer). The meniscus will disappear when the liquid expands to fill the tube. Evidently there is some degree of filling at which the thermal expansion and evaporation effects will cancel and the meniscus *will not move at all*. Under this condition, as the tube is heated, the liquid density decreases because of thermal expansion and the gas density increases because more material is being evaporated into a fixed volume; when the densities of the two phases become equal, the meniscus will disappear! The ability to observe a phase boundary (meniscus) between two, otherwise identical fluids, is possible only if they have different indices of refraction; when the densities of the two phases are equal, so are the indices of refraction, and no phase boundary can be observed. [The distinction between a liquid and a gas is just that. If two phases can

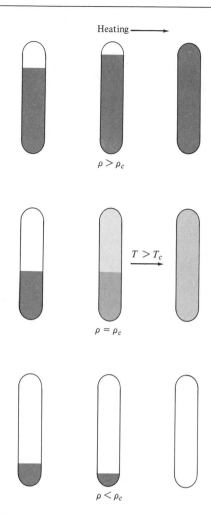

Heating ⟶

$\rho > \rho_c$

$T > T_c$

$\rho = \rho_c$

$\rho < \rho_c$

Figure 1.3   Critical point. A liquid in a closed tube will, upon heating, both expand and evaporate. If the tube is nearly filled, the liquid will expand to fill the tube. If there is only a small amount of liquid, it will evaporate and disappear. At the *critical* filling, the meniscus will not move upon heating but will disappear at the *critical temperature*.

be observed, the denser one (at the bottom of the container, of course) is the liquid. If there is only a single phase, we might call it a liquid if its density is "high" and its compressibility "low," or we might call it a gas if its density is "low" and it is easily compressed, but these are relative terms and it would be easy to find an ambiguous situation. The term "fluid" is neutral.]

The temperature at which the meniscus disappears is the *critical temperature* (T_c). The pressure in the tube at that point is the *critical pressure* (P_c), and the mass of

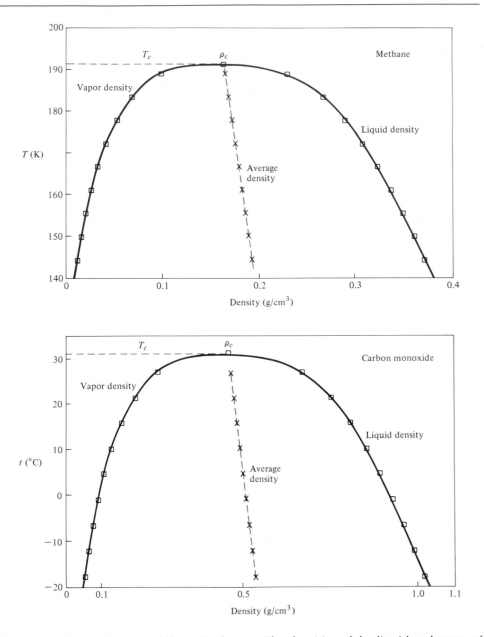

Figure 1.4   Determination of the critical point. The densities of the liquid and vapor of a two-phase fluid are plotted vs. temperature; the point at which they become equal is the *critical point*. A graph of the average density of the two phases is a straight line (the law of rectilinear diameters), and the intersection of this line with the liquid-vapor density line shows the location of the critical point.

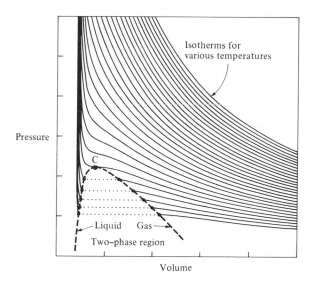

Figure 1.5 The isotherms of a gas. At high temperatures, these are similar to the hyperbolae predicted by the ideal gas law. At the critical temperature, the isotherm has a horizontal inflection at the critical point (C). Below the critical temperature, the isotherms have a horizontal segment (shown by the dotted lines) in the two-phase region, where both liquid and vapor are present. (Cf. Figure 1.7.)

material in the tube divided by the volume of the tube is the *critical density* (ρ_c). The *critical molar volume* (V_c) can be calculated from this density:

$$V_c = \frac{M}{\rho_c}$$

Of course, the experiment as described is a bit idealized; it would be a rare stroke of luck to hit upon the critical amount of material exactly. In an actual experiment, the liquid and vapor densities are both measured at various temperatures and plotted as in Fig. 1.4. The critical point is the point at which the liquid and vapor densities become equal. As can be seen from Fig. 1.4, the shape of the density-vs.-temperature curve determines that the critical density cannot be measured as accurately as the critical temperature. The *law of rectilinear diameters* states that the average density,

$$\rho_{ave} = \tfrac{1}{2}(\rho_{liq} + \rho_{vap})$$

will be a linear function of temperature; the data of Fig. 1.4 demonstrate that this law is correct for the cases shown. The intersection of the density curve and the average-density straight line will determine the location of the critical point.

Another way by which critical behavior can be observed is in graphs of *P* vs. *V* for a gas at constant temperature (isotherms). A plot of this type is shown in Fig. 1.5. (Actual data for CO_2 are given in Fig. 1.6.) At temperatures well above the critical temperature the isotherms of a gas are smooth, looking qualitatively like the hyper-

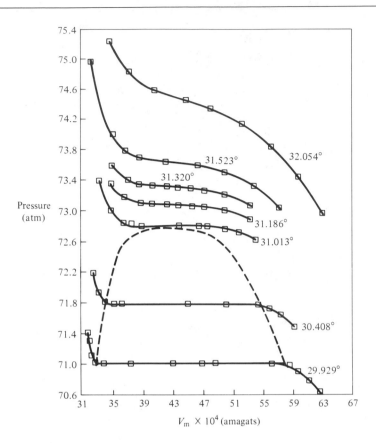

Figure 1.6 Critical point of carbon dioxide. Graph of experimental PVT data for carbon dioxide near the critical point. [Data from Michels, Blaisse, and Michels, *Proc. Roy. Soc. (London)*, A160, 358 (1937). From J. O. Hirschfelder, C. F. Curtiss, and R. B. Bird, *Kinetic Theory of Gases*, 1954: New York, John Wiley & Sons, Fig. 5.2-2a]

bolae predicted by Boyle's law (*PV* = constant). Near the critical temperature the isotherms have a nearly horizontal section; at the critical temperature this becomes an inflection point for which the slope is zero at some point — namely, the critical point (P_c, V_c). Below the critical temperature there are two phases, so the volumes of both the liquid and vapor must be plotted; these are shown in Fig. 1.5 connected by horizontal lines (tie lines). The value of the pressure at which, for a given temperature, a liquid and vapor phase are in equilibrium (i.e., the horizontal lines) is called the *vapor pressure*. The meaning of the horizontal part of the isotherms below the critical temperature, is that the total volume of the fluid (liquid plus vapor) can be decreased without a change in pressure (Fig. 1.7); if there are two phases in the container, compression will cause more vapor to condense into the higher-density liquid, but the pressure (which is the vapor pressure of the liquid at that temperature) will not be

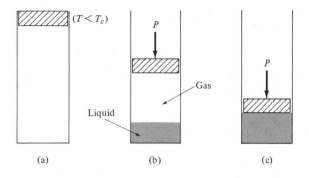

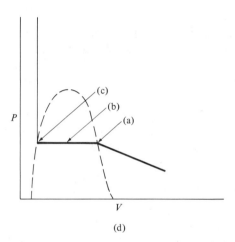

Figure 1.7 Liquefaction of a gas. The isothermal compression of a gas that is below its critical temperature will, at the PV point marked "a", cause the condensation of a liquid. Further reduction of volume will cause more liquid to condense, but the pressure will not change; this is the *vapor pressure* of the liquid at this temperature. At some point ("c" on the diagram) all the gas will have condensed to a liquid, and further reduction of volume requires the compression of the liquid; the sharp rise of the isotherm illustrates the fact that liquids are relatively incompressible.

altered. When the vapor is all gone (the left-hand side of the horizontal lines), the isotherms rise steeply, indicating the incompressibility of the liquid. Also note that at, and above, the critical temperature there is no phase separation but a smooth transition from a low density (gas) to a high density (liquid) as the gas is isothermally compressed. Similarly, above the critical pressure, if the substance were heated at constant pressure, the fluid would go smoothly from high to low density ("liquid" to "vapor") without developing two phases; that is, it does not show the "boiling" behavior we ordinarily expect from "liquids."

20   *Properties of Matter*

[The unit of volume in Fig. 1.6 is *amagats*, which is the ratio of the actual molar volume to the ideal gas molar volume at STP. From the figure we can read $V_c = 43 \times 10^{-4}$ amagats, which translates to:

$$V_c = 43 \times 10^{-4}(22,414 \text{ cm}^3/\text{mole}) = 96 \text{ cm}^3/\text{mole}$$

The term amagat is also applied to density; again, density (amagats) = (actual density)/(ideal density at STP).]

Gas Laws in the Critical Region

An important criterion for the validity of a gas law is its ability to reflect, at least qualitatively, the behavior of a gas at the critical point. The ideal gas law cannot do so; its isotherms (Fig. 1.8) are simple hyperbolae at any temperature whatever. On the other hand, all the two-constant equations of state given earlier [Eqs. (1.3), (1.5), and (1.6)] will show the required inflection in their isotherms at some temperature. Aside from demonstrating that these are realistic equations of state, this inflection provides a convenient method for evaluating the constants (a, b) from the experimentally measured critical constants $(T_c, P_c, \text{and } V_c)$.

Figure 1.8 shows the isotherms of ammonia as calculated by the van der Waals and Dieterici equations. As can be seen, the isotherms at and above the critical temperature are very realistic; below T_c, however, the isotherms show an undulation which is not at all like the observed behavior. What is the meaning of this undulation? Earlier, we saw that the van der Waals equation was cubic in volume [Eq. (1.8)]. A well-known theorem from algebra states that a cubic equation must have three roots; this means that, for any choice of T and P, there will be three values of the volume which will, mathematically, satisfy the van der Waals equation. Above the critical temperature, only one of the roots of Eq. (1.8) is real; the other two roots are complex and clearly have no physical significance. At the critical temperature, all three roots of Eq. (1.8) are real and equal; below T_c the three roots are real and unequal.

Do the roots have any physical significance in the two-phase region? Often the larger root is interpreted as the volume of the gas phase and the smaller as the volume of the liquid; the middle root is presumed to be "unrealistic." Also, if a horizontal line is drawn through the undulation with equal areas above and below, this is interpreted as the vapor pressure of the liquid; Fig. 1.8 shows such a line for 350 K (the dashed line); the van der Waals line predicts a vapor pressure of 62 atm, and the Dieterici equation predicts 79 atm. The actual vapor pressure at that temperature is approximately 38 atm. The fact is that simple, semiempirical gas laws such as the van der Waals or Dieterici equations are not realistic, accurate, or reliable in the critical region. [The van der Waals even predicts *negative* pressures for some cases; note that the Dieterici equation (Fig. 1.8) fails less disastrously than van der Waals in the critical region.]

Considering the complications of critical behavior and phase transitions, the ability of such a simple equation to tell us anything about the critical region is remarkable — rather like Samuel Johnson's comment about a dog's walking on its hind legs, "It is not done well, but you are surprised to find it done at all." Nonetheless, the fact that

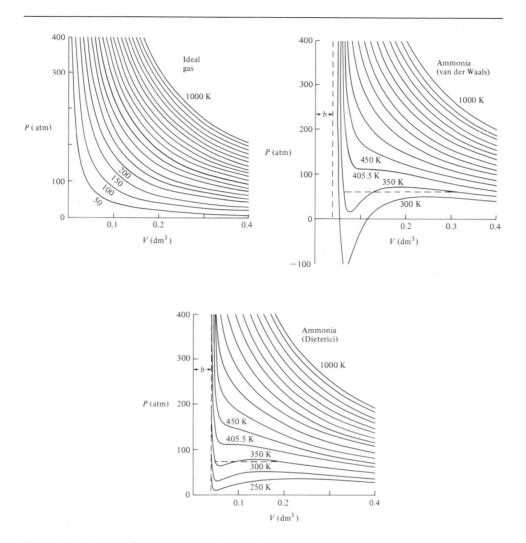

Figure 1.8 Gas isotherms. Comparison of the isotherms of an ideal gas to those predicted by the van der Waals and Dieterici equations for ammonia. An ideal gas will not condense at any temperature. The other two laws predict the location of the critical point, but their behavior in the two-phase region is unrealistic. Note that the van der Waals equation for ammonia at 300 K predicts *negative* pressures.

these equations do show an inflection at some temperature is the source of some useful relationships, which we shall now derive.

Evaluation of Gas Constants from Critical Data

According to calculus, at a horizontal inflection point the first and second derivatives of a function are zero. Using the van der Waals equation to give P as a function of

V (at a given T, considered constant), we can find its derivatives in a straightforward manner; then, at the critical point (P_c, V_c, T_c), we have the following relationships:

$$P_c = \frac{RT_c}{V_c - b} - \frac{a}{V_c^2}$$

$$\frac{dP}{dV} = -\frac{RT_c}{(V_c - b)^2} + \frac{2a}{V_c^3} = 0$$

$$\frac{d^2P}{dV^2} = \frac{2RT_c}{(V_c - b)^3} - \frac{6a}{V_c^4} = 0$$

With a bit of algebra, you should be able to solve these equations to get T_c, V_c, and P_c as related to the constants a, b, P; namely:

$$T_c = \frac{8a}{27bR}, \qquad V_c = 3b, \qquad P_c = \frac{a}{27b^2} \qquad \textbf{(1.19)}$$

Thus van der Waals constants can be used to calculate the critical constants and, historically, this was done. Today, however, the situation is the reverse — the van der Waals constants tabulated for most gases are chosen specifically to make Eqs. (1.19) true. The critical constants of a gas are among its more easily measured characteristics; for that reason it will be convenient to rearrange Eqs. (1.19) so as to permit us to calculate the constants (a, b) from the measured values of the critical constants. When these three equations (1.19) are solved for two constants, some redundancy is present; it is best to avoid relationships involving the critical volume, since it is usually the least accurate of the critical constants. The results are given in Table 1.4; note that the simple and obvious relationship of Eq. (1.19), $b = V_c/3$, has been avoided. This method can also be used to relate the Berthelot and Dieterici constants to the critical constants, and these results are also given in Table 1.4.

The redundancy of Eqs. (1.19) when solved for a and b can be expressed in several ways; one way is to calculate the critical compressibility factor. For the van der Waals equation, this can be shown to give:

$$z_c = \frac{P_c V_c}{RT_c} = \frac{3}{8} = 0.375$$

Table 1.4   Equation-of-state parameters from critical constants

$(e = 2.71828\ldots)$

	van der Waals	Berthelot	Dieterici
a	$\dfrac{27R^2T_c^2}{64P_c}$	$\dfrac{27R^2T_c^3}{64P_c}$	$\dfrac{4R^2T_c^2}{e^2P_c}$
b	$\dfrac{RT_c}{8P_c}$	$\dfrac{RT_c}{8P_c}$	$\dfrac{RT_c}{e^2P_c}$
$\dfrac{P_cV_c}{RT_c}$	$\dfrac{3}{8}$	$\dfrac{3}{8}$	$\dfrac{2}{e^2}$

That is, the critical compressibility factor is predicted to be 0.375 for all gases. As a matter of fact, the critical compressibility factor is reasonably the same for most gases, but the value is usually found to be in the range 0.25 to 0.30. (Try the calculation for a few examples with data from Table 1.1.) The Dieterici equation, on the other hand, predicts a critical compressibility of 0.27 and thus may be a better equation of state to use near the critical point.

1.4   The Law of Corresponding States

The fact, which was noted in the last section, that the ratio P_cV_c/RT_c is nearly the same for all gases is a clue to an important empirical generalization called the *law of corresponding states*. For example, the gases He and CO_2 are very different in their behavior at any given temperature and pressure, but if we compare them, each at its critical point, their compressibility factors are nearly the same:

$$\text{He:}\quad T_c = 5.3 \text{ K},\ P_c = 2.26 \text{ atm},\ V_c = 57.7 \text{ cm}^3/\text{mol};\ \frac{P_cV_c}{RT_c} = 0.300$$

$$CO_2\text{:}\quad T_c = 304.2 \text{ K},\ P_c = 73.0 \text{ atm},\ V_c = 95.6 \text{ cm}^3/\text{mol};\ \frac{P_cV_c}{RT_c} = 0.280$$

We could carry this reasoning further and suspect that the compressibility factor of He at twice its critical temperature and pressure (10.6 K and 4.52 atm) would be nearly the same as that of CO_2 at 608 K, 146 atm. The law of corresponding states states that the compressibility factor (and many other intensive properties) of any gas can be written as a *universal* function of the reduced variables:

$$T_r = \frac{T}{T_c}, \qquad P_r = \frac{P}{P_c}, \qquad V_r = \frac{V}{V_c} \tag{1.20}$$

(See Fig. 1.9.)

Exercise (optional): Use the one-term virial equation:

$$PV_m = RT + BP$$

together with the Berthelot form of the virial coefficient [Eq. (1.17)] to derive the equation:

$$z = 1 + \frac{9P_r}{128T_r}\left(1 - \frac{6}{T_r^2}\right) \tag{1.21}$$

This equation, often called the **second Berthelot equation** to distinguish it from Eq. (1.5), has been widely used to correct for small deviations of gases from ideality.   ∎

Hougen and Watson (ref. 4) have constructed a set of very useful curves for z as a function of P_r and T_r, using the average behavior of a number of gases (N_2, H_2, CO_2, NH_3, CH_4, C_3H_8, and C_5H_{12} for Fig. 1.10). These curves (Fig. 1.10) are the most reliable way of predicting gas behavior under extreme conditions, particularly near the critical region (see Table 1.5).

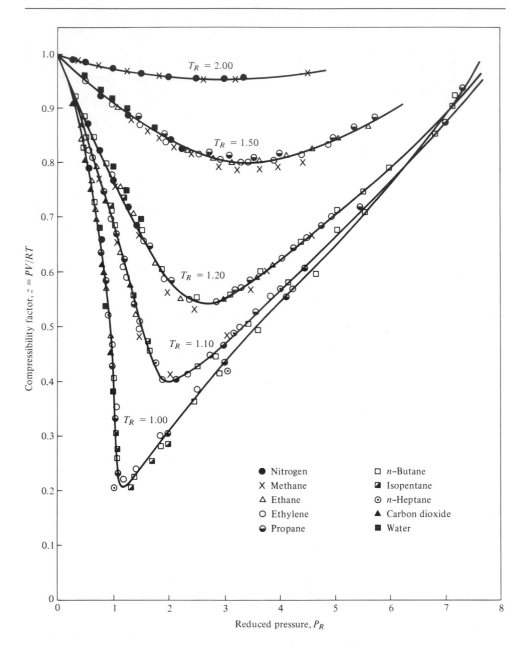

Figure 1.9 Compressibility factors for a variety of gases plotted versus reduced pressure. Note that, at a given ratio of temperature to that gas's critical temperature (the reduced temperature), all gases fall on or near the same isotherm. This is an illustration of the *law of corresponding states*. [From Gouq-Jen Su, *Ind. Eng. Chem.*, 38, 803 (1946)]

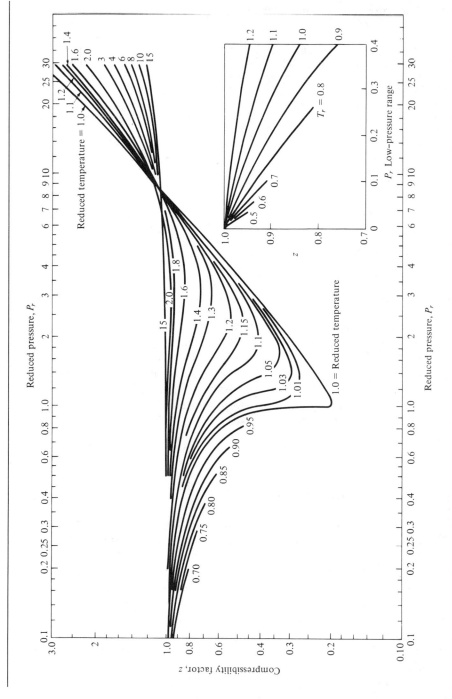

Figure 1.10 Compressibility factor. The compressibility factor as a function of reduced pressure and temperature; curves were constructed from average data of a variety of gases. (From O. A. Hougen and K. M. Watson, *Chemical Process Principles, Part II*, 1947: New York, John Wiley & Sons)

Table 1.5   Calculated and experimental compressibility factors

P (atm)	z (obs.)	One-term virial	van der Waals	Dieterici	Corresponding states (Fig. 1.10)
CO_2 at 313.15 K:					
1	0.995	0.995	0.996	0.995	—
10	0.953	0.957	0.961	0.947	0.95
50	0.739	0.753	0.767	0.695	0.67
100	0.270	0.550	0.350	0.224	0.24
500	0.856	—	1.16	0.922	0.79
1100	1.557	—	2.30	2.00	1.60
CH_4 at 203 K:					
1	0.994	0.994	0.994	0.992	—
10	0.937	0.942	0.942	0.921	0.94
50	0.594	0.711	0.593	0.464	0.58
100	0.377	0.423	0.474	0.313	0.38
500	1.324	—	1.70	1.41	1.3
1000	2.368	—	3.11	2.80	2.4

Another aspect of the law of corresponding states is that the ratio of the second virial coefficient to the critical volume, B/V_c, should be a universal function of T_r; Fig. 1.11 shows this clearly. The average curve of Fig. 1.11 provides another method for estimating $B(T)$ for gases for which the critical constants have been measured.

[The sample calculations of Table 1.5 are worthy of careful examination. In particular, note how well the Dieterici equation performs; the constants for this equation were calculated with the formulas of Table 1.4 (see Table 1.1 for other values). Also, see Fig. 1.12.]

1.5   Kinetic Theory of Gases

Now we have had a brief look at some macroscopic properties of gases; next we shall look at the same situation from the microscopic point of view. The kinetic-molecular theory of gases provides us with a picture of a gas which is composed of huge numbers of submicroscopic particles called molecules which are in constant motion at all times and, randomly, in all directions. These molecules are, in fact, very complex assemblages of electrical charge, the positive nuclei (which contain most of the mass) being held together by a negatively charged cloud of electrons. These are not static entities; the nuclear masses are in constant motion within the molecule, and the whole electrical structure can be influenced (polarized) by external electrical interactions, including those which result when two or more molecules are near each other.

Despite the complexity of molecular structure, a great number of the properties of gases can be explained using a very simple model in which the molecule is treated as a *hard sphere*. Furthermore, at very low densities, when molecular collisions are rare and the distance between molecules is very much larger than their size, an even

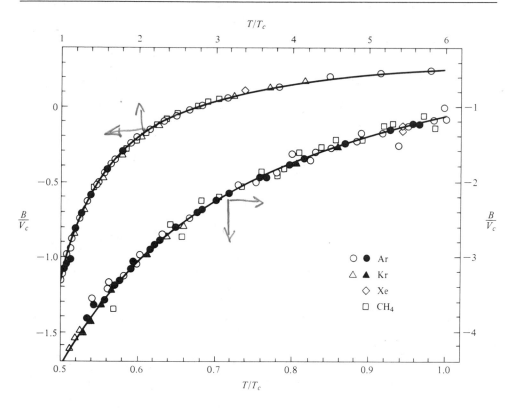

Figure 1.11 Reduced second virial coefficients. Second virial coefficients of argon (circles), krypton (triangles), xenon (diamonds), and methane (squares), when divided by the critical volume and plotted vs. the reduced temperature, fall on the same curve; this is another illustration of the law of corresponding states. The left-hand and upper scales apply to the upper curve, while the lower and right-hand scales apply to the lower curve. (From E. A. Guggenheim, *Thermodynamics*, 1967: Amsterdam, North-Holland Publishing Co., Fig. 3-10, p. 137)

simpler model is usable—the *ideal gas*. We shall examine this model first. It is, in effect, a gas of totally noninteracting and independent point-mass particles; the only property which differentiates one gas from another is the molecular mass (m). An individual molecule is also characterized by its velocity (v), momentum ($p = mv$), or kinetic energy ($mv^2/2$). A collection of such molecules with N molecules in a volume V is characterized by the **number density:**

$$n^* = \frac{N}{V} \tag{1.22a}$$

This number can be calculated from the ideal gas law with $n = N/L$; it should be easy to show:

$$n^* = \frac{PL}{RT} \tag{1.22b}$$

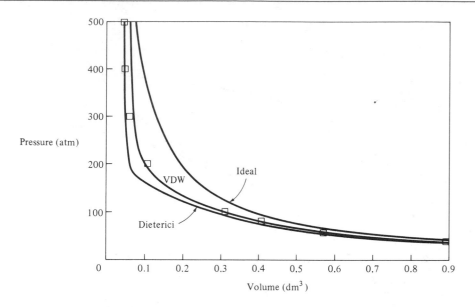

Figure 1.12   Pressure-volume data for ammonia at 473 K. The Dieterici and van der Waals equations follow the experimental data more closely than the ideal gas law, but neither is exact.

Example: Calculate the number density of an ideal gas at STP (standard temperature and pressure, i.e., 0°C and 1 atm).

$$n^* = \frac{(1 \text{ atm})(6.02217 \times 10^{23})}{(82.057 \text{ cm}^3 \text{ atm/K})(273.15 \text{ K})} = 2.6868 \times 10^{19} \text{ cm}^{-3}$$

(This is sometimes called *Loschmidt's number;* this term is also used for what we call Avogadro's number — hence the symbol, L, we have been using.) Alternatively, in SI units:

$$n^* = \frac{(101,325 \text{ Pa})(6.02217 \times 10^{23})}{(8.3143 \text{ J/K})(273.15 \text{ K})} = 2.6868 \times 10^{25} \text{ m}^{-3}$$

Using 10^6 cm^3/m^3, this gives the same answer, $n^* = 2.6868 \times 10^{19}$ cm^{-3}. The reader should check the units of this calculation as an exercise. ■

Velocity Space

The velocity of a particle is a vector quantity (**v**) which specifies the speed of the particle ($v = |\mathbf{v}|$) and its direction of motion. The direction of motion could be given by two angles, for example the polar angle (θ) and the azimuthal angle (ϕ). (Those who are not familiar with polar coordinates and related topics should cover Appendix III at this point.) The velocity of a particle can also be characterized in terms of its rate of motion with respect to the axes of a Cartesian coordinate system:

$$v_x = \frac{dx}{dt}, \qquad v_y = \frac{dy}{dt}, \qquad v_z = \frac{dz}{dt}$$

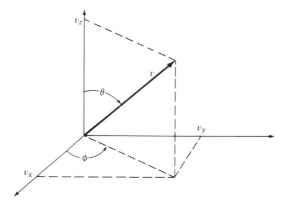

Figure 1.13   Velocity space. All possible velocities are represented by a point in this space. A vector from the origin to the point has a length that is equal to the speed of the particle and a direction that indicates the direction in which the particle is traveling.

These Cartesian velocity components are related to the speed and direction of motion (Appendix III) as:

$$v = \sqrt{v_x^2 + v_y^2 + v_z^2}$$

$$v_x = v \sin \theta \cos \phi$$

$$v_y = v \sin \theta \sin \phi$$

$$v_z = v \cos \theta$$

Note that the Cartesian components of the velocity may be positive or negative, depending on the direction of motion, while the speed (v) is always a positive number. (The Cartesian velocity could, in fact, be zero; this need not mean the molecule is not moving but may simply mean it is moving in a direction perpendicular to that axis.)

Imagine that, for a large collection of molecules, at some point in time we take a "snapshot" measurement of the speed and direction of travel of each molecule. We graph this set of numbers in a Cartesian system with axes labeled v_x, v_y, and v_z by plotting a point at the appropriate location. This is *velocity space* (Fig. 1.13). The distance of a given point from the origin represents its speed, and the polar coordinates (θ, ϕ) represent its direction of travel.

After making this point graph in velocity space, we could divide the space into small elements of volume with size $dv_x\, dv_y\, dv_z$ located at a point (v_x, v_y, v_z). The density of points in this element of volume represents the number of particles having velocities in the range v_x to $v_x + dv_x$, v_y to $v_y + dv_y$, v_z to $v_z + dv_z$. This set of numbers is a three-dimensional function $N(v_x, v_y, v_z)$ which can be used to define a **probability distribution function:**

$$F(v_x, v_y, v_z) = \frac{N(v_x, v_y, v_z)}{N(\text{total})} \tag{1.23}$$

If, overall, the gas is not moving, then the net velocity of the assembly must be zero. This means that if we stood at the origin in velocity space, the density of points, at any given distance, would be the same in any direction. In such a case, the motion is said to be *isotropic,* and the distribution function must be a function only of the speed and independent of θ and ϕ.

If we took another "snapshot" at a later time, any particular molecule could have an entirely different velocity because of collisions with the wall of the container or other molecules, but the distribution of points in velocity space would be the same. Thus, the distribution function is *stationary* — that is, independent of time. (This need not be true, for example, during an explosion; but if the gas is macroscopically uniform in density and temperature, it will be true for the average of a large enough number of particles.)

Collisions with a Wall

Consider the collisions of particles with a wall in the xy plane of a Cartesian space. The rate of progress of a molecule toward this wall is given by the z component of its velocity, $|v_z|$; if the molecule is above the wall, it must have a negative velocity ($v_z < 0$) if it is to hit the wall, otherwise it is going the wrong way.

Now consider a parallelepiped (Fig. 1.14) above the wall with height $|v_z| \Delta t$ and cross-sectional area A. If a molecule having the velocity $-|v_z|$ is anywhere in this volume, it will strike the wall sometime in the interval Δt. The size of this volume is, of course, different for each of the various molecules. However, if the density of particles is n^*, $n^*/2$ molecules per unit volume will be going toward the wall; furthermore, the volume of the parallelepiped is $A|v_z| \Delta t$. Defining $f(v_z)$ as the probability that a molecule has a velocity v_z (i.e., the fraction of the molecules in the gas

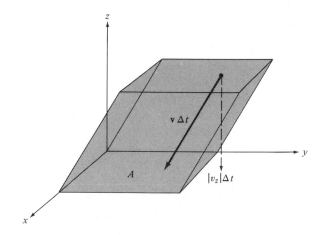

Figure 1.14   Collisions with a wall. All molecules in a parallelepiped with height $|v_z|\Delta t$ will strike the wall of area A in time Δt.

with that velocity), the number of collisions with the wall by molecules with the velocity v_z (i.e., those in the parallelepiped) is:

$$\# \text{ collisions with velocity } v_z = \frac{n^*}{2}(A|v_z|\Delta t)f(v_z) \tag{1.24}$$

What we want to know is the total number of collisions; to calculate this we must know the distribution function $f(v_z)$. Before doing this, we shall examine the implications of Eq. (1.24) regarding the pressure of the gas on the wall due to these collisions. For now, all we need is to realize that, for isotropic motion, the distribution for positive and negative velocities is the same; that is, $f(v_z)$ is an *even* function:

$$f(v_z) = f(-v_z)$$

The Pressure of a Gas

Pressure (P) is defined as the force (F) per unit area; according to Newton's law:

$$F = (\text{mass})(\text{acceleration})$$

$$= m\frac{dv}{dt} = \frac{dp}{dt}$$

where $p = mv$ is the momentum. A molecule heading toward the wall of Fig. 1.14 will have a momentum *before* collision:

$$p_x = mv_x, \qquad p_y = mv_y, \qquad p_z = -m|v_z|$$

and, after collision:

$$p_x = mv_x, \qquad p_y = mv_y, \qquad p_z = m|v_z|$$

so that the change in momentum for each collision is $2m|v_z|$. The total change of momentum for molecules of a given velocity $|v_z|$ is, using Eq. (1.24):

$$\Delta p = 2m|v_z|\left(\frac{n^*}{2}\right)A|v_z|\,\Delta t\,f(v_z)$$

$$= mv_z^2 f(v_z)n^* A\,\Delta t$$

This expression must be summed over all possible velocities for $-\infty < v_z < \infty$; the result involves a quantity which we shall call:

$$\langle v_z^2 \rangle = \sum_{-\infty}^{\infty} v_z^2 f(v_z) \tag{1.25}$$

(we shall soon identify this quantity as the *average value* of the squared velocity component). The pressure is then calculated from

$$P = \frac{1}{A}\frac{\Delta p}{\Delta t}$$

giving:

$$P = n*m\langle v_z^2\rangle \tag{1.26}$$

According to Eq. (1.22), the number density depends directly on the pressure, and, eliminating P between Eqs. (1.26) and (1.22), we get:

$$mL\langle v_z^2\rangle = RT$$

We now define $M = mL$ as the mass of one mole of particles — that is, the molecular weight — to get:

$$\langle v_z^2\rangle = \frac{RT}{M} \tag{1.27}$$

This is a very important result in that it shows a direct relationship between the average of a microscopic quantity, the velocity, and the macroscopic variable, temperature.

In the preceding derivation, our choice of the xy plane for the wall was arbitrary. Because of the isotropy of the molecular motion, the same derivation and the same result [Eq. (1.27)] would apply to the other velocity components, v_x and v_y. This, in turn, allows us to calculate the *speed*, whose square is:

$$v^2 = v_x^2 + v_y^2 + v_z^2$$

whose average will be:

$$\langle v^2\rangle = \langle v_x^2\rangle + \langle v_y^2\rangle + \langle v_z^2\rangle = 3\left(\frac{RT}{M}\right)$$

The square root of this quantity is the *root-mean-square* (rms) speed, for which we shall use the symbol u:

$$u \equiv \langle v^2\rangle^{1/2}$$

$$u = \sqrt{\frac{3RT}{M}} \tag{1.28}$$

This is the first of several answers we shall find to the question, "How fast are molecules moving?"

Example: Calculate the rms speed of N_2 molecules ($M = 0.028$ kg) at 298 K.

$$u = \left[\frac{3(8.3143 \text{ J/K})(298 \text{ K})}{0.028 \text{ kg}}\right]^{1/2} = 515 \text{ m/s}$$

For perspective, this is approximately the speed of sound in air at room temperature, and about 1000 miles per hour. The reader should check the units of this calculation as an exercise. ∎

The Average Kinetic Energy

The kinetic energy of a molecule is:

$$\frac{mv^2}{2} = \frac{m(v_x^2 + v_y^2 + v_z^2)}{2}$$

The average will be:

$$\text{ave. kinetic energy} = \frac{m\langle v^2 \rangle}{2} = \left(\frac{m}{2}\right)\left(\frac{3RT}{M}\right)$$

We will now introduce *Boltzmann's constant:*

$$k = \frac{R}{L} = 1.38062 \times 10^{-23} \text{ J/K}$$

and use $M = mL$ to get

$$\text{ave. kinetic energy} = \tfrac{3}{2}kT \qquad\qquad \textbf{(1.29)}$$

This is another important result: the translational kinetic energy is directly proportional to temperature and independent of mass. Note, however, that real molecules have many other types of energy, including internal (the motions of the electrons and nuclear masses *within* the molecule) and the energy due to intermolecular interactions.

1.6 The Maxwell Distribution Law

In Section 1.5 we referred several times to the probability distribution function for velocities—for example, in Eqs. (1.23) and (1.24). In this section we shall discover exactly what this function is. First, let's look at the general topic of probabilities and averages.

Probability and Averages

Consider a population of objects with some characteristic (denoted x); these could be, for example, a collection of sticks, with x being their length. If N_1 sticks have length x_1, N_2 have length x_2, and so on, the average length, $\langle x \rangle$, of the sticks is

$$\langle x \rangle = \frac{\sum_i N_i x_i}{N}$$

where $N = \sum_i N_i$ is the total number of sticks. The probability of having a length x_i is the fraction of sticks of that length:

$$f(x_i) = \frac{N_i}{N}$$

If this is substituted into the definition of the average value (above), we get:

$$\langle x \rangle = \sum_i x_i f(x_i)$$

Now, if x is a continuous variable, we can define a *probability distribution function,* $f(x)\,dx$, as the fraction of the population with the characteristic in the range x

to $x + dx$. Then, the sum in the definition of average value is replaced by an integral, giving:

$$\langle x \rangle = \int x f(x)\, dx$$

with

$$\int f(x)\, dx = 1$$

where these integrals are over the allowed range of x. The last equation simply states that all members have some characteristic in the range; it is often called the *nor-malization condition*, and such a distribution function is said to be *normalized*.

Now, consider a population with two characteristics — for example, a collection of boards having a length (x) and breadth (y). The number of boards with a particular length x_i *and* a particular breadth y_j is $N(x_i y_j)$, and the fraction is:

$$f(x_i, y_j) = \frac{N(x_i y_j)}{N(\text{total})}$$

This is a *joint* probability distribution function. The total probability of a particular x can be found by adding up all possibilities of y with that length:

$$f(x_i) = \sum_j f(x_i, y_j)$$

Then, the average length is

$$\langle x \rangle = \sum_i \sum_j x_i f(x_i, y_j)$$

The probabilities of the two characteristics are *independently distributed* if:

$$f(x_i, y_j) = f(x_i) f(y_j)$$

For example, suppose we had a population that was 40% female and 10% left-handed. If these characteristics are independent, we would expect to find $(0.40) \times (0.10) = 4\%$ of the population with the joint characteristic (female, left-handed); likewise we would anticipate 6% left-handed males, 36% right-handed females, and 54% right-handed males (total 100%, of course). On the other hand, if the population were 10% blond and 10% blue-eyed, we might find more than 1% blue-eyed blondes, since these characteristics are often linked, possibly through a particular ethnic origin.

For continuous variables we define $f(x, y)\, dx\, dy$ as the fraction with characteristics x to $x + dx$ *and* y to $y + dy$. Then the one-dimensional distribution functions are:

$$f(x)\, dx = \int_y f(x, y)\, dy\, dx, \qquad f(y)\, dy = \int_x f(x, y)\, dx\, dy$$

If the characteristics are independent:

$$f(x, y) = f(x) f(y)$$

Maxwell's Derivation

The velocity distribution function can be derived rigorously from quantum mechanics (the particle in a box, Chapter 11) and statistical mechanics (Chapters 5 and 15). It also follows directly from the Boltzmann law, which will be derived in Chapter 5. Here we shall follow a very simple derivation due to Maxwell, which starts with the presumption that the velocity components (v_x, v_y, v_z) are statistically independent in an isotropic gas (in the absence of external fields). This means that the joint distribution function of Eq. (1.23) can be written:

$$F(v_x, v_y, v_z) = f(v_x)f(v_y)f(v_z)$$

Taking the logarithm of this expression:

$$\ln F = \ln f(v_x) + \ln f(v_y) + \ln f(v_z) \tag{1.30}$$

Since v_x, v_y, and v_z are independent variables, we can differentiate this with respect to v_x, holding v_y and v_z constant:

$$\frac{d \ln F}{dv_x} = \frac{d \ln f(v_x)}{dv_x} \tag{1.31}$$

The squared velocity is:

$$v^2 = v_x^2 + v_y^2 + v_z^2$$

If this relationship is differentiated at constant v_y and v_z, we get:

$$2v\,dv = 2v_x\,dv_x$$

Combining this result with Eq. (1.31) gives:

$$\frac{d \ln F}{v\,dv} = \frac{d \ln f(v_x)}{v_x\,dv_x}$$

This derivation could have been done for any of the velocity components with the same result; therefore:

$$\frac{d \ln f(v_x)}{v_x\,dv_x} - \frac{d \ln f(v_y)}{v_y\,dv_y} = \frac{d \ln f(v_z)}{v_z\,dv_z} = \text{constant}$$

If we call this constant $-2b$ (thereby defining b), we get:

$$d \ln f(v_x) = -2bv_x\,dv_x$$

which integrates to:

$$\ln f(v_x) = \ln a - bv_x^2$$

in which $\ln a$ is a constant of integration. Therefore, the distribution function for v_x is:

$$f(v_x) = ae^{-bv_x^2}$$

(and similarly for v_y or v_z). We can eliminate one of the constants by requiring that the distribution function be normalized:

$$\int_{-\infty}^{\infty} f(v_x)\, dv_x = 1$$

$$a \int_{-\infty}^{\infty} e^{-bv_x^2}\, dv_x = 1$$

This is an integral which can be found on most standard tabulations of integrals (including that on the back cover of this book); the result is:

$$a \sqrt{\frac{\pi}{b}} = 1$$

Therefore, the constant a is related to b as

$$a = \sqrt{\frac{b}{\pi}}$$

and

$$f(v_x) = \sqrt{\frac{b}{\pi}}\, e^{-bv_x^2}$$

The constant b can now be evaluated using the results we obtained earlier [Eqs. (1.25) and (1.27)], which are, of course, the same for v_x as for v_z. The average value of the squared velocity was, from Eq. (1.25):

$$\langle v_x^2 \rangle = \int_{-\infty}^{\infty} v_x^2 f(v_x)\, dv_x$$

This can now be seen to be:

$$\langle v_x^2 \rangle = \sqrt{\frac{b}{\pi}} \int_{-\infty}^{\infty} v_x^2 e^{-bv_x^2}\, dv_x$$

This is, again, an integral found on most standard integral tables, so the relationship of the constant b to the mean squared velocity is readily shown to be:

$$\langle v_x^2 \rangle = \frac{1}{2b}$$

According to Eq. (1.27), $\langle v_x^2 \rangle = RT/M$, so we find that the constants are:

$$b = \frac{M}{2RT} = \frac{m}{2kT} \tag{1.32a}$$

$$a = \sqrt{M/2\pi RT} = \sqrt{m/2\pi kT} \tag{1.32b}$$

where $k = R/L$ is Boltzmann's constant and $m = M/L$ is the mass of one molecule.

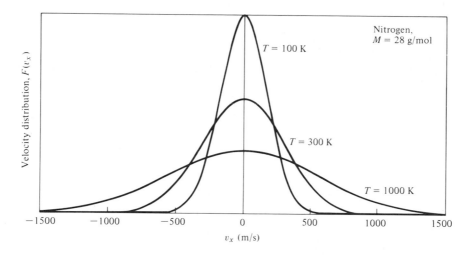

Figure 1.15 The distribution of Cartesian velocities for nitrogen at various temperatures. The most probable velocity is zero, but that does not necessarily indicate that the molecule is stationary; it may merely indicate that the molecule is moving perpendicular to that Cartesian axis so that its velocity projected onto that axis is not changing.

Finally, the desired distribution function is:

$$f(v_x)\, dv_x = \sqrt{\frac{m}{2\pi kT}}\, e^{-mv_x^2/2kT}\, dv_x \qquad (1.32c)$$

A graph of this function is shown in Fig. 1.15.

Collisions with a Wall: Knudsen Flow

As the first application of Eq. (1.32), we shall return to Eq. (1.24) and evaluate the number of collisions with a stationary object such as a wall. The result is important for a number of applications, including surface catalysis, where a molecule must hit and be adsorbed on the surface of the catalyst in order to react.

We define the collision frequency per unit area as:

$$Z_{\text{wall}} = \frac{\#\ \text{collisions}}{A\ \Delta t}$$

which, from Eq. (1.24), is (using v_x rather than v_z):

$$Z_{\text{wall}} = \frac{n^*}{2}\langle |v_x| \rangle$$

The average magnitude of the velocity is

$$\langle |v_x| \rangle = \int_{-\infty}^{\infty} |v_x| f(v_x)\, dv_x$$

with $|v_x| = v_x$ in the positive region and $|v_x| = -v_x$ in the negative region:

$$\langle |v_x| \rangle = \int_0^\infty v_x f(v_x) \, dv_x + \int_{-\infty}^0 (-v_x) f(v_x) \, dv_x$$

Since $f(v_x)$ is an even function of v_x, the second integral can be negated and its limits reversed to give:

$$\langle |v_x| \rangle = 2 \int_0^\infty v_x f(v_x) \, dv_x$$

$$Z_{\text{wall}} = n* \int_0^\infty v_x f(v_x) \, dv_x$$

The reader should evaluate this integral using Eq. (1.32) and a table of integrals to get:

$$Z_{\text{wall}} = n* \left(\frac{RT}{2\pi M} \right)^{1/2} \tag{1.33}$$

Example: A fingernail has an area of about one cm^2. Calculate the number of collisions it suffers in a second by the molecules of the air at room temperature (presumed to be 298 K). Air is a mixture of (roughly) 21% oxygen and 79% nitrogen, but we can treat it as a pure gas with an average molecule weight:

$$\overline{M} = 0.21(32) + 0.79(28) = 28.8 \text{ g}$$

The number density is:

$$n* = \frac{(1 \text{ atm}) (6.022 \times 10^{23})}{(82.057 \text{ cm}^3 \text{ atm/K}) (298 \text{ K})} = 2.46 \times 10^{19} \text{ cm}^{-3}$$

Since we want the answer in centimeters, it may be easier to use cgs units with $R = 8.3143 \times 10^7$ erg/K and erg $= $ g cm^2/s^2:

$$Z_{\text{wall}} = (2.46 \times 10^{19} \text{ cm}^{-3}) \left[\frac{(8.314 \times 10^7 \text{ erg/K}) (298 \text{ K})}{2\pi (28.8 \text{ g})} \right]^{1/2}$$

$$= 2.88 \times 10^{23} \text{ cm}^{-2} \text{ s}^{-1}$$

(Check units!) This is nearly half a mole of collisions in each second.   ■

Equation (1.33) can be easily verified experimentally. If the area of the wall is, in fact, a hole, all molecules which "collide" with the hole will escape, and the rate at which they escape provides a direct measure of the collision frequency. There are limitations on this experiment; if the hole is too large, the gas will whoosh out. What is required is a situation called molecular flow (or *Knudsen flow*) in which the molecule will leave the container only if, in the course of its wanderings, it happens to "collide" with the missing part of the wall. This requires that the *mean free path* of the gas (a term which will be defined in Chapter 9) be large compared to the size of the hole. Practical limitations on the size of the hole which can be drilled will limit the pressure in the container to ~ 1 torr or less.

This effect is utilized in the Knudsen method for determining vapor pressures. An amount of a substance of relatively low volatility is placed into a container which is covered with a plate in which a small hole has been drilled. This container is placed into another which is then evacuated with a vacuum pump. The mass loss (ΔW) of the container in time Δt is used to calculate a quantity

$$\mu \equiv \frac{\Delta W}{A \, \Delta t}$$

in which A is the area of the hole; this is then related to the vapor pressure of the material.

Exercise (optional): Show that the vapor pressure is related to the weight loss in the Knudsen experiment as:

$$P = \mu \left(\frac{2\pi RT}{M}\right)^{1/2} \tag{1.34}$$

∎

The Distribution Function for Speeds

We now know how the individual components of the velocity are distributed, but it is of somewhat more general interest to ask how the speed (v) is distributed. We have, for the density of points in velocity space, Eq. (1.23), with Eq. (1.32):

$$F(v_x, v_y, v_z) = a^3 e^{-bv_x^2} e^{-bv_y^2} e^{-bv_z^2}$$

$$= a^3 e^{-bv^2} \tag{1.35a}$$

As anticipated, the point density is independent of θ and ϕ. The probability of finding a velocity in the range v_x to $v_x + dv_x$, v_y to $v_y + dv_y$, and v_z to $v_z + dv_z$ is

$$F(v_x, v_y, v_z) \, dv_x \, dv_y \, dv_z = a^3 e^{-bv^2} \, dv_x \, dv_y \, dv_z \tag{1.35b}$$

What we really want to know is the probability of a speed between v and $v + dv$. In velocity space, these will be the points in the annulus between a sphere of radius v and that of radius $v + dv$. The volume of this annulus is:

$$\text{ann. vol.} = \frac{4\pi}{3}[(v + dv)^3 - v^3]$$

$$= \frac{4\pi}{3}[v^3 + 3v^2 \, dv + 3v(dv)^2 + (dv)^3 - v^3]$$

Neglecting the higher-order terms in the infinitesimal dv, this becomes:

$$\text{ann. vol.} = \frac{4\pi}{3}(3v^2 \, dv) = 4\pi v^2 \, dv$$

Multiplying this by the point density function, Eq. (1.35), we get the distribution function for speeds between v and $v + dv$:

$$F(v) \, dv = 4\pi a^3 e^{-bv^2} v^2 \, dv$$

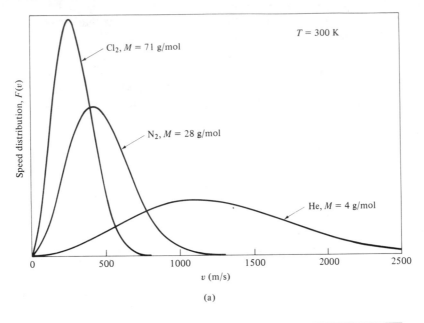

(a)

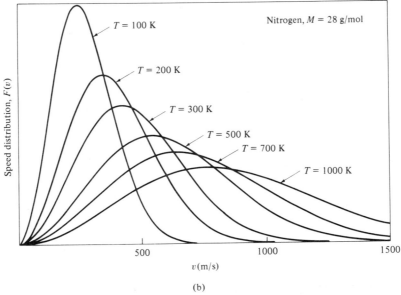

(b)

Figure 1.16 Speed distributions. (a) The distribution of speeds for several gases at 300 K. At a given temperature, the lighter molecules are moving faster. (b) The distribution of speeds for nitrogen at various temperatures. All speeds are found at all temperatures, but at the higher temperature more molecules will be found to be traveling at the faster speeds.

Using the values of a and b derived earlier [Eq. (1.32)]:

$$F(v) \, dv = 4\pi \left(\frac{m}{2\pi kT} \right)^{3/2} e^{-mv^2/2kT} v^2 \, dv \qquad (1.36)$$

This function is shown in Fig. 1.16.

An alternative derivation of Eq. (1.36) proceeds by deriving the relationship between a Cartesian volume element and that in spherical polar coordinates; this is done in Appendix III with the result:

$$dv_x \, dv_y \, dv_z = v^2 \, dv \, d\Omega$$

where the element of solid angle is:

$$d\Omega = \sin \theta \, d\theta \, d\phi$$

If this result is used with Eq. (1.35b) and integrated over all directions in space $(0 < \phi < 2\pi, \, 0 < \theta < \pi)$, Eq. (1.36) results.

Exercise: Show that the distribution function, Eq. (1.36), has a maximum value at the speed $v = v_p$, where v_p is the **most probable speed:**

$$v_p = \sqrt{\frac{2kT}{m}} \qquad (1.37)$$

∎

Exercise: Calculate the **average speed** (denoted $\bar{v}$):

$$\bar{v} = \langle v \rangle = \int_0^\infty v F(v) \, dv$$

The result should be:

$$\bar{v} = \sqrt{\frac{8RT}{\pi M}} \qquad (1.38)$$

∎

Exercise: Show that the ratio of the most probable, average, and rms average speeds is:

$$v_p : \bar{v} : u = 1 : 1.128 : 1.225 \qquad (1.39)$$

∎

From the meaning of a distribution function, we deduce that the fraction of molecules having speeds in a finite interval v_1 to v_2 is:

$$P(v_1 \text{ to } v_2) = \int_{v_1}^{v_2} F(v) \, dv$$

$$= 4\pi \left(\frac{m}{2\pi kT} \right)^{3/2} \int_{v_1}^{v_2} e^{-mv^2/2kT} v^2 \, dv \qquad (1.40a)$$

This is an integral which, between finite limits, must be calculated numerically (see Appendix I). For this reason it is a good idea to rewrite it in terms of the unitless variable:

$$w = \frac{v}{v_p}$$

with

$$v_p = \sqrt{\frac{2kT}{m}}$$

The reader should be able to show that, in terms of w, Eq. (1.40a) can be written:

$$P(v_1 \text{ to } v_2) = \frac{4}{\sqrt{\pi}} \int_{w_1}^{w_2} e^{-w^2} w^2 \, dw \tag{1.40b}$$

Example: Calculate the fraction of molecules with speeds between v_p and $2v_p$. From Eq. (1.40), with $v_1 = v_p$ ($w_1 = 1$), $v_2 = 2v_p$ ($w_2 = 2$):

$$P = \frac{4}{\sqrt{\pi}} \int_{1}^{2} e^{-w^2} w^2 \, dw$$

We shall integrate this with the trapezoidal rule and just four intervals:

w	$e^{-w^2} w^2$
1.00	0.367879
1.25	0.327578
1.50	0.237148
1.75	0.143235
2.00	0.073263

$$\text{Area} = \frac{4}{\sqrt{\pi}} \left(\frac{0.367879}{2} + 0.327578 + 0.237148 + 0.143235 + \frac{0.073263}{2} \right) \times (0.25)$$

$$= 0.523834$$

The accurate answer (obtained with a ten-step Simpson's rule) is 0.526393. The meaning is that 52.6% of the molecules have speeds in this interval. (Our first estimate, 52.4%, is close enough for most purposes.) ■

Conclusion

This ends, for the time being, our discussion of kinetic theory; the topic will be discussed further in Chapter 9. Several important ideas have been introduced which will be of more immediate utility: (1) The pressure exerted by a gas is due to the collisions of individual molecules with the walls of the container. We assumed that the molecules acted independently, and this assumption, strictly true for an ideal gas, must be modified for real gases. (2) The average kinetic energy for the translation of a molecule is $\frac{3}{2} kT$ [Eq. (1.29)]; it is proportional to temperature but independent of the molecular mass.

1.7   Intermolecular Forces

Picking up again our discussion of gas laws, in this section we shall discuss inter-molecular forces in more detail and learn how they can be related to the behavior of real gases. A number of properties can be related to intermolecular forces, including viscosity and other transport properties; we shall focus on PVT behavior as expressed in the virial coefficients.

The interaction of two molecules is usually discussed in terms of a *pairwise poten-tial function, U(r)*, which gives the potential energy of the pair as a function of the distance (*r*) between their centers. The *force* between the molecules is related to this potential as:

$$F = \frac{-dU}{dr}$$

—that is, it is related to the slope of the potential curve. At long range there is a weak attractive force between the molecules which obeys some sort of inverse power law; i.e., $F \propto -1/r^n$. At short range, generally when the electron clouds of the molecules begin to overlap, the force becomes very strongly repulsive. The potential function for a typical interaction is shown in Fig. 1.17.

Figure 1.17 also shows two molecular parameters which are utilized in the dis-cussion of intermolecular potentials, the well depth (ε) and the "diameter" σ. The latter parameter corresponds roughly to the hard-sphere diameter; the actual collision

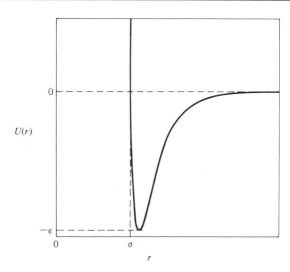

Figure 1.17   The intermolecular potential $U(r)$. The potential is a function of the dis-tance between the centers of the molecules (*r*). This figure illustrates two parameters found in all model potentials: the molecular diameter parameter (σ), which is the dis-tance between the centers of the molecules where the potential curve crosses zero, and the well-depth parameter (ε), which is the depth of the potential well at its minimum.

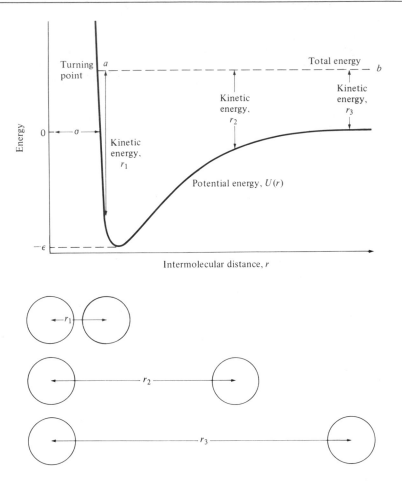

Figure 1.18 Energetics of a collision. The total energy (kinetic plus potential) remains constant, but as the potential energy drops when the molecules approach each other, the kinetic energy (that is, their relative velocity) increases; the velocity of approach is maximum where the potential is minimum, but decreases rapidly upon closer approach. The turning point is where the potential energy is equal to the total energy, and the relative velocity is zero; at this point the molecules reverse their velocity and fly apart. This demonstrates that the collision diameter, the distance of closest approach, will depend on the relative velocity of the colliding molecules.

diameter depends on the relative velocity of the molecules. We can readily see that if we threw a rubber ball against a rigid wall, the minimum distance between the center of the ball and the wall upon impact would depend on how hard the ball was thrown. A similar thing happens with molecules; consider a collision with a total energy, kinetic plus potential, as shown by the line *a-b* in Fig. 1.18. At $r = \infty$, this energy is all relative kinetic energy but, as they approach, the attractive forces accelerate them toward each other; as the potential energy decreases, the kinetic energy (relative velocity) increases such that the total is constant. The relative velocity is a maximum when $U(r)$ is a minimum, but at shorter distances the repulsive forces slow their approach. At point a in Fig. 1.18 the potential and total energy are the same, so the relative kinetic energy must be zero; this is the *turning point*, the brief pause before the molecules fly apart again. This is why, even for spherical objects such as argon atoms, the molecular "diameter" is a concept which must be interpreted carefully.

In most cases we would expect the potential to depend not only on the distance between the molecules, but on their relative orientation as well. For example, at a given distance between their centers of mass, the potential energy of interaction of a polar molecule such as HF would be quite different for the orientations:

$$HF \cdots HF, \qquad HF \cdots FH, \qquad HF \cdots \begin{matrix} H \\ F \end{matrix}, \qquad FH \cdots \begin{matrix} H \\ F \end{matrix}$$

The potentials which we shall discuss depend only on r. Such potentials are exact for He, Ar, Ne, and so on and a good approximation for CH_4, CCl_4, H_2, N_2, CO_2, CH_3CH_3, $CH_2{=}CH_2$, and others. They work less well for distinctly unsymmetrical molecules such as octane, or for polar molecules such as HF, H_2O or HCN.

Quantum mechanics and the theory of atomic and molecular structure can tell us a great deal about these intermolecular forces; this is an active area of research. However, the potential itself is not an experimental observable; rather, a form for the potential is used to calculate observables such as viscosity or virial coefficients as a function of temperature and is thereby compared to experiment. The ability of a proposed potential to fit a variety of data over a range of temperatures with the same set of parameters (e.g., σ and ε of Fig. 1.17) is an indicator of its generality. A number of semiempirical potentials have been proposed and used; we shall discuss a few of the simpler ones.

Most of the calculations referred to above are beyond the scope of this chapter; however, statistical mechanics provides a relatively simple formula for the calculation of the second virial coefficient from a spherically symmetric potential $U(r)$:

$$B(T) = 2\pi L \int_0^\infty (1 - e^{-U/kT}) r^2 \, dr \qquad \textbf{(1.41)}$$

(k is Boltzmann's constant.)

Example: Use Eq. (1.41) to calculate the second virial coefficient of a hard-sphere gas. The potential function is (Fig. 1.19):

$$U = 0 \quad \text{for} \quad r > \sigma, \qquad U = \infty \quad \text{for} \quad r \le \sigma$$

This potential assumes that the molecules know nothing about each other until they collide. Upon collision ($r = \sigma$) the potential jumps suddenly and the repulsive force, the slope of the potential, is infinite. This is clearly an unrealistic limiting case; even billiard balls have some elasticity. Nonetheless the results are revealing.

With a discontinuous potential, Eq. (1.41) must be evaluated in segments. First:

$$0 < r < \sigma, \qquad U = \infty, \qquad 1 - e^{-U/kT} = 1 - 0 = 1$$

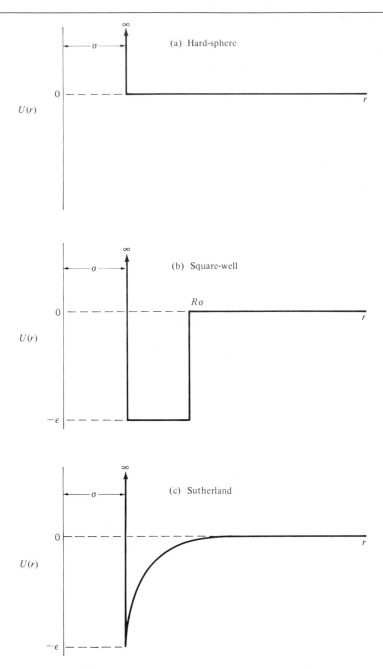

Figure 1.19 Some simple potential functions. (a) The hard-sphere potential is that between two perfectly inelastic spheres with radius σ. (b) The square-well potential gives a rough representation of attractive forces, but its major virtue is its mathematical simplicity. (c) The Sutherland potential combines mathematical simplicity with some degree of realism. The molecules are assumed to be hard spheres with an attractive force that is inversely proportional to some power of the intermolecular distance.

For this region, the integral of Eq. (1.41) is:

$$2\pi L \int_0^\sigma r^2\, dr = \frac{2\pi L\sigma^3}{3}$$

For $\sigma < r < \infty$, $U = 0$ and the integrand is zero:

$$1 - e^0 = 0$$

Therefore, this part of the integral makes no contribution to B. Finally, the second virial coefficient of a hard-sphere gas is:

$$B = \frac{2\pi L\sigma^3}{3} \tag{1.42}$$

Referring to Eq. (1.4), we see that this is just the van der Waals b constant. Also, remembering Eq. (1.18) for the second virial coefficient of a van der Waals gas, we see that Eq. (1.42) is the high-temperature limit of that expression. ∎

The Lennard-Jones Potential

A commonly used and reasonably realistic potential is that proposed by J. E. Lennard-Jones:

$$U(r) = 4\varepsilon\left[\left(\frac{\sigma}{r}\right)^{12} - \left(\frac{\sigma}{r}\right)^{6}\right] \tag{1.43}$$

The parameters in this potential (σ and ε) are those illustrated by Fig. 1.17. It has been quite successful in interpreting virial coefficients, viscosities, and other properties of nonpolar, reasonably spherical molecules. (However, the parameters which work best for virial coefficients differ somewhat from those which work well for viscosity and other transport properties.) Figure 1.20 shows some examples of this potential.

The integral required to calculate $B(T)$ with Eq. (1.41) can be done analytically, but the solution is complicated and in the form of an infinite series. It is a helpful simplification to switch to unitless variables; we shall do this with Eq. (1.41). First we define a unitless distance as the ratio of r to σ:

$$\rho = \frac{r}{\sigma}$$

Equation (1.41) then becomes:

$$B(T) = 2\pi L\sigma^3 \int_0^\infty (1 - e^{-U/kT})\rho^2\, d\rho$$

The result which this equation gave for the hard sphere [Eq. (1.42)] suggests that the collection of constants in front of this integral be used to define a parameter:

$$b_0 = \frac{2\pi L\sigma^3}{3} \tag{1.44}$$

Obviously, this looks exactly like the van der Waals b of Eq. (1.4); the distinction is a fine one, but very important. Equation (1.4) defines a quantity σ called the van der

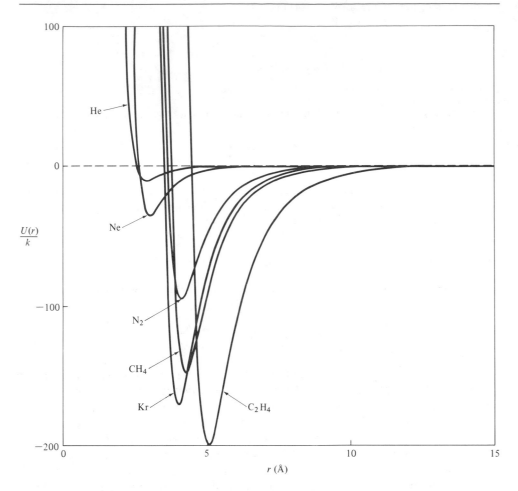

Figure 1.20 The Lennard-Jones potentials for a variety of molecules. This potential function gives a reasonable representation of the potential between closed-shell, non-polar, reasonably symmetrical molecules. The vertical scale is in kelvins (K); these may be compared to the figures in Chapter 13 that show the interaction potential of interaction between open-shell atoms; note that the depth of the open-shell potential is several orders of magnitude greater (1 cm^{-1} = 1.4388 K, 1 eV = 11,604 K).

Waals diameter; it is calculated from the constant b, which is chosen to make Eq. (1.3) work best for a given gas. Equation (1.44) uses the σ which is the value of a parameter that fits a proposed potential [e.g., Eq. (1.43)] to the experimental data. The meaning of σ is always the same, but the actual numerical value will depend on the manner by which it is measured. Table 1.6 gives some values of the "molecular diameter" σ as calculated by various models.

With this definition, Eq. (1.41) looks like:

$$B = 3b_0 \int_0^\infty (1 - e^{-U/kT})\rho^2 \, d\rho$$

Table 1.6 Molecular diameter parameter (σ) from various models
(units: nm)

	van der Waals	Square-well	Lennard-Jones second virial coefficient	Lennard-Jones viscosity
Ne	0.2384	0.2382	0.2749	0.279
Ar	0.2944	0.3162	0.3405	0.342
Kr	0.3160	0.3362	0.360	0.361

For the Lennard-Jones (LJ) potential, the exponent in the integral is:

$$\frac{U}{kT} = \frac{4(\varepsilon/k)}{T}[\rho^{-12} - \rho^{-6}]$$

We note that ε/k has units of temperature, i.e., kelvins. This suggests the use of two *reduced* parameters:

$$B^* = \frac{B}{b_0}, \qquad T^* = \frac{T}{(\varepsilon/k)} \tag{1.45}$$

With these definitions, Eq. (1.41) with the LJ potential is:

$$B^* = 3 \int_0^{\infty} \left\{ 1 - \exp\left[\frac{4(\rho^{-12} - \rho^{-6})}{T^*}\right] \right\} \rho^2 \, d\rho \tag{1.46}$$

From this it can be seen that the reduced virial coefficient is a universal function of the reduced temperature. This function has been calculated, and the results are shown graphically in Fig. 1.21.

Example: A simplified example of how the LJ constants can be calculated from experimental data will be given for the case of N_2. From Fig. 1.21 it can be seen that $B^* = 0$ (therefore $B = 0$) when T^* is about 3.5; this is the Boyle temperature. Actually, the exact Boyle temperature of a Lennard-Jones gas is $T^* = 3.42$; therefore:

$$\frac{\varepsilon}{k} \cong \frac{T_B}{3.42} \tag{1.47a}$$

For N_2, the Boyle temperature (Table 2.5) is 324 K, so:

$$\frac{\varepsilon}{k} \cong \frac{324}{3.42} = 94.7$$

Also, it can be shown (ref. 2, p. 244) that the critical temperature of a Lennard-Jones gas is $T^* \cong 1.30$; therefore, another estimate of the well depth is:

$$\frac{\varepsilon}{k} \cong \frac{T_c}{1.30} \tag{1.47b}$$

For N_2 (Table 1.1):

$$\frac{\varepsilon}{k} \cong \frac{126}{1.30} = 97 \text{ K}$$

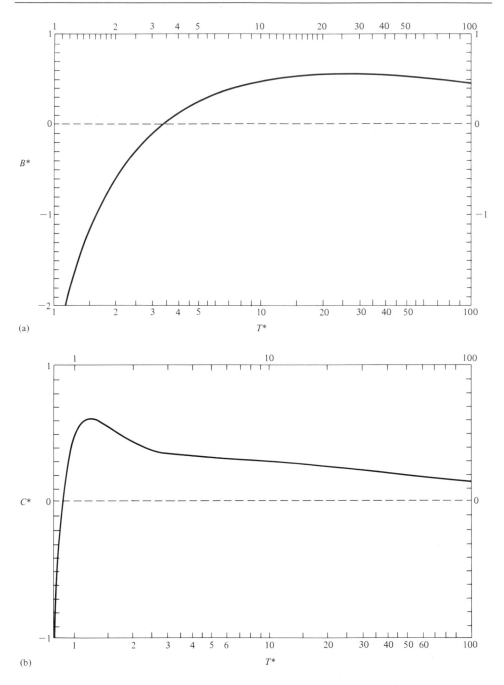

Figure 1.21 Lennard-Jones virial coefficients. (a) The second virial coefficient of a Lennard-Jones gas; compare this to Figures 1.2 and 1.11. (b) The third virial coefficient of a Lennard-Jones gas. These graphs will be useful for estimating second and third virial coefficients for gases whose Lennard-Jones parameters have to be measured.

Table 1.7   Intermolecular potential constants

	Lennard-Jones			Square-well			
	ε/k (K)	σ (nm)	b_0 (cm^3)	ε/k (K)	σ (nm)	b_0 (cm^3)	R
He	10.8	0.263	22.9	—	—	—	—
Ne	35.6	0.2749	26.20	19.5	0.2382	17.05	1.87
Ar	119.8	0.3405	49.79	69.4	0.3162	39.87	1.85
Kr	171	0.360	58.8	98.3	0.3362	47.93	1.85
H$_2$	29.2	0.287	29.8	—	—	—	—
N$_2$	95.0	0.3698	63.78	53.7	0.3299	45.29	1.87
O$_2$	117.5	0.358	57.9	—	—	—	—
CO	100.2	0.3763	67.21	—	—	—	—
CO$_2$	189	0.4486	113.9	119	0.3917	75.80	1.83
CH$_4$	148.2	0.3817	70.14	—	—	—	—
C$_2$H$_6$	243	0.3954	77.97	224	0.3535	55.72	1.652
C$_2$H$_4$	200	0.452	116	222	0.3347	47.29	1.667
C$_6$H$_6$	440	0.527	185	—	—	—	—

These estimates are reasonably close (they are not the same because the LJ potential is not exact); we shall use the average value and take ε/k = 96 K.

Table 1.3 gives two values for the second virial coefficient of N$_2$, one each above and below the Boyle temperature. At T = 223 K, T^* = 223/96 = 2.32 and B = -28.79 cm^3. From Fig. 1.21 we find B^* = $-.42$ at T^* = 2.32, so:

$$b_0 = \frac{B}{B^*} = \frac{-28.79}{-.42} = 68.5 \text{ cm}^3$$

At T = 473 K, T^* = 473/96 = 4.93 and B = 14.763 cm^3. Again from Fig. 1.21, B^* = 0.26 and

$$b_0 = \frac{14.763}{0.26} = 56.8 \text{ cm}^3$$

Taking the average of these two values gives b_0 = 63 cm^3 and [from Eq. (1.44)] σ = 3.7 × 10^{-8} cm = 0.37 nm. In summary, we find:

$$\frac{\varepsilon}{k} = 96 \text{ K}, \qquad \sigma = 0.37 \text{ nm}, \qquad b_0 = 63 \text{ cm}^3$$

These can be seen to be reasonably close to the accepted values (Table 1.7), which were derived using a larger data set of $B(T)$. ∎

Third Virial Coefficients

Third virial coefficients are much more difficult a problem from both the experimental and theoretical points of view. Experimentally, they are difficult to measure, and very few reliable data are available. From the theoretical point of view, $C(T)$ is much more sensitive to the assymmetry of the potential; that is, the assumption that $U(r)$ is independent of θ and ϕ causes much more error. Numerical calculations (which assume additivity of pairwise potentials) have been done for the LJ potential, and the results are shown on Fig. 1.21 for the reduced third virial coefficient:

$$C* = \frac{C}{b_0^2} \tag{1.48}$$

This graph is very useful for estimating $C(T)$ once the LJ parameters have been determined [for example, from $B(T)$ data]. Note that above the critical temperature ($T* = 1.30$), the third virial coefficient is positive and reasonably constant. Remember that the van der Waals equation gave $C = b^2$ [Eq. (1.18)]; from Fig. 1.21 we see that C is reasonably constant when $T > T_c$ ($0.2 < C* < 0.6$), so the van der Waals estimate is not far off. Recalling the earlier discussion of orders of approximation may make it clear that, for many purposes, an order-of-magnitude estimate for C is sufficient.

Example: Estimate $C(T)$ for N_2 at 223 and 473 K using the LJ parameters $\varepsilon/k = 95.0$ K and $b_0 = 63.78$ cm^3 (Table 1.7):

(a) $T = 223$, $T* = \dfrac{233}{95.0} = 2.35$, $C* = 0.40$ (Fig. 1.22)

$$C = 0.40(63.78)^2 = 1.6 \times 10^3 \text{ cm}^6$$

(b) $T = 473$, $T* = \dfrac{473}{95.0} = 4.98$, $C* = 0.32$

$$C = 0.32(63.78)^2 = 1.3 \times 10^3 \text{ cm}^6$$

(Compare, Table 1.3.)   ∎

Other Simple Potential Functions

Several other potential functions are used primarily because of their simplicity. One of these is the square-well potential (Fig. 1.19), which is defined as follows:

$$U = \infty \qquad \text{for } 0 < r < \sigma$$

$$U = -\varepsilon \qquad \text{for } \sigma < r < R\sigma$$

$$U = 0 \qquad \text{for } R\sigma < r < \infty$$

This is a three-parameter model, including a unitless constant R which sets the width of the well in multiples of σ. Some square-well constants are given in Table 1.7; note that R does not vary a lot and it is often given some reasonable value (e.g., 1.8) to make the square-well a two-constant potential. [If $R = 1.8$ is presumed, a well-depth parameter (ε) which is 0.56 times the Lennard-Jones parameter ε will give similar results for $B(T)$.]

Without too much difficulty it can be shown (Problem 1.32) that the second virial coefficient of a square-well is:

$$B(T) = b_0[1 - (R^3 - 1)(e^{\varepsilon/kT} - 1)] \tag{1.49}$$

[b_0 is as defined by Eq. (1.44).]

Another commonly used, simple potential function is the Sutherland potential (Fig. 1.19):

$$U = \infty \qquad \text{for } 0 < r < \sigma$$

$$U = -\varepsilon \left(\frac{\sigma}{r}\right)^n \qquad \text{for } \sigma < r < \infty$$

where n is a constant (usually with a value between 4 and 8). This potential maintains some of the simplicity of the square-well but permits a more realistic long-range attractive potential. (This function is used in Problem 1.31.)

Many other potential functions have been used, some for specialized cases — for example, the Corner potential for long molecules and the Stockmayer potential for polar molecules (ref. 2).

1.8   Mixed Gases

At this point we have the PVT properties of pure gases pretty well under control. However, mixtures of gases are a common occurrence; gaseous fuels (natural gas, synthetic gas, etc.) are generally mixtures, and any gases which are to be chemically reacted must be mixed at some point. If there are a thousand pure gases of interest, it is not impractical to determine empirical constants (e.g., van der Waals constants) for each of them; but these can form a million binary mixtures with all sorts of compositions, so the task of empirical study of mixtures rapidly becomes hopeless (although much work has been done on some important pairs, such as $H_2 + CO$). Therefore there is a big premium for being able to calculate properties of mixtures from those of the pure gases, and this is where the theory of the earlier sections becomes invaluable. As with pure gases, a variety of properties may be of interest — for example, viscosity or thermal conductivity — but we shall only discuss PVT properties.

The simplest assumption is that of an *ideal mixture*. In effect, the ideal mixing approximation assumes that the properties of a mixture are either the sum or the mean of the properties the pure gases would have under the same conditions. For the calculation of the PVT behavior of gas mixtures, the ideal mixing assumption can take two forms called Dalton's law and Amagat's law.

Dalton's Law

The pressure of a gas mixture will be the sum of the *partial pressures* (P_i) — that is, the pressures exerted by the individual gases:

$$P = \sum_i P_i \tag{1.50}$$

The partial pressure of a single gas in a mixture is not directly measurable; however, if we make the reasonable assumption (which follows directly from the kinetic theory) that the pressure of a gas is in proportion to the numbers of molecules present, then:

$$P_i = X_i P \tag{1.51}$$

where X_i is the *mole fraction;* for component 1 of a mixture:

$$X_1 = \frac{n_1}{n_1 + n_2 + n_3 + \cdots} \qquad (1.52)$$

Since, by definition,

$$\sum_i X_i = 1$$

substitution of Eq. (1.51) into (1.50) produces an identity.

Equation 1.50 [or (1.51)] is often referred to as Dalton's law; this is not quite correct, because Eq. (1.51) is really a definition of the partial pressure. Dalton's law refers strictly to how the partial pressure is calculated; it states that the partial pressure of a gas in a mixture is that which it would exert if it were *alone* in the container. For n_i moles at temperature T in a container with volume V, Dalton's law tells us to calculate (for each gas):

$$P_i = f(n_i, V, T)$$

with some convenient equation of state, ignoring the presence of the other gas. Then the total pressure is calculated with Eq. (1.50).

Amagat's Law

Amagat's law states that the volume of a gas mixture at some temperature T and (total) pressure P is the sum of the volumes which the gases would occupy if they were *pure* at the same (T, P). For n_i moles of one of the gases at (T, P) we calculate:

$$V_i = f(n_i, T, P)$$

with some convenient equation of state; then the total volume is:

$$V = \sum_i V_i \qquad (1.53)$$

Both of these laws are clearly exact for ideal gases; how they differ for real gases is not at all obvious. This will be developed shortly, but first the use of these laws will be illustrated with an example.

Example: We shall calculate the pressure exerted by a mixture of $n_1 = 4.00$ moles of NH_3 and $n_2 = 2.00$ moles of N_2 in a container with $V = 2.40$ dm^3 at 500 K, first by the ideal gas, then Dalton's, then Amagat's law.

(a) *Ideal gas.* The total moles are $n = 2 + 4 = 6$ and

$$P = \frac{nRT}{V} = \frac{(6)(0.08206)(500)}{2.40} = 103 \text{ atm}$$

(b) *Dalton's law.* If the NH_3 were alone in the container, it would have a molar volume, $V_m = 2.4/4 = 0.6$ dm^3. Using the van der Waals equation (constants, Table 1.1):

$$P_{NH_3} = \frac{0.08206(500)}{0.60 - 0.0371} - \frac{4.17}{(0.60)^2} = 61.3 \text{ atm}$$

For N_2, $V_m = 2.4/2 = 1.2$ dm^3;

$$P_{N_2} = \frac{0.08206(500)}{1.20 - 0.0391} - \frac{1.39}{(1.20)^2} = 34.4 \text{ atm}$$

The total pressure is therefore:

$$P = 61.3 + 34.4 = 95.7 \text{ atm}$$

(c) *Amagat's law*. Imagine that the container is divided by a partition, and the 4 moles of NH_3 occupy two-thirds of the volume (1.60 dm^3; the molar volume would be $V_m = 1.60/4 = 0.40$) while the 2 moles of N_2 occupy the other third ($V = 0.80$ dm^3, $V_m = 0.8/2 = 0.40$). The reader should use the van der Waals equation to calculate the pressure in each section; the results should be:

$$P(NH_3, \text{ alone}) = 87.0 \text{ atm}$$

$$P(N_2, \text{ alone}) = 105.0 \text{ atm}$$

If the partition were now removed, we would expect the resulting pressure to be a weighted average of these two:

$$P = \tfrac{2}{3}(87.0) + \tfrac{1}{3}(105.0) = 93.0 \text{ atm}$$

Note that the pressures calculated above are *not the partial pressures*. By Eq. (1.51), we can calculate the *partial pressure* of NH_3 as:

$$P_{NH_3} = \tfrac{2}{3}(93.0) = 62.0 \text{ atm}$$

This can be contrasted to our earlier estimate [part (b)] of 61.3 atm. The actual partial pressure of NH_3 is two-thirds of the actual total pressure, but we don't know what that is without doing the experiment.

We shall return to this example later and calculate the pressure by yet another method. ∎

Virial Coefficients of Mixtures

The statistical theory of intermolecular interactions which led to the virial equation can also be applied to mixtures. For a two-component mixture with molecules of types A and B, the calculation of the second virial coefficient must consider three types of pairwise interactions: A-A, A-B and B-B. The result is:

$$B = X_1^2 B_1 + X_2^2 B_2 + 2X_1 X_2 B_{12} \tag{1.54}$$

where B_1 and B_2 are the second virial coefficients of the pure gases, and B_{12} is the second virial coefficient calculated with the potential function U_{12} for interactions between the unlike molecules. Relatively few data are available concerning the interactions of unlike molecules, so the potential constants are generally estimated from those of the like-to-like interactions as follows:

$$\sigma_{12} = \frac{\sigma_1 + \sigma_2}{2}$$

$$b_{012} = \frac{(b_{01}^{1/3} + b_{02}^{1/3})^3}{8}$$

$$\varepsilon_{12} = \sqrt{\varepsilon_1 \varepsilon_2} \tag{1.55}$$

With this set of constants, the cross-virial coefficient B_{12} can be calculated using any of the methods used in the previous section for pure gases. The Lennard-Jones model (using Fig. 1.21) and the square-well [Eq. (1.49)] are convenient. (For square-well, the arithmetic mean of R can be used.)

Example: Calculate B_{12} for mixtures of He and O_2 at 300 K using the Lennard-Jones potential. With constants from Table 1.7:

$$\frac{\varepsilon_{12}}{k} = [10.8(117.5)]^{1/2} = 35.62$$

$$T^* = \frac{300}{35.62} = 8.42$$

From Fig. 1.21 we read $B^* = 0.37$
   Also with data from Table 1.7:

$$\sigma_{12} = \frac{0.263 + 0.358}{2} = 0.3105 \text{ nm} = 3.105 \times 10^{-8} \text{ cm}$$

$$b_0 = \frac{2\pi(3.105 \times 10^{-8} \text{ cm})^3(6.022 \times 10^{23})}{3} = 37.76 \text{ cm}^3$$

Therefore:

$$B_{12} = b_0 B^* = (37.76 \text{ cm}^3)(0.37) = 14 \text{ cm}^3 \qquad \blacksquare$$

Example: Show that Dalton's law corresponds to assuming $B_{12} = 0$. From the virial equation (1.11), neglecting all but the B term, we get for each gas (alone in volume V):

$$P_1 V = n_1 RT + \frac{n_1^2 RT B_1}{V}$$

$$P_2 V = n_2 RT + \frac{n_2^2 RT B_2}{V}$$

With $P = P_1 + P_2$:

$$PV = (n_1 + n_2)RT + \frac{RT}{V}(n_1^2 B_1 + n_2^2 B_2)$$

Dividing by $(n_1 + n_2)$ and defining $V_m = V/(n_1 + n_2)$:

$$\frac{PV_m}{RT} = 1 + (X_1^2 B_1 + X_2^2 B_2)\frac{1}{V_m}$$

Comparison to Eq. (1.54) shows that this is the result if $B_{12} = 0$; in effect, Dalton's law ignores interactions of unlike molecules. $\qquad \blacksquare$

Example: Show that Amagat's law corresponds to assuming that the forces between unlike molecules are similar to those between like molecules, and that the cross-virial coefficient is the *mean*:

$$B_{12} = \frac{B_1 + B_2}{2}$$

To prove this we start with Eq. (1.13) (remembering that β and B are the same and neglecting higher-order terms); the volumes of the pure gases are:

$$V_1 = \frac{n_1 RT}{P} + n_1 B_1$$

$$V_2 = \frac{n_2 RT}{P} + n_2 B_2$$

From Eq. (1.53), the volume of the gases together is:

$$V = V_1 + V_2 = \frac{(n_1 + n_2)RT}{P} + n_1 B_1 + n_2 B_2$$

$$V_m = \frac{RT}{P} + (X_1 B_1 + X_2 B_2)$$

The virial coefficient of the mixture is therefore:

$$B = X_1 B_1 + X_2 B_2$$

If, in Eq. (1.54), we use $B_{12} = (B_1 + B_2)/2$, we get:

$$B = X_1^2 B_1 + X_2^2 B_2 + X_1 X_2 B_1 + X_1 X_2 B_2$$
$$= X_1(X_1 + X_2)B_1 + X_2(X_1 + X_2)B_2$$

Since $X_1 + X_2 = 1$, this is the same result as Amagat's law.   ∎

The significance of the preceding examples is that, in a mixture of gases (red and green), Dalton's law calculates the properties of a gas (red) assuming the other gas (green) is not there, whereas Amagat's law calculates the properties of the red molecules assuming that the green molecules are there and interact with the reds just as do other red molecules. Put another way, the ideal gas law assumes that molecules are blind, they "see" no other molecules; Dalton's law assumes that the molecules are selectively blind, they see only molecules of their own "color"; Amagat's law assumes that the molecules are colorblind, they "see" all molecules but think they are the same color.

A good way to keep these laws straight is to remember that (referring to properties of pure gases) Dalton's law calculates intensive properties (e.g., pressure) as a sum and extensive properties as a weighted mean; Amagat's law calculates extensive properties [e.g., volume, Eq. (1.53)] as a sum and intensive properties as a weighted mean.

van der Waals Equation

Another approach to the calculation of PVT properties is to use the van der Waals equation with, for two components:

$$n = n_1 + n_2 \quad \text{or} \quad V_m = \frac{V}{n_1 + n_2}$$

and constants:

$$a = (a_1^{1/2}X_1 + a_2^{1/2}X_2)^2$$
$$b = b_1 X_1^2 + b_2 X_2^2 + 2b_{12} X_1 X_2 \tag{1.56}$$

where $a_1, b_1, \ldots$ are the van der Waals constants of the pure gases and:

$$b_{12} = \frac{(b_1^{1/3} + b_2^{1/3})^3}{8} \tag{1.57}$$

Example: Again, picking up the earlier example with 4 moles of NH_3 and 2 moles of N_2 in $V = 2.40$ dm^3 at 500 K, we calculate:

$$a = (\tfrac{2}{3}\sqrt{4.17} + \tfrac{1}{3}\sqrt{1.39})^2 = 3.08 \text{ dm}^6 \text{ atm}$$
$$b_{12} = \frac{(0.0371^{1/3} + 0.0391^{1/3})^3}{8} = 0.0381$$
$$b = \tfrac{4}{9}(0.0371) + \tfrac{1}{9}(0.0391) + 2(\tfrac{2}{9})(0.0381) = 0.0378$$

From Eq. (1.3) with $n = 6$:

$$P = \frac{6(0.08206)(500)}{2.40 - 6(0.0378)} - \frac{36(3.08)}{(2.40)^2} = 94.0 \text{ atm}$$

To summarize, the answers by the techniques used are:

ideal gas:	102.6
Dalton's law:	95.7
Amagat's law:	93.0
van der Waals (mixed):	94.0

■

Example: The compressibility factors of a 2:1 mixture of H_2 and CO have been measured [Townsend and Bhatt, *Proc. Roy. Soc. (London)*, A, 134, 502 (1931)]; at 25°C and 300 atm, $z = 1.2968$. From this we can calculate the volume occupied by 2 moles of H_2 mixed with 1 mole of CO at 300 atm (total) as: $z = PV_m/RT$, $V_m = zRT/P$:

$$V = \frac{nzRT}{P} = \frac{(3)(1.2968)(0.082057)(298.15)}{300} = 0.3173 \text{ dm}^3$$

Now let's calculate this volume with Amagat's law, Dalton's law, and the van der Waals equation for mixed gases.

Amagat's law. We calculate the volume which each gas would occupy if the total pressure were due to that gas. For CO ($a = 1.49$, $b = 0.0399$) Eq. (1.9) (solved as before by successive approximations) with $n = 1$, $P = 300$, $T = 298.15$ K gives $V = 0.0908$ dm^3. For H_2 ($a = 0.244$, $b = 0.0266$) Eq. (1.9) with $n = 2$, $P = 300$, $T = 298.15$ K gives $V = 0.2045$ dm^3. The total volume is, from Eq. (1.53):

$$V = 0.0908 + 0.2045 = 0.2953 \text{ dm}^3 \text{ (error, } -7\%)$$

Dalton's law. The partial pressure of H_2 is 200 atm. If it were alone in the container, its volume calculated with Eq. (1.9) using $n = 2$, $P = 200$ atm, $T = 298.15$ K is $V = 0.2838$

dm^3. The partial pressure of CO is 100 atm. If it were alone, its volume, calculated with Eq. (1.9) using $n = 1, P = 100$ atm, $T = 298.15$, would be $V = 0.2312$. The weighted mean of these volumes is:

$$V = \tfrac{2}{3}(0.2828) + \tfrac{1}{3}(0.2312) = 0.2663 \text{ dm}^3 \text{ (error, } -16\%)$$

Note that this is less accurate than Amagat's law, a consequence of its less realistic assumptions.

van der Waals. For the mixed van der Waals, Eqs. (1.56) and (1.57) give:

$$b_{12} = 0.0328, \quad a = 0.542, \quad b = 0.0308$$

With these constants:

$$V_m = \frac{0.082057(298.15)}{300 + \dfrac{0.542}{V_m^2}} + 0.0308$$

This can be solved by successive approximations to give:

$$V_m = 0.09984 \text{ dm}^3, \quad V = 0.2995 \text{ dm}^3 \text{ (error, } -6\%) \qquad \blacksquare$$

1.9 PVT Behavior in Condensed Phases

Although a good bit is known concerning equations of state for dense fluids, the approach which we shall use for calculating the relationships among the variables P, V, and T in dense fluids (liquids) will be an empirical one which is equally useful for solids.

For any substance, the volume $V(T, P)$ can be analyzed in terms of the slope formula (Appendix II):

$$dV = \left(\frac{\partial V}{\partial T}\right)_P dT + \left(\frac{\partial V}{\partial P}\right)_T dP \qquad \textbf{(1.58)}$$

The slopes in this formula will themselves be functions of T and P, but for small changes of (T, P) in condensed phases, they are usually sufficiently constant to be considered a parameter. The change in volume, dV, is an extensive property like the volume, so it is often better to discuss the fractional volume change, dV/V:

$$\frac{dV}{V} = \frac{1}{V}\left(\frac{\partial V}{\partial T}\right)_P dT + \frac{1}{V}\left(\frac{\partial V}{\partial P}\right)_T dP$$

The slopes in this formula are called, respectively, the *coefficient of thermal expansion:*

$$\alpha \equiv \frac{1}{V}\left(\frac{\partial V}{\partial T}\right)_P \qquad \textbf{(1.59)}$$

and the *isothermal compressibility:*

$$\kappa_T = -\frac{1}{V}\left(\frac{\partial V}{\partial P}\right)_T \qquad \textbf{(1.60)}$$

(The minus sign is included in this definition so that the coefficients, κ_T, will be positive numbers. The *compressibility* [Eq. (1.60)] should not be confused with the *compressibility factor* introduced earlier [z, Eq. (1.10)]; the use of the same word for two different concepts is regrettable but unavoidable.)

Exercise: Although they are most useful for condensed phases, Eqs. (1.59) and (1.60) apply equally to gases. Show that, for an ideal gas:

$$\kappa_T = \frac{1}{P}, \qquad \alpha = \frac{1}{T} \qquad \textbf{(1.61)}$$

■

Solids and liquids are very incompressible compared to gases. The change in volume with pressure (at constant T) can be written as a Taylor series about some point $P_0 V_0$ at which the volume is known (normally $P_0 = 1$ atm):

$$V(P) = V_0 + \left(\frac{\partial V}{\partial P}\right)_0 (P - P_0) + \frac{1}{2}\left(\frac{\partial^2 V}{\partial P^2}\right)_0 (P - P_0)^2 + \cdots$$

Keeping only the first-order term, this gives:

$$V(P) = V_0[1 - \kappa_T(P - P_0)] \qquad \textbf{(1.62)}$$

Because changes in volume with pressure are so small, this equation is reasonably accurate up to very high pressures.

Example: With data from Table 1.8, estimate the molar volume of CCl_4 at 100 atm. From Eq. (1.62):

$$V = 97[1 - 91.0 \times 10^{-6}(99)] = 96 \text{ cm}^3 \qquad ■$$

Because these changes are so small, we shall frequently be able to assume that liquids and solids are *incompressible*—that is, that their volume will not change with pressure.

The coefficient of thermal expansion (α) is, strictly, the coefficient of *bulk* expansion; for fluids and isotropic solids it is equal to three times the coefficient of linear expansion. This coefficient may change quite rapidly with temperature in some cases—for example, for water near its freezing point (Table 1.9). For small temperature ranges in which α can be considered to be constant, the Taylor series can be used about any known point $V_0 T_0$ to give:

$$V(T) = V_0[1 + \alpha(T - T_0)] \qquad \textbf{(1.63)}$$

Another important slope on the PVT surface can be calculated using the cyclic rule for partial derivatives (Appendix II); for the variables PVT, this is:

$$\left(\frac{\partial P}{\partial T}\right)_V = -\frac{\left(\dfrac{\partial V}{\partial T}\right)_P}{\left(\dfrac{\partial V}{\partial P}\right)_T}$$

Table 1.8   Coefficients of thermal expansion and compressibility*

(at 20°C unless otherwise specified)

	$10^3\alpha$ (K^{-1})	$10^6\kappa_T$ (atm^{-1})	V_m (cm^3)	C_{pm} (J K^{-1})	C_{vm} (J K^{-1})
Benzene	1.237	63.5	89.	134	92.0
CCl$_4$	1.236	91.0	97.	132	89.5
CCl$_3$H	1.273	90.0	80.2	117	—
Mercury	0.18	3.9	14.8	27.8	23.4
Ag	0.0567	0.98	10.3	25.3	
Cu (-190°C)	0.0270	0.71	—	13.3	—
(20°C)	0.0492	0.76	7.1	24.4	—
(500°C)	0.0600	0.91	—	27.5	—
Zn	0.0893	0.15	7.1	25.0	24.0
H$_2$O (0°)	-0.0547	47	18	76.0	—
(20°)	0.188	45	18	75.3	—
Ice (0°)	0.11		8.8	37.1	—
Acetone	1.487	112.0	45.9	128.0	—
Ethyl ether	1.656	169(13°)	54.5(25°)	169(30°)	—

*Data from N. A. Lange, *Handbook of Chemistry,* 9th Ed., Handbook Publishers, Inc., 1956, Sandusky, Ohio, where more values can be found.

Table 1.9   Specific volume and coefficient of thermal expansion of water

t (°C)	v (cm^3/g)	α (K^{-1})
0	1.0002	-5.47×10^{-5}
4	1.0001	-3.95×10^{-7}
5	1.0001	1.27×10^{-5}
10	1.0003	7.54×10^{-5}
15	1.0009	1.34×10^{-4}
20	1.0017	1.88×10^{-4}
25	1.0027	2.39×10^{-4}
30	1.0041	2.86×10^{-4}
35	1.0056	3.30×10^{-4}
37	1.0063	3.47×10^{-4}
40	1.0074	3.71×10^{-4}
50	1.0115	4.46×10^{-4}
60	1.0164	5.11×10^{-4}
70	1.0219	5.67×10^{-4}
80	1.0279	6.17×10^{-4}
90	1.0345	6.60×10^{-4}
100	1.0416	6.98×10^{-4}

From the definitions (1.59) and (1.60) we see:

$$\left(\frac{\partial P}{\partial T}\right)_V = \frac{\alpha}{\kappa_T} \qquad \textbf{(1.64)}$$

Example: When a mercury-in-glass thermometer is overheated, the top will break off because of the expansion of the mercury. A thick, well-tempered glass tube may be able to withstand ~50 atm pressure without breaking. How far can a thermometer be heated past the temperature at which the capillary is filled until the pressure becomes this large?

Assuming α and κ_T to be constant, Eq. (1.64) can be written for finite change at constant volume as:

$$\Delta T = \left(\frac{\kappa_T}{\alpha}\right) \Delta P$$

With data from Table 1.8:

$$\Delta T = (3.9 \times 10^{-6} \text{ atm}^{-1})(50 \text{ atm})/(0.18 \times 10^{-3} \text{ K}^{-1}) = 1.1 \text{ K} \qquad \blacksquare$$

Postscript

In this chapter, we have seen that the physical properties of matter are mathematical functions; specifically, for pure substances, any intensive property will be a function of two variables, for example temperature and pressure. We have also seen that these properties are the result of the properties and actions of the constituent molecules, although only gases were discussed. The imperfections of gas behavior, their deviations from ideal gas behavior, are due to intermolecular forces. In the next several chapters, many other properties of matter will be introduced and discussed. Then, in Chapter 5, we shall examine in more detail how macroscopic properties are related to molecular properties.

Reference 5 gives a very interesting discussion of the fluid phases of matter, including a brief historical sketch and a discussion of recent research in this field. Reference 1 is a more detailed but relatively simple discussion of many of the topics in this chapter, particularly gas laws and kinetic theory. Reference 3 is more advanced but has a good introduction to probability and statistics. Reference 2 is the last word in the field, but it is not for beginners.

References

1. W. Kauzmann, *Kinetic Theory of Gases,* 1966: New York, W. A. Benjamin, Inc.
2. J. O. Hirschfelder, C. F. Curtiss, and R. B. Bird, *Molecular Theory of Gases and Liquids,* 1954: New York, John Wiley & Sons, Inc.
3. F. C. Andrews, *Equilibrium Statistical Mechanics,* 1963: New York, John Wiley & Sons, Inc.
4. O. A. Hougen and K. M. Watson, *Chemical Process Principles* (Charts), 1947: New York, John Wiley & Sons, Inc.
5. J. A. Barker and D. Henderson, "The Fluid Phases of Matter," *Scientific American,* November 1981, p. 130.

Problems

1.1 Dry air is roughly 79% N_2 and 21% O_2. Calculate its average molecular weight and density at STP using the ideal gas law.

1.2 Use the van der Waals equation to calculate the pressure exerted by SO_2 at 500 K if the density is 100 g/dm^3.

1.3 Calculate the molecular diameter (σ) of CO_2 from its van der Waals constant.

1.4 Calculate the pressure exerted by 3.00 moles of CO_2 in a 9.00-dm^3 container at 400 K using (a) ideal gas, (b) van der Waals. (c) Repeat the calculation for a volume of 2.00 dm^3.

1.5 Use the Dieterici equation to calculate the pressure exerted by 3.00 moles of CO_2 at 400 K in a container with volume (a) 9.00 dm^3, (b) 2.00 dm^3.

1.6 Calculate the molar volume of CH_4 at 298 K and 10.0 atm using (a) ideal gas, (b) van der Waals equation of state.

1.7 Use the van der Waals equation to calculate the volume occupied by 5.00 moles of NH_3 at 300 K, 7.00 atm.

1.8 Show that the Dieterici equation is nearly identical to the van der Waals equation at high temperatures or low densities (i.e., when $a/RTV_m \ll 1$).

1.9 The data below for acetylene at 25°C give the PV product divided by P_0V_0 (at 0°C and 1 atm). Use a graphical or least-squares method to determine P_0V_0 and the second virial coefficient at this temperature.

P (atm)	(PV/P_0V_0)	P (atm)	(PV/P_0V_0)
0.5	1.0989	6	1.0531
1	1.0937	8	1.0385
2	1.0841	10	1.0255
4	1.0684	12	1.0139

1.10 Use the measured compressibility factors given below for methane at 203 K to calculate the second virial coefficient. You could also get an estimate of the third virial coefficient (C) from these data.

P (atm)	z
1	0.9940
10	0.9370
20	0.8683
30	0.7928

1.11 The virial coefficients [for Eq. (1.13)] of hydrogen at 223 K are given below (PV in dm^3 atm); use them to calculate the compressibility of this gas at 50 atm.

$$B = 1.2027 \times 10^{-2}, \qquad \delta = -1.741 \times 10^{-8}$$

$$\gamma = 1.164 \times 10^{-5}, \qquad \varepsilon = 1.022 \times 10^{-11}$$

1.12 Calculate the second virial coefficient of N_2 at 473.15 K using (a) van der Waals, (b) Beattie-Bridgeman, (c) Berthelot equations. The observed value is 14.76 cm^3.

1.13   Calculate the Boyle temperature of argon using (a) van der Waals and (b) Berthelot forms of the second virial coefficient. (c) Use the Beattie-Bridgeman form of the second virial coefficient to calculate the Boyle temperature of argon. The actual value is 410 K.

1.14   Calculate the Boyle temperature of He using the (a) van der Waals, (b) Berthelot, (c) Beattie-Bridgeman forms for $B(T)$.

1.15   Derive an expression for the second virial coefficient of a gas in terms of the Dieterici constants from Eq. (1.6).

1.16   Use the data below to determine the critical constants of Cl_2.

t (°C)	Liquid density (g/cm^3)	Vapor density (g/cm^3)
98.9	1.115	0.124
104.4	1.087	0.139
110.0	1.057	0.156
115.6	1.025	0.179
121.1	0.989	0.203
126.7	0.949	0.231
132.2	0.894	0.268
137.8	0.814	0.321
143.3	0.599	0.523

1.17   Use the Berthelot equation (1.5) at the critical point to derive relationships between the critical constants and the constants a and b.

1.18   Use the law of corresponding states (Fig. 1.10) to calculate the molar volume of NO at 165 K and 19.5 atm.

1.19   Silicon tetrafluoride (SiF_4) has a critical temperature 259.1 K and critical pressure 36.7 atm. Calculate the van der Waals constants; then use the van der Waals equation to calculate the vapor density of this gas at STP.

1.20   (a) Calculate the Dieterici constants of methane from the critical constants.
(b) Use these constants to calculate the pressure of methane when $T = 270$ K, $V_m = 0.1$ dm^3.
(c) Use successive approximations and the Dieterici equation to calculate V_m when $P = 10$ atm and $T = 270$ K.

1.21   Calculate the number density $n*$ of an ideal gas at 298 K and 1 atm.

1.22   Calculate the number of collisions which hydrogen molecules would make with 1 cm^2 of a wall in one second at 150 K and a pressure of (a) 1 torr, (b) 1 atm.

1.23   In a Knudsen experiment, a substance with a molecular weight of 0.210 kg is placed in a cell with a hole of area 3×10^{-5} m^2. At 500 K, the weight lost in 10 minutes is 30 mg. Calculate the vapor pressure of this substance.

1.24   Derive a formula for the speed distribution of a two-dimensional gas. This has application in the study of adsorbed species which may have freedom of motion about the surface of the adsorbent.

1.25   Calculate the average and rms speeds of N_2 at 500°C.

1.26   Calculate the average, rms, and most probable speeds for CO_2 at 300 K.

1.27   Calculate the fraction of molecules in a gas which have velocities greater than $3v_p$. The required integral can be done numerically (Appendix I); it will be sufficiently accurate to use five steps for $3 < (v/v_p) < 3.5$ and five steps for $3.5 < (v/v_p) < 4.5$; above 4.5 the contribution is insignificant.

1.28 Calculate the fraction of molecules in a gas which has a kinetic energy greater than $10kT$.

1.29 Find the distance r/σ at which the Lennard-Jones potential $U(r)$ is a minimum. Show that the value of $U(r)$ at the minimum is $-\varepsilon$.

1.30 Calculate the third virial coefficient (C) of CO at 300 K from its Lennard-Jones constants (Table 1.7).

1.31 Derive a formula for the second virial coefficient from the Sutherland potential with $n = 6$. Assume $\varepsilon/kT < 1$ so that the exponential can be expanded with $e^x = 1 + x + x^2/2 + x^3/6 + \cdots$ (keep exactly that number of terms).

1.32 Derive Eq. (1.49) for the second virial coefficient of a gas with a square-well potential.

1.33 Construct a graph, analogous to Fig. 1.21, of B^* vs. T^* for the square-well potential. Use the same scale as Fig. 1.21 so you can compare the square-well directly to the LJ results.

1.34 (a) For the square-well potential, find a relationship between the well depth (ε) and the Boyle temperature.
(b) Calculate the Boyle temperature of argon using the potential constants of Table 1.7. Compare this to the value predicted by the Lennard-Jones potential.

1.35 (a) Use the square-well potential constants for nitrogen (Table 1.7) to calculate its second virial coefficients at 223.15 and 473.15 K. These may be compared to the results in Table 1.3.
(b) With the same constants, estimate the Boyle temperature of nitrogen.

1.36 A mixture of hydrogen and ammonia has a volume at STP of 153.2 cm^3. The mixture is chilled with liquid nitrogen, the remaining gas drawn off; the volume then (again at STP) was 98.7 cm^3. Calculate the mole fraction of ammonia in the original mixture using Amagat's law.

1.37 Calculate the partial pressure of oxygen in air that has been compressed to 32 atm.

1.38 Use Dalton's law to calculate the pressure exerted by a mixture of 7.0 moles of H_2, 3.0 moles of CO_2, and 2.00 moles of NH_3 at $T = 300$ K if the total volume is 0.84 dm^3. (Use the van der Waals equation for the pure gases.)

1.39 Calculate the volume occupied by 7.00 moles of CO and 5.00 moles of H_2O at 700 K and 1000 atm using Amagat's law (and the van der Waals equation for the pure gases).

1.40 Use the square-well potential constants:

$$CCl_3F: \quad b_0 = 117.6 \text{ cm}^3, \quad \frac{\varepsilon}{k} = 339 \text{ K}, \quad R = 1.545$$

$$CCl_2F_2: \quad b_0 = 140.6 \text{ cm}^3, \quad \frac{\varepsilon}{k} = 345 \text{ K}, \quad R = 1.394$$

to calculate the second virial coefficient of a 50/50 mixture of these gases at 273.15 K and, from that, the compressibility factor of this mixture at STP.

1.41 Use the Lennard-Jones potential to calculate the second virial coefficient of air at 25°C.

1.42 (a) Use the van der Waals equation for mixed gases to calculate the pressure when 6 moles of CO and 4 moles of H_2 are together in a 5-dm^3 container at 500 K.
(b) Calculate the partial pressures of CO and H_2.

1.43 (a) Use the van der Waals equation for mixed gases to calculate the volume occupied by 3 moles of H_2 and 5 moles of N_2 at 398 K, 10 atm.
(b) What is the partial pressure of H_2?

1.44 The specific volume of H_2O (liq.) is given by the following empirical formula:

$$\ln v = -6.70781 + 1.012566 \ln T + \frac{280.663}{T}$$

Derive from this a formula for the coefficient of thermal expansion of water.

1.45 Derive an expression for the coefficient of thermal expansion of a van der Waals gas.

1.46 Show that the coefficient of thermal expansion of a gas that obeys the equation of state:

$$V_m = \frac{RT}{P} + B(T)$$

is

$$\alpha = \frac{1}{T} \left\{ \frac{1 + \dfrac{B'P}{R}}{1 + \dfrac{BP}{RT}} \right\}$$

where $B' = dB/dT$.

1.47 The volume of water at 500 atm (relative to its volume at STP) is:

40°C:	0.9867
60°C:	0.9967
80°C:	1.0071

Estimate the coefficient of thermal expansion of water at 60°C, 500 atm.

1.48 The volume of water (relative to STP) at 40°C is 0.9924 at 1 atm and 0.9867 at 500 atm. Estimate the isothermal compressibility of water at 40°C.

1.49 A tube of CCl_4 is warmed until the liquid fills the entire volume. Calculate the pressure exerted on the walls of the tube if the temperature is raised 1°C more.

1.50 Prove that the coefficient of bulk expansion (α) is three times the coefficient of linear expansion:

$$k = \frac{1}{l} \left(\frac{\partial l}{\partial T} \right)_P$$

1.51 Estimate the molar volume of benzene at 1000 atm from data in Table 1.8.

1.52 Use the five-point differential formula of Appendix 1 and data from Table 1.9 to calculate the coefficient of thermal expansion of water at 30°C.

Thermodynamics has charms
To sooth a savage breast.

–(with apologies to William Congreve)

2

The First Law of Thermodynamics

Thermodynamics is the study of heat and related quantities. It takes fairly loose concepts based on everyday experience, such as heat, temperature, work, and energy, and quantifies them by defining more abstract and rigorous quantities such as internal energy, enthalpy, and entropy. Its importance for the application of physical theory to chemistry is indicated by the fact that, in a first course in physical chemistry, up to half the course may be devoted to this subject. This is only the first of seven chapters of this book with thermodynamics as the major theme.

Thermodynamics, at its purest, is totally macroscopic and phenomenological and is valid independent of any reference to hidden entities or events (such as molecules and molecular collisions). However, the existence of molecules will be no surprise to the modern student, and nonthermodynamic explanations are very useful for placing the subject into the larger context of physical theory. Therefore, while generally adhering to the phenomenological approach, we shall, from time to time, provide molecular "explanations" for various phenomena. At the start, these will be qualitative arguments, because we lack the necessary formalism (statistical mechanics) to argue rigorously. Nonetheless, to keep our priorities straight, what we observe directly is the macroscopic world, and the microscopic picture that we develop is an interpretation of this experience.

2.1 Temperature

Is it necessary to define "temperature" in an age in which thermometers are ubiquitous and the word is bandied about by TV announcers? Perhaps not, but it can't hurt to examine what this term means in a phenomenological sense.

The idea of temperature comes from our qualitative feelings of "hot" and "cold." Temperature is the property of our surroundings that determines whether we wear an overcoat or a swimming suit. To quantify this concept, we can define a temperature scale by measuring a property of some material that changes with temperature. For example, if we seal a column of mercury in a glass tube and measure its length (L), we can use this length to define a temperature scale. The length L_0 in ice water defines the temperature $t = 0$, and the length L_{100} in boiling water (1 atm) defines $t = 100$; then the scale can be defined as:

$$t = \frac{100(L - L_0)}{L_{100} - L_0}$$

We could define such a scale with any material (alcohol is often used in thermometers), but if we checked carefully, these scales would not agree precisely at any temperature other than the defined points. It would be nice to have a scale that does not depend on the property of some particular material (such as mercury or alcohol). We can define such a scale using the properties of an ideal gas.

The PV product of a gas changes with temperature and we can use it to define a temperature scale. Taking the same two fixed points as above, we have:

$$t = 100 \left[\frac{(PV)_t - (PV)_0}{(PV)_{100} - (PV)_0} \right]$$

Even so, precise measurements will show variations among materials, since real gases obey the ideal gas law only approximately. However, if we extrapolate to $P = 0$ (where $V \rightarrow \infty$), the PV product is the same for all gases at a given temperature and is, hence, a property independent of the nature of the material used. In fact we can define an *absolute* scale by the relationship:

$$T = k \lim_{P \rightarrow 0} (PV)$$

The proportionality constant k [seen to be $1/nR$ by comparison to Eq. (1.1)] for the *Kelvin scale* is chosen to make the triple point of water equal to 273.16 K; this has the effect of making the ice point equal to 273.15 to ± 0.02 kelvins. This choice of scale makes the *size* of the Kelvin degree equal to the size of the Celsius degree (ice point = zero) and (denoting the Celsius temperature as t):

$$T(K) = t\,(°C) + 273.15$$

Aside from the fact that this scale is independent of the particular material used, we shall discover in Chapter 3 that the absolute temperature scale has a deeper significance.

2.2 Heat and Work

The development of physics could be described in terms of the evolution of conservation laws of increasing generality. Newton's first law states that the velocity is conserved in magnitude and direction in the absence of an external unbalanced force. This means that the kinetic energy:

$$E_k = \tfrac{1}{2}mv^2$$

is likewise conserved. If a force (F) is applied to the body, the velocity (v) will change according to Newton's second law:

$$F = m\frac{dv}{dt}$$

If a body moves a distance dx through a force field $F(x)$, the change of the kinetic energy is:

$$\int_{x_1}^{x_2} F(x)\,dx = m\int_{x_1}^{x_2} \frac{dv}{dt}\,dx$$

With $dx/dt = v$, this becomes:

$$m\int_{v_1}^{v_2} v\,dv = \tfrac{1}{2}m(v_2^2 - v_1^2)$$

so:

$$\int_{x_1}^{x_2} F(x)\,dx = \tfrac{1}{2}mv_2^2 - \tfrac{1}{2}mv_1^2 = E_{k2} - E_{k1} \tag{2.1}$$

This change is the **work** done on the body by outside forces:

$$w = \int_{x_1}^{x_2} F(x)\, dx \qquad (2.2)$$

If the force is a function of x alone, and not of time, the force can be related to a potential energy function $U(x)$:

$$F = -\frac{dU(x)}{dx}$$

so the integral of Eq. (2.2) can be written as:

$$\int_{x_1}^{x_2} F(x)\, dx = -[U(x_2) - U(x_1)]$$

If this is equated to the change of the kinetic energy, Eq. (2.1), we get:

$$E_{k2} + U(x_2) = E_{k1} + U(x_1) \qquad (2.3)$$

That is, the *total energy*, the sum of the kinetic energy plus the potential energy, is conserved. A useful point of view is that the potential energy is defined so that the law of conservation of energy would hold in the presence of an external force.

But in the real world Eq. (2.3) does not really work. Mechanical energy (motion) will be eventually stopped by friction, and, as we know from experience, friction causes heat. Can conservation be saved by including heat in our definition of energy? That's what the first law is all about.

Actually, the development of the concept of heat came from quite a different direction than mechanics. Heat is a far more intuitive idea than work but more difficult to define quantitatively. Originally scientists thought of heat as a fluid, *caloric;* this was the "substance" which flowed from a hot body to a cold body in order to equalize their temperatures. This idea is correct to a degree; the laws of heat conduction, which treat heat as a conserved fluid, are correct within their domain. The problem is that, although heat may be a fluid, it is not conserved. It can be created—for example, in a chemical reaction or by friction; the latter will be affirmed by any Boy Scout with two sticks.

These two lines of development, mechanics and caloric theory, came together in James Prescott Joule. In a series of careful experiments Joule supplied mechanical energy (via a stirrer) to water and measured a rise in temperature. In modern terms, he did no more than measure the heat capacity of water, but historically the work was of great moment, since it established unequivocally that heat was simply another form of energy. If the concept of energy is broadened to include heat, we may wish to include "chemical" energy as well; then energy will be *conserved* in any process. (We could go even farther and include, for example, nuclear binding energy or mass energy, Einstein's mc^2, but this is not necessary. For present purposes we need to include only those types of energy which *change* during chemical processes.)

The work of Joule (and others) established the first law of thermodynamics: heat is a form of energy and the total energy of an isolated system is conserved. This will be elaborated upon shortly.

2.3   Heat Capacity

Heat and temperature are not the same thing, although in popular parlance they are often confused. To make this obvious, try the following experiment: place a thermometer into a container with ice and water; it should read 0°C (or 32°F). Heat the ice water (you could use a burner). Heat will enter the ice water, and ice will melt, but the temperature will not change as long as any ice is present. A useful distinction is that temperature is an *intensive* property while heat is *extensive* (if you doubled the amount of ice, you'd need twice the heat to melt it all, but the temperature would be the same).

In modern studies of the thermal properties of matter, heat is usually supplied electrically to the material being studied. If an electrical heater with resistance R is supplied with a current I for a time t, it will give off a quantity of heat q:

$$q = I^2 Rt \qquad (2.4)$$

This is **Joule's law** of electrical heating. Because the quantity of electrical heat can be measured very accurately, Eq. (2.4) has become the primary standard for thermal measurements.

When a quantity of heat (q) is supplied to a material (where the material is all in the same phase, unlike the mixture of ice and water in the example above), the temperature of the material will increase; if the temperature change is ΔT, the ratio:

$$C = \frac{q}{\Delta T}$$

is called the **heat capacity.** For exact work, a problem arises with respect to this definition. If a specific quantity of heat, 10 joules for example, is sufficient to raise the temperature of a substance from 20° to 30°, the provision of 10 more joules may not result in another 10° rise. That is, the heat capacity of a substance, in general, is not a constant but will change with temperature. For that reason it is better to define the heat capacity as a function of temperature, $C(T)$, for the addition of an infinitesimal amount of heat dq which results in an increase of temperature from T to $T + dT$:

$$C(T) = \frac{dq}{dT}$$

Another complication arises because the amount of heat required to produce a given temperature rise depends on whether the material is allowed to expand. If the material can expand against its surroundings, some of the heat is used to furnish the required work; this effect is particularly noticeable for gases but is present for solids and liquids as well. We shall, in due course, deal with this complication; for now let's just define it away by assuming that the material being heated is in a rigid container so that its volume is constant. This then defines the **heat capacity at constant volume:**

$$C_v = \left(\frac{dq}{dT}\right)_v \qquad (2.5)$$

James Prescott Joule (1818–1889)
The caloric theory of heat, which held that heat was a conserved material, incapable of being created or destroyed (in effect, a state function), was a mass of contradictions and patches which had been devised to explain such phenomena as the cooling of gases on expansion and the heating of metal when it is "worked" or machined. The first law of thermodynamics, which states that heat and mechanical work are equivalent and interconvertible forms of energy, rests largely on the experiments of Joule.

Joule first experimented with electrical heating and established that the heating effect was proportional to the resistance of the heater and the square of the electrical current. He then devised an electrical stirrer and measured the heating effect of this mechanical action; this led to the concept of heat as a form of dynamical motion. Then, by carefully measuring the temperature increase caused by mechanical work, he measured the "mechanical equivalence of heat" and firmly established that heat and work were interconvertible. (In this, he was approximately contemporary with Julius Mayer, who is also credited with measuring the mechanical equivalence of heat.)

By his careful and painstaking experimental work, Joule laid the foundations of thermodynamics. But he participated little in subsequent developments, apparently because of his lack of mathematical training. Joule was a "nonacademic" scientist in the tradition of Michael Faraday; indeed, his first publication of his experiments was in the local newspaper.

This quantity is a well-defined property of the material. It is, of course, an extensive property, since the heat required to cause a given temperature increase will be proportional to the amount of material which is heated. It is often convenient to define intensive properties by dividing by the moles (*n*) to get the **molar heat capacity:**

$$C_{vm} = \frac{1}{n}\left(\frac{dq}{dT}\right)_V \tag{2.6a}$$

or by the mass (*W*) to get the **specific heat:**

$$c_v = \frac{1}{W}\left(\frac{dq}{dT}\right)_V \tag{2.6b}$$

Molecular Motion and Heat Capacity

How big is the heat capacity, and why does it change with temperature? We can answer these questions most easily for gases, and we shall do so thoroughly in Chapter 5. For present purposes the qualitative discussion that follows may be helpful.

In Chapter 1 it was demonstrated that the average translational kinetic energy of a molecule of mass m is:

$$\tfrac{1}{2}m\langle v^2 \rangle = \tfrac{3}{2}kT$$

(where k is Boltzmann's constant). If thermal energy is supplied to the gas (at constant volume), one effect will be to increase the molecular velocities and, hence, the temperature. If this were all that happened, the heat required to increase the temperature of one mole of gas (L molecules) would be equal to the change in energy:

$$q = L(\tfrac{3}{2})kT_2 - L(\tfrac{3}{2})kT_1$$
$$= \tfrac{3}{2}Lk(T_2 - T_1)$$

Therefore, the heat capacity for L molecules would be:

$$C_{vm} = \tfrac{3}{2}Lk = \tfrac{3}{2}R = 12.47 \text{ J/K}$$

(where k = Boltzmann's constant, $Lk = R$, the gas constant).

Among the common gases, this value is reasonably accurate in just *five* cases: He, Ne, Ar, Kr, and Xe. These, of course, are all monatomic species, and the same value will hold for some other monatomic gases (at least those with a closed-shell electronic structure). Furthermore, the heat capacity of such gases varies very little with temperature — for example, the heat capacity of Ar at $-180°C$ and $15°C$ is

$$C_{vm}(-180°C) = 12.60 \text{ J/K}, \qquad C_{vm} = (15°C) = 12.58 \text{ J/K}$$

The slight changes of the heat capacities of monatomic gases with temperature are largely due to gas nonidealities — that is, to intermolecular forces.

Diatomic Molecules

To understand the heat capacity of molecular gases, we need to understand the types of energies which molecules may have. We shall look in detail at the diatomic molecule; the conclusions are readily generalized to more complex cases.

A diatomic molecule (Figure 2.1) is accurately modeled as two masses (the atoms with masses m_1 and m_2) attached by a stiff spring (the chemical bond). The specification of the position of two masses requires six coordinates, and the description of their movement requires six velocities, which could be the rate of change of those six coordinates. These coordinates could be the three Cartesian coordinates for each mass, but the situation is better understood in terms of the center-of-mass (COM) coordinate system. This topic is discussed in Appendix III; a brief summary will be given here.

For **translation**, the molecule can be considered as a point mass, $m = m_1 + m_2$, located at the COM. The rate of change at the Cartesian coordinates (x, y, z) of the COM in space produces velocities:

$$\dot{x} = \frac{dx}{dt}, \qquad \dot{y} = \frac{dy}{dt}, \qquad \dot{z} = \frac{dz}{dt}$$

and a kinetic energy:

$$\varepsilon_{tr} = \tfrac{1}{2}m(\dot{x}^2 + \dot{y}^2 + \dot{z}^2)$$

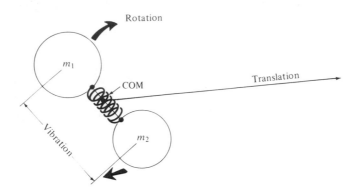

Figure 2.1 Motions of a diatomic molecule. The diatomic molecule can be accurately modeled as two masses connected by a stiff spring. The energy of such a molecule can be divided into contributions due to translation of the center-of-mass (COM), the rotation of the diatom about the COM, and the stretching of the spring (vibration).

These are just the velocities v_x, v_y, v_z that were discussed in Chapter 1, where all types of molecules were treated as point masses.

The relative position of the two masses in the COM system is given by three coordinates; in the spherical polar coordinate system (discussed in Appendix III) these are the bond length R and the polar angles θ and ϕ, which give the orientation of the diatomic vector in space. The changes of these coordinates when the molecule moves can be expressed with three velocities:

$$\dot{\theta} = \frac{d\theta}{dt}, \qquad \dot{\phi} = \frac{d\phi}{dt}, \qquad \dot{R} = \frac{dR}{dt}$$

The first two of these are **rotations** of the diatom about the COM; the kinetic energy for rotation can be shown to be:

$$\varepsilon_{\text{rot}} = \tfrac{1}{2}I(\dot{\theta}^2 + \sin^2\theta \; \dot{\phi}^2)$$

where I is the moment of inertia; in terms of the reduced mass

$$\mu = m_1 m_2/(m_1 + m_2)$$

the moment of inertia of a diatom is

$$I = \mu R^2$$

(This equation is not derived here or in Appendix III; if a derivation is required, any good physics text on classical mechanics should provide it.)

The other velocity, $\dot{R}$, is called a **vibration;** it differs from the other velocities in that it is resisted by the restoring force of the spring — that is, the chemical bond. The restoring force (F) of a spring for small extensions from the equilibrium length (denoted here as R_e, the distance at which the spring exerts no force) can be approximated by Hooke's law:

$$F = -k(R - R_e)$$

where k is the force constant of the spring (or, in this case, the chemical bond). This force can be related to a potential energy $U(R)$, with $F = -dU(R)/dR$; it is then readily shown that

$$U(R) = \tfrac{1}{2}k(R - R_e)^2$$

and the vibrational energy is:

$$\varepsilon_{\text{vib}} = \tfrac{1}{2}\mu\dot{R}^2 + \tfrac{1}{2}k(R - R_e)^2$$

Is there any other type of energy that we must consider? We have ignored the interaction of molecules with each other — that is, the intermolecular forces that were discussed in Chapter 1; this contribution to the energy will be negligible at low densities. We have also ignored the energy of the electrons within the molecule; if we limit our discussion to stable diatomic molecules with closed electronic shells — in effect, molecules for which a Lewis structure can be drawn with no unpaired electrons — the electronic energy will be of no concern except at very high temperatures.

Equipartition of Energy

With the exceptions noted above, the total energy of a diatomic molecule can be written as a sum of translational, rotational, and vibrational energies:

$$\varepsilon = \tfrac{1}{2}m(\dot{x}^2 + \dot{y}^2 + \dot{z}^2) + \tfrac{1}{2}I(\dot{\theta}^2 + \sin^2\theta \, \dot{\phi}^2)$$
$$+ \tfrac{1}{2}\mu\dot{R}^2 + \tfrac{1}{2}k(R - R_e)^2 \tag{2.7}$$

This expression contains six kinetic-energy terms and one potential-energy term. If the molecule is supplied with thermal energy, how will that energy distribute itself among these various types of energy? The principle of *equipartition of energy* states that the average molecular energy will be equally distributed among these terms. The first three terms (translation) were discussed in Chapter 1, where it was shown that each has an average energy of $\tfrac{1}{2}kT$; equipartition then tells us that each term of Eq. (2.7) has an average energy of $\tfrac{1}{2}kT$ and:

$$\langle \varepsilon \rangle = \tfrac{7}{2}kT$$

If the temperature increases by dT, the average energy will increase by $\tfrac{7}{2}k\,dT$ per molecule; for a mole (L molecules) the increase will be $\tfrac{7}{2}Lk\,dT = \tfrac{7}{2}R\,dT$, so the heat capacity should be:

$$C_{vm} = \tfrac{7}{2}R = 29.10 \text{ J/K}$$

Table 2.1 lists experimental heat capacities (at 15°C) for six diatomic molecules; five of these (H_2, N_2, O_2, CO, HCl) have $C_{vm} \cong 20$ J/K, more like $\tfrac{5}{2}R$ (20.79 K) than $\tfrac{7}{2}R$; one, Cl_2, has a larger heat capacity, 25.18 J/K. Figure 2.2 shows the heat capacities of Cl_2, N_2, and H_2 as a function of temperature (with other gases included for comparison); it can be seen — and this is quite typical — that the heat capacity of these diatomic molecules varies from approximately $\tfrac{5}{2}R$ at low temperature and approaches

Table 2.1   Heat capacity of gases

(units: J K^{-1} mol^{-1} at 15°C, 1 atm, except as indicated)

Gas	C_{pm}	C_{vm}	γ	$\dfrac{C_{pm} - C_{vm}}{R}$
He	20.88	12.55	1.664	1.002
Ar	20.93	12.59	1.668	1.003
H$_2$	28.58	20.26	1.410	1.000
N$_2$	29.04	20.69	1.404	1.004
O$_2$	29.16	20.81	1.401	1.003
Cl$_2$	34.13	25.18	1.355	1.076
CO	29.05	20.70	1.404	1.004
HCl	29.59	20.98	1.410	1.035
H$_2$S	36.11	27.36	1.320	1.053
CO$_2$	36.62	28.09	1.304	1.026
SO$_2$	40.64	31.50	1.290	1.100
N$_2$O	36.91	28.33	1.303	1.032
NH$_3$	37.29	28.47	1.310	1.060
CH$_4$ (15°)	35.46	27.07	1.31	1.01
CH$_4$ (−74°)	33.4	24.8	1.35	1.03
CH$_4$ (−115°)	30.2	21.4	1.41	1.06
C$_2$H$_2$ (acetylene)	41.73	31.16	1.260	1.037
C$_2$H$_4$ (ethylene)	42.17	33.56	1.255	1.036
C$_2$H$_6$ (ethane)	48.58	39.46	1.22	1.096

the equipartition value, $\frac{7}{2}R$, only at high temperatures. Why? The answer lies in quantum theory.

According to the quantum theory, to be discussed in detail in Chapters 11 through 14, molecules may accept energy only in certain minimum-sized packages called *quanta*. For translational and rotational energy the minimum quanta are sufficiently small compared to the thermal energy available ($\sim kT$ or 4×10^{-21} J at room temperature) that (excepting H$_2$) these quantum effects can be ignored. Such is not ordinarily the case for vibrations; N$_2$, for example, requires a minimum of 47×10^{-21} joules, more than ten times the average thermal energy available at 300 K, to excite its vibration to a higher energy. At room temperature few collisions have a kinetic energy this large, so the vibration is not "active"—that is, it does not contribute much to the heat capacity; on the other hand, at 3000 K, $kT \sim 40 \times 10^{-21}$ J and the vibration will contribute fully to the heat capacity. The upshot is that the vibrational contribution will vary from nothing at low temperatures to R per mole at high temperatures.

The size of the vibrational quantum depends on the molecule and is related to the atomic masses and the bond stiffness (through the force constant k). According to quantum theory (Chapter 11), the size of the vibrational quantum is:

$$\frac{h}{2\pi} \sqrt{\frac{k}{\mu}}$$

where h is Planck's constant (and μ is the reduced mass of the atoms, Appendix III). The molecule I$_2$, for example, has a minimum quantum requirement for its vibration

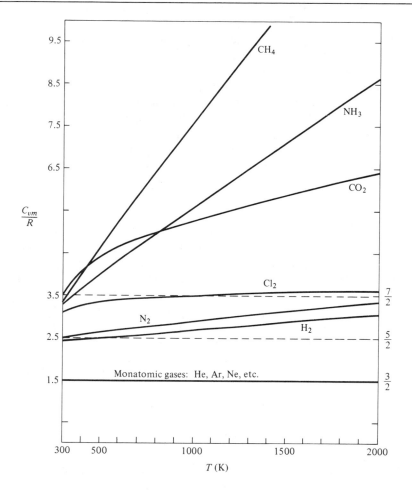

Figure 2.2 Ideal gas heat capacities. The heat capacities shown are for constant-volume heating of a variety of gases at low pressure (where they are reasonably ideal in their behavior) vs. temperature. The increase in heat capacity with increasing temperature is primarily due to the vibrational motions of the molecule; note that the constant-volume heat capacity of monatomic gases is independent of temperature.

of only 4×10^{-21} J, so its vibration will be active at room temperature. As a result, at any particular temperature, diatomic molecules will have different values for their heat capacities, usually in the range 21 to 29 J/K per mole.

Other Substances

Polyatomic molecules have more vibrational degrees of freedom, and generally the heat capacity at any given temperature will be larger the larger the molecule; a study of Table 2.1 will demonstrate this trend. (See also Figure 2.2.) The heat capacities of polyatomic molecules will also tend to vary more extensively with temperature than do those of diatomics.

The heat capacity of liquids is a much more complex topic. The molecules are very densely packed, so that intermolecular interactions cannot be ignored. Some liquids, such as water, are very highly structured owing to hydrogen bonding or other types of specific interactions, and this structure affects the heat capacity. A great deal of theoretical work has been done on liquids, but none of it is simple enough to describe readily (or, if simple, it lacks sufficient generality). We shall regard liquid properties as purely empirical; even on that level there are few simple generalities. Liquids exist only over a limited range of temperature (melting point to boiling point), so it is usually sufficiently accurate to assume that their heat capacity is independent of T.

Solids

Some fairly general and simple theories have been worked out for the heat capacity of atomic solids. An atomic solid consists of atoms or ions occupying lattice points, held there by fairly nonspecific repulsive forces of their neighbors. Most metals and ionic crystals, such as NaCl, fall into this class. In contrast, in a molecular solid complex molecules (atoms held together by strong covalent bonds) occupy the lattice points; examples include CO_2, methane, and benzene. In an atomic solid, heat has the effect of increasing the amplitudes of motion (vibration) about the equilibrium positions. We can think of the atom as being held in a "potential well" formed by the repulsive forces of its neighbors; inside the well it can rattle about with a vigor that increases with temperature. Molecular solids have, in addition, heat-capacity contributions due to intramolecular vibrations and, sometimes, rotations of the molecules about their lattice positions. The conduction electrons of metals will also contribute to the heat capacity.

Heat capacities of solids vary dramatically with temperature, with C_v approaching zero as $T \rightarrow 0$. The high-temperature limit of the molar heat capacity for atomic crystals is usually about $3R$ (or 24.9 J/K). This is an empirical generalization first noted a century ago and called the law of Dulong and Petit. This value is readily explained as a result of the equipartition principle. The N atoms of the crystal each can vibrate in three orthogonal directions, and each of these vibrations has an average energy of kT ($\frac{1}{2}kT$ for the kinetic energy and $\frac{1}{2}kT$ for the potential energy). The arguments used above for diatomic gases then suggest that the heat capacity for N atoms should be $3Nk$, or, per mole, $C_{vm} = 3R$. As for gases, the equipartition principle gives the high-temperature limit, and the principles of quantum theory are required for a full and correct explanation; this is discussed further in Chapter 15.

The Dulong-Petit law had some historical significance in establishing atomic weights; accurate *combining* weights of metals with oxygen can be determined, but this gives the atomic weight only as a multiple of an integer (is the oxide MO, MO_2, M_2O, M_2O_3?), so the determination of atomic weights from heat capacities could be helpful without having to be very precise.

Figure 2.3 shows the heat capacity of water in all three phases; note the discontinuity at the phase transitions. (This figure actually shows the constant-pressure

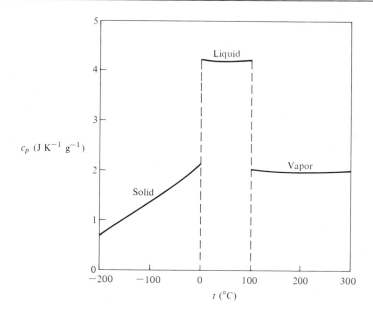

Figure 2.3 Heat capacity of water. The constant-pressure specific heat of water at 1 atm is shown at various temperatures. Note the discontinuities at the melting and boiling points.

heat capacity; C_v would behave in a similar fashion, but it is more difficult to measure. The relationship of the constant-pressure and constant-volume heat capacities will be discussed later in this chapter.)

2.4 The Internal Energy Function

Before quantifying the first law, we shall have to define some terms.

A **system** is some part of the world in which we are interested and plan to describe thermodynamically. This may be, for example, a piece of matter, a reaction vessel and its contents, or an engine. In many thermodynamic problems, the beginning of wisdom is the proper definition of the system.

Surroundings is the rest of the world outside the system. It need not be as broad as the universe. For example, if a reaction vessel is in a thermostat, we may, with proper precautions, be able to limit surroundings to the thermostat. Note that energy is conserved only for the universe; the system and surroundings may exchange energy with each other.

We specify the **state** of a pure homogeneous material by giving values to any two intensive variables. In this chapter and Chapter 3 we shall consider only closed systems — systems containing a constant amount of material. For this reason we may also use extensive properties such as volume to define the state of a system.

A **state function** is a physical property that has a specific value once the state is defined; a number of these were mentioned in Chapter 1, including the volume $V(T, P)$, whose relationship to T and P was given by the equation of state.

The Internal Energy

We quantify the first law of thermodynamics by introducing a new state variable called the **internal energy function,** U. [This is a fairly conventional choice of symbol, although E is used in some texts. Be careful not to confuse it with the potential energies $U(x)$ or $U(R)$ used in this chapter or the intermolecular potential $U(r)$ used in Chapter 1. Alas, there are only 26 letters in the English alphabet!]

If some process is applied to the system to change it from state 1 (with variables T_1, P_1, V_1, and U_1) to state 2 (T_2, P_2, V_2, and U_2), the change in U is the sum of the heat (q) and work (w) supplied to the system:

$$\Delta U = U_2 - U_1 = q + w \tag{2.8}$$

Equation (2.8) is the mathematical statement of the first law of thermodynamics. In this book we shall use the convention that work done *on* the system or heat supplied *to* the system is a positive quantity that increases the internal energy of the system; work done *by* the system or heat removed *from* the system is negative.

The symbol Δ has a very definite meaning in thermodynamics; it means the quantity in the final state minus the quantity in the initial state. For example:

$$\Delta U \equiv U(\text{final state}) - U(\text{initial state})$$

By contrast, heat and work depend on the process by which the state is changed and are not uniquely determined by the initial and final states.

The *process* is what is done to the system or by the system to cause a change of state. For example:

1. A gas expands against a piston, which moves a connecting rod, which turns a wheel.
2. A mixture of H_2 and O_2 react to form water.
3. Tap water at 15°C is heated and boiled.
4. An ice cube at 0°C is melted into water at 0°C.
5. Gin and vermouth are mixed together.

For these processes the system could be defined as (1) the engine, (2) the reaction vessel and its contents, (3) the water (but not the source of heat), (4) the ice cube, and (5) the martini. The definition of the system is obviously somewhat flexible — for example, in (1) it could be just the gas or the whole engine — and, consequently, must be made carefully.

In this chapter and the next we shall be concerned primarily with processes for which the PVT variables of a pure, homogeneous substance are changed. In other chapters we shall consider other types of processes in detail, including phase changes (Chapter 4), chemical reactions (Chapter 6), and mixing (Chapter 7).

A precaution: the symbol q means the heat that *enters* or *leaves* the system. Consider a chemical reaction that is carried out in a container with insulated walls such that heat cannot pass into or out of the system; such a process is called *adiabatic*. If the reaction is exothermic, the contents of the vessel may become very hot, but, since no heat leaves the system, $q = 0$.

Work

The work may be supplied in many forms. The only work of immediate interest is that required to compress a substance, or that done by a substance on its surroundings if it expands. Consider a gas expanding against a piston; if the piston moves from a position x_1 to x_2 against an opposing force F, the work will be:

$$w = -\int_{x_1}^{x_2} F\, dx$$

The negative sign in this equation conforms to our stated convention that work done by the gas is negative. The opposing pressure (Figure 2.4), which we shall denote P_{ex} to distinguish it from the pressure of the gas that is expanded (denoted with an unadorned P), is the force per unit area; if the area of the cylinder is A, then $P_{ex} = F/A$. Also, the volume change will be $dV = A\, dx$, and the work is, therefore:

$$w = -\int_{x_1}^{x_2} \frac{F}{A} A\, dx$$

$$w = -\int_{V_1}^{V_2} P_{ex}\, dV \tag{2.9}$$

The distinction between the pressure of the gas (P) and the pressure applied to the gas (P_{ex}) is an important one. If the pressure of the gas is greater than the opposing pressure, $P > P_{ex}$, the gas will expand; if $P < P_{ex}$, the gas will be compressed. At equilibrium, when no further change in the volume of the gas is apparent, $P = P_{ex}$.

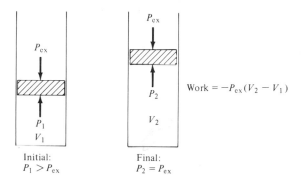

Figure 2.4 Work of gas expansion. The work of a gas expansion depends on the opposing pressure against which it must expand; the pressure of the expanding gas must be greater than the opposing pressure, or no expansion would occur; the expansion stops when these pressures are equal and there is no net force on the piston.

Heat Capacity at Constant Volume

The internal energy of a pure homogeneous substance is a function of two variables; several choices are possible for the independent variables, but it is most convenient to choose T and V and consider $U(T, V)$. The slope formula of Appendix II can then be used to give the change in $U(dU)$ for an infinitesimal change in T and V as:

$$dU = \left(\frac{\partial U}{\partial T}\right)_V dT + \left(\frac{\partial U}{\partial V}\right)_T dV \qquad (2.10)$$

If the state of the system is changed infinitesimally by adding heat dq, and only PV work (that is, work of expansion) is possible, Eq. (2.8) together with Eq. (2.9) gives:

$$dU = dq - P_{ex} dV \qquad (2.11)$$

If, furthermore, this heat is added at constant volume, dV will be zero, and, from Eqs. (2.10) and (2.11):

$$dU = \left(\frac{\partial U}{\partial T}\right)_V dT = dq$$

This demonstrates that the first slope of Eq. (2.10) is simply the heat capacity at constant volume [Eq. (2.5)], which can now be defined as:

$$C_v = \left(\frac{\partial U}{\partial T}\right)_V \qquad (2.12)$$

Example: Calculate the change in the internal energy of 1 kg of argon when it is heated from 25°C to 100°C, at constant volume. From Table 2.1 we get $C_{vm} = 12.59$ J/K. From Eq. (2.10) with $dV = 0$ [and Eq. (2.12)] we get:

$$\int_{U_1}^{U_2} dU = \int_{T_1}^{T_2} C_v dT = n \int_{T_1}^{T_2} C_{vm} dT$$

Since the heat capacity of argon is constant, this equation can be integrated immediately to give:

$$\Delta U = nC_{vm}(T_2 - T_1)$$

The number of moles is:

$$n = \frac{1000}{39.948}$$

Finally:

$$\Delta U = \left(\frac{1000}{39.948}\right)(12.59)(75) = 23.64 \times 10^3 \text{ J}$$

Since $\Delta U = q$ at constant volume, this quantity is also the heat for this process.   ∎

The Internal Pressure

What is the significance of the second slope, $(\partial U/\partial V)_T$? In Chapter 1 we developed the idea that the energy of a gas on the microscopic scale consists of kinetic energy from the molecular translational motion and a potential energy due to intermolecular interactions; in the last section we extended this to include the internal motions, rotation and vibration, of the molecules. For an ideal gas the average kinetic and internal energies of a molecule depend on *temperature alone*. On the other hand, the intermolecular potential, $U(r)$, depends inversely on the distance (r) between the molecules; the average value of this distance, therefore the energy, will depend on the concentration (hence volume) as well as the temperature. This suggests that, for an ideal gas (and for real gases at low pressures), the internal energy (U) is a function of temperature only, and $(\partial U/\partial V)_T = 0$. In general, for solids, liquids, and real gases at high pressure this quantity cannot be ignored.

It should be understood that this slope, $(\partial U/\partial V)_T$, is just another *property* of the material, just as the slope $(\partial U/\partial T)_V$ was a property (the heat capacity); it doesn't have a special symbol (such as C_v for the other slope) but it does have a name — it is called the **internal pressure.** Its relationship to the other properties of the material can be derived only by using the second law of thermodynamics; this will be done in Chapter 3. However, this relationship is so useful it will be given now without proof:

$$\left(\frac{\partial U}{\partial V}\right)_T = T\left(\frac{\partial P}{\partial T}\right)_V - P \tag{2.13}$$

Exercise: Use Eq. (2.13) to prove that $(\partial U/\partial V)_T$ of an ideal gas is zero; that is, evaluate the right-hand side using $P = nRT/V$. ∎

For solids and liquids, Eq. (2.13) can be used to calculate the internal pressure from the coefficients of thermal expansion and isothermal compressibility introduced in Chapter 1. From Eq. (1.64), substituted into Eq. (2.13):

$$\left(\frac{\partial U}{\partial V}\right)_T = T\left(\frac{\alpha}{\kappa_T}\right) - P \tag{2.14}$$

For gases, Eq. (2.13) can be evaluated with any convenient equation of state — for example, the van der Waals equation:

$$P = \frac{RT}{V_m - b} - \frac{a}{V_m^2}$$

$$\left(\frac{\partial P}{\partial T}\right)_V = \frac{R}{V_m - b}$$

$$T\left(\frac{\partial P}{\partial T}\right)_V - P = \frac{RT}{V_m - b} - \frac{RT}{V_m - b} + \frac{a}{V_m^2} = \frac{a}{V_m^2}$$

Therefore:

$$\text{(van der Waals)} \qquad \left(\frac{\partial U}{\partial V}\right)_T = \frac{a}{V_m^2} \tag{2.15}$$

The van der Waals constant a is directly related to the attractive forces between the molecules, as is the internal pressure.

Example: Calculate the internal pressure of H_2O(liquid) and CO_2(gas) at 20°C, 1 atm. For H_2O, using Eq. (2.14) with data from Table 1.8:

$$\left(\frac{\partial U}{\partial V}\right)_T = (293 \text{ K})\left(\frac{0.188 \times 10^{-3} \text{ K}^{-1}}{45 \times 10^{-6} \text{ atm}^{-1}}\right) - 1 \text{ atm} = 1.22 \times 10^3 \text{ atm}$$

Since U/V is an energy/volume, this makes more sense in SI units, for which $Pa = J/m^3$:

$$\left(\frac{\partial U}{\partial V}\right)_T = (1.22 \times 10^3 \text{ atm})(101,325 \text{ Pa/atm}) = 124 \text{ MPa}$$

For CO_2 we use the van der Waals equation with (Table 1.1) $a = 3.59 \text{ dm}^6$ atm and $b = 0.0427 \text{ dm}^3$ to calculate $V_m = 29.95 \text{ dm}^3$. Then, from Eq. (2.15):

$$\left(\frac{\partial U}{\partial V}\right)_T = \frac{3.59 \text{ dm}^6 \text{ atm } (101,325 \text{ Pa/atm})}{(29.95 \text{ dm}^3)^2} = 406 \text{ Pa} \qquad\blacksquare$$

Equation of State for $U(T, V)$

The combination of Eqs. (2.10), (2.12), and (2.13) gives, for changes of U with T and V:

$$dU = nC_{vm} dT + \left[T\left(\frac{\partial P}{\partial T}\right)_V - P\right]dV \qquad (2.16)$$

For changes at constant volume, of course, only the first term is needed. If volume is changing (for example, when the substance is heated at constant pressure), the first term is still, usually, the more important; the second term is often small for gases because the coefficient $[(\partial U/\partial V)_T]$ is small (in fact, it is zero for an ideal gas); for solids and liquids the coefficient of the second term is large, but dV will generally be small for a given temperature change. Therefore, in many practical cases, only the first term is needed.

Equation (2.16) is another example of an *equation of state,* giving the relationship of the state function U to the independent variables T and V. The equation:

$$\Delta U = C_v \Delta T \qquad (2.17)$$

which results from Eq. (2.16) if *either* $dV = 0$ (the process is constant-volume) *or* $(\partial U/\partial V)_T = 0$ (for example, for an ideal gas) *and* C_v is presumed constant, is often useful. This equation is easy to use and often accurate enough, but its limitations must be appreciated. These are:

1. The heat capacity is independent of T. If this condition does not hold, the equation may still be reasonably accurate for small ΔT.
2. (a) The substance is an ideal gas *or* (b) the process is constant-volume ($dV = 0$).
3. There is no phase change in the range ΔT.

Note—and this point is commonly misunderstood—there is no connection between conditions 1 and 2(a). We have seen in the preceding sections (and Chapter 1) that "ideality" neglects intermolecular forces; on the other hand, the temperature dependence of the heat capacity is, for the most part, due to the quantized nature of molecular vibrational energy.

Example: Calculate the heat required to change the temperature of 1 kg of argon from 25°C to 100°C at constant $P = 1$ atm, assuming ideal gas.

The ΔU for this process is, by Eq. (2.16), the same as it was for the previous example (constant-volume heating), since $(\partial U / \partial V)_T = 0$ for an ideal gas; $\Delta U = 23.64 \times 10^3$ J. But the heat is different:

$$q = \Delta U - w = \Delta U + \int_{V_1}^{V_2} P_{ex} \, dV$$

With pressure constant, $P = P_{ex} = 1$ atm, the work term is (using the ideal gas law):

$$w = -P(V_2 - V_1) = -nR(T_2 - T_1)$$

$$= -\left(\frac{1000}{39.948}\right)(8.314)(75) = -15.61 \times 10^3 \text{ J}$$

Then the heat required is:

$$q = \Delta U - w = (23.64 + 15.61) \times 10^3 = 39.25 \times 10^3 \text{ J} \qquad \blacksquare$$

2.5 The Enthalpy Function

In many cases it is more convenient to use temperature and pressure, rather than temperature and volume, as independent variables. This is largely a practical matter; temperature and pressure are more easily measured and controlled than some other variables. For example, chemical reactions are easily carried out at constant pressure by opening the system to the atmosphere (protected perhaps by a reflux condenser). Reactions at constant volume must be carried out in a rigid container called a bomb; the name is ominous, and the effect of miscalculation of the amount of pressure which will be developed by the reaction can be drastic.

There is, mathematically, no reason why we could not treat U as a function of T and P, $U(T, P)$, and write:

$$dU = \left(\frac{\partial U}{\partial T}\right)_P dT + \left(\frac{\partial U}{\partial P}\right)_T dP$$

However, dU (constant P) is not heat and $(\partial U/\partial T)_P$ is not a heat capacity. It proves to be more convenient to define a new function called the **enthalpy** (H):

$$H \equiv U + PV \qquad (2.18)$$

Since U, P, and V are all state functions, it should be obvious that H is a state function as well. Therefore, the change in the value of the enthalpy function when the system is changed from state 1 $(P_1, T_1, \ldots)$ to state 2 $(P_2, T_2, \ldots)$ is:

$$H_2 - H_1 = U_2 - U_1 + (P_2 V_2 - P_1 V_1) \qquad (2.19)$$

The justification for introducing this new variable is in the results; it will greatly simplify our discussion of many processes, especially those which occur at constant pressure. As a first example, consider a process whereby a quantity of heat (q) is added to a system at constant pressure. The change in the internal energy [assuming only PV work, hence, Eq. (2.11)] is:

$$U_2 - U_1 = q - \int_{V_1}^{V_2} P_{ex}\, dV$$

$$= q - P_{ex}(V_2 - V_1)$$

The pressure of the material is unchanged and P_{ex} is the pressure of the gas (P); therefore, from Eq. (2.19):

$$H_2 - H_1 = q - P(V_2 - V_1) + P(V_2 - V_1)$$

$$\Delta H = (q)_P \tag{2.20}$$

The change in the enthalpy at constant pressure (with only PV work) is the heat of the process. [The qualification "only PV work" is important and needs clarification. Electrochemical cells, which we shall examine in detail in Chapter 8, are devices by which the energy of a chemical reaction can be obtained directly as electrical energy. If the cell is discharged at constant pressure so as to do work (for example, by turning an electrical motor), heat may be exchanged with its surroundings, but the ΔH of the process is *not* the heat. Keep this in the back of your mind; for most of the examples we shall encounter before Chapter 8, Eq. (2.20) will be true. There are other types of work as well; we shall discuss these as we encounter them.]

Equation of State for $H(T,P)$

The convenience of the enthalpy function becomes more apparent when we consider it as a function of T and $P, H(T,P)$. The slope formula gives:

$$dH = \left(\frac{\partial H}{\partial T}\right)_P dT + \left(\frac{\partial H}{\partial P}\right)_T dP \tag{2.21}$$

If this equation is taken at constant pressure $(dP = 0)$ and compared to Eq. (2.20) (for an infinitesimal amount of heat dq), we get:

$$dH = (dq)_P = \left(\frac{\partial H}{\partial T}\right)_P dT$$

From this, we see that the first slope of Eq. (2.21) has a very simple physical significance; it is the **heat capacity at constant pressure:**

$$C_p = \left(\frac{\partial H}{\partial T}\right)_P \tag{2.22}$$

The evaluation of the second slope of Eq. (2.21) requires, again, theoretical developments that we shall not be able to justify until Chapter 3. The result will be presented here, for we shall find it very useful:

$$\left(\frac{\partial H}{\partial P}\right)_T = V - T\left(\frac{\partial V}{\partial T}\right)_P \tag{2.23}$$

Putting these together (with $C_{pm} \equiv C_p/n$), we get:

$$dH = nC_{pm}\, dT + n\left[V_m - T\left(\frac{\partial V_m}{\partial T}\right)_P\right] dP \tag{2.24}$$

This is an equation of state for the enthalpy in terms of a readily measured quantity, C_{pm}, and a coefficient that can be either calculated from a PVT equation of state or related to other measurable parameters.

Exercise: Prove that, for an ideal gas, $(\partial H / \partial P)_T = 0$. This means that the enthalpy of an ideal gas is a function of temperature only, $H(T)$, and the enthalpy change for heating an *ideal gas* from T_1 to T_2 is:

$$dH = n \int_{T_1}^{T_2} C_{pm} \, dT \tag{2.25}$$

Note that Eq. (2.25) is also valid for *any* material if $dP = 0$ (and the material remains homogeneous — that is, there is no phase change between T_1 and T_2). ∎

Changes of Enthalpy with Temperature (Constant *P*)

The evaluation of Eq. (2.25) for changes in enthalpy at constant pressure requires some knowledge of how C_p depends on temperature. Although there are theoretical results on this question for gases (Chapter 5) and solids (Chapter 15), heat capacities are largely empirical. There are several methods for various circumstances.

1. If C_P is constant (monatomic gases at low pressure):

$$\Delta H = n C_{pm} (T_2 - T_1) \tag{2.26a}$$

   This may also be used in other cases if (a) insufficient C_p data are available for a better method, (b) and ΔT is small, and/or (c) a rough answer for ΔH is all that is needed.
2. If experimental heat capacities are available between T_1 and T_2, graphical or numerical integration of Eq. (2.25) can be done using the trapezoidal rule (Appendix I). (See Figure 2.5.)
3. When a simple formula is needed, the definition of the mean value provided by calculus (Appendix I) can be used to give:

$$\Delta H = n \overline{C}_{pm} (T_2 - T_1) \tag{2.26b}$$

   $\overline{C}_{pm}$ is (by definition) the mean value of the heat capacity in the interval T_1 to T_2. It may be estimated conveniently as the average of the values at T_1 and T_2 [then Eq. (2.26b) is a one-step trapezoidal rule] or as the heat capacity at the center of the interval — that is, at the mean temperature.
4. For many substances, empirical formulas are available for $C_{pm}(T)$, usually in the form of a power series — for example:

$$C_{pm} = a + bT + \frac{c}{T^2} \tag{2.27a}$$

   or

$$C_{pm} = a' + b'T + c'T^2 + d'T^3 \tag{2.27b}$$

   These constants (see Tables 2.2 and 2.3 for examples) are totally empirical and are valid for only a limited range of temperatures (300 to 2000 K is typical, assuming

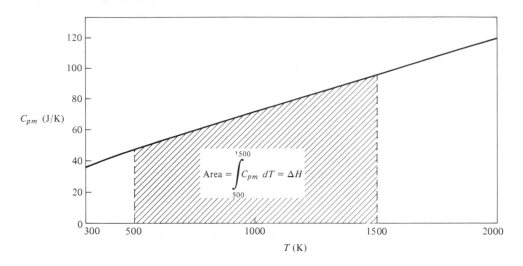

Figure 2.5 Enthalpy change on heating. The enthalpy change for heating one mole of methane from 500 K to 1500 K at constant pressure is the area under the constant-pressure heat capacity between those two temperatures.

the substance has no phase changes in this interval). If such data are available, Eq. (2.25) can be integrated directly.

Example: Calculate ΔH for heating one mole Al_2O_3 from 300 to 500 K. A reference book lists $C_{pm} = 78.99$ J/K for this material. With Eq. (2.26a):

$$\Delta H = 78.99(200) = 15.80 \text{ kJ}$$

But this can be done more accurately. Table 2.2 gives:

$$C_{pm} = 114.8 + (12.8 \times 10^{-3})T - \frac{35.4 \times 10^5}{T^2}$$

At the mean temperature (400 K) this formula gives $C_{pm} = 97.80$ J/K. With Eq. (2.26b):

$$\Delta H = 97.80(200) = 19.56 \text{ kJ}$$

Of course, in this case there is no reason not to calculate the integral exactly:

$$\Delta H = n \int_{300}^{500} C_{pm} \, dT$$

$$= \left[114.8(500 - 300) + (12.8 \times 10^{-3}) \left(\tfrac{1}{2}\right) [(500)^2 - (300)^2] \right.$$

$$\left. + (35.4 \times 10^5) \left(\frac{1}{500} - \frac{1}{300} \right) \right]$$

$$= 19.26 \text{ J/K}$$

Comparison of these results will give some idea of the accuracy of the various methods. If ΔT had been smaller, the difference would have been less significant. ■

Table 2.2   Molar heat capacities at constant pressure

(units: $JK^{-1}\ mol^{-1}$)

$$C^{\theta}_{pm} = a + bT + \frac{c}{T^2}$$

Gases[a]	a	$10^3 b$	$10^{-5} c$	Liquids[b]	a	$10^3 b$	$10^{-5} c$
H_2	27.28	3.26	0.50	I_2	80.33	—	—
O_2	29.96	4.18	−1.67	H_2O	75.48	—	—
N_2	28.58	3.77	−0.50	$C_{10}H_8$	79.50	407.5	—
CO	28.41	4.10	−0.46				
Cl_2	37.03	0.67	−2.85	Solids[c]			
Br_2	37.32	0.50	−1.26				
I_2	37.40	0.59	−0.71	C (graphite)	16.86	4.77	−8.54
CO_2	44.23	8.79	−8.62	Al	20.67	12.38	—
H_2O	30.54	10.29	—	Cu	22.64	6.28	—
H_2S	32.68	12.38	−1.92	Pb	22.13	11.72	0.96
NH_3	29.75	25.10	−1.55	I_2	40.12	49.79	—
CH_4	23.67	47.86	−1.92	$C_{10}H_8$	−116	937	—
				Al_2O_3	114.8	12.8	−35.4
				CuO	38.8	20.1	—
				Ag	21.3	8.54	−1.5
				AgCl	62.26	4.18	−11.3

[a]For gases the constants apply from 298 K to 2000 K for the ideal gas standard state.
[b]For liquids the constants apply at 1 atm from melting point to boiling point.
[c]For solids constants apply at 1 atm from 298 K to melting point or 2000 K.

Source: G. N. Lewis, M. Randall, K. Pitzer, and L. Brewer, *Thermodynamics,* 1961: New York, McGraw-Hill Book Co. (and miscellaneous others).

Table 2.3   Heat capacities of gases

(units: $J\ K^{-1}\ mol^{-1}$; range, bp, or 300 K to 1500 K)

$$C^{\theta}_{pm} = a' + b'T + c'T^2 + d'T^3 \quad (J/K)$$

Gas	a'	$b' \times 10^3$	$c' \times 10^6$	$d' \times 10^9$
Acetone	8.468	269.45	−143.45	29.63
n-Butane	−0.050	387.04	−200.82	40.61
Ethane	5.351	177.67	−68.70	8.514
Ethanol	14.97	208.56	−71.090	—
Ethylene	11.32	122.01	−39.90	—
Methane	14.15	75.496	−17.99	—
Propane	−5.058	308.50	−161.78	33.31
Sulfur dioxide	25.72	57.923	−38.09	8.606
Sulfur trioxide	15.10	151.92	−120.62	36.19

Source: Tables of Physical and Chemical Constants, 13th ed., 1966: New York, John Wiley & Sons.

Changes of Enthalpy with Pressure (Constant *T*)

For solids or liquids, the coefficient of thermal expansion (α) can be used with Eq. (2.23) to give:

$$\left(\frac{\partial H}{\partial P}\right)_T = V - TV\alpha = V(1 - \alpha T)$$

Then Eq. (2.24) becomes:

$$dH = nC_{pm}\,dT + nV_m(1 - \alpha T)\,dP \tag{2.28}$$

This equation could also be used for gases (with the appropriate formula for α), but a better approach for moderate pressures is to use a one-term virial equation:

$$V_m = \frac{RT}{P} + B(T)$$

The derivative of *V* with respect to *T* is then:

$$\left(\frac{\partial V_m}{\partial T}\right)_P = \frac{R}{P} + B'$$

where:

$$B' = \frac{dB}{dT}$$

is the derivative of $B(T)$. Then Eq. (2.23) gives:

$$\left(\frac{\partial H}{\partial P}\right)_T = n[V_m - T(R/P + B')]$$

$$= \frac{nRT}{P} + nB - \frac{nRT}{P} - nB'T$$

$$= n(B - TB')$$

Equation (2.24) then becomes:

$$dH = nC_{pm}\,dT + n(B - TB')\,dP \tag{2.29}$$

Exercise: Use the Berthelot second virial coefficient [Eq. (1.17)] to show:

$$\frac{dB}{dT} = \frac{27RT_c^3}{32P_cT^3} \tag{2.30}$$

and, therefore:

$$\left(\frac{\partial H_m}{\partial P}\right)_T = \frac{9RT_c}{128P_c}\left(1 - \frac{18T_c^2}{T^2}\right) \tag{2.31}$$

∎

Example: Calculate ΔH for increasing the pressure on one mole of (a) benzene liquid (use data from Table 1.8) and (b) CH_4(gas) (use data from Table 1.1), from 1 atm to 100 atm at 20°C.

For benzene, $\alpha - 1.237 \times 10^{-3} \text{ K}^{-1}$, $V_m = 89 \text{ cm}^3$:

$$\Delta H = 89[1 - 1.237 \times 10^{-3}(293)] \int_1^{100} dP$$

$$= 5.62 \times 10^3 \text{ cm}^3 \text{ atm} = 0.57 \text{ kJ}$$

For CH_4, $T_c = 190.6 \text{ K}$, $P_c = 45.8$ atm:

$$\Delta H = \frac{9(8.314)(190.6)}{128(45.8)}\left[1 - 18\left(\frac{190.6}{293}\right)^2\right](99 \text{ atm}) = -1.59 \text{ kJ}$$

(Check units and numerical results for both of these examples.) ∎

The example above illustrates that the change in H with moderate pressures changes is relatively small. As we found for U, it is the temperature dependence of H which is most significant in most cases, and the first term of Eq. (2.24) is often used alone. The second term of Eq. (2.24) is rigorously zero for ideal gas [for which $(\partial H/\partial P)_T = 0$] and for all processes at constant pressure ($dP = 0$).

Changing Both *T* and *P*

If the pressure term of Eq. (2.24) [or (2.28), or (2.29)] is to be integrated for a process for which the temperature is changing, what temperature should be used to calculate the coefficient? There is also a problem in integrating the T term when P is variable, since C_p will generally depend on pressure (a fact not mentioned until now). The resolution of these problems is in the realization that H is a state variable whose change will depend only on the initial and final states and not on the details of the intervening process. Thus, the ΔH for changing:

$$\text{state 1 } (T_1, P_1) \longrightarrow \text{state 2 } (T_2, P_2)$$

can, regardless of how the actual process occurs, be broken into two steps — either:

$$(T_1, P_1) \xrightarrow{\text{step A}} (T_2, P_1) \xrightarrow{\text{step B}} (T_2, P_2)$$

or:

$$(T_1, P_1) \xrightarrow{\text{step C}} (T_1, P_2) \xrightarrow{\text{step D}} (T_2, P_2)$$

Since the final state is the same, the ΔH calculated by either of these paths must be the same.

Before doing an example, we need to find the effect of pressure on C_p. The heat capacity of an ideal gas (which we denote C_p^θ) will not depend on pressure, and the heat capacity of a real gas will approach this value as $P \to 0$. The heat capacity at some finite pressure can be related to the value at zero pressure with a Taylor series:

$$C_p(P) = C_p^\theta + \left(\frac{\partial C_p}{\partial P}\right)_{P=0} P + \cdots \tag{2.32}$$

(The same formula would apply to solids or liquids, but then C_p^θ would be the heat capacity at 1 atm — this would not differ significantly from the value at $P = 0$.) The slope of C_p vs. P is [using the definition, Eq. (2.22)]:

$$\frac{\partial C_p}{\partial P} = \frac{\partial}{\partial P}\left(\frac{\partial H}{\partial T}\right) = \frac{\partial}{\partial T}\left(\frac{\partial H}{\partial P}\right)$$

From Eq. (2.23) we get:

$$\frac{\partial C_p}{\partial P} = \frac{\partial}{\partial T}\left(V - T\frac{\partial V}{\partial T}\right) = \frac{\partial V}{\partial T} - \frac{\partial V}{\partial T} - T\frac{\partial^2 V}{\partial T^2}$$

Therefore:

$$\left(\frac{\partial C_p}{\partial P}\right)_T = -T\left(\frac{\partial^2 V}{\partial T^2}\right)_P \qquad (2.33)$$

For gases, we could use the one-term virial equation:

$$V_m = \frac{RT}{P} + B$$

to give:

$$\left(\frac{\partial V_m}{\partial T}\right)_P = \frac{R}{P} + B'$$

$$\left(\frac{\partial^2 V_m}{\partial T^2}\right)_P = B''$$

where $B'' = d^2B/dT^2$. Then using Eqs. (2.33) and (2.32) (and dividing by n):

$$C_{pm}(P) = C_{pm}^\theta - TB''P \qquad (2.34a)$$

If C_p is known at 1 atm, the same equation can be used with $(P - 1 \text{ atm})$ in place of P:

$$C_{pm}(P) = C_{pm}(1 \text{ atm}) - TB''(P - 1 \text{ atm}) \qquad (2.34b)$$

[Be careful with units when using Eqs. (2.34).]

Exercise: Show that the Berthelot second virial coefficient gives:

$$C_{pm}(T, P) = C_{pm}^\theta(T) + \frac{81R}{32}\left(\frac{T_c}{T}\right)^3 \frac{P}{P_c} \qquad (2.35)$$

The reliability of this equation is questionable. Generally speaking, an approximate relationship such as the Berthelot second virial coefficient will be significantly less accurate for the derivative [for example, Eq. (2.30)] and less yet for the second derivative [as in Eq. (2.35)]. A common rule of thumb is that one order of magnitude is lost in accuracy for each derivative; thus, if an equation has 1% accuracy for the quantity, the first derivative may be in error by 10% and the second derivative by 100%. ∎

Example: We shall calculate the enthalpy change for one mole of argon between $T_1 = 300$ K, $P_1 = 1$ atm and $T_2 = 400$ K, $P_2 = 20$ atm. For Ar (Table 2.1), $C_{pm} = 20.93$ J/K at 1 atm; this number will be reasonably independent of temperature. (This calculation is diagrammed in Figure 2.6. The reader should fill in the missing steps of this example as an exercise.)

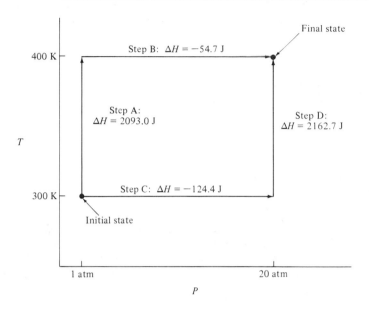

Figure 2.6 The path integral of a state function. The enthalpy change of a process depends only on the initial and final states and is independent of the path. This diagram illustrates two paths for changing the temperature and pressure of argon from 300 K, 1 atm to 400 K, 20 atm; the enthalpy change along either path (or any other path) between these two states is the same.

Step A: Heat from 300 to 400 K at constant $P = 1$ atm:

$$\Delta H_A = 20.93(100) = 2093 \text{ J}$$

Step B: Expand from 1 to 20 atm at constant $T = 400$ K. Equation (2.31) gives (using $T_c = 151$ K, $P_c = 48$ atm):

$$\left(\frac{\partial H}{\partial P}\right)_T = 1.839\left[1 - 18\left(\frac{151}{T}\right)^2\right] \quad \text{(J/atm)}$$

For $T = 400$, $(\partial H/\partial P)_t - -2.88$ J/atm. Therefore:

$$\Delta H_B = -(2.88 \text{ J/atm})(20 - 1) \text{ atm} = -54.7 \text{ J}$$

For the total process:

$$\Delta H = \Delta H_A + \Delta H_B = 2038.3 \text{ J}$$

We shall carry an extra significant figure to show that the two paths give precisely the same result. (As always, we carry extra significant figures in intermediate answers to avoid cumulative round-off errors.)

The calculation by the other path is as follows:

Step C: Expand from 1 to 20 atm at constant $T = 300$ K. The calculation, as in step B but with $T = 300$, gives $(\partial H/\partial P)_T = -6.547$ J/atm.

$$\Delta H_C - -6.547(20 - 1) = -124.4 \text{ J}$$

Step D: Heat from 300 to 400 K at constant $P = 20$ atm. Equation (2.35) with $T_c = 151$ K, $P_c = 48$ atm, gives (with $\Delta P = 20 - 1 = 19$ atm, since C_p at 1 atm is being used):

$$C_{pm} = 20.93 + \frac{3.019 \times 10^7}{T^3} \quad \text{(J/K)}$$

$$\Delta H_D = \int_{300}^{400} C_{pm}\, dT$$

$$= 20.93(400 - 300) - (2.868 \times 10^7)(0.5)\left[\frac{1}{(400)^2} - \frac{1}{(300)^2}\right]$$

$$= 2093 + 69.7 = 2162.7 \text{ J}$$

The total for this path is:

$$\Delta H = \Delta H_C + \Delta H_D = 2038.3 \text{ J}$$

This is, of course, exactly the same answer we calculated by the other path. The answers are surely not as accurate as indicated (even with the proper number of significant figures), but the answers are the same because the same approximations entered into both calculations in the same way. Obviously the calculations for one path (A + B) required less effort, and such will usually be the case. ∎

Recapitulation

A great number of formulas relating to energy and enthalpy have been covered so far in this chapter, but many of the effects discussed were found to be quite small for moderate pressure changes. It will probably be useful to restate the most important of the relationships which have been presented, together with their restrictions.

(A) $$\Delta U = n \int C_{vm}\, dT$$

1. Exact for ideal gas (any process).
2. Exact for constant-volume processes (any substance).
3. A reasonable approximation for most processes in which T changes, if V change is moderate.

(B) $$\Delta H = n \int C_{pm}\, dT$$

1. Exact for ideal gas (any process).
2. Exact for constant pressure process (any substance).
3. A reasonable approximation for most processes in which T changes, if P change is moderate.

(C) $$q = \Delta U$$

Exact for any process at constant V.

(D) $$q = \Delta H$$

Exact for any process at constant P with only PV work.

(E) $$q = \Delta U - \int P_{\text{ex}}\, dV$$

Exact for any process with only PV work.

2.6   The Relationship of C_p to C_v

In general, the heat capacity at constant pressure is more easily measured than that at constant volume, particularly for condensed phases. Remembering the example in Chapter 1 (p. 61), liquids and solids heated at constant volume will exert very large pressures on the container that confines them. On the other hand, it is often necessary to know C_v for certain calculations, and theoretical explanations are usually more easily achieved for C_v. Therefore, being able to calculate C_p from C_v or vice versa is an important topic, which we shall now discuss.

We begin with the definitions:

$$C_p = \left(\frac{\partial H}{\partial T}\right)_P, \qquad H = U + PV$$

These give:

$$C_p = \left[\frac{\partial(U + PV)}{\partial T}\right]_P = \left(\frac{\partial U}{\partial T}\right)_P + P\left(\frac{\partial V}{\partial T}\right)_P$$

The slope $(\partial U/\partial T)_P$ is not at all the same as $(\partial U/\partial T)_V$, but these can be related by the formula of Appendix II for changing the "constant" part of a partial derivative; in this case:

$$\left(\frac{\partial U}{\partial T}\right)_P = \left(\frac{\partial U}{\partial T}\right)_V + \left(\frac{\partial U}{\partial V}\right)_T\left(\frac{\partial V}{\partial T}\right)_P$$

Using this in the equation for C_p (above) gives:

$$C_p = \left(\frac{\partial U}{\partial T}\right)_V + \left(\frac{\partial U}{\partial V}\right)_T\left(\frac{\partial V}{\partial T}\right)_P + P\left(\frac{\partial V}{\partial T}\right)_P$$

The first slope on the right-hand side of this equation is, of course, C_v, so:

$$C_p = C_v + \left[P + \left(\frac{\partial U}{\partial V}\right)_T\right]\left(\frac{\partial V}{\partial T}\right)_P \qquad (2.36)$$

Exercise: Earlier it was shown that $(\partial U/\partial V)_T = 0$ for an ideal gas. Also, with $V = nRT/P$:

$$\left(\frac{\partial V}{\partial T}\right)_P = \frac{nR}{P}$$

so, from Eq. (2.36):

$$C_p = C_v + nR$$

This will be more conveniently written in terms of the molar heat capacities. Dividing by n:

(ideal gas) $\qquad C_{pm} = C_{vm} + R$ $\qquad\qquad\qquad$ **(2.37)**

The reader should peruse Table 2.1 to see how well this relationship works for a variety of real gases at 15°C, one atmosphere.

There is a more direct way to derive Eq. (2.37). It was shown earlier that ΔU for heating an ideal gas was (regardless of changes in P):

$$\Delta U = n \int_{T_1}^{T_2} C_{vm} \, dT$$

Similarly, for the enthalpy change of an ideal gas:

$$\Delta H = n \int_{T_1}^{T_2} C_{pm} \, dT$$

Now the relationship between ΔH and ΔU is given by Eq. (2.19):

$$\Delta H = \Delta U + (P_2 V_2 - P_1 V_1)$$

For an ideal gas, $PV = nRT$ and:

$$\Delta H = \Delta U + nR(T_2 - T_1)$$

$$n \int_{T_1}^{T_2} C_{pm} \, dT = n \int_{T_1}^{T_2} C_{vm} \, dT + nR \int_{T_1}^{T_2} dT$$

Therefore, $C_{pm} = C_{vm} + R$ for an ideal gas. $\qquad\qquad\qquad$ ■

A general relationship can be derived by using Eq. (2.13) with Eq. (2.36); this gives:

$$C_{pm} = C_{vm} + T \left(\frac{\partial V_m}{\partial T} \right)_P \left(\frac{\partial P}{\partial T} \right)_V \qquad\qquad \textbf{(2.38)}$$

For solids and liquids, this relation is more conveniently written in terms of the coefficients of thermal expansion and isothermal compressibility with [compare Eq. (1.59)]:

$$\left(\frac{\partial V_m}{\partial T} \right)_P = V_m \alpha , \qquad \left(\frac{\partial P}{\partial T} \right)_V = \frac{\alpha}{\kappa_T}$$

Therefore:

$$C_{pm} = C_{vm} + \frac{T V_m \alpha^2}{\kappa_T} \qquad\qquad \textbf{(2.39)}$$

Equation (2.38) can, of course, also be evaluated with an equation of state. Equations of state are usually explicit in either P or V (excepting the ideal gas law, which can be either), so one or the other of the derivatives needed for Eq. (2.38) is likely to be inconvenient. This can be remedied by use of the cyclic rule for partial derivatives (Appendix II). For equations of state explicit in P we can use:

$$\left(\frac{\partial V}{\partial T}\right)_P = -\frac{\left(\frac{\partial P}{\partial T}\right)_V}{\left(\frac{\partial P}{\partial V}\right)_T}$$

and for equations of state explicit in V:

$$\left(\frac{\partial P}{\partial T}\right)_V = -\frac{\left(\frac{\partial V}{\partial T}\right)_P}{\left(\frac{\partial V}{\partial P}\right)_T}$$

Example: Calculate $C_p - C_v$ using the one-term virial equation:

$$V_m = \frac{RT}{P} + B(T)$$

$$\frac{\partial V_m}{\partial T} = \frac{R}{P} + B'$$

$$\frac{\partial V_m}{\partial P} = -\frac{RT}{P^2}$$

Therefore:

$$\left(\frac{\partial P}{\partial T}\right)_V = -\frac{\dfrac{R}{P} + B'}{-\dfrac{RT}{P^2}} = \frac{P(R + B'P)}{RT}$$

Using Eq. (2.38):

$$C_{pm} = C_{vm} + T\left(\frac{R}{P} + B'\right)\frac{P(R + B'P)}{RT}$$

With a little bit of algebra, you should get the convenient equation:

$$C_{pm} = C_{vm} + R\left(1 + \frac{B'P}{R}\right)^2 \qquad \textbf{(2.40)}$$

∎

Example: Calculate $C_{pm} - C_{vm}$ for CO_2 (gas) and H_2O (liquid) at 20°C, one atmosphere. For CO_2 we shall use the Berthelot second virial coefficient [Eq. (2.30)] and data from Table 1.1:

$$\frac{B'P}{R} = \frac{27}{32}\left(\frac{T_c}{T}\right)^3\frac{P}{P_c}$$

$$= \frac{27}{32}\left(\frac{304.2}{293.15}\right)^3\frac{1}{73}$$

$$= 0.0129$$

$$C_{pm} - C_{vm} = R(1.0129)^2$$

$$= 8.53 \text{ J/K}$$

Recall that the ideal gas law [Eq. (2.37)] predicts $C_{pm} - C_{vm} = 8.31$ J/K; this would be exact at $P = 0$. For water, we use Eq. (2.39) and data from Table 1.8:

$$C_{pm} - C_{vm} = \frac{(293.15 \text{ K})(18 \text{ cm}^3)(0.188 \times 10^{-3}/\text{K})^2}{45 \times 10^{-6}/\text{atm}}$$

$$= 4.14 \text{ cm}^3 \text{ atm/K}$$

$$= 0.42 \text{ J/K}$$

This is clearly a smaller difference than for gases. Often it is sufficiently accurate to assume that the constant-pressure and constant-volume heat capacities of condensed phases are equal.   ∎

2.7   Expansions of Gases

In this section we shall derive several results for expansions of gases under various conditions. Gas expansions are technologically important for the work they provide, for example in a steam engine or an internal combustion engine. Also, the cooling effect when gases expand is the basis for most refrigerators and air conditioners. The equations we derive will be useful in Chapter 3 when we discuss the second law of thermodynamics. And, most important of all, we shall introduce for the first time a new concept, *reversibility*.

When a gas expands, two things can happen: it may do work on its surroundings, and the temperature of the gas may drop. The amount of work it does depends on the opposing pressure, P_{ex}, and the degree of cooling depends on the extent to which heat is able to enter the gas from the surroundings. This is generally idealized in terms of two limiting types of processes: (1) In an *isothermal* expansion, the gas is in thermal contact with its surroundings (presumed isothermal) and sufficient heat flows into the gas to keep the temperature constant. (2) In an *adiabatic* expansion, no heat enters the gas during the expansion; that is, $q = 0$.

Isothermal Reversible Expansion

The work done by a gas expanding from V_1 to V_2 is:

$$-w = \int_{V_1}^{V_2} P_{ex} \, dV$$

Obviously, the larger the opposing pressure (for a given ΔV), the more work will be done. There is a limit to how large P_{ex} can be, of course, since if it is greater than the pressure of the driving gas (P), no expansion will occur at all. How can the work be maximized?

Consider an ideal gas in a cylinder with a movable piston, with pressure $P_1 = 10$ atm and volume $V_1 = 1$ dm^3. Let us assume that the opposing pressure (also 10 atm at the start) consists of 1 atm due to the air outside and 9 atm due to nine weights sitting on the piston (each exerting 1 atm pressure). If all weights are removed at once, the gas will expand against the constant $P_{ex} = 1$ atm (due to the air in the room) until its pressure is 1 atm — that is, until it reaches equilibrium. If the expansion is isothermal, its final volume will be 10 dm³ and the work done is:

$$-w = (1 \text{ atm}) (10 - 1) \text{ dm}^3 = 9 \text{ dm}^3 \text{ atm} = 912 \text{ J}$$

If, instead, we remove the weights in two stages, the process is:

$$
\boxed{\begin{array}{c} 10 \text{ atm} \\ 1 \text{ dm}^3 \end{array}}
\xrightarrow[P_{ex}=5]{\text{Stage 1}}
\boxed{\begin{array}{c} 5 \text{ atm} \\ 2 \text{ dm}^3 \end{array}}
\xrightarrow[P_{ex}=1]{\text{Stage 2}}
\boxed{\begin{array}{c} 1 \text{ atm} \\ 10 \text{ dm}^3 \end{array}}
$$

In the first stage five weights are removed, $P_{ex} = 5$, and the work is:

$$-w = 5(2 - 1) = 5 \text{ dm}^3 \text{ atm}$$

In the second stage the other four weights are removed, leaving $P_{ex} = 1$ and:

$$-w = 1(10 - 2) = 8 \text{ dm}^3 \text{ atm}$$

The total work is then:

$$-w = 13 \text{ dm}^3 \text{ atm} = 1318 \text{ J}$$

Exercise: Demonstrate that, if the weights are removed one at a time, the work of expansion from ($V = 1$, $P = 10$) to ($V = 10$, $P = 1$) is:

$$-w = 1954 \text{ J} \qquad \blacksquare$$

We could carry this process to an extreme by using, in place of the weights, a pile of sand, which is removed one grain at a time. The maximum work will be obtained if, at each stage of the expansion, the pressure of the gas is only infinitesimally greater than P_{ex}, $P = P_{ex} + dP$, and then in the limit $dP \rightarrow 0$ we sum the infinity of steps to get (with $P_{ex} = P = nRT/V$ for an ideal gas):

$$-w_{max} = nRT \int_{V_1}^{V_2} \frac{dV}{V}$$

$$-w_{max} = nRT \ln\left(\frac{V_2}{V_1}\right) \qquad (2.41)$$

For the expansion above, ($P = 10$, $V = 1$) to ($P = 1$, $V = 10$), the maximum work is (with $nRT = PV = 10 \text{ dm}^3 \text{ atm}$):

$$-w_{max} = (10 \ln 10) \text{ dm}^3 \text{ atm} = 2333 \text{ J}$$

This limiting case is an example of a *reversible* process. (See Figure 2.7.)

In thermodynamics, a reversible change is a change which occurs through a succession of equilibrium states. Since "equilibrium" implies that the system has reached a state where no further change is possible, this may seem like a contradiction in terms. The preceding discussion demonstrates how this could be accomplished. The gas in its initial state, $P = P_{ex} = 10$ atm, is at equilibrium; when a weight is removed,

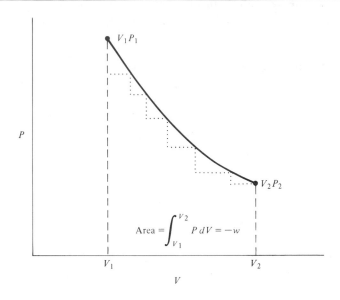

Figure 2.7   Reversible work. The work for an isothermal reversible expansion is the area under the isotherm (solid curve); it can be approximated by a series of constant-pressure expansions (dotted lines), and the approximation becomes closer as the steps become smaller and their number greater.

$P_{ex} < P$, the gas expands to a new equilibrium state. However, during the expansion it is not at equilibrium, so the expansion is *irreversible*. A reversible expansion can be approached if the opposing pressure is reduced gradually in small steps; a truly reversible process occurs in the limit of an infinite number of infinitesimal steps.

If a reversible process is purely hypothetical limiting case, what is its relevance? In the first place, it tells us the *maximum* work that can be obtained from a given process. (Because of the sign convention we are using, this is really the maximum of $-w$.) Also, the concept of reversibility simplifies many thermodynamic calculations by permitting us to use P (the actual pressure of the gas) in place of P_{ex}; state variables such as U and H have changes (ΔU, ΔH) that depend only on the final state and not on the process, so their changes calculated along a reversible path will be the same as the actual change for an irreversible path between the same states. This is a very important point: changes in state variables for any type of process, including irreversible processes, may be calculated along a reversible path, however hypothetical, that goes between the same two states.

For the example we have just discussed, an isothermal expansion of an ideal gas, $\Delta U = 0$ in all cases, because the internal energy of an ideal gas depends only on T — that is, $(\partial U / \partial V)_T = 0$. Therefore, the heat absorbed must always exactly balance the work (since $\Delta U = q + w = 0$):

Isothermal Expansion, $(P = 10, V = 1)$ to $(P = 1, V = 10)$:

1. One-step: $w = -912$ J, $q = 912$ J, $\Delta U = 0$
2. Two-step: $w = -1318$ J, $q = 1318$ J, $\Delta U = 0$
3. Ten-step: $w = -1954$ J, $q = 1954$ J, $\Delta U = 0$
4. Reversible: $w = -2333$ J, $q = 2333$ J, $\Delta U = 0$

This example goes to the heart of *why* thermodynamics prefers to deal with state variables; this advantage will be seen to be even more pronounced in Chapter 3.

Adiabatic Reversible Expansion

For an adiabatic process, $q = 0$, and Eq. (2.11) becomes $dU = dw$ or

$$dU = -P_{ex} dV \qquad (2.42)$$

The expansion begins in a state (T_1, P_1, V_1) and goes to a state (T_2, P_2, V_2) for which only V_2 (or perhaps P_2) is known; the problem is to find the final temperature. To do this, we make two assumptions: (1) The expansion is reversible, so we can replace P_{ex} with $P(T, V)$ of the gas. (2) The gas is ideal, so that (a) $P(T, V)$ can be replaced by nRT/V and (b) the internal energy change [for example, Eq. (2.16)] is simply $C_v dT$. With these assumptions, Eq. (2.42) becomes:

$$nC_{vm} dT = -nRT \frac{dV}{V}$$

The variables T and V are easily separated:

$$C_{vm} \frac{dT}{T} = -R \frac{dV}{V}$$

For a change (T_1, V_1) to (T_2, V_2):

$$\int_{T_1}^{T_2} C_{vm} d(\ln T) = -R \ln \frac{V_2}{V_1} \qquad (2.43)$$

Consideration of the explicit temperature dependence of C_v can result in rather complicated equations, particularly if, as is common, the final temperature is unknown. However, the mean value (as defined in Appendix I) can be used to give the simple result:

$$\overline{C}_{vm} \ln \frac{T_2}{T_1} = -R \ln \frac{V_2}{V_1} \qquad (2.44)$$

This may seem to be small comfort, since we need to know the final temperature to determine $\overline{C}_{vm}$; however, a short series of successive approximations always suffices to give an answer to an accuracy of a few degrees.

Example: Calculate the final temperature when CO_2, $V_m = 2.0$ dm³, $T = 500$ K, is expanded to $V_m = 20$ dm³. Using C_p from Table 2.2, and $C_{vm} = C_{pm} - R$:

$$C_{vm} = (44.23 - 8.314) + (8.79 \times 10^{-3})T - \frac{8.62 \times 10^5}{T^2}$$

At a few sample temperatures, this equation gives:

$$T = 500 \text{ K}: \quad C_{vm} = 36.86 \text{ J/K}$$

$$T = 400 \text{ K}: \quad C_{vm} = 34.04 \text{ J/K}$$

$$T = 300 \text{ K}: \quad C_{vm} = 28.97 \text{ J/K}$$

The final temperature is to be calculated with Eq. (2.44):

$$\ln \frac{T_2}{500} = -\frac{8.314}{\overline{C}_{vm}} \ln 10$$

Not knowing the final temperature, we shall use C_{vm} at 400 K to estimate:

$$\ln \frac{T_2}{500} \cong -\frac{8.314}{34.04} \ln 10$$

$$T_2 \cong 285 \text{ K}$$

Now, if we approximate $\overline{C}_{vm}$ by the heat capacity at the mean temperature:

$$\overline{T} = \frac{500 + 285}{2} = 392.5 \text{ K}$$

We estimate:

$$\overline{C}_{vm} = 33.77 \text{ J/K}$$

This value gives the answer $T_2 = 284$ K, and the successive approximation series can be seen to have converged.

An alternative approximation for $\overline{C}_{vm}$ is the mean heat capacity:

$$\overline{C}_{vm} = \tfrac{1}{2}[C_{vm}(500) + C_{vm}(T_2)]$$

With the first estimate, $T_2 = 285$ K, this gives $\overline{C}_{vm} = 32.34$ J/K, which gives the second approximation $T_2 = 277$ K. Neither of these answers is precise because of the other approximations involved, so another round is not worth doing; in general, we shall use the first method ($\overline{C}$ is the heat capacity at the mean temperature) because it is less work. Either method is more reliable than assuming constant heat capacity at whatever temperature is convenient. (For example, use the heat capacity given in Table 2.1 for CO_2, $C_{vm} = 28.09$ J/K; you should get an answer of $T_2 = 253$ K. On the other hand, if the gas were monatomic, Ar for example, the value of C_{vm} in Table 2.1 could be used without significant error.)   ■

Equation (2.43) or (2.44) is convenient if the volume or temperature of the final state is known. What if only the pressure of the final state is known? As often happens when P is the independent variable, the use of the enthalpy function gives simple results:

$$H = U + PV$$

$$dH = dU + P\,dV + V\,dP$$

$$= dq + dw + P\,dV + V\,dP$$

For a reversible, adiabatic expansion, $dq = 0$ and $dw = -P\,dV$; hence:

$$dH = V\,dP$$

If we assume an ideal gas, Eq. (2.25) can be used for dH, and V can be replaced by nRT/P; therefore:

$$nC_{pm}\,dT = \frac{nRT}{P}\,dP$$

and

$$C_{pm}\frac{dT}{T} = R\frac{dP}{P} \tag{2.45}$$

Upon integration this equation gives:

$$\overline{C}_{pm}\ln\frac{T_2}{T_1} = R\ln\frac{P_2}{P_1} \tag{2.46}$$

Equations (2.44) and (2.46) give two different expressions for the temperature ratio, namely (assuming constant heat capacities):

$$\ln\left(\frac{T_2}{T_1}\right) = -\frac{R}{C_{vm}}\ln\frac{V_2}{V_1} = \frac{R}{C_{pm}}\ln\frac{P_2}{P_1}$$

These can be used to get the PV relations directly:

$$\ln\frac{P_2}{P_1} = -\frac{C_{pm}}{C_{vm}}\ln\frac{V_2}{V_1}$$

The ratio of the heat capacities is given the special symbol:

$$\gamma = \frac{C_p}{C_v} \tag{2.47}$$

This ratio can be measured independently; for example, the velocity of sound (c) in an ideal gas with molecular weight M is:

$$c = \sqrt{\gamma RT/M} \tag{2.48}$$

Using the heat capacity ratio (γ):

$$\ln\frac{P_2}{P_1} = -\gamma\ln\frac{V_2}{V_1} = \ln\left(\frac{V_1}{V_2}\right)^{\gamma}$$

or:

$$P_2V_2^{\gamma} = P_1V_1^{\gamma} \tag{2.49}$$

This equation shows that, during a reversible adiabatic expansion (ideal gas), the product PV^{γ} is a constant. This defines a path on a PV diagram called a **reversible adiabat:**

$$PV^{\gamma} = \text{constant} \tag{2.50}$$

This can be contrasted to the isotherm equation (Boyle's law):

$$PV = \text{constant}$$

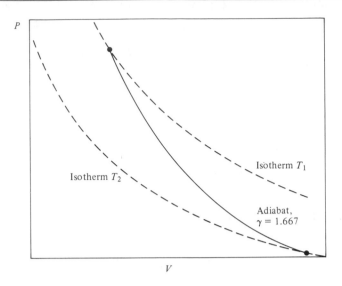

Figure 2.8 Adiabatic reversible expansion. The path of a reversible adiabatic expansion (solid line) as shown here for an ideal, monatomic gas ($\gamma = 1.667$) crosses the isotherms (shown as dashed lines) as the gas cools.

Figure 2.8 contrasts the path of a reversible isothermal and reversible adiabatic expansion between the same volumes, V_1 to V_2. In the next chapter we shall discover that the reversible adiabat has another significance: it is a line of constant entropy (isentrope).

The Joule Expansion

In an adiabatic expansion, the work done by the gas on the surroundings must be obtained at the expense of the internal energy. If the gas is ideal, U is a function of T only, so the loss of internal energy implies a drop in temperature. (This is also, in large part, true for real gases, since the T dependence of U is usually larger than the V dependence.) If the gas expanded and did no work, would the temperature change? For an ideal gas, as we've already implied, the answer is no; but let's look at the situation more closely.

In 1845, J. P. Joule attempted such an experiment. The apparatus (Figure 2.9) allows a gas to expand into a vacuum; since there is no opposing pressure, the work is zero. If, in addition, the gas is expanded adiabatically, $q = 0$ and the internal energy remains constant. This experiment thus measures the change in temperature with volume at constant U: $(\partial T/\partial V)_U$. How can this quantity be related to the gas law? We know [Eq. (2.13)] how to calculate $(\partial U/\partial V)_T$, so this may suggest the use of the cyclic rule for partial derivatives (Appendix II):

$$\left(\frac{\partial T}{\partial V}\right)_U = -\left(\frac{\partial T}{\partial U}\right)_V \left(\frac{\partial U}{\partial V}\right)_T = -\frac{1}{C_v}\left(\frac{\partial U}{\partial V}\right)_T \tag{2.51}$$

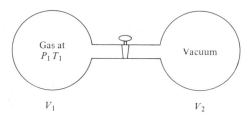

Figure 2.9 The Joule experiment. A gas is expanded into a vacuum; since there is no opposing pressure, the gas does no work in such an expansion. For an ideal gas, the temperature will not change since no work is done; if the experiment were carried out adiabatically with a real gas, the temperature would drop, but the change would be small unless the process started with a liquid or as a gas near the critical region.

We saw earlier that $(\partial U/\partial V)_T$ is zero for an ideal gas. For real gases there would be a small change, but Joule did not observe it because he had his apparatus immersed in a water bath; the whole system, taken as gas plus container, water bath, and thermometers, was reasonably adiabatic, but then the heat capacity in Eq. (2.51) is that of the whole system and is too large for a measurable temperature change to be observed. The experiment would have some chance of succeeding if the gas container were insulated by a vacuum, but then the measurement of the temperature would be a problem. (Because of heat transfer by radiation, the gas container would have to be opaque, so even a tiny thermometer of negligible heat capacity inside the bulb could not be seen.)

The Joule-Thomson Expansion

William Thomson (Lord Kelvin) helped Joule to devise a better experiment (Figure 2.10). A gas at high pressure P_1 is expanded through a throttle valve (originally a porous plug of cotton) into a region of lower pressure, P_2. The apparatus is insulated, so the experiment is adiabatic. Since this is a continuous experiment, with potentially large quantities of gas passing through, the heat capacity of the thermometers, walls, and so on will be negligible; after a while they will equilibrate with the gas and no longer supply heat to it, and a steady temperature reading will be obtained.

The Joule-Thomson expansion is adiabatic, but it is not reversible; nor is any of the variables U, V, T, or P constant. It can be shown, however, that the expansion is *isenthalpic* — that is, H is constant. This can be demonstrated with a schematic version of the experiment (Figure 2.11), in which n moles of gas, initially to the left of the valve at (T_1, V_1) are forced through the valve by a piston at a constant $P_1 = P_{ex}$. The work done on the gas is (since the final volume on the left is zero):

$$w(\text{on ``left"}) = -P_1(0 - V_1) = P_1V_1$$

After it passes through the valve, the gas does work by pushing out against a piston on the right, which resists with a constant pressure $P_2 = P_{ex}$:

$$w(\text{on ``right"}) = -P_2(V_2 - 0) = -P_2V_2$$

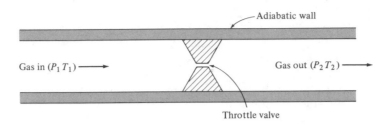

Figure 2.10 The Joule-Thomson experiment. A gas under pressure is expanded through a throttle valve. Such devices are used in most refrigeration equipment. In a refrigerator, the refrigerant gas is compressed by the motor and then allowed to cool by passing it through coils exposed to the room air. It is then expanded through a throttle valve, and the cooled gas is passed through a heat exchanger inside the refrigerator.

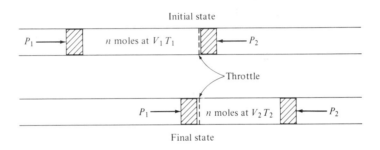

Figure 2.11 Schematic diagram for the Joule-Thomson experiment.

The net work is equal to the change in U, since the expansion is adiabatic, so:

$$U_2 - U_1 = -P_2V_2 + P_1V_1$$
$$U_2 + P_2V_2 = U_1 + P_1V_1$$
$$H_2 = H_1$$

That is, the enthalpy of the gas did not change during the expansion. (cf., Figure 2.12.)

The measurable quantity in this experiment is the change in temperature with pressure at constant enthalpy; this defines the **Joule-Thomson coefficient:**

$$\mu = \left(\frac{\partial T}{\partial P}\right)_H \tag{2.52}$$

The cyclic rule for partial derivatives is again useful:

$$\left(\frac{\partial T}{\partial P}\right)_H = -\left(\frac{\partial T}{\partial H}\right)_P\left(\frac{\partial H}{\partial P}\right)_T$$

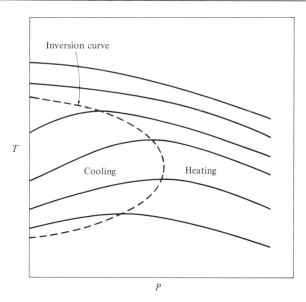

Figure 2.12 Lines of constant enthalpy (isenthalps) of a gas. These lines show the paths of Joule-Thomson expansions. The slope of the line $[(\partial T/\partial P)_H]$ at any point P, T is equal to the Joule-Thomson coefficient; if the slope is positive, the gas will cool upon expansion; if it is negative, the gas will warm upon expansion. (From J. G. Kirkwood and I. Oppenheim, *Chemical Thermodynamics*, 1961: New York, McGraw-Hill Book Co., p. 79)

or:

$$\mu = -\frac{1}{C_p}\left(\frac{\partial H}{\partial P}\right)_T \tag{2.53}$$

Combined with Eq. (2.23), this gives:

$$\left(\frac{\partial H}{\partial P}\right)_T = V - T\left(\frac{\partial V}{\partial T}\right)_P = -\mu C_p$$

Therefore:

$$\mu = \frac{T\left(\frac{\partial V}{\partial T}\right)_P - V}{C_p} \tag{2.54}$$

The Joule-Thomson coefficient is, like the heat capacity, a property of the material that depends on temperature and pressure.

Exercise: Show that the virial series:

$$V_m = \frac{RT}{P} + B + \gamma P + \delta P^2 + \cdots$$

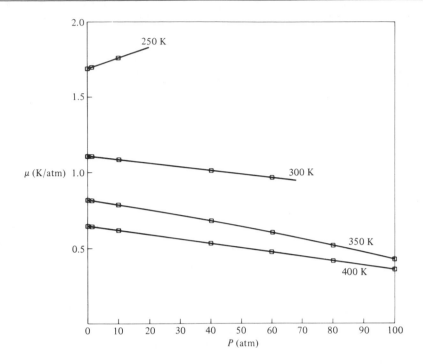

Figure 2.13 The Joule-Thomson coefficient of carbon dioxide vs. *P* at various temperatures. (Data from W. J. Moore, *Physical Chemistry*, 4th ed., 1972: Englewood Cliffs, NJ, Prentice-Hall, Inc.)

gives:

$$T\left(\frac{\partial V_m}{\partial T}\right) - V_m = (TB' - B) + (T\gamma' - \gamma)P + (T\delta' - \delta)P^2 + \cdots$$

[compare Eqs. (2.24) and (2.29)]. Also [compare Eq. (2.34)]:

$$C_{pm}(T, P) = C_{pm}^{\theta} - TB''P - T\gamma''P^2 + \cdots$$

Together, these results give for the Joule-Thomson coefficient:

$$\mu = \frac{(TB' - B) + (T\gamma' - \gamma)P + (T\delta' - \delta)P^2 + \cdots}{C_{pm}^{\theta} - TB''P - T\gamma''P^2 + \cdots} \tag{2.55}$$

$\blacksquare$

Equation 2.55 shows clearly that μ is a function of pressure (explicitly) and temperature (through the virial coefficients and their derivatives) (for example, see the plot of μ for CO_2 in Figure 2.13). Unfortunately, accurate data for effective use of Eq. (2.55) are usually lacking. The low-pressure limit is, however, quite useful and informative:

$$\text{(low pressure)} \qquad \mu C_{pm}^{\theta} = TB' - B \tag{2.56}$$

Table 2.4   Joule-Thomson coefficients at 1 atm

(units: K/atm)

Gas	0°C	100°C
He	−0.0616	−0.0638
H$_2$	−0.013	−0.039
N$_2$	0.2655	0.1291
O$_2$	0.366	0.193
Ar	0.43	0.23
CO$_2$	1.369	0.729
C$_2$H$_5$Cl	5.22	2.43
Air	0.2751	0.1371

Source: M. A. Paul, *Principles of Chemical Thermodynamics*, 1951: New York, McGraw-Hill Book Co.

The discussion and illustrations of Chapter 1 (Figure 1.2, for example) show that the slope $B' = dB/dT$ is generally positive and becomes zero (perhaps slightly negative) at high temperatures. At high temperatures, $B' \sim 0$ and B is positive, therefore μ [Eq. (2.56)] will be negative. A negative Joule-Thomson coefficient means [Eq. (2.52)] that the gas will warm during a Joule-Thomson expansion. At low temperatures, B will be negative while B' is positive, so Eq. (2.56) predicts that μ is positive (the gas cools during a Joule-Thomson expansion). The temperature where the effect changes from cooling ($\mu > 0$) to warming ($\mu < 0$)—that is, when $\mu = 0$—is called the *Joule-Thomson inversion temperature* (T_i). Most common gases, except for He and H$_2$, have positive Joule-Thomson coefficients at room temperature (Table 2.4) and, thus, will cool during a Joule-Thomson expansion that starts at room temperature.

From the data in Table 2.4 it can be seen that Joule-Thomson coefficients are quite small. Despite the small ΔT compared to reversible expansion, this effect finds application in refrigeration and gas liquefaction because of its mechanical simplicity—a simple throttle valve instead of a piston. In refrigerators, the effect is enhanced by compressing the gas until it liquefies, thus taking advantage of the heat of vaporization for cooling.

The Linde process for liquefaction of gases (Figure 2.14) utilizes the Joule-Thomson effect. The smallness of the temperature drop is compensated for by using a cumulative process whereby the cooled gas cools the incoming gas. Some gases, notably helium, must be prechilled to below their inversion temperature so that μ will be positive. Since helium liquefies at 4.2 K, the use of a method with no moving parts is a distinct advantage.

Example: If Eq. (2.30) (for the Berthelot second virial coefficient) is used with Eq. (2.56) [or, alternatively, Eq. (2.31) with Eq. (2.53)], we get:

$$\mu = \frac{9RT_c}{128 P_c C_{pm}} \left(\frac{18T_c^2}{T^2} - 1 \right) \tag{2.57}$$

From this equation we see immediately that the Berthelot equation predicts an inversion temperature:

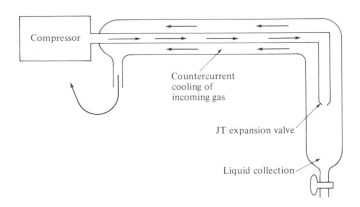

Figure 2.14 The Linde process for liquefying gases (schematic). The gas is run continuously through the throttle valve and used in a countercurrent heat exchanger to cool the incoming gas. When the temperature drops enough, a portion of the gas will liquefy and collect below the valve.

$$T_i = \sqrt{18} T_c$$

Remember that the same equation predicted (Chapter 1) the Boyle temperature to be:

$$T_B = \sqrt{6} T_c$$

These predictions can be tested with data in Table 2.5; the trend is certainly correct, but the accuracy is not, in general, outstanding.

Equation (2.57) can be used to estimate low-pressure Joule-Thomson coefficients. For CO_2 at 273.15 K, the data in Table 2.2 give $C_{pm} = 35.08$ J/K. Data in Table 1.1 then give:

$$\mu = \frac{(9)(8.314 \text{ J/K})(304.2 \text{ K})}{(35.08 \text{ J/K})(128)(73 \text{ atm})} \left[\frac{18(304.2)^2}{(273.15)^2} - 1 \right]$$

$$= 1.48 \text{ K/atm} \qquad \text{(Table 2.4 gives 1.369)}$$

(The reader should check the units of this calculation.)   ■

Table 2.5 Boyle and Joule-Thomson inversion temperatures of some gases

(units: K)

Gas	T_c	T_B	T_i	Gas	T_c	T_B	T_i
He	5.2	24.1	44.8	Ne	44.5	120	231.4
H_2	33.2	107.3	195	CO	134	342	—
N_2	126.0	324	621	O_2	154.3	423	764
Ar	150.8	410	723	CH_4	190.6	509.7	967.8
Air	132.5	347	603	CO_2	304.2	650	~1500

2.8 The Standard State

The absolute values of the state functions U and H cannot be measured; this is of no consequence, since only differences are ever needed. Often enthalpies are tabulated with respect to some arbitrary state; for example, steam tables will tell you the enthalpy of water and steam at various temperatures relative to that of water at 0°C. In Chapter 5 we shall learn how enthalpies can be calculated relative to the enthalpy of the same substance at zero Kelvin. Relative enthalpies of different substances, for example H_2 and O_2, cannot be measured, because there is no process that can convert one to the other (or, at least, no *chemical* process). However, as we shall see in Chapter 6, relative enthalpies for reactions such as:

$$2H_2 + O_2 \longrightarrow 2H_2O$$

for which:

$$\Delta H = 2H_m(H_2O) - 2H_m(H_2) - H_m(O_2)$$

can be measured from the heat of the reaction.

Thermodynamic quantities such as U, H, and C_p are often given for the *standard state* (denoted as a right superscript θ). For solids and liquids the standard state is the substance under 1 atm applied pressure. For gases the standard state is the equivalent ideal gas at 1 atm pressure—that is, the gas at 1 atm, subtracting off the effect of nonideality (intermolecular interactions). Since H, U, C_v, and C_p of an ideal gas do not depend on pressure, we can find the standard values by extrapolating to $P = 0$:

$$H^\theta \text{ (ideal, standard state)} = H \text{ (real gas as } P \to 0)$$

Figure 2.15 diagrams this concept. Equations in this chapter permit calculation of this "nonideality" factor:

$$U_m(T, V) = U_m^\theta(T) - \int_\infty^{V_m} \left[P - T\left(\frac{\partial P}{\partial T}\right)_V \right] dV_m \tag{2.58}$$

$$H_m(T, P) = H_m^\theta(T) + \int_0^P \left[V_m - T\left(\frac{\partial V_m}{\partial T}\right)_P \right] dP \tag{2.59}$$

The usual emphasis is on the enthalpy; however, since these quantities refer to an *ideal* gas, it should be clear that $H_m^\theta = U_m^\theta + RT$.

For solids or liquids, the material is usually assumed to be incompressible (that is, the volume is constant, independent of P), and the result is given in terms of the coefficient of thermal expansion (α):

$$H_m - H_m^\theta = V_m(1 - \alpha T)(P - P^\theta) \tag{2.60}$$

Example: Calculate $H - H^\theta$ for CCl_4(liq.) at 25°C and 2 atm (use data from Table 1.8):

$$H_m - H_m^\theta = (97 \text{ cm}^3)[1 - 298(1.24 \times 10^{-3})](1 \text{ atm})(0.1012 \text{ J/cm}^3 \text{ atm}) = 6 \text{ J} \quad \blacksquare$$

For gases at moderate pressures, the Berthelot equation is convenient. From Eqs. (2.29) and (2.31):

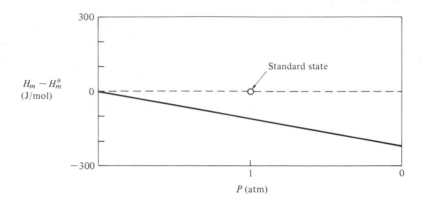

Figure 2.15 The standard-state enthalpy of chlorine. Since the enthalpy of an ideal gas is independent of pressure, and since all gases approach ideal behavior as the pressure approaches zero, the standard enthalpy is the limit of the actual enthalpy (solid line) as the pressure approaches zero.

$$H_m - H_m^\theta = \frac{9RT_c}{128}\left(1 - \frac{18T_c^2}{T^2}\right)\left(\frac{P}{P_c}\right) \tag{2.61}$$

Example: Calculate $H - H^\theta$ for CO_2(gas) at 25°C, 1 atm:

$$H_m - H_m^\theta = \frac{9(8.3143 \text{ J/K})(304.2 \text{ K})}{128}\left[1 - \frac{18(304.2)^2}{(298.2)^2}\right]\frac{1 \text{ atm}}{72.8 \text{ atm}} = -43 \text{ J} \quad \blacksquare$$

These numbers are quite small, so the deviation of H from H^θ will not be important at low to moderate pressures.

Postscript

Of the new ideas introduced in this chapter, the most important by far are state functions and reversible processes. These will be used and, it is hoped, further illuminated in Chapter 3.

In this chapter a number of new properties of materials have been introduced; some have symbols — U, H, C_v, C_p, μ — and some have names (for example, the internal pressure). Some are characterized only as slopes — for example, $(\partial H/\partial P)_T$, $(\partial T/\partial V)_U$ — but they are properties of the material nonetheless. The distinction between total derivatives and partial derivatives should be increasingly evident. The quantity dH/dT is the change of H when T is changed infinitesimally, but $(\partial H/\partial T)_P$ is the heat capacity. For an isothermal process, $dT = 0$; this does not mean that $(\partial T/\partial P)_H$ (the Joule-Thomson coefficient) is zero any more than it means that the heat capacity is infinite. In particular, symbols such as ∂T by themselves have no meaning.

The first law tells us that heat and work are both forms of energy. Work can be turned into heat (via friction) without limit, but the opposite process is usually more interesting. Engines turn heat into work; are there limits to that process? That question is answered by the second law of thermodynamics, the subject of the next chapter.

Problems

2.1 For He(gas), considered as an ideal gas, calculate q, w, and ΔU if one gram is heated from 300 to 400 K (a) at constant volume, (b) at constant pressure (1 atm).

2.2 Calculate ΔU for heating one mole of Cl_2(gas) from 400 to 1600 K at constant volume from the data:

T (K)	C_{vm} (J/K)
400	26.99
600	28.29
800	28.89
1000	29.19
1200	29.42
1400	29.69
1600	29.78

2.3 Calculate the heat required to raise the temperature of 50 g CO_2 from 0°C to 1000°C at constant volume using the data:

t (°C)	C_{vm} (J/K)
0	27.74
100	29.95
400	35.49
1000	42.58

2.4 Use the van der Waals relationship:

$$\left(\frac{\partial U}{\partial V}\right)_T = \frac{a}{V_m^2}$$

to calculate ΔU for an isothermal expansion of one mole of Ar from 50 atm to 1 atm, at $T = 250$ K.

2.5 (a) Calculate the internal pressure, $(\partial U/\partial V)_T$, for CCl_4(liquid) at 293 K, 1 atm.
(b) Repeat this calculation for NH_3(gas) at STP (use van der Waals, and $V_m = 22264$ cm^3).

2.6 (a) Use the one-term virial equation:

$$P = \frac{RT}{V_m}\left(1 + \frac{B}{V_m}\right)$$

to derive the formula:

$$\left(\frac{\partial U}{\partial V}\right)_T = \frac{RT^2 B'}{V_m^2}$$

(b) Evaluate $(\partial U/\partial V)_T$ for CO_2 at STP using the Berthelot virial coefficient; compare this to the van der Waals result [Eq. (2.15)]. (Use $V_m = 22.296$ dm^3 for both cases.)

2.7 To 1 kg of argon (considered as an ideal gas), 2000 J of heat is added at constant $P = 1$ atm. Calculate ΔU, ΔH, ΔT, and ΔV of the gas, and the work. (Use $C_{vm} = 1.5R$, $C_{pm} = 2.5R$.)

2.8 (a) Use the heat capacity from Table 2.2 to calculate ΔH when one mole of Cl_2(gas) is heated from 400 to 800 K at constant pressure.
(b) How much heat must be supplied per gram of the gas to effect this temperature change?

2.9 Calculate ΔH for heating three moles of Al(solid), at constant pressure, from 300 to 600 K. Use heat capacity from Table 2.2.

2.10 Calculate ΔH for heating one mole of copper from 10 to 100 K at constant pressure from the specific heats below:

T (K)	c_p (J K^{-1} g^{-1})	T (K)	c_p (J K^{-1} g^{-1})
10	0.00086	60	0.137
20	0.0077	70	0.173
30	0.027	80	0.205
40	0.060	90	0.232
50	0.099	100	0.254

2.11 From the data table below, calculate ΔH for heating one mole of benzene (solid) from 200 K to its melting point, 278.69 K.

T	C_{pm} (J/K)
200	83.7
240	104.1
260	116.1
278.69	128.7

2.12 Calculate the change in the enthalpy per mole for benzene (liquid) if the pressure is increased by 10 atm at 298 K.

2.13 Calculate the change of the molar enthalpy of argon when it is isothermally compressed at 300 K from 1.0 to 6.0 atm, using $T_c = 151$ K, $P_c = 48.0$ atm.

2.14 (a) Show that the Beattie-Bridgeman second virial coefficient gives:

$$B' = \frac{A_0}{RT^2} + \frac{3c}{T^4}$$

(b) Use this result to derive:

$$\left(\frac{\partial H}{\partial P}\right)_T = B_0 - \frac{2A_0}{RT} - \frac{4c}{T^3}$$

(c) Calculate ΔH for the isothermal compression of one mole of Ar at 300 K from 1.0 to 6.0 atm.

2.15 Calculate the molar enthalpy of NH_3 relative to its enthalpy at 25.0°C, 1.00 atm, when $T = 1000$ K, $P = 100$ atm.

2.16 (a) Calculate C_{pm} of CH_4 at 51.00°C, $P = 0$ (Table 2.2).
(b) Estimate C_{pm} of CH_4 at 51°C, 10 atm, from its critical constants, $T_c = 190.6$ K, $P_c = 45.8$ atm.

2.17 (a) Show that if the Beattie-Bridgeman form of the second virial coefficient is used, the dependence of C_p on pressure is:

$$\left(\frac{\partial C_{pm}}{\partial P}\right)_T = \frac{2A_0}{RT^2} + \frac{12c}{T^4}$$

(b) With constants of Table 1.2, what are the units of this equation?

(c) Calculate C_{pm} of CH_4 at 51°C, 10 atm.

2.18   Prove that:

$$\left(\frac{\partial C_v}{\partial V}\right)_T = T\left(\frac{\partial^2 P}{\partial T^2}\right)_V$$

Evaluate this quantity for a van der Waals gas.

2.19   Acetylene (gas) at 15°C, 1 atm, has $C_{pm} = 41.73$ J/K. Calculate C_{vm} from its critical constant, $T_c = 309$ K, $P_c = 62$ atm (the observed value is 31.16 J/K, Table 2.1). What would C_{vm} be if this gas were ideal?

2.20   (a) Show that for a van der Waals gas:

$$C_{pm} - C_{vm} = \frac{R}{1 - \dfrac{2a(V_m - b)^2}{RTV_m^3}}$$

(b) Evaluate this difference for SO_2 at 15°C, 1 atm ($V_m = 23.83$ dm³; compare Table 2.1).

2.21   Use the square-well potential constants (Table 1.7) for argon at 300 K to estimate $C_{pm} - C_{vm}$ at $P = 1$ and 10 atm [use Eq. (2.40)].

2.22   Calculate the work when an ideal gas, initially at $P = 10$ atm, $V = 2.0$ dm³, $T = 293$ K, is expanded isothermally to $P = 1.0$ atm, if: (a) the expansion occurs vs. $P_{ex} = 1$ atm; (b) the opposing pressure is reduced in nine steps, $P_{ex} = 9, 8, 7, 6, \ldots, 2, 1$; (c) the expansion is done reversibly.

2.23   Calculate the work done by a reversible isothermal expansion of 5.00 moles of an ideal gas at 273 K from 0.025 m³ to 0.112 m³.

2.24   (a) Show that the work for a reversible isothermal expansion of a van der Waals gas is:

$$-w = nRT \ln\left(\frac{V_2 - nb}{V_1 - nb}\right) + n^2a\left(\frac{1}{V_2} - \frac{1}{V_1}\right)$$

(b) Calculate the work of an isothermal, reversible expansion of 6.00 moles of SO_2 from 10.0 dm³ to 150 dm³ at 30°C using the ideal gas and van der Waals equations.

2.25   Calculate the final temperature for an adiabatic reversible expansion from $T_1 = 600$ K, $V_m = 1.00$ dm³ to a final $V_m = 23.0$ dm³, if (a) the gas is argon, (b) the gas is methane. (Assume ideal gas.)

2.26   Adiabatic temperature drops are often a convenient method for measuring heat capacities. A certain gas, when expanded adiabatically and reversibly from 380 K, 3 atm to 1 atm, had a final $T = 278$ K. Calculate its C_{pm}. (Assume ideal gas.)

2.27   A quantity of nitrogen gas at 10 atm is expanded adiabatically and reversibly to 10 times its initial volume. What is the final pressure? (Assume ideal gas.)

2.28   An ideal gas with $C_{pm} = 33.26$ J/K (presumed independent of T) is expanded adiabatically from 400 K, $P = 5.00$ atm to $P = 1.00$ atm. Calculate the final temperature if (a) the expansion is reversible, (b) the expansion is vs. a constant $P_{ex} = 1.00$ atm, (c) the expansion is into a vacuum. [You need to derive an equation for part (b).]

2.29 (a) Show that, for an adiabatic reversible expansion of a van der Waals gas:

$$\overline{C}_{vm} \ln \left(\frac{T_2}{T_1}\right) = -R \ln \left(\frac{V_{2m} - b}{V_{1m} - b}\right)$$

(b) Compare this to the ideal gas result [Eq. (2.43)] for the compression of Ar from $V_m = 20$ dm^3, $T = 500$ K to $V_m = 2.0$ dm^3.

2.30 One mole of an ideal gas with $C_{pm} = 30$ J/K, $C_{vm} = 20.7$ J/K (both considered constant) is expanded from 400 K, 5 atm to P = 1 atm. Calculate the work if (a) the expansion is reversible and isothermal, (b) the expansion is reversible and adiabatic.

2.31 When air is compressed, as with a bicycle pump, the temperature rises. You can readily observe this phenomenon by feeling the coupling at the pump's air outlet. You are pumping air at 22°C, 1 atm, into a bicycle tire with 1 dm^3 volume at a pressure of 65 psig. Calculate the temperature of the air in the tire, assuming the process is adiabatic and reversible.

2.32 Five moles of an ideal gas ($C_{pm} = 22$ J K^{-1}) initially at 300 K, 1 atm, are compressed reversibly to 10 atm. The gas and its container are in thermal contact with a water bath, and the total heat capacity (gas, container, bath) is 9000 J K^{-1}. Calculate the final temperature if the whole system is adiabatic.

2.33 Use the van der Waals relation:

$$\left(\frac{\partial U}{\partial V}\right)_T = \frac{a}{V_m^2}$$

to calculate the temperature change in a Joule expansion ($P_{ex} = 0$, adiabatic) if one mole of CO_2 is expanded from 5 dm^3 to 25 dm^3. (Use $C_{vm} = 28.1$ J/K.)

2.34 Estimate the low-pressure Joule-Thomson coefficient of CO_2 at 400 K from its critical constants. Use this value to estimate the temperature drop if CO_2 at 400 K, 50 atm, is expanded to 1 atm.

2.35 Since the Joule-Thomson coefficient is a function of both T and P, calculation of ΔT for a given ΔP could be complicated. Show that if C_p can be considered constant (independent of P and T), the final temperature of a Joule-Thomson expansion is:

$$T_2 = T_1 - \int_0^P \mu(T_1, P)\, dP$$

2.36 Use the results of Problem 2.35 to calculate the temperature change when CO_2 at 350 K, 100 atm, undergoes a Joule-Thomson expansion to 1 atm, given (for CO_2, 350 K) $\mu = 0.8187 - (3.276 \times 10^{-3})P - (5.66 \times 10^{-6})P^2$ (units: K/atm).

2.37 (a) Show that the van der Waals second virial coefficient, $B = b - a/RT$, gives:

$$\mu C_{pm} = \left(\frac{2a}{RT} - b\right)$$

(b) Use this result to calculate μ of N_2 at 273 K.
(c) Use this result to calculate the inversion temperature of nitrogen. Compare your answers to the actual values, Tables 2.4 and 2.5.

2.38 The gas CHCl$_2$F (Genetron-21) is used as a refrigerant, solvent, and propellant. The following data apply:

$$M = 102.93 \text{ g}, \qquad T_c = 451.7 \text{ K}, \qquad P_c = 51 \text{ atm}$$

$$B(331 \text{ K}) = -528 \text{ cm}^3, \qquad B(339 \text{ K}) = -446 \text{ cm}^3, \qquad B(366 \text{ K}) = -403 \text{ cm}^3$$

(a) Use the $B(T)$ data to calculate B' and μC_p at 339 K. When calculating the derivative, use the three-point formula of Appendix I with $1/T$ as the independent variable.
(b) Calculate the same quantities using the Berthelot virial coefficient.

2.39   Calculate the Joule-Thomson coefficient of ethylene (C_2H_4) at 15°C from its square-well potential constants (Table 1.7). (Use the heat capacity of Table 2.1.)

2.40   (a) Show that, in terms of the Beattie-Bridgeman constants:

$$\mu C_{pm} = -B_0 + \frac{2A_0}{RT} + \frac{4c}{T^3}$$

(b) Calculate μ for N_2 at 0°C. Use $C_{pm} = 29.10$ J/K.
(c) Calculate the inversion temperature of N_2 using this formula.

2.41   (a) Show that for a van der Waals gas:

$$U_m(T, V) = U_m^\theta(T) + \frac{a}{V_m}$$

(b) Calculate this quantity for O_2 when $V_m = 5$ dm^3.
(c) Derive the equivalent expression for the enthalpy.

2.42   The Matheson *Gas Data* book lists the enthalpy of 1-butene (gas) at 140°F and 110.2 psia as 198.4 BTU/lb (relative to 32°F, 1 atm). Calculate H_m^θ relative to the same reference state (0°C, 1 atm) at this temperature in kJ/mole. Data: $T_c = 419.6$ K, $P_c = 39.7$ atm, $M = 56.10$ g, BTU/lb $= 1.80$ cal/g, cal $= 4.184$ J.

"'Entropy,'" repeated Mr. Tompkins. *"I've heard that word before. One of my colleagues once gave a party, and after a few drinks, some chemistry students he'd invited started singing—*

> *'Increases, decreases*
> *Decreases, increases*
> *What the hell do we care*
> *What entropy does?'*

to the tune of 'ach du lieber Augustine.' What is entropy anyway?"

—George Gamow, Mr. Tompkins in Wonderland

3

The Second Law of Thermodynamics

3.1 Heat Engines

3.2 Entropy

3.3 Calculation of Entropy Changes

3.4 Free Energy

3.5 Equations of Thermodynamics

3.6 The Third Law of Thermodynamics

3.7 Entropy of Real Gases: Standard State

3.8 Thermodynamics of Rubberlike Elasticity

It is difficult today to imagine how much glamour was once attached to the science and technology of steam engines — it was the atomic energy, molecular genetics, and microcomputer of the nineteenth century. Certainly, no other invention in history affected the human condition to such an extent. This and related inventions that constituted the industrial revolution were not the work of scientists but of mechanics. However, there was great interest in this device on the part of all, scientists included, and this interest led to the development of the science of thermodynamics. That the science of steam engines can tell us something — and something important — about chemistry (the science of alchemists) is one of the more remarkable developments in the history of science. It is the chemical-applications aspect of thermodynamics that most interests us — but thermodynamics began with engines, and so shall we.

3.1   Heat Engines

A major advance in the understanding of steam engines occurred with the theoretical analysis made by a French military engineer, Sadi Carnot (1824). Carnot reduced the varieties and complexities of steam engines to a simple essence (Figure 3.1): a high-temperature reservoir (the boiler), the engine that does work, and a low-temperature reservoir (the condenser).

Carnot was interested in the theoretical amount of work that an engine could provide from a given amount of heat input. Since the heat input came from a fuel (coal or wood

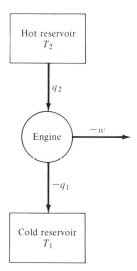

Figure 3.1   The Carnot engine. Carnot's engine is an idealized model for the operation of any heat engines. A quantity of heat (q_2) is withdrawn from the hot reservoir (the boiler), and a quantity of heat ($-q_1$) is discharged to the cold reservoir (the condenser); the work done by the engine is $-w$.

at that time) and economic value arose from the work output, the efficiency of an engine had an obvious practicality. Carnot's analysis was flawed because he knew nothing about the first law of thermodynamics (he was contemporary with Joule) and, hence, assumed that heat was a conserved fluid. He made an analogy between the steam engine and a waterfall in which heat "fell" from the high-temperature reservoir to the low-temperature reservoir, doing work in the process. He then made an analogy between the water head (the height of the fall) and the temperature difference and concluded, correctly, that the amount of work obtainable from a steam engine was proportional to the temperature difference between the hot reservoir (boiler) and the cold reservoir (condenser).

The real genius of Carnot was in the mode of his analysis. This analysis was not perfected until years after his death (he died in 1832, at age 36, of cholera), but it was a strong influence on the subsequent development of engines; the analysis even applies to refrigerators and air conditioners, which were not imagined by Carnot. [Carnot's accomplishments have been reviewed by S. S. Wilson in *Scientific American* (August 1981). Wilson points out that much of what is attributed to Carnot was actually done by later workers, especially Clapeyron and Clausius, who credited Carnot with more than he actually did — or, indeed, could have done without knowing about subsequent developments such as the first law and the absolute temperature scale.]

The Carnot Cycle

All engines work on a cycle. A fluid, the working substance, is heated and expands against a piston, which does work. The fluid is then discharged and the piston returns to its original position for another cycle. The Carnot cycle is an idealization of this process. In order to keep our analysis simple, we shall assume that the working substance is one mole of an ideal gas; this results in no loss of generality.

The Carnot cycle has four steps, which are diagrammed in Figure 3.2.

Step 1. The gas absorbs heat (q_2) from the high-temperature reservoir (temperature T_2) and expands *isothermally* and *reversibly* from V_1 to V_2. For this step:

$$\text{heat} = q_2$$

$$\text{work} = -RT_2 \ln \frac{V_2}{V_1} \quad [\text{Eq. (2.41)}]$$

$$\Delta U = 0 \quad \text{(since isothermal, ideal gas)}$$

Step 2. The gas expands *adiabatically* and *reversibly* from V_2 to V_3; in so doing, its temperature drops from T_2 to T_1. For this step:

$$\text{heat} = \text{zero}$$

$$\text{work} = \Delta U = \int_{T_2}^{T_1} C_v dT$$

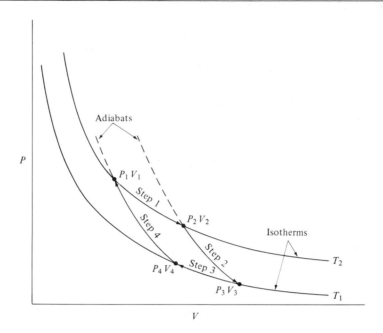

Figure 3.2 The path of a Carnot cycle on a PV diagram. The area inside the cyclic path is the work done by the engine. A Carnot engine uses a permanent (noncondensing) gas as the working fluid and is not at all typical of practical working heat engines. It is used only as an idealized model to derive certain relationships of importance to real engines and to many other areas of science and engineering as well.

Step 3. The gas is compressed from V_3 to V_4 while in thermal contact with the low-temperature reservoir (temperature T_1); this expansion is *reversible* and *isothermal*, with a quantity of heat (q_1 — a negative quantity, since heat is leaving the system) being discharged to the reservoir. For this step:

$$\text{heat} = q_1$$

$$\text{work} = -RT_1 \ln \frac{V_4}{V_3}$$

$$\Delta U = 0$$

Step 4. The gas is compressed *adiabatically* and *reversibly* from V_4 to V_1 (its original volume), warming in the process from T_1 to T_2. For this step:

$$\text{heat} = \text{zero}$$

$$\text{work} = \Delta U = \int_{T_1}^{T_2} C_v dT$$

Adding these together, we get, for the entire cycle:

$$q(\text{cycle}) = q_2 + q_1$$

$$w(\text{cycle}) = -RT_2 \ln \frac{V_2}{V_1} - RT_1 \ln \frac{V_4}{V_3}$$

$$\Delta U(\text{cycle}) = 0$$

The last statement means that, for the entire cycle, $-w = q(\text{net})$ and (remembering that q_1 is negative):

$$-w = q_2 + q_1 = q_2 - |q_1|$$

Also, ΔU was zero for the first and third steps, so we also have for each heat:

$$q_1 = RT_1 \ln \frac{V_4}{V_3}, \qquad q_2 = RT_2 \ln \frac{V_2}{V_1}$$

But the volume ratios $V_1 : V_4$ and $V_2 : V_3$ must be related, because (cf. Figure 3.2) they lie on parallel adiabats between the same pair of temperatures (T_2 and T_1). Using Eq. (2.44) for steps 2 and 4:

$$C_{vm} \ln \frac{T_2}{T_1} = -R \ln \frac{V_2}{V_3} = -R \ln \frac{V_1}{V_4}$$

Therefore $V_2/V_3 = (V_1/V_4)$ or, rearranging slightly, $V_2/V_1 = V_3/V_4$, $\ln (V_2/V_1) = -\ln (V_4/V_3)$.

This, finally, gives a result for the heats and work of a Carnot cycle in terms of a single volume ratio:

$$-w = R(T_2 - T_1) \ln \frac{V_2}{V_1} \qquad \textbf{(3.1a)}$$

$$q_2 = RT_2 \ln \frac{V_2}{V_1} \qquad \textbf{(3.1b)}$$

$$q_1 = -RT_1 \ln \frac{V_2}{V_1} \qquad \textbf{(3.1c)}$$

The efficiency of a Carnot engine is the ratio of the net work obtained ($-w$ with our sign convention) to the fuel burned to provide the heat q_2. From Eqs. (3.1):

$$\text{efficiency} = \frac{-w}{q_2} = \frac{T_2 - T_1}{T_2} \qquad \textbf{(3.2)}$$

This quantity, of course, is optimal for $T_2 \gg T_1$.

Steam Engines and Heat Pumps

It is questionable whether nineteenth-century engineers knew much about Sadi Carnot, but they were certainly aware that higher boiler temperatures produced more power. Unfortunately, for a given working substance (steam), higher boiler temperatures mean higher pressures. In 1880 there were 170 boiler explosions in the United States, injuring 555 and killing 259 persons. (Was it Carnot who sent Casey Jones to the promised land?) The efficiency of an engine can also be improved by lowering the temperature (T_1) of the cold reservoir. For steam electrical generating plants this reservoir is typically a nearby body of water. Carnot's analysis demonstrates that a quantity of heat, q_1, must be discharged into the cold reservoir; it is this which limits the efficiency of the engine. The consequent warming of streams and rivers may cause ecological problems.

Carnot's analysis applies equally to a *heat pump*. Suppose that we supply work to the system, for example by having an electrical motor operate the "engine" (which is now called a compressor); heat q_1 (a positive quantity) will be withdrawn from the cold reservoir and q_2 (negative) will be discharged to the hot reservoir. If the hot reservoir (T_2) is a house, and the cold reservoir (T_1) is the outdoors, the heat pump becomes an efficient method for heating the house. From Eqs. (3.1) we can see that the work (w) required to furnish a heat $(-q_2)$ to the house is (assuming optimal efficiency):

$$w = \left(\frac{T_2 - T_1}{T_2}\right)(-q_2) \tag{3.3}$$

The relative efficiency of a heat pump compared to direct heating (as with a furnace) depends on a number of factors, including the relative cost of various fuels; heat pumps are normally powered by electricity, whereas direct heating can use less expensive forms of energy such as fossil fuels. The advantage [shown by Eq. (3.3)] is that (with typical numbers) you may be able to pump 30 joules of heat into the house with only 1 joule of electrical energy; direct heating, of course, requires 30 joules of energy to get 30 joules of heat (both examples assume perfect efficiencies and, thus, represent lower limits of the energy required). However, the advantage may lie with direct heating if the outside is too cold; why is this? The rate of heat loss of a warm body depends on the temperature difference $(T_2 - T_1)$. Therefore the amount of heat required, and the cost of direct heating, will increase linearly with the temperature difference. But the work required for a heat pump [Eq. (3.3)] also depends on the temperature difference, so that the electrical energy required increases quadratically with the temperature difference; clearly, there is some T_1 below which the heat pump will be less efficient than direct heating.

Looked at from another point of view, the heat pump could be a refrigerator or an air conditioner; it removes a heat q_1 from the cold reservoir (the inside of the refrigerator or the air-conditioned house) to keep it cold. Now we are interested in the ratio of the work required to the desired effect (q_1) and [from Eq. (3.1)]:

$$\frac{w}{q_1} = \frac{T_2 - T_1}{T_1} \tag{3.4}$$

Perpetual Motion

How general is Carnot's analysis? The fact that we assumed reversibility simply means that we calculated the maximum work obtainable from an engine (or the minimum work required to operate a heat pump). Frictional losses and irreversibility

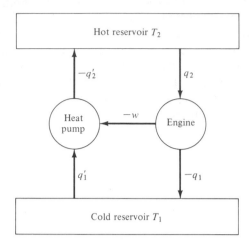

Figure 3.3 Perpetual motion machine of the second kind. Two engines working with the same hot and cold reservoir. One engine does work to operate the other engine as a heat pump. Working at perfect efficiency, the best such a system could accomplish would be no net transfer of heat from cold to hot, with no excess work; any other result would violate the second law of thermodynamics. Such a device is called a perpetual motion machine of the second kind; such devices are hypothetically possible, but only as long as they do no useful work — hence, they are useless as well as practically impossible.

will cause the actual efficiency to be much less than optimum. Also, the optimum efficiency must be the same whether or not an ideal gas is used. Picture two engines (Figure 3.3) operating with the same hot and cold reservoirs. One engine withdraws heat q_2 and does work to operate the other as a heat pump; the heat pump replaces a heat q_2' in the hot reservoir. If $q_2 = q_2'$, and the first engine produced more work than the heat pump required, we could get work with no net input of energy. That seems unlikely. If the net work was zero and $q_2' > q_2$, the hot reservoir would get hotter and the cold colder spontaneously — that is, without any input of energy from the outside world. That this violates a law of nature is implied in our observation that heat always flows *spontaneously* from hot to cold, never the other way around. In such an arrangement *at best* — that is, with a perfectly efficient engine and pump — all we could do is replace the heat, $q_2 = q_2'$ with no net work. A hypothetical machine of this type is called a perpetual motion machine of the second kind; such machines do not violate conservation of energy but, if any useful work were obtained, they would violate the second law of thermodynamics. George Gamow has a picturesque treatment of such machines in his book, *Mr. Tompkins in Wonderland* (available in paperback from Cambridge University Press).

Perpetual motion alone is forbidden by no law of nature so long as such a device is not required to provide work. However, the lure of something for nothing has caused a search for a perpetual motion machine which has continued until this day, even though the French Academy would seem to have said the last word on the subject in 1775:

The construction of a perpetual motion is impossible. Even if the effect of the motive-power were not in the long run destroyed by friction and the resistance of the medium, this power could produce merely an effect equivalent to itself. In order, therefore, to produce a perpetual effect from a finite cause, that effect must be infinitely small in any finite time. Neglecting friction and resistance, a body to which motion has been given will retain it for ever; but only on condition of its not acting on other bodies, and the only perpetual motion possible on this hypothesis (which, besides, cannot occur in nature), would be useless for the object which the devisers of perpetual motions have in view. This species of research has the inconvenience of being costly; it has ruined many a family; and numerous mechanics, who might have done great service, have wasted on it their means, their time, and their talents. These are the principal motives which have led the Academy to its decision. In resolving that it will no longer notice such speculations, it simply declares its opinion of the uselessness of the labors of those who are devoted to them.

Engines operate on a variety of cycles; are these governed by Carnot's equations? An arbitrary reversible cycle, a closed path on a PV diagram (Figure 3.4), can be closely approximated by an alternating series of isothermal and adiabatic steps — that is, a bunch of mini-Carnot cycles. In the limit of infinitesimal steps, the approximation is exact.

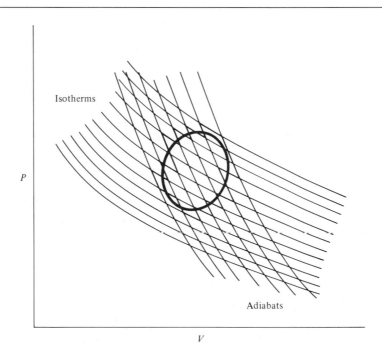

Figure 3.4 A general PV cycle. A general cycle can be approximated by a series of Carnot cycles — that is, a series of alternating isothermal and adiabatic reversible expansions or compressions that end up at the beginning. The area inside the cycle is the reversible work. This is used to show that the conclusions derived for the Carnot cycle are general and apply to other types of engines.

The Carnot cycle is not a practical one for real engines. The cycles actually used, however—for example, the Rankine cycle for steam engines or the Otto cycle for internal combustion engines—are analyzed in much the same way. This is a fascinating topic, but it would carry us too far afield from our principal topic: chemistry. Therefore our discussion of engines ends here.

3.2 Entropy

In the Carnot cycle the net heat, $q_1 + q_2$, is not zero; it, and the work, are not state functions. State functions such as U have no change in a closed cycle, because the final state and initial state are the same. However, from Eqs. (3.1) it should be apparent that q_2/T_2 is the same, with opposite sign, as q_1/T_1, so:

$$\frac{q_2}{T_2} + \frac{q_1}{T_1} = 0$$

In the generalized cycle (Figure 3.4), we have a series of alternating reversible isothermal steps (at temperature T_i) and reversible adiabatic steps. For each adiabatic step, $q = 0$, and for the sum of the isothermal steps about the cycle:

$$\sum_i \frac{q_i}{T_i} = 0$$

Also, ΔU must be zero for the cycle, so the sum of the adiabatic steps is:

$$\sum_i \Delta U_i = 0$$

Now, if we go to the limit of infinitesimal steps with heat dq and energy dU, the integral about the reversible closed path of the cycle is:

$$\oint \frac{dq_{\text{rev}}}{T} = 0, \qquad \oint dU = 0$$

The integral of a state function about a closed path must be zero, for the final state is the same as the initial state. Therefore, the quantity dq_{rev} divided by T must, like U, be a state function; it is called the **entropy**:

$$dS \equiv \frac{dq_{\text{rev}}}{T} \tag{3.5}$$

From this definition, it may appear that entropy is defined only for a reversible process. Not so; entropy is a state function, so, for any type of process:

$$\Delta S = S \text{ (final state)} - S \text{ (initial state)}$$

What Eq. (3.5) tells us is that the change in S must be *calculated* along a reversible path, however hypothetical, that goes between the same initial and final states as the real process. This will be illustrated clearly when we learn to calculate entropy

Rudolf Clausius (1822–1888)

Clausius was responsible for many, if not most, of the major developments in the early history of thermodynamics, including the mathematical formulation of the first law (dq = dU − dw), and the discovery of the internal energy function (U) and the entropy function (S). As is common in science, much of his work was paralleled by the work of others, including that of William Thomson (Lord Kelvin) and William Rankine. Clausius is best remembered for his famous dictum, which summarized the first two laws of thermodynamics: "Die Energie der Welt ist konstant; die entropie der Welt strebt einen Maximum zu."

Clausius became involved in a number of rather chauvinistic controversies involving the relative contributions and priorities of German scientists vis-à-vis the British, including the one between Joule and Mayer regarding the mechanical equivalent of work. These appear to be related to the generally increasing hostility between those countries at the time, which culminated in World War I. One is reminded of recent disputes between American and Russian scientists over the discovery and naming of transuranic elements. Such incidents bear witness that scientists, rather than dwelling in an "ivory tower," tend to be products of the times and societies in which they live.

changes. The reversible heat is not a state function but becomes one when divided by the absolute temperature; T is thus an *integrating factor* for the reversible heat.

Principle of Clausius

A convenient statement of the second law is the principle of Clausius: *the entropy of an isolated system will always increase in a spontaneous process*. Clausius said it more expansively: "The energy of the universe is constant; the entropy of the universe is always increasing to a maximum." Insofar as it applies to cosmology, the first statement is unprovable, hence the second is irrelevant. Nonetheless, Clausius' statement caused a stir among nineteenth-century philosophers, perhaps because it suggested a definite end to the world without overt intervention of any supernatural being. Learned texts were written on the "heat death" of the world due to entropy. It all seems less a concern in the twentieth century, perhaps because more imminent and plausible scenarios for the end of the world have become evident.

As applied to an isolated, closed system the statement of Clausius is exactly correct. The experimental proof of the second law is in our experience that certain processes go spontaneously and irreversibly in one direction; they can be reversed only by intervention from the outside world — that is, by providing energy to the system: Heat flows from hot to cold. Ice added to boiling water produces lukewarm water. In Joule's (first) experiment the gas expanded into the vacuum; "nature abhors a vacuum" (Spinoza). None of these processes has ever been observed to go spontaneously in the opposite direction, and all can (and will, in the next section) be shown to involve an increase in entropy.

An isolated system would have constant volume, be adiabatic, and could exchange no other type of work or energy with its surroundings; that is, it has a constant internal energy U. The mathematical statement of the principle of Clausius is, therefore:

$$(dS)_{U,V} \geq 0 \tag{3.6}$$

It cannot be emphasized too strongly that, as a criterion for what changes may occur spontaneously, the increase of entropy applies only to isolated systems that cannot exchange energy, either work or heat, with any other system. A rigid-walled, insulated container is isolated, and whatever occurs spontaneously within that container will increase the entropy of the contents. If the system can receive energy from the outside, nothing in the second law says that its entropy could not decrease spontaneously. In any limited portion of the universe — for example, the planet Earth — entropy can decrease, provided that the entropy of another portion of the universe (for example, the sun) increases by a greater amount.

The Inequality of Clausius

To discuss systems that can exchange heat or work with the outside world, we must begin with the first law:

$$dU = dq - P_{ex}\, dV$$

For a reversible process, $P_{ex} = P$ and [from Eq. (3.5)] $dq = T\, dS$, so:

$$dU = T\, dS - P\, dV \tag{3.7}$$

Note that every quantity in this equation is a state variable; therefore Eq. (3.7) is true for any type of process, reversible or not. However, if the process is not reversible, then $T\, dS$ is not the heat and $-P\, dV$ is not the work.

Consider a general process with $dU = dq + dw$; the change in U for this process is also given by Eq. (3.7). Therefore:

$$T\, dS - P\, dV = dq - P_{ex}\, dV$$

This can also be written as:

$$T\, dS = dq + (P - P_{ex})\, dV$$

If $P > P_{ex}$, the system will expand spontaneously, $dV > 0$, and the second term on the right-hand side is positive. If $P < P_{ex}$, the system will contract ($dV < 0$), so, again, the second term is positive. For a reversible change, $P = P_{ex}$, and that term is zero. Therefore we conclude:

$$T\, dS \geq dq \tag{3.8}$$

The equality applies at equilibrium or for a reversible process [compare Eq. (3.5)]. (For purposes of understanding thermodynamics, it is useful to regard the terms reversible/equilibrium and irreversible/spontaneous as practical synonyms. The expansion when $P > P_{ex}$ is spontaneous and irreversible, as was stated in Chapter 2; reversible processes are changes in a system at equilibrium.)

Equation (3.8) is an alternative statement of the second law that is particularly suited to applications in nonisolated systems. Before pursuing this topic, let's look at how entropy changes are calculated for a variety of processes.

3.3 Calculation of Entropy Changes

Entropy is an important and useful concept for understanding things such as spontaneity and equilibrium (although, for chemical applications, it is less useful than the free energy, which will be introduced in the next section). It will be a while, perhaps a *long* while, before you will fully appreciate the significance of entropy, but in the meantime you can learn to calculate entropy changes, for that is much less difficult. In Chapter 2 we mentioned the processes heating/cooling, mixing, expansion/compression, phase changes (melting, vaporization, and so on), and chemical reaction. Except for the last, each of these will be discussed in turn.

Entropy Changes with Temperature

The entropy change of a material is readily calculated from Eq. (3.7) if the volume is held constant; with $dV = 0$, $dU = C_v dT$ and

$$dS = \frac{C_v}{T} dT$$

$$\text{(constant } V) \qquad S(T_2, V) - S(T_1, V) = n \int_{T_1}^{T_2} \frac{C_{vm}}{T} dT \qquad \textbf{(3.9)}$$

For changes of entropy with heating at constant pressure, the enthalpy function is, as usual, the most convenient approach. Substitution of $H = U + PV$ into Eq. (3.7) gives:

$$dU = dH - P dV - V dP \quad \text{and} \quad dU = T dS - P dV$$

$$dH = T dS + V dP \qquad \textbf{(3.10)}$$

This equation, like Eq. (3.7), is one of the fundamental relationships of thermodynamics; note that only state variables are involved, so it applies to any type of process.

For a change in temperature at constant pressure, $dH = C_p dT$ and Eq. (3.10) becomes:

$$C_p dT = T dS \qquad \text{(if } dP = 0)$$

$$\text{(constant } P) \qquad S(T_2, P) - S(T_1, P) = n \int_{T_1}^{T_2} \frac{C_{pm}}{T} dT \qquad \textbf{(3.11a)}$$

The considerations involved for the evaluation of Eqs. (3.9) and (3.11) are the same, so the discussion to follow will focus on the latter; all statements will apply to Eq. (3.9) with a simple substitution of V for P.

For purposes of graphical or numerical integration, Eq. (3.11) represents the area of the function C_p/T vs. T; this is illustrated in Figure 3.5 (top). However, since $dT/T = d(\ln T)$, Eq. (3.11) can be written equally well as:

$$\Delta S = n \int_{\ln T_1}^{\ln T_2} C_{pm}\, d(\ln T) \tag{3.11b}$$

Thus, the entropy change upon heating at constant P is the area under the function C_p vs. $\ln T$; this is illustrated by Fig. 3.5 (bottom). Generally C_p vs. $\ln T$ is a smoother function than (C_p/T) vs. T and, therefore, more suitable for graphical or numerical analysis. In particular, the mean value can be used with Eq. (3.11) to give:

$$\text{(constant } P) \qquad \Delta S \cong n\overline{C}_{pm} \ln \frac{T_2}{T_1} \tag{3.12}$$

$$\text{(constant } V) \qquad \Delta S \cong n\overline{C}_{vm} \ln \frac{T_2}{T_1} \tag{3.13}$$

If an empirical formula for C_p as a function of T is available (for example, Tables 2.2 and 2.3), then the evaluation of the integral of C_p/T vs. T can be done directly.

Example: Two identical blocks of a metal have the same heat capacity, 15 J/K (presumed independent of T), but different temperatures, 300 K and 400 K, respectively. Calculate the final temperature and the entropy change if the blocks are in thermal contact and reach equilibrium. The two blocks together are an isolated system, so the heat (ΔH, at constant pressure) that leaves one enters the other, and:

$$\Delta H(\text{system}) = \Delta H(\text{block 1}) + \Delta H(\text{block 2}) = 0$$

If the final temperature is T:

$$(15 \text{ J/K})(T - 300) + (15 \text{ J/K})(T - 400) = 0$$

Since the blocks are the same size, the final temperature is 350 K; no surprise. The entropy change is:

$$\Delta S(\text{system}) = \Delta S(\text{block 1}) + \Delta S(\text{block 2})$$

$$= (15 \text{ J/K}) \ln \frac{350}{300} + (15 \text{ J/K}) \ln \frac{350}{400}$$

$$= +2.312 \text{ J/K} - 2.003 \text{ J/K} = +0.309 \text{ J/K}$$

It should be obvious that this change is spontaneous, and, as Clausius said, the entropy increased. ∎

Example: Calculate the entropy change when 1 kg of lead is heated from 315 to 450 K. From Table 2.2 we get:

$$C_{pm} = 22.13 + (11.72 \times 10^{-3})T + \frac{0.96 \times 10^5}{T^2}$$

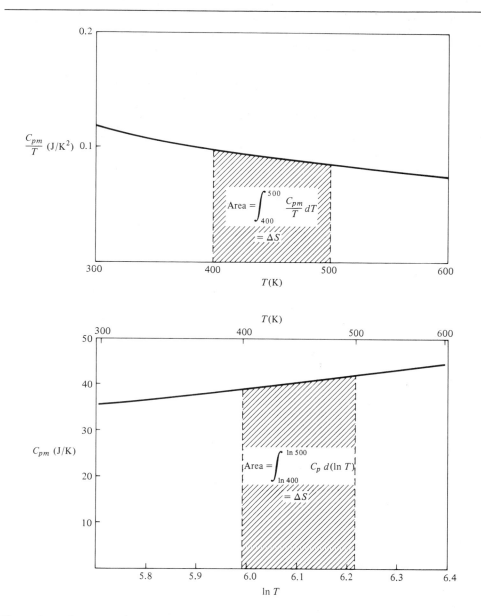

Figure 3.5 Entropy change with temperature. The entropy change for heating at constant pressure is the area under a graph of C_{pm}/T vs. T (top) or, alternatively, the area under a graph of C_{pm} vs. ln (T) (bottom). Ordinarily the second function changes more smoothly (as shown here) and consequently is more suitable for numerical computation.

This, of course, can be integrated directly, but first let's use the mean-value method to see how well it works. The heat capacity at the mean temperature (382.5) can be calculated with this formula: $\overline{C}_{pm} = 27.27$ J/K. The number of moles is $n = 1000/207.19 = 4.8265$.

$$\Delta S = n\overline{C}_{pm} \ln \frac{T_2}{T_1}$$

$$S(450) - S(314) = (4.8265)(27.27) \ln \frac{450}{315}$$

$$= 46.94 \text{ J/K}$$

The exact calculation is:

$$S(450) - S(315) = n \int_{315}^{450} \left[\frac{22.13}{T} + (11.72 \times 10^{-3}) + \frac{0.96 \times 10^5}{T^3} \right] dT$$

$$= n \left\{ 22.13 \ln \frac{450}{315} + 11.72 \times 10^{-3}(450 - 315) \right.$$

$$\left. - \frac{1}{2}(0.96 \times 10^5) \times \left[\frac{1}{(450)^2} - \frac{1}{(315)^2} \right] \right\}$$

$$= 4.8265(7.8932 + 1.5822 + 0.2467)$$

$$= 46.92 \text{ J/K} \qquad\qquad ■$$

Isothermal Expansion

It will be a while before we can treat properly the effect of pressure and volume on the entropy of real gases and other substances, but for an ideal gas the situation is quite simple.

From Eq. (3.7), with $dU = 0$ for an isothermal change of an ideal gas:

$$dS = \frac{P}{T} dV = nR \frac{dV}{V}$$

$$\text{(const. } T\text{, ideal gas)} \qquad S(V_2) - S(V_1) = nR \ln \frac{V_2}{V_1} \qquad\qquad \textbf{(3.14a)}$$

Alternatively, using Boyle's law in Eq. (3.14a), or using Eq. (3.11):

$$\text{(const. } T\text{, ideal gas)} \qquad S(P_2) - S(P_1) = -nR \ln \frac{P_2}{P_1} \qquad\qquad \textbf{(3.14b)}$$

Example: One mole of an ideal gas is expanded reversibly and isothermally from 5 to 22 dm³ at 273 K:

$$\Delta S = 8.3143 \ln \frac{22}{5} = 12.32 \text{ J/K}$$

We could also have used Eq. (2.41) to calculate:

$$-w = 8.3143(273) \ln \frac{22}{5} = 3363 \text{ J}$$

Then, since $\Delta U = 0$ for an isothermal ideal gas, $q = -w = 3363$ J. Since the expansion was reversible, Eq. (3.5) gives:

$$\Delta S = \frac{3363}{273} = 12.32 \text{ J/K}$$

That this is a less general method for calculating entropy changes will be apparent in the next example. ∎

Example: One mole of an ideal gas with $T = 273$ K, $V = 5$ dm^3, is expanded into an evacuated container with $V = 17$ dm^3 (cf. the Joule expansion, Chapter 2). Since the opposing pressure is zero, $w = 0$; also $q = 0$ and the temperature is constant. With a final volume of $5 + 17 = 22$, Eq. (3.14) gives:

$$\Delta S = R \ln \frac{22}{5} = 12.32 \text{ J/K}$$

This is exactly the same as for the previous example; we should have expected that, since the initial and final states are the same. In contrast to that example, this is an irreversible expansion with $q = 0$ [less than $T\Delta S$ as required by Eq. (3.8)]. The expansion is also obviously spontaneous, and, since the gas plus both containers is an isolated system, the increase in entropy is in accord with the principle of Clausius. ∎

Entropy of Mixing (Ideal Gas)

Consider two gases in separate containers that are connected by a tube with a stopcock; gas A, n_A moles, is in a volume V_1 and gas B, n_B moles, is in a volume V_2, at the same temperature and pressure. When the stopcock is opened, the two gases will mix spontaneously and irreversibly. If we assume that they are ideal gases, the entropy change of each is just that due to its increased volume as given by Eq. (3.14a):

$$\Delta S_A = n_A R \ln \left(\frac{V_1 + V_2}{V_1} \right)$$

$$\Delta S_B = n_B R \ln \left(\frac{V_1 + V_2}{V_2} \right)$$

Since the gases were originally at the same T and P, Avogadro's law says that the number of moles of each is proportional to its original volume. For gas A:

$$\frac{n_A}{n_A + n_B} = \frac{V_1}{V_1 + V_2} = X_A$$

This is just the mole fraction (X_A) of gas A. The total entropy change of the two gases is then:

$$\Delta S_{\text{mix}} = -R(n_A \ln X_A + n_B \ln X_B) \tag{3.15}$$

Since the mole fractions are, by definition, smaller than 1 (therefore $\ln X$ is negative), the entropy of mixing is clearly positive, which, by Clausius' principle, it must be for a spontaneous process. Because ideal gases ignore each other, having no inter-molecular forces, this process was, in effect, two Joule expansions, which result in the mixed gas.

Phase Changes

Ice and water, under an applied pressure of 1 atm, can coexist in equilibrium at 0°C. The addition of an amount of heat will melt some ice, the removal of some heat will freeze some water, but the equilibrium is not disturbed so long as there is some of each phase present. Therefore such an addition of heat is a reversible process — moving among equilibrium states. The heat required to melt a quantity of a solid is the heat of fusion; since the system is at equilibrium, and the pressure is constant, we can equate this addition of heat to an enthalpy change, $\Delta H_{fus} = H_{liq} - H_{solid}$, and the entropy change for any *equilibrium* phase change is:

$$\Delta S_\phi = \Delta H_\phi / T_\phi \qquad (3.16)$$

where ϕ indicates a phase change such as melting or vaporization.

Example: The heat of fusion of water is 333 J/g or 6.00 kJ/mole. Calculate ΔS for melting 10 g of ice.

$$\Delta S_{fus} = \frac{10(333)}{273.15 \text{ K}} = 12.2 \text{ J K}^{-1} \qquad \blacksquare$$

It is important to note that equations that calculate changes of H or S or other extensive properties with T — for example,

$$\Delta H = \int_{T_1}^{T_2} C_p \, dT \quad \text{or} \quad \Delta S = \int_{T_1}^{T_2} C_p \frac{dT}{T}$$

— imply that there is *no phase change between T_1 and T_2*. Phase changes will be discussed in detail in the next chapter; for now it is sufficient to realize that phase changes such as melting or evaporation are associated with discontinuous changes in heat capacity (cf. Figure 2.3), enthalpy, entropy and other properties — the enthalpy change is often called the *latent heat*. Calculations in such cases must be made *up to* transition, the "latent" quantity added, and the calculation continued *after* the phase change. For example, referring to Figure 2.3, the area under the "solid" curve is the ΔH for heating 1 g of ice from −200°C to 0°C; the area under the "liquid" curve is ΔH for heating 1 g of water from 0°C to 100°C; the area under the "vapor" curve is ΔH for heating steam from 100°C to 300°C. But the total of these areas is not ΔH for changing 1 g of ice at −200°C to steam at 300°C; to get this quantity the heat of fusion (at 0°C) and the heat of vaporization (at 100°C) must be included. (Cf. Problem 4.24.)

Example: Calculate ΔH and ΔS for heating ice (−10°C) to water (+10°C) at $P = 1$ atm. We use $C_{pm}(\text{ice}) = 37$ J/K and $C_{pm}(\text{water}) = 76$ J/K and assume both heat capacities are independent of temperature. (See Figure 3.6.) For one mole:

$$\Delta H = 37(273.15 - 263.15) + 6000 + 76(283.15 - 273.15) = 7130 \text{ J}$$

$$\Delta S = 37 \ln \frac{273.15}{263.15} + \frac{6000}{273.15} + 76 \ln \frac{283.15}{273.15} = 26 \text{ J/K} \qquad \blacksquare$$

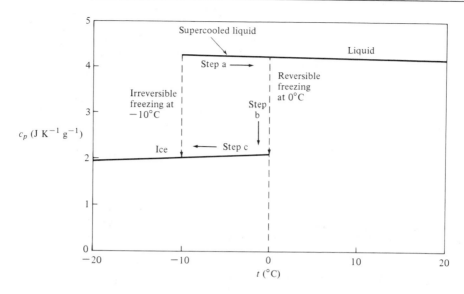

Figure 3.6 The specific heat of water and ice near the freezing point. Liquids can be supercooled below their freezing points and irreversibly frozen, but the entropy change of such a process must be calculated along a reversible path in which the supercooled liquid is warmed to the normal freezing point, frozen reversibly, and then cooled to the original temperature.

With care, liquids can be supercooled to below their normal freezing point; this requires slow cooling in a container free of dust or rough surfaces that could be sites for nucleation of crystals. When the supercooled liquid freezes (which it can be made to do by addition of a tiny crystal to start the process), it freezes instantly, spontaneously, and irreversibly. There is at no point an equilibrium between the initial and final states. The calculation of ΔS for such a process must be done along a reversible path. (The heat capacity of ice and water in this region is illustrated in Figure 3.6; however, for the example to follow, we shall assume a constant heat capacity.)

Example: Calculate ΔS for freezing one mole of water at $-10°C$:

$$H_2O(\text{liq}, -10°C) \longrightarrow H_2O(\text{solid}, -10°C)$$

given $C_{pm}(\text{ice}) = 37$ J/K, $C_{pm}(\text{water}) = 76$ J/K, and the heat of fusion, $\Delta H_f = 6000$ J/mol at $0°C$. The reversible path is:

1. Heat water to $0°C$:

$$\Delta S_a = 76 \ln \frac{273.15}{263.15} = 2.83 \text{ J/K}$$

2. Freeze at 0°:

$$\Delta S_b = \frac{-6000}{273.15} = -21.97 \text{ J/K}$$

3. Cool ice to $-10°C$:

$$\Delta S_c = 37 \ln \frac{263.15}{273.15} = -1.38 \text{ J/K}$$

The desired quantity must be the sum of these steps, so:

$$\Delta S(\text{freeze at } -10°C) = 2.83 - 21.97 - 1.38 = -20.5 \text{ J/K} \qquad \blacksquare$$

Spontaneous or Not

Why is ΔS negative for the preceding example if the process is spontaneous? That example was isothermal, not isolated; heat was lost to the surroundings. By the same path we can calculate ΔH at 263 K for the system. This calculation proceeds as follows:

$$\Delta H(\text{freeze at 263 K}) = \Delta H(\text{heat liq. } 263 \longrightarrow 273 \text{ K}) + \Delta H(\text{freezing at 273 K})$$

$$+ \Delta H(\text{cool solid } 273 \longrightarrow 263 \text{ K})$$

$$\Delta H_f(263) = (76 \text{ J K}^{-1} \text{ mol})(273 \text{ K} - 263 \text{ K}) - 6000 \text{ J/mol}$$

$$+ (37 \text{ J K}^{-1} \text{ mol}^{-1})(263 \text{ K} - 273 \text{ K})$$

$$= -5610 \text{ J/mol}$$

This is the heat released by the system when one mole of supercooled water freezes at $-10°C$. The surroundings must absorb a heat $q(\text{surr.}) = 5610$ J, and the entropy change of the surroundings must be:

$$\Delta S(\text{surr.}) = \frac{5610}{263.15} = 21.3 \text{ J/K}$$

The system plus the surroundings (presumed to have infinite heat capacity and, hence, isothermal at 263 K) are together an isolated system. Therefore:

$$\Delta S(\text{isolated}) = \Delta S(\text{system}) + \Delta S(\text{surr.}) = -20.6 + 21.3 = 0.7 \text{ J/K}$$

This is positive, as it must be for a spontaneous process.

It is perhaps evident that, while entropy can tell us the direction of spontaneous change, unless we are working with isolated systems or willing to deal with such circumlocutions as $\Delta S(\text{surr.})$, it is not very convenient. A dozen more examples, which we shall not pursue, might suggest an easier method. We will always get the quantity:

$$\Delta S(\text{isolated}) = \Delta S(\text{sys.}) + \frac{q(\text{surr.})}{T(\text{surr.})}$$

For an isothermal change at constant pressure, $q(\text{surr.}) = -\Delta H(\text{sys.})$, so:

$$\Delta S(\text{isolated}) = \Delta S(\text{sys.}) - \frac{\Delta H(\text{sys.})}{T}$$

The right-hand side uses quantities that depend only on the system. We could call it the "free entropy" and go from there, but the historical development of the subject took a different path. A similar quantity,

$$\Delta H(\text{system}) - T\Delta S(\text{system})$$

was defined and called the *free energy*. A little thought may convince you that this quantity will spontaneously *decrease* at constant T, P. Thanks to free energy, and in large part to J. Willard Gibbs, we shall shortly be able to forget about the surroundings and deal exclusively with the system.

3.4 Free Energy

What determines the direction of spontaneous change? Ordinary experience — the book falls from the table to the floor, not the opposite — can be summarized as a tendency to seek the minimum energy. On the molecular scale this is not sufficient — the spontaneous mixing of ideal gases involved no energy change — but, for isolated systems, we now know that the entropy will be maximized for a spontaneous change. The tendencies of energy to a minimum and entropy to a maximum will, when both can change, determine the direction of spontaneity and the position of equilibrium (where change stops).

These ideas can be stated more conveniently by defining a new state function called the **Helmholtz free energy:**

$$A \equiv U - TS \tag{3.17a}$$

For an infinitesimal change, this is:

$$dA = dU - T\,dS - S\,dT \tag{3.17b}$$

into which we can substitute the value of dU from the first law (assuming only PV work):

$$dA = dq - P_{\text{ex}}\,dV - T\,dS - S\,dT$$

For a differential change at constant V and T:

$$(dA)_{T,V} = dq - T\,dS$$

The second law, Eq. (3.8), shows that the right-hand side of this relationship must be *negative* for a spontaneous change; therefore the criterion for a spontaneous change at constant T, V is:

$$(dA)_{T,V} \leq 0 \tag{3.18}$$

A convenient statement of the second law for systems at constant T, V is: the Helmholtz free energy will decrease in any spontaneous change; at equilibrium, A is a minimum $(dA = 0)$.

Generally, it is more convenient to work at constant P rather than constant V; as usual when we want P to be an independent variable, the enthalpy function is convenient. The **Gibbs free energy** is defined as:

$$G \equiv H - TS \tag{3.19a}$$

For a differential change:

$$dG = dH - T\,dS - S\,dT \tag{3.19b}$$

If, in this equation, we use:

$$dH = dU + P\,dV + V\,dP$$

and (assuming only PV work) $dU = dq - P_{ex}\,dV$, we get:

$$dG = dq - P_{ex}\,dV + P\,dV + V\,dP - T\,dS - S\,dT$$

For a change at constant T and constant $P = P_{ex}$:

$$(dG)_{T,P} = (dq - T\,dS)$$

Once again the second law, Eq. (3.8), tells us that the right-hand side will be negative for a spontaneous change; therefore, the criterion for a spontaneous change at constant T, P is:

$$(dG)_{T,P} \leq 0 \tag{3.20}$$

The Gibbs free energy is a function that will decrease spontaneously in any system at constant T, P; at equilibrium, G is a minimum ($dG = 0$).

Calculating Free-Energy Changes

If Eq. (3.7) is substituted into Eq. (3.17b) we get:

$$dA = T\,dS - P\,dV - T\,dS - S\,dT$$
$$dA = -S\,dT - P\,dV \tag{3.21}$$

Similarly, if Eq. (3.11) is used in Eq. (3.19b):

$$dG = -S\,dT + V\,dP \tag{3.22}$$

Equations (3.21) and (3.22) will be very useful, although we shall not exploit them fully until later chapters (particularly Chapters 4 and 6, where we shall discuss equilibrium more thoroughly). These equations are very general; note that only state variables appear in them. However, two major restrictions, stated earlier, bear repetition: (1) the system can do work only by expansion—that is, $dw = -P\,dV$ (excluding, for example, electrical work); (2) the system is *closed;* the composition and the amount of material in the system may not change.

Free Energy and Work

Equation (3.17b) with $dU = dq + dw$ is:

$$dA = (dq - T\,dS) + dw - S\,dT$$

For a reversible process, $dq = T\,dS$ and the work is a maximum; the change in A at constant temperature is

$$(dA)_T = dw_{max}$$

The maximum work *by* the system is, because of our sign convention, $-w$; we now can see that the maximum work that can be done by the system in any isothermal process is related to its change in Helmholtz free energy:

$$(-\Delta A)_T = -w_{max} \qquad (3.23)$$

The symbol A, in fact, comes from the German word Arbeit (work) and, in some older texts, A has been called the *work function*.

Sometimes the PV work is largely irrelevant. For example, when a battery is charged or discharged, the volume of the reactants may change, but this work done on or by the atmosphere won't start your automobile; it is the electrical work that the battery reaction can provide that turns the starter motor. We divide work into PV work and "other" work called w':

$$dw = dw' - P_{ex}\, dV$$

If this is used in Eq. (3.19b) with $P = P_{ex}$:

$$dG = dH - T\, dS - S\, dT$$
$$= dU + P\, dV + V\, dP - T\, dS - S\, dT$$
$$= dq + dw' - P\, dV + P\, dV + V\, dP - T\, dS - S\, dT$$

Now, for a reversible process ($dq = T\, dS$) at constant P and T:

$$(dG)_{T,P} = dw'_{max}$$

The maximum "other" work that can be done *by* the system ($-w'$) in any isothermal process is

$$(-\Delta G)_{T,P} = -w'_{max} \qquad (3.24)$$

A book falling from a table will (assume it is attached to a cord on a pulley) be able to do work by lifting another weight; the work it can do is limited (by conservation of energy) to its loss of energy ($mg\, \Delta h$). Equations (3.23) and (3.24) are generalizations of this idea to situations in which the entropy can change; the maximum work that can be done by any isothermal process is limited to the change of the *free energy*. (The choice of G or A depends on whether or not you wish to include PV work.) Equation (3.24) will be especially important in Chapter 8, where we discuss electrochemical cells.

3.5 Equations of Thermodynamics

The numbers of equations which thermodynamics can produce is mind-boggling. Even for closed systems with only PV work, we have eight basic state functions P, V, T, U, H, S, A, G and, among these, 336 possible slopes such as $(\partial H/\partial T)_P$ (that is, C_p), $(\partial T/\partial P)_H$ (the Joule-Thompson coefficient) and $(\partial A/\partial S)_H$. Each of these state functions or slopes can be considered as a function of any two variables, and so on. Allowing changes in composition, surface area, electric, magnetic, or gravitational fields adds another batch of relationships. The number of possible equations has been estimated by P. W. Bridgman as being of the order of 10^{11}. The only sensible course seems to be to learn how to do the derivations; that's the easy part (though you may not think it easy); the hard part is knowing which of the 10^{11} equations you need.

Most derivations start with the *basic equations* of thermodynamics, Eqs. (3.7), (3.10), (3.21), and (3.22); these are listed on Table 3.1 for your convenience, together with some related definitions. We have noted already that these equations are very general; the only limitations are that we have a closed system with only PV work. (These restrictions are easily removed, and will be when the need arises.) In a closed system, we may write any of the variables T, V, P, U, H, S, A, or G as a function of any two others. Any choice is valid, but the most useful tend to be (a) independent (T, V), dependent (U, S, A); (b) independent (T, P), dependent (H, S, G). Also, thermodynamic derivations generally utilize a number of "tricks" involving partial derivatives; these are summarized in Appendix II. This is about all that can be said in general about derivations; beyond this you need intuition, experience, and, often, luck. These can be developed only by practice, so let's try some examples.

Maxwell Relationships: $A(T, V)$

First, consider $A(T, V)$. The slope formula gives:

$$dA = \left(\frac{\partial A}{\partial T}\right)_V dT + \left(\frac{\partial A}{\partial V}\right)_T dV$$

Comparison with the basic equation [Table 3.1(b)] shows immediately:

$$\left(\frac{\partial A}{\partial T}\right)_V = -S, \qquad \left(\frac{\partial A}{\partial V}\right)_T = -P \tag{3.25}$$

Further, if we differentiate Eqs. (3.25) with respect to the opposite variable:

$$\left[\frac{\partial}{\partial V}\left(\frac{\partial A}{\partial T}\right)_V\right]_T = -\left(\frac{\partial S}{\partial V}\right)_T, \qquad \left[\frac{\partial}{\partial T}\left(\frac{\partial A}{\partial V}\right)_T\right]_V = -\left(\frac{\partial P}{\partial T}\right)_V$$

The second derivatives of a state function are independent of the order of differentiation, so these two must be equal:

$$\left(\frac{\partial S}{\partial V}\right)_T = \left(\frac{\partial P}{\partial T}\right)_V \tag{3.26}$$

This is one of the **Maxwell relationships** given in Table 3.1.

Exercise: Starting with $G(T, P)$, prove:

$$\left(\frac{\partial G}{\partial T}\right)_P = -S, \qquad \left(\frac{\partial G}{\partial P}\right)_T = V, \qquad \left(\frac{\partial S}{\partial P}\right)_T = -\left(\frac{\partial V}{\partial T}\right)_P \tag{3.27}$$

∎

Enthalpy

If we wish to obtain a relationship of enthalpy to the independent variables T and P, we start with:

Table 3.1 The fundamental equations of thermodynamics

Definitions:	$H = U + PV$	$A = U - TS$	$G = H - TS$

Properties of matter:

$$C_v = \left(\frac{\partial U}{\partial T}\right)_v \qquad C_p = \left(\frac{\partial H}{\partial T}\right)_P \qquad \mu = \left(\frac{\partial T}{\partial P}\right)_H$$

$$\alpha = \frac{1}{V}\left(\frac{\partial V}{\partial T}\right)_P \qquad \kappa_T = -\frac{1}{V}\left(\frac{\partial V}{\partial P}\right)_T \qquad \kappa_S = -\frac{1}{V}\left(\frac{\partial V}{\partial P}\right)_S$$

Basic equations	Maxwell relationships	Working equations
(a) $dU = T\,dS - P\,dV$	$\left(\frac{\partial T}{\partial V}\right)_S = -\left(\frac{\partial P}{\partial S}\right)_V$	$dU = C_v dT - \left[P - T\left(\frac{\partial P}{\partial T}\right)_v\right]dV$
(b) $dA = -S\,dT - P\,dV$	$\left(\frac{\partial S}{\partial V}\right)_T = \left(\frac{\partial P}{\partial T}\right)_V$	$dS = \frac{C_v}{T}dT + \left(\frac{\partial P}{\partial T}\right)_V dV$
(c) $dH = T\,dS + V\,dP$	$\left(\frac{\partial T}{\partial P}\right)_S = \left(\frac{\partial V}{\partial S}\right)_P$	$dH = C_p dT + \left[V - T\left(\frac{\partial V}{\partial T}\right)_P\right]dP$
(d) $dG = -S\,dT + V\,dP$	$\left(\frac{\partial S}{\partial P}\right)_T = -\left(\frac{\partial V}{\partial T}\right)_P$	$dS = \frac{C_p}{T}dT - \left(\frac{\partial V}{\partial T}\right)_P dP$

Some derived relationships:

$$C_v = T\left(\frac{\partial S}{\partial T}\right)_V \qquad \left(\frac{\partial G}{\partial T}\right)_P = -S \qquad \left(\frac{\partial G}{\partial P}\right)_T = V$$

$$C_p = T\left(\frac{\partial S}{\partial T}\right)_P \qquad \left(\frac{\partial A}{\partial T}\right)_V = -S \qquad \left(\frac{\partial A}{\partial V}\right)_T = -P$$

$$dH = \left(\frac{\partial H}{\partial T}\right)_P dT + \left(\frac{\partial H}{\partial P}\right)_T dP$$

The first coefficient is a measurable parameter, the heat capacity C_p, so nothing need be done there. The second coefficient can be simplified using $dH = T\,dS + V\,dP$, which gives:

$$\frac{dH}{dP} = T\frac{dS}{dP} + V$$

The partial derivative $(\partial H/\partial P)_T$ is the same as the total derivative along a direction in which T is constant, hence:

$$\left(\frac{\partial H}{\partial P}\right)_T = T\left(\frac{\partial S}{\partial P}\right)_T + V$$

If we now use the Maxwell relationship [Eq. (3.27)], we get this coefficient in terms of PVT only:

$$\left(\frac{\partial H}{\partial P}\right)_T = V - T\left(\frac{\partial V}{\partial T}\right)_P \tag{3.28}$$

This is a practical relationship, because V and $(\partial V/\partial T)_P$ are readily available parameters. We found abundant use for this equation in Chapter 2.

Exercise: Derive the following:

$$\left(\frac{\partial U}{\partial V}\right)_T = T\left(\frac{\partial P}{\partial T}\right)_V - P \tag{3.29}$$

Derivations, for this purpose and for the problem set, must always proceed from the basic equations and definitions to the desired result. The derivation of Eq. (3.29) parallels that of Eq. (3.28). ∎

Entropy

Next, we wish to derive an equation for $S(T, V)$ in terms of commonly measured parameters. We start with:

$$dS = \left(\frac{\partial S}{\partial T}\right)_V dT + \left(\frac{\partial S}{\partial V}\right)_T dV$$

and the basic equation, $dU = T\,dS - P\,dV$. At constant volume:

$$(dU)_V = C_v dT = (T\,dS)_V$$

From this, we recognize the first coefficient for $S(T, V)$ as:

$$\left(\frac{\partial S}{\partial T}\right)_V = \frac{C_v}{T}, \qquad C_v = T\left(\frac{\partial S}{\partial T}\right)_V \tag{3.30a}$$

The second coefficient is given in terms of equation-of-state variables (P, V, T) as a Maxwell relation [Table 3.1(b)]. Finally:

$$dS = \frac{C_v}{T}\,dT + \left(\frac{\partial P}{\partial T}\right)_V dV \tag{3.30b}$$

Exercise: Prove that

$$C_p = T\left(\frac{\partial S}{\partial T}\right)_P \tag{3.31a}$$

and

$$dS = \frac{C_p}{T}\,dT - \left(\frac{\partial V}{\partial T}\right)_P dP \tag{3.31b}$$

∎

Adiabatic Expansions

We wish to derive an equation for the temperature change in an adiabatic reversible expansion in terms of the volume. First, let's formulate the problem mathematically; we wish to find $T(V)$ at constant $S(q_{\text{rev}} = 0)$, so it makes sense to start with $T(V, S)$:

$$dT = \left(\frac{\partial T}{\partial V}\right)_S dV + \left(\frac{\partial T}{\partial S}\right)_V dS$$

Letting $dS = 0$, we need to know the slope $(\partial T/\partial V)_S$. It may be helpful to rewrite this partial derivative using a cyclic rule (Appendix II) as:

$$\left(\frac{\partial T}{\partial V}\right)_S = -\left(\frac{\partial T}{\partial S}\right)_V\left(\frac{\partial S}{\partial V}\right)_T$$

Why do this? (1) The desired derivative is not in Table 3.1. (2) It can't hurt; if it doesn't help, try something else. As it happens, it does help (you knew it would, textbooks are always omniscient); the first factor is the reciprocal of Eq. (3.30a) and the second is a Maxwell relation [Table 3.1(b)]; therefore:

$$\left(\frac{\partial T}{\partial V}\right)_S = -\frac{T}{C_v}\left(\frac{\partial P}{\partial T}\right)_V \tag{3.32}$$

Example: Find $T(V)$ for an adiabatic reversible expansion of an ideal gas using Eq. (3.32). For an ideal gas:

$$P = \frac{nRT}{V} \quad \text{and} \quad \left(\frac{\partial P}{\partial T}\right)_V = \frac{nR}{V}$$

Then, using Eq. (3.32):

$$\left(\frac{\partial T}{\partial V}\right)_S = \frac{-nRT}{C_v}$$

Separation of variables gives:

$$C_{vm}\frac{dT}{T} = -R\frac{dV}{V}$$

which leads directly to Eq. (2.44). ∎

Exercise: Derive the following equation for the adiabatic reversible expansion of a van der Waals gas.

$$\int_{T_1}^{T_2} C_{vm}\frac{dT}{T} = R\,\ln\left(\frac{V_2 - nb}{V_1 - nb}\right) \qquad ∎$$

Among the properties listed in Table 3.1 we have included, for the sake of completeness, the *adiabatic compressibility*, κ_S, as contrasted to the isothermal compressibility, κ_T, introduced earlier. It should be a challenge for the reader to prove that:

$$\frac{\kappa_T}{\kappa_S} = \frac{C_p}{C_v} = \gamma \tag{3.33}$$

The adiabatic compressibility of a fluid can be directly measured from the velocity of sound (c). The propagation of sound in a fluid is by means of compressional waves; a segment of the fluid is alternately compressed and decompressed so fast it does not have time to gain or lose heat, therefore its temperature will change adiabatically. The velocity is related to the density and adiabatic compressibility by:

$$c = \frac{1}{\sqrt{\rho\kappa_S}} \tag{3.34}$$

Example: Derive Eq. (2.48) for the velocity of sound (c) in an ideal gas. For an ideal gas we showed earlier (Chapter 1, p. 60) that $\kappa_T = 1/P$, so $\kappa_S = 1/\gamma P$; also $\rho = PM/RT$, and therefore:

$$c^2 = \frac{\gamma RT}{M}$$ ∎

Exercise: In Chapter 2 we showed that:

$$C_p - C_v = \frac{TV\alpha^2}{\kappa_T}$$

Show that this, combined with Eq. (3.33), gives:

$$\kappa_S = \kappa_T - \frac{TV\alpha^2}{C_p} \tag{3.35}$$ ∎

Example: Calculate the velocity of sound in water at 20°C. For use in Eq. (3.35) the data of Table 1.8 are best converted to SI units; for water:

$$\kappa_T = \frac{45 \times 10^{-6} \text{ atm}^{-1}}{101,325 \text{ Pa/atm}^{-1}} = 4.44 \times 10^{-10} \text{ Pa}^{-1}$$

(As usual, we carry an extra significant figure.) Also, $V_m = 18 \times 10^{-6}$ m^3.

$$\kappa_S = 4.44 \times 10^{-10} \text{ Pa}^{-1} - \frac{(293 \text{ K})(18 \times 10^{-6} \text{ m}^3)(0.21 \times 10^{-3} \text{ K}^{-1})^2}{(75.3 \text{ J/K})}$$

$$= 4.41 \times 10^{-10} \text{ Pa}^{-1}$$

$$\rho = \frac{M}{V_m} = 10^3 \text{ kg/m}^3$$

$$c = \sqrt{\frac{1}{\rho\kappa_S}} = 1.5 \times 10^3 \text{ m/s}$$

(Check the units for this example; remember Pa = J/m^3.) ∎

Heat-Capacity Difference

Not the least of the treasures available from Table 3.1 is an easy derivation of Eq. (2.38) for $C_p - C_v$. By now we should recognize that the slopes for S vs. T at constant V and P are:

$$\left(\frac{\partial S}{\partial T}\right)_V = \frac{C_v}{T}, \qquad \left(\frac{\partial S}{\partial T}\right)_P = \frac{C_p}{T}$$

The relationship between the slopes ($\partial S/\partial T$) at constant V and P follows immediately from the formula of Appendix II for changing the variable held constant in a partial derivative; in this case:

$$\left(\frac{\partial S}{\partial T}\right)_P = \left(\frac{\partial S}{\partial T}\right)_V + \left(\frac{\partial S}{\partial V}\right)_T\left(\frac{\partial V}{\partial T}\right)_P$$

Now, using the Maxwell relationship (b), we get Eq. (2.38) directly:

$$C_p = C_v + T\left(\frac{\partial P}{\partial T}\right)_V \left(\frac{\partial V}{\partial T}\right)_P$$

3.6 The Third Law of Thermodynamics

The third law was a latecomer to thermodynamics, being stated for the first time in its modern definitive form by G. N. Lewis and M. Randall in 1923. Its experimental basis arose from the study of the properties of materials at very low temperatures, in particular the heat capacity, and its upshot is to provide additional insight into the meaning of entropy and thermodynamics.

All the equations we have used so far have dealt with differences in the thermodynamic functions — for example, ΔU, ΔH, or ΔS. It was mentioned earlier that only relative values for U and H could be measured. For example, the energy of a solid in its standard state at temperature T could be given as:

$$U^{\theta}(T) = U_0^{\theta} + \int_0^T C_v^{\theta} dT$$

but the value of the energy at $T = 0$ is an unknown constant, which could be chosen arbitrarily for each elemental substance. (The constant for compounds can be determined relative to the elements via heats of reaction, as we shall see in Chapter 6.) The same statement applies also to free energies (since they include U or H in their definition), but not to entropies; the third law permits the measurement of absolute values for the entropy of any material.

Entropy at Zero Kelvin

More than the other two, the third law is very closely connected to the quantization of energy and the kinetic theory of heat. The statistical approach to entropy, introduced by Ludwig Boltzmann, will not be discussed in detail until Chapter 5; for now it is sufficient to state that an increase in entropy is associated with an increasing disorder in matter. The examples we have done can be used to illustrate this notion: entropy increases with temperature because of the increased degree of kinetic motion; the entropy of a gas increases with volume because of the increased space available for kinetic movement; entropy increases when substances are mixed; entropy increases when solids are melted and the molecules can move more freely than when they were tied to a lattice position in the crystal; similarly, entropy increases upon vaporization. All these examples tell us that the entropy of a material goes toward a minimum as $T \rightarrow 0$. The entropy of a crystalline substance at some finite temperature T is related to the entropy at zero Kelvin by:

$$S(T) = S_0 + \int_0^T \frac{C_p}{T} dT \tag{3.36}$$

In order for the integral of Eq. (3.36) to remain finite, it is necessary that $C_p \rightarrow 0$ as $T \rightarrow 0$ in such a way that C_p/T will remain finite; otherwise, the symbol S_0 would have no meaning. This is found experimentally to be true, and this fact is closely connected to the quantum nature of energy. Like the diatomic vibration discussed in Chapter 2, all types of motion have some minimum quantum size by which their degree must be increased. At some sufficiently low temperature, the thermal energies available are so small that all modes of motion become "inactive" and the heat capacity approaches zero. (Figure 3.7 shows this for KCl.)

Furthermore, it appears from experimental results that the entropy of all pure crystalline substances approaches the same limit (S_0) as $T \to 0$. This can be illustrated with an example. Sulfur exists in two crystalline forms, rhombic (rh) and monoclinic (mc). The stable form at $P = 1$ atm below 368.5 K is rhombic. The entropy of rhombic sulfur was determined relative to its entropy at zero Kelvin by heat-capacity measurements which were used to evaluate Eq. (3.36) (how this is done in practice will be illustrated shortly). The result was:

$$S_m(\text{rh}, 368.5 \text{ K}) = S_0(\text{rh}) + 8.810 \text{ cal/K}$$

The phase transition at $T = 368.5$ to monoclinic requires a heat (ΔH_ϕ) of 96.0 cal/mol, so [using Eq. (3.16)]:

$$\Delta S_\phi = \frac{96.0}{368.5} = 0.261 \text{ cal K}^{-1} \text{mol}^{-1}$$

From this, the entropy of monoclinic sulfur is:

$$S_m(\text{mc}, 368.5) = S_0(\text{rh}) + 9.071 \text{ cal/K}$$

Because of the slowness of solid-solid phase transitions, metastable monoclinic sulfur exists below 368.5 K, and its entropy can be determined from heat-capacity measurements and Eq. (3.36):

$$S_m(\text{mc}, 368.5) = S_0(\text{mc}) + 9.04 \text{ cal/K}$$

Within the experimental error, the entropies of monoclinic and rhombic sulfur are the same at $T = 0$.

The third law of thermodynamics states that the entropy of all pure crystalline substances is the same at $T = 0$; the single unknown constant, S_0, is given the value *zero*. From the statistical point of view this seems reasonable; if entropy is disorder, all substances at absolute zero Kelvin will have the same degree of "nondisorder"; that is, they are all perfectly ordered.

Creating Low Temperatures: Adiabatic Demagnetization

One proof of the third law is that "third-law entropies," calculated with Eq. (3.36) with $S_0 = 0$, agree with entropies calculated by statistical theory (Chapter 5); another is a consequence of what happens when we attempt to cool materials to absolute zero temperature.

Considerable effort has been devoted to the achievement of low temperatures. As lower temperatures were reached, they approached zero in sequences such as 0.1, 0.01, 0.001, ... (The lowest temperature achieved in a laboratory to date is $T = 5 \times 10^{-7}$ K by A. Abragam, M. Chapellier, M. Goldman, and Vu Hoang Chau

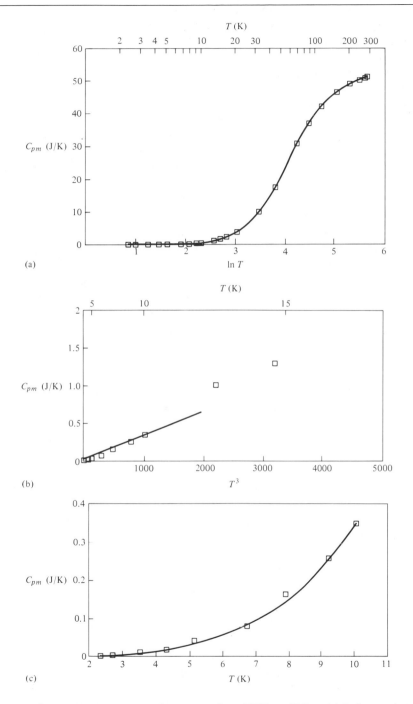

Figure 3.7 The constant-pressure heat capacity of KCl vs. *T*. Panel (c) shows that the *T*³ law is approximately obeyed below 10 K.

at the Centre d'Etudes Nucleaires, Saclay, France, 1969.) It became apparent that absolute zero is approached asymptotically.

An alternate statement of the third law is: absolute zero temperature is unattainable in a finite series of operations. To see how this relates to the other statement, we shall look briefly at how low temperatures are created.

Many gases, including nitrogen, can be cooled and liquefied by adiabatic expansion; the Joule-Thomson expansion is convenient for this purpose. Nitrogen liquefies at 77 K and can be used to cool hydrogen to below its inversion temperature (195 K); from there, hydrogen can be cooled by Joule-Thomson expansion until it liquefies at 20 K. Liquid hydrogen can be used to cool helium to below its inversion temperature (44.8 K); then the Joule-Thomson effect can be used to liquefy it at 4.2 K. A liquid helium bath at 4.2 K is the lowest temperature that can be conveniently maintained in the laboratory; if helium is boiled under reduced pressure, temperatures down to around 1 K can be achieved.

To get lower temperatures, a technique pioneered by William Giauque, adiabatic demagnetization, is used. Certain salts such as gadolinium sulfate have unpaired electrons that behave like little magnets. The salt is placed into a magnet in thermal contact with a helium bath. When the magnet is turned on, the salt is isothermally magnetized as the electrons align themselves with the field. The thermal contact with the helium is via helium gas in an intervening space; if the gas is pumped out, the space becomes a vacuum and the sample becomes adiabatic. The magnet is then turned off and the sample demagnetizes itself adiabatically as the electron spins resume their random alignments; this causes the temperature to drop. Why does this happen?

Compare this to what happens in an ideal gas. If the gas is isothermally compressed, heat must leave the system to balance the work and keep U (hence T) constant. If the gas is then adiabatically expanded, the work done must come at the expense of the energy (hence T), since no heat can enter. In Giauque's experiment, the magnet does work on the spins to magnetize them, and the temperature is maintained constant by heat contributed by the helium bath (some of the helium evaporates). In the adiabatic step, the work required to disalign the spins comes at the expense of temperature.

This process is incapable of reaching $T = 0$, because at that temperature the spins would be aligned whether or not the magnet was on; the entropies of the two states, magnet on and magnet off, are the same at $T = 0$ in accordance with the first statement of the third law.

The Measurements of Third-Law Entropies

Given that the entropy of a material is zero at $T = 0$, the entropy at another temperature can be calculated with Eq. (3.36), taking into account that there are latent entropy changes and discontinuities in the heat capacity at phase transitions. This calculation is generally done by numerical or graphical integration of heat-capacity data. Table 3.2 gives the entropies of various materials at 25°C.

In practice, heat capacities cannot be measured down to zero Kelvin and usually are not measured below 1 K. The gap between zero and the lowest temperature used is filled by theory. Peter Debye showed theoretically that the heat capacity of ionic or atomic solids at low temperature was given by:

Table 3.2   Standard entropies at 298.15 K

(units: J/K)

Solids	S_m^θ		Gases	S_m^θ	Liquids	S_m^θ
Ag	42.68		O	160.96	Br$_2$	152.30
AgCl	96.11		O$_2$	205.06	H$_2$O	70.00
AgBr	107.1		I	180.67	Hg	76.02
AgI	114		I$_2$	260.58	CH$_3$OH	127
Al	28.33		H	114.60	C$_2$H$_5$OH	161
Au	47.36		H$_2$	130.58	C$_6$H$_6$	173
C	5.694	(graphite)	N$_2$	191.50		
C	2.44	(diamond)	Cl$_2$	222.97		
Cd	51.5		F$_2$	202.71		
Cu	33.3		CO	197.48		
Zn	41.6		CO$_2$	213.72		
I$_2$	116.5		HCl	186.77		
Hg$_2$Cl$_2$	196.		NO	210.62		
K	64.18		H$_2$S	205.65		
Na	51.21		CH$_4$	186.06		
Pb	63.97		C$_2$H$_2$	200.8		
PbSO$_4$	147.		C$_2$H$_4$	219.5		
PbCl$_2$	136		C$_2$H$_6$	229.5		
PbO$_2$	75.6		NH$_3$	192.5		
PbO	69.5					
S	31.9	(rhombic)				
S	32.6	(monoclinic)				

$$C_{pm} = aT^3 \qquad (3.37)$$

where a is a characteristic constant for the material. (Cf. Fig. 3.7.) Empirically, this formula is found to work also for molecular solids (but not for metals; the conduction electrons contribute a term to the heat capacity that is linear in T; also, certain solids such as graphite which form two-dimensional sheets obey a T^2 law as $T \to 0$). Debye's theory is discussed in Chapter 15.

The constant in Debye's law can be determined empirically. Let us call the lowest temperature at which the heat capacity was measured T^* and the heat capacity measured at that temperature C_{pm}^*. Then, if T^* is in the range for which Eq. (3.37) is valid:

$$a = \frac{C_{pm}^*}{(T^*)^3}$$

The entropy at this temperature is, from Eq. (3.36):

$$S_m(T^*) = \int_0^{T^*} C_{pm} \frac{dT}{T} = a \int_0^{T^*} T^2 \, dT$$

$$= \frac{a}{3} (T^*)^3$$

With the value of the constant a (above), this is:

$$S_m(T^*) = \frac{C^*_{pm}}{3} \tag{3.38}$$

Above this temperature the heat capacity is integrated numerically to the melting point. At the melting point the ΔS of fusion is added in and the heat-capacity integration is continued for the liquid to the boiling point, where the ΔS of vaporization must be included. If the substance is a gas at the temperature at which you want to calculate S, corrections for nonideality are generally applied to give S^θ; the manner in which this is done will be explained after an example.

Example: The measured heat capacity of benzene (Figure 3.8) and its heat of fusion are given in Table 3.3; these are (as in the original source) in cal/K; at the end we shall have to convert the answer using 4.1840 J/cal. The entropy calculation proceeds as follows:

1. From 0 to 13 K, Eq. (3.38) is used to give $S(13) = 0.685/3 = 0.228$ cal/K.
2. Numerical integration of the heat capacity from 13 K to the melting point can be done using Eq. (3.12). For the first interval:

$$\Delta S(13 \to 14 \text{ K}) = \frac{0.685 + 0.830}{2} \ln \frac{14}{13} = 0.0561 \text{ cal/K}$$

This calculation is repeated for each interval and the results summed, giving:

$$\Delta S(13 \to 278.69 \text{ K}) = 30.6235 \text{ cal/K}$$

[A more accurate evaluation of this integral, using cubic interpolation and a 500-step trapezoidal rule, gives 30.5772 cal/K for this answer. This integral could also be done graphically as the area under C_p/T vs. T (Figure 3.8a) or C_p vs. $\ln T$ (Figure 3.8b); it is apparent from this figure that the latter plot is smoother and, for this case at least, C_p vs. $\ln T$ is more suitable for graphical or numerical integration.]

3. Melting at 278.68 K: $\Delta S = 2358.1/278.69 = 8.461$ cal/K.

Table 3.3   Heat capacity of benzene

(units: cal/K)

T	C_{pm}	T	C_{pm}	T	C_{pm}
13	0.685	80	10.750	278.69	30.760
14	0.830	90	11.430	melts	$\Delta H = 2358.1$ cal
15	0.995	100	12.050	278.69	31.52
20	2.000	120	13.310	298.16	32.52
25	3.145	140	14.700		
30	4.300	160	16.230		
35	5.385	180	18.020		
40	6.340	200	20.010		
50	7.885	220	22.320		
60	9.065	240	24.880		
70	9.975	260	27.760		

Source: G. D. Oliver, M. Eaton, and H. M. Huffman, *J. Am. Chem. Soc.*, **70**, 1502 (1948).

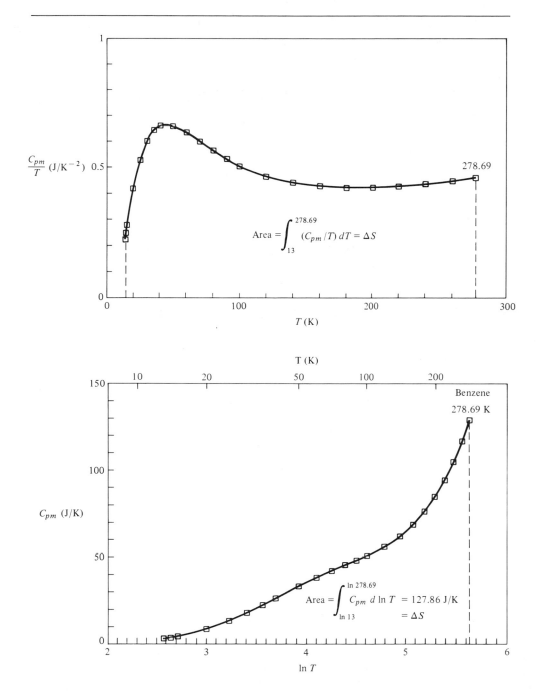

Figure 3.8 The heat capacity of solid benzene. This is used to illustrate the measurement of entropy using the third law of thermodynamics (cf. Table 3.3). As pointed out in Figure 3.6, graphs of C_{pm} vs. ln (T) are usually more suitable for such uses.

4. Heating liquid from melting point to 298.15 K; we can use Eq. (3.12):

$$\overline{C}_p = \frac{32.52 + 31.52}{2} = 32.02$$

$$\Delta S = 32.02 \ln \frac{298.15}{278.69} = 2.161 \text{ cal/K}$$

Finally:

$$S^{\theta}_{298.15} = (0.028 + 30.577 + 8.461 + 2.161)(4.184) = 172.5 \text{ J/K}$$

(Compare Table 3.2.)   ■

3.7   Entropy of Real Gases: Standard State

Entropies are generally listed for the material in the standard state, as in Table 3.2. For gases, this is the entropy which the gas would have if it were an ideal gas at 1 atm. The properties of real gases approach ideality as $P \to 0$, but the simple extrapolation that was used for H^{θ} and U^{θ} in Chapter 2 cannot be used for entropy because, while $S \to S(\text{ideal})$ as $P \to 0$, both approach negative infinity (Figure 3.9); a more subtle approach is needed.

First, the pressure dependence of S for a real gas can be found using the Maxwell equation, Table 3.1(d):

$$\left(\frac{\partial S}{\partial P}\right)_T = -\left(\frac{\partial V}{\partial T}\right)_P$$

This can be evaluated with the virial equation explicit in V:

$$V_m = \frac{RT}{P} + B + \gamma P + \delta P^2 + \cdots$$

$$\left(\frac{\partial S_m}{\partial P}\right)_T = -\left(\frac{\partial V_m}{\partial T}\right)_P = -\frac{R}{P} - B' - \gamma'P - \delta'P^2 - \cdots$$

Keeping only the B' term, this equation can be integrated to give:

$$S_m(P_2) - S_m(P_1) = -R \ln\left(\frac{P_2}{P_1}\right) - B'(P_2 - P_1) \tag{3.39}$$

For an ideal gas, only the first term applies [thus giving Eq. (3.14b)]; the difference between the entropy of a real and ideal gas (denoted S^{id}) is:

$$[S_m(P_2) - S^{\text{id}}_m(P_2)] - [S_m(P_1) - S^{\text{id}}_m(P_1)] = -B'(P_2 - P_1)$$

We now let $P_1 \to 0$ so $S_m(P_1) \to S^{\text{id}}_m(P_1)$ and:

$$S_m(P_2) = S^{\text{id}}_m(P_2) - B'P_2$$

At the standard pressure, $P_2 = P^{\theta} = 1$ atm, $S^{\theta}_m = S^{\text{id}}_m(1 \text{ atm})$:

$$S_m(1 \text{ atm}) = S^{\theta}_m - B'P^{\theta} \tag{3.40}$$

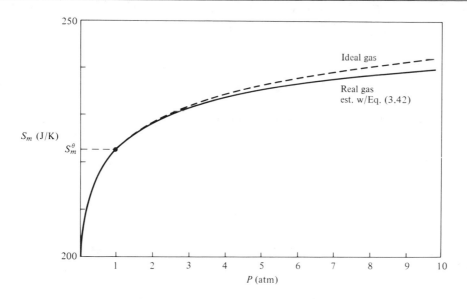

Figure 3.9 Standard entropy. The entropy of chlorine as an ideal gas (dashed curve) and as a real gas [solid curve, estimated using Eq. (3.42)] is shown. As the pressure approaches zero, the real and ideal curves approach each other, but both go to negative infinity. The standard-state entropy is the point on the ideal curve at 1 atm (which is not distinguishable from the real curve on this scale).

Exercise: Third-law entropies of gases, calculated as described in the previous section, are generally for $P = 1$ atm. This value can be used to calculate the standard entropy, using Eq. (3.40). The Berthelot virial coefficient is convenient for this purpose; the reader should demonstrate that the required correction is:

$$S_m^\theta = S_m(1 \text{ atm}) + \frac{27R}{32}\left(\frac{T_c}{T}\right)^3 \frac{P^\theta}{P_c} \qquad \textbf{(3.41)}$$

(In this equation, T is usually 298.15 K, but this is not required; the standard state specifies only pressure, not temperature.) ∎

The entropy at any pressure P is conveniently measured with respect to the standard entropy at that temperature:

$$S(T, P) = S^\theta(T) + (\text{function of } P, T)$$

This can be obtained from Eq. (3.39):

$$S_m(T, P) = S_m^{id}(T, P) - B'P$$

$$S_m(T, P) = S_m^\theta(T) - R \ln \frac{P}{P^\theta} - B'P \qquad \textbf{(3.42)}$$

Equation (3.42) is adequate at low to moderate pressure; if a more accurate equation is needed, or if it is desired to use a different gas law, the exact equation is:

$$S_m(T,P) = S_m^\theta(T) - R \ln \frac{P}{P^\theta} - \int_0^P \left[\left(\frac{\partial V_m}{\partial T} \right)_P - \frac{R}{P} \right] dP \qquad (3.43)$$

Exercise: Show that Eq. (3.43) leads to Eq. (3.42) if $V_m = RT/P + B$. ∎

Changes of entropy with pressure of condensed phases are usually quite small. The Maxwell equation (d) can be used with the coefficient of thermal expansion to give:

$$\left(\frac{\partial S}{\partial P} \right)_T = - \left(\frac{\partial V}{\partial T} \right)_P = -V\alpha$$

For one mole, typically $V \sim 10^2$ cm^3, $\alpha \sim 10^{-3}$ K^{-1}:

$$V\alpha \sim (10^{-1} \text{ cm}^3/\text{K})(8.314 \text{ J/K})/(82.06 \text{ cm}^3 \text{ atm/K}) \cong (0.01 \text{ J/K}) \text{ atm}^{-1}$$

(It can be seen from Table 3.2 that 0.01 J/K is as accurate as entropies are usually measured.)

3.8   Thermodynamics of Rubberlike Elasticity

Thermodynamics applies to more than gases; in fact it can tell us a lot of interesting things about that everyday household object — the rubber band.

Rubbers, like plastics, are long-chain molecules (polymers) formed from monomer units hooked together. Natural rubber is poly(*cis*-isoprene) with a monomer unit (Figure 3.10a):

$$\begin{array}{ccc} \text{CH}_3 & & \text{H} \\ \diagdown & & \diagup \\ & \text{C}=\text{C} & \\ \diagup & & \diagdown \\ -\text{CH}_2 & & \text{CH}_2- \end{array}$$

In 1839, Charles Goodyear discovered that this substance, when cross-linked with sulfur, became a useful elastomer. When cross-linked (typically every 100 or so monomer units) the rubber becomes a *network* of chain strands attached at various points.

It is necessary to modify the thermodynamic equations derived earlier to include the work of stretching (or compressing) the elastomer:

$$w = \int f \, dl \qquad (3.44)$$

where f is the restoring force. According to Hooke's law, the restoring force should be proportional to the displacement from the length, l_0, at equilibrium:

$$f = k(l - l_0) \qquad (3.45)$$

where k is called the *force constant*.

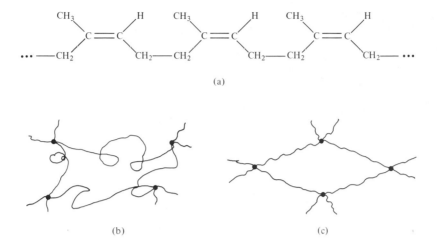

Figure 3.10   The structure of rubber. (a) The structure of poly(*cis*-isoprene), or natural rubber. (b) Rubber in its unstretched state is a tangle of cross-linked polymer chains lying in random directions. (c) When the rubber is stretched, the network of polymer chains becomes more ordered.

Exercise: Show that the work required to stretch from l_0 to l a substance that obeys Hooke's law is:

$$w = \tfrac{1}{2} k (l - l_0)^2 \qquad \blacksquare$$

Rubbers do not obey Hooke's law except for very short displacements, but a force constant can still be defined as:

$$k = \left(\frac{\partial f}{\partial l} \right)_T \qquad (3.46)$$

It should be anticipated that this force "constant" is not really constant but will change with extension. Our sign convention requires that work done *on* the system be positive. If we stretch a rubber band, dl is positive, $l > l_0$, and w is positive. Therefore k, as defined by Eq. (3.46), is a positive quantity.

It can be expected that the force required to stretch a rubber band will depend on the thickness — that is, force is proportional to the cross-sectional area. The **stress** is defined as the force per unit area:

$$\sigma = \frac{f}{A_0} \qquad (3.47)$$

where A_0 is the cross-sectional area of the unstretched material. The **strain** is defined as

$$\varepsilon = \frac{l - l_0}{l_0} \qquad (3.48)$$

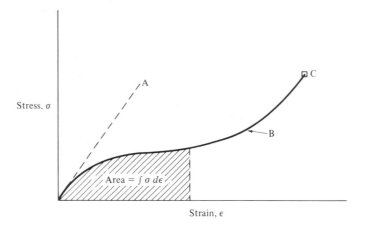

Figure 3.11　Stress-strain curve of a rubber (schematic). The dashed line (A) illustrates Hooke's law (restoring force is proportional to extension); its slope is the initial modulus (also called Young's modulus). Point B is where the rubber begins to stiffen. Point C is the breaking point. The area under this curve is the work per unit volume required to stretch the rubber to a given extension.

In these terms the work of elongation can be written as:

$$w = \int f\,dl = V_0 \int \sigma\,d\varepsilon \tag{3.49}$$

with $V_0 = l_0 A_0$, the volume of the unstretched material. The elastic properties of rubber, and other materials as well, are often described in terms of a stress vs. strain diagram—cf. Figures 3.11 and 3.12. Note that the area under this curve, $\int \sigma\,d\varepsilon$, is the work per unit volume [Eq. (3.49)].

Thermodynamics of Rubber Extension

When a rubber is stretched, the volume does not change much—it gets longer and thinner—so the PV work can be neglected; this gives, for the internal energy:

$$dU = dq + dw = T\,dS + f\,dl \tag{3.50}$$

(When the rubber is stretched, the restoring force is equal to the applied force, so it can be assumed that the process is reversible.) If, in Eq. (3.50), U is replaced by $(A + TS)$, it is readily shown that:

$$dA = -S\,dT + f\,dl \tag{3.51}$$

Exercise:　Show that Eq. (3.51) gives a Maxwell relationship:

$$\left(\frac{\partial f}{\partial T}\right)_l = -\left(\frac{\partial S}{\partial l}\right)_T \tag{3.52}$$

∎

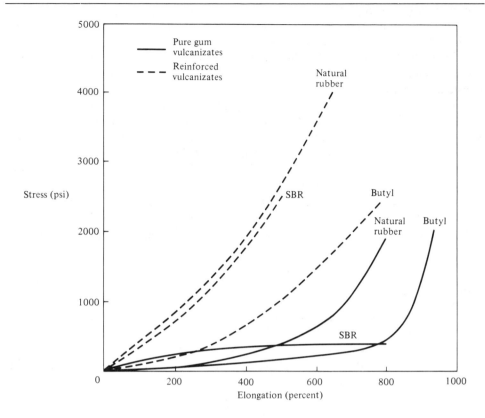

Figure 3.12 Some typical stress-strain curves for vulcanized and reinforced elastomers. (1000 psi = 6.89 MPa) (From A. X. Schmidt and C. A. Marlies, *Principles of High Polymer Theory and Practice*, 1948: New York, McGraw-Hill Book Co., Fig. 9-3)

Also from Eq. (3.51), we can see that the force can be expressed in terms of the free energy as:

$$f = \left(\frac{\partial A}{\partial l}\right)_T$$

and, using $A = U - TS$,

$$f = \left(\frac{\partial U}{\partial l}\right)_T - T\left(\frac{\partial S}{\partial l}\right)_T \tag{3.53}$$

This equation demonstrates that the restoring force has two components — one from energy changes with extension and the other from entropy changes. A rubber network in the unstretched state (Figure 3.10b) can be thought of as a fishing net lying in a jumbled heap. A difference is that, on a molecular scale, thermal motions will cause the net to writhe like a jumble of worms. When the net is stretched, the network strands move with respect to each other. Up to the stiffening point (Figure 3.11) (where you begin to pull against the covalent cross-links) the only resistance is the intermolecular forces between the strands. Forces between hydrocarbon molecules are relatively small — witness the low viscosity and boiling points (relative to molecular weight) of hydrocarbons. Also the volume — hence density — is relatively constant during stretching, so we do not expect the energy to change much with stretching at constant temperature.

On the other hand, the entropy is changing a lot. The stretched net will have its links more often parallel with each other than not and so be in a more ordered state; the statistical interpretation of entropy tells us that greater order means less entropy, so S should decrease with extension. The tendency of S to increase spontaneously is largely responsible for the restoring force — when you pull a rubber band, it is entropy that pulls back! On the macroscopic scale — the fishing net — such considerations are negligible so the net does not "pull back"; however, anyone who has tried to untangle a net has encountered entropy in its alternate incarnation — frustration.

If we incorporate Eq. (3.52) into Eq. (3.53), we get an equation that permits us to measure the extension energy:

$$\left(\frac{\partial U}{\partial l}\right)_T = f - T\left(\frac{\partial f}{\partial T}\right)_l \tag{3.54}$$

By measuring the force required to maintain a given extension as a function of T, we can determine the quantities in Eq. (3.54). [Reference 2 describes a simple experiment to do this. Also, note the close similarity between Eqs. (3.29) and (3.54).] Experimentally, it is found that $(\partial U/\partial l)_T$ is very small, so it becomes worthwhile to define an **ideal rubber** for which

$$\text{(ideal)} \qquad \left(\frac{\partial U}{\partial l}\right)_T = 0$$

Determining the Entropy Change upon Extension

This is by no means an experimental text, but there are two simple experiments that anyone can do with readily available equipment:

Experiment 1. Choose a sturdy rubber band with no nicks, cuts, or obvious signs of deterioration. Hold about 1 cm between your fingers and stretch it tautly (500% at least). Touch it to your lip; it may feel warm. Allow it to equilibrate and then let it contract sharply — you should feel the temperature drop. The effect is small and difficult to measure (how would you attach a thermometer?), but it should be unmistakable.

Experiment 2. Hang a sturdy rubber band (about 1 × 3 mm) from a fixed object and suspend from it a weight sufficient to cause a 50% to 100% extension (a few hundred grams). Now heat the rubber band with a hair dryer — watch carefully to see if the weight moves up or down. (It may be necessary to "temper" the rubber band first. Heat it, allow it to cool, then heat it again. A powerful heat source may be required.)

Both of these experiments tell us the same thing, but we need thermodynamics to connect them. In the first experiment the temperature change with extension is measured; if it is done fast enough, the experiment is adiabatic (constant S) and the coefficient measured is $(\partial T/\partial l)_S$. The second experiment measures the change of length with temperature at constant force, so the coefficient measured is $(\partial l/\partial T)_f$.

For the first experiment we can use the cyclic rule for partial derivatives to derive the relationship:

$$\left(\frac{\partial T}{\partial l}\right)_S = -\left(\frac{\partial T}{\partial S}\right)_l \left(\frac{\partial S}{\partial l}\right)_T = -\frac{(\partial S/\partial l)_T}{(\partial S/\partial T)_l}$$

The denominator should be recognizable as being related to a heat capacity [compare Eqs. (3.30) and (3.31)]:

$$T\left(\frac{\partial S}{\partial T}\right)_l = C_l \tag{3.55}$$

This is the heat capacity at constant extension, but it will not be much different from C_v (Problem 3.40), so we shall use the latter symbol. (We could also use C_p for rough calculations, since it differs from C_v only by approximately 10%.) Our "experimental result" shows that the quantity:

$$\left(\frac{\partial T}{\partial l}\right)_S = -\frac{T}{C_v}\left(\frac{\partial S}{\partial l}\right)_T \tag{3.56}$$

is positive (that is, T decreases when l decreases and vice versa); therefore the quantity $(\partial S/\partial l)_T$ must be *negative* — entropy decreases with length at constant temperature. This was, of course, the conclusion we reached from the net model, which is now confirmed experimentally.

Example: Estimate the magnitude of the temperature change in Experiment 1. As we did for adiabatic expansions of gases, we use the first law with $q = 0$, $dU = dw$. For an ideal rubber, $dU = C_v dT$, so:

$$C_v dT = dw, \qquad \Delta T \cong \frac{w}{C_v}$$

Let us assume a 1-cm segment (unstretched) of a 1×3 mm rubber band; the volume is thus $V_0 = 0.03 \text{ cm}^3$. From Table 3.4 we see that a typical density is about 1 g/cm^3 and a typical specific heat is about 2 J K^{-1} g^{-1}; thus the segment weighs 0.03 g and has $C_v \cong 0.06$ J/K (or 6×10^5 erg/K). To stretch this segment to 6 cm requires about one pound of force (~ 500 g), so the work is:

$$w \cong (500 \text{ g})(980 \text{ cm/s}^2)(5 \text{ cm}) = 2.5 \times 10^6 \text{ ergs}$$

Therefore:

$$\Delta T \cong \frac{2.5 \times 10^6 \text{ ergs}}{6 \times 10^5 \text{ erg/K}} = 4 \text{ K}$$

Table 3.4 Properties of rubber[a]

(T = 298 K. Units: Pa = J m^{-3} = 10 dynes cm^{-2} = 10 ergs cm^{-3})

	Poly(isoprene) (Natural rubber)	Butadiene-styrene (SBR, 23.5% S)	Chloroprene (Neoprene™)	Butene-isoprene (Butyl rubber)
ρ (g cm^{-3})	0.970	0.980	1.320	0.933
α (K^{-1})	6.6×10^{-4}	6.6×10^{-4}	$(6.1–7.2) \times 10^{-4}$	5.6×10^{-4}
c_p (J K^{-1} g^{-1})	1.828	1.83	2.1–2.2	1.85
κ_T (Pa^{-1})	5.14×10^{-10}	5.10×10^{-10}	4.40×10^{-10}	5.08×10^{-10}
ε max (%)[b]	750–850	400–600	800–1000	750–950
Tensile strength (Pa)[b]	$(17–25) \times 10^6$	$(1.4–3.0) \times 10^6$	$(25–38) \times 10^6$	$(18–21) \times 10^6$
Initial modulus (Pa)[c]	1.3×10^6	1.6×10^6	1.6×10^6	1.0×10^6

[a]Properties listed are for vulcanized gum stock without fillers (ref. 4).
[b]Break point.
[c]Initial slope of stress-strain curve.

This is only a rough estimate, but we see immediately that (a) it is detectable by your lip, (b) it won't burn you. However, like Joule's experiment (free expansion of gas), this experiment is difficult to quantitate. ∎

In Experiment 2 we measured $(\partial l/\partial T)_f$. Using the cyclic rule:

$$\left(\frac{\partial l}{\partial T}\right)_f = -\left(\frac{\partial l}{\partial f}\right)_T \left(\frac{\partial f}{\partial T}\right)_l$$

The first slope on the right-hand side can be identified by Eq. (3.46) as the reciprocal of the force constant k. The second term can be replaced by the Maxwell relationship, Eq. (3.52). Therefore:

$$\left(\frac{\partial l}{\partial T}\right)_f = \frac{1}{k}\left(\frac{\partial S}{\partial l}\right)_T \tag{3.57}$$

Since the force constant is positive and (as has been proved already) $(\partial S/\partial l)_T$ is negative, we now see that the length (at constant force) of the rubber band must *decrease* as it is heated. This fact surprises most people; somehow we expect it to get soft and weak when hot. Plastics, which also are long-chain polymers but lack the cross-linked feature, will behave in the expected manner (above the glass transition temperature). This conclusion illustrates the power of thermodynamics to predict the result of one experiment from another, even when the result is "unexpected."

Example: Estimate the degree of shortening of the rubber band in Experiment 2. For an ideal rubber [Eq. (3.53) with $(\partial U/\partial l)_T = 0$]:

$$\left(\frac{\partial S}{\partial l}\right)_T = \frac{-f}{T}$$

We also use Hooke's law, $f = k(l - l_o)$, to give:

$$\left(\frac{\partial S}{\partial l}\right)_T = \frac{-k(l - l_o)}{T}$$

Then, from Eq. (3.57):

$$\left(\frac{\partial l}{\partial T}\right)_f = \frac{1}{k}\left(\frac{\partial S}{\partial l}\right)_l = -\frac{l - l_o}{T}$$

$$\frac{dl}{l - l_o} = -\frac{dT}{T}$$

The reader should integrate this (l_1, T_1) to (l_2, T_2) to get:

$$l_2 - l_o = (l_1 - l_o)\frac{T_1}{T_2} \tag{3.58}$$

Taking, $l_o = 10$ cm, $l_1 = 18$ cm, $T_1 = 300$ K, $T_2 = 400$ K, this result can be used to calculate $l_2 = 16$ cm; this change, 18 cm to 16 cm, should be readily observed, but you may need a powerful hair dryer to attain a 100° rise in temperature. ∎

Postscript

This chapter has laid the formal basis of thermodynamics and derived a large number of useful equations. In particular, we have seen how the various properties of materials and their dependence on temperature, pressure, or other variables can be related to each other. The discussion in this chapter has been limited to closed systems, systems in which the amount and composition may not change, with only PV work; in subsequent chapters, these restrictions will be lifted and more general equations will be developed. New thermodynamic functions, the entropy and the free energies, have been introduced. The entropy function is a necessary addition to the energy functions (U and H) in order for the thermal properties of matter to be understood. The free energies, on the other hand, are more "functions of convenience" which are defined to make easier the understanding of equilibrium and change. Little has been said to this point about the free energy, but in future chapters this will be seen to be the key concept in chemical thermodynamics. The concept of the standard state, introduced in Chapter 2, has been further developed in this chapter; it will be seen, in due course, to be the key to understanding chemical equilibrium and properties of solutions.

Chapter 4 will use the concepts and equations developed in this chapter for the discussion of the properties and equilibria of pure materials; these applications should improve and clarify your understanding of this material.

References

1. G. N. Lewis, M. Randall, K. Pitzer, and L. Brewer, *Thermodynamics,* 1961: New York, McGraw-Hill Book Co.
2. F. Daniels, J. W. Williams, R. A. Alberty, and C. D. Cornwell, *Experimental Physical Chemistry,* 6th ed., 1962: New York, McGraw-Hill Book Co. (Later editions describe a more sophisticated version of this experiment, but this one is somewhat easier to understand.)
3. F. T. Wall, *Chemical Thermodynamics,* 1958: San Francisco, W. H. Freeman Pub.
4. J. Brandrup and E. H. Immergut (eds.), *Polymer Handbook,* 1975: New York, John Wiley & Sons.
5. L. R. G. Treloar, *The Physics of Rubberlike Elasticity,* 1958: Oxford, Clarendon Press.

Problems

3.1   The vapor pressure of water (steam pressure) is 1 atm at 100°C, but 15.3 atm (210 psig) at 200°C. Compare the work obtainable by a steam engine from fuel providing 10^6 J of heat at these temperatures. Assume a condensing temperature of 27°C.

3.2   Calculate the efficiency of an engine operating at 1 atm pressure if the working substance is (a) H_2O, (b) Hg, (c) Na. In parts (a) and (b), assume a condenser temperature of 30°C; use 100°C for (c). (Why wouldn't you use 30°C for Na?)

3.3 Assume a house loses (or gains) heat at a rate of 500 J s^{-1} for each degree Celsius temperature difference. If the outside temperature is 30°C, calculate the power required for an air conditioner to keep the house at (a) 25°C, (b) 20°C. The actual power would be several times this, but the ratios are reasonable.

3.4 (a) The rate of heat loss from a warm body is proportional to the temperature gradient. Show that the power required for a heat pump is proportional to the *square* of the temperature difference; that is, it takes four times as much power to keep a house at 20°C when the outside is 0°C than when the outside is 10°C. (b) Calculate the power needed for an "ideal" heat pump to heat a house with 1000 m^2 of wall area if the inside temperature is 20°C and the outside is 10° or 0°C. Assume that the walls are 10 cm thick and have a thermal conductivity of 0.05 J K^{-1}m^{-1}s^{-1}. The actual power required would be much larger.

3.5 Draw a diagram of a Carnot cycle on a *T* vs. *S* plot. What is the significance of the area inside a Carnot-cycle PV plot (Figure 3.2) and the plot (TS) you have just made?

3.6 Calculate the entropy change when 1 kg of water is heated at constant pressure from 10°C to 60°C.

3.7 Calculate the entropy change when one mole of CO_2(gas) is heated at constant pressure from 300 to 1000 K. Use the heat capacity of Table 2.2.

3.8 Use the heat capacities for copper given below to calculate ΔH and ΔS when 1 kg of copper is heated from 25 K to 200 K at constant *P*.

T (K)	C_{pm} (J/K)
25	1.00
50	6.28
100	16.15
150	20.54
200	22.68

3.9 (a) A mole of an ideal gas is heated at constant pressure from 310 to 420 K. Assume C_{pm} = 30 J/K and calculate q, w, ΔU, ΔH, and ΔS for this process. (b) Do the same for a process of heating (310 to 420 K) at constant volume.

3.10 One kg of iron (specific heat c_p = 0.47 J K^{-1}g^{-1}) at 100°C is placed in 1 kg of water (c_p = 4.19 J K^{-1}g^{-1}) at 0°C. Calculate the final temperature and the entropy change.

3.11 Calculate the entropy change when one mole of an ideal gas is compressed isothermally from 1 atm to 2.5 atm.

3.12 Calculate the entropy of mixing for 1 kg of air (O_2, 21%; N_2, 79% mole).

3.13 Chlorine has two isotopes, mass 35 (75.53%) and mass 37 (24.47%). Calculate the contribution to the molar entropy of Cl_2 from the mixing of the various isotopic species.

3.14 An ideal monatomic gas expands adiabatically from 800 K, 8 atm, to 1 atm final pressure. Calculate ΔS, ΔU, and q for the process if it is (a) reversible, (b) irreversible, doing 3000 J of work, (c) irreversible against zero pressure.

3.15 One mole of an ideal monatomic gas initially at STP is put through each of the reversible steps below, in each case starting at STP. Calculate w, q, ΔU, ΔH, and ΔS for each case.
(a) Cooling at constant volume to -100°C.
(b) Isothermal compression to 100 atm.
(c) Constant pressure heating to 100°C.
(d) Adiabatic expansion to 0.1 atm.

3.16   Tin melts at 231.9°C with a heat of fusion of 7070 J mol^{-1}. The heat capacities are $C_{pm}(\text{solid}) = 28.1$ J K^{-1}, $C_{pm}(\text{liq}) = 30.2$ J K^{-1}.
(a) Calculate ΔS of fusion at 231.9°C.
(b) Calculate ΔS when tin, supercooled 55°C below the normal mp, is frozen.
(c) Calculate $\Delta S(\text{surr.})$ and $S(\text{isolated})$ for (b).

3.17   (a) Show that the entropy change of a van der Waals gas for an isothermal change $V_1 \to V_2$ is:

$$\Delta S = nR \ln\left(\frac{V_2 - nb}{V_1 - nb}\right)$$

(b) Calculate ΔS for expanding one mole of NH_3 from 2 dm^3 to 20 dm^3. Compare this to the ideal gas result.

3.18   Calculate ΔA and ΔG for the isothermal compression of one mole of an ideal gas from 1 atm to 2.5 atm at 27°C.

3.19   Calculate ΔA and ΔG for compressing one mole of NH_3 from 1 atm to 2.5 atm at 27°C using the van der Waals equation. The volume changes from 24.497 dm^3 to 9.718 dm^3.

3.20   A useful objective is to write thermodynamic variables (H, U, S, and so on) in terms of measurable parameters. We have seen (Ch. 2) that the Joule-Thomson coefficient is measurable. Derive the following equation from the basic equations and definitions:

$$dH = C_p \, dT - \mu C_p \, dP$$

3.21   Derive the equation:

$$\frac{\kappa_T}{\kappa_S} = \frac{C_p}{C_v}$$

from the basic equations (Table 3.1) and definitions.

3.22   Some properties of liquids are relatively easy to measure — for example, C_p, α, and κ_S (from the velocity of sound). Other properties are more difficult to measure because liquids are not very compressible — for example, C_v and κ_T.
(a) Derive the equation:

$$\kappa_T = \kappa_S + \frac{TV\alpha^2}{C_p}$$

(b) Calculate κ_S, κ_T, γ, and C_{vm} for benzene at 25°C from these data: $C_{pm} = 134$ J K^{-1}, $V_m = 89.8$ cm^3, $M = 78.11$ g/mole, $c = 1295$ m s^{-1}, $\alpha = 1.24 \times 10^{-3}$ K^{-1}.

3.23   Derive the equation:

$$\left(\frac{\partial H}{\partial T}\right)_V = C_v + \frac{V\alpha}{\kappa_T}$$

3.24   (a) Derive the equation:

$$C_v = C_p - \left(\frac{\partial P}{\partial T}\right)_V (\mu C_p + V)$$

(b) Use this equation to calculate C_{vm} of CO_2 (gas) at 300 K and 1 atm. Use experimental data when possible, van der Waals otherwise. The van der Waals volume is $V_m = 24513.6$ cm^3 under these conditions. (Assume $C_{pm} = C_{pm}^\theta$.)

3.25 Show that the temperature change on adiabatic compression is:

$$\left(\frac{\partial T}{\partial P}\right)_S = \frac{\alpha V_m T}{C_{pm}}$$

Calculate the temperature change when 10 atm pressure is applied adiabatically to benzene at 25°C. Assume incompressibility.

3.26 Aluminum ($V_m = 10$ cm^3, $C_{pm} = 25.4$ J K^{-1}, $\alpha = 2.6 \times 10^{-5}$ K^{-1}) initially at 25°C, is struck by a force of 10 tons (U.S.) on one square inch. Calculate the maximum temperature rise (use the equation derived in the previous problem).

3.27 Calculate the standard entropy of H$_2$ (gas) at 500 K from the value at 25°C given in Table 3.2.

3.28 Use the data below to calculate the standard entropy of ammonia at 298.15 K (units: cal or cal/K, all for one mole).

T	C_{pm}	T	C_{pm}	T	C_{pm}
15	0.175	80	4.954	150	9.272
20	0.368	90	5.612	160	9.846
30	1.033	100	6.246	170	10.42
40	1.841	110	6.877	180	11.03
50	2.663	120	7.497	190	11.71
60	3.474	130	8.120	195.42	11.98
70	4.232	140	8.699		

Melts 195.42, $\Delta H_f = 1351.6$, C_{pm}(liq.) = 17.89 (ave.), boils 239.74, $\Delta H_v = 5581$.

T	C_{pm} (gas)
239.74	8.36
298.15	8.49

3.29 The refrigerant gas, CHCl$_2$F, has a virial coefficient $B = -354$ cm^3 and $B' = 1.66$ cm^3 K^{-1} at 394 K. Calculate ΔS for isothermal compression of this gas at 394 K from 1 atm to 2.5 atm.

3.30 Calculate the entropy of NH$_3$ (gas) at 25°C, 10 atm pressure, assuming (a) ideal gas, (b) the Berthelot virial coefficient.

3.31 Calculate the entropy of benzene (liq.) under 1000 atm pressure at 25°C.

3.32 Propane (gas) has a standard etnropy at 25°C of 270 J K^{-1}. (a) Calculate ΔS^θ when propane at 25°C is burned to form products at 25°C:

$$C_3H_8 + 5 O_2 = 3 CO_2 + 4 H_2O \text{ (liq.)}$$

(b) When burned, a mole of propane provides 2.108×10^6 J of heat (that is, the ΔH^θ for the above reaction is -2.108×10^6 J). Calculate the maximum work available if this heat is used in an engine operating between 300 and 450 K.
(c) A fuel cell is a device that produces an electrical current from a reaction. Calculate the maximum electrical work available if propane is reacted in fuel cell.

3.33 (a) The statistical theory of rubber (refs. 3 and 5) gives the following formula for the stress (σ) of a rubber as a function of the length:

$$\sigma = \frac{\rho RT}{zM}\left(\frac{l}{l_o} - \frac{l_o^2}{l^2}\right)$$

where ρ = density, M = molecular weight of the monomer unit, and z = number of monomer units between cross-links. Show that the work per unit volume for stretching the rubber from l_0 ($\varepsilon = 0$) to $l(\varepsilon)$ is:

$$\frac{w}{V} = \int_{l_0}^{l} \sigma \, d\varepsilon = \frac{\rho RT}{zM} \frac{\varepsilon^2(\varepsilon + 3)}{2(\varepsilon + 1)}$$

(b) Calculate the work for stretching a 10 cm × (0.05 cm²) rubber band to 50 cm. Assume the rubber is isoprene, that it is an ideal rubber, 1.25% cross-linking, and 25°C.

3.34   Calculate the specific heat at constant volume of natural rubber from its specific heat at constant pressure.

3.35   (a) Use the statistical formula for σ (Problem 3.33) to show that the initial modulus of the rubber (also called Young's modulus, the initial slope of the stress-strain curve) is:

$$\left(\frac{d\sigma}{d\varepsilon}\right)_{l_0} = \frac{3\rho RT}{zM}$$

(b) Calculate the percent cross-linking of chloroprene from the initial modulus listed in Table 3.4. (Note that Table 3.4 lists a "typical" value, and this obviously could vary.)

3.36   Calculate ΔT when a 10 cm × (2 mm × 3 mm) rubber band is adiabatically stretched to 20 cm. Assume that the rubber is SBR and use Hooke's law with $\sigma = (3 \times 10^6 \, \varepsilon)$ Pa⁻¹.

3.37   In our discussion of the effect of heat on the length of a rubber band under tension, we neglected the effect of thermal expansion. Assume the coefficient of linear expansion is one-third the volume coefficient (α) and calculate this effect for a 15-cm band and $\Delta T = 100$°C.

3.38   The stress (σ) on vulcanized natural rubber was measured at constant elongation as a function of temperature; the result was linear with (370% stretching) $\sigma = 1.47 \times 10^6 + 1.09 \times 10^4 \, t$ (t in °C, σ in Pa). Calculate $(\partial U/\partial l)_T$ (per unit volume) for this rubber at 0°C.

3.39   Prove:

$$\left(\frac{\partial S}{\partial T}\right)_l = \frac{C_l}{T}$$

where C_l is the heat capacity at constant length.

3.40   Prove that c_l, the specific heat at constant length, is equal to c_v for an ideal rubber.

3.41   (a) Assuming bond lengths (C=C) 1.07×10^{-8} cm and (C—C) 1.09×10^{-8} cm and a C=C—C bond angle of 120°, calculate the length of an isoprene monomer unit. (*Ans:* 3.25×10^{-8} cm.)
(b) In Chapter 9 it is shown that a random walk of N steps of length L will have an rms length of $\sqrt{N} L$. Assuming a random walk, calculate the mean length of an 80-unit isoprene chain. Assume the "links" are rigid units with the length calculated in (a).
(c) Assume the isoprene links to be arranged in a three-dimensional cube, estimate the density of rubber from the answer in (b).

3.42   So, there you are — in the middle of the desert with a warm can of beer and a powerful rubber band. You stretch the rubber band and let it equilibrate in the air, then release it adiabatically around the can. How much will the temperature drop, and how many times must you do this to make the beer drinkable (assuming you are *not* English)? Use V_0(rubber) $= 10^{-5}$ m³, C_p(rubber) $= 20$ J K⁻¹, C_p(can) $= 2000$ J K⁻¹. Assume Hooke's law with $\sigma = (2 \times 10^6 \, \varepsilon)$ Pa⁻¹ and $\varepsilon = 8$ for the stretch (you have powerful arms as well).

"Will you allow me as a chemist to thank you most sincerely for the very wonderful work you have done for us who, without the aid of mathematics, must needs grope so much in the dark?"

—M. M. Pattison Muir, letter to
J. Willard Gibbs (1880)

4

Equilibrium in Pure Substances

4.1 **Chemical Potential**

4.2 **Phase Equilibrium**

4.3 **Chemical Potential and Pressure**

4.4 **Chemical Potential and Temperature**

4.5 **Equilibria Involving Vapors**

4.6 **Surface Tension**

4.7 **Equilibria of Condensed Phases**

4.8 **Triple Points and Phase Diagrams**

4.9 **Non-First-Order Transitions**

In the last chapter we developed a variety of powerful tools which we shall now use to investigate the simplest type of chemical equilibrium, that of a pure substance among its various phases. This application — indeed, most applications of thermo-dynamics to chemistry — followed from the landmark papers by J. Willard Gibbs, *On the Equilibrium of Heterogeneous Substances,* published, unexpectedly, in an obscure provincial journal [*Trans. Conn. Acad., 3,* 108–248 (1876), 343–524 (1878)]. The concepts that Gibbs introduced — in particular, the free energy and the chemical potential — provided the light by which the science of steam engines illuminated chemistry. In this chapter we shall see only a limited number of these applications; the topic will be continued in Chapter 6.

4.1   Chemical Potential

Although we shall continue to limit our discussion to pure materials and systems that are, overall, closed, we now must account for the possibility that the material in the system is not homogeneous but may be distributed among several phases such as solid, liquid, or vapor. Since the amount of material in a particular phase may change, we must first examine how the thermodynamic functions depend on the amount of material.

Extensive properties, such as the Helmholtz free energy A, depend on two variables of state (such as temperature and volume) and on the amount of material, conveniently expressed as the number of moles, n: $A(T, V, n)$. An extension of the partial derivative slope formula (Appendix II) to a function of three variables gives:

$$dA = \left(\frac{\partial A}{\partial T}\right)_{V, n} dT + \left(\frac{\partial A}{\partial V}\right)_{T, n} dV + \left(\frac{\partial A}{\partial n}\right)_{T, V} dn$$

The first two of these slopes were identified in Chapter 3 (where constant n was an implicit condition, since the discussion was limited to closed systems) as $-S$ and $-P$, respectively. The third slope, $(\partial A/\partial n)_{T, V}$, is called the **chemical potential** and denoted by the symbol μ; thus:

$$dA = -S\, dT - P\, dV + \mu\, dn \tag{4.1}$$

If it is preferred to use T and P as independent variables, the Gibbs free energy is generally more convenient; since $G = H - TS$ and $H = U + PV$:

$$G = U + PV - TS = A + PV$$

$$dG = dA + P\, dV + V\, dP$$

Substitution into Eq. (4.1) gives:

$$dG = -S\, dT + V\, dP + \mu\, dn \tag{4.2}$$

This shows that the chemical potential is equally well defined as a slope of G at constant T, P or of A at constant T, V:

$$\mu = \left(\frac{\partial A}{\partial n}\right)_{T, V} = \left(\frac{\partial G}{\partial n}\right)_{T, P} \tag{4.3}$$

Table 4.1 Potentials and displacements

Potential	Displacement	Contribution to A
Pressure, P	Volume, V	$-\int P \, dV$
Stress, σ	Strain, ε	$V_0 \int \sigma \, d\varepsilon$
Surface tension, γ	Surface area, $\mathscr{A}$	$\int \gamma \, d\mathscr{A}$
Voltage, $\mathscr{E}$	Charge, Q	$\int \mathscr{E} \, dQ$
Magnetic field, $\mathscr{H}$	Magnetization, M	$\int \mathscr{H} \, dM$
Chemical potential, μ	Moles, n	$\int \mu \, dn$
Gravitational force, Wg	Height, h	$\int Wg \, dh \; (W = \text{mass})$

What is this "chemical potential"? That is the key question; the diligent student may have a thorough understanding of it by the end of Chapter 8. In this chapter's limited context of a chemically pure material, the answer is almost too trivial. Any extensive property can be made intensive by dividing by the amount of material; like $V_m = V/n$, we can define the free energy per mole as $G_m = G/n$, where $G_m(T, P)$ is an intensive property. Then $G(n, T, P) = nG_m(T, P)$ and $(\partial G/\partial n)_{T, P} = G_m$; for a pure material, the chemical potential is just the free energy per mole. Then why make things complicated by introducing another greek letter and a new name? There are two good reasons:

1. When we get around to discussing mixtures, in particular the liquid mixtures called solutions, the chemical potential will take on a deeper and more complex significance; in particular, the simple interpretation $\mu = G_m$ is no longer true. We shall not discuss this topic in detail until Chapter 7, but, until then, the chemical potential will lend a unity to our discussion of various types of equilibria.

2. It is a good idea to think of the various contributions to the free energy in terms of potential/displacement pairs; thus:

$$dA = -S \, dT + \sum (\text{potential})(\text{displacement}) \qquad (4.4)$$

For example, we may regard pressure as a potential for changing volume; if two portions of a system have different pressures, their volumes will change until the potential difference is zero—that is, until their pressures are equal. Later in this chapter we shall introduce the surface tension (γ) as the potential for changing surface area ($\mathscr{A}$). In Chapter 8 we shall consider the electrical voltage ($\mathscr{E}$) as a potential for moving electrical charge (Q). Analogously, μ is the potential for moving material. If the chemical potential in one region of a system is different than in another region, material will be transferred until the potential difference is zero. Table 4.1 displays these and some other potential/displacement pairs.

Example: An interesting application of the free-energy equation [Eq. (4.4)] arises if the gravitational potential (Table 4.1) is included; then

$$dG = -S \, dT + V \, dP + \mu \, dn + Wg \, dh$$

Consider a column of air (total mass, W) in equilibrium at constant temperature; the amount of material is constant, so $dn = 0$, and:

$$dG = V \, dP + Wg \, dh$$

At equilibrium, the free energy must be at a minimum, so $dG = 0$. Also the volume (V) and mass (W) are related by the density; assuming the ideal gas law:

$$\rho = \frac{W}{V} = \frac{PM}{RT}$$

Therefore ($dG = 0$, $V\,dP = -Wg\,dh$):

$$\frac{dP}{P} = -\frac{Mg}{RT}\,dh$$

Upon integration this gives:

$$\ln\frac{P_2}{P_1} = -\frac{Mg}{RT}(h_2 - h_1) \tag{4.5}$$

Equation (4.5) gives the variation in pressure with altitude of a column of gas at constant temperature. It will approximate the variation of air pressure with altitude in the earth's atmosphere, but only crudely, since the atmosphere is not isothermal. ∎

4.2   Phase Equilibrium

First we shall examine a system divided into two phases (denoted α and β) at constant volume and temperature. This could be a liquid and vapor phase in a rigid container, in thermal contact with a constant-temperature bath; the bath will provide whatever heat is needed for the system to reach equilibrium. Suppose that an amount of material (dn) is transferred isothermally from phase α to phase β (some of the liquid vaporizes or some of the vapor condenses); the free-energy changes are, from Eq. (4.1) (with $dT = 0$):

$$(dA^\alpha)_T = -P\,dV^\alpha - \mu^\alpha\,dn$$
$$(dA^\beta)_T = -P\,dV^\beta + \mu^\beta\,dn$$

(V^α and V^β are the volumes of the respective phases, and μ^α and μ^β are the chemical potentials of the material in that phase.)

The total free energy of the system will be the sum of $\alpha + \beta$:

$$(dA)_{\text{total}} = -P(dV^\alpha + dV^\beta) + (\mu^\beta - \mu^\alpha)\,dn$$

But the total volume, $V^\alpha + V^\beta$, is constant, so:

$$(dA)_{T,V} = (\mu^\beta - \mu^\alpha)\,dn$$

As material evaporates or condenses within the container, the chemical potentials (which are functions of T and P) will change; the second law of thermodynamics, Eq. (3.18), tells us that the spontaneous direction of change (dn positive or negative) will be such that A decreases and, at equilibrium, $(dA)_{T,V} = 0$. It can be seen now that $dA = 0$ when the chemical potentials of the phases are equal; thus, the criterion for equilibrium between phases α and β is:

$$\mu^\alpha = \mu^\beta \tag{4.6}$$

For a given constant temperature, this equality will occur at some particular value of P called the *vapor pressure*.

Now let's look at two phases in equilibrium at constant temperature and pressure. This could be, for example, a mixture of ice and water open to the atmospheric pressure of the room. Suppose that we now add heat to the system, causing a portion of the ice to melt (phase α, $dn^\alpha = -dn$); this increases the amount of liquid (phase β, $dn^\beta = dn$). From Eq. (4.2) with $dP = 0$, $dT = 0$:

$$d(G^\alpha + G^\beta)_{T,P} = (\mu^\beta - \mu^\alpha)\, dn$$

Again, Eq. (3.20) tells us that if the system is to remain in equilibrium, the free-energy change $(dG)_{T,P}$ is zero, and, of course Eq. (4.6) tells us this *will* be true, since the chemical potentials of the two phases must be equal.

In the first example, the position of equilibrium was independent of the volume; if the volume were changed to another value, the amounts of material in each phase would be altered, but the final pressure (for a given temperature) would be the same because of the condition that the chemical potentials be equal [Eq. (4.6)].

In the second example, if some pressure other than atmospheric had been applied to the system, the temperature would be required to change in order that the equilibrium condition, Eq. (4.6), would continue to apply. This tells us that there is a relationship between the temperature and pressure for states that are in equilibrium; to discover what this relationship is, we must explore how μ varies with T and P.

Variation of the Chemical Potential with *T, P*

We start with Eq. (4.2) for either phase:

$$dG = -S\, dT + V\, dP + \mu\, dn$$

Now $S = -(\partial G/\partial T)_{P,n}$ and:

$$\left(\frac{\partial S}{\partial n}\right)_{T,P} = -\frac{\partial}{\partial n}\frac{\partial G}{\partial T} = -\frac{\partial}{\partial T}\frac{\partial G}{\partial n} = -\left(\frac{\partial \mu}{\partial T}\right)_P$$

But $S = nS_m(T,P)$, where S_m is the molar entropy, so $(\partial S/\partial n)_{T,P} = S_m$ and:

$$\left(\frac{\partial \mu}{\partial T}\right)_P = -S_m \tag{4.7}$$

The same procedure also tells us:

$$\left(\frac{\partial \mu}{\partial P}\right)_T = V_m \tag{4.8}$$

Now, $\mu(T,P)$ can be written using the slope formula:

$$d\mu = \left(\frac{\partial \mu}{\partial T}\right)_P dT + \left(\frac{\partial \mu}{\partial P}\right)_T dP$$

$$d\mu = -S_m\, dT + V_m\, dP \tag{4.9}$$

[This equation could have been derived more simply by using the fact that, for a pure material, $\mu = G/n$ and dividing the basic equation, Table 3.1(d), by the number of moles; the procedure used here is a bit more general and could be used, with slight modification, for mixtures.]

Suppose that two phases are in equilibrium and a change is made in T or P. In order to remain at equilibrium, the changes in the chemical potential must be equal:

$$d\mu^\alpha = d\mu^\beta$$

Then, using Eq. (4.9) for each phase:

$$-S_m^\alpha \, dT + V_m^\alpha \, dP = -S_m^\beta \, dT + V_m^\beta \, dP$$

This equation is readily solved to give:

$$\frac{dP}{dT} = \frac{S_m^\alpha - S_m^\beta}{V_m^\alpha - V_m^\beta} \qquad (4.10)$$

Equation (4.10) provides a relationship between the T and P changes that may occur while keeping the phases in equilibrium. For a phase change $\alpha \longrightarrow \beta$ we define:

$$\Delta S_\phi = S_m^\beta - S_m^\alpha$$

$$\Delta V_\phi = V_m^\beta - V_m^\alpha$$

If a phase change occurs at constant T, P, the heat required (for example to melt a solid or vaporize a liquid) is the enthalpy change; for one mole:

$$\Delta H_\phi = H_m^\beta - H_m^\alpha$$

(This is also called the **latent heat** of the transition.) Since we are at an equilibrium point, as shown in Chapter 3 the entropy change must be:

$$\Delta S_\phi = \frac{\Delta H_\phi}{T}$$

With this, Eq. (4.10) becomes:

$$\frac{dP}{dT} = \frac{\Delta H_\phi}{T \, \Delta V_\phi} \qquad (4.11)$$

This is known as the **Clapeyron equation.**

Equation (4.11) is somewhat more convenient than (4.10) because latent heats for phase transitions (that is, ΔH_ϕ) are easily measured and widely tabulated (see Table 4.2); also, enthalpies tend to change less with temperature than the entropies. Applications of this relationship will occupy us for most of the remainder of this chapter, but first we must cover a few more topics.

4.3   Chemical Potential and Pressure

As with other thermodynamic properties (H and U in Chapter 2, S in Chapter 3), we wish to relate $\mu(T, P)$ to its value in a standard state:

Table 4.2   Heats of vaporization and fusion

[Values are at normal (1 Atm) boiling or freezing points]

Substance	T_f (K)	ΔH_f (kJ/mol)	T_b (K)	ΔH_v (kJ/mol)	ΔS_v (J K^{-1} mol^{-1})
He	—	—	4.21	0.084	20
H_2	13.95	0.12	20.38	0.904	44.4
N_2	63.14	0.720	77.33	5.577	72.12
O_2	54.39	0.444	90.18	6.820	75.63
SO_2	197.67	7.40	263.13	24.92	94.71
CO_2		(Sublimes 194.65 K, $\Delta H \cong 25$ kJ/mole)			
NH_3	195.39	5.653	239.72	23.33	97.32
$CHClF_2$	113	—	232.4	20.23	87.05
$CClF_3$	92	—	191.8	15.50	80.81
CCl_2F_2	115	—	243.36	19.97	82.06
CCl_3F	162	—	296.92	25.00	84.20
$CHCl_3$	209.7	9.2	334.4	29.4	87.8
CCl_4	250.3	2.5	349.9	30.0	85.7
CH_4	90.67	0.941	111.66	8.18	73.3
C_2H_6	89.88	2.86	184.52	14.72	79.8
C_3H_8	85.44	3.52	231.03	18.78	81.3
n-C_4H_{10}	134.80	4.66	272.65	22.39	82.1
n-C_6H_{14}	177.80	13.03	341.89	28.85	84.4
n-C_8H_{18}	216.36	20.74	398.81	34.98	87.7
CH_3OH	175.4	3.17	337.9	35.27	104
C_2H_5OH	156	5.02	351.7	38.58	110
H_2O	273.15	6.01	373.15	40.66	109

$$\mu(T, P) = \mu^\theta(T) + (\text{function of } T \text{ and } P)$$

The standard chemical potential, μ^θ, is the value of the chemical potential at the particular pressure $P^\theta = 1$ atm. The pressure dependence of μ is given by Eq. (4.8); therefore:

$$\mu(T, P) = \mu^\theta(T) + \int_{P^\theta}^{P} V_m \, dP \tag{4.12}$$

Equation (4.12) is easily evaluated for two cases, an ideal gas or an incompressible solid or liquid. For an ideal gas $V_m = RT/P$, and Eq. (4.12) can be integrated to give:

$$\mu = \mu^\theta + RT \ln \frac{P}{P^\theta} \tag{4.13}$$

For an incompressible substance, V_m can be assumed constant, and Eq. (4.12) integrates to:

$$\mu = \mu^\theta + V_m(P - P^\theta) \tag{4.14}$$

The evaluation of Eq. (4.12) for real gases is more involved, but the form of Eq. (4.13) is so convenient that a quantity called the **activity** is defined:

$$a \equiv \exp \frac{\mu - \mu^\theta}{RT}$$

$$\mu = \mu^\theta + RT \ln a \tag{4.15}$$

In effect, the activity as defined in Eq. (4.15) expresses the pressure dependence of the chemical potential in a way that will be seen (ultimately) to be very convenient. For the two cases above, it should be apparent that:

$$a \text{ (ideal gas)} = \frac{P}{P^\theta} \tag{4.16}$$

$$a \text{ (incompressible solid or liquid)} = \exp\left[\frac{V_m(P - P^\theta)}{RT}\right] \tag{4.17}$$

The full convenience of the activity concept may not become apparent until Chapter 6; for this chapter you need only appreciate the distinction between the chemical potential (μ) and its standard-state value (μ^θ).

Example: Calculate the activity of water at a pressure of 10 atm and $T = 298$ K. For water $V_m = 18.01$ cm^3.

$$\ln a = \frac{(18.01 \text{ cm}^3)(10 - 1) \text{ atm}}{(82.06 \text{ cm}^3 \text{ atm/K})(298 \text{ K})} = 0.006628, \qquad a = 1.0066$$

This example demonstrates that, for most solids and liquids at moderate pressure, the activity will be nearly equal to 1; in such cases we shall often assume $a = 1$ and, hence, $\mu = \mu^\theta$. ∎

4.4 Chemical Potential and Temperature

One of the equations derived in Chapter 3 (cf. Table 3.1) is:

$$\left(\frac{\partial G}{\partial T}\right)_P = -S$$

Generally it is more convenient to discuss the temperature dependence of the free energy in terms of the enthalpy rather than entropy. To perform this switch, we start with $G = H - TS$ and:

$$\frac{G}{T} = \frac{H}{T} - S$$

$$\frac{\partial}{\partial T}\left(\frac{G}{T}\right)_P = \frac{1}{T}\left(\frac{\partial H}{\partial T}\right)_P - \frac{H}{T^2} - \left(\frac{\partial S}{\partial T}\right)_P$$

In Chapter 3, two definitions were given for C_p:

$$C_p = \left(\frac{\partial H}{\partial T}\right)_P = T\left(\frac{\partial S}{\partial T}\right)_P$$

These demonstrate that the first and third terms of $\partial(G/T)/\partial T$ (above) are the same. Finally:

$$\left(\frac{\partial(G/T)}{\partial T}\right)_P = \frac{-H}{T^2} \tag{4.18a}$$

This equation can be divided by the number of moles (n) to give a similar equation for the chemical potential.

$$\left(\frac{\partial(\mu/T)}{\partial T}\right)_P = \frac{-H_m}{T^2} \tag{4.18b}$$

$$\frac{d(\mu^\theta/T)}{dT} = \frac{-H_m^\theta}{T^2} \tag{4.18c}$$

Note that the derivative for μ^θ can be written as a total derivative, since this quantity is not a function of pressure [cf. Eq. (4.12)].

It is often more useful to write these equations in terms of the variable $1/T$ rather than T; if $f(T)$ is a function of T, the chain rule gives:

$$\frac{df(T)}{d(1/T)} = \frac{df}{dT}\frac{dT}{d(1/T)} = -T^2\frac{df}{dT}$$

Using this with the equations above gives:

$$\left(\frac{\partial(G/T)}{\partial(1/T)}\right)_P = H \tag{4.18d}$$

$$\left(\frac{\partial(\mu/T)}{\partial(1/T)}\right)_P = H_m \tag{4.18e}$$

$$\frac{d(\mu^\theta/T)}{d(1/T)} = H_m^\theta \tag{4.18f}$$

Exercise: Prove that the temperature dependence of the activity [Eq. (4.15)] is:

$$R\left(\frac{\partial \ln a}{\partial(1/T)}\right)_P = H_m - H_m^\theta \tag{4.19}$$

The last section of Chapter 2 showed that H does not change very much with pressure, so for moderate pressures $H \cong H^\theta$ and the activity will be nearly independent of T. ∎

4.5 Equilibria Involving Vapors

Both solids and liquids have a vapor pressure, although in the case of some solids (granite, for example) it may be unmeasurably small. For liquids, this phenomenon is fairly obvious: the odor of organic solvents, the "humidity" due to water vapor in the air, and so on. The vapor pressure increases with temperature, and, when it equals the pressure of the environment, rapid evaporation, "boiling," occurs. (Boiling actually occurs at a temperature slightly higher than the equilibrium temperature because of surface effects.) Vapor pressure of solids, more often called the *sublimation pressure,* is less obvious. Iodine crystals in a closed container show a purple vapor phase — this is more evident if they are warmed a bit. Wet laundry on a clothes line will dry even if the weather is freezing — albeit slowly. Ice cubes left in a freezer too long will shrink, owing to evaporation. Carbon dioxide gas at 1 atm will, if chilled to $-78.5°C$, freeze directly to a solid; liquid CO_2 exists only when $P > 5$ atm. Solid CO_2, "dry ice," sublimes directly at 1 atm, $-78°C$, to a gas and is widely used for keeping things cold.

If the effect of surface tension is ignored, there is no difference (mathematically) between liquid-vapor or solid-vapor equilibria. For the remainder of this section we shall discuss only liquids, but the formulas derived apply to solids as well with only trivial changes in the symbols used (for example, vap $\longrightarrow$ sub, liq $\longrightarrow$ sol). The effect of surface tension on the vapor pressure of liquids will be discussed in the next section.

From Eq. (4.13) for the chemical potential of the vapor (assumed to be an ideal gas) and (4.14) for the liquid, the condition for equilibrium [Eq. (4.6)] gives:

$$\mu_{liq} = \mu_{vap}$$

$$\mu_{liq}^{\theta} + V_m(P° - P^{\theta}) = \mu_{vap}^{\theta} + RT \ln\left(\frac{P°}{P^{\theta}}\right) \qquad \textbf{(4.20a)}$$

In this equation, $P°$ is the equilibrium vapor pressure and $P^{\theta} = 1$ atm. The second term on the left-hand side is generally negligible; this can be shown by introducing the activity of the liquid [Eq. (4.17)], giving:

$$\mu_{liq}^{\theta} = \mu_{vap}^{\theta} + RT\left(\ln \frac{P°}{P^{\theta}} - \ln a_{liq}\right)$$

An earlier example demonstrated that the activity of liquids was approximately 1 at low pressure, therefore $\ln a \cong 0$. Dropping this term, we get:

$$R \ln \frac{P°}{P^{\theta}} = \frac{\mu_{liq}^{\theta}}{T} - \frac{\mu_{vap}^{\theta}}{T} = \frac{-\Delta G_v^{\theta}}{T} \qquad \textbf{(4.20b)}$$

(ΔG_v^{θ} is the standard free-energy change for the vaporization of one mole of the liquid.) Then, Eq. (4.18f) for the derivative of μ^{θ}/T, applied to Eq. (4.20), gives:

$$R\frac{d \ln(P°/P^{\theta})}{d(1/T)} = H_{liq}^{\theta} - H_{gas}^{\theta}$$

$$R\frac{d \ln(P°/P^{\theta})}{d(1/T)} = -\Delta H_v^{\theta}$$

(ΔH_v^{θ} is the molar heat of vaporization.) Since P^{θ} is a constant, $d \ln P^{\theta}/d(1/T) = 0$, so this can also be written:

$$\frac{d \ln P°}{d(1/T)} = \frac{-\Delta H_v^{\theta}}{R} \qquad \textbf{(4.21a)}$$

This is known as the **Clausius-Clapeyron equation.** If the variable T is preferred to $1/T$, this can be readily changed to:

$$\frac{d \ln P^\circ}{dT} = \frac{\Delta H_v^\theta}{RT^2} \tag{4.21b}$$

This equation also follows directly from Eq. (4.11):

$$\frac{dP}{dT} = \frac{\Delta H_v}{T \Delta V}$$

For the vaporization of a liquid:

$$\Delta V = V_{m,\,vap} - V_{m,\,liq} = \frac{RT}{P} - V_{m,\,liq}$$

Neglecting the volume of the liquid (this is equivalent to setting its $a = 1$) gives:

$$\frac{dP}{dT} = \frac{P \,\Delta H_v}{RT^2}$$

Then, with $dP/P = d(\ln P)$, Eq. (4.21b) follows immediately.

Equations (4.21) are generally useful from several points of view. Plots of $\ln P^\circ$ vs. $1/T$ can be seen, by Eq. (4.21a), to have a slope that is the heat of vaporization divided by R. Such plots are often found to be nearly linear, especially over small temperature ranges, indicating that the heat of vaporization does not vary rapidly with temperature. Hence, $\ln P^\circ$ and $(1/T)$ are useful variables for interpolating vapor-pressure data; if we know P_1° at T_1 and P_2° at T_2, $\ln P^\circ$ and $(1/T)$ may be linearly interpolated for temperatures between T_1 and T_2.

Figure 4.1 shows the variation of the vapor pressure of $CHClF_2$ with temperature. The graph of $\ln P$ vs. $1/T$ is nearly linear, indicating that its slope $(-\Delta H_v/R)$ is nearly constant in this temperature range. [Close examination of Figure 4.1 will reveal a slight curvature; the line drawn through the points is a least-squares fit of all points, and its slope corresponds to the mean heat of vaporization in this temperature range (200–320 K).]

Equations (4.21a) and (4.21b) apply as well to sublimation. Figure 4.2 shows the vapor pressure of solid and liquid carbon dioxide on a $\ln P$ vs. $1/T$ plot; note the difference in the slopes, which indicates that the heats of vaporization and sublimation are not the same (cf. Problem 4.29).

Equation (4.21) can be also written in integral form between two pairs of equilibrium points (P_2°, T_2) and (P_1°, T_1):

$$\ln \frac{P_2^\circ}{P_1^\circ} = \int_{T_1}^{T_2} \frac{\Delta H_v^\theta}{RT^2} \, dT \tag{4.22}$$

If the temperature dependence of ΔH_v is not known (in any case, it may not change very much over a small range of temperature), Eq. (4.22) can be integrated, assuming constant ΔH_v, to give:

$$\ln \frac{P_2^\circ}{P_1^\circ} = -\frac{\Delta H_v^\theta}{R}\left(\frac{1}{T_2} - \frac{1}{T_1}\right) \tag{4.23}$$

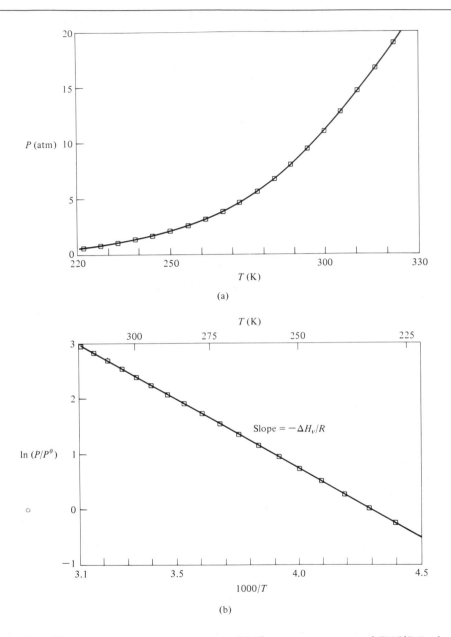

Figure 4.1   Vapor pressure vs. temperature. (a) The vapor pressure of $CHClF_2$ is shown vs. temperature. (b) A graph of ln (P) vs. the reciprocal of the absolute temperature is nearly linear, indicating that the heat of vaporization does not vary rapidly with temperature.

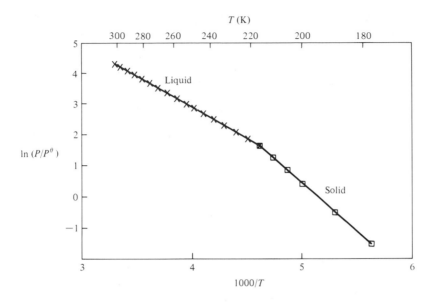

Figure 4.2 Vapor pressure of liquid and solid carbon dioxide. Note the discontinuity of the slope at the phase transition, which indicates that the heat of vaporization (related to the slope of the "liquid" line) is different than the heat of sublimation (related to the slope of the "solid" line).

This is a particularly convenient equation for interpolating vapor-pressure data or for estimating heats of vaporization. According to the differential mean-value theorem (Appendix I), the ΔH_v that appears in Eq. (4.23) is, in effect, the mean value in the interval (T_1, T_2).

Example: The vapor pressure of water is 23.756 torr at 25°C and 760 torr at 100°C (the normal boiling point). From this we can estimate the heat of vaporization using Eq. (4.23):

$$\ln \frac{760}{23.756} = \frac{-\Delta H_v}{8.3143}\left(\frac{1}{373.15} - \frac{1}{298.15}\right)$$

$$\Delta H_v = 42.74 \text{ kJ/mol}$$

What does this number mean? Heats of vaporization of water have been accurately measured over a wide temperature range because of its obvious importance (for example, in steam engines); at 25°, $\Delta H_v = 44.80$ and at 100°C, $\Delta H_v = 40.65$ kJ/mole. The average of these two values is:

$$\Delta \overline{H}_v = \tfrac{1}{2}(44.00 + 40.65) = 42.32 \text{ kJ/mol}$$

This value is reasonably close to the value calculated using Eq. (4.23). Also at 60°C [close to the mean temperature, $(100 + 25)/2 = 62.5°C$], the heat of vaporization is 42.48 kJ/mol; again, this value is very close to the calculated value. ∎

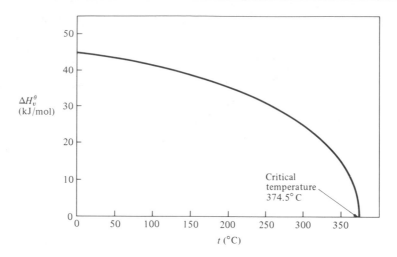

Figure 4.3   Heat of vaporization of water vs. temperature. Note the sharp dropoff near the critical temperature. At the critical temperature, the heat of vaporization goes to zero, as it must. (Cf. Figure 4.4.)

The variation of the heat of vaporization of water with temperature is shown by Figure 4.3. Notice that near the critical temperature the heat of vaporization drops rapidly to zero. That this must occur is demonstrated by Figure 4.4, which shows the relative enthalpy of nitrogen plotted as isotherms vs. P. Below the critical temperature, isothermal compression of the gas will cause condensation and a consequent discontinuity in H (the vertical lines in the two-phase region); this is the enthalpy of vaporization. Above the critical temperature, no phase separation occurs and the isotherms are continuous.

The Effect of Hydrostatic Pressure on Vapor Pressure

Up to this point we have assumed that the pressure of the system is the vapor pressure of the gas. If other gases are present, as there would be if the liquid were under air, the pressure applied to the liquid will not be the same as its vapor pressure.

Assume that there is some hydrostatic pressure P_x applied to the liquid in addition to the vapor pressure. This pressure could be applied mechanically, perhaps via a membrane that is permeable to the vapor, but it is most conveniently supplied by an inert gas. To be inert, a gas must meet two requirements: (1) it must not interact strongly with the vapor so as to cause the ideal gas approximation to fail; (2) it must not dissolve appreciably in the liquid. In Chapter 7 we shall discuss the effect of dissolved gases (and other solutes) on the vapor pressure of liquids and find that this effect is relatively small for gases such as nitrogen or oxygen (that is, air) in water.

Assuming a vapor phase that is an ideal gas, and an incompressible liquid phase, the condition for equilibrium in the presence of an excess pressure P_x is:

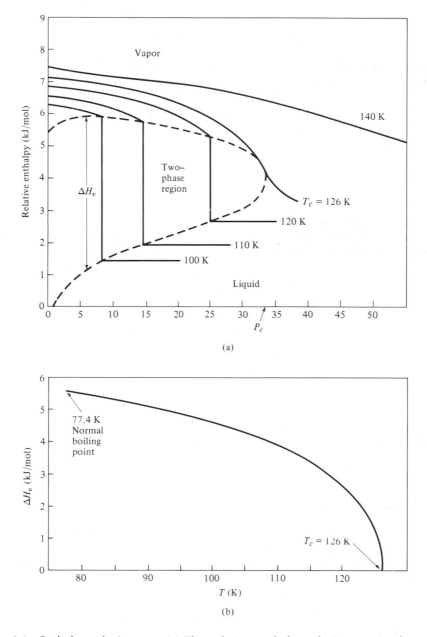

Figure 4.4 Enthalpy of nitrogen. (a) The relative enthalpy of nitrogen is plotted vs. pressure for various temperatures (isotherms). The discontinuous isothermal change of the enthalpy in the two-phase region is the "heat of vaporization" at that temperature. From this figure it can be seen that the heat of vaporization must go to zero at the critical point. (The isotherms in the liquid region are somewhat speculative.) (b) The enthalpy of vaporization of nitrogen vs. T.

$$\mu_{\text{liq}} = \mu_{\text{vap}}$$

$$\mu^{\theta}_{\text{liq}} + V_m(P_x + P^{\circ} - P^{\theta}) = \mu^{\theta}_{\text{vap}} + RT \ln \frac{P}{P^{\theta}}$$

The normal vapor pressure (P°) is given by Eq. (4.20a) when the pressure on the liquid is only the vapor pressure:

$$\mu^{\theta}_{\text{liq}} + V_m(P^{\circ} - P^{\theta}) = \mu^{\theta}_{\text{vap}} + RT \ln \frac{P^{\circ}}{P^{\theta}}$$

Subtracting these two equations gives:

$$\ln \frac{P}{P^{\circ}} = \frac{V_m P_x}{RT} \tag{4.24}$$

Example: Lange's *Handbook of Chemistry* lists the vapor pressure of water at 25°C *under its own vapor* as 23.756 torr. From this, calculate the vapor pressure of water under air (assumed "inert") at $P = 1$ atm (total) and 100 atm.

$$P_x = \frac{760 - 23.756}{760} = 0.9687 \text{ atm}$$

$$\ln \frac{P}{23.756} = \frac{(18.06 \text{ cm}^3)(0.9687 \text{ atm})}{(82.057 \text{ cm}^3 \text{ atm/K})(298.15 \text{ K})}$$

$$= 0.0007152$$

$$P = 23.773 \text{ torr}$$

This is a very small correction, although it may be necessary for very accurate work. The reader should show that when $P_x = 100$ atm, the vapor pressure is 25.58 torr. In Chapter 7 (p. 334) we shall return to this example and determine the effect of dissolved air on the vapor pressure; the correct answer (at 100 atm) will be shown to be 25.54 torr. ∎

4.6   Surface Tension

As the properties of molecules in a dense phase (liquid) differ from those in a vapor phase, so do the properties of those molecules at the interface between those two phases, the surface, differ from those in the interior. To increase the surface area of a liquid, thereby bringing more molecules from the bulk phase to the surface, requires work. The work required to increase the surface area ($\mathscr{A}$) is

$$dw = \gamma \, d\mathscr{A} \tag{4.25}$$

where γ is called the **surface tension.** (Surface tension can also be defined for solids, and this property is related to tensile strength; but that is another story, and this discussion will be limited to liquids.) Inclusion of this work in the expression for the change of the internal energy gives:

$$dU = T \, dS - P \, dV + \gamma \, d\mathscr{A} \tag{4.26}$$

Josiah Willard Gibbs (1839–1903)

Science is a hothouse flower that generally flourishes in established, stable civilizations; developing countries are, for good reason, concerned with more practical matters. In that context it is not surprising that Gibbs, one of the greatest mathematical physicists ever and unquestionably the greatest that the United States produced in the nineteenth century, began as an inventor. His 1866 patent for "An Improved Railway Brake" anticipated some features introduced by George Westinghouse, the inventor of the railway air brake; however, Westinghouse became rich and Gibbs became a professor.

  Gibbs made important contributions to mathematics and optics and monumental contributions to statistical mechanics and thermodynamics, including the invention of the free-energy function. His papers were so advanced and mathematical that scarcely a handful of scientists worldwide could appreciate them, and none of these lived in North America. The springing of such a person, unheralded, from the "provinces" is hardly less astonishing than the finding of a coloratura soprano singing Rossini arias on a mountain in Tibet. Although the Nobel prize was established two years before Gibbs' death, he was never nominated because he was too obscure. Even today he is relatively unknown and unappreciated among his countrymen. But no class in physical chemistry should let his birthday (February 11) pass without paying homage to him as the father of us all.

Using $A = U - TS$ gives:

$$dA = -S\,dT - P\,dV + \gamma\,d\mathscr{A} \tag{4.27}$$

Then, with $G = A + PV$:

$$dG = -S\,dT + V\,dP + \gamma\,d\mathscr{A} \tag{4.28}$$

These equations show three equivalent definitions for the surface tension:

$$\gamma = \left(\frac{\partial G}{\partial \mathscr{A}}\right)_{T,P} = \left(\frac{\partial A}{\partial \mathscr{A}}\right)_{T,V} = \left(\frac{\partial U}{\partial \mathscr{A}}\right)_{S,V} \tag{4.29}$$

If we take the derivative of $\gamma = (\partial A/\partial \mathscr{A})_{T,V}$ with respect to T:

$$\left(\frac{\partial \gamma}{\partial T}\right)_V = \frac{\partial}{\partial T}\frac{\partial A}{\partial \mathscr{A}} = \frac{\partial}{\partial \mathscr{A}}\left(\frac{\partial A}{\partial T}\right)_V$$

But from Eq. (4.27) we see that $(\partial A/\partial T)_V = -S$, so:

$$S_s \equiv \left(\frac{\partial S}{\partial \mathscr{A}}\right)_{T,V} = -\left(\frac{\partial \gamma}{\partial T}\right)_{V,\mathscr{A}} \tag{4.30}$$

Table 4.3   Surface tensions of some liquids
(units: dyne cm^{-1} = 10^{-3} N m^{-1})

Substance	°C	γ	Substance	°C	γ
Platinum	2273	1900	Water	0	75.7
Copper	1404	1100	Water	20	72.75
Aluminum	700	840	Water	25	72.0
Lead	350	453	Water	40	69.6
Mercury	20	472	Water	60	66.2
Acetone	20	23.7	Water	80	62.6
Benzene	20	28.88	Water	100	58.8
Chloroform	20	27.14	Oxygen	−183	13.1
Ethanol	10	23.6	Oxygen	−203	18.3
Ethanol	20	22.8	Argon	−188	13.2
Ethanol	30	21.9	Argon	−183	11.9
Methanol	0	24.5	Nitrogen	−203	10.5
Methanol	20	22.6	Nitrogen	−193	8.3
Methanol	50	20.2	Nitrogen	−183	6.2

The quantity defined by Eq. (4.30) (S_s) is called the **surface entropy** of the liquid; it is the entropy per unit area due to the surface.

The surface energy $U_s \equiv (\partial U/\partial \mathscr{A})_{T,V}$, which is the internal energy per unit area due to the surface, can be computed as follows:

$$A = U - TS$$

$$\left(\frac{\partial A}{\partial \mathscr{A}}\right)_{T,V} = \left(\frac{\partial U}{\partial \mathscr{A}}\right)_{T,V} - T\left(\frac{\partial S}{\partial \mathscr{A}}\right)_{T,V}$$

$$A_s = U_s - TS_s$$

But from Eq. (4.29), the left-hand side is just γ; also, Eq. (4.30) gives the surface entropy, so:

$$\gamma = U_s + T\left(\frac{\partial \gamma}{\partial T}\right)_V$$

$$U_s = \gamma - T\left(\frac{\partial \gamma}{\partial T}\right)_V \tag{4.31}$$

The unit relationships in this equation derive from the fact that force/length = energy/area; in cgs units, dyne/cm = erg/cm^2 and in SI units N/m = J/m^2.

What are the physical effects of surface tension? Equation (4.27) shows that, at constant T, V, a liquid can minimize its free energy by decreasing its surface area. We have not, of course, included the gravitational force in our equations, but if that is negligible (as it will be for small droplets), the droplet will assume the shape of a sphere, the shape that has the minimum surface area for a given volume. A quantity of liquid dispersed into fine droplets (an aerosol) may have a very large surface area; 1 cm^3 dispersed as 10 nm droplets has about 300 square meters of surface area. Two droplets that collide can lower their combined surface area and, hence, free energy by merging into a single droplet—a phenomenon of obvious interest in meteorology. This spontaneous combination of droplets can be observed with droplets on a non-adsorbent surface; if you push them together (with a nonadsorbent finger) until they touch, they will combine to minimize the surface area. This effect is clearly demonstrated using small mercury droplets on glass, but be cautious, mercury is somewhat toxic. (The shape of droplets on a surface is rather complicated, involving the surface tension to the air and to the surface as well as the force of gravity. Clearly they are not simple spheres.)

Bubble Pressure

The effect of surface tension is also observable as bubbles. Before looking at this, we need to obtain a relationship between the volume and surface area. A sphere of radius r has volume (V) and surface area $(\mathscr{A})$:

$$V = \frac{4\pi r^3}{3}, \qquad \mathscr{A} = 4\pi r^2$$

A curved surface with radius of curvature r can be considered to be a section of a sphere with infinitesimal elements:

$$dV = 4\pi r^2\, dr, \qquad d\mathscr{A} = 8\pi r\, dr$$

Putting these together, we get:

$$d\mathscr{A} = \frac{2}{r}\, dV$$

This can be used to write Eq. (4.27) for the free energy as:

$$dA = -S\, dT - P\, dV + \frac{2\gamma}{r}\, dV \qquad \textbf{(4.32)}$$

The quantity $2\gamma/r$ has units of pressure and is often called the *surface pressure*. This is a force per unit area on the surface in addition to that exerted by the pressure of the vapor phase. The force is directed toward the center of curvature, inward for a concave surface and outward for a convex surface. If the radius of curvature is infinite—that is, a flat surface—the surface pressure is zero.

This force is demonstrated in a commonly observed phenomenon, the formation of bubbles. A liquid in the shape of a hollow sphere can minimize its surface area, and hence its free energy, by contracting. Were it not for the presence of gas inside the bubble, it would contract to a spherical droplet of minimum surface area. However, as it contracts, the gas inside is compressed and an excess pressure, ΔP, exerts an outward force to counteract the contraction. At equilibrium the inward force ($4\gamma/r$, since there are two surfaces) exactly balances the pressure difference, and the excess pressure inside the bubble must be:

$$\Delta P = \frac{4\gamma}{r} \qquad \textbf{(4.33)}$$

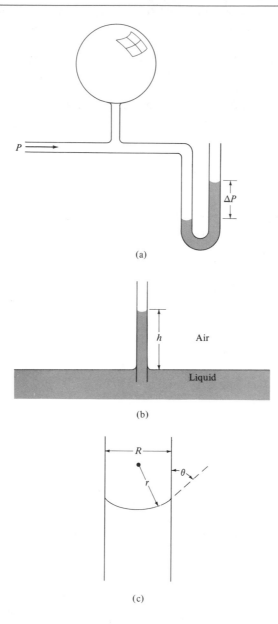

Figure 4.5 Surface tension. (a) The formation of a bubble requires an excess pressure inside the bubble to counteract the surface tension, which would cause the bubble to contract and minimize its surface area. (b) The rise of a liquid in a capillary is used to measure surface tension. (c) The parameters used to analyze capillary rise: R is the radius of the capillary, r is the radius of curvature of the liquid surface, and θ is the contact angle between the surface and the side of the capillary. If the surface of the liquid is concave (negative radius of curvature), the liquid level in the capillary will be depressed.

When someone "breaks your bubble," the excess pressure inside is released and the liquid contracts rapidly to a droplet. Figure 4.5(a) shows, schematically, an apparatus that could measure this pressure (and, from it, the surface tension).

Capillary Rise

Another method for measuring surface tension arises from the phenomenon of capillarity. If a capillary tube (typically glass) is placed in a liquid, the surface of the liquid inside the capillary will be curved. If the liquid "wets" (that is, adheres to) the glass, the surface will be concave and the liquid column rises into the tube — for example, water in glass. If the liquid does not wet the glass, the surface will be convex and depressed with respect to the bulk liquid surface — for example, mercury in glass [Figure 4.5(b)].

We can analyze the capillary rise by examining the balance of forces that places the column of liquid in equilibrium. If the column of liquid is approximated as a cylinder with cross section $\mathscr{A}$ and height h, the mass of liquid in the column is $m = \rho h \mathscr{A}$ where ρ is the density. The force of gravity is then $mg = \rho g h \mathscr{A}$. This is counterbalanced by the surface pressure $2\gamma/r$ per unit area. If we assume that these two areas, the surface area and the cross-sectional area of the tube, are approximately the same, then, at equilibrium:

$$\rho g h \mathscr{A} = \frac{2\gamma}{r} \mathscr{A}$$

$$\gamma = \frac{1}{2} \rho g h r \tag{4.34a}$$

In this equation, r is the radius of curvature, not necessarily the radius of the tube. A more accurate formula must take into account the fact that the top of the column of liquid is not flat. With R = radius of the tube, $\rho_l - \rho_v$ the difference of the liquid and vapor densities (to account for Archimedes' principle), and θ the contact angle of the liquid surface with the side of the tube [Fig. 4.5(c)], a more accurate formula is:

$$\gamma = \left(h + \frac{R \cos \theta}{3} \right) (\rho_l - \rho_v) \frac{Rg}{2 \cos \theta} \tag{4.34b}$$

Effect of Surface Tension on Vapor Pressure

The effect of surface tension on the vapor pressure of a liquid, neglected earlier, can now be discussed. The total free energy [Eq. (4.2) with $dT = 0$ and the surface-pressure term of Eq. (4.32) added] is:

$$dA_{\text{liq}} = -P \, dV_{\text{liq}} + \frac{2\gamma}{r} \, dV_{\text{liq}} + \mu_{\text{liq}} \, dn_{\text{liq}}$$

For the vapor phase we have only:

$$dA_{\text{vap}} = -P \, dV_{\text{vap}} + \mu_{\text{vap}} \, dn_{\text{vap}}$$

If the liquid and vapor are at equilibrium in a container of fixed volume, material balance requires $dV_{\text{liq}} = -dV_{\text{vap}}$, $dn_{\text{liq}} = -dn_{\text{vap}}$, and the condition for equilibrium is $d(A_{\text{liq}} + A_{\text{vap}}) = 0$; this then requires:

$$d(A_{\text{liq}} + A_{\text{vap}}) = (\mu_{\text{liq}} - \mu_{\text{vap}}) \, dn_{\text{liq}} + \frac{2\gamma}{r} \, dV_{\text{liq}} = 0$$

The change in volume and moles of the liquid phase are related via the density:

$$dn = \frac{\rho}{M} \, dV$$

Then the equilibrium condition [Eq. (4.6)] becomes:

$$\mu_{\text{vap}} = \mu_{\text{liq}} + \frac{2\gamma M}{\rho r} \tag{4.35}$$

As before, we assume $\mu = \mu^{\theta}$ for the liquid and $\mu = \mu^{\theta} + RT \ln (P/P^{\theta})$ for the vapor, giving:

$$\mu_{\text{vap}}^{\theta} + RT \ln P = \mu_{\text{liq}}^{\theta} + \frac{2\gamma M}{\rho r}$$

The standard free-energy difference is related to the "normal" vapor pressure P° by Eq. (4.20); this is, in effect, the vapor pressure when the radius of curvature is infinite (a plane surface). Using Eq. (4.20) in the equation above gives:

$$\ln \frac{P}{P^{\circ}} = \frac{2\gamma M}{\rho r RT} \tag{4.36}$$

Example: At 25°C, water has $P^{\circ} = 23.76$ torr. Calculate the vapor pressure of water if it is dispersed into droplets with $r = 100$ μm, 1 μm, or 1 nm. With $M = 0.01801$ kg, $r = 1 \times 10^{-4}$ m, surface tension (Table 4.3) $\gamma = 72.0$ dyne/cm $= 72.0 \times 10^{-3}$ N/m, $\rho = 1$ g/cm^3 $= 1000$ kg/m^3:

$$\ln \frac{P}{23.76} = \frac{2(72.0 \times 10^{-3} \text{ N/m})(0.01801 \text{ kg})}{(1000 \text{ kg/m}^3)(10^{-4} \text{ m})(8.3143 \text{ J/K})(298.15 \text{ K})}$$

$$= 1.0462 \times 10^{-5}$$

$$P = 23.76 \text{ torr}$$

Obviously the effect is negligible for $r = 100$ μm. For $r = 1 \times 10^{-6}$ m, the reader should show that $P = 23.78$ torr, still a pretty small change. However when $r = 1 \times 10^{-9}$ m, $P = 67.64$ torr. The last case is somewhat questionable, since the laws of thermodynamics apply only to matter on the macroscopic scale, and 10^{-9}-m is close to the size of a molecule. As an exercise, calculate the number of water molecules in a 10^{-9}-m droplet, assuming it has the normal liquid density. [*Answer:* 140.]   ∎

Since a condensing liquid must begin with very small droplets, the phenomenon of elevated droplet vapor pressure makes it questionable whether a vapor, with no dust particles around for nucleation sites, would ever condense. In fact, dust-free vapors do become readily supersaturated. In the Wilson cloud chamber a supersaturated vapor is used to detect charged particles; the passing of a charged particle provides a "trigger" for the condensation of the vapor along its path and makes that path visible to the eye. Provision of nuclei to condense vapors is the principle of cloud seeding to "make" rain. Dust due to atmospheric pollution has been blamed for excessive rainfall over urban centers.

4.7 Equilibria of Condensed Phases

The most obvious example of an equilibrium between condensed phases is that between a solid and liquid at the melting point (fusion); but examples of solid/solid phase transitions abound. Many solids exist in more than one crystalline form (allotropes) — for example: sulfur (rhombic and monoclinic), $CaCO_3$ (calcite and aragonite), tin (grey and white), and carbon (graphite and diamond). Phase changes in solids often occur very slowly, especially at low temperatures, so it is not uncommon to find metastable phases. Diamond at 1 atm and 298 K is metastable with respect to graphite; if this "spontaneous" change occurred readily, diamonds would have little value. On the other hand, the Wentorf process makes diamonds from graphite by applying high pressures (high temperatures and a catalyst are used to accelerate the rate of the conversion); the diamonds so made are not of gem quality but are useful for industrial applications. The high-pressure research of P. W. Bridgman has shown that ice has numerous allotropic forms at very high pressures; one of these (ice VII) melts at 100°C!

If we assume that two phases (α and β) are incompressible, Eqs. (4.6) and (4.14) give:

$$\mu_\alpha^\theta + V_{m\alpha}(P - P^\theta) = \mu_\beta^\theta + V_{m\beta}(P - P^\theta) \tag{4.37}$$

Since the standard chemical potential depends only on temperature, this relation demonstrates that, for a given temperature, there will be only one pressure at which both phases can coexist in equilibrium. At any other pressure, one or the other of the phases is metastable or nonexistent.

As with vapor equilibrium, a number of (T, P) pairs satisfy the equilibrium condition. The relationship of these pairs is given by [Eq. (4.11)]:

$$\frac{dP}{dT} = \frac{H_{m\beta} - H_{m\alpha}}{T(V_{m\beta} - V_{m\alpha})} \tag{4.38}$$

A common application of Eq. (4.38) is to find the dependence of the melting temperature (T_f) on pressure. If we assume that the heat of fusion is independent of temperature, Eq. (4.38) integrates readily to:

$$\ln \frac{T_f}{T_f^\theta} = \frac{\Delta V_f(P - P^\theta)}{\Delta H_f} \tag{4.39}$$

where

$$\Delta V_f = V_{m,\text{liq}} - V_{m,\text{solid}}$$

(Since the melting point does not change much with pressure, the variation of the heat of fusion with temperature is usually not a significant concern.)

The direction in which T_f changes with P depends on the relative densities of the phases. Water is rather unusual, in that the density of the solid is less than that of the liquid at the melting point (demonstrated by the fact that ice floats on water); this means that ΔV_f is negative and T_f decreases with increasing pressure. The melting of ice under pressure at temperatures below $0°C$ has been alleged to be responsible for making ice skating easier. Apparently, no one has tried to skate on a substance with a positive ΔV_f, but anyone who has watched small children (who create negligible pressure on the ice) playing hockey outdoors at $-20°C$ would be convinced that this effect, if not negligible, is certainly unnecessary below the age of 15. (There is an undeniable effect of temperature on the coefficient of friction, but this is probably due to a thin layer of water on the surface and has nothing to do with the applied pressure.)

Example: Calculate the fp lowering of ice due to a 315-lb person on a $\frac{1}{8}''$ by $8''$ ice skate. From Table 4.2, $\Delta H_f = 6010$ J/mol.

$$P - P^\theta = 300 \text{ psi} = 2.07 \times 10^6 \text{ Pa}$$

$$\rho(\text{ice}) = 0.915 \text{ g cm}^{-3}$$

$$\rho(\text{water, } 0°) = 0.999841 \text{ g cm}^{-3}$$

$$\Delta V_f = 18.01 \left(\frac{1}{0.999841} - \frac{1}{0.915} \right) = -1.67 \text{ cm}^3 \text{ mol}^{-1}$$

$$\ln \frac{T_f}{273.15} = \frac{(-1.67 \times 10^{-6} \text{ m}^3)(2.07 \times 10^6 \text{ Pa})}{6010 \text{ J}} = -5.75 \times 10^{-4}$$

$$T_f = 272.99$$

$$\Delta T = -0.16°C = -0.3°F \qquad \blacksquare$$

Exercise: For small ΔT, use the expansion $\ln(1 + x) = x$ with $x = \Delta T / T_f^\theta$ to rewrite Eq. (4.39) as:

$$\Delta T = \frac{T_f^\theta \Delta V_f (P - P^\theta)}{\Delta H_f} \qquad (4.40)$$

Use this equation to calculate the result of the preceding example; you should get exactly the same answer. $\qquad \blacksquare$

4.8 Triple Points and Phase Diagrams

Consider the following experiment. A material is frozen in a container and the vapor space is evacuated and the container sealed. The material is then allowed to warm up and melt under its own vapor pressure. At this melting point there are three phases, solid, liquid, and vapor, all in equilibrium; this is called a triple point. At this point, the chemical potentials in all three phases are equal:

$$\mu_{\text{solid}} = \mu_{\text{liq}} = \mu_{\text{vap}} \qquad (4.41)$$

The requirement of two-phase equilibrium, as shown in previous section, gives rise to a functional relationship between the T and P at which these may be in equilibrium — for example, Eq. (4.21) for liquid/vapor or solid/vapor equilibrium or Eq. (4.39) for solid/liquid or solid/solid equilibrium. A function of two variables, such as the chemical potential, is a surface in (μ, T, P) space; an equality for two such functions, such as $\mu_\alpha(T, P) = \mu_\beta(T, P)$, is the intersection of the two surfaces (which is a line). For a simple solid/liquid/vapor system, there are three such lines: sublimation pressure vs. temperature for solid/vapor, vapor pressure vs. temperature for liquid/vapor, and melting point vs. pressure for solid/liquid. These lines must intersect at a point for which Eq. 4.41 will apply — the triple point.

When these (P, T) relationships are plotted — as, for example, in Figure 4.6 — the result is called a *phase diagram*. The lines of such a diagram represent the (P, T) pairs at which equilibrium among the phases is possible.

Phase diagrams are particularly useful when other phases are possible. Figure 4.7, for example, shows the equilibria of sulfur, which has two solid phases — monoclinic and rhombic; note that there are three triple points. [The dashed lines show the paths of metastable equilibrium; the slowness of solid-solid phase transitions and the reluctance of vapors to condense (mentioned in Section 4.6) are the causes of these metastable situations.]

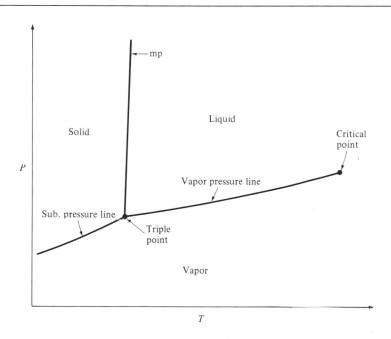

Figure 4.6 Phase diagram for a pure material. The three equilibrium lines, melting point vs. pressure, sublimation pressure vs. temperature, and vapor pressure vs. temperature, intersect at the triple point. Note that the vapor-pressure line does not extend beyond the critical point, since at that point there is no discontinuous phase change in a fluid. (Recall Figures 4.3 and 4.4.)

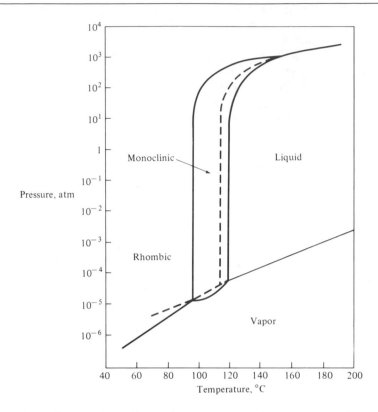

Figure 4.7 Phase diagram for sulfur. Sulfur has two solid forms, monoclinic and rhombic, and therefore three triple points. Dashed lines show metastable states. (From Guggenheim, *Thermodynamics,* 1967: North-Holland Publishing Co., Amsterdam, p. 126.)

Figure 4.8 shows the phase diagram of carbon. As mentioned earlier, graphite can be converted to diamond at high pressures (the Wentorf process). It also should be evident that, at ordinary temperatures and pressures, diamond is unstable with respect to graphite. This fact will have very little effect on the diamond market and illustrates a major limitation to phase diagrams and, in effect, to all thermodynamics; thermodynamics tells you only what would happen *if* anything happens, not that something must occur in a finite time. In the long run diamonds will turn to graphite, mountains will wash into the sea, and the sun will burn out; but the run is very long and, in the interim, there are more practical matters to which thermodynamics can be applied.

4.9   Non-First-Order Transitions

All the phase transitions discussed or mentioned to this point are of a type called first-order. There are transitions (most commonly solid/solid transitions) that differ from these in several significant respects. To see what these differences are, we must first review the distinctive features of first-order transitions.

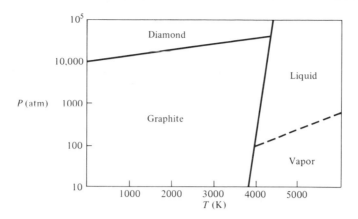

Figure 4.8 Phase diagram for carbon. Graphite is the stable form of carbon under ordinary conditions, with diamond being formed only at very high pressures.

First-order transitions are characterized by a latent heat. This means that, at the transition temperature, the enthalpy changes discontinuously (Figs. 4.9a and 4.2); the entropy and volume behave in a similar manner. The free energy (or chemical potential), on the other hand, is continuous, but its slope $[-S = (\partial G/\partial T)_P]$ is not. Figure 4.9(a) shows the free-energy lines for two phases α and β; at the point of their intersection, $\mu_\alpha = \mu_\beta$, the system switches phases to keep its free energy at the required minimum. (The dashed lines represent metastable states.) [Also recall Figure 4.2 for the vapor pressure of CO_2; the vapor-pressure curves ($\ln P$ vs. $1/T$) are continuous and meet at the triple point; this means [compare Eq. (4.13)] that the free energy is continuous. However, there is a discontinuity in the slope of the curves, meaning that the enthalpy change for vaporization changes discontinuously when solid CO_2 melts.]

In one sense, the heat capacity also takes a finite jump; for example, at 0°C water (liquid) has $C_{pm} = 75.99$ J/K while ice has $C_{pm} = 37.08$ J/K. Looked at from another point of view, the heat capacity of water at 0°C is *infinite;* that requires some explanation. Suppose that we place an ice cube into a "black box," together with an electrical heater and a thermocouple for measuring temperature. The box is closed, with the leads to the heater and thermocouple available, and is given to an unknowing experimenter. This person then supplies electrical heat to the box, measures the temperature rise, and calculates $C_p = q/\Delta T$. Everything is fine until the temperature reaches 0°C; then, when a finite quantity of heat is supplied, ΔT is zero and the experimenter dutifully calculates "$C_p = q/0 = \infty$." This is, admittedly, something of a circumlocution; in effect the existence of a latent heat for the transition is being attributed to a point infinity in the heat capacity.

By way of contrast, a second-order transition [Fig. 4-9(b)] has no latent heat and has a finite discontinuity in the heat capacity. The entropy and enthalpy are continuous functions, but their first derivatives $[C_p = (\partial H/\partial T)_P, C_p = T(\partial S/\partial T)_P]$ are not. The second derivative of G is, accordingly, discontinuous, but the first derivative is continuous. Examples of second-order transitions include:

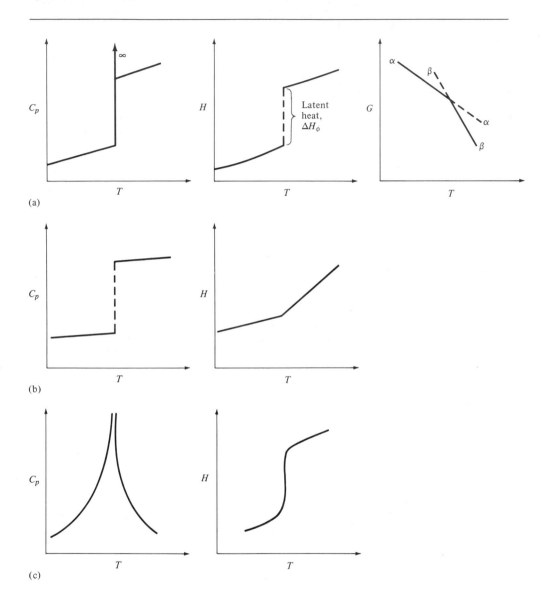

Figure 4.9   Thermodynamics of phase transitions. Changes of the thermodynamic functions at various types of phase transitions: (a) first-order, (b) second-order, and (c) lambda transitions.

1. NH_4Cl, $\sim -30°C$, the onset of free rotation of the NH_4^+ group.
2. NaCN, $\sim 150°C$, the onset of CN^- rotation about its lattice position.
3. MnO, alignment of spins of unpaired electrons (Curie temperature).
4. Ti_2O_4, semiconductor-to-metal transition.
5. Cu_3Au, an order-disorder transition in which the Cu and Au atoms are ordered at lower temperature but mixed randomly higher temperatures.

Ehrenfest suggested a convenient classification scheme for phase transitions in terms of the lowest-order derivative of G that is discontinuous: first-order, $(\partial G/\partial T)_P$; second-order, $(\partial^2 G/\partial T^2)_P$; third-order, $(\partial^3 G/\partial T^3)_P$; and so on. However, not all higher-order transitions fit into this scheme in an unambiguous fashion.

A common type of higher-order transition, illustrated by Figure 9(c), is called a lambda transition; the heat capacity becomes infinite at the lambda point in a manner that resembles the Greek letter lambda (λ), hence the name. The area under this infinity in C_p (the enthalpy) is finite; the enthalpy is continuous but has a vertical inflection point. As with all non-first-order transitions, there is no true latent heat; however, the "jump" in the enthalpy at the transition temperature [Figure 9(c)] is sometimes called the ΔH of the transition.

A fascinating example of a lambda transition occurs in the behavior of ^4He at low temperatures (Figure 4.10). (^4He is the common isotope; the other isotope, ^3He, which is immiscible with ^4He at low temperatures, behaves quite differently; cf. ref. 1, p. 125.) Uniquely, helium remains a liquid down to the lowest temperature achieved and presumably would not freeze even at absolute zero temperature if $P < 25$ atm. It is also unusual in that there are two liquid phases, He I and He II. The low-temperature form, He II, is a very unusual material, which has been characterized as a *superfluid;* it has an anomalously large heat conductivity and a viscosity that is virtually zero. The transition between He I and He II is a lambda transition. It has no latent heat, but the heat capacity increases apparently without limit as $T \to T_\lambda$; the heat capacity has been measured as close as 10^{-6} degrees to T_λ and still found to be increasing. Note [Figure 4.10(c)] that the volume of helium is also continuous at T_λ, but its slope $(\partial V/\partial T)_P$ [the coefficient of thermal expansion (α)] is apparently infinite at T_λ, just like C_p.

Pippard (ref. 1) has an extended discussion of higher-order phase transitions.

Postscript

This chapter has discussed equilibrium in pure materials. The important new concepts of activity and chemical potential have been introduced, and the manner by which the free energy functions are used in the discussion of equilibrium has been illustrated. This topic will be continued in Chapter 6 where chemical equilibrium is discussed, and in Chapters 7 and 8 where solution equilibria are covered. Aside from the principles involved, an understanding of the phase equilibria of pure materials is important for the understanding it gives us of how materials behave, and how their properties change with temperature, pressure, and other variables such as surface area.

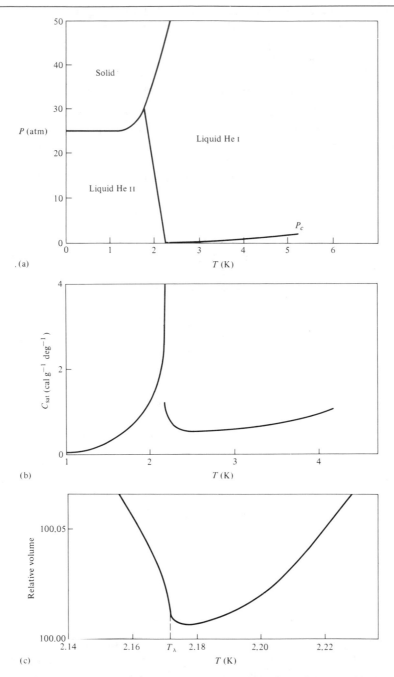

Figure 4.10 Phase diagram for ^{4}He. The transition from liquid He I to He II, panel (a), is a lambda transition, as shown by the heat-capacity curve, panel (b). Panel (c) shows the volume change at this transition, which has an apparent vertical inflection point, indicating that the coefficient of thermal expansion (α) is infinite at that point. (From A. B. Pippard, *Elements of Classical Thermodynamics*, 1966: Cambridge University Press, London.)

Before continuing our discussion of equilibrium, we shall, in the next chapter, take a look at how thermodynamic functions may be calculated theoretically.

References

1. A. B. Pippard, *Elements of Classical Thermodynamics*, 1966: Cambridge, U.K., at the University Press.
2. A. W. Adamson, *The Physical Chemistry of Surfaces*, 1967: New York, Wiley (Interscience).

Problems

4.1 Calculate the pressure of air above the earth at 10 km, assuming the atmosphere to be isothermal. The mean temperature at this altitude is 230.8 K compared to 288.0 K at sea level; use the average of these two values.

4.2 For some purposes, barometric readings are corrected to sea level. If a barometer at 20°C in Denver (altitude 1638 m, $g = 979.594$ cm/s^2) reads 620 torr, what would be the sea-level reading under the same conditions?

4.3 From data given in an example in Section 3.3, the ΔG of the process:

$$\text{water}(-10°C) \longrightarrow \text{ice}(-10°C)$$

can be calculated as:

$$\Delta G = \Delta H - T\,\Delta S = -5610 - 263.15(-20.5) = -215 \text{ J/mole}$$

At $-10°C$, ice has a vapor pressure of 1.950 torr, while super-cooled water has a vapor pressure of 2.149 torr. Use these data to calculate ΔG for freezing water at $-10°C$.

4.4 Use data in Table 4.2 to estimate the vapor pressure of CCl_2F_2 at 200 K.

4.5 Methyl mercaptan (CH_3SH) has vapor pressures $P° = 100$ torr at $-34.8°C$, $P° = 400$ torr at $-7.9°C$.
(a) Calculate the heat of vaporization.
(b) Estimate the normal boiling point.

4.6 The vapor pressure of water is 149.38 torr at 60°C and 233.7 torr at 70°C. Calculate the heat of vaporization.

4.7 Use the vapor pressures of ClF_3 given below to calculate the heat of vaporization. Use a graphical method or linear regression.

t (°C)	$P°$ (torr)	t (°C)	$P°$ (torr)
-46.97	29.06	-33.14	74.31
-41.51	42.81	-30.75	86.43
-35.59	63.59	-27.17	107.66

4.8 Use the vapor-pressure data below to calculate the heat of vaporization of 1-butene.

T (K)	$P°$ (atm)
273.15	1.268
277.60	1.490
283.15	1.810

4.9   Vapor pressures of Cl_2 are given below. Make a graph of $\ln P$ vs. $1/T$ and determine the heat of vaporization by either a graphical or a least-squares method.

T (K)	P° (atm)	T (K)	P° (atm)
227.6	0.585	283.15	4.934
238.7	0.982	294.3	6.807
249.8	1.566	305.4	9.173
260.9	2.388	316.5	12.105
272.0	3.483	327.6	15.676

4.10   Use the vapor pressures of ice given below to calculate the enthalpy of sublimation at $-30°C$.

t (°C)	P° (torr)
-28	0.351
-30	0.2859
-32	0.2318

4.11   The vapor pressure of carbonyl sulfide (OCS) is given by the following empirical formula (for 162–224 K):

$$\log_{10} (P^\circ/\text{torr}) = \frac{-1318.260}{T} + 10.15309 - (1.4778 \times 10^{-2})T + (1.8838 \times 10^{-5})T^2$$

(a) Determine the vapor pressure and heat of vaporization of this substance at 200 K.
(b) Calculate the temperature for which $P^\circ = 50$ torr.

4.12   Liquid nitrogen, normal boiling point 77.33 K, is a convenient cryoscopic bath for use in the laboratory. If a lower temperature is required, it can be obtained by reducing the pressure over the boiling nitrogen. Use data from Table 4.2 to estimate the temperature of boiling N_2 when $P = 100$ torr.

4.13   The heat of vaporization of N_2 has been given as (units: J K^{-1} mol^{-1}):

$$\Delta H_v = 8070 - 32.07T$$

(a) Derive a formula for the vapor pressure of liquid nitrogen as a function of T; the normal bp is 77.33 K.
(b) Calculate the boiling temperature for $P = 100$ torr; contrast your answer to that from the previous problem, which was calculated by a less accurate method.

4.14   The following data apply to $CHClF_2$ (Freon-22) at 300 K. $\Delta H_v = 15.65$ kJ/mol, $P^\circ = 10.86$ atm, $C_{pm}(\text{liq}) = 121.2$ J/K, $C_{pm}(\text{vap}) = 55.0$ J/K.
(a) Derive the following formula for the heat of vaporization:

$$\Delta H_v = R(4270 - 7.96T)$$

(State any assumptions explicitly.)
(b) Use this result to derive the formula for the vapor pressure:

$$\ln P^\circ = 62.02 - 7.96 \ln T - \frac{4270}{T}$$

(c) Estimate the normal boiling temperature of this gas using the results of part (b).

4.15   The heat of vaporization of Zn(nbp 907°C) is given by (units: J):

$$\Delta H_v = 1.286 \times 10^5 - 8.87T - 2.41 \times 10^{-3} T^2$$

Zinc (gas) is monatomic with $C_{pm} = 2.5R$. Use these data to calculate the heat capacity of Zn (liq) at the normal bp.

4.16   Ammonia (NH_3) at 0°C has a vapor pressure of 4.2380 atm and a specific volume of 1.5660 cm^3/g. Calculate the vapor pressure of ammonia at this temperature under an excess pressure of 100 atm.

4.17   The normal vapor pressure of thallium at 1200 K is 8.26 torr; its density at that temperature is approximately 10.4 g/cm^3. Calculate its vapor pressure under 1000 atm of an inert gas.

4.18   Use data in Table 4.3 to calculate the surface free energy, surface energy, and surface entropy of ethanol at 20°C.

4.19   Use data in Table 4.3 to calculate the surface energy and entropy of water at 60°C. (Use the 5-point differentiation method of Appendix I.)

4.20   Calculate the surface area if 5 grams of water (25°C) is dispersed into droplets with a radius of 20 nm. Calculate the work required to create this dispersion.

4.21   (a) From the data below calculate the surface energy of acetone at 20°C:

t (°C)	γ (dynes cm^{-1})
0°	26.2
20°	23.7
40°	21.2

(b) Calculate the added energy when 1 cm^3 of acetone is dispersed as droplets of 1 micron (10^{-6} m) radius.

4.22   Methanol has vapor pressure of 1.000 atm at 64.7°C; the density of the liquid is 0.7510 g/cm^3. Calculate the vapor pressure of methanol if it is dispersed as 200-nm diameter droplets at this temperature.

4.23   The normal melting point of lead is 327.3°C, where its densities are ρ(liq) = 10.51 g/cm^3, ρ(solid) = 11.23 g/cm^3. Calculate its melting point at a pressure of 1000 atm. ($\Delta H_f = 5.10$ kJ/mol. Note that you must assume no phase change for the solid when the pressure is applied.)

4.24   Calculate the enthalpy change for heating 1 g of ice at -200°C to steam at 300°C. Use the heat capacities of Figure 2.3.

4.25   Carbon dioxide at its triple point ($T = 216.5$ K, $P = 5.11$ atm) has molar volumes 29.1 cm^3 (solid), 37.4 cm^3 (liquid). Estimate the melting point of CO_2 at 100 atm. (Use $\Delta H_f = 7950$ J/mol.)

4.26   The standard free energy per mole of diamond is greater than that of graphite by 2.87 kJ at 25°C. (How this could be determined will be discussed in Chapter 6.) The density of graphite is 2.260 g/cm^3 and that of diamond 3.513 g/cm^3. Estimate the pressure required to convert graphite to diamond at 25°C.

4.27   Use the fact that enthalpy is a state variable to derive a relationship among the enthalpies for vaporization, sublimation, and fusion at the triple point.

4.28   The heat of sublimation of metals is usually determined indirectly because the vapor pressures are so low. The vapor pressure of liquid Cd is:

$$\ln\left(\frac{P°}{torr}\right) = 28.292 - \frac{1.340 \times 10^4}{T} - 1.2572 \ln T$$

The heat of fusion at the triple point (593 K) is 6138 J/mol. Calculate the heats of vaporization and sublimation at 594 K.

4.29   The vapor pressure of ice is 1.950 torr at $-10°C$ and 4.579 torr at $0°C$. The vapor pressure of water is 9.209 torr at $10°C$, 4.579 torr at $0°C$. Calculate the enthalpies of vaporization, sublimation, and fusion for water at $0°C$. (The triple point is actually $0.01°C$.)

4.30   The *Handbook of Chemistry and Physics* (CRC Publishing Co.) gives the following empirical formulas for NO: sublimation pressure:

$$\ln P° = 23.136 - \frac{1975}{T}$$

and vapor pressure:

$$\ln P° = 19.434 - \frac{1568}{T}$$

(In both cases, $P°$ is in *torr*.) Calculate (a) $\Delta H(\text{vap})$, (b) $\Delta H(\text{sub})$, (c) $\Delta H(\text{fus})$, and (d) the temperature at the triple point for this substance.

4.31   Use the data in Figure 4.2 to determine the heats of vaporization and sublimation of carbon dioxide. From these values, calculate the heat of fusion.

4.32   The Clapyron equation:

$$\frac{dP}{dT} = \frac{\Delta H}{T \Delta V}$$

was derived in the text by using the continuity of the chemical potential at the phase transition point. It does not apply to a second-order transition, since, for such a case, $\Delta H = 0$ and $\Delta V = 0$. For a second-order transition, use the continuity of the entropy to derive the analogous equation:

$$\frac{dP}{dT} = \frac{\Delta C_p}{TV \Delta \alpha}$$

where α is the coefficient of thermal expansion.

4.33   Use the continuity of V at a second-order phase transition to derive the Ehrenfest equation:

$$\frac{dP}{dT} = \frac{\Delta \alpha}{\Delta \kappa_T}$$

(See the discussion in the previous problem.)

Has matter more than motion? Has it thought
Judgement and genius? is it deeply learn'd
In mathematics?...

     –Edward Young, Night Thoughts *(1745)*

I feel like a fugitive from th' law of averages.

     –Bill Mauldin, Up Front *(1944)*

5

Statistical Thermodynamics

Thermodynamics, as stated earlier, deals with macroscopic, measurable properties of matter and is valid independent of the existence of any substructure of invisible particles or unobserved events. Yet, there is clear evidence for a substructure of atoms and molecules, and the properties of these entities are readily measured. Is it possible to calculate macroscopic properties from the known properties of atoms and molecules?

The behavior of a set of mutually interacting bodies is calculable from the laws of mechanics if the forces are known, but the difficulty of the calculations increases rapidly with the number of significant interactions involved. The orbits of the planets can be precisely calculated from the laws of Newton and Kepler, but this calculation involves, perhaps, a few dozen major interactions. The tiniest, most insignificant piece of matter contains such a huge number of atomic-level particles that such a direct calculation would be totally impractical. This huge number, however, makes the use of statistical methods very attractive. In a collection of 6×10^{14} particles (a nanomole) it is not possible to calculate the behavior of each particle individually, but the *average* behavior of the group can be predicted with great certainty once the mechanical laws and forces are specified with sufficient accuracy.

The procedure by which these calculations are done is called *statistical mechanics*. As it turns out, in many cases thermodynamic properties can be calculated with about the same accuracy with which they can be measured; this often makes computation an attractive alternative to direct measurement. It will be admitted in advance that the statistical method is practical only for relatively simple cases; the only applications that we shall discuss in this chapter are ideal gases. If you needed to know the heat capacity of octane at 25°C, you would probably do the thermal measurement; the calculation would be relatively difficult. On the other hand, if you needed to know the entropy of methyl radicals ($CH_3\cdot$) at 4000 K, the statistical calculation would be the only practical method. The value of the statistical calculations of thermodynamic properties is well illustrated by the following quotation by G. V. Mock (1964) from the preface to the first edition of the JANAF Tables (ref. 5):

> Beginning in the mid-1950's, when elements other than the conventional carbon, hydrogen, oxygen, nitrogen, chlorine, and fluorine came into consideration as rocket propellant ingredients, formidable difficulties were encountered in conducting rigorous theoretical performance calculations for these new propellants. The first major problem was calculational techniques. The second was the lack of accurate thermodynamic data.
>
> By the end of 1959, the calculational technique problem had been substantially resolved by applying the method of minimization of free energy to large, high-speed digital computers. At this point the calculations became as accurate as the thermodynamic data upon which they were based. However, serious gaps were present in the available data. For propellant ingredients, only the standard heat of formation is required to conduct a performance calculation. However, this must be known to a high degree of accuracy. For combustion products, the enthalpy and entropy must be known, as a function of temperature, in addition to the standard heat of formation.

Statistical calculations using spectroscopic data played a major role in the compilation of the needed data. In this connection it should be noted that, in this modern age, microcomputers nearly as powerful as the computers of the 1950s are available for scarcely more than what a good mechanical calculator cost at that time.

The laws that govern the behavior of atoms and molecules are those of quantum mechanics; this topic will not be covered until Chapters 11–14. This chapter presumes no background in quantum mechanics and, as a result, will have to use the results of this theory without proof. These results will be fully justified later. [Those who do not prefer this method can, with small loss, proceed to Chapter 6 and cover this chapter later, together with Chapter 15. Readers who already have sufficient background in quantum mechanics should cover the first four sections of this chapter, followed by Chapter 15, returning to this chapter for the practical applications (pp. 223 ff).]

5.1 Probability and Entropy

The basic formula for the statistical interpretation of entropy is that proposed by Ludwig Boltzmann:

$$S = k \ln W \tag{5.1}$$

In this formula, W is the statistical probability of the state whose entropy is S. This formula is the basis for the interpretation of thermodynamics in terms of the atomic/molecular hypothesis. At the time it was proposed (1896) there was virtually no direct evidence for the existence of atoms and molecules, so Boltzmann's views were the cause of great controversy. Possibly as a result of this controversy, Boltzmann committed suicide in 1906, just as his views were about to prevail. His famous formula is carved on his tomb in the Central Cemetery in Vienna.

Free Expansion of an Ideal Gas

As an example, we shall consider Joule's experiment, the free expansion of an ideal gas, whose thermodynamics was treated earlier. It is an important simplification that this process is both adiabatic and isothermal, for then $\Delta U = 0$ and only the entropy will change. The process (Figure 5.1) starts with N molecules ($n = N/L$ moles) in a volume V_1 that is connected to a vacuum with volume V_2. Experience tells us that the gas will expand spontaneously into the vacuum; the entropy change, calculated as in Chapter 3 [Eq. (3.14)], will be:

$$\Delta S = nR \ln \frac{V(\text{final})}{V(\text{initial})} = nR \ln\left(\frac{V_1 + V_2}{V_1}\right) \tag{5.2}$$

For simplicity we shall assume that the volumes V_1 and V_2 are equal, so the final volume of a gas is just twice the initial volume and the entropy change is:

$$\Delta S = nR \ln 2$$

Next we shall examine this same process from the statistical point of view, using Boltzmann's equation. To do this, we must determine the probabilities for having various numbers of molecules in each half of the container—that is, in V_1 or V_2.

If there were only one molecule, the probability of its being in one or the other of the containers is just $\frac{1}{2}$. If there were two molecules, the probability that both would be in V_1 is $(\frac{1}{2}) (\frac{1}{2}) = \frac{1}{4}$; for three molecules, this probability is $(\frac{1}{2}) (\frac{1}{2}) (\frac{1}{2}) = \frac{1}{8}$. For N

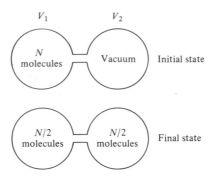

Figure 5.1 The Joule expansion of a gas. The spontaneous expansion of an ideal gas into a vacuum is entirely statistical in origin, reflecting the greater probability of states with equal (or nearly equal) numbers of molecules in each half of the container compared to the initial state where all molecules are in one half of the container.

molecules, the probability of the initial state, all molecules in V_1, is just:

$$W(\text{initial}) = (\tfrac{1}{2})^N \tag{5.3}$$

Of course, if the containers are connected, it is intuitively unlikely that that all would be in the same container. Suppose there are four molecules, which for convenience we shall label A, B, C, and D. The probability that all four are in V_1 is, as discussed above, $(\tfrac{1}{2})^4 = \tfrac{1}{16}$. However, if only three molecules are in V_1, there are four ways these three molecules could be chosen (ABC, ABD, ACD, BCD), so the probability of this state is greater. The probability of any particular result—for example A, B, and C in V_1, D in V_2—is still $\tfrac{1}{16}$ ($\tfrac{1}{2}$ for each event), but there are four results that put three molecules in V_1, so the total probability is $\tfrac{4}{16}$. The most probable result is, of course, two molecules in each container; there are six ways to choose which will be in V_1 (AB, AC, AD, BC, BD, CD) so the probability of this state is $\tfrac{6}{16}$.

The general result, provided by the theory of permutations and combinations, is as follows: given N trials with two equally probable outcomes, the probability that the distribution of results is N_1, N_2 (with $N = N_1 + N_2$) is:

$$W(N_1, N_2) = \frac{1}{2^N} \frac{N!}{N_1! N_2!} \tag{5.4}$$

This formula would apply to, for example, the results of flipping coins; N_1 could be the number of heads and N_2 the number of tails in N trials. For example, if you flip a coin 10 times, the probability of obtaining 3 heads and 7 tails (or vice versa) is:

$$W(3, 7) = W(7, 3) = \left(\frac{1}{2}\right)^{10} \frac{10!}{7! 3!} = 0.117188$$

The more likely result, 5 heads and 5 tails, has:

$$W(5, 5) = \left(\frac{1}{2}\right)^{10} \frac{10!}{5! 5!} = 0.246094$$

Exercise: Calculate the following probabilities for 10 coin flips:

$$W(0, 10) = W(10, 0) = 0.000977$$

$$W(1, 9) = W(9, 1) = 0.009766$$

$$W(2, 8) = W(8, 2) = 0.043945$$

$$W(4, 6) = W(6, 4) = 0.205078$$

Show that the sum for all possible outcomes is 1. (Because of round-off, the sum may be slightly different than 1.)   ■

Equation (5.4) also applies to the case of N molecules distributed between two equal volumes. The most probable result is that half the molecules will be in each container, so $N_1 = N_2 = N/2$, and the probability of the final state for the free expansion of an ideal gas is:

$$W(\text{final}) = \left(\frac{1}{2}\right)^N \frac{N!}{\left(\frac{N}{2}\right)!\left(\frac{N}{2}\right)!} \tag{5.5}$$

This result—equal numbers in each container—is probably intuitively obvious; however, it is derived rigorously in Section 9.2 under the theory of random walks. This, of course, is only the most probable result, and there will be a range of results that have a finite probability. It can be shown (Section 9.2) that the standard deviation of the distribution of Eq. (5.5) is equal to $\sqrt{N}$. Suppose $N = 2 \times 10^{16}$; the most probable number of molecules in either container is 1×10^{16}, but this number may fluctuate by an amount of the order of $\sqrt{N} \sim 10^8$. This fluctuation is quite insignificant for such a large number of molecules; the number of molecules in either container may range from

$$1.00000001 \times 10^{16} \quad \text{to} \quad 0.99999999 \times 10^{16}$$

In other words, the most probable result is the only result we are likely to observe for the numbers of particles generally encountered with macroscopic amounts of matter. (Note that if the container volume were 1 cm³ and the temperature were 0°C, the pressure for this number of particles would be 0.284 torr.)

The entropy change for the free expansion of an ideal gas can now be calculated with Eq. (5.1) using the results of Eqs. (5.3) and (5.5) [note that the former equation also follows from Eq. (5.4) with $N_1 = N$ and $N_2 = 0$—recall that $0! = 1$]:

$$\Delta S = k \ln \frac{W(\text{final})}{W(\text{initial})} = k \ln \frac{N!}{(\frac{1}{2}N)!(\frac{1}{2}N)!} \tag{5.6}$$

Stirling's Approximation

The evaluation of factorials of large numbers is commonly encountered in statistics and is usually done using Stirling's approximation. The logarithm of $N!$ is:

$$\ln N! = \ln 1 + \ln 2 + \ln 3 + \cdots + \ln N = \sum_{x=1}^{N} \ln x$$

Table 5.1 Test of Stirling's approximation

N	N!	ln N! Exact	ln N! Eq. (5.8)	ln N! $N \ln N - N$
10	3,628,800	15.104 413	15.104 415	13.025 851
100	9.3326×10^{157}	—	363.739 376	360.517 019
1000	4.0239×10^{2567}	—	5912.128 178	5907.755 279
1×10^{10}	$10^{10^{11}}$	—	$2.202\,585 \times 10^{11}$	$2.202\,585 \times 10^{11}$

$$L! \cong 10^{14\,000\,000\,000\,000\,000\,000\,000\,000}$$

This sum can be approximated as an integral:

$$\ln N! \cong \int_1^N \ln x \, dx = (x \ln x - x)_1^N = N \ln N - N + 1$$

Neglecting the 1, we get:

$$\ln N! \cong N \ln N - N \tag{5.7}$$

A more accurate formula is sometimes needed; it is

$$\ln N! = \frac{1}{2} \ln 2\pi + \left(N + \frac{1}{2}\right) \ln N - N + \frac{1}{12N} - \mathbb{O}\left(\frac{1}{N^3}\right) \tag{5.8}$$

Table 5.1 compares these formulas; for the large values of N in a macroscopic system, Eq. (5.7) is sufficiently accurate.

Using the simple form of Stirling's approximation, Eq. (5.6) becomes:

$$\Delta S = k\left(\ln N! - 2 \ln \frac{N}{2}!\right) = k\left(N \ln N - N - N \ln \frac{N}{2} + N\right)$$

$$\Delta S = Nk \ln 2 \tag{5.9}$$

This will be exactly the thermodynamic result derived earlier if Boltzmann's constant is identified as:

$$k = \frac{R}{L} = 1.38062 \times 10^{-23} \text{ J K}^{-1} \tag{5.10}$$

Fluctuations from Equilibrium

Boltzmann's formula can also be used to calculate the probability (p) of finding some state other than the most probable (equilibrium) state. If W is the probability of some state, and W_{eq} is the probability for the equilibrium state, then:

$$p = \frac{W}{W_{eq}} = e^{\Delta S/k} \tag{5.11}$$

This formula can be used with the thermodynamic results for ΔS (Chapter 3) to calculate the probability of various fluctuations.

Example: Two samples of water, each 10 g with $C_p = 42$ J K^{-1}, are in thermal contact at 300 K. What is the chance that, when measured, one sample will have $T = 301$ K while the other has $T = 299$ K? The ΔS for this process is [by Eq. (3.12)]:

$$\Delta S = 42\left(\ln \frac{301}{300} + \ln \frac{299}{300}\right) = -4.667 \times 10^{-4} \text{ J K}^{-1}$$

$$\frac{\Delta S}{k} = \frac{-4.667 \times 10^{-4} \text{ J/K}}{1.38 \times 10^{-23} \text{ J/K}} = -3.38 \times 10^{19}$$

$$p = e^{-3.38 \times 10^{19}}$$

What if the samples were only 10 μg and the temperature fluctuation 0.001 K? (This is as small a change in T as can be readily measured.) Then the heat capacity of the sample is 42×10^{-6} J/K, and:

$$\Delta S = 42 \times 10^{-6}\left(\ln \frac{300.001}{300.000} + \ln \frac{299.999}{300.000}\right) = -1.73 \times 10^{-15} \text{ J/K}$$

$$\frac{\Delta S}{k} = \frac{-1.73 \times 10^{-15} \text{ J/K}}{1.38 \times 10^{-23} \text{ J/K}} = -1.25 \times 10^{8}$$

The probability is still very small:

$$p = \frac{1}{10^{54\,000\,000}}$$

For both of these calculations the probabilities were very small; this demonstrates that the chance of observing a random fluctuation from thermodynamic equilibrium in a macroscopic system, while finite, is very small. ■

5.2 Configurations and Gibbs' Paradox

The probability of a state is proportional to the number of distinct configurations that can lead to that state. If we remove a new deck of playing cards from its box, we will nearly always find the deck to be perfectly ordered. If we then throw the cards into a pile and pick them up randomly, what is the probability that we will get the same sequence? The first card can be chosen 52 ways, the second 51, the third 50, and so on. There will be:

$$52 \times 51 \times 50 \times \cdots = 52! = 8.0658 \times 10^{67}$$

possible sequences; only one of these sequences would be perfectly ordered ($4! = 24$ if we don't care about the order of the suits), so it is very unlikely that we will get an ordered sequence. Each possible sequence (configuration) is equally probable, but those which we would call disordered greatly outnumber those which we would call ordered and are correspondingly more probable. This is the basic postulate of statistics, whether we are concerned with cards or molecules: the probability of an event is proportional to the number of ways in which it can occur. The various ways to make up a state at random are its *configurations*.

A gas has an entropy due to the fact that its molecules could be, randomly, in any part of the container at any given time; this is called the configurational entropy. To count the number of configurations a gas may have, we suppose that the container (volume V) is divided into a large number of tiny cells (volume v); these cells can be very small and their number very large (but not infinite). The number of configurations possible for a single molecule is, then, simply the number of cells, V/v:

$$(\text{\# configurations for one molecule}) = \frac{V}{v}$$

For N particles, each with V/v configurations, the total number of configurations is:

$$(\text{\# } N\text{-particle configurations}) = \left(\frac{V}{v}\right)^N$$

[This is somewhat oversimplified, since it ignores the possibility that cells are multiply occupied. This possibility can be ignored without significant error because, for a molecular gas, the density of cells is much greater than the density of molecules, so that multiply occupied cells will be a small fraction of the whole. We shall examine this question more carefully in Chapter 15.]

If we assume that this number of N-particle configurations is the W of Eq. (5.1), the configurational entropy will be:

$$S^{\text{conf}} = k \ln\left(\frac{V}{v}\right)^N = Nk \ln\left(\frac{V}{v}\right)$$

This equation may appear to be reasonable — the logarithmic dependence of S on V is as expected from thermodynamics — *but it is wrong*.

To see that the equation above must be incorrect, let us suppose that the gas (N molecules in volume V) is arbitrarily divided into halves by an imaginary partition. Each half has $N/2$ molecules in a volume $V/2$ and a configurational entropy (according to the result above):

$$S(\text{half}) = \frac{N}{2} k \ln \frac{V}{2v}$$

Now, suppose the barrier is removed; the entropy change is:

$$S(\text{whole}) - 2 \times S(\text{half}) = Nk \ln \frac{V}{v} - Nk \ln \frac{V}{2v} = Nk \ln 2$$

But this is an entropy change for an *imaginary* process; this problem was first discussed by J. Willard Gibbs and is called Gibbs' paradox.

This paradox arises because we have overcounted the number of configurations available to N particles. The molecules of a gas are indistinguishable, so that when we count a configuration such as:

(molecule 1 in cell 13, molecule 2 in cell 71, molecule 3 in cell 100, ...)

this is not really distinguishable from the configuration:

(molecule 2 in cell 13, molecule 1 in cell 71, molecule 3 in cell 100, ...)

All that really matters is that there is *a molecule* in cells 13, 71, 100, ... For this case—three molecules in cells 13, 71, 100—there are 3! = 6 ways to permute them among their cells; for N molecules there are $N!$ ways.

Now, the number of configurations of N indistinguishable particles is:

$$W^{\text{conf}} = \left(\frac{V}{v}\right)^N \frac{1}{N!} \tag{5.12}$$

and

$$S^{\text{conf}} = k \ln \frac{1}{N!} \left(\frac{V}{v}\right)^N$$

With Stirling's approximation:

$$\ln N! = N \ln N - N$$

the configurational entropy is readily shown to be:

$$S^{\text{conf}} = Nk\left(\ln \frac{V}{N} - \ln v - 1\right) \tag{5.13}$$

Now it should be evident that the entropy of half the gas ($N/2$ molecules in volume $V/2$) is exactly half the entropy of Eq. (5.13), so Gibbs' paradox has disappeared.

Entropy and the Third Law of Thermodynamics

The third law of thermodynamics, as discussed in Section 3.6, states that the entropy of all perfect crystalline substances is the same at $T = 0$; this value (S_0) is chosen to be zero. This, we can now see, is in accord with Boltzmann's hypothesis; a perfectly ordered crystal (like an unshuffled deck of cards) has only one configuration at $T = 0$, so $W = 1$ and [from Eq. (5.1)] $S = 0$.

Some substances apparently violate the third law in that their entropy extrapolated to $T = 0$ is not zero. That this is so can be determined by comparing their third-law entropies (measured as described in Chapter 3, assuming $S_0 = 0$) and their statistical entropies (calculated as described later in this chapter). Many of these apparent exceptions occur because some sort of disorder has been frozen in and persists at the lowest temperature to which the heat capacity was measured.

Carbon monoxide is a good example. If a crystal of carbon monoxide were perfectly ordered—either:

$$...C\equiv O \ C\equiv O \ C\equiv O \ C\equiv O \ C\equiv O \ C\equiv O ...$$

or:

$$...C\equiv O \ O\equiv C \ C\equiv O \ O\equiv C \ C\equiv O \ O\equiv C ...$$

its entropy would be zero at $T = 0$. However, $C\equiv O..C\equiv O$ and $C\equiv O..O\equiv C$ have nearly the same energy, so when this substance freezes, a random mixture of the two possible $C\equiv O$ orientations results. With two orientations possible, the number of configurations possible for N molecules follows immediately from Eq. (5.4):

$$W(\text{random}) = \frac{1}{2^N} \frac{N!}{\left(\dfrac{N}{2}\right)!\left(\dfrac{N}{2}\right)!}$$

The same reasoning as used earlier will show that the residual entropy at $T = 0$ due to this disorder is (for one mole, $N = L$, $R = Lk$)

$$S_0 = R \ln 2$$

Other substances show such anomalies. For example, NNO has, like CO, a residual entropy of $\sim R \ln 2$. Phosgene:

$$O=C\underset{\diagdown\ Cl}{\overset{\diagup\ Cl}{}}$$

can crystallize in three nearly equivalent configurations and has a residual entropy $\sim R \ln 3$. Note that tabulated entropies (for example, Table 3.2) always are corrected for this effect and, thus, correspond to the statistical entropy rather than the third-law values, regardless of how they were actually measured.

Configurations of Polymer Chains

Contrary to what has been implied so far, counting configurations is not the principal method by which statistics is applied to thermodynamics. Before we go on to the practical aspects of the subject, however, it may be useful to look at another area for which the idea of configurations is fruitful.

In Chapter 3 it was pointed out that the elasticity of rubber is largely due to entropy. The polymer chain between the cross-links of a rubber network can be modeled as a number (N) of rigid segments (length λ) connected by universal joints. If the cross-links are separated by some distance (L), there will be a large but finite number of chain configurations that can reach between those two points; Figure 5.2(a) shows one of these. If the rubber is stretched so that the distance between the cross-links is large [as in Figure 5.2(b)], there will be fewer configurations that reach between the ends; in fact, if $L = N\lambda$, there is just one such configuration. Also, if the rubber is compressed, the cross-link points will be close together [Figure 5-2(c)] and the number of chain configurations will also be fewer than at the intermediate configuration. There will be some length at which the number of configurations and, hence, the entropy, will be a maximum. For an ideal rubber, this will be the equilibrium length—the length at which there is no restoring force. (The theory of random walks, Section 9.2, suggests that this point of maximum entropy will be when $L \sim \sqrt{N}\lambda$. Also, see Problem 5.3.) Thus the entropy change, which, as noted in Chapter 3, is largely the

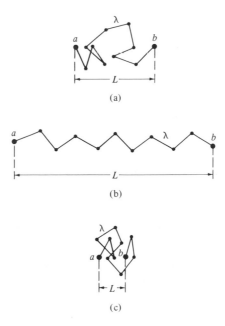

Figure 5.2 Configurations of a flexible chain. The number of configurations is the number of distinct paths that the links can take between the ends of the chain. There are fewer configurations for the stretched state [panel (b)] or the compressed state [panel (c)] than at some intermediate state [panel (a)]. The most probable distance between the ends (L) is approximately $\sqrt{N}$ times the length of a single link (λ).

cause of the restoring force of a rubber, can be viewed as a result of the number of possible chain configurations between the cross-links. There are, of course, a number of other effects that are operational in the phenomenon of elasticity, but this example will illustrate that the concept of configurations together with Boltzmann's formula can provide easily pictured mechanisms that can explain the macroscopic behavior of matter.

5.3 Boltzmann's Distribution Law

Thus far, we have considered only entropy changes that macroscopically are isothermal and microscopically involve no more than shuffling molecules among configurations having the same energy. We now wish to establish a connection between the temperature and the energies of the particles and, from that, information about the temperature dependence of the entropy and other thermodynamic quantities.

Consider a set of N identical molecules; each molecule has available to it a set of energy levels, $\{\varepsilon_i\}$. This set represents every possible energy that a molecule may have. These values can be calculated, given the mechanical laws that specify the motions of the molecules and the nuclei and electrons within the molecules. Even at thermal equilibrium, all molecules will not be in the same energy state; recall from Chapter 1 that the molecular velocities of a gas were distributed over a wide range. If, however, the gas is an isolated system (in the thermodynamic sense — that is, constant n, U, V), then the total energy (E) and the total number of particles (N) will be constant. If n_1 molecules have energy ε_1, n_2 have energy ε_2, and so on, the total energy of the set of N molecules is:

$$E = \sum_i n_i \varepsilon_i \tag{5.14}$$

The total number of particles is:

$$N = \sum_i n_i \tag{5.15}$$

The *state* of the system is specified by giving the set of *population numbers:*

$$\{n_i\} = (n_1, n_2, n_3, \ldots)$$

The number of configurations corresponding to a given population set, $\{n_i\}$, will be the number of ways of placing N objects into boxes so that there are n_1 in box 1, n_2 in box 2, n_3 in box 3, and so on. The theory of permutations and combinations gives this factor as:

$$W = \frac{N!}{n_1! n_2! n_3! \ldots} \tag{5.16}$$

(Remember that $0! = 1$ by definition, so that unoccupied boxes make no contribution to W.)

The maximum probability, hence the maximum entropy, will occur when each particle is in a different box; all $n_i = 1$ (or zero) and $W = N!$. However, this assumes that there is no restriction on the energy; minimization of E would argue for putting all N particles into the same box, that of lowest energy. What we want to find is the population set that gives a maximum value of W consistent with a given total energy.

Example: If we have three boxes labeled A, L, M containing three objects, there are $3! = 6$ ways to get a population $(1, 1, 1)$; these are just the permutations of the letters: ALM, AML, LAM, MAL, MLA, LMA. The population $(1, 2, 0)$ has $3!/1!2!0! = 3$ configurations — namely, LLA, LAL, and ALL.   ∎

Example: Let us examine the possible states of N particles if each particle has only three possible values for its energy. While not particularly realistic (molecules have large numbers of energy states available), it is convenient because of its simplicity. With the total number of particles:

$$N = n_1 + n_2 + n_3$$

and the total energy:

$$E = n_1 \varepsilon_1 + n_2 \varepsilon_2 + n_3 \varepsilon_3$$

constant, only one of the populations is an independent variable; this means that each state of the system is specified by giving the population of any one of the energy levels.

Ludwig Boltzmann (1844–1906)

As late as 1907, Wilhelm Ostwald, the most prominent physical chemist of the time, could write a treatise on chemistry without mentioning the atomic-molecular hypothesis, except to disparage it as "the so-called structural theory." Ostwald was also a violent opponent of the kinetic theory developed by Boltzmann and Maxwell, since he felt that atoms had no physical reality except, perhaps, as models. Boltzmann, who (along with Gibbs) also developed statistical mechanics, knew the truth and answered his critics effectively; however, direct physical evidence for the existence of atoms and molecules was scant at that time. (The evidence that was available, such as atomic spectra, had not yet been interpreted.)

Boltzmann's suicide has often been ascribed to his despair over the prospects of the kinetic-molecular theory, but solid evidence for this is lacking. He certainly despaired of success, writing, "It would be a great tragedy for science if the theory of gases were temporarily thrown into oblivion," and "I am conscious of being only an individual struggling weakly against the stream of time." But if he was opposed, he was never alone, for many scientists favored his views. Also, he had been subject to fits of depression and had attempted suicide before. In any case, it is ironic that he died just as his views, supported by new experimental evidence, were about to prevail with finality.

It should be noted that organic chemists had accepted the atomic theory 40 years earlier and had made immense strides in developing a modern and sophisticated science; they did not wait for the physical chemists to vote. There must be a moral in that somewhere.

Suppose, for example, we choose $\varepsilon_1 = 0$, $\varepsilon_2 = 1$, and $\varepsilon_3 = 3$ (arbitrary units), $N = 20,000$, $E = 10,000$. Then if n_3 is used as the independent variable, the other two populations are fixed by the conditions:

$$n_2 = 10,000 - 3n_3$$

$$n_1 = 20,000 - n_2 - n_3$$

Then the probability W (actually $\ln W$) can be calculated using Eq. (5.16) (using Stirling's approximation for factorials too large to be calculated directly — for example, $>69!$ on a calculator, roughly $>30!$ on a microcomputer.)

How many distinguishable states are there for this system? A population cannot be negative, so n_3 can range upward from zero; however, n_3 cannot be greater than 3333, for otherwise n_2 must (according to the total-energy restriction above) become negative. There are therefore 3333 possible states for 20,000 particles, which have a total energy of 10,000 (and the three energy levels specified); each state is specified by the population number for level 3: $0 < n_3 < 3333$.

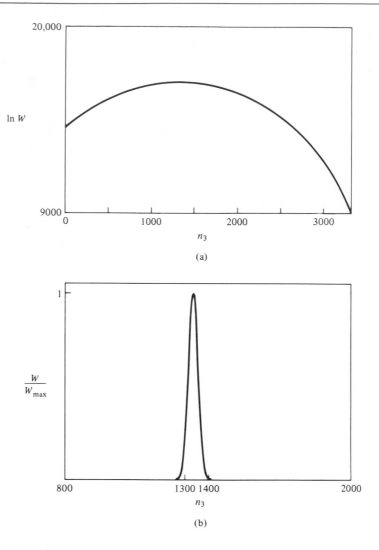

Figure 5.3 Distributions of particles among energy levels. The distributions of 20,000 particles among three energy levels (with energies of 0, 1, and 3 units, respectively), that have a total energy $E = 10,000$ units, can be characterized by giving the population of any one level (here n_3). The probabilities of the various distributions (W) are shown here for all possibilities. The most probable population for level 3 (n_3) is that for which W is a maximum. The logarithmic plot of panel (a) is somewhat deceptive because of the "leveling effect" of logarithms. Panel (b) shows that W is quite sharply peaked about the most probable value.

The results of this calculation are displayed in Figure 5.3. The graph of ln W vs. n_3 shows clearly that this quantity has a maximum in the vicinity of $n_3 \cong 1300$ (the actual maximum is for $n_3 = 1336$). Recall that, according to Eq. (5.1), this function is proportional to the entropy of the system; the tendency of this isolated system to be in the state of maximum probability, therefore entropy, is just another statement of the second law of thermodynamics.

The logarithmic plot of Figure 5.3 shows that many states other than the most probable are possible, but it is deceptive because of the leveling effect of the logarithmic scale. Figure 5.3(b) shows the relative probabilities, $W/W_{\max}$; it can be seen that the distribution of probable states is rather narrow, about 0.3% of the whole. For macroscopic numbers of particles (say 10^{15} or more) the range of permissible states is relatively narrower (that is, a smaller fraction of the whole), so that it is reasonable to discuss the state of the system in terms of a single, most probable distribution for the populations, $\{n_i\}$. For such large numbers, direct calculation is obviously impractical, so it will be useful to find a more general method for determining the most probable population distribution. ∎

Lagrange Method of Undetermined Multipliers

We distinguish between the absolute or global maximum of a function and a constrained maximum — the largest value the function can have subject to the constraint of being on some path. The distinction is that between the highest point in Colorado (Mt. Elbert, 14,423 ft) and the highest point in Colorado, which is on interstate highway 70 (Loveland Pass, 11,992 ft) (Figure 5.4).

For a function of two variables, $f(x, y)$:

$$df = \left(\frac{\partial f}{\partial x}\right)_y dx + \left(\frac{\partial f}{\partial y}\right)_x dy \qquad (5.17)$$

A maximum in f requires $df = 0$, which will be true if both slopes, $(\partial f/\partial x)$ and $(\partial f/\partial y)$, are zero; this is the global maximum.

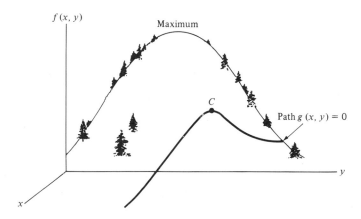

Figure 5.4 A constrained maximum. Point C is the highest point on a particular path that is the intersection of the function $g(x, y)$ with the function $f(x, y)$.

We seek a maximum in the function $f(x, y)$ subject to the constraint of being on some path. This path will be defined by another functional relationship:

$$g(x, y) = 0$$

Since $g(x, y)$ is constant on the path, at all points on the path $dg = 0$ and:

$$dg = \frac{\partial g}{\partial x} dx + \frac{\partial g}{\partial y} dy = 0$$

Therefore the partial derivatives of $g(x, y)$ provide a relationship between the coordinate changes dx and dy that will not leave the path, namely:

$$dx = \frac{(-\partial g/\partial y)}{(\partial g/\partial x)} dy \quad \text{or} \quad dy = \frac{(-\partial g/\partial x)}{(\partial g/\partial y)} dx \qquad \textbf{(5.18)}$$

Now a maximum in $f(x, y)$ requires $df = 0$ and:

$$df = \left(\frac{\partial f}{\partial x}\right) dx + \left(\frac{\partial f}{\partial y}\right) dy = 0$$

The requirement that the maximum be on the path ($g = 0$) can be enforced by using Eq. (5.18) to eliminate either dx or dy from the above equation. If dx is eliminated:

$$df = \left[\frac{\partial f}{\partial y} - \frac{\partial f}{\partial x} \frac{(\partial g/\partial y)}{(\partial g/\partial x)} \right] dy = 0 \qquad \textbf{(5.19a)}$$

If dy is eliminated:

$$df = \left[\frac{\partial f}{\partial x} - \frac{\partial f}{\partial y} \frac{(\partial g/\partial x)}{(\partial g/\partial y)} \right] dx = 0 \qquad \textbf{(5.19b)}$$

In Eqs. (5.19), dx and dy are infinitesimals, small but not zero, so these equations require the following equality:

$$\frac{\partial f}{\partial x} = \frac{\partial f}{\partial y} \frac{(\partial g/\partial x)}{(\partial g/\partial y)} \quad \text{or} \quad \frac{(\partial f/\partial x)}{(\partial g/\partial x)} = \frac{(\partial f/\partial y)}{(\partial g/\partial y)}$$

These equalities are used to define an *undetermined multiplier* (λ):

$$\lambda = \frac{(\partial f/\partial x)}{(\partial g/\partial x)} = \frac{(\partial f/\partial y)}{(\partial g/\partial y)} \qquad \textbf{(5.20)}$$

If this multiplier is introduced into Eqs. (5.19), the result is

$$\frac{\partial f}{\partial x} - \lambda \frac{\partial g}{\partial x} = 0 \quad \text{or} \quad \frac{\partial}{\partial x}(f - \lambda g) = 0 \qquad \textbf{(5.21a)}$$

$$\frac{\partial f}{\partial y} - \lambda \frac{\partial g}{\partial y} = 0 \quad \text{or} \quad \frac{\partial}{\partial y}(f - \lambda g) = 0 \qquad \textbf{(5.21b)}$$

But Eqs. (5.21) are the conditions for finding an *unconstrained* maximum of the function:

$$G = (f - \lambda g)$$

with λ a parameter to be determined.

Example: Find a maximum of $f(x, y) = e^{-(x^2+y^2)}$ subject to the constraint $g(x, y) = x + 4y - 17 = 0$. From the slope formula:

$$df = -2xe^{-(x^2+y^2)} dx - 2ye^{-(x^2+y^2)} dy = 0$$

Clearly this condition is met by the point $x = y = 0$; this is the global maximum and is not on the path for which $x + 4y = 17$. The constrained maximum could be found directly by using this condition to eliminate x or y from the $df = 0$ equation, but for purposes of illustration we shall use Lagrange's method. Let:

$$G(x, y) = f(x, y) - \lambda g(x, y) = e^{-(x^2+y^2)} - \lambda(x + 4y - 17)$$

An unconstrained maximum in G has the requirements:

$$\frac{\partial G}{\partial x} = -2xe^{-(x^2+y^2)} - \lambda = 0, \qquad \lambda = -2xf(x, y)$$

$$\frac{\partial G}{\partial y} = -2ye^{-(x^2+y^2)} - 4\lambda = 0, \qquad \lambda = -\frac{y}{2}f(x, y)$$

This clearly requires $2x = y/2$ or $y = 4x$. The original condition ($g = 0$) then becomes $x + 4(4x) = 17$. Therefore the constrained maximum is at the point $x = 1$, $y = 4$. ∎

If there is a set of coordinates x_i, and two conditions to be satisfied, $g(x_i) = 0$, $h(x_i) = 0$, then a constrained maximum of $f(x_i)$ will be found when the function:

$$G = (f + \alpha g + \beta h)$$

has an unconstrained maximum; there are two undetermined multipliers, α and β.

Most Probable Populations

Back to the distribution of Eq. (5.16). We seek a maximum in $\ln W$ subject to two constraints; the total number of particles [N, Eq. (5.15)] and the total energy [E, Eq. (5.14)] are constant. These conditions can be expressed as:

$$N - \sum_i n_i = 0 \quad \text{and} \quad E - \sum_i n_i \varepsilon_i = 0 \tag{5.22}$$

With undetermined multipliers α and β, we need to maximize

$$G(n_i) = \ln W + \alpha \left(N - \sum_i n_i \right) + \beta \left(E - \sum_i n_i \varepsilon_i \right) \tag{5.23a}$$

Using W from Eq. (5.16):

$$\ln W = \ln N! - \sum_j \ln n_j!$$

and Stirling's approximation [Eq. (5.7)] for each term of this expression gives:

$$\ln W = N \ln N - N - \sum_j n_j \ln n_j + \sum_j n_j$$

The second and last terms on the right-hand side are, of course, the same and will cancel; therefore:

$$\ln W = N \ln N - \sum_j n_j \ln n_j \tag{5.24}$$

If this is substituted into Eq. (5.23a), we get:

$$G(n_i) = N \ln N - \sum n_j \ln n_j + \alpha\left(N - \sum n_j\right) + \beta\left(E - \sum n_j \varepsilon_j\right) \tag{5.23b}$$

The undetermined multipliers will guarantee that conditions (5.22) are met, so we now can vary the numbers n_i in an arbitrary fashion. For example, we can take the derivative of Eq. (5.23) with respect to any one of the populations (n_i), holding all others constant, and this derivative must be zero at the maximum:

$$\frac{\partial G}{\partial n_i} = -\ln n_i - 1 - \alpha - \beta \varepsilon_i = 0$$

Therefore, the condition required of any population n_i to make W a maximum (for a given E and N) is:

$$\ln n_i = -(1 + \alpha) - \beta \varepsilon_i$$

or

$$n_i = e^{-(1+\alpha)} e^{-\beta \varepsilon_i} \tag{5.25}$$

The constant α can be eliminated by using the fact that the sum of all n_i is just N, the total number of particles:

$$N = \sum_i n_i = e^{-(1+\alpha)} \sum_i e^{-\beta \varepsilon_i}$$

Then in Eq. (5.25) we can replace:

$$e^{-(1+\alpha)} = \frac{N}{\sum_j e^{-\beta \varepsilon_j}}$$

to get:

$$n_i = \frac{N e^{-\beta \varepsilon_i}}{\sum e^{-\beta \varepsilon_j}} \tag{5.26}$$

(The constant β is still undetermined.) This equation is also usefully written for the ratio of the populations of two levels (n_i and n_j) as:

$$\frac{n_i}{n_j} = \frac{e^{-\beta \varepsilon_i}}{e^{-\beta \varepsilon_j}} = e^{-\beta(\varepsilon_i - \varepsilon_j)} \tag{5.27}$$

Equations (5.26) and (5.27) are called **Boltzmann's law** and provide (once β and ε_i are specified) a relationship that gives the most probable distribution of particles among a set of energy levels at constant E, N; these are called the *Boltzmann populations.*

Example: Consider a set of 111 particles distributed among three equally spaced energy levels ($\varepsilon_1 = 0$, $\varepsilon_2 = a$, $\varepsilon_3 = 2a$, where a is some constant). Three possible population distributions are:

Level	Set A	Set B	Set C
$\varepsilon_3 = 2a$	$n_3 = 1$	2	0
$\varepsilon_2 = a$	$n_2 = 10$	8	12
$\varepsilon_1 = 0$	$n_1 = 100$	101	99

Each of these distributions has the same energy—for example, set A:

$$E = (1)(2a) + 10(a) + 0 = 12a$$

The reader should demonstrate that sets B and C have the same energy. The first of these sets is a Boltzmann distribution; from Eq. (5.27), the population ratios are:

$$\frac{n_2}{n_1} = \frac{e^{-\beta a}}{e^0} = e^{-\beta a}$$

$$\frac{n_3}{n_2} = \frac{e^{-2\beta a}}{e^{-\beta a}} = e^{-\beta a}$$

That is, for equally spaced levels, the population ratios for adjacent levels are equal; clearly this is true for set A, but not for the others.

The reader should check the following calculations:

$$W_A = \frac{111!}{100!10!1!} = 5.20564 \times 10^{15}$$

$$W_B = \frac{111!}{101!8!2!} = 2.31934 \times 10^{15}$$

$$W_C = \frac{111!}{99!12!0!} = 3.94366 \times 10^{15}$$

Evidently the Boltzmann distribution is the more probable.   ∎

5.4   The Partition Function

The sum that entered so innocuously into Eq. (5.26) is far more important than it may appear; it is called the **partition function** with symbol z:

$$z \equiv \sum_{\text{states}} e^{-\beta \varepsilon_i} \tag{5.28}$$

(The symbol q is also used; z comes from the German *Zustandsumme* = state sum, a far more descriptive term than "partition function.")

It often happens that a number of states have the same energy; this is called a *degeneracy*. If the number of states having an energy ε_i is g_i, the partition function can be written as a sum over *energy levels:*

$$z = \sum_{\text{levels}} g_i e^{-\beta \varepsilon_i} \tag{5.29}$$

The second form of z is often more convenient, but understand that Eqs. (5.28) and (5.29) are simply different ways to count the states of a particle.

There is a good example of degeneracy with which the reader may be already familiar — the hydrogen atom. The states of the electron in the hydrogen atom are characterized by four quantum numbers: principal (n), azimuthal (l), magnetic (m), and spin (m_s). Each permissible combination of these four numbers is a separate state; however the energy depends only on n, so all states with various l, m, m_s (for a given n) have the same energy — they are *degenerate*. For example, if $n = 2$, the following states have the same energy:

$$(2s) \quad \begin{cases} n = 2, \ l = 0, \ m_s = \tfrac{1}{2} \\ n = 2, \ l = 0, \ m_s = -\tfrac{1}{2} \end{cases}$$

$$(2p) \quad \begin{cases} n = 2, \ l = 1, \ m = 1, \quad m_s = \pm\tfrac{1}{2} \\ n = 2, \ l = 1, \ m = 0, \quad m_s = \pm\tfrac{1}{2} \\ n = 2, \ l = 1, \ m = -1, \ m_s = \pm\tfrac{1}{2} \end{cases}$$

The total degeneracy for the $n = 2$ state is therefore $g = 8$.

The Boltzmann Temperature

Next, we need to identify the constant β. Without further developments this cannot be done rigorously, so the relationship will simply be stated, and proven later:

$$\beta = \frac{1}{kT} \tag{5.30}$$

(k = Boltzmann's constant, T = temperature in kelvins). Equation (5.30) need not be as arbitrary as it appears; if you prefer, Eq. (5.30) can be taken to be the definition of the *Boltzmann temperature (T)*, which will later be shown to be identical to the ideal gas temperature we have been using all along.

Energy and Enthalpy

The Boltzmann population of an energy level (with degeneracy g_i) will be [from Eq. (5.26), counting levels rather than states]:

$$n_i = \frac{N g_i e^{-\beta \varepsilon_i}}{\sum_j g_j e^{-\beta \varepsilon_j}} = \frac{N g_i e^{-\beta \varepsilon_i}}{z}$$

This can be used to calculate the total energy:

$$E = \sum n_i \varepsilon_i = \frac{N \sum g_i \varepsilon_i e^{-\beta \varepsilon_i}}{z}$$

Note the following: the derivative of z [Eq. (5.29)] with respect to β (at constant V, so the ε_i will be constant) is:

$$\frac{\partial z}{\partial \beta} = \frac{\partial}{\partial \beta} \sum g_i e^{-\beta \varepsilon_i} = -\sum g_i \varepsilon_i e^{-\beta \varepsilon_i}$$

But this is exactly the sum needed to calculate E (above), so:

$$E = -\frac{N}{z}\frac{\partial z}{\partial \beta} = -N\frac{\partial \ln z}{\partial \beta} \tag{5.31a}$$

In terms of $T = 1/k\beta$, this becomes:

$$E = NkT^2\left(\frac{\partial \ln z}{\partial T}\right)_V \tag{5.31b}$$

The student encountering Eq. (5.31) for the first time would be perfectly justified in thinking its derivation to be circular, if not outright chicanery. The utility of this equation arises from the manner in which the partition function concept will be developed. The energy levels (ε_i) of a particle are independent of T, but they do depend on the volume of the container in which the particle is located. The partition function is then a function of T (via $\beta = 1/kT$) and of V (via the ε_i): $z(T,V)$. This can be viewed as some new state function from which, provided $z(T,V)$ can be evaluated, other state functions can be derived. We shall, in due course, develop the form of this function (but only for ideal gases), but our immediate objective is to show how $z(T,V)$ can be used to calculate the thermodynamic state functions.

What is the meaning of E? If we choose our energy scale so that the lowest energy (ε_0) is equal to zero, then:

$$z = g_0 + g_1 e^{-\varepsilon_1/kT} + g_2 e^{-\varepsilon_2/kT} + \cdots$$

At $T = 0$, z is constant ($z = g_0$) and E, according to Eq. (5.31), must be equal to zero. If T is now increased by adding heat at constant volume, E will increase (as the higher energy levels are occupied). But since no work is done (V is constant), this added heat must equal the change in the internal energy (U); therefore:

$$E = U - U_0$$

where U_0 is the internal energy at $T = 0$. (All of the practical equations of this chapter will be limited to ideal gases, so U_0 will normally mean the internal energy of an ideal gas at $T = 0$. That is a pretty hypothetical state, so it is probably best to view U_0 as some unknown constant; in the end, it is fixed to satisfy the thermodynamic conventions for energy.)

The internal energy is therefore related to the partition function as:

$$U - U_0 = NkT^2\left(\frac{\partial \ln z}{\partial T}\right)_V \tag{5.32a}$$

We shall more often want to evaluate U for one mole ($N = L$, $R = Lk$) of an ideal gas in the standard state ($P^\theta = 1$ atm $= 101,325$ Pa); for this case, Eq. (5.32a) becomes:

$$U_m^\theta - U_0 = RT^2\left(\frac{\partial \ln z^\theta}{\partial T}\right)_V \tag{5.32b}$$

[Since U_0 is simply some unknown constant, it seems unnecessary to complicate the notation by defining a new symbol for "an unknown constant per mole"; if you like, in Eq. (5.32b), U_0 is the molar internal energy of an ideal gas at $T = 0$, $P^\theta = 1$ atm, but this is, as stated earlier, a very hypothetical state.]

The enthalpy, $H = U + PV$, is given by $H_m = U_m + RT$ for an ideal gas, so:

$$H_m^\theta - H_0 = RT^2\left(\frac{\partial \ln z^\theta}{\partial T}\right)_V + RT \tag{5.33}$$

[At $T = 0$ for an ideal gas, $H = U$, so the unknown constant U_0 can also be written as H_0; this more common notation is used in Eq. (5.33).]

Entropy and Free Energy

The entropy can be calculated from Eq. (5.1) by assuming the particles are in their most probable state as given by Boltzmann's law [Eq. (5.26)]. From Eq. (5.24):

$$\ln W = N \ln N - \sum_i n_i \ln n_i$$

But Boltzmann's law requires:

$$\ln n_i = \ln N - \ln z - \frac{\varepsilon_i}{kT}$$

Together, these give:

$$\ln W = N \ln N - \sum_i n_i \ln N + \sum_i n_i \ln z + \frac{\sum_i n_i \varepsilon_i}{kT}$$

$$= N \ln z + \frac{\sum n_i \varepsilon_i}{kT}$$

But in the second term on the right-hand side, $\sum n_i \varepsilon_i$ is just E (or $U - U_0$), so:

$$\ln W = N \ln z + \frac{U - U_0}{kT}$$

If this is used in Eq. (5.1), the entropy is:

$$\text{(distinguishable)} \qquad S = Nk \ln z + \frac{U - U_0}{T} \tag{5.34}$$

The Helmholtz free energy, $A = U - TS$, follows immediately:

$$\text{(distinguishable)} \qquad A - U_0 = -NkT \ln z \tag{5.35}$$

For reasons that will not be immediately apparent, Eqs. (5.34) and (5.35) are correct *only for distinguishable particles*. This will be explained in the next section.

Equations (5.32) to (5.35) show that the evaluation of the partition function is surely a worthwhile task, since from it come all of the thermodynamics of the system; the evaluation of the partition function for molecular gases and the use of these results to calculate the thermodynamic functions for gases will occupy us for the remainder of this chapter.

5.5 Translational Motions

The energy of an atom or molecule can be rigorously separated into contributions from the translation of the molecule as a whole and from the internal motions of the molecule. The internal energy levels of a molecule are a consequence of its electronic energy, the vibrations of the nuclei with respect to each other, and the rotation of the molecule. An atom has only translational and electronic energies.

We write the energy of an atom or molecule as a sum of translational and internal contributions:

$$\varepsilon = \varepsilon_{tr} + \varepsilon_{int} \tag{5.36}$$

From this we deduce that the partition function is a product; since $e^{a+b} = e^a e^b$, we can write each term of z as:

$$\exp \frac{-\varepsilon}{kT} = \exp \frac{-\varepsilon_{tr}}{kT} \exp \frac{-\varepsilon_{int}}{kT}$$

Therefore, if we define:

$$z_{tr} = \sum e^{-\varepsilon_{tr}/kT} \tag{5.37}$$

$$z_{int} = \sum e^{-\varepsilon_{int}/kT} \tag{5.38}$$

then it is readily shown that the partition function can be written as a product:

$$z = z_{tr} z_{int} \tag{5.39}$$

All the thermodynamic equations, (5.32) through (5.35), involve the logarithm of z; taking the logarithm of Eq. (5.39) gives:

$$\ln z = \ln z_{tr} + \ln z_{int}$$

Therefore all the thermodynamic functions will be sums of internal and translational contributions — for example:

$$S = S(tr) + S(int)$$

$$U = U(tr) + U(int)$$

In this section we shall calculate the translational part of these functions.

The Evaluation of the Translational Partition Function

The translational energy of an ideal gas molecule is purely kinetic, so:

$$\varepsilon = \frac{p^2}{2m} \tag{5.40}$$

where $p = mv$ is the momentum of the particle. To evaluate the partition function, we must evaluate the sum:

$$z_{tr} = \sum e^{-(p^2/2m)/kT} \tag{5.41}$$

over all possible states of the particle. This could be done using either classical or quantum mechanics; both methods give the same result, so we shall use the classical method (the quantum derivation is given in Chapter 15).

In classical mechanics, the state of a particle is specified by giving its position (x, y, z) and momentum (p_x, p_y, p_z) at any point in time. Then, if the force (F) acting on the particle is specified, Newton's law:

$$F = \frac{dp}{dt}$$

will give the course of the particle for all future times. Such a state can be envisioned as a set of points in a six-dimensional space with coordinates x, y, z, p_x, p_y, and p_z; such a space, with both momenta and ordinary distances as coordinates, is called *phase space*. Given the difficulties of imagining a six-dimensional space, it will certainly be better to work with the simpler case of motion in one direction with coordinate x and momentum $p = p_x$. The results will be easily generalized to three dimensions. Since:

$$p^2 = p_x^2 + p_y^2 + p_z^2 \tag{5.42}$$

the three-dimensional partition function will be [from Eq. (5.41.)]:

$$z = z_x z_y z_z \tag{5.43}$$

where z_x, z_y, and z_z are the partition functions for one-dimensional translation.

In the one-dimensional case, phase space can be represented as a graph of p vs. x, as shown by Figure 5.5. Each point in this space represents a state of the particle; but

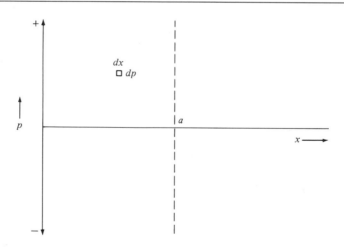

Figure 5.5 Phase space. The classical state of a particle moving in one dimension is specified by giving its position (x) and momentum (p); each state corresponds to a point on a graph of p vs. x.

there are an infinity of mathematical points in any part of this plane, and that is unacceptable, for, as we shall see, it would make the entropy infinite. Instead, we suppose that the plane is divided into small cells with size $(dp\,dx)$; each cell, whose size is small but not zero, represents the state of the particle. If the dimension of the container is (in the x direction) equal to a, then all possible states have $0 < x < a$. In the momentum direction, the coordinate values possible are $-\infty < p < \infty$. Since the cells are small and numerous, the sum of Eq. (5.41) can be approximated by an integral, giving:

$$z_x = \frac{1}{h} \int_0^a dx \int_{-\infty}^{\infty} e^{-p^2/2mkT}\, dp \qquad (5.44)$$

A mysterious factor, h, has crept into Eq. (5.44); what is that? Some such constant is required by simple unit analysis; z is clearly unitless, yet the product $(dp\,dx)$ has units of $(mass)(length)^2(time)^{-1}$, and so must h. In the SI system the units of h are $kg\ m^2\ s^{-1} = J\ s$. In effect, h is the size of a unit cell in phase space.

The integral of Eq. (5.44) is readily evaluated (at least with the help of an integral table), giving:

$$z_x = \sqrt{\pi}\ \sqrt{2mkT}\ \frac{a}{h}$$

The results in the y direction, with $0 < y < b$, and in the z direction ($0 < z < c$) will be obtained in the same manner, giving, from Eq. (5.43):

$$z = \frac{(2\pi mkT)^{3/2}V}{h^3} \qquad (5.45)$$

In Eq. (5.45) we have introduced the volume of the container, $V = abc$.

Equation (5.45) is precisely the same as that obtained by the quantum-mechanical calculation (Chapter 15), but quantum mechanics adds one vital piece of information — the constant h is **Planck's constant:**

$$h = 6.62602 \times 10^{-34}\ J\ s = 6.62602 \times 10^{-27}\ erg\ s$$

In classical mechanics, the momentum and the position of an object can, in principle, be specified with any degree of accuracy. There is therefore no mathematical reason why the cell size in phase space could not be made arbitrarily small. According to Heisenberg's uncertainty principle, if the position of a particle is specified to an accuracy δx, then the momentum cannot be specified more accurately than δp with

$$\delta x\ \delta p \gtrsim h \qquad (5.46)$$

Planck's constant is so small that this uncertainty is not significant for macroscopic objects; for atom-sized particles, however, it is quite important. [The question of the size of a cell in phase space was a serious problem for classical mechanics and, along with the black-body radiation and other problems, was a harbinger of the quantum revolution. Several persons including Planck himself (1921) deduced the relationship of this cell size to Planck's constant, but a rigorous derivation was possible only after the advent of quantum mechanics (1925).]

If Eq. (5.45) were used with Eq. (5.34) to calculate the entropy, the result would be of the form:

$$S(\text{tr}) = Nk(\text{function of T}) + Nk \ln V$$

This equation would give Gibbs' paradox all over again. The reason is that, when evaluating z, we counted the states of one particle; when generalizing to N particles it is not permissible to replace $\ln z$ by $N \ln z$ (assuming, in effect, that the partition function of N particles is z^N) because this fails to take into account the indistinguishability of the particles. The same argument used earlier, that there are $N!$ ways to permute N particles among their cells, suggests that the partition function must be divided by $N!$, giving, for the entropy:

$$S(\text{tr}) = k \ln \frac{z^N}{N!} + \frac{U - U_0}{T}$$

With Stirling's approximation, this becomes:

$$\text{(indistinguishable)} \quad S = Nk \ln \frac{z}{N} + Nk + \frac{U - U_0}{T} \qquad \textbf{(5.47)}$$

The free energy is then:

$$\text{(indistinguishable)} \quad A - U_0 = -NkT \ln \frac{z}{N} - NkT \qquad \textbf{(5.48)}$$

[Compare Eqs. (5.34) and (5.35).] Note that the energy [Eq. (5.32)] is not affected if z is divided by a constant, so this relationship is unchanged.

The Ideal Gas Law

One of the many equations in Table 3.1 is:

$$P = -\left(\frac{\partial A}{\partial V}\right)_T$$

From Eq. (5.48):

$$P = NkT \frac{\partial}{\partial V}(\ln z_{\text{tr}} - \ln N - NkT)$$

$$= \frac{NkT}{z_{\text{tr}}}\left(\frac{\partial z_{\text{tr}}}{\partial V}\right)_T$$

Using Eq. (5.45) for z, and introducing the number of moles, $n = N/L$, and the gas constant $R = kL$, it is readily shown that this equation becomes:

$$P = \frac{nRT}{V} \qquad \textbf{(5.49)}$$

That was a lot of work to derive the ideal gas law, but the derivation tells us two things: (1) the volume dependence of the partition function as derived is in accord with experimental facts; (2) the Boltzmann temperature that was defined in Eq. (5.30) is, in fact, the ideal gas temperature.

[Partition functions for real gases require the inclusion in the molecular energy, Eq. (5.40), of terms for the potential energy of interaction between molecules. This makes the integrals much more difficult — see refs. 2 and 4. If only pairwise inter-actions are included, Eq. (1.41) for the second virial coefficient will result.]

Energy and Heat Capacity

To calculate the internal energy U we need to evaluate:

$$\left(\frac{\partial \ln z_{tr}}{\partial T}\right)_V = \frac{1}{z_{tr}}\left(\frac{\partial z_{tr}}{\partial T}\right)_V$$

With Eq. (5.45), it is readily shown that this is:

$$\left(\frac{\partial \ln z_{tr}}{\partial T}\right)_V = \frac{3}{2}\frac{1}{T}$$

Then, from Eq. (5.32), the internal energy is (for one mole):

$$U_m^\theta(\text{tr}) - U_0 = \frac{3}{2}RT \tag{5.50}$$

This result is exactly that predicted by equipartition theory. The translational heat capacity follows directly, as $C_v = (\partial U/\partial T)_V$:

$$C_{vm}^\theta(\text{tr}) = \frac{3}{2}R \tag{5.51}$$

For many monatomic gases, translation is the only contribution to U and C_v, so Eqs. (5.50) and (5.51) are complete as they stand. This statement applies to gases of atoms with closed electronic shells, such as He, Ne, Ar, and Hg; it applies to some open-shell atoms (for example, Na) but not others (for example, Cl). Without a better understanding of atomic energy levels (Chapter 12) one should be cautious in using these results for open-shell atoms.

Entropy

It will be appropriate to evaluate S for the standard state, since changes in entropy with pressure are easily calculated (as in Chapter 3). Using Eqs. (5.45) and (5.47) and the ideal gas law (for V) gives:

$$S_m^\theta(\text{tr}) = R \ln\left[\frac{(2\pi M)^{3/2}R^{5/2}T^{5/2}}{L^4 P^\theta h^3}\right] + \frac{5}{2}R \tag{5.52}$$

(where $M = Lm$ is the molecular weight). This is called the **Sackur-Tetrode equation.**

Exercise: Derive Eq. (5.52). Then insert values for the constants to get the convenient equation:

$$S_m^\theta(\text{tr}) = \tfrac{3}{2}R \ln M + \tfrac{5}{2}R \ln T - 1.1649R \tag{5.53}$$

Equation (5.53) requires the molecular weight to be in grams, so be careful of units in this calculation. (If SI units are used, $P^\theta = 101{,}325$ Pa, and a factor of 1000 g/kg must be included.) ∎

Exercise: Equation (5.53) should give the entire entropy for closed-shell monatomic gases. Calculate the standard entropy of Ne at 298.15 K and compare your answer to the third-law value, 146.5 ± 0.4 J K^{-1}. (This is the first result that has used the fact that h is Planck's constant. That the result is correct is one of the experimental proofs of the correctness of the quantum theory.) ∎

5.6   Internal Motions

Only quantum statistical mechanics can correctly interpret the internal energies of atoms and molecules; for that reason we will have to use certain theoretical formulas and empirical facts that cannot be fully justified until later (Chapters 11 through 13).

The key to the understanding of molecular energies is the Born-Oppenheimer approximation (Chapter 13); this permits the separation of a molecule's internal energy into separate contributions from the electrons, the vibrations of the nuclei, and the overall rotation of the molecule:

$$\varepsilon_{\text{int}} = \varepsilon_{\text{elec}} + \varepsilon_{\text{vib}} + \varepsilon_{\text{rot}} \tag{5.54}$$

This separation is not nearly as rigorous as the separation, which we did earlier, of the translation and internal motions [Eq. (5.36)], but it is accurate enough for most purposes.

Equations (5.36) and (5.54) mean that the molecular partition function can be written as a product:

$$z = z_{\text{tr}} z_{\text{vib}} z_{\text{rot}} z_{\text{elec}} \tag{5.55}$$

This, in turn, means that the thermodynamic functions will be sums—for example:

$$U = U(\text{tr}) + U(\text{vib}) + U(\text{rot}) + U(\text{elec})$$

Units and Molecular Constants

Looking ahead a bit, the energy levels of atoms and molecules are quantized—that is, only certain values of energy are permitted. These energy levels are measured by spectroscopy by determining the wavelengths of light emitted or absorbed by the molecules; according to a formula first given by Bohr:

$$\text{energy of photon} = \text{change in molecule's energy}$$

$$h\nu = \Delta\varepsilon$$

If the wavelength (λ) of the photon is measured rather than its frequency (ν), these energies may be given as reciprocal wavelengths (wave numbers):

$$\tilde{\nu} = \frac{1}{\lambda} = \frac{\Delta\varepsilon}{hc}$$

These measured energies are generally written in terms of a set of molecular constants, which are tabulated in a variety of places; the units for these constants are usually given, as measured, in cm^{-1}, so unit conversions will be necessary. In cgs units:

$$\text{energy in ergs} = hc(\text{energy in } cm^{-1})$$

with $h = 6.62602 \times 10^{-27}$ erg s and $c = 2.9979 \times 10^{10}$ cm s^{-1}.

Since we shall frequently encounter quantities such as ε/kT, it will be very convenient to convert these constants into kelvins:

$$\text{energy in kelvins} = \frac{hc}{k}(\text{energy in } cm^{-1})$$

with $hc/k = 1.4388$ K cm. This will usually be done by defining a "characteristic temperature," which is generally given the symbol θ.

Vibrations

The vibrations of a diatomic molecule can be modeled by considering two masses (the nuclei) connected by a spring (the bond). The vibrational energy can be calculated with reasonable accuracy by assuming that the spring obeys Hooke's law (the restoring force is proportional to the displacement from the equilibrium length); this is the *harmonic oscillator* approximation. Quantum mechanics (Chapter 11) shows that such an oscillator will have a set of equally spaced energy levels with energy:

$$\varepsilon_v = (v + \tfrac{1}{2})hc\omega_e \tag{5.56}$$

In Eq. (5.56), v is the vibrational quantum number and has values $v = 0, 1, 2, \ldots, \infty$; ω_e is called the *vibrational constant*. The vibrational constant has units of cm^{-1} and its value is related to the masses of the nuclei and the force constant of the "spring." Some values are given in Table 13.1. Characteristic temperatures for vibration:

$$\theta_v = \frac{hc\omega_e}{k} \tag{5.57}$$

are listed in Table 5.2.

It will be convenient to shift our zero of energy so that the energy of the $v = 0$ state is zero and to calculate the partition function in terms of the differences:

$$\varepsilon_v - \varepsilon_0 = vhc\omega_e$$

$$z_{\text{vib}} = \sum_{v=0}^{\infty} e^{-vhc\omega_e/kT} = \sum_{v=0}^{\infty} e^{-v\theta_v/T} \tag{5.58}$$

As can be seen from an inspection of Table 5.2, vibrational constants tend to be quite large, so, at ordinary temperatures, only a few terms contribute to the sum in this partition function.

Table 5.2   Molecular constants for rotation and vibration[a]

Diatomics	θ_r (K)	θ_v (K)	Polyatomics	θ_v (degeneracy in parentheses)
H_2	87.5	6323.8	H_2O	5254, 2295, 5404
HF	30.127	5954.5	CH_2	4270, 2078, 4316
DF	15.837	4313.9	SO_2	1656, 754, 1958
HCl	15.238	4301.6	CO_2	1997, 3380, 960(2)
HBr	12.19	3812	HCN	3006, 4765, 1024(2)
HI	9.43	3323	NH_3	4801, 1367, 4912(2), 2342(2)
N_2	2.89	3395	CH_4	4193, 2196(2), 4345(3), 1879(3)
CO	2.78	3122	SiF_4	1151, 374(2), 1470(3), 604(3)
NO	2.50	2739		
O_2	2.08	2274		
Cl_2	0.351	813		
Br_2	0.116	470		
ICl	0.164	553		

[a]These constants are calculated using $\tilde{B}_e$ and ω_e. For greater accuracy use (Table 13.1):

$$\theta_r = 1.4388\left(\tilde{B}_e - \frac{\alpha_e}{2}\right), \qquad \theta_v = 1.4388(\omega_e - 2\omega_e x_e)$$

Example: The vibrational constant of N_2 is $\theta_v = 3395$ K. At 300 K:

$$z_{\text{vib}} = 1 + e^{-3395/300} + e^{-2(3395/300)} + \cdots$$

If we define $X = e^{-3395/300} = 1.2168 \times 10^{-5}$, then:

$$z_{\text{vib}} = 1 + X + X^2 + X^3 + \cdots = 1.0000122$$

At 1000 K, $X = 0.03354055$ and

$$z_{\text{vib}} = 1 + 0.03354 + 0.00112 + 0.00004 + 0.00000 = 1.0347$$

The vibrational contribution to the thermodynamic functions will not be significant when $z \cong 1$ ($\ln z \cong 0$) — that is, until $T \sim \theta_v$.   ∎

In the long run it will be better to get an explicit expression for the partition function as a function of T. This can be done for vibrations by using the Taylor series expansion (Appendix I):

$$\frac{1}{1 - X} = 1 + X + X^2 + X^3 + \cdots = \sum_{n=0}^{\infty} X^n$$

From this it may be evident that Eq. (5.58) will be:

$$z_{\text{vib}} = \frac{1}{1 - e^{-\theta_v/T}} \tag{5.59}$$

The thermodynamic formulas for vibration can be derived from this partition function using Eq. (5.32) for U and (5.35) for S. (Note that indistinguishability is not an issue here; it was taken care of under translation, and in any case each molecule has its own private set of vibrational states.)

Exercise: Use Eq. (5.59) to calculate the partition functions of the previous example, N_2 at 300 K and 1000 K. You should get exactly the same answers.   ■

Example: Derive formulas for U, C_v, and S for vibrations. Define $u = \theta/T$, so:

$$z = (1 - e^{-u})^{-1}$$

$$\frac{\partial z}{\partial T} = \frac{\partial z}{\partial u}\frac{\partial u}{\partial T} = \frac{\theta_v}{T^2}(1 - e^{-u})^{-2}e^{-u}$$

$$U_m - U_0 = \frac{RT^2}{z}\frac{\partial z}{\partial T} = R\theta_v e^{-u}(1 - e^{-u})^{-1}$$

$$U_m(\text{vib}) - U_0 = \frac{R\theta_v}{e^u - 1} \qquad (5.60)$$

The reader should show that $C_v = \partial U/\partial T$ is:

$$C_{vm}(\text{vib}) = \frac{Ru^2 e^u}{(e^u - 1)^2} \qquad (5.61a)$$

Equation (5.61) and other equations of this section are written in terms of e^u, as they are in nearly all texts. For purposes of computation, however, Eq. (5.61a) is better written as:

$$C_{vm}(\text{vib}) = \frac{Ru^2 e^{-u}}{(1 - e^{-u})^2} \qquad (5.61b)$$

These equations are mathematically identical, but for practical purposes the second form is superior; if you attempt to calculate e^u for $u > 230$ (on most calculators) or $u > 85$ (on a typical microcomputer) an "OVERFLOW" error message will result. On the other hand, e^{-u} is usually set to zero when u is too large for the computer to evaluate it meaningfully.

From Eq. (5.34):

$$S_m = R \ln z + \frac{U_m - U_0}{T}$$

Using the partition function of Eq. (5.59) and the formula already derived for U [Eq. (5.60)], the reader should show that:

$$S_m(\text{vib}) = \frac{Ru}{e^u - 1} - R \ln (1 - e^{-u}) \qquad (5.62)$$

The free energy is, from Eq. (5.35):

$$A_m(\text{vib}) - U_0 = RT \ln (1 - e^{-u}) \qquad (5.63)$$
■

The characterization of the vibrations of polyatomic molecules requires a number of vibrational constants. A linear molecule with N nuclei will have $3N - 5$ vibrational constants, while a nonlinear molecule has $3N - 6$. Often, two or more of these vibrations may have the same constant; this is called a degeneracy: $g \equiv$ number of vibrations having the same vibrational constant. Equations (5.60) through (5.63) apply to each of these vibrations; to get the total they must be summed over $3N - 5$ (linear) or $3N - 6$ (nonlinear) values of the vibrational constant.

Example: The nonlinear triatomic molecule H_2O has $3N - 6 = 3$ vibrations with $\theta_v = 5254, 2295, 5404$ K (Table 5.2). Calculate the vibrational contribution to the internal energy at 1000 K.

(1) $\qquad u = 5.254, \qquad U_m - U_0 = \dfrac{8.3143(5254)}{e^u - 1} = 229.5$ J

(2) $\qquad u = 2.295, \qquad U_m - U_0 = \dfrac{8.3143(2295)}{e^u - 1} = 2138.1$ J

(3) $\qquad u = 5.404, \qquad U_m - U_0 = 203.0$ J

The total is: $U_m(\text{vib}) - U_0 = 2571$ J.   ∎

Exercise: The linear triatomic molecule CO_2 has $3N - 5 = 4$ vibrations (Table 5.2): $\theta_v = 1997, 3380$, and 960 (twice). Calculate the internal energy due to vibration at 1000 K. [*Answer:* 13,503 J]   ∎

Rotations

Molecular rotations can be treated with good accuracy by the rigid-rotor approximation; this works because the vibrational "springs" are, in fact, usually quite stiff, and the bond lengths are therefore not too much distorted by the centrifugal force of the rotation. The theory is simplest for linear molecules, and it is only this case that we shall discuss in detail.

A collection of masses m_i (the nuclei) arranged along an axis (z) will have a moment of inertia:

$$I = \sum_i m_i (z_i - z_0)^2 \tag{5.64}$$

where z_0 is the location of the center of mass (COM), which is determined by the requirement:

$$\sum_i m_i (z_i - z_0) = 0 \tag{5.65}$$

The rotational energy of a rigid linear rotor is quantized; the allowed values of the energy are (Chapter 13):

$$\varepsilon_J = J(J + 1) \frac{h^2}{8\pi^2 I} \tag{5.66}$$

where J is the rotational quantum number; $J = 0, 1, 2, 3, \ldots, \infty$. Each of these energy levels has a degeneracy $g_J = 2J + 1$.

The quantity that is usually tabulated is the **rotational constant** (units: cm^{-1}):

$$\tilde{B}_e = \frac{h}{8\pi^2 Ic} \qquad (5.67)$$

We shall use the **rotational characteristic temperature:**

$$\theta_r = \frac{hc\tilde{B}_e}{k} \qquad (5.68)$$

which is tabulated in Table 5.2.

With these definitions, the partition function is:

$$\text{(linear)} \qquad z_{\text{rot}} = \sum_{J=0}^{\infty} (2J + 1)e^{-J(J+1)\theta_r/T} \qquad (5.69)$$

Equation (5.69) is correct for unsymmetrical molecules, but there is a difficulty with molecules that have a center of symmetry. A heteronuclear diatomic molecule AX must rotate by 360° to return to its original position:

but a homonuclear diatomic need only rotate by 180° to arrive at a position indistinguishable from where it would be after 360°:

For this reason an oxygen molecule may have only *odd* values of J ($J = 1, 3, 5, 7, \ldots$), and the sum of Eq. (5.69) would be roughly twice as large as the actual partition function. The situation is similar, but rather more complicated, for other molecules. (Oxygen is simple because the $^{16}_{8}O$ nucleus has no spin.) This topic cannot be explained better without more quantum theory, so it will be delayed until Chapter 15. When $T \gg \theta_r$, the partition function can accurately be corrected by dividing the sum of Eq. (5.69) by a *symmetry number* (σ): $\sigma = 1$ for unsymmetrical linear molecules such as HCl and HCN; $\sigma = 2$ for symmetrical linear molecules such as O_2 and CO_2 (that is, OCO).

For most molecules at temperatures at or above room temperature (indeed, any temperature above their boiling point, even if below room temperature) it can be seen (Table 5.2) that $T \gg \theta_r$; H_2 is a major exception, so we shall exclude this single molecule from further discussion until Chapter 15. When $T \gg \theta_r$, many rotational states will be populated, and the sum in the partition function can be approximated by an integral:

$$z_{\text{rot}} = \frac{1}{\sigma} \int_0^{\infty} (2J + 1)e^{-J(J+1)\theta_r/T} \, dJ$$

This integral is easily evaluated by a change in variable:

$$y \equiv J(J + 1), \qquad dy = (2J + 1) \, dJ$$

$$z_{\text{rot}} = \frac{1}{\sigma} \int_0^{\infty} e^{-(\theta_r/T)y} \, dy$$

The result, valid for all linear molecules when $T \gg \theta_r$, is:

$$z_{\text{rot}} = \frac{T}{\sigma \theta_r} \qquad (5.70)$$

Exercise: Use the rotational partition function [Eq. (5.70)] to derive the following (for linear molecules):

$$U_m(\text{rot}) - U_0 = RT \qquad (5.71)$$

$$C_{vm}(\text{rot}) = R \qquad (5.72)$$

$$S_m(\text{rot}) = R \ln \frac{T}{\sigma \theta_r} + R \qquad (5.73)$$

Note that the heat capacity is exactly that predicted by the equipartition theorem. These formulas apply only for $T \gg \theta_r$ (see Figure 5.6 for the exact form of U and C_v at lower T), but, as indicated, this will be the practical case for all molecules except H_2 at, and above, their boiling points (Figure 5.6). ∎

The rotations of a nonlinear molecule are described using three moments of inertia about a set of orthogonal axes called the *principal* axes (Chapter 14). The partition function is:

$$(\text{nonlinear}) \qquad z_{\text{rot}} = \frac{8\pi^2}{\sigma h^3}(2\pi kT)^{3/2}(I_x I_y I_z)^{1/2} \qquad (5.74)$$

Exercise: Prove that the contribution of rotation to the internal energy of a nonlinear molecule is $\frac{3}{2}RT$. Contrast this to the linear case, which gave RT. ∎

Finding the symmetry number of a nonlinear molecule can be difficult. The symmetry number (σ) is defined as the number of indistinguishable positions in space that can be reached by rigid rotations. For example: H_2O ($\sigma = 2$), NF_3 ($\sigma = 3$), BF_3 ($\sigma = 6$), CH_4 ($\sigma = 12$), SF_6 ($\sigma = 24$). The first of these should be fairly obvious; the last two emphatically are not. Let's look more closely at NF_3 and BF_3.

Example: The symmetry number of NF_3 (pyramidal):

(top view)

The top view is easier to visualize, but *remember that N is not in the same plane as F.* The fluorines must be numbered so we can keep track of the "indistinguishable" positions:

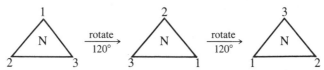

There are three indistinguishable positions, therefore $\sigma = 3$. ∎

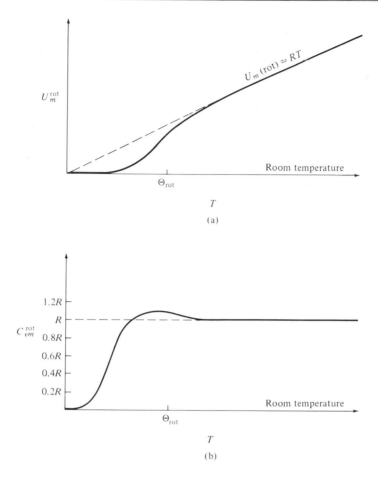

Figure 5.6 Rotational energy and heat capacity. The rotational contributions to (a) the molar energy and (b) the molar heat capacity are shown for various temperatures near the rotational characteristic temperature (Θ_{rot}). The equipartition values ($U = RT$ and $C_v = R$) are correct at temperatures well above the rotational characteristic temperature. (From F. C. Andrews, *Equilibrium Statistical Mechanics*, 1963: New York, John Wiley & Sons, Inc., Fig. 19-1, p. 95)

Example: The symmetry number of BF_3 (planar):

$$\begin{matrix} F & & F \\ & \diagdown \; \diagup & \\ & B & \\ & | & \\ & F & \end{matrix}$$ (all atoms in the same plane)

Once again, we number the fluorines and rotate the molecule by 120° about the axis perpendicular to the molecular plane; but the molecule can also be rotated about the BF bonds by 180°:

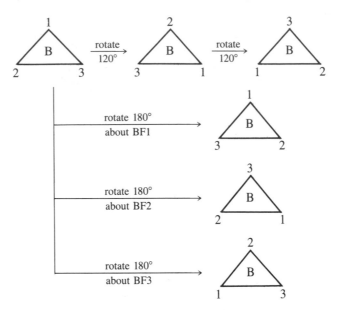

Therefore, there are six equivalent positions, and $\sigma = 6$.   ■

Electronic Energy Levels

The electrons of an atom or molecule have numerous energy levels available to them. The reader may be familiar with the hydrogen atom whose electron is characterized by a quantum number n and that is permitted to have energies:

$$E_n = -\frac{4\pi^2 me^4}{n^2 h^2}$$

for $n = 1, 2, 3, \ldots, \infty$. The electronic levels of other atoms and molecules cannot be represented by such a simple formula, but those levels are also quantized and their energies and degeneracies can usually be determined from spectroscopy.

In most cases, only a few electronic levels will contribute to the partition function. Setting the ground state energy to be zero, $\varepsilon_0 = 0$, the partition function is:

$$z_{\text{elec}} = g_0 + g_1 e^{-\varepsilon_1/kT} + g_2 e^{-\varepsilon_2/kT} + \cdots \tag{5.75}$$

These energies are usually tabulated (for example, in Table 5.3) in units of cm^{-1} (that is, as ε/hc).

Stable closed-shell atoms or molecules rarely have any excited states below 10,000 cm^{-1} ($\theta \sim 14{,}000$ K), so at $T < 5000$ K the partition function is constant:

$$z_{\text{elec}} = g_0$$

In this case, the electrons will contribute nothing to the internal energy (U); also, in many cases $g_0 = 1$, so there will also be no contribution to the entropy. In effect, if $g_0 = 1$ and there are no excited states below $\varepsilon \sim kT$, the electronic contributions to thermodynamic functions are negligible. There are some exceptions; oxygen (O_2),

Table 5.3   Electronic energy levels

	g_0	ε/hc (cm^{-1}) (degeneracy in parentheses)
Atoms		
C	1	16.40(3), 43.40(5), 10,193(5)
N	4	19,224(6), 19,233(4), 28,839(6)
O	5	158.265(3), 226.997(1), 15,868(5)
F	4	404.1(2)
Cl	4	882.36(2)
Br	4	3685.24(2)
I	4	7603.15(2)
Molecules		
O_2	3	7918.1(2), 13,195(1)
NO	2	121.1(2)
CH	2	17.9(2), 4500(?)(4), 23150(2)
CH_3	2	46,205(2)
OH	2	139.7(2)

for example, has two unpaired electrons in its ground state, which causes $g_0 = 3$ (a "triplet," see Chapter 13). Table 5.3 lists some excited electronic states and degeneracies.

Exercise: Show that a ground-state degeneracy adds $R \ln g_0$ to the molar entropy.   ■

Example: Calculate S_m^θ of O_2 at 298.15 K. (Data are in Tables 5.2 and 5.3.)

1. Translation:

$$S_m^\theta = R(2.5 \ln 298.15 + 1.5 \ln 32 - 1.1649) = 151.967 \text{ J K}^{-1}$$

2. Vibration: $u = 2274/298.15 = 7.627$:

$$S_m = \frac{8.3143(2274)}{298.15(e^u - 1)} - 8.3143 \ln (1 - e^{-u}) = 0.035 \text{ J K}^{-1}$$

3. Rotation: $\sigma = 2$, $\theta = 2.080$, $U - U_0 = RT$:

$$S_m = R \ln \frac{298.15}{2(2.080)} + R = 43.834 \text{ J K}^{-1}$$

4. Electronic:

$$S_m = R \ln 3 = 9.134 \text{ J K}^{-1}$$

TOTAL: $S_m^\theta(298.15) = 204.97 \text{ J K}^{-1}$. (Table 3.2 lists 205.06.)   ■

When it is necessary to include excited electronic states in the calculation of thermodynamic properties, it is usually best to use Eq. (5.75) directly with Eqs. (5.32), (5.34) and (5.35) [computing C_v as $(\partial U/\partial T)$]. The two-level case is relatively simple: denoting the ground-state degeneracy as g_0, and assuming only one excited state (g_1, ε_1), it can be shown that $(x = \varepsilon_1/kT)$:

$$U_m(\text{elec}) - U_0 = \frac{Rg_1(\varepsilon_1/k)}{g_0 e^x + g_1} \tag{5.76}$$

$$C_{vm}(\text{elec}) = \frac{Rg_0 g_1 x^2 e^x}{(g_0 e^x + g_1)^2} \tag{5.77}$$

Note that at high T $(x << 1)$ the electronic contribution to the energy becomes constant:

$$U_m\left(T >> \frac{\varepsilon_1}{k}\right) - U_0 = \frac{g_1}{g_0 + g_1} R\varepsilon_1/k$$

$$= \frac{L\varepsilon_1 g_1}{g_0 + g_1} \tag{5.78}$$

and the heat capacity of Eq. (5.77) is zero.

Summary

Table 5.4 gives a summary of the formulas for calculating thermodynamic properties of gases; these formulas are restricted to $T >> \theta_r$ and to the calculation of ideal gas properties (or the standard-state properties of real gases). The fact that these formulas

Table 5.4   Thermodynamic properties of ideal gases

	Translation[a]	Rotation[b] $(T >> \theta_r)$		Vibration[c]
C_{vm}^θ	$1.5R$	R		$\dfrac{Ru^2 e^u}{(e^u - 1)^2}$
		$1.5R$		
$U_m^\theta - U_0$	$1.5RT$	RT		$\dfrac{R\theta_v}{e^u - 1}$
		$1.5RT$		
S_m^θ	$1.5R \ln M$ $+2.5R \ln T$ $-1.1649R$	$R \ln (T/\sigma\theta_r) + R$		$\dfrac{Ru}{e^u - 1}$ $-R \ln (1 - e^{-u})$
		$R \ln (AT^{3/2}/\sigma) + 1.5R$		
ϕ°	$1.5R \ln M$ $+2.5R \ln T$ $-3.6649R$	$R \ln (T/\sigma\theta_r)$		$-R \ln (1 - e^{-u})$
		$R \ln (AT^{3/2}/\sigma)$		

[a]M = molecular weight in grams.
[b]Linear above dotted line, nonlinear below. $A = 8\pi^2(2\pi k)^{3/2}(I_x I_y I_z)^{1/2}/h^3$.
[c]$u = \theta_v/T$. Sum over all vibrations.

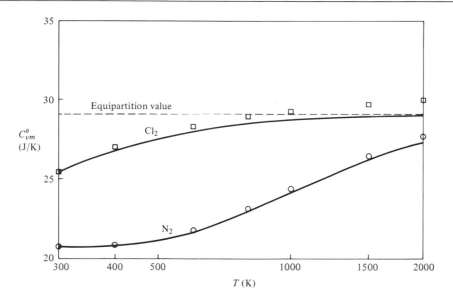

Figure 5.7 Heat capacities of chlorine and nitrogen vs. temperature. The solid lines are calculated as described in the text (formulas of Table 5.4). The experimental values are as given by Walter J. Moore, *Physical Chemistry*, 4th ed., 1972: Englewood Cliffs, NJ, Prentice-Hall, Inc. The deviations at higher temperatures are due to approximations made in deriving the formulas.

are limited to ideal gases is less of a problem than it may, at first, appear. They can be used to calculate *standard state* properties of real gases and, as we saw earlier (sections 2.8 and 3.7), the thermodynamic properties of real gases are easily calculated from standard state properties, given an equation of state. Also, as we shall learn in the next chapter, standard state properties can be used to calculate equilibrium constants. Figure 5.7 shows sample calculations for the heat capacity of Cl_2 and N_2. The errors at high temperatures are largely due to the approximations in the quantum-mechanical formulas we have used; for example, the formulas neglected vibrational anharmonicity, vibration-rotation interactions, and centrifugal stretching. (These terms are explained in Chapter 13; ref. 3 gives more accurate formulas.)

The formulas of Table 5.4 can be used for polyatomic molecules by simply summing the vibrational contributions over all normal modes. However some polyatomic molecules present additional complications such as internal rotations (for example, rotation about the C-C bond in ethane) or "umbrella" inversions (as in ammonia); these are discussed in advanced texts.

Electronic contributions (not included on Table 5.4) can be neglected for most stable molecules. One exception to this statement is NO, which has a low-lying excited electronic state with $\varepsilon_1/hc = 121.1$ cm^{-1} ($g_1 = 2$, $g_0 = 2$, cf. Table 5.3). The effect of this state on the heat capacity of NO is demonstrated by Figure 5.8; note that when $T \gg \varepsilon_1/k$, the electronic contribution to the heat capacity is *zero*.

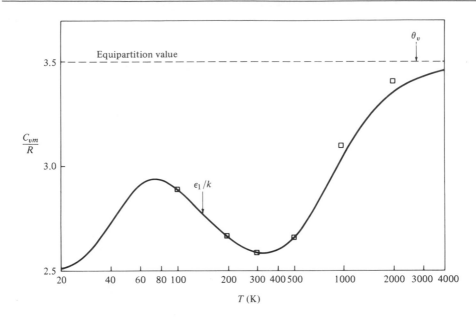

Figure 5.8 Heat capacity of nitrous oxide (NO) vs. temperature. (Experimental values from Ref. 5.) The peak near 60 K is due to low-lying electronic excited states. The deviations between the calculated and observed values at high temperature are primarily due to approximations in the formulas used for the calculation (Table 5.4 and Eq. 5.77). The low-lying electronic states do not contribute significantly to the heat capacity above about 400 K, but they will contribute to the entropy at all temperatures.

5.7 The Free-Energy Function

In Chapter 6 we shall discover (you may know it already) that the standard Gibbs' free-energy change of a chemical reaction is related to the equilibrium constant (K) of that reaction as:

$$\Delta G^\theta = -RT \ln K$$

Thus, equilibrium constants can be calculated (in some cases) from spectroscopic data, but this is done rather indirectly.

The free energy of a substance is [Eq. (5.48) for translation and Eq. (5.35) for the internal motions]:

$$A_m^\theta - U_0 = -RT \ln \frac{z^\theta}{L} - RT \tag{5.79}$$

with

$$z^\theta = z_{\text{tr}}^\theta z_{\text{vib}} z_{\text{rot}} z_{\text{elec}}$$

The Gibbs' free energy of a gas is then:

$$G = A + PV$$

$$G_m^\theta = A_m^\theta + RT$$

Therefore:

$$G_m^\theta - H_0 = -RT \ln \frac{z^\theta}{L} \tag{5.80}$$

[The zero Kelvin constants H_0 and U_0 are exactly the same; the change in notation between Eqs. (5.79) and (5.80) is in accord with standard usage.]

Equation (5.80) is used to define the **free-energy function** $\phi°$:

$$\phi° \equiv \frac{-(G_m^\theta - H_0)}{T} \tag{5.81}$$

(There is, apparently, no standard symbol for this quantity, so this symbol will not be found in other texts or references.) This quantity is, of course, directly calculable from the partition function; Eq. (5.80) gives:

$$\phi° \equiv R \ln \frac{z^\theta}{L} \tag{5.82}$$

Exercise: Combine the results derived earlier to show that, for a gas of linear molecules with $T \gg \theta_r$:

$$\phi° = R[3.5 \ln T + 1.5 \ln M - 3.6649 - \ln (\sigma\theta_r) - \sum \ln (1 - e^{-u}) + \ln z_{\text{elec}}] \tag{5.83}$$

(where M = molecular weight in grams; $u = \theta_v/T$; the sum is over all vibrations). Figure 5.9 displays several sample calculations. ∎

The free energy can be calculated from this function as.

$$G_m^\theta = H_0 - T\phi°$$

and the change for a chemical reaction is

$$\Delta G^\theta = \Delta H_0 - T\Delta\phi° = -RT \ln K$$

Thus, to relate partition functions to equilibrium constants, it is necessary to evaluate ΔH_0; this is the heat of the ideal gas reaction at $T = 0$. This quantity can be calculated from either thermal or (sometimes) spectroscopic data. (A simple example: For a diatomic dissociation such as $N_2 = 2N$, ΔH_0 is simply the bond energy per mole.)

Free-Energy Function Based on 298 K

Practical use of Eq. (5.82) for chemical reactions requires that one be able to evaluate the partition function for each of the reactants and products; this is not always practical (so far we know how to do it only for simple gases). For that reason, it is convenient to define another free-energy function based on enthalpies at 25°C (a temperature for

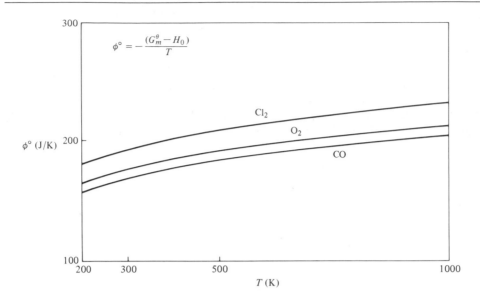

Figure 5.9   The free-energy function (0 K based) for chlorine, oxygen, and carbon monoxide. The free-energy function is often tabulated in reference books at 100-K intervals; this graph demonstrates that the function behaves smoothly over such a range (always presuming that there are no phase transitions in the range) and can therefore be interpolated accurately with a low-order polynomial.

which heats of reaction are often measured and commonly tabulated). We shall call this function ϕ':

$$\phi' \equiv \frac{-[G_m^\theta - H_m^\theta(298.15 \text{ K})]}{T} \tag{5.84}$$

This free-energy function is the one most commonly tabulated (see, for example, ref. 5 or 3. Some values will be given in Chapter 6.)

The relationship of the two free-energy functions is, from Eqs. (5.81) and (5.84):

$$\phi' = \phi^\circ + \frac{H_m^\theta(298.15) - H_0}{T} \tag{5.85}$$

The required enthalpy difference can be calculated for gases from spectroscopic data.

Exercise: Combine earlier results (use $H_m^\theta = U_m^\theta + RT$) to show:

$$(\text{linear}) \qquad H_m^\theta(T) - H_0 = 3.5RT + \frac{\sum R\theta_v}{e^u - 1} + [\text{electronic}] \tag{5.86}$$

$$(\text{nonlinear}) \qquad H_m^\theta(T) - H_0 = 4RT + \frac{\sum R\theta_v}{e^u - 1} + [\text{electronic}] \tag{5.87}$$

∎

Example: For Cl_2 gas Eq. (5.86) can be used to calculate the relationship between the two free-energy functions; with $T = 298.15$ K ($\theta_v = 813$ K, $u = 813/298.15 = 2.727$):

$$H_m^\theta(298.15) - H_0 = 3.5R(298.15) + \frac{R(813)}{e^{2.727} - 1}$$

$$= 9149 \text{ J}$$

Therefore at any temperature T:

$$\phi'(T) = \phi°(T) + \frac{9149}{T}$$

■

The implication of Eqs. (5.85) to (5.87) is that ϕ', as well as $\phi°$, can be calculated from spectroscopic data if the substance is a gas. The real advantage of ϕ' over $\phi°$ is that it can be calculated for any substance from thermodynamic data. Using:

$$G = H - TS$$

in Eq. (5.84) gives

$$\phi'(T) = \frac{TS_m^\theta - H_m^\theta(T) + H_m^\theta(298.15)}{T}$$

$$\phi'(T) = S_m^\theta(T) - \frac{H_m^\theta(T) - H_m^\theta(298.15)}{T} \qquad \textbf{(5.88)}$$

Entropies at 25°C are commonly tabulated (as in Table 3.2, for example), and the results of Chapter 3 give:

$$S_m^\theta(T) = S_m^\theta(298.15) + \int_{298.15}^{T} C_{pm}^\theta \, d(\ln T)$$

Also the enthalpy difference of Eq. (5.88) can be calculated from the same heat capacities:

$$H_m^\theta(T) - H_m^\theta(298.15) = \int_{298.15}^{T} C_{pm}^\theta \, dT$$

Therefore, Eq. (5.87) can be evaluated using readily measured thermodynamic properties:

$$\phi'(T) = S_m^\theta(298.15) + \int_{298.15}^{T} C_{pm}^\theta \, d(\ln T) - \frac{1}{T} \int_{298.15}^{T} C_{pm}^\theta \, dT \qquad \textbf{(5.89)}$$

Thus, this free-energy function can be calculated from either spectroscopic or thermal data.

Example: Although Eq. (5.89) and the equations of Table 5.4 (for gases) are generally useful in different circumstances, we shall use them both to calculate the free-energy function (ϕ') of N_2 at 100 K; this will serve to illustrate the two types of calculation. From Table 2.1:

$$C_{pm}^\theta = 28.58 + 3.77 \times 10^{-3}T - \frac{5 \times 10^4}{T^2}$$

(Remember that this empirical equation is valid only for $298 < T < 2000$ and therefore cannot be used to calculate quantities such as $H_m^\theta(298.15) - H_0$.) From Table 3.2:

$$S_m^\theta(298.15) = 191.50 \text{ J K}^{-1}$$

(a) $$\int_{298.15}^{1000} C_{pm}\, dT = 28.58(1000 - 298.15) + \frac{1}{2}(3.77 \times 10^{-3})[(1000)^2$$

$$- (298.15)^2] + (5 \times 10^4)\left(\frac{1}{1000} - \frac{1}{298.15}\right) = 21659 \text{ J}$$

(b) $$\int_{298.15}^{1000} \frac{C_{pm}}{T}\, dT = 28.58 \ln \frac{1000}{298.15} + 3.77 \times 10^{-3}(1000 - 298.15)$$

$$+ \frac{1}{2}(2 \times 10^4)\left[\frac{1}{(1000)^2} - \frac{1}{(298.15)^2}\right] = 36.98 \text{ J K}^{-1}$$

From Eq. (5.89):

$$\phi' = 191.50 + 36.98 - \frac{21659}{1000} = 206.82 \text{ J K}^{-1}$$

Next we shall calculate the same quantity using spectroscopic data. From Table 5.2: $\theta_r = 2.89$, $\theta_v = 3395$; also $\sigma = 2$, $g_0 = 1$, $M = 28.01$ g. From Eq. (5.83) with $u = 3.395$:

$$\phi^\circ = 8.3143[3.5 \ln (1000) + 1.5 \ln (28.01) - 3.6649$$

$$- \ln [(2)(2.89)] - \ln (1 - e^{-3.395})]$$

$$= 197.80 \text{ J K}^{-1}$$

From Eq. (5.86) with $u = 3395/298.15 = 11.387$:

$$H_m^\theta(298.15) - H_0 = (3.5)(8.3143)(298.15) + \frac{(8.3143)(3395)}{e^{11.387} - 1}$$

$$= 8677 \text{ J}$$

Then, using Eq. (5.85):

$$\phi'(1000) = 197.80 + \frac{8677}{1000} = 206.48 \text{ J K}^{-1}$$

This value differs from the thermodynamic result by 0.16%; ref. 3 gives 206.65 for this quantity; ref. 5 gives 206.60.   ∎

Postscript

Statistical mechanics makes two important contributions to thermodynamics. First, it provides valuable insight into the molecular-level phenomena that are responsible for macroscopic properties. Second, it provides useful and accurate formulas for calculating certain thermodynamic properties. It is especially important that the spectroscopic constants involved in the statistical calculations are independent of temperature; the formulas that use them are valid over a wide range of temperatures and are particularly valuable at very high temperatures. (However, the formulas of Table 5.4 are inaccurate at high temperature because of the approximations used. Even the accurate formulas are primarily useful for gases of simple molecules, but this is somewhat self-compensating, since, at a sufficiently high temperature, where the formulas are most valuable, simple molecules are all you are likely to have.)

For those who seek to know more about this topic, refs. 1 and 2 present extended treatments at about the same level as this chapter. Reference 3 gives more details on practical calculations of thermodynamic properties. Reference 4 is a concise but more rigorous introduction to the general areas of probability, kinetic theory, and statistical mechanics. (Other references follow Chapter 15.)

References

1. Leonard K. Nash, *Elements of Statistical Mechanics*, 1971: Reading, Mass., Addison-Wesley Publishing Co.
2. Walter Kauzmann, *Thermodynamics and Statistics*, 1967: New York, W. A. Benjamin, Inc.
3. G. N. Lewis, M. Randall, K. S. Pitzer, L. Brewer, *Thermodynamics*, 1961: New York, McGraw-Hill Book Co.
4. Frank C. Andrews, *Equilibrium Statistical Mechanics*, 1963: New York, John Wiley & Sons, Inc.
5. D. R. Stull and H. Prophet (Project Directors), *JANAF Thermochemical Tables*, 2d ed., 1971, U.S. Department of Commerce, National Bureau of Standards (NSRDS-NBS 37).

Problems

5.1   Compute the number of distinct permutations of the letters in the words LEAK, LEEK, MISSISSIPPI. Confirm your calculation for the first two by writing out all the possibilities.

5.2   In a SCRABBLE™ set there are (of 98 tiles), 2 P, 2 C, 2 H, 12 E, 2 M.
(a) If you drew 5 tiles at random, what is the chance that they would spell (in order) PCHEM?
(b) Is ORGANIC more probable? (8 O, 6 R, 3 G, 9 A, 6 N, 9 I, 2 C.)
(c) What is the probability of drawing the letters P, C, H, E, M in any sequence?

5.3   A flexible polymer chain can be modeled as a set of rigid links (length λ). It can be shown that for N links, the number of configurations is:

$$W = AL^2 \exp\left(-\frac{L^2}{N\lambda^2}\right)$$

where L is the end-to-end distance and A is a constant:
(a) Prove that S is a maximum when $L = \sqrt{N}\,\lambda$.
(b) Calculate ΔS for stretching the chain by 10%. Assume $N = 100$ and a mole of chains.

5.4   Comparison of statistical and third-law entropies of CH_3D reveals a discrepancy of 11.6 J K^{-1} mol^{-1}. Explain. What would you anticipate the discrepancy in S to be for CH_2D_2?

5.5   A system containing 38 particles has three equally spaced energy levels available. Two population distributions are A: $(18, 12, 8)$, B: $(17, 14, 7)$. Show that both distributions have the same energy. Is either of these a Boltzmann distribution? Calculate W for each. Which is more probable? If the energy-level spacing ($\Delta\varepsilon$) is 10^{-22} J, what is T?

5.6   A molecule has three equally spaced energy levels. At equilibrium the number of molecules (out of 5550) in each level is 5000, 500, 50. Confirm that this is a Boltzmann distribution. Use Stirling's approximation to calculate $\ln W$ for this distribution and for the distribution (5001, 498, 51) (with the same total energy) to show that the Boltzmann distribution is the more probable.

5.7   What is the probability that two blocks of a material (with $C_p = 10$ J K^{-1} for each) in thermal contact will differ in temperature by 0.003 K at 300 K (when they have reached equilibrium)?

5.8   A population of 1,000,000 particles has two energy states available to it with $\Delta\varepsilon = 4.14 \times 10^{-21}$ J.
(a) Calculate the populations of the levels for $T = 10$ K, 300 K, 3000 K.
(b) What will be the populations as $T \rightarrow \infty$?

5.9   Prove that:

$$C_v = Nk\beta^2 \frac{\partial^2 \ln z}{\partial \beta^2}$$

5.10   Prove that a shift in the zero of energy for a set of energy levels does not affect either C_v or S. (The effect on U will be to add a constant $N \Delta\varepsilon$.)

5.11   If the volume of a cell (that is, state) in phase space is h^3, estimate the number of such states for a He atom in a 1-cm^3 box having $\varepsilon < kT$ at 4 K.

5.12   Calculate the standard entropy of Ar at 298.15 K. This can be compared to the third-law value of 154.6 ± 0.8 J K^{-1}.

5.13   Use Eq. (5.60) to prove that when $T \gg \theta_v$, the vibrational energy will have the equipartition value, RT per mole.

5.14   Calculate the exact rotational partition function of [Eq. (5.69)] CO gas at $T = 13.9$ K and compare the result to the approximate value given by Eq. (5.70).

5.15   Calculate the rotational partition function of O_2 gas for $T = 20.8$ K; do this exactly [Eq. (5.69) with $J = 1, 3, 5, 7, \ldots$] and compare this result to the approximation of Eq. (5.70).

5.16   Calculate the fraction of molecules in each of the $J = 0, 3, 10$ rotational states of HCl at 304.76 K. If a computer or programmable calculator is available, do all J's and graph your results.

5.17   Derive the formula for the rotational contribution to the internal energy (per mole) (a) for linear molecules, (b) for nonlinear molecules.

5.18   For the rotation of a linear molecule, prove:

$$S_m^\theta(\text{rot}) = R\left(\ln \frac{T}{\sigma\theta} + 1\right)$$

5.19   The product of the moments of inertia of SF_4 is:

$$I_x I_y I_z = 6.721 \times 10^{-114} \text{ g}^3 \text{ cm}^6$$

(a) Calculate the rotational partition function at 298.15 K ($\sigma = 2$).
(b) Calculate $S(\text{rot})$ for this case.

5.20 List the symmetry numbers of the following molecules: $H\!-\!C\!\equiv\!N$, $S\!=\!C\!=\!S$ (both linear), $H\!-\!O\!-\!H$, $H\!-\!O\!-\!D$ (both bent), CH_3Cl, CH_2Cl_2.

5.21 Give symmetry numbers for the following substituted benzene molecules:

5.22 (a) Show that the rotational characteristic temperature has the value:

$$\theta_r = \frac{h^2}{8\pi^2 Ik} = \frac{4.0275 \times 10^{-39}}{I}$$

(with I in g cm^2).

(b) For a nonlinear molecule, three rotational constants can be defined, one for each moment of inertia. Show that the rotational partition function [Eq. (5.74)] can be written:

$$z_{\text{rot}} = \frac{\sqrt{\pi}}{\sigma} \frac{T^{3/2}}{(\theta_x \theta_y \theta_z)^{1/2}}$$

5.23 It was stated that the high-temperature limit for the rotational partition function was valid for all gases above their normal boiling points, except for H_2, which has a very small moment of inertia (large θ_r) and a very low boiling point. Methane also has a very small moment of inertia (the carbon is at the center of mass):

$$I = 5.313 \times 10^{-40} \text{ g cm}^2$$

and a low boiling point (111.7 K). Calculate the rotational characteristic temperature of methane to see if it is indeed an order of magnitude smaller than the boiling temperature, as required (use the results of problem 5.22).

5.24 Calculate the electronic partition function for a nitrogen atom at 10,000 K.

5.25 Atomic sodium has a doubly degenerate ground state, and no low-lying excited electronic states. Calculate its molar entropy at 1800 K.

5.26 Starting with Eq. (5.76) prove that the contribution to C_v from a low-lying electronic state (g_1, ε_1) is:

$$C_{vm}(\text{elec}) = \frac{Rg_1 g_0 x^2 e^x}{(g_0 e^x + g_1)^2}$$

(where $x = \varepsilon_1/kT$). Show that this function has a maximum.

5.27 Atomic chlorine has a ground electronic state degeneracy of 4 and a low excited state with $\varepsilon_1 = 881$ cm^{-1}, $g_1 = 2$. Calculate C_{vm} (total) at 300 K, 600 K, 1000 K (use result of problem 5.26).

5.28 Calculate the internal energy $U_m^\theta - U_0$ of CO at 500 K.

5.29 Calculate $H_m^\theta - H_0^\theta$ at 298.15 K for H_2O (ideal gas).

5.30 Calculate the standard entropy of N_2 at 298.15 K and 1000 K from spectroscopic data. (Compare Table 3.2.)

5.31 Calculate the total entropy of O_2 at 298.15 K and 1000 K from its spectroscopic constants. (Compare Table 3.2.)

5.32 In some texts, $C_{pm} = \frac{7}{2}R$ is stated as if it represented *truth*. For which of the following gases will this statement be reasonably accurate (5%) at 25°C?

$$HCl, \quad Br_2(gas), \quad H_2O(gas), \quad CH_4, \quad SiF_4$$

5.33 Calculate the heat capacity (at constant P) of Cl_2 at 298.15, 500, and 1000 K. What is the equipartition value?

5.34 Calculate C_{vm}^θ of HCN at 300 K.

5.35 Calculate C_{pm}^θ of SiF_4 at 300 K.

5.36 Calculate the free-energy functions (ϕ° and ϕ') for O_2 at 1000 K. (The JANAF tables give $\phi' = 220.77$ J.)

5.37 HF is a difficult substance to handle in the lab, so statistical calculations are a very attractive alternative to experimentation. For this molecule, calculate $S_m^\theta(298.15 \text{ K})$, $H_m^\theta(298.15) - H_0$ and ϕ' at 500 K.

5.38 Calculate the free-energy function (ϕ') of $H_2O(liq)$ at 100°C from its entropy (Table 3.2) and heat capacity ($C_{pm}^\theta = 75$ J K^{-1}, assumed independent of T).

5.39 Calculate the free-energy function $\phi'(T)$ for HI at 1000 K from the thermodynamic data below.

$$S_m^\theta(298.15) = 49.351 \text{ cal K}^{-1}$$

T	C_{pm} (cal K^{-1})	T	C_{pm} (cal K^{-1})
298.15	6.969	700	7.424
400	7.010	800	7.600
500	7.107	900	7.767
600	7.253	1000	7.920

5.40 If an atom or molecule has a low-lying excited electronic state with $\varepsilon_1 \ll kT$ at some temperature, show that the contribution of the electronic state to the free energy is $\ln (g_0 + g_1)$ and to the molar energy, $Lg_1\varepsilon_1/(g_0 + g_1)$. [Since $L\varepsilon_1 \ll RT$, this permits us to ignore the small energy and approximate the situation as a single state of degeneracy $g_0 + g_1$. In other words, very small splitting of ground state degeneracies can be ignored.]

5.41 The CH radical is not stable under ordinary conditions but will be an important factor in reactions at high temperatures. It has a low-lying excited state (Table 5.3), but at high temperature this will only add $R \ln 4$ to the free-energy function (ϕ°). Calculate ϕ' at 1000 K. You will need to add the electronic contribution of $L\varepsilon/2 = 107$ J/mol to $H_m(298) - H_0$. The value listed by the JANAF tables is 197.95 J K^{-1}. (Use $\theta_v = 3932$ K, $\theta_r = 20.79$ K.)

5.42 Gas-phase ionizations can be treated in the same fashion as other chemical reactions:

$$M \longrightarrow M^+ + e$$

To this end, one must know the thermodynamic properties of the product — an electron gas. Calculate $S_m^\theta(298.15)$ for a mole of electrons. This can be compared to an early estimate by Lewis and Randall (1922), 3.28 cal K^{-1}.

5.43 Starting with Eq. (5.82) and the appropriate partition function, show that the free-energy function of a monatomic species is:

$$\phi^\circ = R(2.5 \ln T + 1.5 \ln M - 3.665 + \ln z_{\text{elec}})$$

where M is the atomic weight in grams.

We must have energy!
    – T. S. Garp

6
Chemical Reactions

The chemical reaction is the central phenomenon of chemistry; if thermodynamics did nothing else, its contributions to our understanding of chemical reactions would of itself justify teaching the subject in a chemistry course. Now, the preliminaries are over and the main event begins.

Thermodynamics treats two important facets of chemical reactions: in which direction will the reaction go spontaneously, and how much heat will be released. But that's not the whole story. Thermodynamics deals only in potentialities; it tells you what will happen *if* it happens. If hydrogen and oxygen are mixed at room temperature and react, there is no question that they will spontaneously form water. But if no spark is applied and no catalyst is present, they could remain together for years without reacting. How do reactions occur? How fast? These are not thermodynamic questions; they not answered by and really are of no concern to thermodynamics. Kinetics, Chapter 10, will be needed to complete the picture.

Reactions may be classified as gas phase, solution, or heterogeneous (for example, the reaction of a gas with a solid). The solution process, even without reaction, involves complications and concepts with which we are not yet ready to deal. For that reason we shall exclude solutions for the present and, in this chapter, deal only with reactions involving ideal or near ideal gases and pure solids.

6.1 The Reaction Process

The symbol Δ in thermodynamics is used to indicate the difference of some property between a final and initial state. In Chapters 2 and 3 we discussed processes that changed the T, P, V state of pure homogeneous substances. In Chapter 4 we discussed processes that changed the phase of pure materials. In this chapter we discuss chemical reactions as a process in which the chemical identity of the material is altered with consequent changes in energy, entropy, and so on:

$$\boxed{\begin{array}{c} \text{Initial State:} \\ \text{Reactants, } T_1 \; P_1 \end{array}} \xrightarrow{\text{Reaction}} \boxed{\begin{array}{c} \text{Final State:} \\ \text{Products, } T_2 \; P_2 \end{array}}$$

Changes in state functions do not depend on the details of the process but only on the initial and final states; this great simplification permits us to deal with all sorts of reactions—explosions, burning of coal, corrosion of iron, and so on—with the same facility.

For any extensive property, V, U, H, S, and so on, we can define the change for reaction as:

$$\Delta(\text{property})_{\text{rxn}} = (\text{property of products}) - (\text{property of reactants})$$

For ideal or near ideal gases we can reasonably assume that the property of the substance in a mixture, either the reactant mixture or the product mixture, is the same as that of the pure substance at that temperature and partial pressure. For the hypothetical reaction:

$$a\text{A} + b\text{B} = c\text{C} + d\text{D}$$

the entropy change would be:

$$\Delta S_{rxn} = cS_{m,C} + dS_{m,D} - aS_{m,A} - bS_{m,B}$$

where $S_{m,A}$ is the molar entropy of pure A, and so on.

To simplify the notation we shall define ν_i as the stoichiometric coefficient of substance i such that it is positive for products and negative for reactants. In this notation, the entropy change of a reaction is:

$$\Delta S_{rxn} = \sum_i \nu_i S_{m,i} \qquad \textbf{(6.1a)}$$

This can be done for any extensive property. For example:

$$\Delta V_{rxn} = \sum_i \nu_i V_{m,i} \qquad \textbf{(6.1b)}$$

$$\Delta H_{rxn} = \sum_i \nu_i H_{m,i} \qquad \textbf{(6.1c)}$$

$$\Delta G_{rxn} = \sum_i \nu_i \mu_i \qquad \textbf{(6.1d)}$$

Example: Calculate ΔS^{θ} for the reaction $H_2(g) + Cl_2(g) = 2\,HCl(g)$ from data in Table 3.2 (at 25°C).

$$\Delta S_{rxn}^{\theta} = 2(186.77) - 130.58 - 222.97 = 19.99 \text{ J K}^{-1}$$

Note that the reaction:

$$\tfrac{1}{2}H_2 + \tfrac{1}{2}Cl_2 = HCl$$

has $\Delta S_{rxn}^{\theta} = 9.995$ J K^{-1}, half as much. The reaction and amount must be specified to make Δ(rxn) meaningful. ∎

The symbol Δ(rxn) generally refers to an *isothermal* reaction; the products have the same temperature as did the reactants. Standard state means, in effect, 1 atm pressure. In the example above the process is:

$$\left.\begin{array}{l} 1 \text{ mole } H_2, \text{ ideal gas, 1 atm, 25°C} \\ 1 \text{ mole } Cl_2, \text{ ideal gas, 1 atm, 25°C} \end{array}\right\} 2 \text{ moles HCl, ideal gas, 1 atm, 25°C}$$

For this process $\Delta S_{rxn}^{\theta} = 19.99$ J K^{-1}. This is a very hypothetical process; how it is connected to what really happens is the mystery that thermodynamics will resolve.

6.2 Heats of Reaction

A chemical reaction in which heat is released is called *exothermic*, while one that requires heat to be supplied is *endothermic*. In this section we shall discuss only reactions that go spontaneously to completion; such reactions are usually exothermic. Such cases are common, but, as we shall discover later in this chapter, there are reactions which reach an equilibrium with significant amounts of reactants remaining, and there is no necessary connection between the direction of a spontaneous reaction and exothermicity.

By the first law of thermodynamics, the heat of any process at constant volume must be equal to the change in the internal energy. For a reaction carried out at constant volume:

$$q_V = \Delta U_{rxn} \tag{6.2}$$

If the reaction is carried out at constant pressure, and there is a volume difference between the products and reactants, the system will have to do work:

$$q_P = \Delta U_{rxn} - w$$
$$= \Delta U_{rxn} + P \, \Delta V_{rxn}$$

This is just the enthalpy difference:

$$q_P = (U + PV)_{products} - (U + PV)_{reactants}$$
$$= H_{products} - H_{reactants}$$
$$q_P = \Delta H_{rxn} \tag{6.3}$$

Remembering our convention that heat added to the "system" is positive, we conclude that an exothermic reaction is characterized by a decrease in H or U; that is, the ΔH or ΔU of the reaction would be negative.

Calorimetry

The details of how heats of reaction are measured are beyond the scope of this book, but a general discussion will illuminate what the term means.

The reaction is carried out in thermal contact with a water bath (Figure 6.1) so that the heat released will raise the temperature of the bath. The whole system, reaction plus bath, is adiabatic, so:

$$q_{rxn} = -q_{bath}$$

The heat delivered to the bath can be calculated from the temperature rise if the heat capacity of the bath (C) is known. This quantity (C) can be determined by heating the bath electrically; if a current I is run through a heater of resistance R for a time t:

$$q_{elec} = I^2 Rt = C \, \Delta T$$

Example: A 100-watt electrical heater raises the temperature of a water bath by 1.362°C in 10 minutes. A reaction in this bath gives $\Delta T = 0.754$ K. Calculate the heat of reaction.

$$C = \frac{(100 \text{ watts}) (10 \text{ min}) (60 \text{ sec/min})}{1.362 \text{ K}} = 44.05 \text{ kJ/K}$$

$$q_{rxn} = -(44.05 \text{ kJ/K}) (0.754 \text{ K}) = -33.2 \text{ kJ} \qquad \blacksquare$$

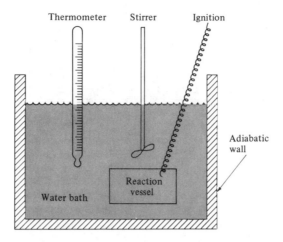

Figure 6.1 Calorimeter (schematic). The heat evolved by the reaction is measured by the temperature rise of the water bath. To relate the temperature rise to the heat evolved, the bath (with all contents) is calibrated by electrical heating.

Heats of reaction are generally reported for the standard state — ΔU^{θ}, ΔH^{θ}. Since H and U do not change a lot with pressure (Chapter 2), corrections to the standard state are usually small.

Heats of Formation

Since the isothermal heat of reaction at constant P or V is related to state variables U [Eq. (6.2)] or H [Eq. (6.3)], it is itself a state function. Hess's law of heat summation states that the heat of reaction is independent of the particular method by which the reaction takes place. This means that we can calculate the heat of a reaction that, perhaps, we cannot conveniently carry out, from the heats of any series of reactions that goes to the same place.

Example: The hydrogenation of acetylene to ethylene:

$$C_2H_2 + H_2 = C_2H_4$$

can be done directly, but not easily. The easiest thing to measure for hydrocarbons is the heat of combustion. If we do this for each of these gases, we get the following (note the sign reversal for the second reaction):

$$C_2H_2 + \tfrac{5}{2} O_2 \longrightarrow 2\,CO_2 + H_2O(l), \qquad \Delta H = -1299.63 \text{ kJ}$$

$$2\,CO_2 + 2\,H_2O(liq) \longleftarrow C_2H_4 + 3\,O_2, \qquad \Delta H = +1410.97 \text{ kJ}$$

$$H_2 + \tfrac{1}{2} O_2 \longrightarrow H_2O(liq), \qquad \Delta H = -285.830 \text{ kJ}$$

Table 6.1 Thermodynamic properties at 298.15 K

Substance	ΔH_f^{θ} (kJ)	ΔG_f^{θ} (kJ)	C_{pm} (J/K)	Substance	ΔH_f^{θ} (kJ)	ΔG_f^{θ} (kJ)	C_{pm} (J/K)
$H_2O(g)$	−241.826	−228.593	33.58	$N_2O_4(g)$	9.661	98.286	79.1
$H_2O(liq)$	−285.830	−237.191	83.66	$NH_3(g)$	−46.19	−16.64	35.66
$CO_2(g)$	−393.513	−394.383	37.13	$SO_2(g)$	−296.9	−300.4	39.8
$CO(g)$	−110.523	−137.268	29.14	$SO_3(g)$	−395.2	−370.4	50.6
Graphite	0	0	8.64	$H_2S(g)$	−20.15	−33.02	34.0
Diamond	1.896	2.866	6.06	$HF(g)$	−272.5	−274.6	29.14
$COCl_2(g)$	−223.0	−210.5	—	$HCl(g)$	−92.312	−95.265	29.6
$CH_4(g)$	−74.848	−50.794	35.71	$HBr(g)$	−36.2	−53.22	29.1
$CH_2O(g)$	−116	−110	35.4	$Br(g)$	111.8	82.38	20.79
$CH_3OH(g)$	−201.3	−161.9	—	$Br_2(g)$	30.7	3.14	36.0
$CH_3OH(liq)$	−238.64	−166.3	81.6	$HI(g)$	25.9	1.3	29.1
$C_2H_2(g)$	226.75	209.20	43.93	$PbCl_2(s)$	−359.2	−314.0	—
$C_2H_4(g)$	52.283	68.124	43.56	$PbSO_4(s)$	−918.34	−811.24	—
$C_2H_6(g)$	−84.667	−32.89	52.66	$Hg_2Cl_2(s)$	−264.9	−210.66	—
$C_2H_4O(g)$[a]	−51.00	−11.7	48.16	$Hg_2Br_2(s)$	−206.8	−178.72	—
$C_3H_8(g)$	−103.85	−23.49	73.09	$HgO(s, red)$	−90.71	−58.534	—
$C_3H_6(g)$	20.41	62.72	—	$AgCl(s)$	−127.04	−109.72	—
$C_3H_4(g)$[b]	185.43	193.77	—	$AgBr(s)$	−99.50	−93.68	—
$C_6H_6(g)$[c]	82.927	129.67	81.67	$CaO(s)$[d]	−635.5	−604.2	—
$C_6H_6(liq)$[c]	49.04	124.5	—	$CaCO_3$[d]	−1206.9	−1128.8	—
$NO(g)$	90.374	86.688	29.86	$CaCO_3$[e]	−1207.0	−1127.7	—
$NO_2(g)$	33.85	51.840	37.9	$NaCl(s)$	−411.00	−384.03	—
$N_2O(g)$	81.55	103.6	38.7	$KCl(s)$	−435.868	−408.32	—

[a] Ethylene oxide.
[b] Methyl acetylene.
[c] Benzene.
[d] Calcite.
[e] Aragonite.

Source: Most data are from ref. 3, converted by 4.184 J/cal.

The reactions and heats are added to give:

$$C_2H_2 + H_2 = C_2H_4, \qquad \Delta H_{rxn} = -174.49 \text{ kJ}$$

■

Heats of reaction are, of course, related to the molar enthalpies:

$$\Delta H_{rxn} = \sum \nu_i H_{m,i}$$

Only relative enthalpies can be measured, and only some of them, so this is not a practical equation. But, since all possible compounds can, in principle, be made from the same elements, it should be possible to define a set of reactions, N reactions for N compounds, from which the heat of any conceivable reaction among the N compounds could be calculated. The way this is done is through the *heats of formation*.

The formation reaction of a compound is the reaction by which it is formed from its elements in their standard state. The elements must be in their most stable form at that temperature; this means, for example, at 25°C, $H_2(g)$, not H; $Br_2(liq)$, not gas; $I_2(s)$, not vapor; C(graphite), not diamond; S(rhombic), not monoclinic. The formation reaction need not necessarily be practical, since Hess's law can always be used.

Example: The following are formation reactions for the compounds on the right-hand side:

$$Na(s) + \tfrac{1}{2} Cl_2(g) = NaCl(s)$$

$$K(s) + \tfrac{1}{2} Cl_2(g) + 2 O_2(g) = KClO_4(s)$$

$$H_2(g) + S(s,\text{rhombic}) + 2 O_2(g) = H_2SO_4(liq)$$

$$2C(s,\text{graphite}) + 3 H_2(g) = C_2H_6(g)$$

■

The standard heats of formation, ΔH_f^θ, for some compounds at 25°C are listed in Table 6.1. (Heats of formation are always for one mole of the substance.) The definition of ΔH(rxn), Eq. (6.1c), is not a practical equation, since the absolute molar enthalpies are not measurable. With the definition of the formation reaction, it can be cast into a practical form:

$$\Delta H_{rxn} = \sum_i \nu_i \Delta H_f^\theta(i) \qquad (6.4)$$

Since the reactants and products all could be made from the same elements, the molar enthalpies of the elements will drop out of Eq. (6.4). Note that ΔH_f^θ of the elements is zero at all temperatures *by definition*.

Example: Heats of formation of hydrocarbons are usually determined from the heat of combustion, ΔH_c^θ. For propane (Table 6.2):

$$C_3H_8(g) + 5 O_2(g) = 3 CO_2(g) + 4 H_2O(liq), \qquad \Delta H_c^\theta = -2220.0 \text{ kJ}$$

The heats of formation of CO_2 and H_2O, needed in such calculations, have been measured very accurately (see Table 6.1). For this reaction:

$$\Delta H_c^\theta = 4 \Delta H_f^\theta(H_2O,l) + 3 \Delta H_f(CO_2,g) - \Delta H_f^\theta(C_3H_8)$$

Table 6.2   Heats of combustion at 298.15 K[a]

Substance	Formula	ΔH_c^{θ} (kJ/mol)
Graphite	C	−393.51
Diamond	C	−395.39
Methane	CH_4	−890.4
Ethane	C_2H_6	−1559.8
Propane	C_3H_8	−2220.0
n-Butane	C_4H_{10}	−2878.5
n-Octane	C_8H_{18}	−5452
Methanol	CH_3OH	−726.1
Ethanol	C_2H_5OH	−1367
Benzoic acid	C_6H_5COOH	−3226.7
Sucrose	$C_{12}H_{22}O_{11}$	−5643.8
Thiophene	C_4H_4S	−2805

[a]For one mole of substance with $CO_2(g)$, $SO_2(g)$ and $H_2O(liq)$ as products.

(*Note:* The heat of formation of O_2 is zero by definition.) With data from Table 6.1:

$$\Delta H_c^{\theta} = 4(-285.830) + 3(-393.513) - \Delta H_f^{\theta}(C_3H_8) = -2220.0 \text{ kJ}$$

This can be solved to give:

$$\Delta H_f^{\theta}(C_3H_8) = -103.8 \text{ kJ} \qquad \blacksquare$$

Temperature Dependence of ΔH

Tables such as Table 6.1 give only heats of reaction at the temperature indicated, in this case 25°C. While heats of formation at other temperatures can be found for some substances (for example, from the JANAF tables), it is often necessary to calculate them from other data.

The definition of the ΔH of reaction, Eq. (6.1c), applies at any temperature. For two temperatures:

$$\Delta H_{rxn}(T_2) = \Delta H_{rxn}(T_1) + \sum_i \nu_i [H_{m,i}(T_2) - H_{m,i}(T_1)] \qquad \textbf{(6.5)}$$

For Eq. (6.5), the quantities $H(T_2) - H(T_1)$ could be calculated from spectroscopic data, as explained in Chapter 5; however, this procedure must be practical for every reactant and product.

A convenient method for calculating the change of ΔH with T is obtained by taking the derivative of Eq. (6.1c) with respect to T:

$$\left(\frac{\partial \Delta H_{rxn}}{\partial T}\right)_P = \sum_i \nu_i \left(\frac{\partial H_{m,i}}{\partial T}\right)_P$$

The derivatives in the sum are the heat capacities, so, by analogy with Eq. (6.1), we define:

$$\Delta C_p = \sum_i \nu_i C_{pm}(i) \qquad \textbf{(6.6)}$$

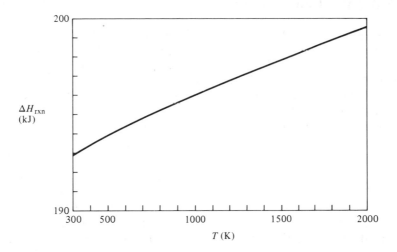

Figure 6.2 Enthalpy change (per mole) for dissociation of bromine gas. Note that this quantity changes by only a few percent over a thousand degrees. The frequently used approximation that enthalpies of reaction are constant is often valid over limited temperature ranges. (Compare Figures 6.5 and 6.6.)

This gives:

$$\left(\frac{\partial \Delta H_{\text{rxn}}}{\partial T}\right)_P = \Delta C_p \tag{6.7}$$

or:

$$\Delta H_{\text{rxn}}(T_2) = \Delta H_{\text{rxn}}(T_1) + \int_{T_1}^{T_2} \Delta C_p \, dT \tag{6.8}$$

The evaluation of integrals such as this was discussed in Chapter 2. For small temperature ranges, it may be sufficiently accurate to assume that the heat capacities are constant. (As an example, the enthalpy change for the bromine dissociation reaction is shown in Figure 6.2.)

Example: Calculate the heat of reaction at 100°C for:

$$\tfrac{1}{2} \text{H}_2 + \tfrac{1}{2} \text{Cl}_2 = \text{HCl(g)}$$

This is the formation reaction, so Table 6.1 gives directly $\Delta H(298) = -92\,312$ J. Using heat capacities (at 15°C) from Table 2.1:

$$\Delta C_p = (29.59) - 0.5(28.58 + 34.13) = -1.765$$
$$\Delta H(T_2) = \Delta H(T_1) + \Delta C_p(T_2 - T_1)$$
$$\Delta H(373) = -92\,312 - 1.756(373 - 298) = -92\,444 \text{ J}$$

The use of 15°C heat capacities is justified only as a convenience. For greater accuracy (or for a larger change in T), the temperature-dependent heat capacities of Table 2.2 or 2.3 should be used. ∎

6.3   Adiabatic Flame Temperature

The isothermal reaction is not the common experience. If methane is burned, the heat will not escape instantly and the product gases will be very hot. (Indeed, for kinetic reasons, the reaction may not even occur at 25°C.) In the short term — for example, at the burning edge of a flame — the reaction is essentially adiabatic; all the "heat" released by the reaction is used to raise the temperature of the product gases. The final temperature of an adiabatic reaction is called the *adiabatic flame temperature*. Even if a reaction is not adiabatic, the concept is useful, since the adiabatic temperature represents the maximum temperature the reaction can achieve.

Adiabatic flames are used to study combustion reactions of fuels, propellants, and the like. Such experiments are usually done in a burner not unlike the Bunsen or Meker burner commonly used in the chemistry laboratory. The temperature and chemical constituents of the flame can be measured spectroscopically (see ref. 1); some idea of how temperature is measured spectroscopically will be obtained from Chapter 13.

The calculation of the flame temperature from thermodynamics is diagrammed in Figure 6.3. Assuming constant pressure, the process can be analyzed in two steps:

1. Isothermal reaction at T_1:

$$q_1 = \Delta H_{rxn}$$

2. Heating of products:

$$q_2 = \int_{T_1}^{T_2} C_p^{prod} \, dT$$

The net process is adiabatic, so $q_1 + q_2 = 0$, and:

$$-\Delta H_{rxn} = \int_{T_1}^{T_2} C_p^{prod} \, dT \tag{6.9}$$

By "products" we mean everything in the product stream, including, possibly, reactants (if one was in excess or the reaction was not complete) and inert gases (for example, nitrogen, if a combustion is done with air); T_1 would typically be 298 K.

Because the final temperature will be quite high (Table 6.3), the heat capacities of the products must be known at high temperatures. Since the common products include only H_2O, CO_2, CO, and N_2, the heat capacities of Table 2.2 are sufficient; however, since T_2 is unknown, the evaluation of Eq. 6.9 can be cumbersome. It is easiest to define $\overline{C}_p$ as the mean heat capacity of the products:

$$\overline{C}_p = \tfrac{1}{2}[C_p^{prod}(T_2) + C_p^{prod}(T_1)] \tag{6.10}$$

Then, Eq. (6.9) becomes:

$$\Delta T = -\frac{\Delta H_{rxn}}{\overline{C}_p} \tag{6.11}$$

Equations (6.10) and (6.11) can be solved by a short series of successive approximations. It may be more convenient to approximate $\overline{C}_p$ as the heat capacity of the products at the mean temperature $\overline{T}$, particularly since the heat-capacity formulas of Table 2.2 are not accurate for temperatures above 2000 K.

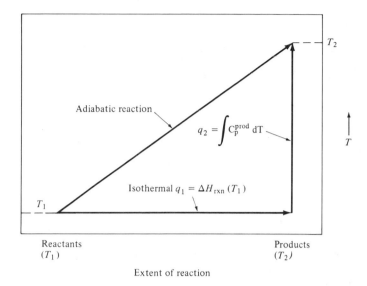

Figure 6.3 Adiabatic reaction. In an adiabatic reaction, no heat is permitted to escape from the reaction vessel; consequently, the temperature rises. Because enthalpies are state functions, an adiabatic, constant-pressure reaction ($\Delta H = 0$) can be analyzed into two steps: (1) an isothermal, constant-pressure reaction, and (2) heating of the reaction products at constant pressure up to the final temperature (T_2) of the adiabatic reaction.

Table 6.3   Adiabatic temperatures of various combustion reactions

Fuel	Reaction	T (K)[a]	q (J/g)[b]
Hydrogen-oxygen	$2\,H_2 + O_2 \longrightarrow 2\,H_2O$	3100	$-24\,000$
Methane-oxygen	$CH_4 + 2\,O_2 \longrightarrow CO_2 + 2\,H_2O$	3000	$-10\,000$
Methane-air	$CH_4 + air \longrightarrow CO_2 + 2\,H_2O(+8\,N_2)$	2200	$-2\,700$
Octane-oxygen	$2\,C_8H_{10} + 25\,O_2 \longrightarrow 16\,CO_2 + 18\,H_2O$	3100	$-9\,900$
Acetylene-oxygen	$2\,C_2H_2 + 5\,O_2 \longrightarrow 4\,CO_2 + 2\,H_2O$	3300	$-18\,800$
Cyanogen-oxygen	$C_2N_2 + O_2 \longrightarrow 2\,CO + N_2$	4800	$-6\,300$
Producer gas-air	$2\,CO + 4\,H_2 + air \longrightarrow$		
	$2\,CO_2 + 4\,H_2O\,(+12\,N_2)$	2400	$-4\,100$
Methylhydrazine-nitrogen tetroxide	$CH_4N_2 + N_2O_4 \longrightarrow 2\,H_2O + CO_2 + 2\,N_2$	3000	$-7\,500$

[a]Flame temperature at low pressure. At higher pressures the dissociation of product molecules would be suppressed, and consequently the flame temperature would be higher.

[b]Heat released per gram of reactants including fuel, oxidizer, and (in the case of air) inert components, after the products have cooled to the original temperature.

Source: William C. Gardiner, Jr., *Scientific American,* February 1982.

Example: Estimate the adiabatic flame temperature of methane burning in air. (We shall assume air to be 80% nitrogen, so $N_2 : O_2 = 4 : 1$ in the reactant mixture.) Assuming a stoichiometric amount of oxygen:

$$CH_4 + 2\,O_2 + 8\,N_2 = CO_2 + 2\,H_2O + 8\,N_2$$

From Table 2.2 (units: J/K):

$$C_p(CO_2) = \left(44.23 + 8.79 \times 10^{-3}T - \frac{8.62 \times 10^5}{T^2}\right)$$

$$C_p(H_2O) = 2(30.54 + 10.29 \times 10^{-3}T)$$

$$C_p(N_2) = 8\left(28.58 + 3.77 \times 10^{-3}T - \frac{0.5 \times 10^5}{T^2}\right)$$

These are added to get the heat capacity of the product mixture:

$$C_p^{prod} = 333.95 + 0.05953T - \frac{1.262 \times 10^6}{T^2}$$

As a first approximation (no more than a guess) we shall assume a mean $T = 1000$ K and, from that, estimate $\overline{C}_p = 392.22$ J/K.

The heat of combustion of methane is given by Table 6.2 as -890.4 kJ (at 298 K), but this is the heat with H_2O(liq) as the product, and in the flame the product will surely be H_2O(g). The heat of vaporization of water is given by Table 4.2 as 40.66 kJ/mol; Hess's law can be used to calculate the heat of combustion of methane to H_2O(g):

$$CH_4 + 2\,O_2 = CO_2 + 2\,H_2O(liq), \qquad \Delta H = -890.4 \text{ kJ}$$

$$2\,H_2O(liq) = 2\,H_2O(g), \qquad \Delta H = 2(40.66) \text{ kJ}$$

The reactions and heats are added to give:

$$CH_4 + 2\,O_2 = CO_2 + 2\,H_2O(g), \qquad \Delta H = -809.1 \text{ kJ}$$

Now the temperature change can be estimated from Eq. (6.11):

$$\Delta T = \frac{809.1 \times 10^3}{392.22} = 2063 \text{ K}, \qquad T_2 = 298 + 2063 = 2361 \text{ K}$$

With this estimate the mean temperature will be:

$$\overline{T} = \frac{2361 + 298}{2} = 1329.5$$

With this temperature, $\overline{C}_p$ is calculated to be 412.38 J/K. The reader should demonstrate that the next round of approximation gives $T_2 = 2260$ K. Because of other approximations involved (see below), it is not worthwhile to do another round in the successive approximation. ∎

From Table 6.3 it can be seen that the observed flame temperature of methane in air is 2200 K. The method used above will generally overestimate the temperature, because it fails to take into account the dissociation of the products at such high temperatures. Some of the dissociation reactions that are important include:

$$H_2O = \tfrac{1}{2}\,H_2 + OH$$

$$CO_2 = CO + \tfrac{1}{2}\,O_2$$

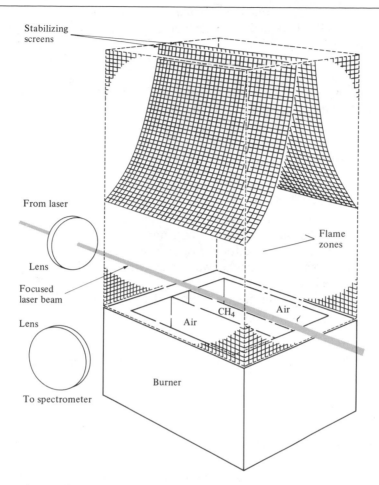

Figure 6.4 Chemical analysis by laser-induced fluorescence. The chemical species present in a flame can be identified by laser-induced fluorescence. A gaseous fuel flows through the center slot, and oxidant (air or oxygen in most cases) flows through slots on either side of the fuel. The stabilizing screen prevents turbulence in the flame. A laser beam is focused on the reaction zone where the fuel and oxidant streams meet, and scattered light, which includes light absorbed and re-emitted by atoms or molecules in the flame (fluorescence), is detected by a spectrometer at right angles to the direction of the laser beam. (From W. C. Gardiner, Jr., *Scientific American*, February 1982)

and the atomic dissociations such as:

$$H_2 = 2\,H, \qquad O_2 = 2\,O, \qquad N_2 = 2\,N$$

All of these are endothermic reactions, and their heat must be deducted from the heat $(-\Delta H_{rxn})$ available to raise the temperature. We shall discuss such equilibria later; for now we will only note that, if the degree of dissociation is α, then an amount of heat equal to α times the heat of dissociation must be subtracted from the heat of combustion. Since α also depends on T, this makes accurate calculation of flame temperatures very complicated.

The study of flame reactions is also of interest from the point of view of kinetics. Reference 1 provides an interesting discussion of this fascinating subject. The equipment used for analysis of chemical species present in a flame by means of laser-induced fluorescence is illustrated in Figure 6.4.

6.4   Reversible Reactions

Why do some chemicals react, while others do not? What determines the spontaneous direction of a reaction? Why are certain combinations of elements preferred? What is the source of *chemical affinity?* These were major concerns for nineteenth-century chemists; in fact, early studies of heats of reaction were motivated by the belief that this energy was the driving force for chemical reaction. The discovery of reversible reactions, reactions that could go either way depending on conditions, showed that this idea was an oversimplification.

The law of mass action, formulated in its modern form by Van't Hoff (ca. 1877), states that a reaction — for example:

$$a\mathrm{A} + b\mathrm{B} = c\mathrm{C} + d\mathrm{D}$$

will go spontaneously in whichever direction which will create a ratio of concentrations:

$$K = \frac{[\mathrm{C}]^c[\mathrm{D}]^d}{[\mathrm{A}]^a[\mathrm{B}]^b} \qquad (6.12)$$

that is constant at any given temperature.

Example: One of the earliest investigations of chemical equilibrium was that of Berthelot and St. Gilles [*Ann. Chim. Phys.,* 68, 225 (1863)] on the esterification reaction of ethyl alcohol and acetic acid:

$$C_2H_5OH + CH_3COOH = CH_3COOC_2H_5 + H_2O$$

When 1 mole of alcohol was mixed with 2 moles of acid, 0.86 moles of ester (and the same amount of water) were formed. The mass-action equilibrium constant is then:

$$K = \frac{[\text{ester}][H_2O]}{[\text{alcohol}][\text{acid}]} = \frac{(0.86)^2}{(1-0.86)(2-0.86)} = 4.6$$

When the proportions were reversed, 2 moles of alcohol to 1 mole of acid, 0.83 moles of ester were formed, and:

$$K = \frac{(0.83)^2}{(2-0.83)(1-0.83)} = 3.5$$

With one mole each of reactants, 0.665 moles of ester were formed, and:

$$K = \frac{(0.665)^2}{(1-0.665)^2} = 3.9$$

Claude Louis Berthollet (1748–1822)

Napoleon's military expedition to Egypt in 1799 was accompanied by an impressive corps of scholars. Among the important discoveries made was the Rosetta Stone, a tablet with a parallel text in Greek, ancient Egyptian hieroglyphics, and demotic characters; this was to provide the key for the translation of ancient Egyptian manuscripts.

Another important discovery was made by the chemist Berthollet on the shores of Lake Natron. He found there that sodium carbonate had been formed by reaction of the excess sodium chloride in the area with the calcium carbonate in the limestone. This reaction was known in the laboratory, but in the opposite direction: calcium chloride reacts with sodium carbonate to form calcium carbonate as an insoluble precipitate. Thus Berthollet made the first recorded observation of a reversible chemical reaction, and he formulated an early version of the principle of mass action and discussed the effect of temperature and concentration on chemical affinity. Unfortunately, Berthollet's opposition to Dalton's law of definite proportions brought discredit to his theories; he based his opposition on faulty data and on analyses of metal oxides having varying degrees of hydration. He never denied that the law of definite proportions worked, but he felt that there were cases where the amounts of reactants could affect the composition of the compound. Like so many other scientific discoveries, the law of mass action had to be rediscovered—a half century later.

Nonstoichiometric compounds that do not obey Dalton's law are now known and are called berthollides, *as opposed to the more common* daltonides, *which obey the law of definite proportions.*

These ratios are reasonably constant, considering the fourfold change in the reactant ratio; for reasons that we shall discuss in Chapter 7, this ratio cannot be expected to be precisely constant. ∎

Thermodynamics can greatly clarify our discussion of equilibrium. In Chapter 3 it was shown that the requirement for a spontaneous change at constant temperature and pressure is:

$$(\Delta G)_{T,P} \leq 0$$

If the process is a reaction:

$$\Delta G_{\text{rxn}} = G(\text{products}) - G(\text{reactants})$$

$$= \sum_i \nu_i \mu_i$$

the concentrations of the reactants and products will change in the direction that decreases the difference between G(products) and G(reactants); at equilibrium, $\Delta G_{rxn} = 0$. (As in Chapter 4, we use the chemical potential μ in place of the molar free energy G_m. For ideal or near ideal gases, the distinction is not significant; in solutions, as we shall discover in Chapter 7, the distinction is quite major.)

In Chapter 4 the *activity* was defined by:

$$\mu_i(T, P) = \mu_i^\theta(T) + RT \ln a_i \tag{6.13}$$

In terms of activity, the ΔG of reaction becomes:

$$\Delta G_{rxn} = \sum_i \nu_i \mu_i^\theta + RT \sum_i \nu_i \ln a_i$$

We shall denote the first term on the right-hand side as:

$$\Delta G_{rxn}^\theta = \sum_i \nu_i \mu_i^\theta \tag{6.14}$$

We note that this quantity is a function of temperature, but is independent of pressure/concentration. Equation (6.14) gives the free-energy change for the (possibly hypothetical) reaction:

The second term will be used to define the activity quotient Q:

$$\sum_i \nu_i \ln a_i = \ln Q, \qquad Q = \prod_i a_i^{\nu_i} \tag{6.15}$$

$$\Delta G_{rxn} = \Delta G_{rxn}^\theta + RT \ln Q \tag{6.16}$$

At equilibrium, $\Delta G_{rxn} = 0$ and:

$$\Delta G_{rxn}^\theta = -RT \ln K_a \tag{6.17}$$

K_a is the thermodynamic equilibrium constant; it has the same definition as Q [Eq. (6.15)] except that the activities have their equilibrium values. K_a, like ΔG^θ, depends on temperature but is independent of pressure. We will be well into Chapter 8 before we return to this point and discuss the measurement of activities and ΔG_{rxn} in nonequilibrium situations. In the meantime it is important to realize that *it is ΔG, not ΔG^θ, that determines the spontaneous direction of a reaction*.

For the general reaction:

$$a\text{A} + b\text{B} = c\text{C} + d\text{D}$$

$$K_a = \frac{a_\text{C}^c a_\text{D}^d}{a_\text{A}^a a_\text{B}^b} \tag{6.18}$$

Activities were briefly discussed in Chapter 4; they will be discussed in greater detail in Chapters 7 and 8. All we need for the moment are the results of Chapter 4 for low pressures:

$$a_i(\text{pure solid}) = 1, \qquad a_i(\text{ideal gas}) = \frac{P_i}{P^\theta} \tag{6.19}$$

(P_i is the partial pressure of gas i.) Modifications required at higher pressures (roughly $P > 10$ atm) will be discussed later in this chapter.

If the activities of Eq. (6.19) are used in the definition of K_a, the result is an equilibrium constant called K_p. If all the components of the general reaction [Eq. (6.18)] are gases, the result is:

$$K_p = \frac{P_C^c P_D^d}{P_A^a P_B^b} (P^\theta)^{-\Delta \nu_g} \tag{6.20}$$

where the difference between the stoichiometric coefficients of the product gases and the reactant gases has been denoted:

$$\Delta \nu_g = c + d - a - b$$

For heterogeneous reactions involving pure solids, K_p is derived from K_a by setting the activities of the pure solids equal to 1.

At low to moderate pressures, where the ideal gas law is reasonably valid, K_p is independent of the total pressure and can be used in place of K_a for equations such as Eq. (6.17).

In many texts, K_p is defined with units—that is, as in Eq. (6.20) but without the P^θ term. Such equilibrium constants cannot be used directly in the thermodynamic equations unless the pressures are in atmospheres (in which case $P^\theta = 1$). In fact, for the remainder of this chapter we shall omit the P^θ part of Eq. (6.20) with the understanding that the partial pressures must be in atmospheres.

Example: The K_p's of the following reactions are (P in atm):

$$\tfrac{1}{2} N_2(g) + \tfrac{3}{2} H_2(g) = NH_3(g), \qquad\qquad K_p = \frac{P_{NH_3}}{P_{N_2}^{1/2} P_{H_2}^{3/2}}$$

$$N_2(g) + 3 H_2(g) = 2 NH_3(g), \qquad\qquad K_p = \frac{P_{NH_3}^2}{P_{N_2} P_{H_2}^3}$$

(Note that this is the square of the K_p of the previous reaction.)

$$2 NO(g) + O_2(g) = 2 NO_2(g), \qquad\qquad K_p = \frac{P_{NO_2}^2}{P_{NO}^2 P_{O_2}}$$

$$NO(g) + \tfrac{1}{2} O_2(g) = NO_2(g), \qquad\qquad K_p = \frac{P_{NO_2}}{P_{NO} P_{O_2}^{1/2}}$$

(Note that this is the square root of the K_p of the previous reaction.)

$$Fe_2O_3(s) + CO(g) = CO_2(g) + 2 FeO(s), \qquad K_p = P_{CO_2}/P_{CO}$$

$$Na_2SO_4 \cdot 10H_2O(s) = Na_2SO_4(s) + 10 H_2O(g), \qquad K_p = P_{H_2O}^{10} \qquad\blacksquare$$

Example: The reaction:

$$\tfrac{3}{2}\,H_2(g) + \tfrac{1}{2}\,N_2(g) = NH_3(g)$$

was carried out at $T = 620$ K and $P = 10$ atm. With an initial mixture $H_2 : N_2 = 3:1$, the mixture at equilibrium was 7.35% ammonia. From Dalton's law:

$$P_{NH_3} = 0.0735(10) = 0.735 \text{ atm}$$

The remaining pressure must be three-fourths hydrogen and one-fourth nitrogen:

$$P_{H_2} = 0.75(10 - 0.735) = 6.949, \qquad P_{N_2} = 0.25(10 - 0.735) = 2.316 \text{ atm}$$

$$K_p = \frac{(0.735)}{(2.316)^{1/2}(6.949)^{3/2}} = 0.0264$$

Note that the sum of the partial pressures $(0.735 + 6.949 + 2.316)$ is equal to the total pressure (10 atm) as required; this is a useful cross-check.   ∎

Exercise: The same reaction was carried out with a total $P = 50$ atm, and the equilibrium mixture was 25.11% ammonia. Calculate K_p. [*Answer:* $K_p = 0.0276$; this value differs slightly from that at 10 atm because of the failure of the ideal gas approximation.]   ∎

Example: The total pressure of the reaction:

$$NH_4HSe(s) = NH_3(g) + H_2Se(g)$$

is 23.1 torr at 30.1°C. Calculate ΔG^θ. There are equal moles of each gas, so their partial pressures are each equal to one-half the total pressure:

$$P_{NH_3} = P_{H_2Se} = 0.5\left(\frac{23.1}{760}\right) = 1.52 \times 10^{-2} \text{ atm}$$

$$K_p = P_{NH_3} P_{H_2Se} = (1.52 \times 10^{-2})^2 = 2.31 \times 10^{-4}$$

$$\Delta G^\theta_{rxn} = -(8.3143)(303.2)\ln(2.31 \times 10^{-4}) = 21,108 \text{ J}$$ ∎

6.5   Calculation of Equilibrium Constants

Free Energies of Formation

The free energy is, like the enthalpy, a state variable, so standard free energies of reaction will be additive in the same manner as are enthalpies; this is just a simple extension of Hess's law (which historically was formulated for heats of reaction) to other state functions. It is therefore possible to make a tabulation of standard free energies of formation (ΔG_f^θ) for each compound and, from these, to calculate the equilibrium constant for any reaction among those compounds:

$$\Delta G^\theta_{rxn} = \sum_i \nu_i \Delta G_f^\theta(i) = -RT \ln K_a \qquad \textbf{(6.21)}$$

A number of standard free energies of formation can be found in Table 6.1 (for 25°C).

Standard free energies of formation may, in some cases, be calculated directly from the equilibrium constant of the formation reaction, provided this equilibrium constant is directly measurable:

$$\Delta G_f^\theta = -RT \ln K_a(\text{formation}) \tag{6.22}$$

Otherwise, free energies of formation are calculated indirectly using Hess's law in a manner exactly as for enthalpies of formation.

Standard free energies of formation may also be calculated from enthalpies of formation (measured as described earlier) if ΔS^θ for the formation can be calculated from either third-law or statistical entropies:

$$\Delta G_f^\theta = \Delta H_f^\theta - T\Delta S_f^\theta \tag{6.23}$$

Example: The formation reaction for benzene:

$$6\,C(s, \text{graphite}) + 3\,H_2(g) = C_6H_6(\text{liq})$$

is an unlikely candidate for a direct equilibrium study. However, the heat of formation is readily calculated from the heat of combustion, as described earlier; its value is given by Table 6.1 as 49.04 kJ/mol. The calculation of the standard entropy of benzene was discussed as an example in Chapter 3, and that of hydrogen could be calculated as described in Chapter 5. Using data from Table 3.2 (at 25°C):

$$\Delta S_f^\theta = S_m^\theta(C_6H_6, \text{liq}) - 3\,S_m^\theta(H_2, g) - 6\,S_m^\theta(C, \text{graphite})$$

$$= 173 - 3(130.58) - 6(5.694) = -252.9 \text{ J K}^{-1}$$

Therefore, the standard free energy of formation of benzene at 25°C is:

$$\Delta G_f^\theta = 49\,040 - (298.15)(-252.9) = 124.4 \text{ kJ} \qquad \blacksquare$$

Example: The Boudouard reaction:

$$C(\text{graphite}) + CO_2(g) = 2\,CO(g)$$

was studied by Rhead and Wheeler [*J. Chem. Soc.*, **97**, 2178 (1910)] between 1123 and 1473 K. At 1273 K, 1 atm, they found that the gas phase was 99.41% CO:

$$K_p = \frac{P_{CO}^2}{P_{CO_2}} = \frac{(0.9941)^2}{0.0059} = 167.5$$

$$\Delta G_{rxn}^\theta = -8.314(1273) \ln (167.5) = -54.2 \text{ kJ} \qquad (\text{at } 1273 \text{ K})$$

Data were also collected for the oxidation of CO to CO_2, and extrapolation of these results (collected for temperatures between 1100 and 1800 K) to 25°C gave the following:

(1) $\qquad\qquad C(s) + CO_2(g) = 2\,CO(g), \qquad \Delta G^\theta(298.15) = 122.3 \text{ kJ}$

(2) $\qquad\qquad CO(g) + \tfrac{1}{2} O_2(g) = CO_2(g), \qquad \Delta G^\theta(298.15) = -258.4 \text{ kJ}$

If these reactions are added, the result is:

$$C(s) + \tfrac{1}{2} O_2(g) = CO(g), \qquad \Delta G^\theta = 122.3 - 258.4 = -136.1 \text{ kJ}$$

This is, of course, the formation reaction for CO (compare Table 6.1, which gives -137.268 kJ for this quantity). If reaction (1) is added to twice reaction (2), the result is:

$$C + O_2 = CO_2, \qquad \Delta G^\theta = 122.3 - 2(258.4) = -394.5 \text{ kJ}$$

This is the formation reaction for CO_2 (compare Table 6.1). $\qquad\qquad\qquad \blacksquare$

Free-Energy Functions

Calculation of equilibrium constants from spectroscopic data is best done through the free-energy function defined in Chapter 5:

$$\phi° = -\frac{(G_m^\theta - H_0^\theta)}{T}$$

(See Table 6.4.) We define the change of this function upon reaction as:

$$\Delta\phi° = \sum_i \nu_i \phi°(i) \qquad \qquad \textbf{(6.24)}$$

Table 6.4   Free-energy functions (based on 0 K)

	$\phi° = -\dfrac{G_m^\theta - H_0}{T}$ (J/K)					$H_m^\theta(298.15) - H_0$ (kJ)	$\Delta H_{0,f}^\theta$ (kJ)
	298.15	500	1000	1500	2000		
Graphite	2.2	4.85	11.6	17.5	22.5	1.050	0
H(g)	93.81	104.6	114.8	127.4	133.4	6.197	216.0
H$_2$(g)	102.2	116.9	137.0	148.9	157.6	8.468	0
H$_2$O(g)	155.5	172.8	196.7	211.7	223.1	9.908	−238.94
O(g)	138.4	150.0	165.1	173.8	179.8	6.724	246.2
O$_2$(g)	176.0	191.0	212.1	225.1	234.7	8.661	0
O$_3$(g)	204.1	227.9	251.8	270.7	284.5	10.36	145
N(g)	132.4	143.2	157.6	166.0	172.0	6.197	470.87
N$_2$(g)	162.4	177.5	197.9	210.4	219.6	8.669	0
NO(g)	179.8	195.6	217.0	230.0	239.5	9.180	89.87
N$_2$O(g)	187.8	205.5	233.3	252.2	—	9.586	84.98
NO$_2$(g)	205.8	224.3	252.0	270.2	284.0	10.31	36.32
CO(g)	168.4	183.5	204.1	216.6	225.9	8.673	−113.81
CO$_2$(g)	182.3	199.5	226.4	224.7	258.8	9.364	−393.17
CH$_4$(g)	152.5	170.5	199.4	221.1	239	10.03	−66.90
CH$_2$O(g)	185.1	203.1	230.6	250.6	266.0	10.01	−112
CH$_3$OH(g)	201.4	222.3	257.7	—	—	11.43	−190.2
C$_2$H$_5$OH(g)	235.1	262.8	315.0	356.3	—	14.2	−219.3
CH$_3$CHO(g)	221.1	245.5	288.8	—	—	12.84	−155.4
CH$_3$COOH(g)	236.4	264.6	317.6	357.1	—	13.8	−420.49
C$_2$H$_2$(g)	167.3	186.2	230.2	239.5	256.6	10.01	227.3
C$_2$H$_4$(g)	184.0	203.9	239.7	267.5	290.6	10.56	60.75
C$_2$H$_6$(g)	189.4	212.4	255.7	290.6	—	11.95	−69.12
NH$_3$(g)	159.0	176.9	203.5	221.9	236.6	9.92	−39.2
Cl$_2$(g)	192.2	208.6	231.9	246.2	256.6	9.2	0
COCl$_2$(g)	240.6	266.2	304.6	331.1	351.1	12.87	−217.8
CH$_3$Cl(g)	198.5	217.8	251.1	274.2	—	10.41	−74.1
CH$_2$Cl$_2$(g)	230.5	252.5	291.1	318.17	—	11.86	−79
CHCl$_3$(g)	248.1	275.3	321.2	353.0	—	14.18	−96
C$_3$H$_6$(g)	221.5	248.2	299.4	340.7	—	13.54	35.4
C$_3$H$_8$(g)	220.6	250.2	310.0	359.2	—	14.69	−81.50

Source: Ref. 2.

This gives:

$$T\Delta\phi° = -\Delta G^\theta_{\text{rxn}} + \Delta H^\theta_0$$

$$= RT \ln K_a + \Delta H^\theta_0$$

Finally:

$$R \ln K_a = \Delta\phi° - \frac{\Delta H^\theta_0}{T} \tag{6.25}$$

ΔH^θ_0 is the heat of the reaction at zero kelvin. This quantity can be measured spectroscopically for some simple cases (mostly those involving diatomic molecules); more often, ΔH^θ_0 will be calculated from heats of formation at zero kelvin, $\Delta H^\theta_{0,f}$ which are, in turn, calculated from heats of formation at 298.15 by methods described in Chapter 5:

$$\Delta H^\theta_{0,f} = \Delta H^\theta_f(298.15) - \sum_i \nu_i[H^\theta_m(298.15) - H^\theta_0] \tag{6.26}$$

The free-energy function is simply related to the partition function through Eq. (5.82); combined with Eq. (6.25), this gives:

$$K_a = \prod \left(\frac{z^\theta(i)}{L}\right)^{\nu_i} e^{-\Delta H^\theta_0/RT} \tag{6.27}$$

Equation (6.27) is interesting in that it shows a clear relationship between the partition function and the activity [compare Eq. (6.15)] and, hence, relates activity to molecular properties; however, it is usually less convenient than Eq. (6.25) for computation.

Often the free-energy function based on 298 K will be more convenient:

$$\phi' = -\frac{G^\theta_m - H^\theta_m(298.15)}{T}$$

(See Table 6.5.) A derivation similar to that for Eq. (6.25) gives:

$$R \ln K_a = \Delta\phi' - \frac{\Delta H^\theta(298.15)}{T} \tag{6.28}$$

As discussed in Chapter 5, this method is most convenient when spectroscopic data must be used together with thermal data.

Example: Calculate the equilibrium constant for the dissociation of oxygen, $O_2 = 2O$, at 298.15 and 2000 K. Data are from Table 6.4:

$$\Delta H^\theta_0 = 2(246.2 \times 10^3) = 4.924 \times 10^5 \text{ J}$$

At 298.15 K:

$$\Delta\phi° = 2(138.4) - 176.0 = 100.8 \text{ J K}^{-1}$$

Table 6.5   Free-energy functions (based on 298.15 K)

$$\phi' = -\frac{G_m^\theta - H_m^\theta(298.15)}{T} \quad (\text{J/K})$$

	298.15 K	500 K	1000 K		1500 K	$\Delta H_f^\theta(298.15)$ (kJ)
$Fe_3C(s)$	101	114.1	160.1		196.2	+21
$Cr_{23}C_6(s)$	610.0	689.5	979.1		1128	−411
$Fe(s)$	27.2	30.2	42.1		53.8	0
$Cr(s)$	23.8	26.5	37.0		46.1	0
$Cu(s)$	33.3	36.1	46.4	melts	55.8	0
$CuO(s)$	42.6	47.8	67.3		(84.0)	−157
$Fe_3O_4(s)$	146	165.5	246.7		306.3	−1120
$Fe_2O_3(s)$	87.4	100.4	152.4		196.3	−823.4
$Ti(s)$	30.5	33.5	44.3		54.2	0
$TiC(s)$	24.2	28.7	46.1		60.8	−185
$TiN(s)$	30.3	35.0	53.2		68.3	−338
$N_2(g)$	191.5	194.8	206.6		216.2	0
$O_2(g)$	205.1	208.5	220.8		230.9	0
$O_3(g)$	238.8	243.6	262.1		277.8	143
$S(s, rh.)$	31.93	34.97	46.85		58.07	0
$S_2(g)$	228.1	231.9	245.7		256.7	129.0
$S_8(g)$	430.2	448.6	516.1		569.8	101.3
$H_2S(g)$	205.6	209.6	224.5		237.3	−20.4
$SO_2(g)$	248.1	252.9	271.2		286.7	−296.8
$SO_3(g)$	256.7	262.9	267.7		308.9	−396.8

Source: Ref. 2.

Now, Eq. (6.25) gives:

$$\ln K_a = \frac{100.8}{8.3143} - \frac{4.924 \times 10^5}{8.3143(298.15)} = -186.5$$

$$K_a = 1.0 \times 10^{-81}$$

Note that such a small equilibrium constant has little meaning, except in a statistical sense. The earth's entire atmosphere contains only 3.5×10^{19} moles of oxygen, so, if it were at equilibrium at 25°C, this equilibrium constant would predict that there are approximately 600–700 dissociated O atoms (10^{-21} moles) in all the earth. (Actually there probably are more because of photodissociation in the upper atmosphere.) Of course this computational method works as well at higher temperatures, where dissociation is more significant. At 2000 K:

$$\Delta\phi° = 2(179.8) - 234.7 = 124.9 \text{ J K}^{-1}$$

$$\ln K_a = \frac{124.9}{8.3143} - \frac{4.924 \times 10^5}{8.3143(2000)} = -14.59, \quad K_a = 4.61 \times 10^{-7}$$

Clearly this constant, very important in high-temperature reactions such as flames and explosions, would be difficult to measure otherwise. In Chapter 13 we shall discover that ΔH_0^θ of this reaction is simply the O-O *bond energy*.   ∎

Temperature Dependence of Equilibrium Constants

Equilibrium constants will change with temperature, sometimes quite drastically (a point illustrated by the example above). Figures 6.5 and 6.6 show typical examples. Thermodynamics explains this change, beginning with Eq. (4.18):

$$\left(\frac{\partial(G/T)}{\partial(1/T)}\right)_P = H$$

This is applied to each term of

$$\Delta G_{rxn}^{\theta} = \sum_i \nu_i G_m^{\theta}(i)$$

$$\frac{d(\Delta G^{\theta}/T)}{d(1/T)} = \sum_i \nu_i \frac{d(G^{\theta}/T)}{d(1/T)} = \sum_i \nu_i H_m^{\theta}(i)$$

The partial derivative at constant P was replaced by a total derivative because G^{θ} and H^{θ}, being by definition the free energy and enthalpy at some particular pressure, do not depend on pressure. The right-hand side of this relation is ΔH_{rxn}^{θ}, so:

$$\frac{d \ln K}{d(1/T)} = -\frac{\Delta H_{rxn}^{\theta}}{R} \tag{6.29}$$

$$\frac{d \ln K}{dT} = \frac{\Delta H_{rxn}^{\theta}}{RT^2} \tag{6.30}$$

These equations are simply a quantification of Le Chatelier's principle, which states that the position of equilibrium for an exothermic reaction (ΔH negative) will be shifted to the left (K is decreased) when the temperature is increased, and vice versa.

Equation (6.29) can be used to calculate heats of reaction in situations for which the calorimetric method is not practical — namely, reversible reactions. If $\ln K$ is plotted vs. $1/T$, the slope of the plot will be $-\Delta H_{rxn}/R$; such plots are often nearly linear, particularly over limited temperature ranges, indicating that ΔH does not change as rapidly with temperature as does K. (Contrast Figures 6.5 and 6.6.) If the plot is not linear, Eq. 6.29 applies to the slope of the tangent line at a particular temperature.

Equation (6.29) or (6.30) can also be used to calculate K at some new temperature from a value at another temperature and the heat of reaction:

$$\ln K(T_2) = \ln K(T_1) + \frac{1}{R} \int_{T_1}^{T_2} \frac{\Delta H_{rxn}^{\theta}}{T^2} dT \tag{6.31}$$

For small temperature ranges, it may be sufficiently accurate to integrate Eq. (6.31), assuming ΔH is independent of temperature:

$$\ln K(T_2) = \ln K(T_1) - \frac{\Delta H_{rxn}^{\theta}}{R}\left(\frac{1}{T_2} - \frac{1}{T_1}\right) \tag{6.32}$$

More accurately, the ΔH of Eq. (6.32) is the mean value in the range T_1 to T_2.

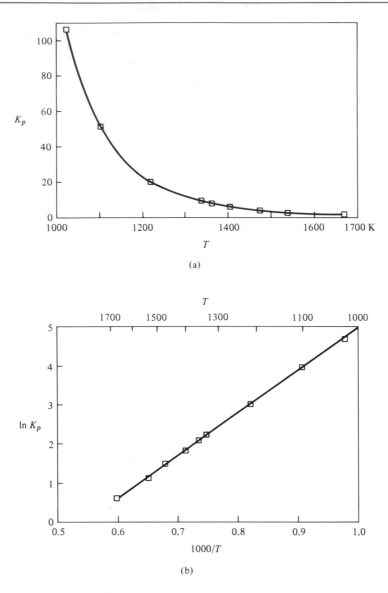

(a)

(b)

Figure 6.5 Equilibrium constant and temperature. (a) The equilibrium constant for the reaction $H_2(g) + \frac{1}{2}S_2(g) = H_2S(g)$ is shown as a function of temperature. (b) A graph of the logarithm of the equilibrium vs. reciprocal temperature is nearly linear; that is, the enthalpy change of this reaction does not vary much with temperature over this range. (Data from ref. 2.)

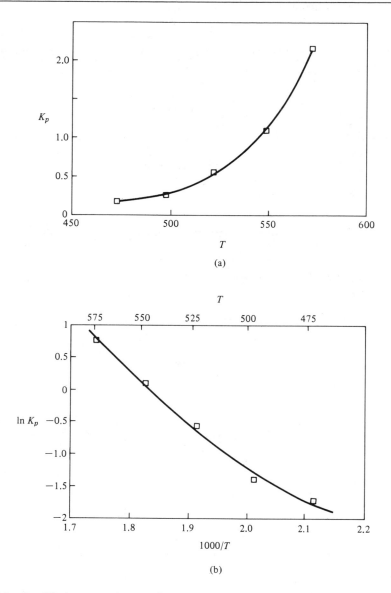

Figure 6.6 Equilibrium constant and temperature. (a) The equilibrium constant for the reaction of ethanol to form acetaldehyde and hydrogen vs. temperature. (b) The graph of the logarithm of the equilibrium constant vs. reciprocal temperature shows a distinct curvature, because the enthalpy change of this reaction varies quite a bit with temperature over this range. (Data from A. H. Lubberly and M. B. Mueller, *J. Am. Chem. Soc.*, 1149 (1946)

Example: The equilibrium constant of the reaction

$$C(s) + CO_2(g) = 2\ CO(g)$$

is 14.11 at 1123 K and 43.07 at 1173 K. Calculate ΔH of this reaction. From Eq. (6.32):

$$8.3143 \ln \frac{43.07}{14.11} = -\Delta H_{rxn}\left(\frac{1}{1173} - \frac{1}{1123}\right)$$

$$\Delta H_{rxn}^{\theta} = 244\ kJ$$

This value will be correct at $T \cong 1150$ K. It can be compared to the heat of this reaction at 298 K, which can be calculated from heats of formation (Table 6.1). The result, $\Delta H_{rxn}^{\theta}(298\ K) = 172$ kJ, clearly shows that the heat of this reaction does change significantly with temperature, but the change between 1123 and 1173 K would be so small that Eq. (6.32) can be used without significant error.   ■

Example: Estimate the equilibrium constant of the dissociation:

$$Br_2(g) = 2\ Br(g)$$

at 500, 1000, and 2000 K, using the data at 25°C (Table 6.1) and Eq. (6.32). From Table 6.1 at $T = 298.15$ K:

$$\Delta H_{rxn}^{\theta} = 1.929 \times 10^5\ J$$

$$\Delta G_{rxn}^{\theta} = 1.6162 \times 10^5\ J, \qquad \ln K(298) = -65.198$$

$$\ln K(500) = -65.198 - \frac{1.929 \times 10^5}{8.3143}\left(\frac{1}{500} - \frac{1}{298.15}\right) = -33.78$$

$$K(500) = 2.13 \times 10^{-15}$$

The reader should repeat this calculation at 1000 and 2000 K. The results are given in Table 6.6 (line 1) together with the more accurate estimates calculated from free-energy functions.   ■

Example: Estimate the equilibrium constant of $Br_2 = 2Br$, assuming constant heat capacities. From Table 6.1:

$$\Delta C_p = 2(20.79) - 36.0 = 5.58\ J/K$$

From Eq. (6.8) with $T_1 = 298.15$ K and ΔC_p presumed constant:

$$\Delta H_{rxn}^{\theta}(T) = \Delta H_{rxn}^{\theta}(298.15) + \Delta C_p(T - 298.15)$$

$$= 1.929 \times 10^5 - 5.58(298.15) + 5.58T$$

$$= 1.912 \times 10^5 + 5.58T$$

The integral required for Eq. (6.31) is:

$$\int \frac{\Delta H}{T^2}\, dT = 1.912 \times 10^5 \int \frac{dT}{T^2} + 5.58 \int \frac{dT}{T}$$

Taking the lower limit again at 298 K, Eq. (6.31) becomes:

$$\ln K(T) = \ln K(298) - \frac{1.912 \times 10^5}{8.3143}\left(\frac{1}{T} - \frac{1}{298.15}\right) + \frac{5.58}{8.3143} \ln \frac{T}{298.15}$$

Table 6.6 Calculated equilibrium constants for $Br_2 = 2Br$

Method	500 K	1000 K	2000 K
1. Constant ΔH [Eq. (6.32)]	2.13×10^{-15}	2.54×10^{-5}	2.77
2. Constant ΔC_p	2.28×10^{-15}	3.53×10^{-5}	5.54
3. $\Delta C_p(T)$ (Table 2.2)	2.28×10^{-15}	3.38×10^{-5}	4.83
4. Free-energy functions (ref. 2)	2.24×10^{-15}	3.26×10^{-5}	4.94
5. Free-energy functions (JANAF)	2.18×10^{-15}	3.23×10^{-5}	5.10

The reader should fill in the missing steps of this derivation and check the calculations of Table 6.6 (line 2). This is a fortuitous example, because the heat capacities of diatomic and monatomic species are not large and do not change rapidly with temperature; the good results of this method at temperatures as high as 2000 K cannot be counted upon in general. ∎

Exercise (optional): Using the heat capacity of Table 2.2 for Br_2 and $C_p = \frac{5}{2}R$ for Br (assumed independent of T), derive the following results:

$$\Delta C_p = 4.25 - 5.0 \times 10^{-4}T + \frac{1.26 \times 10^5}{T^2}$$

$$\Delta H^\theta_{rxn} = 1.9208 \times 10^5 + 4.25T - 2.5 \times 10^{-4}T^2 - \frac{1.26 \times 10^5}{T}$$

This function is graphed in Figure 6.2.

$$R \ln K = 77.315 - \frac{1.9208 \times 10^5}{T} + 4.25 \ln T - 2.5 \times 10^{-4}T + \frac{6.3 \times 10^4}{T^2}$$

From this result, line 3 of Table 6.6 can be calculated.

From Table 5.3 we can see that Br atoms have an electronic state at 3685 cm^{-1}, so for $T > 2000$ K, the heat capacity of the atoms may differ somewhat from its presumed value of $\frac{5}{2}R$. (Cf. Problem 6.53.) ∎

6.6 Activity of Real Gases

It is the equilibrium constant K_p that is of practical interest, since it will tell us the extent of reaction. But only the thermodynamic equilibrium constant K_a is truly "constant"—that is, independent of pressure at any given temperature. At low pressure, $K_p \cong K_a$; in this section we shall seek a more general relationship.

The logarithmic dependence of the chemical potential of an ideal gas on pressure:

$$\mu(\text{ideal gas}) = (\text{const.}) + RT \ln P$$

suggests the definition of a quantity called the *fugacity* such that:

$$\mu = (\text{const.}) + RT \ln f$$

Clearly $f(\text{ideal gas}) = P$. Since only relative chemical potentials can be measured, we must define some *standard state* with:

$$\mu^\theta = (\text{const.}) + RT \ln f^\theta$$

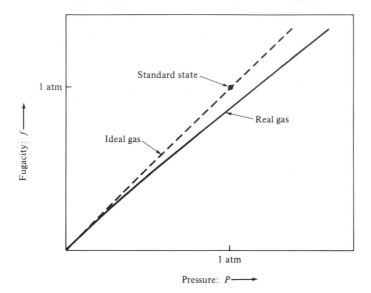

Figure 6.7 Standard state for gases. The most common standard state used for gases is the hypothetical point on the ideal gas curve at 1 atm. Although real gases at one atmosphere are not in the standard state, the correction is usually small enough that it can be calculated accurately. (Compare Figure 6.8.)

Then:

$$\mu - \mu^\theta = RT \ln \frac{f}{f^\theta} \tag{6.33}$$

Comparison to the definition of activity used earlier [Eq. (6.13)] shows:

$$a = \frac{f}{f^\theta} \tag{6.34}$$

Fugacity has the units of pressure, and for gases the standard state is defined so that $f^\theta = 1$ atm; with this unit, the fugacity and activity are numerically the same (if $a = 20, f = 20$ atm), so the terms are often used interchangeably. Although deviations from ideality of gases at 1 atm are small, the standard state of a gas is defined as the hypothetical point (Fig. 6.7) at 1 atm pressure for an ideal gas; a real gas will not have $f = 1$ atm at $P = 1$ atm but will approach the ideal curve as $P \to 0$. (Figure 6.8 is a plot for carbon monoxide.)

From Chapter 3, the change in the free energy at constant T is $dG = V\,dP$. This gives (letting $f^\theta = 1$ atm):

$$d\mu = V_m\,dP$$

$$RT\,d(\ln f) = V_m\,dP \tag{6.35}$$

Since $f \to P$ as $P \to 0$, this expression must be integrated from $P = 0$, since that

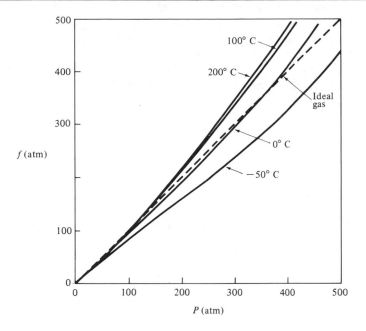

Figure 6.8 Fugacity of carbon monoxide vs. pressure at various temperatures. Note that ideal behavior ($f = P$) is always approached in the low-pressure limit.

is where we know what f is. This procedure is a bit awkward, since the logarithm of zero is infinite; for that reason it will be convenient to define a fugacity coefficient γ such that:

$$f = \gamma P \qquad (6.36)$$

(ideal gas, $\gamma = 1$; $\gamma \rightarrow 1$ as $P \rightarrow 0$). Substitution of this definition into Eq. (6.35) gives:

$$RT[d(\ln P) + d(\ln \gamma)] = V_m \, dP$$

$$RT \, d(\ln \gamma) = V_m \, dP - RT \, d(\ln P)$$

Since $d(\ln P) = dP/P$, we get, finally:

$$RT \, d(\ln \gamma) = \left(V_m - \frac{RT}{P} \right) dP$$

This expression is easily evaluated with $P = 0$ as a lower limit, since, at $P = 0$, $\gamma = 1$ and $\ln \gamma = 0$:

$$\ln \gamma = \frac{1}{RT} \int_0^P \left(V_m - \frac{RT}{P} \right) dP \qquad (6.37)$$

This equation is the basis for the evaluations of fugacity coefficients, hence fugacities, at high pressures.

Fugacities of gases can be calculated in three ways: (1) Gas laws can be used to write the integrand of Eq. (6.37) as a function of P; integration then gives γ as a function of P and T. This method tends to fail at very high pressures, just when things are getting interesting. (2) Measured values of V_m vs. P can be used with Eq. (6.37) to calculate γ by numerical integration. (3) Graphs constructed by R. H. Newton with average data from a variety of gases (evaluated by method 2) (see Figure 6.9) can be used to determine γ from the reduced temperature and pressure:

$$T_r = \frac{T}{T_c}, \qquad P_r = \frac{P}{P_c}$$

Use of these graphs is the most convenient method for finding fugacities at very high pressures. To use such a graph, the reduced pressure, P/P_c is calculated and located as the abscissa; this point is then traced up to the isotherm for the appropriate $T_r = T/T_c$ (interpolating as necessary) and the fugacity coefficient is read from the ordinate.

Example: We have often used a one-term virial series to estimate small deviations from ideality:

$$PV_m = RT + BP$$

$$V_m - \frac{RT}{P} = B$$

Using this in Eq. (6.37) gives a simple relation:

$$\gamma = e^{BP/RT} \qquad\qquad (6.38)$$

Methods for estimating the second virial coefficient $B(T)$ were discussed in Chapter 1. Equation (6.38) will be tested for CO_2 at 373.15 K. The Beattie-Bridgeman equation gives (using data from Table 1.2):

$$B = B_0 - \frac{A_0}{RT} - \frac{c}{T^3}$$

$$= 0.10476 - \frac{5.0065}{0.08206(373.15)} - \frac{66 \times 10^4}{(373.15)^3}$$

$$B = -0.07144 \text{ dm}^3, \qquad \frac{B}{RT} = -0.002333 \text{ atm}$$

Then, Eq. (6.38) gives $\gamma = \exp(-0.002333P)$; also, $f = \gamma P$:

$$P = \quad 1 \text{ atm}; \quad \gamma = 0.998, \quad f = \ 0.998 \text{ atm}$$

$$P = \quad 5 \text{ atm}; \quad \gamma = 0.988, \quad f = \ 4.94 \quad\text{atm}$$

$$P = \quad 50 \text{ atm}; \quad \gamma = 0.890, \quad f = 44.5 \quad\ \text{atm}$$

$$P = 1000 \text{ atm}; \quad \gamma = 0.0970, \quad f = 97 \qquad \text{atm (?)}$$

This calculation shows that our use of the ideal gas approximation ($\gamma = 1$) at low pressure was quite reasonable. We shall see shortly that the calculation at 1000 atm is absurd.   ∎

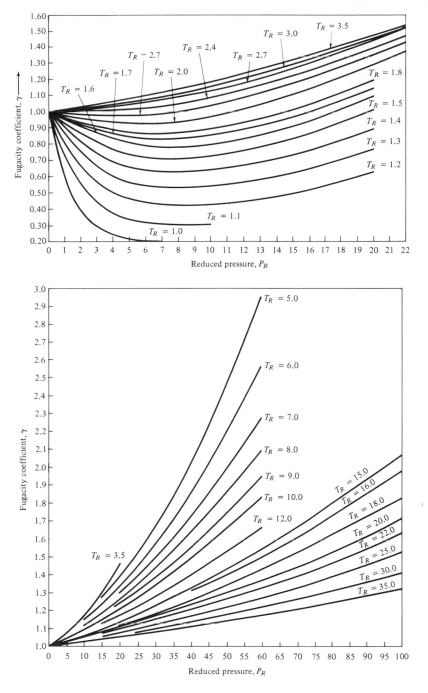

Figure 6.9 Corresponding state graphs for fugacity coefficients. [From R. H. Newton, *Ind. Eng. Chem.*, 27, 302 (1935)]

Table 6.7   Compressibility and fugacity coefficients of CO_2 at 373.15 K

P (atm)	z	$\dfrac{1000(z-1)}{P}$	Integral[a]	Cumulative[b]	γ
0	—	-2.334[c]		—	—
			-0.120		
50	0.8772	-2.456		-0.120	0.887
			-0.124		
100	0.7489	-2.511		-0.244	0.787
			-0.123		
150	0.6384	-2.411		-0.367	0.693
			-0.111		
200	0.5922	-2.039		-0.478	0.620
			-0.090		
250	0.6075	-1.570		-0.568	0.566
			-0.069		
300	0.6471	-1.176		-0.637	0.529
			-0.089		
400	0.7551	-0.612		-0.727	0.484
			-0.043		
500	0.8729	-0.254		-0.770	0.463
			-0.013		
600	0.9928	-0.012		-0.783	0.457
			0.007		
700	1.1113	0.159		-0.776	0.460
			0.022		
800	1.2280	0.285		-0.754	0.471
			0.033		
900	1.3422	0.380		-0.720	0.487
			0.042		
1000	1.4534	0.453		-0.679	0.507

[a]Trapezoidal rule from P(above) to P(below); see Appendix I.
[b]Trapezoidal rule from 0 to P.
[c]Estimated as B/RT; see text.

Example: Fugacities can be estimated from Newton's graphs, Figure 6.9. For CO_2 at 373 K and 1000 atm (data from Table 1.1):

$$T_r = \frac{373}{304} = 1.23, \qquad P_r = \frac{1000}{73} = 13.7$$

Interpolating Figure 6.9 gives $\gamma = 0.51$, $f = 510$ atm (compare above).   ■

Example: For purposes of numerical integration, it will be convenient to introduce the compressibility factor (z) into Eq. (6.37):

$$z = \frac{PV_m}{RT}$$

$$\ln \gamma = \int_0^P \frac{z-1}{P}\, dP \tag{6.39}$$

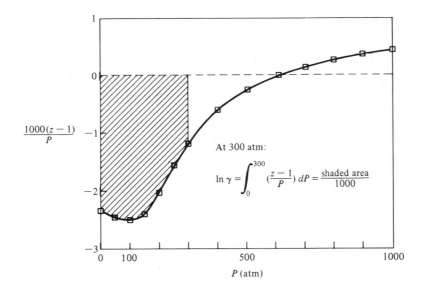

Figure 6.10 Fugacity coefficient of carbon dioxide. The evaluation of the fugacity coefficient is illustrated for carbon dioxide ($T = 373.15$ K, cf. Table 6.7) from experimental compressibility factors (z).

A slight complication exists at the lower limit of Eq. (6.39); since $z = 1$ at $P = 0$, the integrand is equal to $0/0$ and is indeterminate. The limit can be found by using the virial equation:

$$PV_m = RT + BP + \gamma P^2 + \cdots$$

$$\frac{z - 1}{P} = \frac{B}{RT} + \frac{\gamma}{RT} P + \cdots$$

Therefore:

$$\lim_{P \to 0} \left(\frac{z - 1}{P} \right) = \frac{B}{RT} \tag{6.40}$$

Table 6.7 shows a sample calculation for CO_2 at 373 K; the reader should work through this example. Also, compare the results at 50 and 1000 atm to the results of the previous examples. The graphical integration of Eq. (6.39) is illustrated by Figure 6.10. ∎

Equilibrium Constants at High Pressure

Using the definition of the fugacity coefficient [Eq. (6.36)], we can write the thermodynamic equilibrium constant for a general gas-phase reaction:

$$a\text{A} + b\text{B} = c\text{C} + d\text{D}$$

$$K_a = K_p K_\gamma \tag{6.41}$$

$$K_\gamma = \frac{\gamma_C^c \gamma_D^d}{\gamma_A^a \gamma_B^b} \tag{6.42}$$

The fugacity coefficients of Eq. (6.42) are, strictly, those for the gases in the equilibrium mixture (as opposed to those of pure gases, as were calculated above). Since there is no easy method for calculating fugacity coefficients of gases in mixtures, it is usually assumed that the fugacity of a gas (i) in the mixture in which its partial pressure is P_i is $f_i = \gamma_i P_i$, with γ_i being the same as the fugacity coefficient of the pure gas at the same *total* pressure. (Cf. Amagat's law, Chapter 1.)

Example: Calculate the fugacities of CO and CO_2 in a mixture 40% CO at 400 K and 500 atm total pressure (that is, 200 atm CO, 300 atm CO_2) using Newton's graphs (Figure 6.9). The reduced pressures are calculated as the ratio of the total pressure (500 atm) to the critical pressure (Table 1.1):

CO: $T_c = 134$, $P_c = 35$, $T_r = 3.0$, $P_r = 14.3$;   Figure 6.9: $\gamma = 1.28$

CO_2: $T_c = 304$, $P_c = 73$, $T_r = 1.3$, $P_r = 6.8$;   Figure 6.9: $\gamma = 0.57$

$$f_{CO} = 1.28(200) = 256 \text{ atm}, \qquad f_{CO_2} = 0.57(300) = 171 \text{ atm} \qquad \blacksquare$$

Example: Nitrogen is an important nutrient for plants, but only a few of them can use it directly as N_2 from the air. An important process for making atmospheric nitrogen available as fertilizer is the Haber process for making ammonia, which has been in use since early in the century:

$$\tfrac{3}{2} H_2(g) + \tfrac{1}{2} N_2(g) = NH_3(g)$$

For kinetic reasons this reaction is carried out at high temperature, although, as we shall see, this affects the equilibrium unfavorably. Le Chatelier's principle tells us that increased pressure will shift the equilibrium of such a reaction to the right, and, for that reason, the reaction is usually carried out at high pressures. Because of its commercial importance, the reaction has been intensively studied at high temperature and pressure; some data are given in Table 6.8.

Table 6.8   Equilibrium constants for ammonia synthesis

$K_p = \dfrac{P_{NH_3}}{P_{N_2}^{1/2} P_{H_2}^{3/2}}$

P (atm)	673 K	723 K	773 K
10	0.0129	0.00659	0.00381
30	0.0129	0.00676	0.00386
50	0.0130	0.00690	0.00388
100	0.0137	0.00725	0.00402
300	—	0.00884	0.00498
600	—	0.01294	0.00651
1000	—	0.02328	—

Source: Data are from A. T. Larson and R. L. Dodge, *J. Am. Chem. Soc.*, 45, 2918 (1923); 46, 367 (1924).

Extrapolation of the data at 773 K to $P = 0$ gives us the thermodynamic equilibrium constant, $K_a = 0.00377$. This is the quantity that can be calculated theoretically by the methods of the previous section. Suppose this were a new reaction and, knowing K_a at 773 K, we wanted to estimate K_p at 600 atm. We could do so by estimating the fugacity coefficients with Figure 6.9:

$$NH_3: \quad T_c = 405.5, \quad P_c = 111.5, \quad T_r = 1.91, \quad P_r = 5.4; \quad \gamma = 0.89$$

$$N_2: \quad T_c = 126, \quad P_c = 33.5, \quad T_r = 6.13, \quad P_r = 18; \quad \gamma = 1.31$$

$$H_2: \quad T_c = 33, \quad P_c = 12.8, \quad T_r = 23.2, \quad P_r = 47; \quad \gamma = 1.26$$

$$K_\gamma = \frac{0.89}{(1.31)^{1/2}(1.26)^{3/2}} = 0.55$$

$$K_p = \frac{K_a}{K_\gamma} = \frac{3.77 \times 10^{-3}}{0.55}$$

$$= 6.85 \times 10^{-3} \quad (6.51 \times 10^{-3} \text{ obs.}) \qquad \blacksquare$$

In a heterogeneous equilibrium, it may be necessary to estimate the effect of high pressure on the activity of solids. Since K_p is constituted by setting $a = 1$ for solid phases, any deviation from this value should be included in K_γ. Since, for pure solids, the activity and activity coefficient are the same, the special symbol Γ is used for this quantity:

$$a(\text{pure solid}) = \Gamma(\text{pure solid})$$

From Eq. (4.17) (replacing a with Γ):

$$RT \, d(\ln \Gamma) = V_m \, dP$$

If it is assumed that the solid is incompressible (V_m independent of P), its activity will be:

$$\Gamma(\text{pure solid}) = \exp\left(\frac{PV_m}{RT}\right) \qquad \textbf{(6.43)}$$

The same arguments could be applied to liquids, but then we must be concerned that the gas will dissolve in the liquid; we will discuss this further in the next chapter. (Gas solubility in solids is not unknown, so even here one must be cautious.)

6.7 Extent of Reaction

Up to this point we have focused on methods for calculating equilibrium constants, but the practical question that equilibrium constants are intended to answer is how much product will be formed under a given set of conditions. To answer this question we have available, first of all, the equilibrium constant, which relates the amounts of products and reactants; after that we must use the various conditions of material balance, the stoichiometry of the reaction, and Dalton's law of partial pressures to find the extent of reaction.

The equilibrium constant K_p is not always the most convenient one to use. Using Dalton's law of partial pressures for all *gases* in the reaction:

$$P_i = X_i P$$

gives the equilibrium constants in terms of the mole fractions:

$$K_p = \prod_i P_i^{\nu_i} = \left(\prod_i X_i^{\nu_i}\right) P^{\Delta \nu_g}$$

where $\Delta \nu_g$ is the sum of the stoichiometric coefficients (negative, as usual, for reactants) of the *gases* only. The product of the mole fractions gives a quantity called K_X, the mole fraction equilibrium constant:

$$K_X = \prod_i X_i^{\nu_i} \tag{6.44a}$$

$$K_X = K_p(P)^{-\Delta \nu_g} \tag{6.44b}$$

For example, the Haber synthesis has:

$$K_X = \frac{X_{NH_3}}{X_{N_2}^{1/2} X_{H_2}^{3/2}}$$

with:

$$X_{NH_3} = \frac{\text{moles } NH_3}{\text{total moles in mixture}}$$

and so on. We note that K_X is not a true constant (independent of P at constant T) unless the number of moles of gases in the products is the same as for the reactants ($\Delta \nu_g = 0$). Equation (6.44) is simply a quantification of Le Chatelier's principle, which says that the equilibrium position of a reaction for which gases are lost ($\Delta \nu_g$ negative) will be shifted to the right (K_X is increased) if the pressure is increased, and vice versa.

Example: Find the effect of pressure on the Boudouard reaction:

$$C(s) + CO_2(g) = 2\, CO(g)$$

for which $K_p = 14.11$ at 1123 K.

$$K_p = \frac{P_{CO}^2}{P_{CO_2}} = \frac{X_{CO}^2}{X_{CO_2}} P$$

Therefore:

$$K_X = \frac{K_p}{P} = \frac{X_{CO}^2}{X_{CO_2}}$$

The material balance is simple in this case, since $X_{CO_2} = 1 - X_{CO}$:

$$X_{CO}^2 - K_X(1 - X_{CO}) = 0$$

This is a quadratic equation, which, of course, has two roots given by:

$$X_{CO} = -\frac{K_X}{2} \pm \frac{1}{2}\sqrt{K_X^2 + 4K_X}$$

Since X_{CO} must be a positive number, the root with the $+$ sign must be the correct one — the other root has no physical significance. Introducing $K_X = K_p/P = 14.11/P$, we can write this as:

$$X_{CO} = \frac{7.055}{P}(\sqrt{1 + aP} - 1) \tag{6.45}$$

with:

$$a = \frac{4}{K_p} = 0.283487$$

The reader should fill in the missing steps of this derivation and check the following calculations (also see Figure 6.11):

P	X_{CO}	P_{CO}
1	0.938	0.938 atm
10	0.676	6.76 atm
100	0.312	31.2 atm

At 10 atm and 1123 K, the mixture will be 67.6% CO. (At 100 atm, the error due to our use of the ideal gas law — that is, the presumption that K_p is independent of P — could be significant.) ■

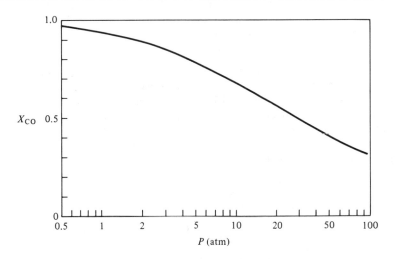

Figure 6.11 Reaction equilibrium. The mole fraction carbon monoxide in the reaction $C(s) + CO_2(g) = 2\ CO(g)$ at 1123 K, vs. pressure. As expected from the Le Chatelier principle, the equilibrium is shifted to the left by increasing pressure.

Two other equilibrium constants are sometimes useful. If we introduce into K_p the ideal gas relationship:

$$n_i = P_i \frac{V}{RT}$$

we can define an equilibrium constant for the number of moles:

$$K_n = \prod_i n_i^{\nu_i} \tag{6.46a}$$

Note the distinction between n_i, the number of moles of i present at equilibrium, and ν_i, the stoichiometric coefficient of i. This is related to K_p as:

$$K_n = K_p \left(\frac{RT}{V} \right)^{-\Delta\nu_g} \tag{6.46b}$$

Similarly, the equilibrium constant can be written in terms of concentrations:

$$C_i = \frac{n_i}{V} = \frac{P_i}{RT}$$

$$K_c = \prod_i C_i^{\nu_i} \tag{6.47a}$$

$$K_c = K_p (RT)^{-\Delta\nu_g} \tag{6.47b}$$

In solving equilibrium problems, the choice among K_p, K_X, K_n, or K_c is a matter of convenience or, perhaps, mere taste. There are usually several effective methods for solving a given problem. For gas reactions, it is usually best to start (at least) with K_p; K_X is apt to be convenient if the final equilibrium pressure is known; K_n may be convenient if the volume is constant (and known).

Example: The calculation of the degree of dissociation for reactions such as $Br_2(g) = 2\,Br(g)$ is a typical problem (equilibrium constants, Table 6.6).

The gas Br_2 has a molecular weight of 159.808 g. If you had a container with 15.9808 g of bromine, you would have, formally, 0.1 moles of Br_2; actually you would have some Br_2 and some Br. The degree of dissociation (α) is defined as the fraction of the formal moles of Br_2 (n_0) which are dissociated. Suppose we have a vessel containing n_0 moles of Br_2 at some (T, P); then the gas is allowed to dissociate (at constant P):

$$Br_2(g) = 2\,Br(g)$$

Initial moles: n_0 0

Final moles: $n_0(1 - \alpha)$ $2\alpha n_0$ Total: $n_T = n_0(1 + \alpha)$

Final mole fraction: $\dfrac{1 - \alpha}{1 + \alpha}$ $\dfrac{2\alpha}{1 + \alpha}$

With the pressure constant it will be best to use the mole fraction equilibrium constant K_X:

$$K_p = \frac{P_{Br}^2}{P_{Br_2}} = \frac{X_{Br}^2 P}{X_{Br_2}}$$

$$K_X = \frac{\dfrac{4\alpha^2}{(1 + \alpha)^2}}{\dfrac{1 - \alpha}{1 + \alpha}} = \frac{4\alpha^2}{(1 - \alpha)(1 + \alpha)}$$

Noting that $(1 + \alpha)(1 - \alpha) = 1 - \alpha^2$, we get:

$$K_X = \frac{4\alpha^2}{1 - \alpha^2} \tag{6.48}$$

At 2000 K, Table 6.6 (last line) gives $K_p = 5.10$. Equation (6.48) with $K_X = K_p/P$ gives (after some algebra):

$$\frac{4\alpha^2}{1 - \alpha^2} = \frac{K_p}{P}, \qquad \alpha = \left(\frac{K_p}{4P + K_p}\right)^{1/2}$$

For example (using $K_p = 5.10$):

$$P = 0.1 \text{ atm}, \qquad \alpha = 0.96$$

$$P = 1 \quad \text{atm}, \qquad \alpha = 0.75$$

$$P = 5 \quad \text{atm}, \qquad \alpha = 0.45$$

This means that, at 2000 K, $P = 1$ atm, 75% of the bromine molecules are dissociated. Figure 6.12 shows more calculations, with K_p calculated as in Problem 6.53.   ∎

Exercise: Show that, for an unsymmetrical dissociation such as:

$$COCl_2(g) = CO(g) + Cl_2(g)$$

the value of K_X is:

$$K_X = \frac{\alpha^2}{1 - \alpha^2} \tag{6.49}$$

∎

Example: The reaction

$$\tfrac{1}{2} N_2(g) + \tfrac{3}{2} H_2(g) = NH_3(g)$$

has $K_p = 0.336$ at 500 K. Calculate the final pressure and percent NH_3 in the product mixture if an initial P of 10 atm N_2 is reacted with (initial) 1 atm H_2 at 673 K at constant volume. To avoid fractional exponents, it will be helpful to double the reaction and square the equilibrium constant:

$$N_2 + 3 H_2 = 2 NH_3$$

$$K_p = \frac{P_{NH_3}^2}{P_{N_2} P_{H_2}^3} = (0.336)^2 = 0.113$$

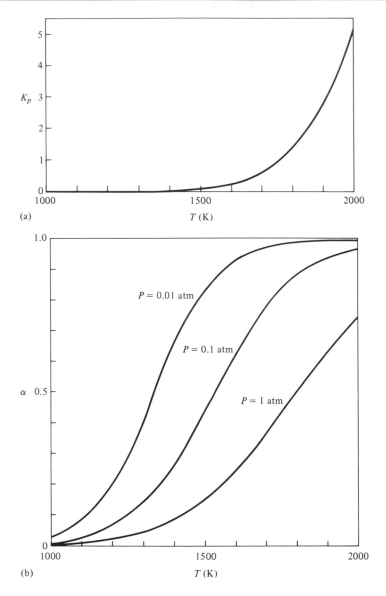

Figure 6.12   The bromine dissociation equilibrium. (a) The ̃equilibrium constant vs. temperature. (This is calculated using the theoretical result of problem 6.53.) (b) The degree of dissociation at various temperatures and pressures. Low pressure always promotes dissociation (Le Chatelier's principle); NASA scientists working on the space shuttle were surprised at the amount of corrosion they found, considering the limited supply of oxygen at the altitudes at which the vehicle had been cruising; they attributed the corrosion to the fact that, at such low oxygen pressures, the oxygen was highly dissociated into the more reactive atoms. The corrosion was also accelerated by the high velocity, which caused the vehicle to encounter more of the reactive atoms than one would naively expect.

Since the final pressure is not known, we will probably find it easiest to work with K_p. To account for the material balance, it will be convenient to introduce a temporary variable (y) as the loss of nitrogen pressure for the reaction; then (because of the stoichiometry) the pressure of ammonia formed must be $2y$ and the loss of hydrogen pressure must be $3y$:

	N_2	$+$	$3 H_2$	$=$	$2 NH_3$	Total P
Initial P:	10		1		0	11
Final P:	$10 - y$		$1 - 3y$		$2y$	$11 - 2y$

In terms of this variable, the equilibrium constant is:

$$K_p = \frac{4y^2}{(10 - y)(1 - 3y)^3} = 0.113$$

To calculate the appropriate value of y, it is necessary to arrange this in the form $f(y) = 0$; for example:

$$f(y) = 4y^2 - K_p(10 - y)(1 - 3y)^3 = 0$$

If a computer or programmable calculator is available, the root of this equation is readily found by the *Regula Falsi* method (Appendix I); otherwise the Newton-Raphson method will be best. It may be convenient (but is hardly necessary) to multiply out the factors in the second term; if this is done, $f(y)$ becomes:

$$f(y) = 4y^2 - K_p(27y^4 - 279y^3 + 279y^2 - 91y + 10)$$

The derivative needed for the Newton-Raphson method is:

$$f' = 8y - K_p(108y^3 - 837y^2 + 558y - 91)$$

Next, we need an initial guess for y. Since the pressure of the ammonia cannot be negative, it is required that $y > 0$. Also, the hydrogen pressure ($1-3\ y$) cannot be zero, so $y < 0.333$. At $y = 0$, it should be apparent that $f(y) (= -10K_p = -1.13)$ is negative. If $y = 0.1$ is used as a trial value, then:

$$y(0.1) = -0.3437$$

This value is negative, so we now know that the root is greater than 0.1. Next we might try $x = 0.2$; the result is:

$$f(0.2) = 0.0891265$$

The function is now positive so the root is $0.1 < y < 0.2$. At this point it is probably worthwhile to use the Newton-Raphson formula to get the next estimate:

$$y(\text{new}) = y(\text{old}) - \frac{f(y)}{f'(y)}$$

For $y = 0.2$ the slope is (from the formula above):

$$y'(0.2) = 2.95781$$

Therefore, the next estimate is:

$$y = 0.2 - \frac{0.0891265}{2.95781} = 0.169867$$

Succeeding calculations are as follows:

$$y = 0.169867, \quad f(y) = -0.015586, \quad f'(y) = 3.60044$$

$$y = 0.174196, \quad f(y) = 5.6 \times 10^{-4}, \quad f'(y) = 3.4983$$

$$y = 0.174$$

The root can be seen to have converged. Now the equilibrium partial pressures can be calculated:

$$P_{NH_3} = \quad\quad 2y = \quad 0.348 \text{ atm}$$

$$P_{N_2} = \quad 10 - y = \quad 9.826 \text{ atm}$$

$$\underline{P_{H_2} = \quad 1 - 3y = \quad 0.478 \text{ atm}}$$

$$P(\text{total}) = 11 - 2y = 10.652 \text{ atm}$$

Since there were numerous opportunities for error in this calculation, it is wise to check back to see if these pressures are consistent with the original K_p:

$$K_p = \frac{P_{NH_3}^2}{P_{H_2}^3 P_{N_2}} = \frac{(0.348)^2}{(9.826)(0.478)^2}$$

$$= 0.113 \quad (\text{compare above})$$

Finally, the mole percent ammonia at equilibrium is:

$$\%NH_3 = 100\frac{P_{NH_3}}{P_{\text{total}}} = 3.27\%$$

■

Example: It is important to realize that, in the example preceding, the pressures specified were initial pressures; the final equilibrium pressure was different because the reaction occurred at constant volume. It is also possible that the reaction could be carried at constant pressure, so the equilibrium pressure would be, like the initial pressure, 11 atm. In such a case a different approach is needed; the fact that P is constant may suggest using the mole-fraction equilibrium constant:

$$K_X = K_p P^2 = (0.113)(11)^2 = 13.7$$

Next, the material balance must be deduced from the initial mole ratio, $N_2 : H_2 = 10 : 1$. The number of moles present initially will depend on the volume of the container — which has not been specified. However, that is irrelevant, since the position of the equilibrium at a given T, P will not depend on the size of the container; therefore we shall simply assume an initial moles of 10 for N_2 and 1 for H_2. Also, we denote the moles nitrogen reacted as x:

	N_2	$+$	$3 H_2$	$=$	$2 NH_3$	Total
Initial moles	10		1		0	11
Final moles	$10 - x$		$1 - 3x$		$2x$	$11 - 2x$
Mole fraction	$\dfrac{10 - x}{11 - 2x}$		$\dfrac{1 - 3x}{11 - 2x}$		$\dfrac{2x}{11 - 2x}$	1

$$K_X = \frac{X_{NH_3}^2}{X_{N_2} X_{H_2}^3} = \frac{4x^2(11 - 2x)^2}{(10 - x)(1 - 3x)^3} = 13.7$$

Using the value calculated above for K_X, the problem can be formulated as finding the root $(0 < x < 0.33)$ of the equation:

$$f(x) = 4x^2(11 - 2x)^2 - 13.7(10 - x)(1 - 3x)^3 = 0$$

The reader should solve this by any convenient method; the answer should be $x = 0.245$. From this, the mole fraction of ammonia is:

$$X_{NH_3} = \frac{2x}{11 - 2x} = 0.0466 \qquad \text{(that is, 4.66\%)}$$

The partial pressure is:

$$P_{NH_3} = X_{NH_3}P = (0.0466)(11) = 0.513 \text{ atm}$$

(Compare above.) ∎

The Effect of Foreign Gases

Discussions of the effect of foreign gases on equilibrium are often unnecessarily confusing; in the ideal gas approximation, the effect of the presence of a gas that does not react with any of the products or reactants is *exactly nothing*. If K_p is used, this should be obvious, since only the partial pressures of the active gases (those which enter into the reaction) are in the definition. If K_X is used, some difficulty results, since in Eq. (6.44) the foreign gas will enter into both the total pressure and the definition of the mole fractions. Actually, if in Eq. (6.44) we interpret "P" as the pressure of the *active gases only* and X as the mole fraction of the active gases only, then K_X can be used just as if the foreign gas were not present.

Example: The Boudouard reaction considered earlier had $K_p = 14.11$ at 1123 K. Suppose we react a mixture 50% CO_2 and 50% N_2 with graphite at a constant $P = 20$ atm and 1123 K. After reaction (at constant P):

$$P_{N_2} + P_{CO} + P_{CO_2} = 20 \text{ atm}$$

the nitrogen pressure is still 10 atm, so:

$$P_{CO} + P_{CO_2} = 10 \text{ atm}$$

$$K_p = \frac{P_{CO}^2}{P_{CO_2}} = \frac{P_{CO}^2}{10 - P_{CO}} = 14.11$$

$$P_{CO}^2 + 14.11 P_{CO} - 10(14.11) = 0$$

This equation is readily solved to give $P_{CO} = 6.76$ atm. This is exactly the result obtained earlier when the pressure was $P = P_{CO} + P_{CO_2} = 10$ atm.
In fact Eq. (6.45) is still usable — provided we interpret, as we did then:

$$"P" = P_{CO} + P_{CO_2} = 10 \text{ atm}$$

$$"X_{CO}" = \frac{n_{CO}}{n_{CO} + n_{CO_2}} = 0.676$$

This "mole fraction" is a significant quantity, since it is the fraction of CO_2 that reacted and, as such, is unaffected by the nitrogen. The actual mole fraction of the CO in the product mixture is, of course:

$$X_{CO} = \frac{n_{CO}}{n_{CO} + n_{CO_2} + n_{N_2}}$$

This can be calculated from Dalton's law:

$$X_{CO} = \frac{P_{CO}}{P} = \frac{6.76}{20} = 0.338$$

The mole fraction of CO at equilibrium is changed, not because the N_2 has shifted the equilibrium, but simply because the nitrogen has diluted the mixture.

Of course, the foreign gas may shift the equilibrium at very high pressures through its effect on the fugacity coefficients. ∎

6.8 Heterogeneous Reactions

Cyclopentene (which we shall call A):

$$A = \langle\!\!\!\bigcirc\!\!\!\rangle$$

will react with iodine to form cyclopentadiene (B)

$$B = \langle\!\!\!\bigcirc\!\!\!\rangle$$

by a gas-phase homogeneous reaction:

$$A(g) + I_2(g) = 2\,HI(g) + B(g)$$

The equilibrium constant is

$$K_p(\text{homo.}) = \frac{P_{HI}^2 P_B}{P_{I_2} P_A} \tag{6.50}$$

But I_2 exists as a solid, so it is also possible to have a heterogeneous reaction:

$$A(g) + I_2(s) = 2\,HI(g) + B(g)$$

$$K_p(\text{hetero.}) = \frac{P_{HI}^2 P_B}{P_A} \tag{6.51}$$

In the second case, of course, there will be some I_2 in the gas phase, as determined by the equilibrium vapor pressure of

$$I_2(s) = I_2(g)$$

Does the heterogeneous reaction occur on the surface or in the gas phase? Kinetically this is an important question, but thermodynamically it is irrelevant. If the $I_2(g)$ reacts, it must be replaced by the evaporation of $I_2(s)$ to maintain the equilibrium vapor pressure, at least as long as there is *any* amount of solid present. The homogeneous equilibrium condition, Eq. (6.50), must be maintained—but P_{I_2} will be *constant* if solid I_2 is present. In fact $P_{I_2} = P_{I_2}^\circ$, the equilibrium vapor pressure of $I_2(s)$ at that temperature. From Eq. (6.50) (for the case that solid is present):

$$\frac{P_{HI}^2 P_B}{P_A} = K_p(\text{homo.})P_{I_2}^\circ$$

But the left-hand side of this relation is just Eq. (6.51), so:

$$K_p(\text{hetero.}) = K_p(\text{homo.})P_{I_2}^\circ$$

If there is no solid I_2 present, then P_{I_2} becomes variable, and the situation is as usual for homogeneous reactions.

A similar argument could be applied to any heterogeneous reaction. In the Boudouard reaction discussed earlier, for example, there would presumably be some vapor pressure of carbon, however small. In this case the reaction is most certainly a surface reaction, but this has no effect on any of the thermodynamic calculations.

Postscript

The Boudouard reaction, mentioned several times in this chapter, is important industrially in a negative sense — it is a nuisance. The Fischer-Tropsch process is used to make liquid fuels from coal; the coal is reacted with steam to form CO and H_2; the CO then reacts with H_2 to form hydrocarbons:

$$n\text{CO} + 2n\,\text{H}_2 \longrightarrow (-\text{CH}_2-)_n + \text{H}_2\text{O}$$

The hydrogen is subsequently regenerated by the water-gas shift reaction

$$\text{CO} + \text{H}_2\text{O} \longrightarrow \text{H}_2 + \text{CO}_2$$

Both of these reactions are catalyzed heterogeneously and complicated by the formation of coke on the surface by the Boudouard reaction:

$$2\,\text{CO} \longrightarrow \text{C(s)} + \text{CO}_2$$

and by the reaction:

$$\text{H}_2 + \text{CO} \longrightarrow \text{C(s)} + \text{H}_2\text{O}$$

The Fischer-Tropsch process was important historically; during World War II this process provided much of the liquid fuels for Germany's tanks and airplanes. During the 1970s, interest in the process revived when oil prices skyrocketed. (*Chemical & Engineering News*, October 26, 1981, gives an overview of this subject.)

If nothing else, this chapter should have impressed you with the immense value of thermodynamic data compilations. They make dry reading to be sure (the heat capacity of NaI is 16.76 cal K^{-1} mol^{-1} at 1900 K, the entropy of LiCl is 72.96 cal K^{-1} mol^{-1} at 3600 K, the heat of formation of NBr is 68.89 kcal mol^{-1} at 5300 K, yawn!), but their value, and the labor required to measure or calculate all those numbers, should not be underestimated.

References

1. W. C. Gardiner, "The Chemistry of Flames", *Scientific American,* February 1982.
2. G. N. Lewis, M. Randall, K. S. Pitzer, and L. Brewer, *Thermodynamics,* 2d ed., 1961: New York, McGraw-Hill Book Co.
3. M. A. Paul, *Principles of Chemical Thermodynamics,* 1951: New York, McGraw-Hill Book Co.

Problems

6.1   Give the formation reactions for the following:

$$C_{12}H_{22}O_{11}(s), \quad PbSO_4(s), \quad NaIO_3(s), \quad C_2H_5OH(liq)$$

6.2   The heat of combustion of acetic acid, $CH_3COOH(l)$, (at 25°C and constant pressure) is $\Delta H_c^\theta = -871.7$ kJ/mol. Use this to determine its heat of formation.

6.3   Calculate the heat of formation of thiophene from its heat of combustion, Table 6.2.

6.4   Calculate the heat of formation of TiN from the data ($T = 298.15$ K):

$$Ti(s) + O_2(g) = TiO_2(s,rutile), \qquad \Delta H = -939.7 \text{ kJ}$$

$$TiN(s) + O_2(g) = TiO_2(s,rutile) + \tfrac{1}{2} N_2(g), \qquad \Delta H = -606.7 \text{ kJ}$$

6.5   The reactions:

$$(1) \quad H_2S(g) + \tfrac{3}{2} O_2(g) = SO_2(g) + H_2O(g)$$

$$(2) \quad SO_2(g) + 2 H_2S(g) = 3 S(s) + 2 H_2O(g)$$

are important in a process for removing H_2S from gas streams. Calculate the heats of these reactions at 298 K.

6.6   Calculate the heat of combustion per kilogram of water gas, which consists of 40% CO, 3% CO_2, 52% H_2, and 5% N_2.

6.7   The heat of vaporization of Hg at 25°C is 31.76 kJ/mole. Calculate ΔH^θ of the reaction:

$$HgO(s,red) = Hg(g) + \tfrac{1}{2} O_2(g)$$

at 25°C from the heat of formation.

6.8   Estimate the heat of the reaction:

$$H_2O(g) + CO(g) = H_2(g) + CO_2(g)$$

at 425°C. Assume constant C_p.

6.9   The reaction:

$$4 H_2(g) + 2 CO(g) = C_2H_4(g) + 2 H_2O(g)$$

is one of those in the Fischer-Tropsch synthesis of hydrocarbons. Derive a formula for the

ΔH of this reaction as a function of T, assuming constant heat capacities (Tables 2.1 and 6.1). Calculate ΔH at 600 K.

6.10   Derive a formula for the heat of the reaction:

$$C(\text{graphite}) + 2\,H_2(g) = CH_4(g)$$

as a function of T. (Use data from Tables 6.1 and 2.1 and assume constant heat capacity.) Calculate ΔH at 500 and 1000 K.

6.11   Estimate the adiabatic flame temperature for burning propane (C_3H_8,g) in air. Assume the reactants start at 298 K; neglect dissociation.

6.12   The heat from chemical reactions cannot be dissipated immediately, so the products of an exothermic reaction can become quite hot; the maximum temperature can be estimated by assuming an adiabatic reaction. Estimate the maximum temperature for the lead-chamber process:

$$NO_2 + SO_2 = NO + SO_3$$

if it is carried out with $T_1 = 373$ K. Assume ΔH_{rxn} and C_p's independent of temperature.

6.13   The reaction:

$$C(s) + S_2(g) = CS_2(g)$$

was equilibrated at 1282 K; the product mixture was found to be 85% CS_2. Calculate K_p.

6.14   The dimerization of acetic acid ($M = 60.05$ g/mole):

$$2\,HAc(g) = (HAc)_2(g)$$

was studied by determining the apparent molecular weight (M_a) from vapor-density measurements. At 132°C, $P = 0.4862$ atm, $M_a = 78.13$ g. Calculate K_p.

6.15   The reaction:

$$2\,NaHCO_3(s) = Na_2CO_3(s) + H_2O(g) + CO_2(g)$$

has a total $P = 0.5451$ atm at 90°C.
(a) Calculate K_p from this pressure.
(b) Calculate the total pressure if the $NaHCO_3$ is placed into a container with 1 atm of CO_2 (before reaction) at 90°C.

6.16   (a) Show that the dissociation pressure (P) of ammonium carbamate:

$$NH_2COONH_4(s) = 2\,NH_3(g) + CO_2(g)$$

is related to the equilibrium constant as $K_p = \frac{4}{27}P^3$.
(b) At 25°C, the total pressure of the above reaction is 0.117 atm. Calculate ΔG_f^θ of ammonium carbamate.

6.17   Calculate ΔG_f^θ at 298.15 K for C_2H_5OH(liq) from its heat of combustion (Table 6.2) and its standard entropy (Table 3.2).

6.18   A standard reaction for coal gasification is:

$$H_2O(g) + \text{coal} = H_2(g) + CO(g)$$

(a) Treating coal as (approximately) graphite, calculate the equilibrium constant of this reaction at 1000 K (use free-energy functions).
(b) Calculate the percent H_2O reacted if the initial H_2O pressure is 10 atm (assume ideal gas).

6.19   Calculate the equilibrium constant for the reaction:

$$Ti(s) + \tfrac{1}{2} N_2(g) = TiN(s)$$

at 1500 K. Will this reaction be spontaneous if $P_{N_2} = 1$ torr?

6.20   Oxygen as an impurity in a gas stream can be removed by passing the gas over hot copper turnings and permitting the reaction:

$$Cu(s) + \tfrac{1}{2} O_2(g) = CuO(s)$$

What would be the concentration of O_2 (per cm^3) in the gas stream following this reaction at 500 K? (Use data from Table 6.5.)

6.21   Calculate the equilibrium constant K_a for the formation of ammonia at 773 K. Use $\Delta H_f^\theta(298) = -10.97 \pm 0.1$ kcal. The free energy functions (298 based) from the JANAF tables are given below (in cal K^{-1}).

	H_2	N_2	NH_3
$\phi'(700)$	33.153	47.731	48.647
$\phi'(800)$	33.715	48.303	49.467

Check the effect of the error in ΔH by repeating the calculation with $\Delta H_f = -10.90$ kcal. Compare your answer to Table 6.8.

6.22   One of the oldest reactions studied in a laboratory is the conversion of mercuric oxide to mercury by heat:

$$2\,HgO(s,red) = 2\,Hg(liq) + O_2$$

Estimate the temperature at which this conversion will occur if HgO is heated in air. For this reaction at 25°C, $\Delta H = 303.67$ kJ, $\Delta G^\theta = 180.60$ kJ. (You may assume ΔH is constant.) (When the vapor is condensed, shining droplets of mercury appear. This seemingly magical transformation of a red powder to a silvery liquid was used by alchemists to impress their research sponsors.)

6.23   The heat-capacity change for a reaction can be expressed (as in Table 2.1):

$$\Delta C_p = \Delta a + \Delta b T + \frac{\Delta c}{T^2}$$

(a) Derive an expression for $\ln K$ in terms of Δa, Δb, Δc, the enthalpy of reaction at 298.15 K, and two constants of integration (I and J) which can be determined if K and ΔH are known at any one temperature.

(b) Derive such a formula for the Haber synthesis and compare its results to Table 6.8.

6.24   (a) Use the results for $\Delta H(T)$ obtained in Problem 6.10 to derive a formula for $\ln K$ as a function of T for the reaction:

$$C(s) + 2\,H_2(g) = CH_4(g)$$

(b) Compare your calculation at 1000 K to that obtained from the free-energy functions, Table 6.4.

6.25   Estimate ΔH^θ of the reaction:

$$H_2(g) + \tfrac{1}{2} S_2(g) = H_2S(g)$$

given $K_p = 106$ at 1023 K, $K_p = 20.2$ at 1218 K.

6.26   Estimate the equilibrium constant at 500 K for:

$$SO_2(g) + 2 H_2S(g) = 3 S(s) + 2 H_2O(g)$$

Assume ΔH is independent of T.

6.27   From the equilibrium constants below for the reaction:

$$V_2O_5(s) + SO_2(g) = V_2O_4(s) + SO_3(g)$$

calculate (a) ΔH^θ_{rxn}, (b) ΔG^θ_{rxn}, (c) the percent of the SO_2 converted to SO_3 at 878 K.

T	K_p
831	0.0154
857	0.0170
878	0.0182
906	0.0202
918	0.0215

6.28   The dissociation pressures of:

$$CaCO_3(s) = CaO(s) + CO_2(g)$$

are given below. Calculate the heat of this reaction. Use either linear regression (Appendix I) or a graphical method.

T	P (atm)	T	P (atm)
1115.4	0.4513	1177.4	1.157
1126.0	0.5245	1179.6	1.151
1127.6	0.5317	1210.1	1.770
1142.0	0.6722		

6.29   In Problem 6.9 the heat of reaction for:

$$4 H_2(g) + 2 CO(g) = C_2H_4(g) + 2 H_2O(g)$$

was estimated as:

$$\Delta H = -191{,}844 - 61.98T$$

Derive a formula for $\ln K$ as a function of T and estimate K at 600 K.

6.30   The equation of state:

$$PV_m = RT(1 + 6.4 \times 10^{-4} P)$$

works for hydrogen gas at 25° up to 1500 atm with a maximum of 0.5% error. Use it to calculate the fugacity of hydrogen at 500 and 1000 atm.

6.31   Show that the fugacity of a van der Waals gas is given by:

$$f = \left(\frac{RT}{V_m - b}\right) \exp\left(\frac{b}{V_m - b} - \frac{2a}{RTV_m}\right)$$

This is best done by integrating Eq. 6.35 by parts from a low-pressure point (P^*, V^*) at which the ideal gas law is valid $(P^*V^* = RT, f^* = P^*)$; then find the limit $P^* \to 0$, $V^* \to \infty$.

6.32   Calculate the fugacity coefficients of methane between 100, 500, and 1000 atm from the data below ($T = 203$ K).

P (atm)	z	P (atm)	z
1	0.9940	160	0.5252
10	0.9370	180	0.5752
20	0.8683	200	0.6246
30	0.7928	250	0.7468
40	0.7034	300	0.8663
50	0.5936	400	1.0980
60	0.4515	500	1.3236
80	0.3429	600	1.5409
100	0.3767	800	1.9626
120	0.4259	1000	2.3684
140	0.4753		

6.33   Calculate the fugacity coefficients of ammonia between 100 and 1000 atm from the data below ($T = 473$ K).

P (atm)	z	P (atm)	z
10	0.9805	200	0.5505
20	0.9611	300	0.4615
30	0.9418	400	0.4948
40	0.9219	500	0.5567
60	0.8821	600	0.6212
80	0.8411	800	0.7545
100	0.8008	1000	0.8914

6.34   (a) Derive a formula for the fugacity of a gas from the equation of state:

$$PV_m = RT + \beta P + \gamma P^2 + \delta P^3 + \varepsilon P^4$$

(b) For hydrogen at 0°C, $\beta = 1.3638 \times 10^{-2}$, $\gamma = 7.851 \times 10^{-6}$, $\delta = -1.206 \times 10^{-8}$, $\varepsilon = 7.354 \times 10^{-12}$ (P in atm, V in dm^3). Calculate the fugacity of hydrogen at $P = 500$ atm, 0°C.

6.35   Zinc has (20°C) $V_m = 7.1$ cm^3 and (Table 1.8):

$$\kappa_T = 1.5 \times 10^{-7} \text{ atm}^{-1}$$

Estimate the activity of zinc at 298 K and $P = 1000$ atm, assuming (a) it is incompressible and (b) Eq. (1.62) applies.

6.36   A mixture containing 200 atm methane and 400 atm carbon dioxide is heated to 900 K. Use Newton's graphs (Figure 6.9) to calculate the fugacity of each gas.

6.37   The reaction:

$$C(s) + 2\,H_2(g) = CH_4(g)$$

has $K_a = 0.46$ at 873 K. Estimate K_p and the percent methane in the equilibrium mixture when the equilibrium total pressure is 1000 atm. You may assume that graphite is incompressible with density 2.26 g/cm^3.

6.38   The equilibrium:

$$N_2O_4(g) = 2\,NO_2(g)$$

was studied by equilibrating the gas at 37.0°C in a container with $V = 2042$ cm^3. The pressure was adjusted to 750 torr (by allowing gas to escape) and the container weighed; it contained 5.80 g of gas. Calculate the degree of dissociation and K_p at this temperature.

6.39 Calculate the degree of dissociation for:

$$N_2O_4 = 2\ NO_2$$

using data from Table 6.1 at $P = 1$ atm, $T = 298$ K.

6.40 The equilibrium constant for the dissociation of phosgene:

$$COCl_2(g) = CO(g) + Cl_2(g)$$

is 22.5 at 668 K. Calculate the degree of dissociation at $P = 2$ atm.

6.41 Use the calculated equilibrium constant (Table 6.6) for $Br_2 = 2$ Br to calculate the degree of dissociation at 2000 K, 1 atm, for each of the estimated values.

6.42 The lead-chamber process for making sulfuric acid uses the reaction:

$$NO_2(g) + SO_2(g) = NO(g) + SO_3(g)$$

At 373 K, $K_p = 15.8 \times 10^3$. Calculate the percent conversion of SO_2 to SO_3 if the NO_2 and SO_2 are mixed in equal proportions.

6.43 Urea can react with water vapor to form ammonia:

$$CO(NH_2)_2(s) + H_2O(g) = CO_2(g) + 2\ NH_3(g)$$

with $K_p = 1.63$ at 25°C. If urea is stored in a closed container under moist air with an initial water pressure of 20 torr, what will be the pressure of ammonia at equilibrium?

6.44 The reaction:

$$CO(g) + H_2O(g) = CO_2(g) + H_2(g)$$

has $K_p = 1.374$ at 1000 K. If a mixture of CO (40%) and H_2O (60%) at 10 atm pressure is reacted at 1000 K, what will be the percent H_2 in the equilibrium mixture?

6.45 If hydrogen is equilibrated over solid iodine:

$$I_2(s) + H_2(g) = 2\ HI(g)$$

at 25°C and a total (constant) pressure of 3 atm, what will be the mole fraction of HI in the vapor phase?

6.46 The reaction:

$$C(s) + 2\ H_2(g) = CH_4(g)$$

has $K_a = 1.38$ at 800 K. Calculate the percent methane in the equilibrium mixture for final (equilibrium) pressures of 50 and 500 atm (assume ideal gas).

6.47 Sulfur exists in the vapor phase as either S_2 or S_8. Calculate the mole fraction of S_8 in sulfur vapor at 1000 and 1500 K when the total pressure is 3 atm.

6.48 The reaction:

$$2\ SO_2(g) + O_2\ (g) = 2SO_3(g)$$

has $K_p = 3.46$ at 1000 K. A mixture of 10% SO_2 in air (which is initially 21% oxygen) is reacted (at constant V) with an initial total pressure of 1 atm. What percent of the SO_2 will be converted to SO_3?

6.49   The reaction (A = cyclopentene, B = cyclopentadiene):

$$A(g) + I_2(g) = 2\ HI(g) + B(g)$$

has $K_p = 0.30$ at 600 K. If an equimolar mixture of A and I_2 is reacted, and the equilibrium pressure is 2 atm, what will be the percent conversion of A to B?

6.50   For the reaction of Problem 6.29, $K_p = 7.79$ at 600 K. If a mixture of hydrogen and carbon monoxide with (initial) pressures $P_{H_2} = 4$ atm, $P_{CO} = 1$ atm, is reacted at constant V, T, calculate the partial pressure of ethylene and the total pressure in the final equilibrium mixture.

6.51   The reaction:

$$2\ H_2(g) + S_2(g) = 2\ H_2S(g)$$

has $K_p = 408$ at 1218 K. A reaction mixture with initial pressures of 2 atm for H_2 and 1 atm for S_2 is reacted at constant V. Calculate the final partial pressure of H_2S.

6.52   Titanium metal is to be used for a high-temperature process in which coking is possible. If elemental carbon is deposited, a carbide can form:

$$C(s) + Ti(s) = TiC(s)$$

Determine whether this reaction is spontaneous at 1500 K.

6.53   (a) Use spectroscopic data from Chapter 5 for the reaction $Br_2(g) = 2\ Br(g)$ (using $\Delta H_0^\theta = 1.901 \times 10^5$ J) to derive the equation:

$$\ln K = 0.4057 + 1.5 \ln T + \ln(1 - e^{-470/T}) - \frac{22{,}864}{T} + 2 \ln(4 + 2e^{-5302/T})$$

(b) Calculate the degree of dissociation at 1 atm, 2200 K.

6.54   (a) Calculate the equilibrium constant at 623 K for the ionization of cesium vapor:

$$Cs(g) = Cs^+(g) + e^-$$

from the ionization potential of Cs (3.89405 eV) and the mass of an electron (9.1091×10^{-28} g). (*Note:* 1 eV/molecule = 96,487 J/mol; Cs has an electronic degeneracy $g_0 = 2$ while Cs^+ has $g_0 = 1$; neither has any low-lying excited states; for the electron, use $g_0 = 2$.)

(b) At 623 K, the vapor pressure of Cs is 7.8×10^{-3} atm. Calculate the number of electrons per cubic centimeter in the vapor at this temperature.

6.55   The table below lists the standard free energies of formation of several hydrates of $MgCl_2$ at 25°C. Which hydrate will be the stable form in air at 25° if the relative humidity is 80%? (The vapor pressure of water is 23.76 torr at 25°C.)

$MgCl_2$	$\Delta G_f^\theta = \ -592.33$ kJ
$MgCl_2 \cdot H_2O$	-862.36
$MgCl_2 \cdot 2H_2O$	-1118.5
$MgCl_2 \cdot 4H_2O$	-1633.8
$MgCl_2 \cdot 6H_2O$	-1278.8

6.56   What would be the minimum pressure of carbon dioxide required to change calcium oxide to calcium carbonate (calcite) at 25°C?

6.57   The formation of nitrogen oxides (collectively called NO$_x$) in internal combustion engines that use air is a significant factor in air pollution. Ignoring other reactions, determine the partial pressure of NO in a combustion chamber at 2000 K if there is 80 atm nitrogen and 10 atm oxygen present (before reaction).

6.58   Show that the change of the Helmholtz free energy A of a reaction — ΔA_{rxn} — is related to K_c as:

$$\Delta A_{rxn}^{\theta} = -RT \ln K_c$$

What approximation or assumptions are implicit in this relationship? Read problem 6.59 for background.

6.59   Standard free energies of gases are (as in this chapter) usually for an ideal gas state at 1 atm. For some purposes it is more convenient to use a standard state of unit concentration (for example, 1 mol/dm^3). Calculate the free energy of formation for CO at 25°C for a standard state of 1 mol/dm^3.

6.60   In Problem 4.26 the pressure required to convert graphite to diamond at 298 K was estimated. For kinetic reasons this process is usually carried out at a higher temperature. Estimate the pressure required at 500 K. (You may assume enthalpies and densities to be independent of T.)

*It is better to know some of the
questions than all of the answers.*

– James Thurber

7

Solutions

Λ great number of chemical reactions occur in solution, including nearly all the biological processes that, together, constitute life. To understand these processes we must learn how thermodynamics is applied to mixtures. This topic, solution thermodynamics, is probably the most difficult that must be taught in a first course in physical chemistry; in the end you may find that more questions have been raised than answers provided. But, in this case as many others, just understanding the question is half the victory.

For gas-phase reactions we were able to get by with an important simplifying assumption. We assumed, often implicitly, that, for example, the thermodynamic properties of a gas that was 20% of a mixture at 1 atm were the same as those of the pure gas at 0.2 atm. This was never completely true but, for gases at moderate pressures, was nearly always good enough. At high densities — in liquids or in solution — it is only rarely a valid assumption.

The key concepts are *chemical potential, activity,* and the *standard states*. How these are defined, measured, and used in solutions is our primary topic (although others will be addressed). In this chapter we shall, for the most part, limit discussion to nonelectrolytes. When the solute can ionize, several new twists are added; these will be covered in Chapter 8.

7.1 Measures of Composition

The mass-action principle [Eq. (6.12)] was originally formulated for solutions in terms of the concentrations of the reactants and products; comparison to the thermodynamic formulation of the equilibrium constant [for example, Eq. (6.18)] should make it clear that activity must be closely related to concentration. There are numerous ways to express the composition of a solution; we shall briefly review a few of the more important ones.

Most of the time we shall consider mixtures with two components — a solvent (n_1 moles) and a solute (n_2 moles). (In some cases, such as acetone + water mixtures, either component could be called the solvent or solute, so an arbitrary choice can be made.) The simplest measure of composition is the **mole ratio:**

$$r = \frac{n_2}{n_1} \tag{7.1}$$

We shall distinguish between extensive properties such as volume and free energy, which depend on n_1 and n_2, and intensive properties such as density and chemical potential, which depend only on the relative amounts as expressed in the mole ratio (r) or some other measure of composition. An equally fundamental measure of composition is the **mole fraction:**

$$X_1 = \frac{n_1}{n_1 + n_2} = \frac{1}{1 + r}$$

$$X_2 = \frac{n_2}{n_1 + n_2} = \frac{r}{1 + r} \tag{7.2}$$

Obviously $X_1 = 1 - X_2$, so one of these is redundant.

In some situations it is more convenient to express composition as **molality** (*m*): moles of solute per kg of solvent. The relationship of molality to the mole fraction can be derived by considering a solution containing $n_2 = m$ moles of solute in 1 kg of solvent (molecular weight M_1), so:

$$n_1 = \frac{1000}{M_1}$$

Then it is easily shown from Eq. (7.2) that:

$$X_2 = \frac{m}{m + \dfrac{1000}{M_1}} \tag{7.3}$$

$$m = \frac{1000X_2}{M_1(1 - X_2)} \tag{7.4}$$

Another frequently used measure of composition is the **concentration** (*c*): moles of solute per dm³ of solution (also called the **molarity** — this term is not preferred). This measure is less popular than molality in solution thermodynamics because it will change with temperature. The relationship of *c* to the mole fraction can be derived by considering 1 dm³ of solution containing $n_2 = c$ moles of solute. If ρ is the density of the solution (in g/cm³), then the total mass is 1000ρ, of which $n_2 M_2$ grams are solute, and the remainder is solvent.

$$\text{mass of solvent} = 1000\rho - n_2 M_2$$

$$n_1 = \frac{1000\rho}{M_1} - n_2 \frac{M_2}{M_1}$$

The reader should now use Eq. (7.2) to derive:

$$X_2 = \frac{cM_1}{1000\rho + c(M_1 - M_2)} \tag{7.5}$$

Exercise: Determine the mole fractions for a 1-molal solution of a solute in water ($M_1 = 18.01$). (*Answer:* $X_1 = 0.9823$, $X_2 = 1.77 \times 10^{-2}$.)   ∎

7.2 Partial Molar Quantities

In previous chapters we examined thermodynamic changes for a variety of processes including changes of *T* or *P*, phase changes, and chemical reaction. Now, in the same fashion, we consider the mixing process with:

$$\Delta(\text{mixing}) = (\text{property of solution}) - \sum (\text{properties of pure substance})$$

For example, the volume change on mixing:

$$\Delta V_{\text{mix}} = V(\text{solution}) - V(\text{solvent}) - V(\text{solute}) \tag{7.6}$$

Table 7.1   Symbols for thermodynamic properties of solutions

Material	Symbols			
Pure liquid i	$V_i^{\bullet}$	$H_i^{\bullet}$	$S_i^{\bullet}$	$G_i^{\bullet}$
Pure liquid i per mole	$V_{mi}^{\bullet}$	$H_{mi}^{\bullet}$	$S_{mi}^{\bullet}$	$\mu_i^{\bullet}$
Whole solution	V	H	S	G
Solution/(total moles of all components)[a]	V_m	H_m	S_m	G_m
Partial molar of i in solution	$\overline{V}_i$	$\overline{H}_i$	$\overline{S}_i$	μ_i
Apparent molar (of solute)	$^{\phi}V$	$^{\phi}H$		
Reference state, per mole	V_i^{θ}	H_i^{θ}	S_i^{θ}	μ_i^{θ}

[a]For example, if a solution containing n_1 moles of solvent and n_2 moles of solute has a volume V, then $V_m = V/(n_1 + n_2)$.

The total volume of the solution (two components) will depend on four variables, T, P, n_1, and n_2. A generalization of the slope formula gives:

$$dV = \left(\frac{\partial V}{\partial T}\right)_{P,n_1,n_2} dT + \left(\frac{\partial V}{\partial P}\right)_{T,n_1,n_2} dP + \left(\frac{\partial V}{\partial n_1}\right)_{T,P,n_2} dn_1 + \left(\frac{\partial V}{\partial n_2}\right)_{T,P,n_1} dn_2 \quad (7.7)$$

The coefficient of dT is related to the coefficient of thermal expansion [α, Eq. (1.59)] of the solution, while the coefficient of dP is related to the isothermal compressibility [κ_T, Eq. (1.60)]. The other coefficients are called the **partial molar volumes,** denoted $\overline{V}_i$:

$$\overline{V}_i \equiv \left(\frac{\partial V}{\partial n_i}\right)_{T,P,n_{j \neq i}} \quad (7.8)$$

(There is no uniformity of notation for partial molar quantities; a summary of the notation used in this chapter is given in Table 7.1.)

A similar equation could be written for any extensive property — for example, the free energy:

$$dG = \left(\frac{\partial G}{\partial T}\right)_{P,n_1,n_2} dT + \left(\frac{\partial G}{\partial P}\right)_{T,n_1,n_2} dP + \left(\frac{\partial G}{\partial n_1}\right)_{T,P,n_2} dn_1 + \left(\frac{\partial G}{\partial n_2}\right)_{T,P,n_1} dn_2 \quad (7.9)$$

This generalization of Eqs. (3.22) and (4.2) defines the partial molar free energy, also called the **chemical potential:**

$$\mu_i \equiv \left(\frac{\partial G}{\partial n_i}\right)_{T,P,n_{j \neq i}} \quad (7.10)$$

The other slopes were given previously (Table 3.1); together these give:

$$dG = -S\,dT + V\,dP + \mu_1\,dn_1 + \mu_2\,dn_2 \quad (7.11)$$

Similar definitions can be made for any extensive property including the **partial molar enthalpy:**

$$\overline{H}_i \equiv \left(\frac{\partial H}{\partial n_i}\right)_{T,P,n_{j \neq i}} \quad (7.12)$$

and the **partial molar entropy:**

$$\overline{S}_i \equiv \left(\frac{\partial S}{\partial n_i}\right)_{T,P,n_{j \neq i}} \tag{7.13}$$

These partial molar quantities are the key to understanding solutions, and their significance is by no means readily apparent. To illustrate some applications of this concept we will primarily utilize volume, since it is easiest to visualize; in the end, however, it is the chemical potential that will be most important.

Partial Molar Volumes

The change in volume of a solution due to adding more material, dn_1 or dn_2, at constant T and P is, from Eqs. (7.7) and (7.8):

$$dV = \overline{V}_1 \, dn_1 + \overline{V}_2 \, dn_2 \tag{7.14}$$

The partial molar volumes depend on concentration and not on the amount of solution, so Eq. (7.14) can be integrated along a path of constant concentration, along which the partial molar volumes are constant. Suppose the moles of each component increased by an amount (dn) that is in proportion to the amount already there:

$$dn_1 = n_1 \, d\lambda, \qquad dn_2 = n_2 \, d\lambda$$

The increase in the volume of the solution will be $dV = V \, d\lambda$, so Eq. (7.14) becomes:

$$V \, d\lambda = \overline{V}_1 n_1 \, d\lambda + \overline{V}_2 n_2 \, d\lambda$$

Then, dividing by $d\lambda$, we get:

$$V = n_1 \overline{V}_1 + n_2 \overline{V}_2 \tag{7.15}$$

The volume change on mixing, the difference between the solution volume (V) and the total volume of the separated components, is:

$$\Delta V_{\text{mix}} = V - (n_1 V_{m1}^{\bullet} + n_2 V_{m2}^{\bullet})$$

where $V_{mi}^{\bullet}$ denotes the molar volume of the pure liquid (cf. Table 7.1). Using Eq. (7.15) gives:

$$\Delta V_{\text{mix}} = n_1(\overline{V}_1 - V_{m1}^{\bullet}) + n_2(\overline{V}_2 - V_{m2}^{\bullet}) \tag{7.16}$$

Another quantity frequently reported is the **apparent molar volume** $({}^{\phi}V)$ of the solute; this is the change in the volume of the solution when n_2 moles of solute is added to n_1 moles of pure solvent, per mole of solute:

$$^{\phi}V = \frac{V - n_1 V_{m1}^{\bullet}}{n_2} \tag{7.17}$$

Equations (7.15) and (7.16) suggest that the partial molar volume is something like an "effective" molar volume of the substance in the solution, but this idea must be used with care, since partial molar properties of a solute depend on the solvent as much as the solute. Molar volumes, apparent molar volumes, and partial molar volumes are

Table 7.2   Volumes of solutes in 1-Molal solutions in N-methylacetamide (CH₃CONHCH₃)

(Units: cm³/mol)

Solute	$V^{\bullet}_{m2}$	$^{\phi}V$	$\overline{V}_2$
n-Octane	164.8	165.9	166.0
n-Nonane	180.8	182.8	182.8
n-Decane	196.9	199.75	199.5
3,3-Diethylpentane	172.0	173.6	173.7

Source: Q. D. Craft and R. H. Wood, *J. Solution Chem.*, 6, 525 (1977).

closely related, particularly in dilute solution (see Table 7.2). However, in some situations they may differ, drastically. For example, solutions of highly charged salts in water often show apparent molar volumes that are *negative* because of the *electrostriction effect* (see Figure 7.1). The electrostriction effect refers to the tendency of highly charged solutes to break down the structure of water; thus, the negative apparent volume of the solute is actually due to a change in the structure of the solvent.

Why shouldn't molar volume and partial molar volume be the same? If we mixed a bushel of oranges with a bushel of apples, we would most likely get two bushels of fruit; the ΔV_{mix} is zero and the volume occupied by an orange in the mixture would be the same as the volume needed in a basket of oranges. On the other hand, if we mixed a bushel of sand with a bushel of coal (large chunks) we would get a great deal less than two bushels; the ΔV_{mix} is negative. In fact the first sand added would not increase the volume at all and would have an apparent volume of zero. If you had two boxes, one filled with books and the other with baseballs, and attempted to repack

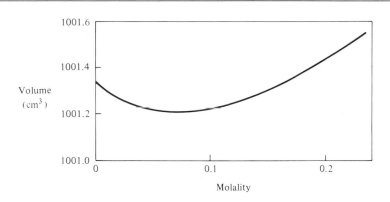

Figure 7.1   Volume of solutions of magnesium sulfate in 1000 g water. The shrinkage of an aqueous solution when highly charged ionic solutes are added is a common phenomenon and is caused by the structure-disrupting effects of the ions (electrostriction effect). (From G. N. Lewis, M. Randall, K. S. Pitzer, and L. Brewer, *Thermodynamics*, 1961: New York, McGraw-Hill Book Co., Fig. 17.1)

them with a random mixture, they wouldn't fit; in this case the ΔV_{mix} is positive. Solutions have these types of effects (called "packing" effects) and more; specifically, molecules have electrical interactions that may be quite strong at short range.

Measurement of Partial Molar Volumes

If the volume of a solution containing n_1 moles of component 1 and n_2 moles of component 2 is measured, the molar volume of the solution can be calculated as:

$$V_m = \frac{V}{n_1 + n_2}$$

A typical graph of V_m vs. X_1 is shown in Figure 7.2. Note that the intercepts of this curve are the molar volumes of the pure liquids. We wish to determine the partial molar volumes at some particular concentration X' (point c in Figure 7.2) at which the molar volume has some value (V_m') (point d). A tangent line is drawn to the curve at the point (c, d). The equation of this tangent line in terms of its intercepts, point a when $X_1 = 1$ and point b when $X_2 = 1$, is:

$$y = b + (a - b)X_1$$

At the point where the tangent line touches the curve, $y = V_m'$ and:

$$V_m' = b + (a - b)X_1'$$
$$= aX_1' + b(1 - X_1')$$
$$= aX_1' + bX_2'$$

If Eq. (7.15) is divided by $n_1 + n_2$, it becomes:

$$V_m = X_1\overline{V}_1 + X_2\overline{V}_2$$

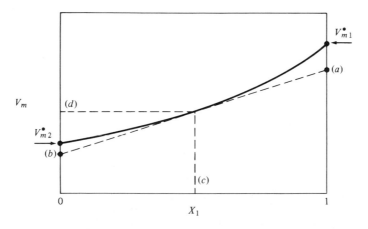

Figure 7.2 The measurement of partial molar volume. The molar volume of the solution is plotted vs. mole fraction. A tangent line at some concentration (c) has intercepts that are (a) the partial molar volume of component 1 and (b) the partial molar volume of component 2 at that concentration.

When we compare this to the equation before, it should be apparent that the intercepts of a tangent line at any mole fraction are just the partial molar volumes at that concentration:

$$a = \overline{V}_1, \qquad b = \overline{V}_2$$

It should also be evident from Figure 7.2 that the partial molar volumes will, in general, depend on concentration (some results are shown in Figure 7.3). More practically, partial molar volumes can be measured by mixing weighted portions of the

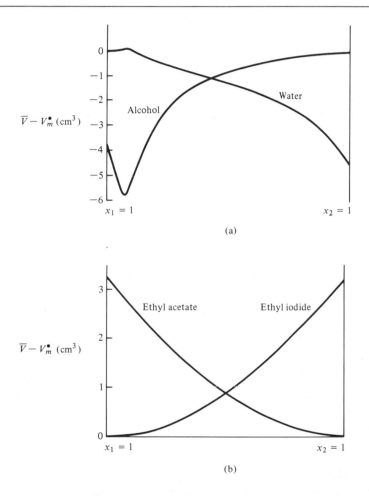

(a)

(b)

Figure 7.3 Partial molar volumes. (a) The difference between the partial molar volumes and molar volumes of water (component 1) and ethanol (component 2) at various concentrations of their solutions. (b) The same graph for solutions of ethyl acetate (component 2) and ethyl iodide (component 1). In the first case the solution volume will be somewhat smaller than the sum of the volumes mixed to make the solution; in the second case the volume is larger than the sum of those of the pure components. (From Lewis et al., *Thermodynamics*, Figs. 17.5, 17.6)

components and measuring the resulting volumes. The weights (W) of a solution are related to the moles through the molecular weights (M):

$$W_1 = n_1 M_1, \qquad W_2 = n_2 M_2$$

We denote the weight fraction as:

$$X_1^w = \frac{W_1}{W_1 + W_2}$$

and the specific volume of the solution as:

$$v_s = \frac{V}{W_1 + W_2}$$

A similar derivation to that above will give:

$$v_s = X_1^w \frac{\overline{V}_1}{M_1} + X_2^w \frac{\overline{V}_2}{M_2}$$

Therefore, plots of V_s vs. weight fraction will have tangent lines whose intercepts (a' at $X_2^w = 0$, b' at $X_1^w = 0$) are the **partial specific volumes,** (v_s) which are related to the partial molar volumes as:

$$\overline{V}_1 = a' M_1, \qquad \overline{V}_2 = b' M_2$$

7.3   Gibbs' Phase Rule

Many of the methods for measuring activities in solution involve phase equilibria — for example, the equilibrium between a solution and its vapor or a solid (at the freezing point). In this section we shall consider the phase equilibria of multicomponent mixtures; this will be a generalization of the discussion in Chapter 4, which covered only pure substances. Before starting, we must define three terms.

Degrees of freedom (F) is the number of intensive variables that can be independently specified for a thermodynamic system. For pure homogeneous materials this number is two, typically T and P.

Number of components (c) is the number of chemically distinct constituents of a thermodynamic system whose amounts can be independently varied. Often this is obvious; solutions of acetone-chloroform or sucrose-water clearly have two components. A solution of NaCl in water will be dissociated into Na^+ and Cl^- ions, but still $c = 2$, since the requirement of electrical neutrality prevents the concentrations $[Na^+]$ and $[Cl^-]$ from being independent. Acetic acid (HAc) partially dissociates, so an aqueous solution may contain H_2O, HAc, H^+, and Ac^-; still $c = 2$, since the equilibrium requirement for the dissociation fixes the ratio $[H^+][Ac^-]/(HAc)$ at a given (T, P).

In more complex mixtures, the determination of c can be very complicated. A practical definition is to take c to be the *minimum* number of pure compounds from which each phase of the mixture could be created; there may be several choices as to how these pure phases could be specified, but the minimum number is always the same.

Another practical definition is that c is the number of distinct chemical species present minus the total number of constraints (such as material balance or electrical neutrality) and the number of chemical equilibria that must be satisfied.

Example: Determine the number of components in the system $NH_4Cl(s) = NH_3(g) + HCl(g)$. There are three distinct species present, but (if no excess NH_3 or HCl is added), material balance requires that the amount of NH_3 is equal to the amount of HCl. Also, the equilibrium constant determines the pressure of these gases, so there are two constraints, and $c = 1$. ∎

Example: How many components are there in an aqueous solution of NaCl and KBr? At least five species are present, H_2O, Na^+, Cl^-, K^+, Br^-, but material balance requires $[Na^+] = [Cl^-]$, $[K^+] = [Br^-]$, so there are only three components. (Electrical neutrality is satisfied by the material balance and so is not a separate restriction.)

If to the solution above KCl and NaBr are added, how many components are there? There are still five species (as above), but electrical neutrality requires:

$$[Na^+] + [K^+] = [Cl^-] + [Br^-]$$

so there are four components. ∎

Example: Determine the number of components in a mixture of $CaCO_3(s)$, $CaO(s)$, and $CO_2(g)$. Because of the equilibrium:

$$CaCO_3(s) = CaO(s) + CO_2(g)$$

a mixture with any ratio $CaCO_3/CaO$ could be created by mixing the appropriate amounts of $CaCO_3 + CaO$ or $CaO + CO_2$; in either case $c = 2$. One could argue that the mixture could be made from three substances — for example, $Ca + C + O_2$ — but this is not the *minimum* number. Alternatively, one could argue that there are three distinct species present, and one equilibrium that must be satisfied — therefore, $c = 2$. ∎

Note that the definition of the number of components saves you from error caused by species you may not have thought of. For example, in water the ions may be either

$$H_2O = H^+ + OH^-$$

or

$$H_2O + H_2O = H_3O^+ + OH^-$$

or something more complicated. However, each new species you may suspect is canceled by the new equilibrium required to create it, so, if you never thought of them, your conclusion as to the number of components would be the same.

The number of components is not always unambiguous. At room temperature in the absence of a catalyst, one could create a mixture of $H_2 + O_2 + H_2O$ in any proportions, and $c = 3$ for all practical purposes; however, if the reaction:

$$H_2 + \tfrac{1}{2} O_2 = H_2O$$

occurs at a finite rate, then any possible mixture that is consistent with the equilibrium constant could be created with $H_2 + O_2$ alone; therefore $c = 2$.

Number of phases (*p*): A phase is a quantity of material that is homogeneous throughout down to the molecular level. The state of subdivision will not matter (so long as surface effects can be neglected); a single crystal of NaCl is one phase; if it is ground into a fine powder, it is still a single phase. A system consisting of KCl(solid) + KBr(solid) has two phases regardless of whether they are separated or mixed in powder form. [If the degree of subdivision were very great, we would have to include surface effects in the equation for the free energy; see Eq. (4.28) as an example.] On the other hand, some solids can form solid solutions; we would have to distinguish between a mixture of Cu and Ni powders (*p* = 2) and a solid solution Cu-Ni (*p* = 1), the latter being a random mixture on a molecular scale. Liquids that form two phases when mixed, such as oil + water, are called *immiscible*. The two immiscible phases need not be pure, just different in composition. For example, the system *n*-hexane + nitrobenzene at room temperature forms two phases; one is about 80% hexane and the other about 20% hexane in nitrobenzene.

Again, there is a possibility of ambiguity in defining a phase; micelles are an example. Molecules with a long hydrocarbon chain linked to a hydrophilic head group (for example, sodium dodecylsulphate) may, under proper conditions, form large-scale aggregates called micelles; these are spherical with the hydrophilic heads outward and a hydrocarbon interior. Such a situation could be considered either as a solution, one phase, with a very high degree of association (*N*-mer formation with $N \sim 100$) or as a separate phase with a very high surface area; either point of view works, so long as it is pursued consistently.

Equilibrium in Multicomponent Systems

Consider a system with *p* phases (labeled α, β, and so on) and *c* components (labeled *i*). At equilibrium, the temperature of all phases must be the same, otherwise heat would flow from one to the other until the temperature was the same. Likewise all phases must have the same pressure, otherwise one would expand at the expense of another. For two phases, α and β, at the same *T* and *P*, the Gibbs' free energy is:

$$dG^\alpha = -S^\alpha \, dT + V^\alpha \, dP + \sum_{i=1}^{c} \mu_i^\alpha \, dn_i^\alpha$$

$$dG^\beta = -S^\beta \, dT + V^\beta \, dP + \sum_{i=1}^{c} \mu_i^\beta \, dn_i^\beta$$

If an amount of component *i*, dn_i, is transferred from α to β at constant *T, P*, $-dn_i^\alpha = +dn_i^\beta = dn_i$, and the change in the total free energy is:

$$d(G^\alpha + G^\beta) = \sum_i (\mu_i^\beta - \mu_i^\alpha) \, dn_i$$

If the system is to remain in equilibrium, the total free energy must not change: $d(G^\alpha + G^\beta) = 0$. Therefore, the condition for equilibrium is:

$$\mu_i^\alpha = \mu_i^\beta \tag{7.18}$$

for each component of the system.

The total number of intensive variables for such a system includes $(c - 1)$ concentrations for each of p phases plus two more (which could be, for example, T and P):

$$\text{total variables} = p(c - 1) + 2$$

These are not independent because of the conditions of Eq. (7.18) for equilibrium; there will be $(p - 1)$ such conditions for each of c components:

$$\text{number of constraints} = c(p - 1)$$

The degrees of freedom will be the difference between the total variables and constraints:

$$F = p(c - 1) + 2 - c(p - 1)$$

$$F = c - p + 2 \tag{7.19}$$

In Chapter 4 we considered the phase equilibrium of pure material; Eq. (7.19) with $c = 1$ reflects these results. When there is one phase, $F = 2$; T and P can be varied independently. Where there are two phases, $F = 1$; this means that T and P must be functionally related [for example, Eqs. (4.10) or (4.23)]. When three phases are in equilibrium, the system is invariant $(F = 0)$; this was called a triple point.

If there are two components, the situation as reflected by Eq. (7.19) is somewhat different. Consider the two-component systems $NaCl + H_2O$. There are three independent, intensive variables—for example, T, P, and the concentration of $NaCl$. If the solution is in equilibrium with its vapor phase, there are two phases and $F = 2$; for example, the temperature or concentration can be varied and the vapor pressure will vary accordingly (that is, the vapor pressure will be a function of T and concentration). If $NaCl$ is added until a saturated solution is achieved, there will be three phases (solid $NaCl$, solution, and vapor) so, from Eq. (7.19), $F = 1$. This additional constraint could be expressed as a dependence of the solubility of $NaCl$ in water on temperature. If a dilute solution of $NaCl$ in water is chilled, at some temperature (the freezing point), ice crystals will form. There are now three phases—ice, solution and vapor—and $F = 1$; the freezing temperature will be a function of the concentration of $NaCl$. If, to the freezing mixture, $NaCl$ is added until a saturated solution is achieved, there will be four phases—ice, solid $NaCl$, solution, and vapor—and $F = 0$; that is, there is only one set of the variables, T, P, and $[NaCl]$, at which these four phases can coexist. Obviously this is the maximum number of equilibrium phases possible in a two-component system.

In the preceding example, the pressure was considered to be a totally dependent variable. If the same series of experiments were carried out under an independent external pressure, the results would not be much different if the pressure were modest (~ 1 atm). Condensed-phase equilibria (melting points, solubilities) do not change much with pressure, so it is often useful to define an effective number of degrees of freedom (F'), excluding pressure:

$$F' = c - p + 1 \tag{7.20}$$

If, in the examples above, we had ignored the vapor in counting phases and calculated F' in each case, our conclusions would have been the same. For example, the system ice, solution, solid NaCl under some constant pressure P has $F' = 0$, meaning it is an invariant point so long as P is not changed.

   A practical use of Gibbs' phase rule is in its application to phase diagrams. This topic is covered later in this chapter (Section 7.11).

7.4 Fugacity

The fugacity, discussed briefly in Chapter 6, is a quantity that is logarithmically related to the chemical potential; for component i of a mixture:

$$\mu_i = (\text{constant}) + RT \ln f_i \qquad (7.21)$$

The difference of the chemical potential between two phases is then:

$$\mu_i^\alpha - \mu_i^\beta = RT \ln \frac{f_i^\alpha}{f_i^\beta}$$

Therefore, the condition for equilibrium, Eq. (7.18), can be alternatively expressed, for each component, as:

$$f_i^\alpha = f_i^\beta \qquad (7.22)$$

   If a component of a solution is volatile, Eq. (7.21) provides a method for measuring its fugacity. If the solution is in equilibrium with a vapor phase, for each component:

$$f_i(\text{in solution}) = f_i(\text{vapor})$$

At low pressures, the vapor can be assumed to obey the ideal gas law and [cf. Eq. (6.36)] the fugacity of the vapor will be equal to the vapor pressure:

$$f_i(\text{in solution}) = P_i \qquad (7.23)$$

The fugacity is also called the "escaping tendency," since it is directly measured by the volatility of a component.

   If fugacity and vapor pressure are the same thing, why introduce fugacity at all? In the first place, Eq. (7.23) is correct only at low vapor pressures at which the ideal gas law applies; however, nearly all the examples in this chapter will be in this limit. Also, Eq. (7.23) may not be practical for substances that are not volatile. Considering the equilibrium between solid NaCl and a saturated aqueous solution of NaCl at room temperature, it is true that:

$$f_{\text{NaCl}}(\text{solid}) = f_{\text{NaCl}}(\text{sat'd soln.})$$

This equation is also true for the vapor pressure of NaCl, but at room temperature the vapor pressure is immeasurably small. It is best to regard Eq. (7.23) as a method for measuring fugacity rather than as a defining relationship; Eq. (7.21) is the definition of fugacity.

Example: The vapor pressure of NaCl (solid) can, however, be calculated using methods explained in Chapters 5 and 6. The process is:

$$\text{NaCl(solid)} = \text{NaCl(vapor)}$$

with $K_a = P(\text{NaCl, vapor})$. Reference 4 lists the following properties (all at 298.15 K) for solid NaCl:

$$\phi' = 17.33 \text{ cal/K}, \qquad \Delta H_f^\theta = -98.2 \times 10^3 \text{ cal}$$

Also, for NaCl(gas) the properties listed by ref. 4 (calculated from spectroscopic data) are:

$$\phi° = 47.18 \text{ cal/K}, \qquad \Delta H_f^\theta = -42.7 \times 10^3 \text{ cal}$$

$$H_m^\theta(298.15 \text{ K}) - H_0^\theta = 2295 \text{ cal}$$

Then [using Eq. (5.85)]:

$$\phi'(\text{gas}) = 47.18 + \frac{2295}{298.15} = 54.88 \text{ cal/K}$$

Then, Eq. (6.28) can be used to calculate the equilibrium constant for vaporization (that is, the vapor pressure) with:

$$\Delta\phi' = \phi'(\text{gas}) - \phi'(\text{solid}) = 37.55 \text{ cal/K}$$

$$\Delta H^\theta(298.15) = -42.7 \times 10^3 + 98.2 \times 10^3 = 55.5 \times 10^3 \text{ cal}$$

$$\ln K = \frac{37.55}{1.987} - \frac{55.5 \times 10^3}{(1.987)(298.15)} = -74.79$$

Therefore, the vapor pressure of NaCl at 25°C is:

$$P = 3.3 \times 10^{-33} \text{ atm}$$

This amounts to one *molecule* in $1 \times 10^7 \text{ m}^3$! (*Note:* The spectroscopic measurements from which the vapor properties were calculated were obtained at elevated temperatures.) ∎

Standard States

Since absolute chemical potentials cannot be measured, it is usual to measure them relative to a *standard* or *reference* state; from Eq. (7.21):

$$\mu_i - \mu_i^\theta = RT \ln \frac{f_i}{f_i^\theta} \tag{7.24}$$

This relationship defines the **relative activity:**

$$a_i = \frac{f_i}{f_i^\theta} \tag{7.25}$$

Another way to discuss relative fugacities is with respect to some *ideal* case for which the fugacity (f^{id}) can be calculated or measured (often in some limiting case); this is used to define an **activity coefficient:**

$$\gamma_i = \frac{f_i}{f_i^{\text{id}}} = \frac{a_i}{a_i^{\text{id}}} \tag{7.26}$$

For gases, these relationships were used in Chapter 6. The ideal case was then, of course, the ideal gas, and the reference state was the ideal gas at 1 atm. Molecules behave ideally at low pressure — that is, when they are near no other molecules. There

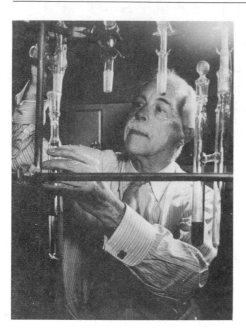

Gilbert N. Lewis (1875–1946)

Lewis is well memorialized in chemistry by Lewis acids and bases; Lewis pairing and Lewis structures are still taught as the simplest way to understand chemical bonding and structure. However, his somewhat more prosaic contributions in the area of thermodynamics were equally important and far-reaching in their influence on chemistry.

The concepts of free energy and chemical potential introduced by Gibbs held the key to chemical thermodynamics, but few chemists at the time (ca. 1900) understood or appreciated this work. Lewis expanded upon and applied these concepts to chemical problems and, in the process, invented fugacity, activity, and partial molar quantities and derived the exact form for Nernst's equation for electrochemical potential (Chapter 8). He also did a great deal of work in collecting, correcting, and systematizing thermodynamic data on free energies. His text, Thermodynamics, *written with Merle Randall in 1923 set the pattern for physical chemical education for the next several generations.*

In later years Lewis became interested in photochemistry, and two years before his death he made the suggestion that led to the discovery of the triplet state by Melvin Calvin and Michael Kasha. A chemist to the end, he died in the laboratory while conducting an experiment.

is no reason that this reference state could not be used for solutions, but for practical reasons it usually is not.

There are two commonly used reference states for solutions: (1) The *solvent* (or *Raoult's law*) standard state. In this case, a molecule will behave ideally when it is completely surrounded by other molecules of the same type as far as it can "see." In other words, the reference state is the pure liquid. (2) The *solute* (or *Henry's law*) standard state. In this case, a solute molecule behaves ideally when it is completely surrounded by solvent molecules as far as it can "see"—that is, in the dilute-solution limit.

How far can a molecule "see"? This depends on the long-range portion of the intermolecular potential. Neutral, nonpolar molecules are well represented by the Lennard-Jones potential [Eq. (1.43)], so their energy of interaction will drop off as r^{-6}. Polar molecules have potentials that, at long range, tend to drop off as r^{-4}; also, the intermolecular forces at a given distance are stronger, so polar molecules can "see" farther. Charged species obey Coulomb's law, with the potential dropping as r^{-1}; this is a very long-range interaction, and very strong, so ions can "see" very far. Thus, "dilute solution" is a term whose meaning depends on the nature of the solute. [The situation is quite analogous to that in gases (Chapter 1), where the deviation from ideality at a given pressure depended on the strength of the intermolecular forces.]

As mentioned earlier, in certain mixtures (water-acetone, benzene-chloroform) either component could be called the solvent. In such a case it would be reasonable to use the solvent standard state for both components. This is the first case we shall discuss under "Raoult's law."

Fugacity, activity, chemical potential—these are by no means obvious concepts, so it will not be surprising if the reader is a bit baffled at this point. With any luck, they will all become clearer as we discuss the specifics in subsequent sections. We shall see many equations for fugacity, activity, and chemical potential; the reason for starting as we did is to emphasize that the relations of this section are *defining relations,* which will be true in all situations that will be encountered.

7.5 Raoult's Law

If the number of moles in a two-component solution is altered at constant T and P, Eq. (7.11) becomes:

$$dG = \mu_1 \, dn_1 + \mu_2 \, dn_2 \tag{7.27}$$

The identical reasoning that led from Eq. (7.14) to Eq. (7.15) will demonstrate that the total free energy of a solution containing (n_1, n_2) moles is:

$$G = \mu_1 n_1 + \mu_2 n_2 \tag{7.28}$$

The free energy of mixing is then:

$$\Delta G_{\text{mix}} = G - (n_1 \mu_1^\bullet + n_2 \mu_2^\bullet)$$
$$\Delta G_{\text{mix}} = n_1(\mu_1 - \mu_1^\bullet) + n_2(\mu_2 - \mu_2^\bullet) \tag{7.29}$$

If both components are volatile, the chemical potential of the component i in solution relative to the pure liquid is readily obtained from Eqs. (7.21) and (7.23) (for ideal vapor phase):

$$\mu_i - \mu_i^\bullet = RT \ln \frac{f_1}{f_1^\bullet} \cong RT \ln \frac{P_i}{P_i^\bullet} \tag{7.30}$$

$$\Delta G_{\text{mix}} = RT\left(n_1 \ln \frac{P_1}{P_1^\bullet} + n_2 \ln \frac{P_2}{P_2^\bullet}\right) \tag{7.31}$$

where $P_1^\bullet$ and $P_2^\bullet$ are the vapor pressures of the respective pure liquids at the same temperature as the solution.

If we had a solution with equal amounts of two components, we would expect the vapor phase to be richer in the more volatile constituent; if we had a mixture of two components with equal volatilities, we would expect to find, in the vapor phase, a concentration of either component proportional to that in the solution. In other words, the vapor pressure of a component in a solution will be proportional to its volatility (which can be roughly measured by its vapor pressure when pure) and its concentration. This was first demonstrated by the systematic investigations of F. M. Raoult (1888). These results are usually expressed as **Raoult's law (RL):**

$$P_i = X_i P_i^\bullet \tag{7.32}$$

Actually Raoult's results did not show the strict proportionality of Eq. (7.32); the definitive study of Zawidski (1900) for 13 binary mixtures found only one mixture (CH_2Br-CH_2Br + CH_2Br-$CHBr$-CH_3) that obeyed Eq. (7.32) closely over the full range of concentrations. Evidently this "law" is obeyed more often than not in the breach; Eq. (7.32) can be used quantitatively only for a very limited class of solutions for which the components are very similar in size, shape, and intermolecular forces—for example, benzene + toluene, *ortho*-xylene + *meta*-xylene, butane + pentane. Nonetheless it is a very useful idea.

Before proceeding, let's try to anticipate a common confusion: Eq. (7.32) is often confused with Dalton's law, which relates partial pressures to the mole fraction *in the vapor phase*. Since we already are using the symbol X_i to mean mole fraction of component *i* in the solution, we shall try to avoid confusion by introducing a new symbol, Y_i, for the mole fraction in the vapor phase; Dalton's law is then

$$P_i = Y_i P, \qquad Y_i = \frac{P_i}{P} \tag{7.33}$$

where P is the total pressure of the vapor phase above the solution.

Exercise: Show that, if both components of a binary mixture obey Raoult's law, the mole fraction in the vapor is related to the mole fraction in the liquid by:

$$Y_2 = \frac{X_2 P_2^{\bullet}}{P_1^{\bullet} + X_2(P_2^{\bullet} - P_1^{\bullet})} \tag{7.34}$$

Hint: Start with Eq. (7.33) and $P = P_1 + P_2$.   ∎

The Ideal Solution (RL)

If Eq. (7.32) is used in Eq. (7.30), the free-energy change of mixing for a solution obeying Raoult's law is:

$$\Delta G_{mix}^{id} = RT(n_1 \ln X_1 + n_2 \ln X_2) \tag{7.35}$$

Such a solution will be called an *ideal solution* (RL).

The properties of such a solution can be found by applying the principles of thermodynamics. From Table 3.1 we find:

$$S = -\left(\frac{\partial G}{\partial T}\right)_P$$

Therefore, for the mixing process:

$$\Delta S_{mix} = -\left(\frac{\partial \Delta G_{mix}}{\partial T}\right)_P$$

Now, Eq. (7.35) gives:

$$\Delta S_{mix}^{id} = -R(n_1 \ln X_1 + n_2 \ln X_2) \tag{7.36}$$

This is exactly the result derived in Chapter 3 [Eq. (3.15)] for mixing *ideal gases*. This result works well for gases because the density is so low that intermolecular interactions are relatively small. For it to work for liquids, we must require that the molecules of all constituents be very similar in size and shape and lacking in specific interactions (such as hydrogen bonds) that would cause a molecule to prefer one type of neighbor over another.

Likewise, from Eq. (4.18) the heat of mixing can be calculated:

$$\Delta H_{mix} = \frac{\partial(\Delta G_{mix}/T)}{\partial(1/T)} \tag{7.37}$$

From Eq. (7.35) we see that $\Delta G^{id}_{mix}/T$ does not depend on T, so:

$$\Delta H^{id}_{mix} = 0 \tag{7.38}$$

Again this is quite reasonable for gases, because intermolecular forces are small; at liquid densities, intermolecular forces are never negligible, but it will suffice for Eq. (7.38) to be true that the forces (in an A + B mixture) of A..A, A..B, B..B interactions are similar in magnitude. If the intermolecular interaction energy is denoted as ε, then this requirement is:

$$\varepsilon_{AA} \cong \varepsilon_{AB} \cong \varepsilon_{BB}$$

This is much less restrictive than requiring all these interaction energies to be negligible.

Standard State (RL)

The concept of the ideal solution suggests the use of the pure liquid as the reference state (denoted as "RL") for a component in solution. Then:

$$\mu_i^\theta = \mu_i^\bullet \tag{7.39}$$

$$a_i = \frac{P_i}{P_i^\bullet} \tag{7.40}$$

The behavior of a real solution can be compared to that of an ideal solution of the same composition by use of Raoult's law:

$$f_i^{id} = X_i P_i^\bullet \tag{7.41}$$

which gives the activity coefficient as:

$$\gamma_i = \frac{f_i}{f_i^{id}} = \frac{P_i}{X_i P_i^\bullet} \tag{7.42}$$

The combination of Eqs. (7.39) and (7.41) shows the relationship between activity and the concentration of a component of a solution:

$$a_i = \gamma_i X_i \tag{7.43}$$

Example: Raoult (1890) reported that a solution of aniline in CS_2 ($X_{CS_2} = 0.9088$) at 24° lowered the vapor pressure of CS_2 by 5.00%. In modern terms we would say (for $CS_2 = A$):

$$a_A = \frac{P_A}{P_A^\bullet} = 1 - 0.0500 = 0.9500$$

$$\gamma_A = \frac{a_A}{X_A} = \frac{0.9500}{0.9088} = 1.045$$

He also reported (at 23°C) that a solution of ethyl benzoate in chloroform (A), with $X_A = 0.9244$, lowered the vapor pressure of chloroform by 8.27%:

$$a_A = \frac{P_A}{P_A^\bullet} = 1 - 0.0827 = 0.9173$$

$$\gamma_A = \frac{0.9173}{0.9244} = 0.9923$$

Note that the latter case is very nearly ideal ($a_A = x_A$) at this concentration.   ∎

Deviations from Raoult's Law

Raoult's law behavior is purely statistical; it would apply to mixtures of red balls and green balls, apples and oranges. All chemistry, the properties that give molecules their individual "personalities," is then to be found in the deviations from this purely statistical behavior. These deviations from ideality are discussed in terms of either activity coefficients or the excess thermodynamic functions such as:

$$\Delta G_{mix}^{ex} = \Delta G_{mix} - \Delta G_{mix}^{id}$$

With Eqs. (7.31), (7.35), and (7.42) it is readily shown that:

$$\Delta G_{mix}^{ex} = RT(n_1 \ln \gamma_1 + n_2 \ln \gamma_2) \tag{7.44}$$

The advantage of this approach is that the deviations can be discussed in terms of entropic (ΔS_{mix}^{ex}) and enthalpic ($\Delta H_{mix}^{ex} = \Delta H_{mix}$) contributions using Eqs. (7.36), (7.37), and/or

$$\Delta G = \Delta H - T\Delta S \tag{7.45}$$

which is valid for any isothermal process including mixing.

Negative deviations from Raoult's law ($\gamma < 1$, $\Delta G_{mix}^{ex} < 0$) occur when the solvent and solute molecules decrease each other's escaping tendency. Sometimes this occurs for a specific reason; for example, acetone and chloroform mixtures can form hydrogen bonds:

This is a relatively strong interaction, which is not available to the pure liquids (cf. Figure 7.4).

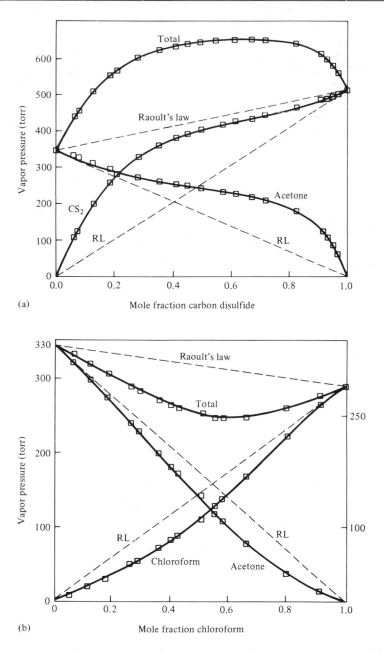

Figure 7.4 Vapor pressure vs. mole fraction diagrams. (a) Carbon disulfide/acetone at 35.17°C, and (b) chloroform/acetone at 35.17°C. The first case shows substantial positive deviations from Raoult's law, while the deviations in the second case are negative. (From J. H. Hildebrand and R. L. Scott, *The Solubility of Nonelectrolytes*, 3d ed., 1950: New York, Van Nostrand Reinhold Company, p. 30, Figs. 1 and 2)

Positive deviations from Raoult's law ($\gamma > 1$, $\Delta G_{mix}^{ex} > 0$) occur when the solute tends to break up strong interactions of the solvent with itself. This nearly always happens with molecules containing OH groups — water, alcohols, organic acids — because these may form strong hydrogen bonds. Figure 7.5 (ethanol + heptane) shows such an example; note how much the volatility of ethanol in the solvent is increased relative to the statistical expectation. For example, at 10% ethanol, Raoult's law would predict a vapor pressure for ethanol of $0.1 \, P^{\bullet}(\text{ethanol}) = 13$ torr; the vapor pressure of ethanol above this solution is actually 92 torr. Correspondingly, the activity of ethanol is much greater than the ideal value.

Positive deviations from Raoult's law result when the molecules "like" their own kind much more than the other; the ultimate expression of this solvent-solute antipathy is for the substances to be immiscible. Figure 7.6 demonstrates this situation for butane + perfluorobutane mixtures. At 296 K these liquids exhibit very strong positive deviations from Raoult's law; these become more pronounced as the temperature is lowered. At 233 K and below these liquids will separate into two immiscible phases, one rich in butane and the other in perfluorobutane; the temperature where this begins is called the *critical solution* or *consolute* temperature. Figure 7.6(d) shows a plot of composition of the two immiscible phases vs. temperature.

Regular Solutions

A great deal of effort has been put into understanding and interpreting solution behavior. One area in which considerable progress has been made is a class of solutions that J. H. Hildebrand called *regular solutions*. Solutions in this class are generally those for which the molecules have no strong specific interactions, such as hydrogen bonding or acid-base association.

Regular-solution theory is simplest for mixtures for which the molecular size is nearly the same. For such cases the chemical potential can be written as:

$$\mu_1 - \mu_1^{\bullet} = RT \ln X_1 + wX_2^2$$
$$\mu_2 - \mu_2^{\bullet} = RT \ln X_2 + wX_1^2 \qquad \textbf{(7.46)}$$

In these equations, w is an empirical parameter (Table 7.3) related to the differences of the interaction energies:

$$w \propto (2\varepsilon_{12} - \varepsilon_{11} - \varepsilon_{22})$$

Table 7.3   Regular-solution constants

Mixture	T	w (joules)	$\dfrac{\partial w}{\partial T}$	$\dfrac{\partial^2 w}{\partial T^2}$
CCl_4 + benzene	298	324	−0.368	−0.021
Benzene + cyclohexane	293	1275	−7.03	−0.046
CCl_4 + cyclohexane	313	267	−0.937	
Benzene + toluene	353	−41	−0.62	0.005
CS_2 + acetone	308	4175	−5.36	−0.043

Source: Reference 4, p. 286; converted by 4.184 J/cal.

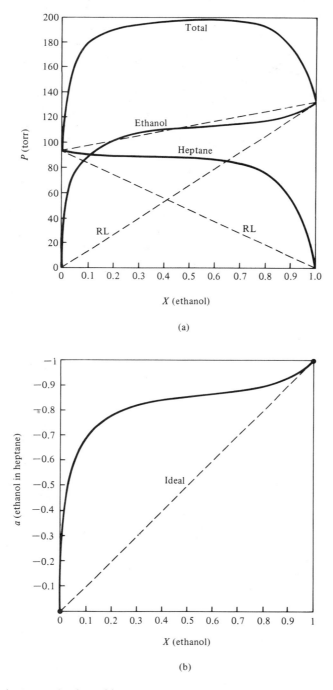

Figure 7.5   Solutions of ethanol/heptane. (a) Vapor pressure vs. composition. (b) Activity of ethanol vs. composition. (Both at 40°C.) [Data from G. A. Ratcliff and K. C. Chao, *Can. J. Chem. Eng.*, 47, 148 (1969)]

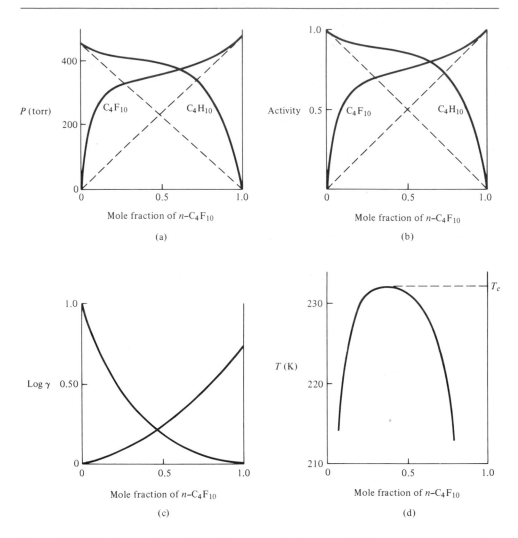

Figure 7.6 Solutions of *n*-perfluorobutane and *n*-butane. (a) Vapor pressure vs. composition. (b) Activity vs. composition. (c) Activity coefficient vs. composition. (All at 296 K.) This system shows large positive deviations from Raoult's law, and the components become immiscible below ca. 232 K; panel (d) is a graph of the compositions of the two immiscible phases vs. temperature. The butane-rich phase (on the left) and the fluorobutane-rich phase become closer in composition as the critical solution temperature (T_c) is approached; above the critical solution temperature the components are totally miscible, and there is only one phase. (From J. H. Hildebrand and R. L. Scott, *Regular Solutions*, 1962: Englewood Cliffs, N.J., Prentice-Hall, Inc., p. 17)

Although the interaction energies can be estimated from heats of vaporization, it is probably best to take w to be an empirical parameter, characteristic of a given binary mixture. It can be shown (Problem 7.17) that the activity coefficients of a regular solution are related to the parameter w by:

$$\ln \gamma_1 = \frac{X_2^2 w}{RT}, \qquad \ln \gamma_2 = \frac{X_1^2 w}{RT} \qquad (7.47)$$

Also, the excess free energy and entropy of mixing is:

$$\Delta G_{mix}^{ex} = (n_1 + n_2) X_1 X_2 w \qquad (7.48)$$

$$\Delta S_{mix}^{ex} = -(n_1 + n_2) X_1 X_2 \frac{\partial w}{\partial T} \qquad (7.49)$$

7.6 Henry's Law

Henry's law (HL) is based on the pioneering research of William Henry (1804) on the solubility of gases in liquids. The results can be expressed as a proportionality between the mole fraction of the solute and the pressure of the gas above the solution:

$$P_2 = k X_2 \qquad (7.50)$$

This law is obeyed very well for gases for which (a) the solubility is sparing, (b) the gas phase is reasonably ideal, and (c) the gas does not react with, nor ionize in, the solvent. Figure 7.7 shows that propane, n-butane and 1-butene in water all obey Henry's law quite well. Figure 7.8, on the other hand, shows that HCl in water does not follow Eq. (7.50) at any pressure. Gas solubilities are given in Table 7.4.

Looked at from another point of view, Eq. (7.50) expresses a proportionality between vapor pressure and concentration; in that sense it is similar to Raoult's law

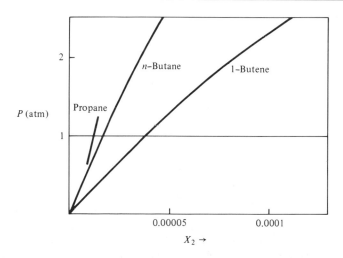

Figure 7.7 Solubility of several hydrocarbon gases in water at 37.8°C. Henry's law, which predicts a linear relationship between solubility and pressure, is reasonably accurate for the gases at and below 1 atm. (From W. Gerrard, *Solubility of Gases and Liquids*, 1976: New York, Plenum Press)

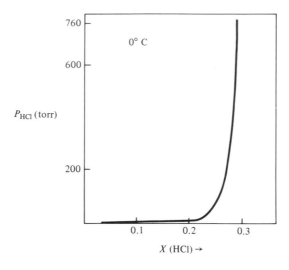

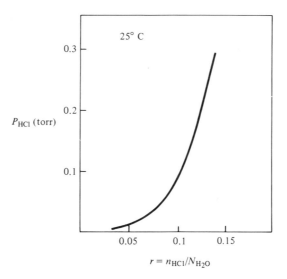

Figure 7.8 Solubility of HCl in water at 0°C and 25°C vs. pressure. Henry's law is not obeyed even approximately in this case, nor for most cases for which the gas ionizes in solution. (From Gerrard, *Solubility of Gases and Liquids*)

(in fact, for a perfectly ideal solution, they are the same, with $k = P_2^\bullet$). Henry's law, however, permits us to choose a constant (k) specifically to fit the empirical behavior in dilute solutions, whereas Raoult's law works for nonideal solutions only at high concentrations — that is, as $X \to 1$. In fact, Henry's law can be used to define an *ideal dilute* solution and a new reference state (denoted "HL"). The Henry's law constant

Table 7.4 Solubility of gases at 1 atm pressure

(25°C unless otherwise specified)

Solvent	Gas	X_2	Solvent	Gas	X_2
Benene	Ar	8.77×10^{-4}	Water	CO_2	7.1×10^{-4}(20°C)
Cyclohexane	Ar	1.49×10^{-3}	Water	CO_2	6.1×10^{-4}(25°C)
n-Hexane	Ar	2.53×10^{-3}	CCl_4	CO_2	1.0×10^{-2}(20°C)
Water	Ar	2.7×10^{-5}	Water	CH_4	2.82×10^{-5}(19.8°C)
Water	N_2	1.2×10^{-5}	Water	C_2H_6	4.00×10^{-5}(19.8°C)
Water	O_2	2.3×10^{-5}	Water	C_3H_8	3.17×10^{-5}(19.8°C)

Source: Most data from ref. 2.

can be precisely defined as the limiting slope of the solute P_2 vs. X_2 curve as $X_2 \rightarrow 0$:

$$k = \lim_{X_2 \rightarrow 0} \left(\frac{P_2}{X_2} \right) \tag{7.51}$$

(Note that HCl in water, Figure 7.8, does not obey Henry's law even in this limiting form; the reason is that, in water, this gas ionizes into H^+ and Cl^-, and Henry's law must be modified for electrolytes. For the time being, we exclude ionizing solutes; we will discuss this subject in Chapter 8.)

Figure 7.9 for solutions of Br_2 in CCl_4 shows the actual vapor pressure of Br_2 along with the two "ideal" cases, Henry's law and Raoult's law. It can be seen that, while RL is best as $X_{Br_2} \rightarrow 1$, HL works best for dilute solutions. In situations where the dilute solution is specifically of interest, it is useful to define an "ideal" state based on Henry's law:

$$f_2^{id} = kX_2 \tag{7.52a}$$

$$\gamma_{2x}(HL) = \frac{f_2}{f_2^{id}} = \frac{f_2}{kX_2} \tag{7.52b}$$

$$a_2(HL) = \gamma_{2x}X_2 = \frac{f_2}{k} \tag{7.52c}$$

Equations (7.52) have been deliberately written in terms of the fugacity rather than the pressure of the solute. If the solute is volatile, then the fugacity in solution can be measured by measuring its fugacity in the vapor phase — that is, its partial pressure. However, the definitions of Eq. (7.52) are useful for nonvolatile solutes as well; in fact, this is the area of its widest application. If one wished to study solutions of substances such as NaCl or sucrose, a standard state (RL) that requires one to know the vapor pressure of the pure liquid (at the same temperature) can be inconvenient, if not impossible, to use — what is the vapor pressure of pure liquid NaCl (mp 800°C) at 25°C? Of course, even with the HL convention, Eq. (7.52) is not practical for *measuring* activities if the solute vapor pressure is vanishingly small; in Section 7.7, we shall learn how activities are measured in such cases.

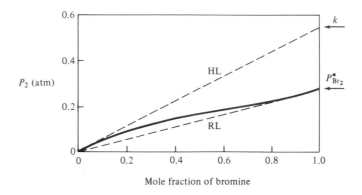

Figure 7.9 Vapor pressure of bromine vs. composition for solutions in carbon tetra-chloride. The reference lines and standard states are shown for Henry's law (HL) and Raoult's law (RL). (From Lewis et al., *Thermodynamics*, Fig. 20.3)

Molality-Scale Henry's Law

Solute concentrations are frequently expressed as molalities rather than mole fraction. A Henry's law constant can be defined in terms of solute molality in a manner similar to Eq. (7.51):

$$k_m = \lim_{m \to 0} \frac{P_2}{m} \qquad (7.53)$$

This is used to define another reference state for which the following equations apply:

$$f^{id} = k_m m \qquad (7.54a)$$

$$\gamma_{2m} = \frac{f_2}{k_m m} \qquad (7.54b)$$

$$a_2(HL) = \gamma_{2m} m \qquad (7.54c)$$

Yet another reference state can be defined if the concentration (c) is used in Eqs. (7.53) and (7.54) in place of molality.

Comparison of RL and HL Activities

We now have four definitions for activity — Eqs. (7.40), (7.52), and (7.54) (for m and c); this should emphasize the point that activity is a *relative* quantity that has no meaning unless the reference state is specified. For a given solution these activities will have entirely different values. Nonetheless, we shall use the same symbol, since there is no conventional terminology to distinguish among them; it will usually be clear from the context which is intended. For example, we shall always use the RL convention for the solvent [Eq. (7.40)]; for a nonvolatile solute, unless otherwise specified, the HL convention with molality scale will be used. These conventions are

Table 7.5   Conventions for solution standard states

| | Activity | | Ideal state |
	Defined	Measured	
Solvent (RL)	$a_1 = \gamma_1 X_1$	$a_1 = \dfrac{P_1}{P_1^{\bullet}}$	$\gamma_1 \to 1$ as $X_1 \to 1$
Solute (HL) (X)	$a_2 = \gamma_{2x} X_2$	$a_2 = \dfrac{P_2}{k}$	
(m)	$a_2 = \gamma_{2m} m$	$a_2 = \dfrac{P_2}{k_m}$	$\gamma_2 \to 1$ as $X_2 \to 0$
(c)	$a_2 = \gamma_{2c} c$	$a_2 = \dfrac{P_2}{k_c}$	

summarized in Table 7.5. [The notation RL or HL will be added where ambiguity is possible. Also, for solutions in water the symbol "aq" is used generally to denote an HL (molality) reference state.]

Note that, in all cases:

$$a = \gamma[\text{concentration}]$$

$$a(\text{ideal}) = [\text{concentration}]$$

with "concentration" being expressed as mol/dm^3 (concentration, strict usage), mol/kg (molality) or mole fraction. The solvent reference state (RL) is the pure liquid, so $\gamma_1 \to 1$ as $X_1 \to 1$. The solute reference state (HL) is a hypothetical state of unit concentration (either $X_2 = 1$, $m = 1$, or $c = 1$) in which the solute behaves as if it were at infinite dilution — that is, obeys Henry's law; then $\gamma_2 \to 1$ as $X_2 \to 0$.

Example: Br_2 in CCl_4 at 25°C and $X_2 = 0.0250$ ($m = 0.167$) has a vapor pressure of 10.27 torr. The Henry's law constant (Problem 7.19) is $k = 391$ torr (mole fraction); on the molality scale (Problem 7.33), $k_m = 61$ torr/molal. The vapor pressure of pure Br_2 at 25°C is $P_2^{\bullet} = 236$ torr. The activities for several reference states are:

$$a_2(\text{RL}) = \frac{P_2}{P_2^{\bullet}} = \frac{10.27}{236} = 0.0435, \qquad \gamma_2 = \frac{a_2}{X_2} = 1.741$$

$$a_2(\text{HL}, X) = \frac{P_2}{k} = \frac{10.27}{391} = 0.0263, \qquad \gamma_{2x} = \frac{a_2}{X_2} = 1.052$$

$$a_2(\text{HL}, m) = \frac{P_2}{k_m} = \frac{10.27}{61} = 0.168, \qquad \gamma_{2m} = \frac{a_2}{m} = 1.006 \qquad \blacksquare$$

7.7   The Gibbs-Duhem Equation

As mentioned earlier, if a material is not volatile, we cannot measure its activity using its vapor pressure. In this section we shall develop a method that will permit us to calculate the activity of one component (the solute) from measurements of the activity of the other component (the solvent).

Equation (7.9), with T and P constant, gave:

$$dG = \mu_1 \, dn_1 + \mu_2 \, dn_2$$

This was used to derive Eq. (7.27) for the free energy of a solution:

$$G = \mu_1 n_1 + \mu_2 n_2$$

Taking the derivative of this equation gives:

$$dG = \mu_1 \, dn_1 + n_1 \, d\mu_1 + \mu_2 \, dn_2 + n_2 \, d\mu_2$$

Comparison of this result to Eq. (7.9) shows that:

$$n_1 \, d\mu_1 + n_2 \, d\mu_2 = 0 \qquad\qquad \textbf{(7.55)}$$

Equation (7.55) means that there must be a functional relationship between the chemical potentials of the two components and the way they change with composition. That this must be true should have been evident from Gibbs' phase rule; when $c = 2$ and T and P are held constant, there is only one degree of freedom, so the specification of one additional intensive variable will fix the values of all the remaining ones. This means that μ_2 must be a function of μ_1; Eq. (7.55), called the **Gibbs-Duhem equation,** expresses this necessary relationship.

The usual mode of application of the Gibbs-Duhem equation is to write it in terms of activities:

$$\mu_i = \mu_i^\theta + RT \ln a_i$$

$$d(\ln a_2) = -\frac{n_1}{n_2} d(\ln a_1) \qquad\qquad \textbf{(7.56)}$$

Osmotic Coefficient

Now there will be some concentration at which a_2 is known; for example, with the RL reference state, $a_2 = 1$ when $X_2 = 1$, or with the HL reference state, $a_2 = m$ in the dilute-solution limit. If the solvent activity (a_1) is measured over a range of concentrations between where a_2 is known (m^*) and the concentration (m) at which a_2 is to be measured, integration of Eq. (7.56) will give the desired result:

$$\ln a_2(m) = \ln a_2(m^*) - \int_{m^*}^{m} \frac{n_1}{n_2} d(\ln a_1)$$

This technique is, as stated, most useful for nonvolatile solutes, for which the HL reference state (molality scale) is generally used. In such cases the evaluation of Eq. (7.56) is done conveniently using the **practical osmotic coefficient:**

$$\phi = -\frac{n_1}{n_2} \ln a_1 \qquad\qquad \textbf{(7.57a)}$$

In a solution of molality m, there are $n_2 = m$ moles of solute per kg of solvent ($1000/M_1$ moles), so:

$$\phi = \frac{(-1000/M_1) \ln a_1}{m} \qquad\qquad \textbf{(7.57b)}$$

It is assumed that a_1 is measurable, for example from solvent vapor pressures or by some other method, so ϕ can then be calculated directly using Eq. (7.57) and is itself an empirical quantity.

Example: Find the limit of ϕ as $n_2 \to 0$. Since when $n_2 = 0$, $a_1 = 1$, $\phi = 0/0$ and is therefore indeterminate. We find the limit from the fact that as $X_2 \to 0$, the solvent becomes ideal and $a_1 \to X_1 = 1 - X_2$. Then, for small X_2:

$$\ln a_1 \cong \ln (1 - X_2) \cong -X_2$$

At small n_2 [from Eq. (7.57)]:

$$\phi \to \frac{n_1}{n_2} X_2 = \frac{n_1 n_2}{n_2(n_1 + n_2)} = X_1$$

Now let $n_2 = 0$ and $X_1 = 1$, so:

$$\lim_{n_2 \to 0} \phi = 1 \qquad \blacksquare$$

Calculation of Activity Coefficients

The relationship of ϕ to the Gibbs-Duhem equation is found by the following — by no means obvious — derivation:

$$m\phi = -(1000/M_1) \ln a_1$$

$$d(m\phi) = m \, d\phi + \phi \, dm$$

$$= (-1000/M_1) \, d(\ln a_1)$$

Now from Eq. (7.56) with $n_2 = m$ and $n_1 = 1000/M_1$ we get:

$$m \, d(\ln a_2) = m \, d\phi + \phi \, dm$$

Because this equation must be integrated from $m = 0$ (where $a_2 = 0$), we will need to introduce the activity coefficient, $a_2 = \gamma_{2m} m$:

$$m \, d(\ln \gamma_{2m}) + m \, d(\ln m) = m \, d\phi + \phi \, dm$$

$$d(\ln \gamma_{2m}) = d\phi + \frac{(\phi - 1) \, dm}{m}$$

This equation is now integrated from a lower limit where $m = 0$, $\phi = 1$, $\gamma_{2m} = 1$, to the molality at which γ_{2m} is to be calculated:

$$\ln (\gamma_{2m}) = (\phi - 1) + \int_0^m \left(\frac{\phi - 1}{m'}\right) dm' \qquad \textbf{(7.58a)}$$

It is common in many texts to introduce a symbol:

$$j = 1 - \phi$$

$$\ln (\gamma_{2m}) = -j - \int_0^m \left(\frac{j}{m'}\right) dm' \qquad \textbf{(7.58b)}$$

The steps required to measure solute activity coefficients at some molality from solvent activities are: (1) The solvent activity is measured at various concentrations from very dilute to the desired molality m. We have noted that this can be done from solvent vapor pressure; in Section 7.8 another method will be described. (2) The osmotic coefficient ϕ is calculated as a function of m using Eq. (7.57b); note that $1000/M_1$ is a solvent characteristic; for example, it is 55.51 for water. (3) Equation (7.63) is used to calculate γ_{2m}; the integral can be evaluated numerically or graphically.

Step 3 sounds straightforward—plot j/m vs. m and measure the area under the curve—until you try it. There is a great deal of scatter in such plots, particularly at low concentrations. At small m, ϕ is near to 1 and, even with the most careful measurements, j has very few significant figures. Actually j/m is approaching 0/0; the limit is in fact finite (see ref. 4, p. 239, for a proof), but the effect of small errors is greatly exaggerated at low concentration, and a great deal of scatter results. The preferred procedure is to fit the j/m data to an empirical equation; for nonelectrolytes this is a power series in m:

$$\frac{j}{m} = a + bm + cm^2 + \cdots \tag{7.59}$$

Exercise: Show that if Eq. (7.59) is truncated after two terms:

$$\frac{j}{m} = a + bm$$

the activity coefficient is given by:

$$\ln \gamma_{2m} = -2am - 1.5bm^2 \qquad\blacksquare$$

Example: Table 7.6 gives a data set for aqueous solutions of sucrose at 25°C. The water activity (a_1) was measured from vapor pressures:

$$a_1 = \frac{\text{vapor pressure of } H_2O \text{ above solution}}{\text{vapor pressure of pure } H_2O}$$

From these data, the osmotic coefficients are calculated;

$$\phi = \frac{-(1000/18.01) \ln a_1}{m}$$

At 1.4 molal:

$$\phi = \frac{-55.51 \ln 0.97193}{1.4} = 1.129$$

Then $j/m = (1 - 1.129)/1.4 = -0.0921$.

Figure 7.10 shows these calculations graphically. The integral of Eq. (7.58) is the area under this curve (actually the negative of the area, since $-j/m$ is plotted) between $m = 0$ and the molality at which γ_{2m} is to be measured.

Table 7.6 Osmotic coefficients and activity coefficients of sucrose at 25°C

m	a_1	ϕ	γ_2	m	a_1	ϕ	γ_2
0.1	0.99819	1.006	1.02	1.6	0.96740	1.150	1.33
0.2	0.99634	1.018	1.03	1.8	0.96280	1.169	—
0.3	0.99448	1.024	—	2.0	0.95807	1.189	1.44
0.4	0.99258	1.033	—	2.5	0.94569	1.240	—
0.5	0.99067	1.041	1.08	3.0	0.93276	1.288	1.75
0.6	0.98872	1.050	—	3.5	0.91933	1.334	—
0.7	0.98672	1.060	—	4.0	0.90567	1.375	2.10
0.8	0.98472	1.068	—	4.5	0.8917	1.414	—
0.9	0.98267	1.079	—	5.0	0.8776	1.450	2.47
1.0	0.98059	1.088	1.19	5.5	0.8634	1.482	—
1.2	0.97634	1.108	—	6.0	0.8493	1.511	2.88
1.4	0.97193	1.129	—	6.18	(saturated solution)		2.965

Source: Data are from R. A. Robinson and R. H. Stokes, *Electrolyte Solutions,* 1955: London, Butterworth's.

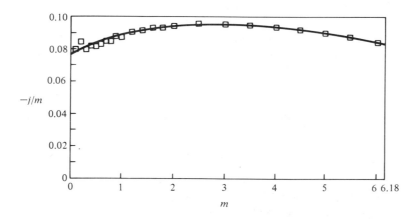

Figure 7.10 Graph of $-j/m$ vs. m for sucrose in water at 25°C. Such graphs are used to calculate activity coefficients using the Gibbs-Duhem equation (cf. Table 7.6).

These data may also be fitted to Eq. (7.59) by polynomial regression with the result:

$$\ln \gamma_{2m} = 0.073367 + 0.019685m - (5.073 \times 10^{-3})m^2 + (3.558 \times 10^{-4})m^3$$

With this result the activity coefficients of Table 7.6 were calculated. Other results are shown in Table 7.7. ■

Table 7.7   Activity coefficients of amino acids in water
(25°C, solute standard state, molality scale)

	\multicolumn{4}{c}{m}				
	0.2	0.5	1.0	2.0	Sat. γ	Sat. m
Alanine	1.00	1.01	1.02	—	1.045	1.862
Alanylalanine	0.98	0.99	1.04	—	—	—
Glycine	0.960	0.908	0.875	0.787	0.729	3.37
Proline	1.02	1.05	1.10	1.21	3.13	14.1
Serine	0.95	0.89	0.81	0.70	0.602	4.02
Alanylglycine	0.93	0.87	0.86	—	—	—

7.8   Colligative Effects

Colligative effects are a group of solution properties that, in dilute solution, depend only on the number of solute molecules present and not on any specific chemical property of the solute. These include: vapor-pressure lowering of the solvent, boiling-point elevation of the solvent by a nonvolatile solute, freezing-point depression, and osmotic pressure. These effects were of great importance historically in establishing the foundations of solution theory. They were also used widely for determining molecular weights. Today, for the most part, these methods have been supplanted by more modern techniques such as mass spectrometry or gel-permeation chromatography.

We shall not discuss the boiling-point elevation; the theory is very similar to that of the freezing-point depression.

Exercise: Show that an ideal dilute solution of a nonvolatile solute will lower the solvent vapor pressure by an amount determined only by the number of moles of solute and solvent present:

$$\frac{P_1^\bullet - P_1}{P_1^\bullet} = X_2 \tag{7.60}$$

∎

Example: In Chapter 4 (page 182) it was calculated that the vapor pressure of water at 25°C and $P = 100$ atm would be increased by 7.62% over the value at 1 atm (23.76 torr). At that time it was mentioned that, if the pressure were due to a gas, the solubility of the gas in the water could cause complications. The data in Table 7.4 for N_2 and O_2 can be used with Henry's law to give (for $P_{N_2} = 79$ atm, $P_{O_2} = 21$ atm) the solubility of air in water at 100 atm:

$$X_2 = 21(2.3 \times 10^{-5}) + 79(1.2 \times 10^{-5}) = 1.4 \times 10^{-3}$$

By Eq. (7.60) the vapor pressure of the water will be:

$$P_1 = X_1 P_1^\bullet = P_1^\bullet(1 - X_2)$$

so the effect of the dissolved air will be to *lower* the vapor pressure of the water by a fraction X_2 or 0.14%. If it is assumed that these two effects—the hydrostatic pressure and the effect of the dissolved gas—are additive, the vapor pressure of water at 25°C under 100 atm of air is calculated as:

$$P_{H_2O} = (23.76)(1 + 0.0762 - 0.0014) = 25.54 \text{ torr}$$

∎

Freezing-Point Depression

Freezing points can be measured accurately and conveniently and so are an attractive alternative to vapor pressure for determining activities. A disadvantage is that this method can determine activities at only one temperature, the freezing point of the solution.

At its freezing point, the chemical potential of the pure solid solvent (μ_1^s) is equal to that of the solvent in the solution:

$$\mu_1(\text{solid}) = \mu_1(\text{solution})$$

$$\mu_1^s = \mu_1^\bullet + RT \ln a_1$$

The activity (RL) of the solvent is (from this):

$$\ln a_1 = \frac{1}{R}\left(\frac{\mu_1^s}{T} - \frac{\mu_1^\bullet}{T}\right)$$

Taking the derivative with respect to T gives:

$$\frac{\partial \ln a_1}{\partial T} = -\frac{1}{RT^2}(H_1^s - H_1^\bullet)$$

But the enthalpy of fusion of the pure solvent is:

$$\Delta H_f = H_1^\bullet - H_1^s$$

so:

$$\frac{\partial \ln a_1}{\partial T} = \frac{\Delta H_f}{RT^2}$$

This equation is integrated with the lower limit $T = T_f^\bullet$, $a_1 = 1$ — that is, pure solvent:

$$\ln a_1 = \int_{T_f^\bullet}^{T_f} \frac{\Delta H_f}{RT^2}\, dT$$

where T_f is the freezing point of the solution whose activity is a_1. If the freezing-point change is small, we can assume that the integrand is constant and replace dT with:

$$\theta \equiv T_f^\bullet - T_f$$

$$\ln a_1 = -\frac{\Delta H_f}{R(T_f^\bullet)^2}\theta \qquad (7.61)$$

Equation (7.61) can now be used to calculate the osmotic coefficient:

$$\phi = \left|\frac{1000\,\Delta H_f}{M_1 R(T_f^\bullet)^2}\right| \frac{\theta}{m}$$

All the factors in the bracket are characteristic of the pure solvent and are used to define the **molal freezing-point-depression constant:**

$$K_f = \left[\frac{M_1 R(T_f^\bullet)^2}{1000\,\Delta H_f}\right] \qquad (7.62)$$

Table 7.8   Freezing-point constants

Solvent	$T_f^\bullet$	K_f	b
Water	273.15	1.860	4.8×10^{-4}
N-methylacetamide	303.66	5.77	1.4×10^{-3}
Benzene	278.6	5.12	—
Camphor	451.5	40.0	—
Cyclohexane	279.6	20.0	—
Ethylene carbonate	309.52	5.32	2.5×10^{-3}
Phenol	315	7.27	—

Then, the osmotic coefficient is:

$$\phi = \frac{\theta}{K_f m} \tag{7.63a}$$

Freezing-point-depression constants for some solvents are listed in Table 7.8. For a dilute solution, $\phi \cong 1$, and:

$$\theta \cong K_f m \tag{7.63b}$$

Exercise: For an ideal dilute solution, $\phi \to 1$ as $m \to 0$. Show that:

$$\lim_{c_2 \to 0} \left(\frac{\theta}{c_2} \right) = \frac{K_f}{M_2} \tag{7.63c}$$

where c_2 is equal to the grams of solute per kg of solvent and M_2 is the molecular weight of the solute. This equation has been used for determining molecular weights from freezing-point depressions.   ∎

A somewhat more accurate formula relating ϕ to the freezing-point depression can be derived by explicitly considering the temperature dependence of the integrand of Eq. (7.61b); the result is:

$$\phi = \frac{\theta}{K_f m}(1 + b\theta)$$

$$b = \frac{1}{T_f^\bullet} - \frac{\Delta C_{pf}}{2 \Delta H_f} \tag{7.64}$$

where $\Delta C_{pf} = C_{pm}(\text{liq}) - C_{pm}(\text{solid})$ is the heat-capacity change on fusion of the pure solvent. Some values for b can be found in Table 7.8.

Example: Q. D. Craft and R. H. Wood [*J. Solution Chem.*, **6**, 525 (1977)] measured the freezing points of N-methylacetamide (NMA) with various solutes. For *n*-octane

Table 7.9 Osmotic coefficients for *n*-octane dissolved in N-methylacetamide*

m	θ	ϕ	j/m	γ_{2m}
0.1242	0.6811	0.9517	0.3889	0.91
0.1233	0.6762	0.9515	0.3933	—
0.2937	1.5031	0.8888	0.3786	0.80
0.2905	1.4884	0.8898	0.3793	—
0.4897	2.3178	0.8229	0.3616	0.70
0.4852	2.2999	0.8243	0.3621	—
0.7402	3.1772	0.7473	0.3368	0.59
0.7314	3.1494	0.7496	0.3424	—

*Q. D. Craft and R. H. Wood, *J. Solution Chem.*, 6, 525 (1977).

at $m = 0.2937$ they found $\theta = 1.5031$. With constants from Table 7.8, the osmotic coefficient is:

$$\phi = \frac{1.5031[1 + 1.4 \times 10^{-3}(1.5031)]}{5.77(0.2937)} = 0.8888$$

$$j = 1 - \phi = 0.1112, \qquad \frac{j}{m} = 0.3785$$

Other results are given in Table 7.9. Linear regression of these data gives:

$$\frac{j}{m} = 0.4025 - 0.0843m$$

with correlation factor $r = -0.9963$. The activity coefficient of *n*-octane in NMA is then (using the results of the exercise on p. 332):

$$\ln \gamma_{2m} = -0.805m + 0.13m^2 \qquad \blacksquare$$

Osmotic Pressure

Certain materials, called semipermeable membranes, will permit the passage of some molecules but not others. Figure 7.11 diagrams an experiment in which pure solvent is separated from a solution by a membrane through which the solvent molecules, but not the solute molecules, may pass freely. If there were no barrier, the solute would flow left while the solvent flowed right until all concentration gradients disappeared. Since only the solvent can flow, its attempts to dilute the solution will result in excess material on the right and the build-up of a hydrostatic pressure (Π) called the osmotic pressure. The effect of pressure on the Gibbs' free energy at constant T is (cf. Table 3.1):

$$dG = V\,dP$$

At equilibrium for the situation shown in Figure 7.11:

$$\mu_1(\text{left}) = \mu_1(\text{right})$$

$$\mu_1^\bullet = \mu_1^\bullet + RT \ln a_1 + \overline{V}_1\Pi$$

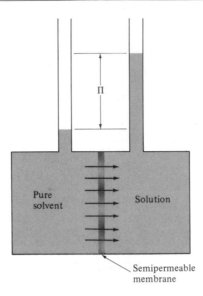

Figure 7.11   Osmotic pressure (schematic). The solute and solution are separated by a semipermeable membrane through which only the solvent can pass. The solvent flows through the membrane to dilute the solution in order to equalize the chemical potential; this results in an excess hydrostatic pressure of the solution side of the membrane, which is called the osmotic pressure.

From this, the activity of the solvent in the solution is seen to be:

$$\ln a_1 = -\frac{\Pi \overline{V}_1}{RT} \tag{7.65}$$

Rigorous application of these equations requires us to know the partial molar volume of the solvent. More often the dilute-solution approximation is made: $\overline{V}_1 = V_{m1}^{\bullet}$, $a_1 = X_1 = 1 - X_2$. Then, using:

$$\ln(1 - X_2) \cong X_2 \cong -\frac{n_2}{n_1}$$

gives:

$$n_2 \cong \frac{\Pi(n_1 V_{m1}^{\bullet})}{RT}$$

But $nV_{m1}^{\bullet}$ is just $V_1^{\bullet}$, the total volume of the solvent in the solution, so:

$$\Pi V_1^{\bullet} \cong n_2 RT$$

Approximating $n_2/V_1^{\bullet}$ by the concentration ($c = n_2/V$) gives:

$$\Pi \cong cRT \tag{7.66}$$

The derivation of Eq. (7.66) made many approximations; the statistical mechanical theory of solutions of W. G. McMillan and J. E. Mayer [*J. Chem. Phys.*, 13, 276 (1945)] demonstrated that it is exact in the dilute-solution limit.

The similarity of Eq. (7.66) to the equation for the pressure of an ideal gas:

$$P = \frac{n}{V} RT = c\, RT$$

cannot be missed. This equation was first obtained by J. H. van't Hoff (1887) by making an analogy between the pressure of a gas and the osmotic pressure of a solute. Also, the McMillan-Mayer theory utilizes a virial expansion very much like that used for gases (Section 1.2); Eq. (7.66) then becomes the leading term of a series:

$$\frac{\Pi V}{nRT} = 1 + Bc + \cdots$$

and the second virial coefficient (B) is calculated from solute pairwise interactions with an equation like Eq. (1.41). However, this analogy is more mathematical than physical; the osmotic pressure does not arise from the solute molecule's striking the membrane (as is the case with a gas, cf. Section 1.5) but, rather, is due to the flow of the *solvent* through the membrane.

If a mechanical pressure is applied to the solution in excess of the osmotic pressure, the solvent will flow out of the solution; this is called *reverse osmosis*. Using membranes that pass water but not salt, reverse osmosis has become a commercial process for water purification. Some have speculated that, if a long pipe with such a selective membrane were sunk into the ocean, a fresh-water fountain would result, with pure water flowing and no net input of energy. This seems unlikely [see O. Levenspiel and N. deNevers, *Science,* 183, 157 (1974)], but compared to distillation, reverse osmosis does require a very small amount of energy input to obtain a given amount of pure water. It was shown in Chapter 3 that the reversible work for a process at constant T and P was just $-\Delta G$; this is a great deal less than the heat required for distillation.

Example: For purifying one mole of water, the work required can be calculated from the osmotic pressure:

$$w = -\Delta G = \mu_1^{\bullet} - \mu_1 = -RT \ln a_1 = \Pi \overline{V}_1$$

The osmotic pressure of the salt water of the ocean is about 23 atm, so for one mole of water:

$$w = (23\ \text{atm})(18\ \text{cm}^3/\text{mol}) = 414\ \text{cm}^3\ \text{atm}/\text{mol} = 42\ \text{J/mol}$$

The heat of vaporization of water (Table 4.2) is 40,660 J/mole; this is the energy required to purify water by distillation. Of course, clever design can recover a lot of the heat in the distillation process; still, the advantage of reverse osmosis is impressive. ■

7.9 Solutions of Macromolecules

In the attempt to study molecular properties, the gas phase is very important, for there we can observe molecules in near isolation. For macromolecules, this is not a realistic path to follow, so investigation of dilute-solution properties has served in its stead.

The nonideality of macromolecular solutions is due, in large part, to the sheer size of the solute; this makes their interpretation relatively straightforward. Because of the ready availability of selective membranes that will pass small molecules (the solvent) but not macromolecules, measurement of osmotic pressure is a popular method for studying the thermodynamics of macromolecular solutions. It has the added advantage of good sensitivity and utility at ambient temperature.

Example: A solute with molecular weight 40,000 g/mole dissolved in water with a concentration of 0.2 mg/cm^3 has a molality of 5×10^{-6} moles/kg. The freezing-point depression, which must in any case be measured near 0°C, is only 9×10^{-6} K; the measurement of temperature to μK accuracy requires great care and very expensive equipment.

The osmotic pressure at 25°C is, by Eq. (7.66), ($n_2 = 5 \times 10^{-9}$, $V_1^{\bullet} = 1$ cm^3):

$$\Pi = \frac{(5 \times 10^{-9})(82 \text{ cm}^3 \text{ atm/K})(298 \text{ K})}{1 \text{ cm}^3}(760 \text{ torr/atm}) = 0.093 \text{ torr}$$

This pressure seems small until you realize that the measurement will be of a *water head;* 1 torr = 13.58 mm H$_2$O, and:

$$\Pi = 1.3 \text{ mm H}_2\text{O}$$

A water head of this size can, with proper care, be easily measured to a few percent accuracy. ∎

The analogy of Eq. (7.66) to the ideal gas law suggests that a virial series, like that used for gases, may be useful. Since the molecular weight is often unknown, it is useful to expand the chemical potential in terms of the weight concentration:

$$c_2 = \frac{\text{grams of solute}}{\text{kg of solvent}}$$

$$\mu_1 - \mu_1^{\bullet} = -RT V_1^{\bullet} c_2 \left(\frac{1}{M_2} + Bc_2 + Cc_2^2 + \cdots \right) \tag{7.67}$$

Using Eq. (7.65), the osmotic pressure follows from this:

$$\Pi = c_2 RT \left(\frac{1}{M_2} + Bc_2 + Cc_2^2 + \cdots \right) \tag{7.68}$$

As with gases, the third virial coefficient is difficult to measure or interpret theoretically. The molecular weight and second virial coefficient are readily measured from plots of (Π/c_2) vs. c_2. (This is analogous to the P/ρ plots used for gases — Chapter 1.)

If you recall the van der Waals formulation of the second virial coefficient [Eq. (1.18)]:

$$B = b - \frac{a}{RT}$$

then it may be evident that there will be two distinct contributions to the nonideality of polymer solutions: the excluded volume (van der Waals' *b*) and the forces of attraction or repulsion between the macromolecule and the solvent or each other (roughly the van der Waals a, but the analogy is breaking down here).

For example, the virial coefficient of solid spheres that do not interact strongly with the solvent can be calculated solely on the basis of excluded volume. The excluded volume is proportional to the molar volume of the solute (more precisely, the partial molar volume); since Eqs. (7.67) and (7.68) are written in terms of a weight concentration (c_2), the specific volume $v_2 = V_m/M_2$ is somewhat more appropriate. Tanford (ref. 6) has shown that the second virial coefficient of solid spheres is:

$$B = \frac{4v_2}{M_2}$$

Since most organic materials have specific volumes ~ 0.75 cm^3/g (cf. Chapter 9), one would expect, for such a case, that:

$$BM_2 \sim 3$$

The data of Table 7.10 for globular proteins show that this expectation is met reasonably well.

Synthetic linear polymers can be treated statistically as a chain of flexible links in a random coil. This theory is rather complicated and will not be discussed here (but see ref. 6); the results predict that a random coil with $M_2 = 100,000$ should have a second virial coefficient of about 5×10^{-4}. This magnitude is borne out (Table 7.10)

Table 7.10 Virial coefficients of macromolecules

	$\overline{M}_n$	B	
Poly(isobutylene) in cyclohexane 30°C	81 400	7.22×10^{-4}	
	169 000	6.62×10^{-4}	
	720 000	5.32×10^{-4}	
	M_2	B	BM_2
Proteins in water (high salt, at isoelectric point)			
β-Lactoglobulin	39 000	8.2×10^{-5}	3.2
Ovalbumin	45 000	3.1×10^{-5}	1.4
Hemoglobin	67 000	5.5×10^{-5}	3.7
Bovine serum albumin	69 000	2.1×10^{-5}	1.5
	M_2	B	
Theoretical			
Sphere	100 000	$3 \ \times 10^{-5}$	
Rod ($l/d = 100$)	100 000	7.5×10^{-4}	
Random coil (good solvent)	100 000	$5 \ \times 10^{-4}$	

for poly(isobutylene) in cyclohexane. (Synthetic polymers generally have a mixture of molecular weights, so the molecular weight determined by osmotic pressure is actually the number-average molecular weight, $\overline{M}_n$.) These theoretical results apply only to polymers in a "good" solvent, where it is a truly random coil; in "poor" solvents the chain tends to associate and be balled up rather than stretched out randomly. For example, poly(isobutylene) ($\overline{M}_n = 720{,}000$) in benzene has $B = 0.71 \times 10^{-5}$, a good bit different than that in cyclohexane.

We have had here only a limited view of a very broad and important area. Those who wish to know more will find ref. 5 to be a good introduction to the subject; ref. 6 gives a more thorough discussion of synthetic macromolecules, and ref. 7 covers biological macromolecules.

7.10   Equilibrium in Solution

The importance of activities and activity coefficients is largely in what they can tell us regarding equilibrium and reactions in solution. The derivation leading up to Eq. (6.17) was perfectly general [nothing was said about gases until Eq. (6.19) *et seq.*]; a general reaction:

$$aA + bB \longrightarrow cC + dD$$

has a standard free-energy change and equilibrium constant as follows:

$$\Delta G^{\theta}_{rxn} = c\mu^{\theta}_C + d\mu^{\theta}_D - a\mu^{\theta}_A - b\mu^{\theta}_B = \sum_i \nu_i \Delta G^{\theta}_f(i)$$

$$K_a = \frac{a^c_C a^d_D}{a^a_A a^b_B}$$

$$\Delta G^{\theta}_{rxn} = -RT \ln K_a \qquad\qquad (7.69)$$

By now it should be clear that activity is a relative quantity, whose value depends on the choice of standard state. Several choices of standard state are possible, and any will work, so long as the same standard state is used for K_a and ΔG^{θ}_{rxn}.

Free Energies of Formation in Solution

Free energies of reaction in solution can, as with gases, be calculated from standard free energies of formation (or vice versa). Figure 7.12 shows the relationship among the various free energies of formation. It is often desirable to be able to calculate $\Delta G^{\theta}_f(\text{soln})$ from ΔG^{θ}_f of the pure material; to do so it is necessary to consider the equilibrium between the two states involved. There are three cases:

1. The pure substance is a solid. The equilibrium between a solid and a saturated solution must be created and the activity of the solute in a saturated solution (a_{sat}) measured as described in this chapter. Neglecting the small effect of pressure on the solid phase:

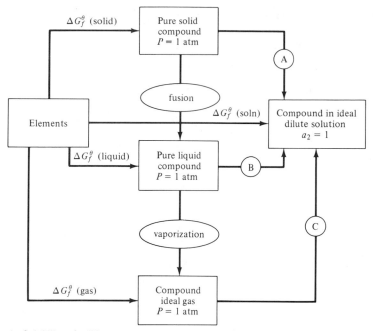

A. Solubility of solid
B. Dilution
C. Henry's law

Figure 7.12   Free energies of formation. The free energy of formation — the free-energy change of the reaction (possibly hypothetical) by which a compound is formed from its elements in their standard states — can be defined for the compound in any phase including in solution. This diagram shows how these various free energies of formation are related. In solution, the reference state must be specified. For the RL reference state, the standard free energy is that of the pure liquid. The standard free energy for the HL reference state (ideal dilute solution) can be calculated from that in the solid using solubilities (A), or from that in the vapor phase using Henry's law (C). If the pure solute is a liquid, and the HL convention is used for the solute, free energies of dilution (B) must be known.

$$\mu_2(\text{pure solid}) = \mu_2(\text{sat'd solution})$$

$$\mu_2^\theta(\text{solid}) = \mu_2^\theta(\text{soln}) + RT \ln a_{\text{sat}}$$

$$\Delta G_f^\theta(\text{soln}) = \Delta G_f^\theta(\text{solid}) - RT \ln a_{\text{sat}} \qquad \textbf{(7.70)}$$

Example: In an earlier example (Table 7.6) it was found for sucrose + water at 25° that the activity coefficient in a saturated solution (m = 6.18 moles/kg) was 2.965, so $a_{\text{sat}} = \gamma_{2m} m = 18.3$. Therefore [from Eq. (7.70)]:

$$\Delta G_f^\theta(\text{soln}) - \Delta G_f^\theta(\text{solid}) = -7.21 \text{ kJ}$$

Table 7.11 gives (for the same quantity) $-1551.8 + 1544.7 = -7.1$ kJ.   ∎

Table 7.11   Free energies of formation in aqueous solution

(25°C; Solute standard state, molality scale; Units: kJ/mole)

	ΔG_f^θ(pure)		ΔG_f^θ(aq)
Acetaldehyde	−133.7	(gas)	−139.7
Acetic acid	−392.3	(liq)	−399.5
Carbon dioxide	−394.4	(gas)	−386.2
Hydrogen sulfide	−33.0	(gas)	−27.4
Ammonia	−16.7	(gas)	−26.7
Ethanol	−174.8	(liq)	−181.5
Methanol	−166.23	(liq)	−175.23
n-Propanol	−172.4	(liq)	−175.8
iso-Propanol	−181.0	(liq)	−185.9
n-Butanol	−169.0	(liq)	−171.8
Formaldehyde	−110.0	(gas)	−130.5
Urea	−197.15	(s)	−203.84
Sucrose	−1554.7	(s)	−1551.8
α-D-Glucose	−910.27	(s)	−917.22
Fructose	—		−915.38
Glycine	−370.7	(s)	−373.0
L-Alanine	−369.9	(s)	−371.3
Alanylglycine (DL)	—		−479.36
Water	−237.191	(liq)	—

Source: Data are from ref. 7, p. 236.

2. The pure substance is a gas. In this case gas solubilities (Henry's law) must be measured.

$$\mu_2(gas) = \mu_2(soln)$$

$$\mu_2^\theta(g) + RT \ln P_2 = \mu_2^\theta(soln) + RT \ln a_2$$

If the molality-scale standard state is used, then Eqs. (7.54) are applicable; for an ideal gas phase, $a_2 = P_2/k_m$ and:

$$\mu_2^\theta(g) = \mu_2^\theta(soln) - RT \ln k_m$$

$$\Delta G_f^\theta(soln) = \Delta G_f^\theta(g) + RT \ln k_m \qquad \textbf{(7.71)}$$

(*Note:* Because of the conventions for the gas standard state, k_m must be in atm/molal.)

Example: Table 7.4 gives the solubility of CO_2 in water at 25°C and 1 atm as $X_2 = 6.1 \times 10^{-4}$. This can be converted to a molality using Eq. (7.4); $m = 3.4 \times 10^{-2}$ moles/kg. If we neglect the hydrolysis of CO_2 in water and assume a dilute ideal solution (HL), we get:

$$k_m = \frac{P_2}{m} = 29.4 \text{ atm/molal}$$

Table 6.1 gives $\Delta G_f^\theta = -394.383$ kJ for CO_2(gas) at 25°C.

$$\Delta G_f^\theta(CO_2, \text{aq}) = -394.383 + \frac{(8.3143)(298.15)}{1000} \ln 29.4 = -386.0 \text{ kJ}$$

This can be compared to the value in Table 7.11, -386.2 kJ. [Here, and elsewhere, "aq" will be used to indicate the solute (HL) standard state in water, molality scale.] ∎

3. The pure substance is a liquid. If the substance is normally a liquid at the temperature at which you are working, it may be desirable to use the solvent (RL) standard state:

$$\mu_i^\theta(\text{RL}) = \mu_i^\bullet$$

in which case the free energy of formation of the pure liquid can be used directly. If vapor phase free energies are known, the vapor pressure of the pure liquid can be used to relate the free energies of the vapor to that of the pure liquid (as in Chapter 4):

$$\mu_i(\text{gas}) = \mu_i^\bullet(\text{liq})$$

$$\mu_i^\theta(g) + RT \ln P_i^\bullet = \mu_i^\bullet(\text{liq})$$

If the HL reference state is preferred, the Henry's law constant must be determined (for example, as in Figure 7.9) and Eq. (7.71) can be used.

Example: Table 6.1 lists two free energies of formation for methanol at 25°C:

$$C(s) + 2\,H_2(g, 1\text{ atm}) + \tfrac{1}{2}O_2(g, 1\text{ atm}) \longrightarrow CH_3OH(g, 1\text{ atm}), \qquad \Delta G_f^\theta = -161.9 \text{ kJ}$$

$$C(s) + 2\,H_2(g, 1\text{ atm}) + \tfrac{1}{2}O_2(g, 1\text{ atm}) \longrightarrow CH_3OH(\text{liq}), \qquad \Delta G_f^\theta = -166.3 \text{ kJ}$$

Adding these equations gives:

$$CH_3OH(\text{liq}) \longrightarrow CH_3OH(g, 1\text{ atm}), \qquad \Delta G^\theta = 4.4 \text{ kJ}$$

$$\mu^\theta(\text{gas}) - \mu^\bullet(\text{liq}) = -RT \ln P^\bullet = 4400 \text{ J}$$

$$\ln P^\bullet = -\frac{4400}{(8.314)(298.15)}, \qquad P^\bullet = 0.17 \text{ atm} \qquad ∎$$

Concentration Equilibrium Constants

For practical purposes the activities in K_a [Eq. (7.69)] must be related to concentration. The choice of reference state is somewhat arbitrary but must be made with absolute consistency. If the reference state for ΔG_f^θ is HL(molality) (generally denoted aq in tables), then the activity is replaced by:

$$a_i = \gamma_i m_i \qquad (\gamma_i \to 1 \text{ in dilute solution})$$

If the value of ΔG_f^θ for the pure liquid is used, then the reference state is RL(mole fraction), and, in the equilibrium constant, the activity is:

$$a_i = \gamma_i X_i \qquad (\gamma_i \to 1 \text{ as } X_i \to 1)$$

The apparent (concentration) equilibrium constant for:

$$a\text{A} + b\text{B} = c\text{C} + d\text{D}$$

is:

$$K' = \frac{[\text{C}]^c[\text{D}]^d}{[\text{A}]^a[\text{B}]^b} \tag{7.72}$$

where [ ], in this equation, denotes some measure of concentration determined by the reference state being used. This is related to the thermodynamic equilibrium constant as:

$$K_a = K'\left(\frac{\gamma_C^c \gamma_D^d}{\gamma_A^a \gamma_B^b}\right) \tag{7.73}$$

The convenience of the HL reference state can now be seen: if the reactants are all solutes, then, as the solution is diluted, all $\gamma_i \to 1$ and K' (the experimental equilibrium constant) approaches K_a.

Example: The distribution of a solute between immiscible solvents can be treated in this fashion — for example, I_2 distributed between H_2O and CCl_4:

$$I_2(\text{soln in } CCl_4) = I_2(\text{soln in } H_2O)$$

$$K' = \frac{c(I_2 \text{ in } H_2O)}{c(I_2 \text{ in } CCl_4)}$$

The following data were measured at 25°C:

c (in H_2O)	c (in CCl_4)	K' (concentration scale)
0.00134	0.1196	0.0112
0.00115	0.1010	0.0114
0.000763	0.0654	0.0117
0.000322	0.02745	0.01173

It is clear that K_c is becoming constant at low concentrations where the ideal dilute-solution laws apply. Evidently $K_a \cong 0.0117$ and $\Delta G^\theta = -11.0$ kJ. Note that if the concentrations had been expressed as molalities or mole fraction, the ΔG^θ would be different. ∎

If the solvent participates in the reaction, it must be referred to the solvent standard state. For example, the hydrolysis of an ester:

$$\text{ester} + H_2O = \text{acid} + \text{alcohol}$$

with water as a solvent will have:

$$K' = \frac{[\text{acid}][\text{alcohol}]}{[\text{ester}][H_2O]}$$

If the acid, alcohol, and ester are referred to the solute standard state and water is referred to the solvent standard state ($[H_2O] = X_{H_2O}$), then the equilibrium constant for this reaction becomes:

$$K_a = \frac{m(\text{acid})m(\text{alcohol})}{m(\text{ester})X_{H_2O}} \frac{\gamma(\text{acid})\gamma(\text{alcohol})}{\gamma(\text{ester})\gamma(H_2O)}$$

As the solution is diluted, $\gamma_i \to 1$ for all solutes, but also $\gamma_{H_2O} \to 1$ as $X_{H_2O} \to 1$; that is, $K' \to K_a$ as the solution is progressively diluted. Also at sufficiently dilute solution, X_{H_2O} is nearly 1 and:

$$K_a \cong K' \cong \frac{m(\text{acid})m(\text{alcohol})}{m(\text{ester})}$$

That is, the solvent can be omitted from the equilibrium expression. Also, note that the standard free energy of this reaction would be calculated using that of pure H_2O:

$$\Delta G^{\theta}_{rxn} = \Delta G^{\theta}_f(\text{acid, aq}) + \Delta G^{\theta}_f(\text{alcohol, aq}) - \Delta G^{\theta}_f(\text{ester, aq}) - \Delta G^{\theta}_f(\text{pure } H_2O, \text{liq})$$

7.11 Phase Diagrams

In Chapter 4 the phase diagram was introduced as a graph of (T, P) points at which two or more phases could coexist in equilibrium. For two-component systems, representation of the various phase equilibria requires three variables (typically T, P, and composition). While it is possible to represent such situations graphically, it is more common to graph two variables with the other constant — for example, vapor pressure vs. composition at constant T, or boiling temperature vs. composition at constant P.

Liquid-Vapor Diagrams

The equilibria of a binary solution with its vapor can be represented as a graph of vapor pressure vs. composition (constant T) — but two compositions are of interest: the vapor-phase composition (for example, the mole fraction Y_2) and the liquid-phase composition (mole fraction X_2). For an ideal solution:

$$P = X_1 P_1^{\bullet} + X_2 P_2^{\bullet}$$

so, using $X_1 = 1 - X_2$, the vapor pressure vs. liquid composition is:

$$X_2 = \frac{P - P_1^{\bullet}}{P_2^{\bullet} - P_1^{\bullet}}$$

Also, the vapor-phase composition was given by Eq. (7.34) as:

$$Y_2 = \frac{X_2 P_2^{\bullet}}{P_1^{\bullet} + X_2(P_2^{\bullet} - P_1^{\bullet})}$$

For each pressure there are two compositions, X_2 and Y_2; Figure 7.13 shows a graph of these compositions vs. P for a hypothetical ideal mixture, with each pair of equilibrium compositions connected by a *tie line*. (The scale on the bottom could be X_2 or Y_2, depending on which line is read.) Figure 7.13 is representative of a two-component liquid-vapor phase diagram. The upper region is labeled "liquid" because the pressure/composition points in that region represent a liquid with no vapor phase. The lower region is labeled "vapor"; the points in that region represent the various vapor mixtures at pressures too low (for that T) to condense any vapor. A point in the two-phase region has no meaning except as a point on a tie line that connects the composition vs. P lines of the two equilibrium phases. Phase diagrams are conventional representations, and this is the first of several important conventions:

Convention: All two-phase regions are regions of horizontal tie lines connecting equilibrium phases. These tie lines are *not* usually drawn on the diagram but can be furnished (or imagined) where needed.

Figure 7.14 illustrates the behavior of a vapor mixture upon isothermal compression; this should be contrasted to Figure 1.7 for a pure vapor. If a vapor at pressure

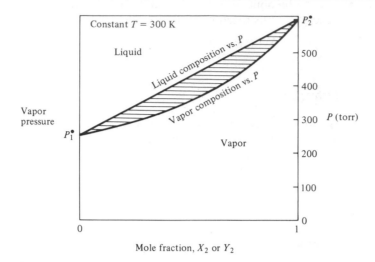

Figure 7.13 Vapor pressure vs. composition diagram for an ideal solution. The composition of the liquid phase (top line) and the vapor phase (bottom line) are both plotted, with *tie lines* connecting pairs of points representing equilibrium phases at a given pressure.

P_1 is compressed, at P_2 some liquid will begin to condense. At P_3, there are two phases with compositions given by points a and b (at the ends of a tie line). As the system is compressed, there are changes in both the amount and composition of the phases, so (in contrast to Figure 1.7) the pressure is not constant. At pressures above P_4, there is only a single phase — the liquid.

Boiling-Point Diagrams

A somewhat more useful way to represent the phase behavior of mixtures is to plot liquid/vapor composition vs. T at constant P. If a liquid mixture is boiled, the vapor usually will not have the same composition as the liquid. If the liquid/vapor compositions are measured and plotted vs. boiling temperature, a diagram such as Figure 7.15 may result. On this diagram (which could be the same hypothetical ideal mixture as Figure 7.13, with typical heats of vaporization) the equilibrium phases are connected by tie lines, but the reader is reminded that these are not normally shown on phase diagrams.

The boiling behavior of a mixture is illustrated by Figure 7.16. A mixture of composition indicated by the vertical dotted line is heated and, at T_2, begins to boil. The vapor phase is richer in the more volatile (lower-boiling) component. At T_3, the compositions of the equilibrium phases are given by the ends of the tie line — point a gives the composition of the vapor phase, point b the composition of the liquid. At T_4, the last liquid disappears; above this temperature there is only a vapor phase.

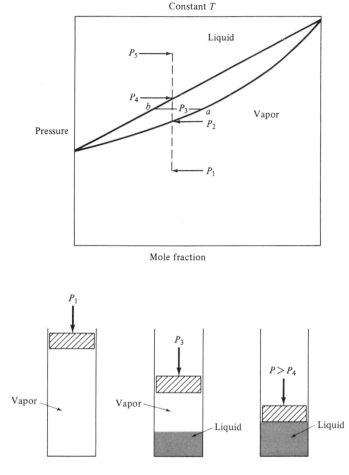

Figure 7.14 The effect of increasing pressure on a mixed vapor. In contrast to the behavior of a pure gas (Figure 1.7), the equilibrium pressure varies during compression when two phases are present (between P_2 and P_4 on this diagram).

Note that, in contrast to the behavior of a pure liquid, the boiling occurs over a range of temperatures.

The phase diagrams for ideal mixtures (Figures 7.13 and 7.15) tell us nothing that could not be calculated from properties of the pure liquids (for example, vapor pressures at one T and heats of vaporization). For real mixtures, phase diagrams provide a convenient representation of the observed boiling behavior.

A mixture that shows positive deviations from Raoult's law will have a higher vapor pressure and, hence, a lower boiling point than predicted by Raoult's law. The phase diagram of a mixture with strong positive deviations may have a minimum in the

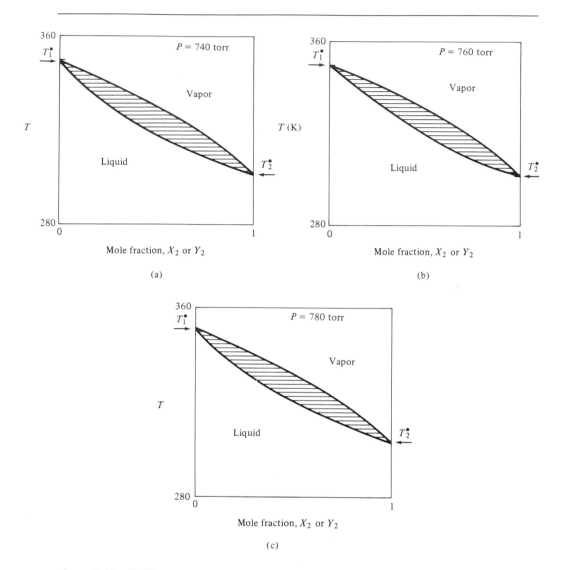

Figure 7.15 Boiling temperature vs. composition for an ideal solution. The horizontal tie lines connect the compositions of liquid (left) and vapor (right) phases that are in equilibrium at that temperature. Note the change in boiling behavior with pressure.

boiling point, as illustrated by Figure 7.17(a). Note that at one point (the minimum) the liquid and vapor compositions are identical; such a mixture will show the constant-T boiling behavior, just like that of a pure substance; it is called an *azeotrope*.

A phase diagram for a mixture that has strong negative deviations from Raoult's law (decreased vapor pressure, increased boiling T) is illustrated by Figure 7.17(b); note the presence of a maximum-boiling azeotrope.

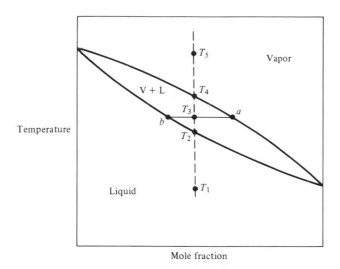

Figure 7.16 The effect of heating a two-component liquid. In contrast to a pure liquid, the liquid boils over a range of temperatures.

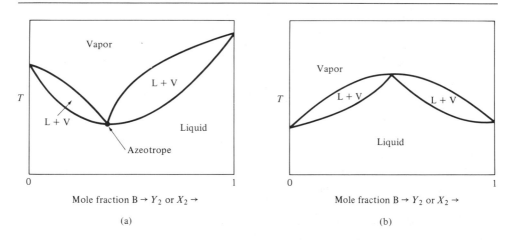

Figure 7.17 Boiling temperature vs. composition diagrams for two cases. (a) A minimum-boiling azeotrope is characteristic of solutions that have large positive deviations from Raoult's law. (b) Maximum-boiling azeotropes occur when solutions have large negative deviations from Raoult's law.

Immiscible Liquid Phases

It was mentioned earlier that liquid mixtures with strong positive derivations from Raoult's law may be (in the extreme case) immiscible. Figure 7.18 shows typical phase diagrams for such cases. In Figure 7.18(a), the liquids are miscible (one liquid

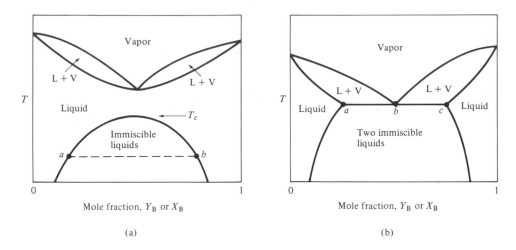

Figure 7.18 Phase diagrams for immiscible liquids. (a) Phase diagram for two components that are immiscible at temperatures below the minimum boiling temperature but miscible at the boiling point. (b) Phase diagram for two components that are immiscible at the boiling point.

phase) at the boiling point. Note that the two-phase region is, as always, a region of tie lines; the one illustrated connects the composition (*a*) of the A-rich phase and the composition (*b*) of the B-rich phase, which are in equilibrium at that temperature. At some temperature (T_c, the consolute or critical solution temperature) the liquids become identical in composition — that is, there is only one liquid phase, so we would say the liquids are miscible.

Figure 7.18(b) illustrates the situation if the liquids remain immiscible up to the boiling temperature. Note that, even with immiscible liquids, the boiling temperature of the mixture is lower than those of the pure substances; this is the basis of *steam distillation*, which is commonly used for the purification of organic compounds that may decompose at their normal boiling temperature. Also note in Figure 7.18(b), a horizontal line is drawn connecting three phases in equilibrium: the A-rich liquid (point *a*), the B-rich liquid (point *c*), and the vapor with composition *b*.

Convention: Horizontal lines on two-component phase diagrams connect the compositions of *three* phases in equilibrium.

Melting Behavior of Mixtures

Figure 7.19(a) illustrates the melting behavior of a mixture (A + B) that is miscible in the liquid phase. The curve descending from the left represents the freezing-point depression of A when B is added; the two-phase region represents changes of the liquid composition (to the right of the invisible tie lines) as more pure A is frozen out. The curve on the right similarly represents the freezing-point depression of B when A is added. The point at which these curves intersect, the minimum melting point, is

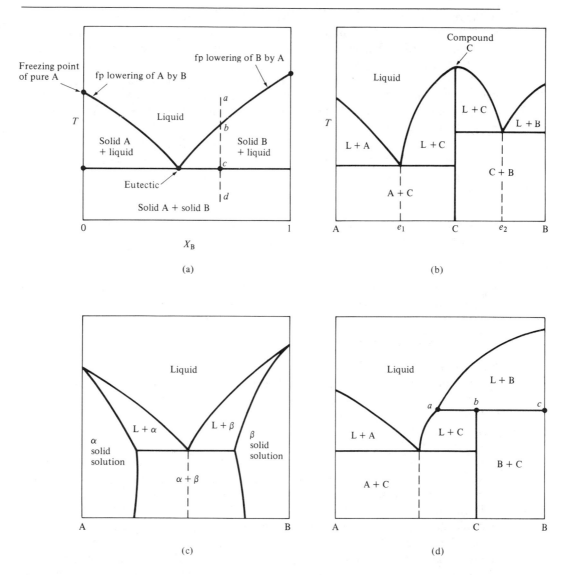

Figure 7.19 Freezing-point diagrams. (a) The simplest case for the freezing of a two-component solution; the components both freeze out as pure materials. (b) Phase diagram for the freezing of a liquid, exhibiting compound formation. (c) Phase diagram for the freezing of a liquid if the solid phases are mixtures (solid solutions). (d) Phase diagram exhibiting compound formation with an incongruent melting point.

called the *eutectic*. The horizontal line at that point represents the equilibrium of three phases, pure A, pure B, and the eutectic liquid. Below this horizontal line there are two solid phases, pure solid A and pure solid B. (Some authors denote the region to the left of the eutectic point as a mixture of pure A and solid "eutectic," and the region to the right as pure B and "eutectic." The eutectic is actually a mixture (as contrasted to a solid solution or compound, see below), but it is a very fine mixture and does have physical properties distinct from those of a coarser mixture; therefore this is a useful idea from the materials-science point of view.)

The melting/freezing behavior of a mixture can be illustrated using the vertical dashed line (*a-d*) of Figure 7.19(a). If a liquid mixture with composition and temperature of point *a* is cooled, at point *b* it will begin to freeze; "freeze" in this instance means that pure solid A appears. At any temperature in the two-phase region (*b* to *c*) a tie line can be drawn (or imagined); the left end of this line will give the composition of the liquid at that temperature. Note that the temperature is changing during the freezing process; this is in sharp contrast to the behavior of a pure substance, which freezes (has an equilibrium between solid and liquid) at only one temperature. The melting range of a material provides a useful criterion for its purity. Note that at the eutectic temperature, *T* is invariant, because there are three phases present; the effective number of degrees of freedom is, from Eq. (7.20):

$$F' = c - p + 1 = 2 - 3 + 1 = 0$$

Also note that a mixture with a composition of the eutectic will melt sharply at a single temperature, just like a pure substance.

Compound Formation

Figure 7.19(b) shows an example of a phase diagram with two eutectics (minima in the freezing point) and a maximum in the freezing temperature (composition C). The composition C is represented by a vertical solid line and is called a compound.

Convention: A vertical line on a two-component phase diagram denotes the formation of a compound.

In a compound (as distinct from a physical mixture) the substances are mixed at the atomic level in a stoichiometric manner. The melting behavior of a compound is exactly like that of a pure substance; that is, the melting point is sharp, and the composition of the solid phase is the same as that of the liquid.

Recall that a eutectic mixture also has a sharp melting point; but it is easy to distinguish this from the behavior of a compound or any other pure material. If [cf. Figure 7.19(a) or (b)] a small amount of the pure phase is added to a eutectic mixture, the freezing point (the temperature at which the first solid appears upon cooling) will *increase*. If any material is added to a compound (or other pure material), the freezing point will be *decreased*.

Note that Figure 7.19(b) is, effectively, two simple phase diagrams like Figure 7.19(a), back-to-back, one for mixtures of A + C, the other for mixtures of C + B. [Does the compound persist in the liquid phase? That is, should the vertical line C of Figure 7.19(b) be extended above the melting point? From the narrow point of view of phase diagrams, the question is irrelevant and cannot be answered; the mixture is still a two-component mixture by the definition given earlier because, in the liquid, there will be an equilibrium A + B ⇌ C.]

Figure 7.20 shows an example of compound formation in the Zn/Mg system.

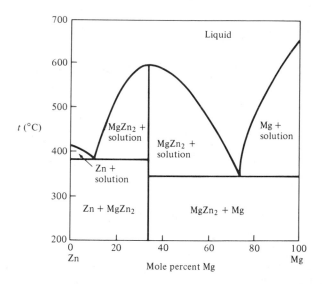

Figure 7.20 Phase diagram for zinc and magnesium. This is an example of compound formation from the melt. (From F. Daniels and R. A. Alberty, *Physical Chemistry*, 3d ed., 1966: New York, John Wiley & Sons, Inc., p. 162, Fig. 5-7)

Solid Solutions

Figure 7.19(c) illustrates yet another type of melting-point diagram. In this case, the solid formed is not a pure material, but a solid solution of variable composition. A solid solution can be contrasted to a compound or eutectic as follows: (1) it is an atomic-level random mixture, (2) its composition is variable over some range, (3) its melting point is not sharp. Note that the regions labeled α and β are single-phase regions. [Also, compare this to the phase diagram for immiscible liquid solutions—for example, Figure 7.18(b).] Figure 7.21 shows an example of such behavior in the Cu/Al system.

Some materials form solid solutions with a full range of compositions, for example the Cu/Ni system shown in Figure 7.22.

Incongruent Melting Points

Some compounds do not melt into a liquid of the same composition but, rather, decompose upon melting into another solid. Such cases are characterized by phase diagrams such as Figure 7.19(d). The melting of the compound C (point *b*) results in the formation of another solid (pure B at point *c*) and a solution (point *a*); note the horizontal line that denotes three phases in equilibrium: solid B, solid C, and liquid (composition *a*). This is called an *incongruent melting point* or a *peritectic*.

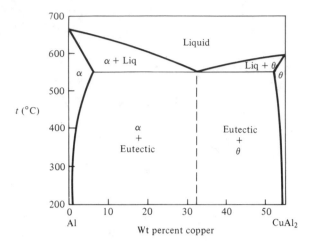

Figure 7.21 Phase diagram for aluminium and copper. This is an example of solid-solution formation. (From W. Moore, *Physical Chemistry*, 4th ed., 1972: Englewood Cliffs, N.J., Prentice-Hall, Inc., p. 266, Fig. 7-22)

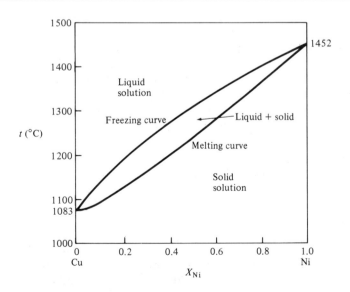

Figure 7.22 Phase diagram for copper and nickel. This system forms solid solutions at all compositions—compare to Figure 7.15. (From Moore, *Physical Chemistry*, p. 266, Fig. 7-21)

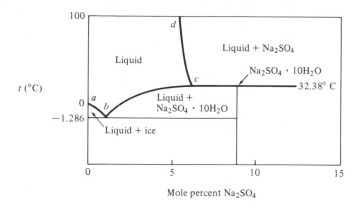

Figure 7.23 Phase diagram for sodium sulfate/water solutions. This is an example of an incongruent melting point. The compound, sodium sulfate decahydrate, when it melts at 32.38°C, decomposes into the anhydrous salt and a saturated solution (point c). Note (line d) the unusual temperature dependence of the solubility of the anhydrous salt—it becomes less soluble as temperature increases. (From Daniels and Alberty, *Physical Chemistry*, p. 164, Fig. 5-18)

The Na_2SO_4/water system (Figure 7.23) is an example of a peritectic. The compound sodium sulfate decahydrate ($Na_2SO_4 \cdot 10H_2O$) is, by all criteria, a pure compound. Yet when it melts (at 32.38°C), three phases form: anhydrous Na_2SO_4, the decahydrate, and the solution (point c on Figure 7.23). (This invariant point is a useful standard for thermometer calibration.) The hydrate does not so much "melt" as dissolve itself in its own water of hydration.

Figure 7.23 can serve as a useful summary of the ideas discussed in this section: On the left, the line a-b represents the freezing-point depression of water by sodium sulfate. The eutectic at point b represents a freezing, saturated solution with ice, solution, and solid $Na_2SO_4 \cdot 10H_2O$ present. Line b-c is the solubility curve (vs. T) for $Na_2SO_4 \cdot 10H_2O$ in water. Line c-d (above the peritectic temperature) is the solubility curve for Na_2SO_4 in water; note the unusual decrease in solubility of this salt with temperature. At the peritectic temperature (point c), the solid phase changes from decahydrate to anhydrous sodium sulfate.

Postscript

As promised at the beginning of this chapter, the thermodynamics of solutions has proved to be a difficult and complex topic. Among the important quantities whose definition and measurement have been discussed in this chapter are: partial molar quantities, fugacity, activity, activity coefficients, and standard states. A major objective has been to obtain a relationship between free energy (hence reactivity) and the concentration of the solute. This was approached by using two limiting ideal cases: Raoult's law and Henry's law. These defined reference states that were used to relate activity to concentration via the activity coefficient. One very important topic has been slighted, namely, heats of solution; the reason for this omission is that this topic would introduce a whole new set of complications to an already complex chapter. However, the student who has mastered the material of this chapter should have no difficulty in reading specialized texts (such as reference 4) which deal with this topic. Two major applications have been introduced—the phase behavior of mixtures and chemical equilibrium in solution. The second of these will be expanded upon in the next chapter where we shall discuss ionic solutions.

References

1. J. H. Hildebrand and R. L. Scott, *The Solubility of Nonelectrolytes,* 3d ed., 1950: New York, Van Nostrand Reinhold Company.
2. W. Gerrard, *Solubility of Gases and Liquids,* 1976: New York, Plenum Press.
3. J. H. Hildebrand and R. L. Scott, *Regular Solutions,* 1962: Englewood Cliffs, NJ, Prentice-Hall, Inc.
4. G. N. Lewis, M. Randall, K. S. Pitzer, and L. Brewer, *Thermodynamics,* 1961: New York, McGraw-Hill Book Co.
5. K. E. Van Holde, *Physical Biochemistry,* 1971: Englewood Cliffs, NJ, Prentice-Hall, Inc.
6. C. Tanford, *Physical Chemistry of Macromolecules,* 1967: New York, John Wiley & Sons, Inc.
7. J. T. Edsall and J. Wyman, *Biophysical Chemistry,* Vol. I, 1958: New York, Academic Press, Inc.

Problems

7.1   Calculate the molality of a solution having $X_2 = 0.132$ in water.

7.2   0.15 mole of a solute is dissolved in 300 g of CCl_4 ($M = 153.823$). Calculate (a) the mole fraction of the solute, (b) the molality.

7.3   Use the data in Table 7.2 for 1-molal solutions of *n*-decane in *N*-methylacetamide ($M_1 = 73.09$ g/mol, $V_{ml}^{\bullet} = 77.03$ cm^3/mole) to calculate $\overline{V}_1$ and ΔV_{mix} for one mole of this solution.

7.4   The partial molar volumes of water and ethanol in a solution $X_{H_2O} = 0.6$ (25°C) are 17 and 57 cm^3 mol^{-1}, respectively. Calculate the volume change on mixing sufficient ethanol with two moles of water to give this composition. Use densities (H_2O) 0.997 g/cm^3 (EtOH) 0.7893 g/cm^3.

7.5   When n_2 moles of NaCl are added to 1 kg of water, the volume of the solution is (in cm^3):

$$V = 1001.38 + 16.6253 n_2 + 1.7738 n_2^{3/2} + 0.1194 n_2^2$$

Calculate the partial molar volumes of NaCl and H_2O at $m = 1.5$ and 0 (that is, at infinite dilution).

7.6   A solution 0.1 molal AgNO$_3$ in water is saturated with AgCl and AgBr (that is, both solids are present). How many components are in this system? How many phases (including vapor)? How many degrees of freedom? If P is fixed, can T vary?

7.7   Solid I$_2$ is added to a mixture of water and CCl$_4$ (immiscible liquids) until saturated (solid present). How many components are there? Phases (including vapor)? Degrees of freedom?

7.8   How many components and phases are there in an equilibrium mixture of:
(a) N$_2$O$_4$(g) = 2 NO$_2$(g).
(b) COCl$_2$(g) = CO(g) + Cl$_2$(g) (no excess CO or Cl$_2$ is added).
(c) COCl$_2$, CO, and Cl$_2$ with arbitrary amounts of CO and Cl$_2$.
(d) HCl, NH$_3$, NH$_4$Cl (solid) with arbitrary amounts of the gases.

7.9   Assume that benzene and toluene form ideal solutions. Pure benzene boils at 80°C; at that temperature, toluene has a vapor pressure of 350 torr.
(a) Calculate the partial and total pressures of a solution at 80° with X(benzene) = 0.2.
(b) What composition of solution would boil at 80°C under a reduced pressure of 500 torr?

7.10   Assuming ideal solution, calculate ΔG, ΔH, and ΔS of mixing 0.25 moles of benzene with 0.5 moles of toluene at 30°C.

7.11   At 39.9°C a solution of ethanol (X_1 = 0.9006, $P_1^\bullet$ = 130.4 torr) and isooctane ($P_2^\bullet$ = 43.9 torr) forms a vapor phase with Y_1 = 0.6667, P = 185.9 torr.
(a) Calculate the activity and activity coefficient of each component.
(b) Calculate the vapor pressure of this solution using Raoult's law.

7.12   At 90°C a solution of n-propanol (X_2 = 0.259, $P_2^\bullet$ = 577.5 torr) and water ($P_1^\bullet$ = 527.76 torr) has a vapor pressure of 820.3 torr; the vapor phase is 39.7% n-propanol.
(a) Calculate the vapor pressures, activities, and activity coefficients of each component.
(b) What should the vapor pressure of this solution be according to Raoult's law?

7.13   Ratcliff and Chao [*Can. J. Chem. Eng.*, 47, 148 (1969)] measured vapor pressures of isopropanol ($P_1^\bullet$ = 1008 torr) and n-decane ($P_2^\bullet$ = 48.3 torr) at 90°C.
(a) Calculate the activity coefficients of n-decane and isopropanol at each concentration.
(b) Make a graph of $\Delta G_{\text{mix}}^{\text{ex}}$ vs. X_2.

	n-Decane	
Pressure (torr)	X_2	Y_2
942.6	0.1312	0.0243
909.6	0.2040	0.0300
883.3	0.2714	0.0342
868.4	0.3360	0.0362
830.2	0.4425	0.0411
786.8	0.5578	0.0451
758.7	0.6036	0.0489

7.14   Below are given activity coefficients (25°C) for acetone (1)–chloroform (2) solutions. Do these data conform reasonably to Eq. (7.47)? If so, calculate w.

X_2	γ_1	γ_2
0.184	0.98	0.59
0.361	0.91	0.69
0.508	0.82	0.77
0.662	0.68	0.88

7.15   Use data from Table 7.3 to calculate w at 310 K for benzene + cyclohexane mixtures.

7.16   Use the regular solution constants (Table 7.3) for CCl_4 + benzene to calculate ΔG, ΔS, and ΔH of mixing for a solution 4 moles CCl_4 with 6 moles of benzene.

7.17   Derive Eqs. (7.47), (7.48), and (7.49) for a solution obeying Eq. (7.46).

7.18   Use the data below for the solubility of 1-butene in benzyl alcohol at 0°C to calculate the Henry's law constant for this gas.

P_2 (torr)	X_2
200	0.040
400	0.087
600	0.151
700	0.193
760	0.226

7.19   G. N. Lewis and H. Storch [*J. Am. Chem. Soc.*, 39, 2544 (1917)] measured the partial pressures of Br_2 above its solutions in CCl_4. Determine the Henry's law constant for Br_2 in CCl_4 from these data.

X_{Br_2}	P_{Br_2} (torr)	X_{Br_2}	P_{Br_2} (torr)
0.00394	1.52	0.0130	5.43
0.00420	1.60	0.0236	9.57
0.00599	2.39	0.0238	9.83
0.0102	4.27	0.025	10.27

7.20   Wood and Delaney [*J. Phys. Chem.*, 72, 4651 (1968)] measured the solubility of He in N-methylacetamide (NMA) at 1 atm and temperatures between 35° and 70°C. The results fit the equation:

$$\ln X_2 = -\frac{1152.5}{T} - 6.0579$$

Assuming ideal solution (HL), calculate ΔG^θ and ΔH^θ for one mole He(gas) $\longrightarrow$ He(dilute solution in NMA) at 310 K.

7.21   The Henry's law constant (mole-fraction scale) for krypton in water is 2.00×10^4 atm at 20°C.
(a) How many grams of Kr would dissolve in 1000 g water at that temperature and $P = 100$ atm?
(b) How much would this solubility depress the vapor pressure of H_2O? Compare this effect to that of the applied pressure [Eq. (4.24)]. (Use $P^\bullet = 17.535$ torr.)

7.22   If we assume an ideal dilute solution, the solvent activity $a_1(RL) = X_1$ and the solute activity (HL, mole fraction scale) $a_2(HL) = X_2$. Find the derivatives:

$$\frac{d(\ln X_1)}{dn_1} \quad \text{and} \quad \frac{d(\ln X_2)}{dn_1}$$

and prove that these conventions are consistent with the Gibbs-Duhem equation.

7.23   Vapor pressures of solvents containing nonvolatile solutes can be measured by bubbling dry nitrogen through the solution. If the exiting gas is saturated with solvent vapor, the vapor pressure can be calculated from the weight loss. 23.50 dm^3 of dry N_2 ($P = 760.0$ torr) is bubbled through an aqueous solution; the solution loses 0.5312 g. Calculate the vapor pressure of the water. (The outlet pressure also was 760.0 torr.)

7.24   Q. Craft and R. H. Wood [*J. Solution Chem.*, **6**, 525 (1977)] measured osmotic coefficients of N-methylacetamide in solution with *n*-nonane by freezing-point depression. The results were fitted by least squares to give:

$$\phi = 1 - 0.5035m + 0.2364m^2 - 0.1206m^3$$

Use this to calculate the activity coefficients of *n*-nonane at 0.01, 0.1, and 0.5 molal.

7.25   Calculate the freezing-point constant K_f for *n*-octane (C_8H_{18}) (use data from Table 4.2).

7.26   Use the data below for freezing point depression by *n*-decane in N-methylacetamide [Craft and Wood, *J. Solution Chem.*, **6**, 525 (1977)] to determine the activity coefficient of *n*-decane when $m = 0.5$ mol/kg.

m	θ	m	θ
0.1753	0.9127	0.4069	1.8677
0.1742	0.2708	0.5699	2.3951
0.2708	1.3387	0.5633	2.3754
0.2697	1.3294		

7.27   Show that the osmotic coefficient is related to the osmotic pressure as:

$$\phi = \frac{n_1 \overline{V}_1 \Pi}{n_2 RT}$$

7.28   Use Eq. (7.66) to calculate the osmotic pressure of 0.20 moles/dm³ solution of sucrose in water at 20°. The observed value is 5.06 atm.

7.29   Twenty milligrams of a protein is dissolved in 10 g water. The osmotic pressure at 25°C was 0.30 torr. Calculate the molecular weight, assuming ideal dilute solution.

7.30   P. J. Flory [*J. Am. Chem. Soc.*, **65**, 372 (1943)] reported the following osmotic pressure data for solutions of polyisobutylene at 25°C.

Concentration (g/dm³)	Π (Pa, in C_6H_{12})	Π (Pa, in C_6H_6)
20.0	1187	210.3
15.0	667	150.4
10.0	306	100.5
7.5	176	
5.0	92	49.5
2.5	35	

Calculate the average molecular weight and second virial coefficients.

7.31   D'Orazio and Wood [*J. Phys. Chem.*, **67**, 1435 (1963)] measured the solubility of hydrazoic acid (HN_3) in water as a function of $P(HN_3)$ and T. Use these results to calculate ΔG^θ, ΔS^θ, and ΔH^θ for $HN_3(g) \longrightarrow HN_3(aq)$ at 24.42°C.

	0°C		24.42°C		49.46°C	
$m(HN_3)$	$P(HN_3)$		m	P	m	P
1.807	33.3 torr		1.651	101.0	1.397	228.0
0.9473	16.8		0.9020	55.6	0.8187	134.8
0.3964	7.0		0.3922	24.5	0.3848	64.9
0.1023	1.8		0.1019	6.3	0.1008	17.7

7.32   Calculate $\Delta G_f^\theta(\text{aq})$ for H_2S in water at 25°C from the gas solubility data below:

$m(H_2S)$	P_{H_2S} (atm)
0.050	0.486
0.101	0.992
0.150	1.474
0.204	2.049
0.254	2.514

7.33   Use the data below (at 25°C) to calculate ΔG_f^θ of Br_2 in CCl_4 (solute standard state, molal scale):

Molality	P_{Br_2} (torr)
0.0257	1.52
0.0392	2.39
0.0670	4.27
0.0856	5.43

7.34   Glycine forms a saturated solution in water (25°C) at $m = 3.3$. At that concentration $\gamma_{2m} = 0.872$. Calculate the ΔG_f^θ of glycine(aq) from that of the pure substance (Table 7.11). You may compare your answer to that given in the table.

7.35   The standard free energy of formation of L-serine (solid) is -508.8 kJ/mole at 25°C. This compound forms a saturated solution in water at 4.02 moles kg^{-1}, at which concentration $\gamma_m = 0.602$. Calculate $\Delta G_f^\theta(\text{aq})$.

7.36   Use the data below for the distribution of H_3BO_3 between H_2O and amyl alcohol to calculate the thermodynamic K_a (water/alcohol) for the distribution equilibrium.

moles/dm³ in H_2O	moles/dm³ in alcohol
0.02602	0.00805
0.05140	0.01545

7.37   Amino acids can polymerize through the formation of peptide bonds:

$$RCOOH + R'NH_2 \longrightarrow R-\overset{\displaystyle O}{\overset{\displaystyle \|}{C}}-\overset{\displaystyle H}{\overset{\displaystyle |}{N}}-R' + H_2O$$

These form the primary structure of proteins. Calculate the concentration of the alanine-glycine dimer at equilibrium in water if alanine and glycine were initially at $0.1m$; assume ideal solution.

7.38   The freezing points (the temperature at which the first solid appears) of various Pb/Sn alloys are listed below. The eutectic temperature is 182°C. The Pb-rich solid was analyzed and found to be a solid solution, 25% Sn in Pb. The Sn-rich solid was 5% Pb in Sn. Draw the phase diagram.

Wt % Pb	T_f	Wt % Pb	T_f
0	232°C	50	216°C
10	212°C	70	236°C
20	197°C	80	254°C
30	190°C	90	280°C
40	182°C	100	327°C

Wherefore waste our elocution
On impossible solution?

—W. S. Gilbert

8

Ionic Solutions

Salt water and sugar water are both solutions, but they differ in several important ways. In the first place, salt water conducts electricity, and very well too (not in the league with copper, but a lot better than pure water). Sugar water conducts electricity only reluctantly; its resistance may be 4 to 6 orders of magnitude greater than that of salt water and is comparable to that of the solvent alone. Also, the colligative properties (osmotic pressure, freezing-point lowering, and so on) of a salt solution are, mole for mole, substantially greater than those of a sugar solution. The reason for both of these differences is that salts are ionized in water (and some other polar solvents). This fact, only very slowly accepted by chemists, was first suggested by the careful research of Svante Arrhenius on the conductivity of solutions and of J. H. van't Hoff on colligative properties, especially osmotic pressure (Figure 8.1).

In this chapter we shall discuss three major aspects of ionic solutions: thermodynamic properties, electrical properties, and, putting these two together, electrochemical cells. In the hope that Chapter 7 is still fresh in your mind, we shall start with an attempt to make sense of the manner by which activities are defined for solutes that ionize in solution.

8.1   Ionic Activities

The fact that certain materials dissociate into ions when dissolved in polar solvents such as water was established historically by experimental results concerning the conductivity and colligative effects of such solutions. Solution conductivity is probably the most graphic proof that solutes dissociate into ions — we shall discuss this subject in Section 8.3 — but we shall begin with a brief discussion of colligative effects, since this will provide valuable clues as to how activities of electrolytes must be defined.

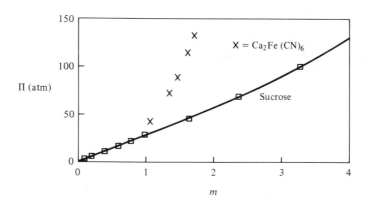

Figure 8.1   Osmotic pressures of aqueous solutions of sucrose (□) and $Ca_2Fe(CN)_6$ (×). This is an example of the radical deviations from ideal solution behavior shown by ionized solutes.

The colligative effects of electrolyte solutions were found to differ greatly from those predicted on the basis of their formal molality (the molality calculated from the formula weights). Van't Hoff expressed these deviations in terms of an "osmotic factor" denoted i. For example, the freezing-point depression and osmotic pressure were written as [compare Eqs. (7.63) and (7.66)]:

$$\theta = iK_f m$$
$$\Pi = icRT \tag{8.1}$$

[This quantity, i, is related to the "osmotic coefficient," ϕ, discussed in Chapter 7. They are not quite the same; the relationship will be derived later. For now we shall take i, as intended, to be simply an empirical factor based on Eq. (8.1).]

It was found, for an electrolyte that could dissociate into ν_+ cations and ν_- anions (total ions $\nu = \nu_+ + \nu_-$), that in the infinite-dilution limit, $i \to \nu$. This is a logical result if the electrolyte is completely dissociated in dilute solution; colligative effects depend on the number of particles, and the "effective molality" of a dissociated solute is just the product of ν and the molality (νm). The question remains as to why i is different from ν at small but finite concentrations; for example, a 0.005-molal solution of HCl in water has $i = 1.98$, respectably close to the expected value, $\nu = 2$. On the other hand, 0.005-molal $CuSO_4$ in water has $i = 1.54$ (ν, of course, is equal to 2). (For the remainder of this chapter, all solutions will be assumed to be in water unless otherwise specified.)

One reasonable explanation would be that electrolytes at finite concentration are not 100% dissociated. If the degree of dissociation is α and the formal molality is m_0, then there will be $m_0(1 - \alpha)$ moles of undissociated solute and $\nu \alpha m_0$ moles of ions; the effective molality of all species is then:

$$\text{effective molality} = m_0(1 - \alpha) + \nu \alpha m_0$$

In such a case, the osmotic factor can be related to the degree of dissociation:

$$i = 1 + (\nu - 1)\alpha$$
$$\alpha = \frac{i - 1}{\nu - 1} \tag{8.2}$$

This explanation works quite well for substances called "weak electrolytes," acetic acid for example. [The "strong/weak" classification of electrolytes is based on their conductivity in a manner that will be discussed in Section 8.3. "Works quite well" means that the degree of dissociation calculated from Eq. (8.2) is the same as that calculated from conductivity.] However, Eq. (8.2) can give misleading, or even absurd, results for strong electrolytes.

Example: HCl ($m = 1$ mole/kg) has $i = 2.12$. Calculate α.

$$\alpha = \frac{2.12 - 1}{2 - 1} = 1.12$$

It's 112% dissociated?

$CuSO_4$ ($m = 1$ mole/kg) has $i = 0.93$:

$$\alpha = \frac{0.93 - 1}{2 - 1} = -0.07$$

Now there is a truly meaningless calculation! ∎

The modern point of view is that strong electrolytes are 100% dissociated at low concentrations and that the observed differences between i and ν are due to solution nonidealities. (Ion pairing and complex formation may occur at higher concentration.) As we shall see shortly, such deviations can be extreme even at very low concentrations (~0.01 molal). The class of strong electrolytes includes nearly all salts, as well as inorganic acids and bases (HCl, NaOH), if the solvent is water or some other very polar liquid.

Activity of Strong Electrolytes

Let us assume a salt (or any other strong electrolyte) that dissociates into ν_+ cations with charge z_+ and ν_- anions with charge z_-; the solution process is:

$$M_{\nu_+}X_{\nu_-}(\text{solid}) \longrightarrow \nu_+ M^{z+}(\text{aq}) + \nu_- X^{z-}(\text{aq})$$

The chemical potential of the solute can reasonably be written as a sum of chemical potentials for the ions:

$$\mu_{\text{salt}} = \nu_+ \mu_+ + \nu_- \mu_- \tag{8.3}$$

For each ion, the activity is defined as:

$$\ln a_i = \frac{\mu_i - \mu_i^\theta}{RT}$$

The activity of the salt as a whole is similarly related to its chemical potential:

$$\ln a_{\text{salt}} = \frac{\mu_{\text{salt}} - \mu_{\text{salt}}^\theta}{RT}$$

Using Eq. (8.3), we see:

$$\nu_+ \ln a_+ + \nu_- \ln a_- = \ln a_{\text{salt}}$$

or

$$a_{\text{salt}} = a_+^{\nu_+} a_-^{\nu_-}$$

Note that activities of salts and the individual ions are always referred to the solute standard state (usually molality scale).

From the colligative effects we know that a mole of salt at infinite dilution behaves more like ν moles of a nonelectrolyte; this suggests that the interesting quantity is μ/ν for the salt. Since:

$$\frac{\mu_{\text{salt}}}{\nu} = \frac{\mu_{\text{salt}}^\theta}{\nu} + RT \ln a_{\text{salt}}^{1/\nu}$$

this suggests that the $(1/\nu)$ power of the salt activity is the significant quantity. Also, activities of individual ions can never be measured—a solution must always contain both cations and anions to ensure electrical neutrality. Therefore, it is convenient to define a quantity called the **mean ionic activity** $(a_\pm)$ as the geometrical mean of a_+ and a_-:

$$a_\pm = (a_+^{\nu_+} a_-^{\nu_-})^{1/\nu}$$

$$a_{\text{salt}} = a_\pm^\nu \tag{8.4}$$

With this quantity, the chemical potential of the salt *per mole of ions* is just:

$$\frac{\mu_{\text{salt}}}{\nu} = \frac{\mu_{\text{salt}}^\theta}{\nu} + RT \ln a_\pm$$

This, again, is suggested by van't Hoff's results, which demonstrate that, in dilute solution at least, a mole of salt acts like ν moles of a nonionized solute.

Ionic Activity Coefficients

An activity coefficient can be defined for each ion:

$$a_+ = \gamma_+ m_+, \qquad a_- = \gamma_- m_-$$

where $m_+ = \nu_+ m$ and $m_- = \nu_- m$ are the molalities of the ions. For the salt, this definition gives

$$a_{\text{salt}} = a_+^{\nu_+} a_-^{\nu_-} = m_+^{\nu_+} m_-^{\nu_-} \gamma_+^{\nu_+} \gamma_-^{\nu_-}$$

This suggests the definition of a **mean ionic activity coefficient** $\gamma_\pm$:

$$\gamma_\pm = (\gamma_+^{\nu_+} \gamma_-^{\nu_-})^{1/\nu} \tag{8.5}$$

The reason for doing this is, again, that γ_+ and γ_- cannot be measured but $\gamma_\pm$ can. Putting it all together:

$$a_{\text{salt}} = a_\pm^\nu = (m_+^{\nu_+} m_-^{\nu_-}) \gamma_\pm^\nu \tag{8.6}$$

Example: Some authors introduce a quantity called the mean ionic molality, $m_\pm$, so that $a_\pm = m_\pm \gamma_\pm$ by analogy with, for example, Eq. (7.54c). This seems to be a needless confusion. Let's work through a specific example to see how it comes out.

The salt Na_2SO_4, with a formal molality m, dissociates into $m_+ = 2m$ sodium ions and $m_- = m$ sulfate ions. From Eq. (8.6), the activity is:

$$a(Na_2SO_4, \text{aq}) = (2m)^2(m)\gamma_\pm^3 = 4m^3\gamma_\pm^3$$

The mean ionic activity is therefore:

$$a_\pm = (4)^{1/3} m \gamma_\pm$$

The appearance of factors such as the $(4)^{1/3}$ in this relationship seems to cause a lot of confusion; think through this example carefully — the factor of $4^{1/3}$ is simply a result of the fact that one mole of Na_2SO_4 produces two moles of Na^+. ∎

Exercise: Show that the mean ionic activity of $La_2(SO_4)_3$ is:

$$a_\pm = (2.55085)m\gamma_\pm$$
∎

Some examples of mean ionic activity coefficients are given in Figure 8.2 and Table 8.1. Comparison to earlier results, Table 7.7 for example, will convince you that the deviations from ideal solution behavior of electrolytes are indeed extreme.

Measurement of Ionic Activity Coefficients

Actually, there are a number of ways to measure ionic activity coefficients; in Section 8.5 we shall see how they can be measured using cell emf's. In this section we shall discuss their measurement using colligative effects and the Gibbs-Duhem equation. The derivations of Chapter 7 will not be repeated, but the steps leading to Eq. (7.58) will be very much the same if a_2 is replaced by $a_\pm^\nu$. It is convenient to redefine the **osmotic coefficient** as:

$$\phi = \frac{-1000/M_1}{\nu m} \ln a_1 \qquad (8.7)$$

[This is the same as the original definition, Eq. (7.57), for $\nu = 1$—that is, for nonelectrolytes. This definition has the advantage that the limit of ϕ as $m \to 0$ is always 1, electrolyte or not.] As before, the solvent activity can be measured by vapor-pressure lowering or freezing-point depression. As a consequence of the new definition of ϕ, Eq. (7.64) for the freezing-point depression is now:

$$\phi = \frac{\theta}{\nu K_f m} (1 + b\theta) \qquad (8.8)$$

Neglecting the b correction, this is just:

$$\theta = \nu \phi K_f m$$

from which it should be apparent that the van't Hoff i factor is just $\nu\phi$, and, since $\phi \to 1$ as $m \to 0$, $i \to \nu$ in the same limit.

Osmotic coefficients could also be measured from solvent vapor pressure, using Eq. (8.7) and $a_1 = P_1/P_1^\bullet$, as was illustrated in Chapter 7.

The solute activity coefficient, which is now $\gamma_\pm$, can be calculated as before from the Gibbs-Duhem equation with the formula ($j = 1 - \phi$):

$$-\ln \gamma_\pm = j + \int_0^m \left(\frac{j}{m'}\right) dm' \qquad (8.9)$$

There is a difference in the manner in which the data are treated for electrolytes. For nonelectrolyes, it was suggested that j/m be graphed vs. m, or expanded in a power series in m [Eq. (7.59)]. Brønsted (1922) noted, on purely empirical grounds, that in dilute solutions, $\ln \gamma_\pm$ was linear in the square root of the molality. This suggests that it would be better to write j as a power series in $\sqrt{m}$ rather than m:

$$j = a\sqrt{m} + bm + cm^{3/2} + \cdots$$

$$\frac{j}{m} = \frac{a}{\sqrt{m}} + b + cm^{1/2} + \cdots \qquad (8.10)$$

Figure 8.3 displays data obtained from the vapor-pressure lowering of aqueous solutions of NaCl. The graph of j/m vs. m clearly demonstrates that data at lower

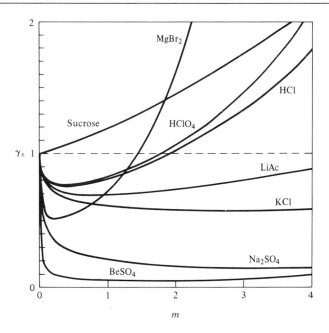

Figure 8.2 Mean ionic activity coefficients for salts and strong acids in water at 25°C. (Sucrose is included for comparison.)

Table 8.1 Mean ionic activity coefficients in water (25°C)

m	KCl	NaCl	HCl	$NaClO_4$	H_2SO_4	$CuSO_4$
0.001	0.9648	0.966	0.966	—	0.830	0.74
0.005	0.927	0.929	0.928	—	0.639	0.53
0.01	0.901	0.904	0.904	—	0.544	0.41
0.02	0.868	0.875	0.875	—	0.453	0.31
0.05	0.816	0.823	0.830	—	0.340	0.21
0.1	0.769	0.778	0.796	0.775	0.265	0.16
0.2	0.718	0.735	0.767	0.729	0.209	0.11
0.5	0.649	0.681	0.757	0.668	0.154	0.068
1.0	0.603	0.673	0.809	0.629	0.130	0.047
4.0	—	0.791	1.74	—	0.172	—

m	$CaCl_2$	$ZnCl_2$	$CdSO_4$	$Cr(NO_3)_3$	$ZnSO_4$	$LaCl_3$
0.005	0.789	0.767	0.476	—	0.477	—
0.01	0.732	0.708	0.383	—	0.387	—
0.02	0.669	0.642	—	—	0.298	—
0.05	0.584	0.556	0.199	—	0.202	—
0.1	0.524	0.502	0.137	0.319	0.148	0.331
0.2	0.491	0.448	—	0.285	0.104	0.298
0.5	0.510	0.376	0.061	0.291	0.063	0.266
1	0.725	0.325	0.042	0.401	0.044	0.481

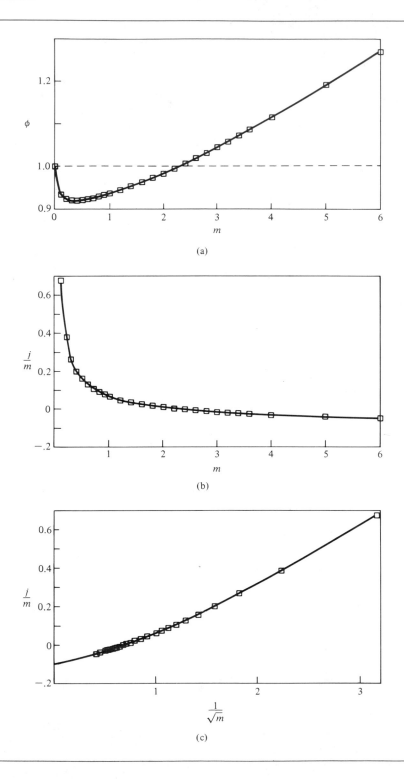

(a)

(b)

(c)

concentrations are needed in order for the graphical integration used in Chapter 7 (Figure 7.10) to be effective. But such data are not easily measured; at the lowest concentration shown ($m = 0.1$ mol/kg) the water vapor pressure is lowered by only 0.3%, and at lower concentration accurate measurement would be difficult. On the other hand, a plot of j/m vs. $1/\sqrt{m}$ [Figure 8.3(c)] is clearly well suited for representation by a low-order polynomial or for other types of numerical analysis.

Exercise: Show that a series of the form of Eq. (8.10) gives the equation:

$$\ln \gamma_\pm = -3a\sqrt{m} - 2bm - \tfrac{5}{3}cm^{3/2} - \cdots \tag{8.11}$$

■

Example: R. H. Wood, R. Kirk Wicker and R. W. Kreis [*J. Phys. Chem.*, 75, 2313 (1971)] measured osmotic coefficients by freezing-point depression for solutions of $NaNO_3$ in N-methylacetamide. Some results:

m	ϕ	$\dfrac{j}{m}$	$\dfrac{1}{\sqrt{m}}$	$\gamma_\pm$(calculated)
0.05	0.979	0.42	4.47	—
0.1	0.972	0.28	3.16	0.971
0.2	0.962	0.19	2.23	0.915
0.3	0.955	0.15	1.82	0.864
0.4	0.948	0.13	1.58	—
0.5	0.942	0.12	1.41	0.833

For dilute solutions, it is sufficiently accurate to use only the first two terms of Eq. (8.10). Linear regression of j/m vs. $1/\sqrt{m}$:

$$\frac{j}{m} = b + a\frac{1}{\sqrt{m}}$$

gives for these data an intercept $b = -0.02738$, and slope $a = 0.09899$ ($r = 0.99899$). From Eq. (8.11) (neglecting the third term) this gives:

$$\ln \gamma_\pm = -0.297\sqrt{m} + 0.055m$$

This equation gives the "calculated" values above. The authors used a more sophisticated, and complicated, method of analysis; in this particular case the simple analysis above gives nearly the same results. ■

Actually the determination of ionic activity coefficients from colligative effects is very complicated; Lewis and Randall (ref. 4) give the details. The more accurate analysis rests on the fact that the limiting slope ["$3a$" in Eq. (8.11)] can be calculated theoretically; that is the next topic.

Figure 8.3 Osmotic coefficient of NaCl in water. (a) The osmotic coefficient plotted vs. molality (25°C). (b) Plot of j/m vs. m for NaCl in water ($j = 1 - \phi$); compare this to the behavior of nonionized solutes as illustrated by sucrose (Figure 7.10). Calculating the numerical integral required to calculate activity coefficients from such a graph would be very inaccurate. (c) Graph of j/m vs. $1/\sqrt{m}$; such a graph is well behaved for ionic solutes and is well-suited for numerical analysis.

Theory of Electrolyte Solutions

The behavior of ionic solutes in dilute solutions was explained theoretically by Debye and Hückel (1923). This theory treats the ions as point charges ($z_i e$) interacting by Coulomb's law; the solvent was treated as a continuous medium characterized by a dielectric constant ε. This is a clear oversimplification; the solvent is a complex, molecular aggregation that is "lumpy" on the scale of ion sizes. Furthermore, its electrical structure can be altered by strong electric fields such as those due to nearby ions.

The theory (discussed in more detail in ref. 4, p. 335) introduced a parameter a_0 as the distance of closest approach, a radius about the ion into which no other ions could penetrate. The theory also showed the importance to ionic theory of the **ionic strength** (I), which had been introduced by Lewis and Randall a few years earlier:

$$I = \tfrac{1}{2} \sum_i z_i^2 m_i \qquad (8.12)$$

After a number of approximations, all of which are rigorous in an infinitely dilute solution, Debye and Hückel obtained the result:

$$\ln \gamma_{\pm} = \frac{-\alpha |z_+ z_-| \sqrt{I}}{1 + B a_0 \sqrt{I}} \qquad (8.13)$$

$$\alpha = \frac{e^3}{(\varepsilon k T)^{3/2}} \left(\frac{2\pi \rho^{\bullet} L}{1000} \right)^{1/2} \qquad (8.14a)$$

$$B = \left(\frac{8\pi L e^2 \rho^{\bullet}}{1000 \varepsilon k T} \right)^{1/2} \qquad (8.14b)$$

where e is the electron charge, $\rho^{\bullet}$ is the density of the pure solvent, ε is the solvent's dielectric constant, k is Boltzmann's constant, and L is Avogadro's number.

The Debye-Hückel theory correctly predicted the dependence of $\ln \gamma_{\pm}$ on $\sqrt{m}$ in dilute solution (when the B term can be neglected) that had been observed empirically by Brønsted. It also provided a numerical value for the limiting slope ["$3a$" in Eq. (8.11)]; this value, $-|z_+ z_-|\alpha$, can be calculated if the dielectric constant and density of the solvent are known (see Table 8.2 for values for H_2O) and has received abundant experimental verification (see Figure 8.4).

Equation (8.13) without the B term is called the **Debye-Hückel limiting law** *(DHLL);* it is useful for calculating limiting slopes as $m \rightarrow 0$. Use of Eq. (8.13) at finite concentrations is limited because of the parameter a_0, whose value is not subject to independent experimental verification. (For large ions a_0 could be taken to be the mean diameter of the ion pair, but for small ions such as K^+ and Cl^- its value is considerably larger than the ionic radii as determined by crystallography; see Problem 8.5.)

Guggenheim (1935) pointed out that a_0 is often ~ 3 Å, so $B a_0$ is usually of the order of magnitude 1; he proposed an empirical equation:

$$\ln \gamma_{\pm} = \frac{-\alpha |z_+ z_-| \sqrt{I}}{1 + \sqrt{I}} + 2 \left(\frac{\nu_+^2 + \nu_-^2}{\nu_+ + \nu_-} \right) \beta m \qquad (8.15)$$

Table 8.2 Constants for the Debye-Hückel formula for water

t (°C)	$\rho^{\bullet}$ (H_2O)	α (molal)	α (molar)	B^*
0	0.999841	1.133	1.133	0.324
10	0.999700	1.149	1.149	—
20	0.998263	1.167	1.168	—
25	0.997044	1.177	1.179	0.329
30	0.995646	1.190	1.195	—
40	0.99221	1.207	1.212	0.332
100	0.95835	1.372	1.401	—

*For a_0 in angstroms $= 10^{-8}$ cm.

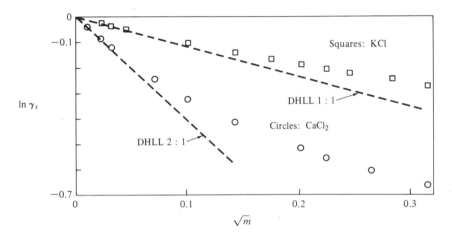

Figure 8.4 Examples of the Debye-Hückel limiting law (DHLL). The importance of this law is not in its accuracy (which is rather poor in most cases) but in its predicting the correct limiting slopes.

where β is an empirical parameter that is specific for the ion pair to which $\gamma_\pm$ applies. (Table 8.3 gives some values.) For brevity we shall call Eq. (8.15) the **DHG equation;** it gives excellent results for calculating $\gamma_\pm$ up to $I \cong 0.1$ if β is known. Even when β is not known, the DHG equation ($\beta = 0$) works reasonably well up to $I \cong 0.1$. (It is not, in either case, reliable at higher ionic strengths.)

As written, Eq. (8.15) applies only to single-electrolyte solutions; there is a generalization for mixed electrolytes [E. A. Guggenheim and J. C. Turgeon, *Trans. Faraday Soc.*, 51, 747 (1955)], but it is complicated and we shall not use it. The effect of other electrolytes on the activity coefficient of a particular pair is accounted for to a large extent by their contributions to the ionic strength, Eq. (8.12); when we need such calculations we shall use the DHG equation without the second (β) term.

Our coverage of the theory of electrolyte activities has brought us to about 1935. A great deal of progress has been made since then; in particular, McMillan and Mayer

Table 8.3   Specific ion interaction constants

Salt	$\beta(0°)$	$\beta(25°)$	Salt	$\beta(0°)$	$\beta(25°)$
HCl	0.25	0.27	KCl	0.04	0.10
HBr	—	0.33	KBr	0.06	0.11
HI	—	0.36	$KClO_4$	−0.55	—
$HClO_4$	—	0.30	KNO_3	−0.28	−0.11
HNO_3	0.16	—	KAc	0.30	0.26
LiCl	0.20	0.22	CsCl	—	0
LiBr	0.30	0.26	$CsNO_3$	—	−0.15
$LiNO_3$	0.23	0.21	$TiNO_3$	—	−0.36
LiAc	0.19	0.18	$BaCl_2$	0.7	—
NaCl	0.11	0.15	$Ba(NO_3)_2$	−0.5	—
NaBr	0.20	0.17	$CoCl_2$	1.1	—
NaAc	0.26	0.23	K_2SO_4	−0.1	—
$AgNO_3$	—	−0.14			

Source: E. A. Guggenheim and J. C. Turgeon, *Trans. Faraday Soc.*, 51, 747 (1955).

[*J. Chem. Phys.*, 13, 276 (1945)] demonstrated how statistical mechanics could be used to calculate solution properties. Reliable methods for calculating thermodynamic properties of concentrated electrolyte solutions are now available, but we shall go no farther.

Example: Calculate the ionic strength of a solution that contains 0.1 mol/kg NaCl and 0.05 mol/kg Na_2SO_4.

$$Na^+: \quad m = 0.1 + 2(0.05) = 0.2, \quad z = 1$$

$$Cl^-: \quad m = 0.1, \quad\quad\quad\quad\quad\quad z = 1$$

$$SO_4^{2-}: \quad m = 0.05, \quad\quad\quad\quad\quad z = 2$$

$$I = \tfrac{1}{2}[0.2 + 0.1 + 4(0.05)] = 0.25 \quad ■$$

Exercise: Show that the ionic strength of a single-electrolyte solution is related to the molality (m) as follows:

$$1:1 \text{ electrolyte (e.g., NaCl):} \quad\quad I = m$$

$$2:2 \text{ electrolyte (e.g., CaSO}_4\text{):} \quad\quad I = 4m$$

$$1:2 \text{ or } 2:1 \text{ (e.g., Na}_2SO_4 \text{ or CaCl}_2\text{):} \quad I = 3m \quad ■$$

Example: Estimate $\gamma_\pm$ for a 0.05 mol/kg solution of KCl at 25°C. The value of β is given in Table 8.3.

$$\ln \gamma_\pm = \frac{-1.177\sqrt{0.05}}{1 + \sqrt{0.05}} + 2(0.10)(0.05) = -0.2051$$

$$\gamma_\pm = 0.815 \quad \text{(Table 8.1 gives 0.816)} \quad ■$$

Peter Debye (1884–1966)

Debye was born in Maastricht, Netherlands, and was trained as an electrical engineer. In 1908 he received his doctorate in physics and in 1911 succeeded Einstein as professor of physics at Zurich. In the next five years he made three major contributions: his theory of specific heats (Chapter 15), the powder method for X-ray crystallography (with Paul Scherrer), and the discovery of permanent dipole moments in molecules and their effect on the dielectric constant of materials (Chapter 13). The unit for measuring dipole moments in called the debye in his honor.

Arrhenius's 1887 theory that salts were ionized in aqueous solution has been called by some the beginning of physical chemistry. As impressive as this theory was, a number of troubling problems remained, that could not be explained away by assuming partial dissociation. In 1923 Debye and Hückel published their theory, which effectively explained the concentration dependence of the conductivity and the thermodynamics of strong electrolytes as a result of the Coulombic forces between the ions.

In the 1930's Debye was at the Kaiser Wilhelm Institute in Berlin (where, again, he succeeded Einstein) but remained a Dutch citizen. In 1939 he was barred from his laboratory and told he could not return unless he became a German citizen. Instead, he emigrated to the United States, ending up at Cornell University.

Example: Estimate $\gamma_\pm$ for 0.01 mol/kg H_2SO_4 solution when the solution also contains 0.02 mol/kg NaCl.

$$I = 0.02 + 3(0.01) = 0.05$$

$$\ln \gamma_\pm = \frac{-1.177 |2| \sqrt{0.05}}{1 + \sqrt{0.05}} = -0.4302, \qquad \gamma_\pm = 0.650 \qquad \blacksquare$$

8.2 Ionic Equilibria

Many important types of chemical reactions in solution involve ions. In this section we shall look at two — solubility and acid-base reactions; later in this chapter we shall discuss some further aspects of ionic reactions, particularly oxidation-reduction reactions.

Solubility Product

The equilibrium between a salt and its saturated solution is:

$$M_{\nu_+} X_{\nu_-}(\text{solid}) = \nu_+ M^{z+} + \nu_- X^{z-}$$

The **solubility product** is defined as:

$$K_{sp} = m_+^{\nu_+} m_-^{\nu_-} \qquad (8.16)$$

This quantity is, in the dilute-solution approximation, a constant that is independent of the concentrations of the individual ions.

Example: AgCl in water has a solubility $S = 1.27 \times 10^{-5}$ mole/kg. From Eq. (8.16), $K_{sp} = S^2 = 1.61 \times 10^{-10}$. If this salt is dissolved in a solution with 0.1 mol/kg NaCl, we can calculate:

$$[Ag^+] = S, \qquad [Cl^-] = 0.1 + S \cong 0.1$$

$$S = \frac{K_{sp}}{0.1} = 1.61 \times 10^{-9} \text{ mol/kg}$$

One may well question the use of the ideal dilute-solution approximation in this case; we'll check that out shortly. ∎

The thermodynamic treatment of this equilibrium gives the true equilibrium *constant* as:

$$K_a = a_+^{\nu_+} a_-^{\nu_-}$$

Then, with:

$$a_+ = m_+ \gamma_+ \quad \text{and} \quad a_- = \gamma_- m_-$$

$$K_a = (m_+^{\nu_+} m_-^{\nu_-})(\gamma_+^{\nu_+} \gamma_-^{\nu_-})$$

The first part of the right-hand side is just the solubility product of Eq. (8.16); the second part is replaced by the mean ionic activity coefficient [Eq. (8.5)]; the result is:

$$K_a = K_{sp} \gamma_\pm^\nu \qquad (8.17)$$

For a very insoluble salt such as AgCl, the difference between K_a and K_{sp} is minor if there are no other ions present. However, for more soluble or more highly charged salts, the difference can be significant, particularly if there are other electrolytes present to contribute to the ionic strength.

Example: Calculate K_a for the solubility of AgCl. For a 1:1 electrolyte, $I = m$, and this is equal to the solubility: $I = S = 1.27 \times 10^{-5}$. From the DHG equation:

$$\ln \gamma_\pm = \frac{-1.177\sqrt{1.27 \times 10^{-5}}}{1 + \sqrt{1.27 \times 10^{-5}}} = -0.00418, \qquad \gamma_\pm = 0.996$$

$$K_a = K_{sp}(0.996)^2 = 1.60 \times 10^{-10}$$

As predicted, the correction is negligible. ∎

Example: Next we shall check the results of the previous example for the solubility of AgCl in 0.1 mol/kg NaCl. If the solubility is S:

$$I = 0.1 + S$$

We can neglect S, the contribution of the AgCl to I, and use the DHG formula (without the β term) to calculate:

$$\ln \gamma_{\pm} = -\frac{1.177\sqrt{0.1}}{1 + \sqrt{0.1}}, \qquad \gamma_{\pm} = 0.754$$

The solubility product will be:

$$K_{sp} = \frac{1.60 \times 10^{-10}}{(0.754)^2} = 2.81 \times 10^{-10}$$

$$= [Ag^+][Cl^-]$$

Again $[Cl^-] = 0.1 + S$, and we can neglect S compared to 0.1 to calculate:

$$S = \frac{2.81 \times 10^{-10}}{0.1} = 2.81 \times 10^{-9}$$

Thus, the solubility is predicted to be 75% greater than that predicted by the ideal solution theory. The calculation is still somewhat suspect on at least two grounds. First, $m = 0.1$ mol/kg is at the upper limit for the validity of the DHG equation. Also, at such a high chloride concentration, complex ions such as $AgCl_2^-$ may form, and this will enhance the solubility even more. ∎

In the example above, the solubility of the salt did not contribute significantly to the ionic strength. For situations where the solubility is a significant part of the ionic strength, a short series of successive approximations must be used. The following procedure is recommended: (1) Estimate S assuming an ideal solution. (2) Calculate I and, from that, $\gamma_{\pm}$ and K_{sp}. (3) Calculate S again. (4) Repeat steps 2 and 3 until S converges.

Example: Calculate the solubility of AgCl in water (25°C) with 1.50×10^{-3} mol/kg $Ba(NO_3)_2$ using $K_a = 1.60 \times 10^{-10}$. The ionic strength is:

$$I = 3(1.5 \times 10^{-3}) + S = 0.0045 + S$$

where the solubility (S) is given by:

$$K_a = \gamma_{\pm}^2 S^2, \qquad S = \frac{\sqrt{K_a}}{\gamma_{\pm}}$$

The ideal solution solubility is 1.26×10^{-5}, so:

$$I = 0.0045 + 1.26 \times 10^{-5} = 0.0045126$$

The activity coefficient for Ag^+Cl^- is:

$$\ln \gamma_{\pm} = \frac{-1.177\sqrt{I}}{1 + \sqrt{I}} = -0.07409, \qquad \gamma_{\pm} = 0.9286$$

With this, the solubility is:

$$S = \frac{1.26 \times 10^{-5}}{0.9286} = 1.36 \times 10^{-5} \text{ mol/kg}$$

(The measured solubility is 1.372×10^{-5} mol/kg.) A second round of approximation is not needed, since the change in I is negligible. However, if this is not obvious, the reader should do the next round.   ∎

Weak Acids and Bases

Weak acids (HA) and bases (B) undergo hydrolysis in water:

$$HA(aq) = H^+(aq) + A^-(aq)$$

$$K_a' = \frac{[H^+][A^-]}{[HA]} \tag{8.18}$$

$$B(aq) + H_2O = BH^+(aq) + OH^-(aq)$$

$$K_b' = \frac{[BH^+][OH^-]}{[B]} \tag{8.19}$$

[The dissociation constant of Eq. (8.18) is traditionally called K_a, but we have been using that symbol to mean the thermodynamic or activity equilibrium constant of any reaction. To avoid confusion, a prime is added to the acid dissociation constant, and to K_b as well for consistency's sake.]

The thermodynamic equilibrium constant for an acid dissociation is:

$$K_a = K_a' \frac{\gamma_\pm^2}{\gamma_{HA}} \tag{8.20}$$

For dilute solutions it is usually assumed that the neutral solute HA behaves ideally ($\gamma_{HA} \cong 1$).

Example: The degree of dissociation of acetic acid, with a formal molality $m_0 = 2.41 \times 10^{-3}$ mol/kg, is $\alpha = 0.0829$. Calculate the thermodynamic equilibrium constant.

$$[H^+] = [Ac^-] = \alpha m_0, \qquad [HAc] = m_0(1 - \alpha)$$

$$K_a' = \frac{\alpha^2 m_0}{1 - \alpha} = \frac{(0.0829)^2(2.41 \times 10^{-3})}{1 - 0.0829} = 1.81 \times 10^{-5}$$

The ionic strength is $I = \alpha m_0 = 2.00 \times 10^{-4}$. Using the DHG equation ($\beta = 0$) gives $\gamma_\pm = 0.984$.

$$K_a = (1.81 \times 10^{-5})(0.984)^2 = 1.75 \times 10^{-5} \qquad ∎$$

Example: Calculate the pH (puissance d'Hydrogen) of a 0.01 mol/kg solution of acetic acid in water (25°C) if 0.05 mol/kg NaCl is added.

First, assuming ideal solution:

$$K_a = \frac{[H^+][Ac^-]}{[HAc]} = \frac{\alpha^2 m_0}{1 - \alpha}$$

$$\alpha^2 = \left(\frac{K_a}{0.01}\right)(1 - \alpha) = 1.75 \times 10^{-3}(1 - \alpha)$$

This can be solved to give $\alpha = 0.0410$. Then:

$$[H^+] = 0.01(0.0410) = 4.10 \times 10^{-4}, \qquad pH \cong -\log_{10}[H^+] = 3.387$$

With the salt added, the ionic strength is:

$$I = 0.05 + 0.01\alpha$$

Neglecting the contribution of the ionized acetic acid to I (the second term), we can use the DHG formula to calculate $\gamma_{\pm} = 0.806$. Then:

$$K_a' = \frac{K_a}{\gamma_{\pm}^2} = 2.691 \times 10^{-5}$$

Solving for α as above:

$$\alpha^2 = (0.00269)(1 - \alpha)$$

gives $\alpha = 0.0505$. The ionic strength is:

$$I = 0.05 + 0.01(0.0505) = 0.0505$$

The reader can demonstrate that this small change is negligible. The hydrogen-ion activity is, strictly, $[H^+]\gamma_+$; since single ion activity coefficients cannot be measured, the practical definition is:

$$a_{H^+} = [H^+]\gamma_{\pm}$$

$$pH = -\log_{10} a_{H^+}$$

The reader should show that this calculation gives:

$$[H^+] = 5.05 \times 10^{-4}$$

$$pH = 3.390$$

Despite a 20% change in the H^+ concentration, the pH is nearly the same as that estimated by the ideal solution calculation; the reason is that the "ideal" calculation underestimated α by 20% and (by assuming the activity coefficient $= 1$) overestimated $\gamma_{\pm}$ by 20%, and these errors cancel for the pH calculation. This will generally be the case when $\alpha \ll 1$. ∎

Free Energies of Formation

As for other types of solutes, a standard free energy of formation can be defined for electrolytes in solution. For example, $\Delta G_f^\theta(HCl, aq)$ is the free energy of the reaction:

$$\tfrac{1}{2} H_2(g, 1 \text{ atm}) + \tfrac{1}{2} Cl_2(g, 1 \text{ atm}) \longrightarrow HCl(aq, a = 1)$$

Recalling Eq. (8.3), it would seem reasonable to write this quantity as a sum of contributions for individual ions; in this case:

$$\Delta G_f^\theta(HCl, aq) = \Delta G_f^\theta(H^+, aq) + \Delta G_f^\theta(Cl^-, aq)$$

But, of course, the free energy of formation of a single ion cannot be measured any more than its activity. It is taken as a convention that the standard free energy of formation of the H^+ is zero; therefore the entire amount of ΔG^θ for the above reaction is assigned to Cl^-:

$$\Delta G_f^\theta(Cl^-, aq) = \Delta G_f^\theta(HCl, aq)$$

If, then, the free energy of formation some other salt is measured, NaCl for example:

$$\Delta G_f^\theta(\text{NaCl}, \text{aq}) = \Delta G_f^\theta(\text{Na}^+, \text{aq}) + \Delta G_f^\theta(\text{Cl}^-, \text{aq})$$

the free energy of formation of $\text{Na}^+(\text{aq})$ can be calculated. Table 8.4 lists values for some common ions.

Free energies of formation of electrolytes can be determined as for any other solute (Section 7.10); in addition, these can be calculated from emf measurements of electrochemical cells (Section 8.5).

Example: The thermodynamic solubility product of AgCl in water was calculated earlier (p. 376): $K_a = 1.60 \times 10^{-10}$. The standard free energy of the reaction:

$$\text{AgCl}(s) = \text{Ag}^+(\text{aq}) + \text{Cl}^-(\text{aq})$$

can be calculated from this number:

$$\Delta G_{\text{rxn}}^\theta = -(8.3143)(298.15)\ln(1.60 \times 10^{-10}) = 55.91 \text{ kJ}$$

This quantity can also be calculated from data on Tables 6.1 and 8.4:

$$\Delta G_{\text{rxn}}^\theta = \Delta G_f^\theta(\text{Ag}^+, \text{aq}) + \Delta G_f^\theta(\text{Cl}^-, \text{aq}) - \Delta G_f^\theta(\text{AgCl}, s)$$

$$= 77.10 - 131.17 - (-109.72) = 55.65 \text{ kJ} \qquad \blacksquare$$

Table 8.4   Thermodynamic properties of ions in solution (water, 25°C)

ion	S_m^θ (J/K)	ΔG_f^θ (kJ)	ΔH_f^θ (kJ)
H^+	0*	0*	0*
Li^+	19.7	−293.8	−278.46
Na^+	58.6	−261.87	−239.66
K^+	103.3	−282.2	−252.7
Ag^+	73.2	77.11	−105.9
Mg^{2+}	−132.2	−456.01	−461.96
Ca^{2+}	−53.1	−553.04	−542.96
Cd^{2+}	−61.1	−77.74	−72.38
Cu^{2+}	−110.9	64.98	+ 64.39
Zn^{2+}	−107.5	−147.21	−152.4
Pb^{2+}	21	−24.3	1.6
F^-	−9.6	−276.5	−329.1
Cl^-	56.5	−131.17	−167.46
Br^-	82.4	−102.82	−120.9
I^-	105.9	−51.67	−55.94
OH^-	10.4	−157.30	−229.9
NO_3^-	125.1	−110.5	−206.57
SO_4^{2-}	18.4	−741.99	−907.51
ClO_4^-	—	−10.3	−131.4

*All values relative to H^+, which is zero by convention.

8.3 Electrochemistry

In this section we shall discuss electrical properties of ionic solutions — in particular, the conductivity. Before beginning, it will no doubt be useful to briefly review some definitions and units relating to electricity.

An electrical current I (SI unit: amperes, A) is a rate of flow of charge Q (SI unit: coulombs, C):

$$I = \frac{dQ}{dt}$$

$$[\text{ampere}] = [\text{coulomb/sec}] \tag{8.21}$$

The electromotive force (emf) is $\mathscr{E}$ (SI unit: volts, V), also called the electrical potential or voltage. The work required to move a charge through an electric potential is:

$$w = \mathscr{E}Q$$

$$[\text{joule}] = [\text{volt-coulomb}] \tag{8.22}$$

The power is the work per unit time:

$$\text{power} = \mathscr{E}I$$

$$[\text{watt}] = [\text{J/s}] = [\text{volt-ampere}] \tag{8.23}$$

The ratio of the emf to the current flowing through any material is called the resistance R (SI unit: ohms, Ω) (**Ohm's law**):

$$R = \frac{\mathscr{E}}{I}$$

$$[\text{ohm}] = [\text{volt/ampere}] \tag{8.24a}$$

The reciprocal of the resistance is called the *conductance* (SI unit: siemens, S); this quantity is often given the symbol L, but since we are already using that symbol for Avogadro's number, we shall simply denote the conductance as R^{-1}:

$$R^{-1} = \frac{I}{\mathscr{E}}$$

$$[\text{S}] = [\Omega^{-1}] = [\text{A/V}] \tag{8.24b}$$

The transport of electricity through a metallic conductor is by motion of electrons, each of which carries a charge $-e$:

$$e = 1.60210 \times 10^{-19} \text{ C}$$

Thus, the passage of one coulomb involves the transport of 6.24181×10^{18} electrons.

The transport of charge through a solution is by ions, each of which carries a charge that is an integer multiple of the electronic charge — $z_i e$. One mole of ions carries a charge:

$$Q_i = z_i \ (Le)$$

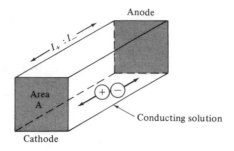

Figure 8.5   The conduction of electricity between electrodes of area A. The solution between the electrodes has the same area as the electrodes—that is, there are no constrictions. The current in an ionic solution has two components, that due to anions moving toward the anode, and that due to cations moving toward the cathode.

The quantity Le, the value of one mole of unit charges, is called **Faraday's constant**:

$$\mathscr{F} = Le = 96\,487 \text{ C}$$

Ion Mobility

The passage of a current through an ionic solution has two components: that due to a flow of cations (I_+) in one direction and that due to a flow of anions (I_-) in the opposite direction (Figure 8.5):

$$I = I_+ + I_- \tag{8.25}$$

We shall now examine the situation of a current flowing through a solution between parallel plate electrodes; each electrode has an area A and they are separated by a distance l. (The intervening solution also has a cross-sectional area A; that is, there is no constriction between the electrodes.) The current due to the flux of a particular ion i is:

$$I_i = \frac{dQ_i}{dt} = |z_i|e\frac{dN_i}{dt}$$

If we assume that the concentration of ions (N_i/V) between the plates is uniform, and the ions are moving at a uniform velocity (v_i), the number of ions arriving at the electrode (area A) in a time interval Δt will be:

$$\Delta N = \frac{N_i}{V}Av_i\,\Delta t$$

and the current will be:

$$I_i = |z_i|e\frac{N_i}{V}Av_i \tag{8.26}$$

According to Coulomb's law, the force exerted on a charge $q = z_i e$ in an electric field $E = d\mathscr{E}/dx$ is:

$$\text{force} = z_i eE$$

By Newton's law, (force) = (mass)(acceleration), the charge should be accelerated toward the oppositely charged electrode with a velocity that increases at a constant rate:

$$m\frac{dv}{dt} = z_i eE$$

This is indeed what would happen if the ion were in a vacuum, but in a solution the motion of the ion will be slowed by a frictional drag; a frictional force, one that is proportional to the velocity, must be subtracted from the electrostatic force, giving a net force:

$$\text{force} = z_i eE - f v_i$$

Because of this, the ion will soon reach a **terminal velocity** at which the net force is zero:

$$v_i = \frac{z_i eE}{f} \tag{8.27}$$

The frictional drag on the ion is primarily a viscous effect. Any object moving through a fluid carries with it the closest layer of the fluid; layers farther out follow along with a velocity that decreases with the distance from the object. The resistance of a fluid to such a differential movement is called viscosity—this topic is discussed more thoroughly in Chapter 9. The friction constant of Eq. (8.27) is proportional to the viscosity (η) of the solvent; for example, **Stokes' law** gives the friction constant (f) for a sphere of radius r as:

$$f = 6\pi\eta r \tag{8.28}$$

Two other effects, involving ion-ion interactions, make the ion's velocity depend on ion concentration. The *electrophoretic effect* is due to the fact that an ion, moving toward its electrode, carries the nearby solvent along. A nearby ion that is attempting to go the other way is, in effect, swimming "upstream" and experiences an enhanced viscous drag. The *relaxation effect* is rather more subtle. An ion in solution is in an electric field due to the ions around it—the "ionic atmosphere." At equilibrium, the ion is in a position at which there is no net force due to its ion atmosphere; but if, under the influence of an external electric field, the ion makes a jump toward the electrode, it will experience a restoring force back toward its original position. By thermal motions, the ion atmosphere will quickly readjust (relax) to the new situation, but in the meantime the ion's progress has been slowed.

The direct dependence of the ion velocity on the electric field strength [Eq. (8.27)] is very straightforward and rather uninteresting. A more fundamental quantity, the **mobility**, is defined as the velocity per unit electric field:

$$u_i \equiv \frac{v_i}{E} \tag{8.29}$$

Now, going back to Eq. (8.26) and the current between parallel plates, we can use the fact that the electric field between parallel plates, which are separated by l, is uniform and related to the emf by:

$$E = \frac{\mathscr{E}}{l}$$

The ion velocity of Eq. (8.26) is replaced by the mobility:

$$v_i = \frac{u_i \mathscr{E}}{l}$$

to give:

$$I_i = |z_i| e \frac{N_i}{V} A u_i \frac{\mathscr{E}}{l}$$

If the solute has a degree of dissociation α and produces ν_+ cations and ν_- anions for each dissociation, then n moles give:

$$N_+ = \alpha \nu_+ L n, \qquad N_- = \alpha \nu_- L n$$

and the current can be written:

$$I_i = \alpha u_i \mathscr{F} \nu_i |z_i| \frac{n}{V} \frac{A \mathscr{E}}{l} \tag{8.30}$$

[For the time being we shall assume that electrolytes are 100% dissociated, $\alpha = 1$; partial dissociation is discussed in detail starting on page 394 with Eq. (8.46).]

Transference Number

We shall return to Eq. (8.30) later; for now we simply note that electrical neutrality requires $\nu_+ |z_+| = \nu_- |z_-|$, and the ratio of the cation current to the anion current will be given by the ratio of the mobilities:

$$\frac{I_+}{I_-} = \frac{u_+}{u_-}$$

Stated simply: the faster ion carries the most current.

The transference (or transport) number (t) is defined as the fraction of the current carried by a given ion; for two ions:

$$t_+ = \frac{I_+}{I_+ + I_-} = \frac{u_+}{u_+ + u_-}$$

$$t_- = 1 - t_+ = \frac{u_-}{u_+ + u_-} \tag{8.31}$$

How this number could be measured is shown schematically in Figure 8.6; this shows an electrolysis cell with an electrolyte (M^+X^-) and metal electrodes. When the electrodes are connected to an external source of electric current, metal

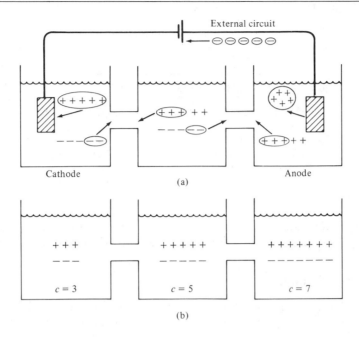

Figure 8.6   Hittorf method for measuring transference numbers (schematic).

ions are reduced at the cathode ($M^+ + e \longrightarrow M$) and oxidized at the anode ($M \longrightarrow M^+ + e$). The current in the cell will consist of cations moving toward the cathode (to be reduced) and anions moving toward the anode (to neutralize the newly produced cations). For analysis, the cell is divided into three compartments. Let us assume that 5 electrons pass through the external circuit; this will result in the removal of 5 cations in the cathode compartment and the placing of 5 cations into solution in the anode compartment. In Figure 8.6 it is assumed that the current in the cell will be 3 cations moving toward the cathode and 2 anions moving toward the anode; that is, $t_+ = \frac{3}{5} = 0.60$, $t_- = \frac{2}{5} = 0.40$. Note that afterward [Figure 8.6(b)] the concentrations of electrolyte in the anode and cathode compartments are changed; the analyses of the concentrations in the anode and cathode compartments provide a way to measure the transference number. Some values are given in Table 8.5. (This is called the Hittorf method; detailed discussions of this and other methods for measuring the transference number are given by ref. 2 and most physical chemistry laboratory texts. In Section 8.6 we shall discuss another method for measuring this quantity.)

In conclusion, the rate at which an ion moves through a solution in an electric field is measured by its mobility (u)—the velocity per unit electric field. The unit of mobility could be, for example:

$$\frac{m/s}{V/m} = m^2 s^{-1} V^{-1}$$

Table 8.5   Cation transference numbers at 25°C

Substance	Concentration (eq/dm³)					
	0.01	0.02	0.05	0.1	0.2	1.0
HCl	0.8251	0.8266	0.8292	0.8314	0.8337	
LiCl	0.3289	0.3261	0.3211	0.3168	0.3112	0.287
NH_4Cl	0.4907	0.4906	0.4905	0.4907	0.4911	
NaCl	0.3918	0.3902	0.3876	0.3854	0.3821	
KCl	0.4902	0.4901	0.4899	0.4898	0.4894	0.4882
KNO_3	0.5084	0.5087	0.5093	0.5103	0.5120	
$AgNO_3$	0.4648	0.4652	0.4664	0.4682	—	
NH_4NO_3				0.4870		

The fraction of the current carried by an ion is called the transference number (t), which depends on the ion's mobility relative to that of the counterion (Eq. 8.31).

Conductance of Solutions

The experiments of Georg Ohm showed that the resistance of a material is proportional to its length (l) and inversely proportional to its cross-sectional area (A):

$$R = \rho \frac{l}{A}$$

The proportionality constant (ρ) is called the **resistivity** or **specific resistance** and is an intensive property of the material. The reciprocal of the resistivity is called the **conductivity** or **specific conductance:**

$$\kappa = \frac{1}{\rho}$$

The conductivity of a solution between parallel plates (area A, separation l) can be calculated from its measured resistance:

$$\kappa = \frac{l}{AR} \tag{8.32}$$

Example: The measured resistance of a solution is 7.85 Ω. The cell plates have $A = 5.75$ cm² and are separated by $l = 1.32$ cm.

$$\kappa = \frac{1.32 \text{ cm}}{(5.75 \text{ cm}^2)(7.85 \text{ }\Omega)} = 2.92 \times 10^{-2} \text{ S/cm}$$

($S = \Omega^{-1}$, siemens; in many older sources the term mho is used in place of Ω^{-1} or S.)   ∎

Returning now to Eq. (8.30), we can see that the conductance ($R^{-1} = I/\mathscr{E}$) is:

$$R_i^{-1} = \alpha u_i \mathscr{F} \frac{A}{l} \nu_i |z_i| \frac{n}{V}$$

and the conductivity is [with Eq. (8.32)]:

$$\kappa_i = \alpha u_i \mathscr{F} \nu_i |z_i| \frac{n}{V}$$

For a salt, the total conductivity is the sum of the cation and anion conductivity, $\kappa_+ + \kappa_- = \kappa$ and, since electrical neutrality requires $\nu_+|z_+| = \nu_-|z_-|$, the conductivity of an ionic solution is:

$$\kappa = \alpha(u_+ + u_-)\mathscr{F}\nu_i|z_i|\frac{n}{V} \qquad (8.33)$$

Now would probably be a good time to talk about units. All the electrical units we have used (Ω, S, and so on) are SI units. To be totally consistent, the cell dimensions A and l should be given in meters; in addition the ion velocity should be in m/s and the electric field in volts/m; this would require the concentration (n/V) of Eq. (8.33) to be given as moles/m^3 — chemists generally prefer to express concentration as mol/dm^3. Since mixed units are inevitable, we shall adopt the traditional system in which the cell dimensions are given in cm and velocities are cm/s; the other units are:

$$\text{mobility:} \qquad \text{cm}^2 \text{ V}^{-1} \text{ s}^{-1}$$

$$\text{conductivity:} \quad \text{S/cm}$$

$$\text{concentration:} \quad \text{moles/dm}^3$$

(1 dm^3 = 1000 cm^3). With these units, the reader should demonstrate that the volume in Eq. (8.33) must be in cm^3; the term n/V can be replaced by $c/1000$, where c is the concentration in mol/dm^3.

We still must account for the $\nu_i z_i$ factor; this is most conveniently done by defining an **equivalent concentration:**

$$\tilde{c} \equiv \nu_i|z_i|c \qquad (8.34)$$

Then Eq. (8.33) for the **conductivity** of a solution becomes:

$$\kappa = \frac{\alpha(u_+ + u_-)\mathscr{F}\tilde{c}}{1000} \qquad (8.35)$$

Example: Na_2SO_4 has a formula weight of 142.04 g. In a solution with 1.4204 g of this solute per dm^3 of solvent, the concentration is:

$$c = 0.01 \text{ mol/dm}^3$$

However, for the purpose of electrochemistry, the significant fact is that there are 0.02 moles of positive charge and 0.02 moles of negative charge per dm^3 (this is the same amount of charge that would be provided by a solution of a 1:1 electrolyte like NaCl if $c = 0.02$ mol/dm^3). For Na_2SO_4 $\nu_i|z_i| = 2$ and $\tilde{c} = 0.02$ eq./dm^3.

Another method of doing this is found in some texts; this method uses the concentration of a salt called "$\frac{1}{2} Na_2SO_4$." The formula weight of this substance is $142.04/2 = 71.02$, so a solution with 1.4204 g/dm^3 of "sodium sulphate" is said to be 0.02 mol/dm^3 in "$\frac{1}{2} Na_2SO_4$." There are two arguments against this system: First, it is not consistent with the usual thermodynamic convention; for example, Table 8.4 gives the entropy and free energy for a mole of

SO_4^{2-}, not for a "mole" of "$\frac{1}{2} SO_4^{2-}$." Also, the potential for confusion in this business is endless and there are many traps for the unwary; the concept of equivalence may be old-fashioned, but the use of Eq. (8.34) and its special symbol can eliminate at least one such trap. ∎

Equivalent Conductivity

Equation (8.35) shows that the conductivity of a solution depends on the concentration of the solute in three ways: the degree of dissociation (α), the mobility (u), and, of course, the direct dependence ($\tilde{c}$). It is very revealing to calculate a quantity called the **equivalent conductivity** (or, in some references, the equivalent conductance):

$$\Lambda = \frac{1000\kappa}{\tilde{c}} \tag{8.36}$$

Values are given in Table 8.6.

Now, from Eq. (8.35), the equivalent conductivity can be seen to be:

$$\Lambda = \alpha(u_+ + u_-)\mathscr{F} \tag{8.37}$$

Exercise: Show that, with the system of units we have been using, the equivalent conductivity will have units cm^2 S/equiv. ∎

Figure 8.7 shows the change of the equivalent conductivity with concentration in two extreme cases; HCl has a very large equivalent conductivity at all concentrations, increasing slightly as $c \to 0$; acetic acid has a very small equivalent conductivity for $c > 0.01$ (note the discontinuity of the vertical scale), but its equivalent conductivity increases very rapidly as $c \to 0$, nearly reaching that of HCl in the $c = 0$ limit. Substances whose equivalent conductivity behaves like that of HCl are called *strong electrolytes;* these substances are considered to be 100% dissociated [$\alpha = 1$ in Eq. (8.37)], and the change of their equivalent conductivity with concentration is assumed to be due to changes in mobility with concentration (for the reasons discussed earlier). Substances whose equivalent conductivity behaves like that of acetic acid are called *weak electrolytes;* we shall discuss strong electrolytes first and return to weak electrolytes in a few pages.

Table 8.6   Equivalent conductivity in water at 25°

(Units: cm^2 S equiv^{-1})

Electrolyte	$\Lambda°$	Concentration (eq/dm³)		
		0.001	0.01	0.1
KCl	149.86	146.95	141.27	128.96
NaCl	126.45	123.74	118.51	106.74
HCl	426.16	421.36	412.00	391.32
AgNO₃	133.36	130.51	124.76	109.14
KNO₃	144.96	141.84	132.82	120.40
NH₄Cl	149.7	—	141.28	128.75
LiCl	115.03	112.40	107.32	95.86

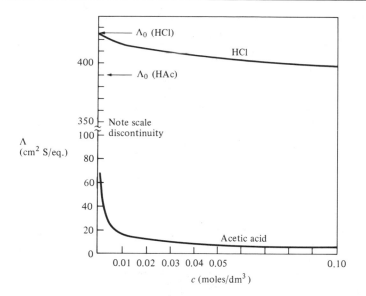

Figure 8.7 Equivalent conductivity of hydrochloric acid and acetic acid vs. concentration. (Note the discontinuity in the vertical scale.)

Strong Electrolytes: Kohlrausch's Law

The extrapolation of the equivalent conductivity to infinite dilution (that is, to $c = 0$) gives a quantity called the limiting ionic conductivity, Λ°. In this limit, the ion mobility will depend only on solvent-ion interactions and not at all on the nature of the counter-ion; this is Kohlrausch's law of independent ion migration. The limiting conductivity of a salt can be written as a sum of ion contributions:

$$\Lambda^\circ = \lambda^\circ_+ + \lambda^\circ_- \qquad \textbf{(8.38a)}$$

where λ°_i is a property of the ion that is the same regardless of what the counter-ion is. (Table 8.7 lists some values.) It should be evident from Eqs. (8.25) and (8.35) that the equivalent conductivity can *always* be written as a sum of ionic contributions:

$$\Lambda = \lambda_+ + \lambda_- \qquad \textbf{(8.38b)}$$

but λ_i will, like u_i, depend on concentration because of the interactions of the ion with the other ions.

Equivalent ion conductivities are related to mobility [Eq. (8.37) with $\alpha = 1$] and can be calculated from transference numbers [Eq. (8.31)]:

$$\lambda_i = u_i \mathscr{F}, \qquad \lambda_i = t_i \Lambda \qquad \textbf{(8.39)}$$

The unique statement in Kohlrausch's law is that the ionic conductivity is a property of the ion alone, and independent of the counter-ion. This is true at infinite dilution, but not exact at finite concentrations. (If it were so, the transport numbers of Table 8.5 would not depend on concentration.) However, this law is often used at finite concentration as an approximation.

Table 8.7   Limiting ionic conductivity ($\lambda°$) in water (cm² S/equiv.)

Cations	18°C	25°C	Anions	18°C	25°C
H^+	315	349.8	OH^-	174	197.6
Li^+	32.55	38.69	F^-	47.6	55.4
Na^+	42.6	50.11	Cl^-	66.3	76.34
K^+	63.65	73.50	Br^-	68.2	78.14
Rb^+	66.3	77.8	I^-	66.8	76.97
Cs^+	66.8	77.3	CN^-	—	82
NH_4^+	63.6	73.4	NO_3^-	62.6	71.44
Ag^+	53.25	61.92	ClO_4^-	59.1	67.4
Tl^+	64.8	74.7	$\frac{1}{2}SO_4^{2-}$	68.7	80
$\frac{1}{2}Mg^{2+}$	44.6	53.06	$\frac{1}{3}Fe(CN)_6^{3-}$	—	99.1
$\frac{1}{2}Ca^{2+}$	50.4	59.50	$\frac{1}{4}Fe(CN)_6^{4-}$	—	111
$\frac{1}{2}Sr^{2+}$	50.6	59.46	Formate$^-$	48	54.6
$\frac{1}{2}Zn^{2+}$	45.0	52.8	Acetate$^-$	35	40.9
$\frac{1}{2}Cd^{2+}$	45.1	54	Chloroacetate$^-$	—	39.8
$\frac{1}{2}Pb^{2+}$	60.5	70	Dichloroacetate$^-$	—	38
$\frac{1}{2}Mn^{2+}$	44.5	53.5	Trichloroacetate$^-$	—	35
$\frac{1}{3}Al^{3+}$	—	63	n-Propionate$^-$	—	35.8
$\frac{1}{3}Co(NH_3)_6^{3+}$	—	99.2	Benzoate$^-$	—	32.3
$N(CH_3)_4^+$	—	44.92	Picrate$^-$	25.14	31.39
$N(C_2H_5)_4^+$	—	32.66			
$N(C_3H_7)_4^+$	—	23			

Source: G. Milazzo, *Electrochemistry,* 1963: Amsterdam, Elsevier Publishing Co.

Example: An important application of Kohlrausch's law is for determining the limiting equivalent conductivities of weak electrolytes. As can be seen in Figure 8.7, such limiting equivalent conductivities cannot be determined accurately as the intercept of Λ vs. c plots because of the extremely steep approach of the curve to the $c = 0$ axis. The salt (MA) of a weak acid (HA) will usually be a strong electrolyte; the limiting equivalent conductivity of this salt, a strong acid (HX), and its salt (MX) are related to the single-ion values as:

$$\Lambda°_{MA} = \lambda°_{M^+} + \lambda°_{A^-}$$

$$\Lambda°_{MX} = \lambda°_{M^+} + \lambda°_{X^-}$$

$$\Lambda°_{HX} = \lambda°_{H^+} + \lambda°_{X^-}$$

Then, the equivalent conductivity of the weak acid:

$$\Lambda°_{HA} = \lambda°_{H^+} + \lambda°_{A^-}$$

can be calculated as:

$$\Lambda°_{HA} = \Lambda°_{HX} - \Lambda°_{MX} + \Lambda°_{MA}$$

If the appropriate ionic conductivities are already known, as in Table 8.7, the calculation of $\Lambda°$ is even easier; for acetic acid (25°C):

$$\Lambda° = \lambda°_{H^+} + \lambda°_{Ac^-} = 349.8 + 40.9 = 390.7 \text{ cm}^2 \text{ S/equiv.} \qquad \blacksquare$$

Example: Data from Tables 8.5 and 8.6 will be used to calculate the ionic conductivity of chloride ion in HCl and KCl. At $c = 0.1$ moles/dm^3 (moles = equivalents for $1:1$ electrolytes):

$$\Lambda_{KCl} = 128.96, \quad t_+ = 0.4898:$$

$$\lambda_{Cl^-} = (1 - 0.4898)(128.96) = 65.80 \text{ cm}^2 \text{ S/equiv.}$$

$$\Lambda_{HCl} = 391.32, \quad t_+ = 0.8314:$$

$$\lambda_{Cl^-} = (1 - 0.8314)(391.32) = 65.97 \text{ cm}^2 \text{ S/equiv.}$$

The difference is small but significant; λ_{Cl^-} is affected by the counter-ion even at 0.1 molar. At $c = 0.01$ moles/dm^3 the reader should repeat this calculation to get:

$$\lambda_{Cl^-} = 72.02 \text{ cm}^2 \text{ S/equiv. in KCl}$$

$$\lambda_{Cl^-} = 72.06 \text{ cm}^2 \text{ S/equiv. in HCl}$$

Note that both are approaching $\lambda° = 76.34$ cm^2 S/equiv. (Table 8.7).

To keep things in perspective, these are small differences; Kohlrausch's law is often useful at small but finite concentrations. It would not be unreasonable to assume $\lambda_{Cl^-} \cong 65.9$ at $c = 0.1$ for any strong $1:1$ electrolyte; but be careful — a more highly charged ion, as in CaCl$_2$, will affect the mobility of the chloride more, even at the same *equivalent* concentration ($\bar{c} = 0.1$, $c = 0.05$ moles/dm^3). ∎

Example: Calculation of ion mobilities and velocities. Using Eq. (8.39) for chloride ion at infinite dilution:

$$u_{Cl^-} = 76.34 \text{ cm}^2 \text{ S}/96\,487 \text{ C} = 7.912 \times 10^{-4} \text{ cm}^2 \text{ S C}^{-1}$$

The units can be worked out from Eqs. (8.24) and (8.21):

$$\text{S C}^{-1} = \text{A V}^{-1} \text{ C}^{-1} = \text{V}^{-1} \text{ s}^{-1}$$

$$u_{Cl^-} = 7.912 \times 10^{-4} \text{ cm}^2 \text{ V}^{-1} \text{ s}^{-1}$$

In a unit field — for example, an emf of 1 volt between plates 1 cm apart, $E = 1$ V/cm [Eq. (8.29)]

$$v_{Cl^-} = Eu_{Cl^-} = 7.912 \times 10^{-4} \text{ cm s}^{-1}$$

The time required for the ion to travel between the electrodes (1 cm) would be 1264 s = 21 minutes.

The motion of the ion from electrode to electrode is not a direct flight but, rather, a uniform drift velocity superimposed on a diffusional random walk. To put our 1-cm-in-21-minutes velocity into perspective, let us estimate how far diffusion would, on the average, move the ion in the same amount of time.

The thermal motions of the ions are very fast, much faster than the drift velocity imposed by the electric field, but because of frequent collisions (with consequent changes in direction) the net progress is quite small. The rms displacement (Δx) in time t is related to the diffusion constant D by the relationship (cf. Chapter 9):

$$\Delta x = \sqrt{2Dt}$$

The diffusion constant of an ion is related to its mobility by an equation derived by Nernst (1888):

$$D_i = \frac{u_i RT}{\mathscr{F}|z_i|} \tag{8.40}$$

Using the mobility of Cl^- (above) gives:

$$D_{Cl^-} = 2.03 \times 10^{-5} \ cm^2 \ s^{-1}$$

Then the rms distance traveled in 1264 s is:

$$\Delta x = [2(2.03 \times 10^{-5})(1264)]^{1/2} = 0.23 \ cm$$

In the same time, of course, the ion "drifted" 1 cm from electrode to electrode.

As an aside, let us note that if the diffusion of an electrolyte (MX) is measured in a concentration gradient, the rate at which one ion can move is limited by the requirement of electrical neutrality. For example, in HCl the H^+ can move very fast, but it must wait for the Cl^- ion to catch up in order to maintain neutrality. Nernst showed that the mean diffusion constant of an ion pair is:

$$\overline{D}_{MX} = \frac{2D_{M^+}D_{X^-}}{D_{M^+} + D_{X^-}} \tag{8.41}$$

∎

Onsager's Theory

The limiting equivalent conductivity is an important quantity whose value must be obtained by extrapolation to $c = 0$. Empirically it is known that equivalent conductances of strong electrolytes at very low concentrations are nearly linear in the square root of the concentration:

$$\Lambda \cong \Lambda° + k\sqrt{\tilde{c}} \tag{8.42}$$

Plots of Λ vs. $\sqrt{\tilde{c}}$ can be smoothly extrapolated to find $\Lambda°$ (Figure 8.8). However, only the points at very low concentrations will be strictly linear; unfortunately these tend to be the least accurate values. The reason is that the measured conductivity (κ) really contains three components: that due to the electrolyte, that due to impurities in the solvent used to make up the solution, and that due to the ionization of the solvent—for example, $H_2O = H^+ + OH^-$. The latter two are usually corrected for by measuring the conductivity of the solvent used, as a control:

$$\kappa = \kappa(measured) - \kappa(control)$$

Water, purified with the greatest care, has $\kappa \sim 10^{-7}$ S/cm; typical "conductivity" water has $\kappa \sim 10^{-6}$ S/cm. Typically $\Lambda \sim 100 \ cm^2$ S, so if $\tilde{c} = 10^{-4}$:

$$\kappa = \frac{\tilde{c}}{1000} \sim 10^{-5} \ S/cm$$

This means that the control value is a significant fraction of the measured value, and additional uncertainty will result below $\tilde{c} \sim 10^{-4}$ unless the greatest care is taken.

A better method for extrapolating equivalent conductivities results from the theo-

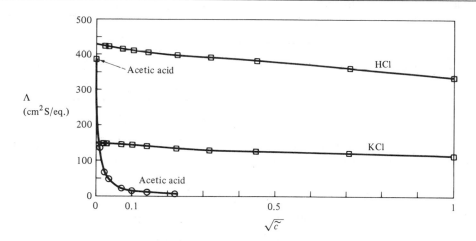

Figure 8.8 Equivalent conductivity vs. $\sqrt{c}$. for three solutes in water. The behavior of acetic acid is characteristic of *weak electrolytes,* which are only partially ionized, with the degree of ionization increasing with decreasing concentration. The behavior of HCl and KCl is characteristic of *strong electrolytes,* which are, presumably, 100% ionized at these concentrations.

retical work of Lars Onsager (1928) on the effect of ion interactions on mobility. This theory treats the solvent as a continuous medium characterized by a dielectric constant (ε) and a viscosity (η) with the ions interacting by Coulomb's law; in that sense it is quite similar to the Debye-Hückel theory with the same types of limitations. The result (ref. 2, p. 327) is fairly simple for the special case of a $1:1$ electrolyte:

$$\Lambda = \Lambda^\circ - (\theta\Lambda^\circ + \sigma)\sqrt{c}$$

$$\theta = \frac{8.147 \times 10^5}{(\varepsilon T)^{3/2}}, \qquad \sigma = \frac{40.93}{\eta(\varepsilon T)^{1/2}} \qquad \textbf{(8.43)}$$

where η is the viscosity of the solvent and ε is its dielectric constant. (Values of these constants at 25°C are given in Table 8.8.)

This result is, like the Debye-Hückel limiting law, strictly valid only in a limiting sense; that is, it correctly predicts the limiting slope of Λ vs. $\sqrt{c}$ plots, but cannot be used directly to calculate accurate equivalent conductivities.

The utility of Onsager's equation is demonstrated by solving Eq. (8.43) for Λ°:

$$\Lambda^\circ \cong \frac{\Lambda + \sigma\sqrt{c}}{1 - \theta\sqrt{c}} \equiv \Lambda' \qquad \textbf{(8.44)}$$

This approximate value of Λ° (denoted Λ') will approach the correct value as $c \to 0$; thus:

$$\Lambda' = \Lambda^\circ + Bc \qquad \textbf{(8.45)}$$

where B is an empirical slope used for the linear extrapolation.

Table 8.8   Constants for the Onsager equation (25°C)

	θ	σ
Water	0.229	60.2
Methanol	0.923	156.1
Ethanol	1.33	89.7

Example: We shall illustrate this method with data for KI in H_2O at 25°C. For each measured value of the equivalent conductivity we calculate:

$$\Lambda' = \frac{\Lambda + 60.2\sqrt{c}}{1 - 0.229\sqrt{c}}$$

$c = 0.005, \quad \Lambda = 144.37: \quad \Lambda' = 151.07$

$c = 0.01, \quad \Lambda = 142.18: \quad \Lambda' = 151.67$

$c = 0.02, \quad \Lambda = 139.45: \quad \Lambda' = 152.92$

$c = 0.05, \quad \Lambda = 134.97: \quad \Lambda' = 156.44$

These values can be seen to be varying smoothly with concentration. Linear regression (Appendix I; graphical extrapolation can also be used) gives $\Lambda° = 150.49 \pm 0.07$ cm^2 S/equiv. ($r = 0.999925$). (Error estimate is for 90% confidence limits; see Appendix I.) ∎

Weak Electrolytes: Degree of Dissociation

Equation (8.36) can be used to calculate the degree of dissociation from measured values of the equivalent conductivity:

$$\alpha = \frac{\Lambda}{\mathscr{F}(u_+ + u_-)} \tag{8.46}$$

if the ion mobilities can be independently measured. For very weak electrolytes it is often assumed that the limiting values of the mobility can be used, and the denominator of Eq. (8.46) is approximated by $\lambda_+^° + \lambda_-^° = \Lambda°$. This is never quite correct, but this approximation can be used to define an *apparent* degree of dissociation:

$$\alpha' = \frac{\Lambda}{\Lambda°} \tag{8.47}$$

from which an apparent dissociation equilibrium constant can be calculated:

$$K'' = \frac{c(\alpha')^2}{1 - \alpha'} \tag{8.48}$$

Compared to the thermodynamic equilibrium constant [see Eqs. (8.18), (8.19), (8.20)] the "constant" of Eq. (8.48) is doubly approximate: it uses concentrations in place of activities, and the degree of dissociation should have been corrected for the change of mobility with concentration. However, if the values so calculated are

Table 8.9   Apparent dissociation constant of acetic acid in water (25°C)

$10^5 c$	Λ	$\alpha' = \Lambda/\Lambda°$	(α, ref. 2)	$10^5 K''$
11.135	127.75	0.3270	(0.3277)	1.7689
21.844	96.493	0.2470	(0.2477)	1.7694
136.340	42.227	0.1081	(0.1086)	1.7856
344.065	27.199	0.0696	(0.0700)	1.7922
599.153	20.962	0.0537	(0.0540)	1.7982
984.21	16.371	0.0419	(0.0422)	1.8036

extrapolated to infinite dilution, both of these approximations become exact:

$$\lim_{c \to 0} K'' = K_a$$

[The K_a so measured is for a concentration standard state (mol/dm³), as opposed to the molality standard state that we have been using up to this point. At infinite dilution, the concentration-scale equilibrium constant can be converted to the usual molality-scale constant by dividing it by the density of the solvent.]

A more accurate method for calculating degrees of dissociation from conductivity uses Kohlrausch's law to estimate the ionic conductivities at the various concentrations; then α is calculated using Eq. (8.46). This method is discussed by ref. 2 (p. 344).

Example: Use measured values of the equivalent conductivity of acetic acid (Table 8.9) to calculate the thermodynamic dissociation constant:

$$K_a = \frac{a_{H^+} a_{Ac^-}}{a_{HAc}}$$

(concentration-scale activities).

For acetic acid, $\Lambda° = 390.7 \text{ cm}^2$ S; this value is used to calculate α' (column 3). (The correct values of α as given by ref. 2 are listed in column 4.) From this number, K'' (column 5) is calculated using Eq. (8.48). These values were extrapolated vs. $\sqrt{c}$ to give:

$$K_a = (1.766 \pm 0.005) \times 10^{-5}$$

(A more accurate extrapolation using third-order polynomial regression in $\sqrt{c}$ gives a value of 1.758×10^{-5}. Reference 3 gives 1.753×10^{-5} as the answer.) ■

8.4   Electrochemical Cells

Electrochemical cells are devices that can make chemical energy available in the very convenient form of electrical energy. They are also useful and flexible devices for studying the thermodynamics of ionic solutions.

As an initial example, we shall look at the Daniell cell (Figure 8.9). If a bar of zinc is placed into a solution of copper sulfate, the chemical reaction:

$$Zn(s) + Cu^{2+}(aq) \longrightarrow Cu(s) + Zn^{2+}(aq) \tag{8.49a}$$

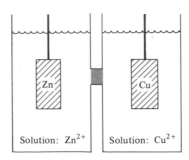

Figure 8.9 The Daniell cell. The solutions are separated by a barrier, such as a glass frit, that permits the passage of electrical current but inhibits the convectional mixing of the solutions.

will occur spontaneously. When done in this way, virtually no work can be obtained and the chemical energy is dissipated as heat. The reaction can also be carried out as indicated in Figure 8.9. Two solutions, $CuSO_4(aq)$ and $ZnSO_4(aq)$, are separated by a porous barrier that permits ions to cross but prevents convective mixing of the solutions. A copper bar in the $CuSO_4$ solution is the site of the reduction:

$$Cu^{2+} + 2\,e \rightarrow Cu(s)$$

and a zinc bar in the $ZnSO_4$ solution is oxidized:

$$Zn(s) \rightarrow Zn^{2+} + 2\,e$$

The electrons exchanged by this redox pair must pass through the external circuit, where they can be made to do work (for example, by driving an electric motor).

This cell was actually the first practical source of electrical power invented (J. F. Daniell, 1836), and, in the days when copper and zinc were cheap and electricity could not be obtained by plugging into a socket, it was widely used; one important application was as a power source for the electrical telegraph invented by S.F.B. Morse in 1844. Electrochemical cells and batteries are still very important sources of electrical power in applications where portability is important, but the major interest in this chapter will be in their use for studying the thermodynamics of chemical reactions, especially redox reactions in solution.

Example: For the cell reaction, Eq. (8.49a), the heats and free energies of formation of Table 8.4 can be used to calculate:

$$\Delta H_{rxn}^{\theta} = -216.8 \text{ kJ}, \qquad \Delta G_{rxn}^{\theta} = -212.2 \text{ kJ}$$

Then:

$$T\,\Delta S_{rxn}^{\theta} = \Delta H_{rxn}^{\theta} - \Delta G_{rxn}^{\theta} = -4.6 \text{ kJ}$$

The irreversible reaction, when Zn is placed directly into the copper sulfate solution, would release a heat approximately equal to ΔH_{rxn}^{θ}, or 217 kJ/mol. (This is not the precise number, since heats of mixing are probably not negligible.) If the reaction could be carried out reversibly in the cell of Figure 8.9, the work obtained [Eq. (3.24)] would be approximately $-\Delta G_{rxn}^{\theta}$ or 212 kJ, while only $T\,\Delta S$ (-4.6 kJ) of heat would be released.

This example makes some rather crude assumptions, some of which will be improved upon in the coming pages, but it should demonstrate that electrochemical cells can make chemical energy available efficiently and in a useful and convenient form.   ■

Cell Diagrams and Conventions

The conventions that will be outlined in this section may seem to be a bit arbitrary, but if they are followed exactly, you will avoid some of the ambiguity that often creeps into this subject. Unfortunately, other conventions are used in other sources, particularly the older ones.

The cell is diagrammed formally by listing its phases, electrode to electrode. We shall use a vertical bar (|) to indicate a phase separation; if a single phase contains several components, they will be separated by commas (excluding solvent). The Daniell cell, for example, is diagrammed:

$$Zn(s)\,|\,ZnSO_4(aq)\,|\,CuSO_4(aq)\,|\,Cu(s) \qquad \textbf{(8.49b)}$$

It will be understood that strong electrolytes in water are ionized, so $CuSO_4(aq)$ *means* $Cu^{2+}(aq) + SO_4^{2-}(aq)$. The phase boundary between the two solutions is called a *liquid junction;* such junctions raise complications when discussing cell thermodynamics, so we shall prefer, when possible, to discuss single-electrolyte cells without liquid junctions. But let's continue with the conventions, using the Daniell cell as an example.

The reaction at the left electrode is always written as an *oxidation,* while that at the right electrode is written as a *reduction*; the sum of these half-reactions is the *cell reaction:*

Left (ox.):	$Zn = Zn^{2+} + 2\,e$
Right (red.):	$Cu^{2+} + 2\,e = Cu$
Cell reaction:	$Zn + Cu^{2+} = Zn^{2+} + Cu$

It is useful to think of an electrode as having an electrical potential of its own. These single-electrode potentials cannot be measured, but all possible electrodes could, in principle, be measured vs. a single electrode chosen as a standard, and relative values of single-electrode potentials thus obtained. The convention is that the emf of the cell is equal to that of the right electrode *minus* that of the left electrode (both being "reduction" potentials):

$$\mathscr{E}_{cell} = \mathscr{E}_R - \mathscr{E}_L \qquad \textbf{(8.50)}$$

Equation (8.50) refers only to electrode processes; a cell such as (8.49), the Daniell cell, has a potential due to the liquid junction ($\mathscr{E}_J$), so its emf must be written:

$$\mathscr{E} = \mathscr{E}_{Cu^{2+}|Cu} - \mathscr{E}_{Zn^{2+}|Zn} + \mathscr{E}_J \qquad \textbf{(8.49c)}$$

There are methods by which liquid-junction potentials can be reduced or eliminated, but even if this were not possible (and it is possible to find such cases), it is useful to consider the existence of cells for which the liquid junction has been eliminated; this will be indicated with a double bar ($\parallel$) — for example:

$$Zn(s)\,|\,ZnSO_4(aq)\,\|\,CuSO_4(aq)\,|\,Cu(s) \tag{8.49d}$$

for which the emf is:

$$\mathscr{E} = \mathscr{E}_{Cu^{2+}|Cu} - \mathscr{E}_{Zn^{2+}|Zn} \tag{8.49e}$$

In a later section we shall discuss how this cell can be approximated using a *salt bridge*. The double vertical bar is often used to indicate a salt bridge, but here it has a more general meaning of a hypothetical cell with no liquid-junction potential.

[It is possible to make too much of liquid junctions; the nineteenth-century telegraphers who used the Daniell cell did not have to worry about it. In any case, there is always a potential between phases, including one between the electrode and the electrolyte. However, electrode-electrolyte contact potentials are, effectively, included in the electrode emf and will cause no problem if the electrode is always made the same way. (Electrochemistry is as much an art — some would say "black art" — as a science, and the last statement glosses over many difficult problems.) However, liquid junctions do cause serious problems for the thermodynamic interpretation of cell emf's.]

For these conventions, "right" and "left" are pure formalities, which depend on how we choose to write down the cell diagram. If we wrote the cell for Eqs. (8.49) with copper on the left and zinc on the right, the cell reaction would be reversed to:

$$Cu + Zn^{2+} = Zn + Cu^{2+}$$

and the calculated emf would have the opposite sign (it would, in fact be negative — cell (8.49b) would have a positive emf). It is a custom in some quarters always to write a cell diagram in such a manner that the emf will be positive; this will result, as we shall soon see, in a cell reaction that is spontaneous in the direction written. However, there is no law against writing a reaction, for example:

$$H_2O \longrightarrow H_2 + \tfrac{1}{2}O_2$$

that is not, under ordinary conditions, spontaneous, and there is likewise no reason that a cell emf could not be a negative number. If the cell as written has a positive emf, the right electrode will be positive and the left electrode negative; otherwise vice versa. This is of some minor concern when the cell is put together in the laboratory; however, a few moments' experimentation would suffice to show which electrode is positive or negative.

Reversible emf and Free Energy

The manner in which cell emf's are measured is illustrated by Figure 8.10. The cell voltage is opposed by another, accurately calibrated, voltage until no current flows through the galvanometer. In this manner we can measure the cell emf while drawing negligible current from the cell; this number is called the *reversible emf*. (Some authorities distinguish between the cell "voltage" or "potential" and the reversible emf; the symbol $\mathscr{E}$ is reserved for the latter. We have assumed, and shall continue to

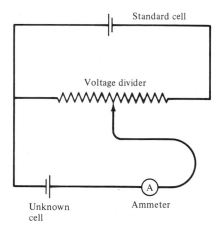

Figure 8.10 Measurement of reversible emf. A voltage is opposed to the cell voltage and adjusted until no current flows through the ammeter (A); at this point, the voltage of the divider (measured in a separate experiment) is equal to the cell voltage.

assume, that all cell emf's are measured *correctly* — that is, reversibly — and use only one symbol.)

By the opposing-emf method, we could imagine permitting the cell reaction to occur with infinite slowness and, by that process, to get the maximum work from the electrons that flow through the external circuit. The electrical work obtained from moving n moles of electrons through a potential is, by Eq. (8.22):

$$w_{max} = n\mathscr{F}\mathscr{E}$$

In Chapter 3 [Eq. (3.24)] we saw that the maximum work (other than PV work) obtainable from any process at constant (T, P) was $-\Delta G$ of that process. Now, the process is the cell reaction, and:

$$\Delta G_{rxn} = -n\mathscr{F}\mathscr{E} \qquad (8.51)$$

This is the fundamental equation for relating cell emf's to thermodynamics. We shall explore the full consequences of Eq. (8.51) in Section 8.5; first let's look at some of the various types of electrodes that can be used to make up cells.

Electrode Types

We have already seen examples of metal electrodes and their reactions. Another useful cell type is a gas electrode, such as:

$$H^+(aq) \,|\, H_2(g), Pt$$

with the electrode reaction:

$$2\,H^+(aq) + 2\,e = H_2(g)$$

In this cell, hydrogen gas is bubbled over a specially prepared platinum electrode, which is immersed in the electrolyte. Electrical contact is made through the platinum. It is likely that the hydrogen that is reacting is dissolved in the platinum, but since at equilibrium the fugacity of the gas is equal to that of the dissolved substance, the cell can be treated as if the reaction were, in fact, directly from the gas phase.

Another very useful type of electrode is made by combining an insoluble salt of a metal with a metal electrode — for example:

$$Cl^-(aq) \,|\, AgCl(s) \,|\, Ag(s)$$

The electrode reaction in this case is:

$$AgCl(s) + e = Ag(s) + Cl^-(aq)$$

The electrical contact is through the silver and both the Ag and AgCl are in contact with the electrolyte. The chloride in the electrolyte must be provided by a soluble salt such as NaCl or KCl. Again, it is at least possible that the actual cell reaction is due to the slight amount of dissolved AgCl, but, as long as the solution is saturated, the fugacities are equal and the reaction can be treated as it if were heterogeneous. (Recall the discussion of gas-phase heterogeneous reactions, Section 6.8.)

An important example of this type of electrode is the calomel electrode, which is commonly used as a secondary standard for electrode potentials (Figure 8.11):

$$KCl(aq, m) \,|\, Hg_2Cl_2(s) \,|\, Hg(liq)$$

The electrode reaction is:

$$Hg_2Cl_2(s) + 2\,e = 2\,Hg(liq) + 2\,Cl^-(aq)$$

Electrical contact is made by a wire inserted into the mercury. [Calomel is the ancient name for mercury (I) chloride. Calomel was once used medicinally as a purgative but is now used as a fungicide, which illustrates how dangerous medicine can be.]

The standard, alluded to earlier, against which single-electrode emf's are measured is the standard hydrogen electrode, which will be discussed in the next section. The calomel electrode has been accurately calibrated vs. this standard with the results (25°C):

$$KCl, m = 0.1 \text{ mol/kg:} \qquad \mathscr{E}_{cal} = 0.3338 \text{ V}$$

$$KCl, m = 1.0 \text{ mol/kg:} \qquad \mathscr{E}_{cal} = 0.2800 \text{ V}$$

$$KCl, \text{ saturated solution:} \qquad \mathscr{E}_{cal} = 0.2415 \text{ V}$$

If a cell is made from an electrode with unknown emf vs. calomel — for example:

$$M \,|\, MX(aq) \,|\, KCl(aq, m) \,|\, Hg_2Cl_2(s) \,|\, Hg$$

the measured emf of the cell is:

$$\mathscr{E}_{cell} = \mathscr{E}_{cal} - \mathscr{E}_{M^+/M} + \mathscr{E}_J$$

If the junction potential can be minimized, eliminated, or calculated, the unknown emf can be calculated from this measurement. (To some extent this junction potential is included in $\mathscr{E}_{cal}$, since it was there when the cell was calibrated.)

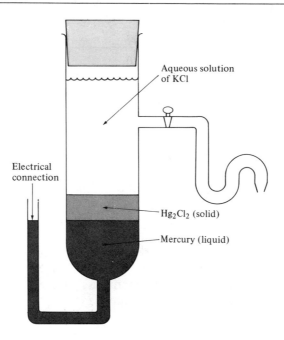

Figure 8.11   The calomel electrode. This electrode is often used as a secondary standard for measuring half-cell emf's. The emf of this cell, which will depend on temperature as well as the KCl electrolyte concentration, has been accurately measured.

8.5   The Standard Emf

The free energy of the cell reaction can be written in terms of the chemical potential as:

$$\Delta G_{rxn} = \sum_i \nu_i \mu_i$$

For each reactant and product the activity is given by:

$$\mu_i = \mu_i^{\theta} + RT \ln a_i$$

Then [Eq. (6.16)]:

$$\Delta G_{rxn} = \Delta G_{rxn}^{\theta} + RT \ln Q$$

with:

$$Q = \prod_i a_i^{\nu_i} \qquad \textbf{(8.52)}$$

The Gibbs' free energy can be related to the cell emf by Eq. (8.51) to give the **Nernst equation:**

$$\mathscr{E} = \mathscr{E}^{\theta} - \frac{RT}{n\mathscr{F}} \ln Q \qquad \textbf{(8.53)}$$

where the **standard emf** is defined as:

$$\mathscr{E}^\theta \equiv -\frac{\Delta G^\theta_{\text{rxn}}}{n\mathscr{F}} \qquad (8.54)$$

Nernst actually presented his equation in terms of concentrations, with (in place of Q):

$$Q' = \prod_i c_i^{\nu_i}$$

and it is still used as such for approximate work. We shall, however, always use Eqs. (8.52) and (8.53), which are thermodynamically correct.

We shall see shortly how the standard emf of a cell is measured; for now we need only note that, like the cell emf of Eq. (8.50), it can be written in terms of half-cell or single-electrode standard emf's as:

$$\mathscr{E}^\theta_{\text{cell}} = \mathscr{E}^\theta_{\text{right}} - \mathscr{E}^\theta_{\text{left}} \qquad (8.55)$$

Since only relative emf's (standard or otherwise) can be measured, it will be a great convenience to choose one half-cell and arbitrarily assign its emf a specific value. The convention is that the standard emf of the electrode:

$$H^+(aq) \,|\, H_2(g), Pt$$

is assigned the value of zero. Table 8.10 lists some standard single-electrode potentials measured relative to this convention.

Example: Calculate the emf of the cell:

$$Zn(s) \,|\, ZnCl_2(aq, m = 1) \,|\, AgCl(s) \,|\, Ag$$

using data from Table 8.10. First we write the cell reaction:

Left (ox.):	$Zn = Zn^{2+} + 2\,e$	$\mathscr{E}^\theta_L = -0.7628$ V
Right (red.):	$2\,AgCl(s) + 2\,e = 2\,Ag + 2\,Cl^-$	$\mathscr{E}^\theta_R = +0.2225$ V
Cell:	$Zn(s) + 2\,AgCl(s) = 2\,Ag(s) + ZnCl_2(aq)$	$\mathscr{E}^\theta = 0.9853$ V

$$Q = \frac{a_{ZnCl_2} a^2_{Ag}}{a_{Zn} a^2_{AgCl}}$$

For pure solids at moderate pressures, $a = 1$. For the electrolyte:

$$a_{ZnCl_2} = m_+ m_-^2 \gamma_\pm^3 = 4m^3 \gamma_\pm^3$$

The activity coefficient can be obtained from Table 8.1 (the ionic strength, $I = 3m$, is too high to use the DHG formula): $\gamma_\pm = 0.325$. Then, from the Nernst equation:

$$\mathscr{E} = 0.9853 - \frac{8.3143 \times 298.15}{2 \times 96,487} \ln\left[(4)(1)(0.325)^3\right]$$

$$= 0.9853 - (-0.0255) = 1.0108 \text{ V}$$

Note: Some texts would present this example with the sign changed for emf of the left electrode; then the cell emf is the sum rather than the difference of the electrode potentials. The

Table 8.10   Standard electrode potentials in water at 25°C

Electrode	Electrode reaction (acid solution)	$\mathscr{E}^{\theta}$ (volts)
$Li^+ \mid Li$	$Li^+ + e = Li$	-3.045
$K^+ \mid K$	$K^+ + e = K$	-2.925
$Ba^{2+} \mid Ba$	$Ba^{2+} + 2\,e = Ba$	-2.906
$Ca^{2+} \mid Ca$	$Ca^{2+} + 2\,e = Ca$	-2.866
$Na^+ \mid Na$	$Na^+ + e = Na$	-2.714
$Zn^{2+} \mid Zn$	$Zn^{2+} + 2\,e = Zn$	-0.7628
$Fe^{2+} \mid Fe$	$Fe^{2+} + 2\,e = Fe$	-0.4402
$Cd^{2+} \mid Cd$	$Cd^{2+} + 2\,e = Cd$	-0.4029
$SO_4^{2-} \mid PbSO_4 \mid Pb$	$PbSO_4 + 2\,e = Pb + SO_4^{2-}$	-0.3546
$I^- \mid AgI \mid Ag$	$AgI + e = Ag + I^-$	-0.1522
$Sn^{2+} \mid Sn$	$Sn^{2+} + 2\,e = Sn$	-0.136
$Pb^{2+} \mid Pb$	$Pb^{2+} + 2\,e = Pb$	-0.126
$Fe^{3+} \mid Fe$	$Fe^{3+} + 3\,e = Fe$	-0.036
$D^+ \mid D_2, Pt$	$2\,D^+ + 2\,e = D_2$	-0.0034
$H^+ \mid H_2, Pt$	$2\,H^+ + 2\,e = H_2$	(zero by convention)
$Br^- \mid AgBr \mid Ag$	$AgBr + e = Ag + Br^-$	0.0711
$Sn^{4+}, Sn^{2+} \mid Pt$	$Sn^{4+} + 2\,e = Sn^{2+}$	0.15
$Cu^{2+}, Cu^+ \mid Pt$	$Cu^{2+} + e = Cu^+$	0.153
$Cl^- \mid AgCl \mid Ag$	$AgCl + e = Ag + Cl^-$	0.2225
$Cl^- \mid Hg_2Cl_2 \mid Hg$	$Hg_2Cl_2 + 2\,e = 2\,Hg + 2\,Cl^-$	0.2680
$Cu^{2+} \mid Cu$	$Cu^{2+} + 2\,e = Cu$	0.337
$I^- \mid I_2 \mid Pt$	$I_2 + 2\,e = 2\,I^-$	0.5355
$Ag^+ \mid Ag$	$Ag^+ + e = Ag$	0.7991
$Hg^{2+} \mid Hg$	$Hg^{2+} + 2\,e = Hg$	0.854
$Hg^+ \mid Hg$	$Hg^+ + e = Hg$	0.92
$Br^- \mid Br_2 \mid Pt$	$Br_2 + 2\,e = 2\,Br^-$	1.0652
$Mn^{2+} \mid H^+ \mid MnO_2 \mid Pt$	$MnO_2 + 4\,H^+ + 2\,e = Mn^{2+} + 2\,H_2O$	1.23
$Cr^{3+}, Cr_2O_7^{2-}, H^+ \mid Pt$	$Cr_2O_7^{2-} + 14\,H^+ + 6\,e = 2\,Cr^{3+} + 7\,H_2O$	1.33
$Cl^- \mid Cl_2, Pt$	$Cl_2 + 2\,e = 2\,Cl^-$	1.3595
	(Basic solutions)	
$OH^- \mid Ca(OH)_2 \mid Ca \mid Pt$	$Ca(OH)_2 + 2\,e = 2\,OH^- + Ca$	-3.02
$ZnO_2^{2-}, OH^- \mid Zn$	$Zn(OH)_4^{2-} + 2\,e = Zn + 4\,OH^-$	-1.215
$OH^- \mid H_2 \mid Pt$	$2\,H_2O + 2\,e = H_2 + 2\,OH^-$	-0.82806
$CO_3^{2-} \mid PbCO_3 \mid Pb$	$PbCO_3 + 2\,e = Pb + CO_3^{2-}$	-0.509
$OH^- \mid HgO \mid Hg$	$HgO + H_2O + 2\,e = Hg + 2\,OH^-$	0.097

point of view represented here, which is that recommended by IUPAC, is that the emf is the potential of the electrode relative to the standard electrode and has a value (including sign) that is independent of how the cell reaction is written or whether it is the "right" or "left" electrode. The measured potential of the cell is the difference between the potentials of the two electrodes with the convention:

$$\mathcal{E}_{\text{cell}} = \mathcal{E}_R - \mathcal{E}_L$$

$$\mathcal{E}_{\text{cell}}^\theta = \mathcal{E}_R^\theta - \mathcal{E}_L^\theta$$

Suppose you had two electrodes. The first, measured vs. the standard hydrogen electrode, had a voltage of 6 V while the other was 4 V. If the voltmeter was placed between these two electrodes, the voltage would be $6 - 4 = 2$ V—the difference, not the sum. ∎

Measurement of Standard emf's

The standard emf is the emf of the cell in which all reactants and products are at unit activity—easier said than measured, for there is no convenient method to measure the activity of an electrolyte except with a cell for which $\mathcal{E}^\theta$ is already known. Measuring the standard emf of each electrode presents unique problems related to the chemical process involved (every number on Table 8.10 is a story); the general procedure will be illustrated with the cell:

$$\text{Pt}, \text{H}_2(\text{g, 1 atm}) \,|\, \text{HCl}(\text{aq}, m) \,|\, \text{AgCl(s)} \,|\, \text{Ag} \tag{8.56a}$$

Example: Determine the cell reaction and Nernst equation for cell (8.56).

L (ox.): $\frac{1}{2} \text{H}_2(\text{g}) = \text{H}^+ + \text{e}$

R (red.): $\text{AgCl(s)} + \text{e} = \text{Ag(s)} + \text{Cl}^-$

Cell: $\frac{1}{2} \text{H}_2(\text{g, 1 atm}) + \text{AgCl(s)} = \text{HCl}(\text{aq}, m) + \text{Ag(s)}$ $\tag{8.56b}$

For solids at moderate pressure, $a = 1$. If the hydrogen is assumed to be an ideal gas, $a_{\text{H}_2} = P_{\text{H}_2}/P^\theta = 1$. Therefore, the Nernst equation is (with $a_{\text{HCl}} = m^2\gamma_\pm^2$):

$$\mathcal{E} = \mathcal{E}^\theta - \frac{RT}{\mathcal{F}} \ln m^2\gamma_\pm^2 \tag{8.56c}$$

∎

The molality of Eq. (8.56c) is a known quantity, but the only thing we know for certain about the activity coefficient is that, by the convention we always use for ionic solutes, $\gamma_\pm \to 1$ as $m \to 0$. If the emf of the cell is measured at many molalities, we could calculate:

$$\mathcal{E}' \equiv \mathcal{E} + \frac{2RT}{\mathcal{F}} \ln m = \mathcal{E}^\theta - \frac{2RT}{\mathcal{F}} \ln \gamma_\pm \tag{8.57}$$

This quantity will approach $\mathcal{E}^\theta$ as $m \to 0$ (so $\gamma_\pm \to 1$ and $\ln \gamma_\pm \to 0$). It is somewhat more accurate to use something else that is known about activity coefficients: the first term of the DHG equation. Substituting Eq. (8.15) into (8.57) and collecting all the known terms on the left-hand side gives (with $I = m$ for a 1:1 electrolyte):

$$\mathcal{E}'' \equiv \mathcal{E} + \frac{2RT}{\mathcal{F}}\left(\ln m - \frac{1.177\sqrt{m}}{1 + \sqrt{m}}\right) = \mathcal{E}^\theta + Bm \tag{8.58}$$

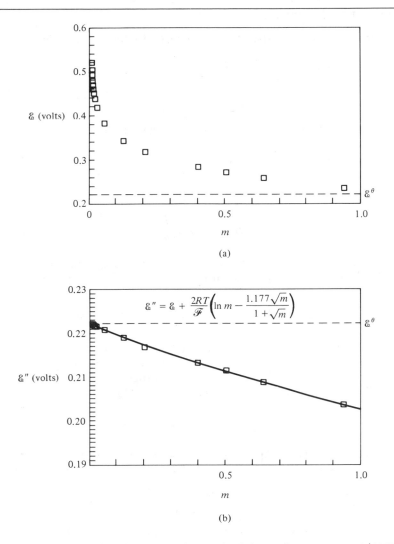

Figure 8.12 Cell emf vs. molality. (a) The emf of the cell Pt, H_2(1 atm) | HCl(aq, m) | AgCl(s) | Ag vs. molality of HCl. Panel (b) shows the graph used to determine standard emf from data such as those shown on panel (a).

The constant B, related of course to Guggenheim's β, is an empirical constant used to extrapolate the left-hand side to $m = 0$, where its value is $\mathscr{E}^\theta$. Equation (8.58) applies only for cells with 1:1 electrolytes; several of the problems will give the reader practice with this and other types of cells. Figure 8.12 illustrates that $\mathscr{E}''$ [as defined by Eq. (8.58)] behaves smoothly up to $m = 1$ mol/kg and is thus very suitable for numerical or graphical extrapolation. The easier method [plotting $\mathscr{E}'$ of Eq. (8.57) vs. $\sqrt{m}$] cannot effectively utilize data for $m > 0.1$, and the extrapolation is not as accurate.

Thermodynamics from Cells

Cell emf's may, of course, be used to calculate ΔG^{θ} for the cell reaction by Eq. (8.54);

$$\Delta G^{\theta}_{rxn} = -n\mathscr{F}\mathscr{E}^{\theta}$$

$$= \sum_i \nu_i \Delta G^{\theta}_{fi} \tag{8.59}$$

Taking the derivative of ΔG^{θ}_{rxn} with respect to T and using:

$$S = -\left(\frac{\partial G}{\partial T}\right)_P$$

gives the standard entropy change of the cell reaction:

$$\Delta S^{\theta}_{rxn} = n\mathscr{F}\frac{d\mathscr{E}^{\theta}}{dT} \tag{8.60}$$

Then, since $\Delta H = \Delta G + T\,\Delta S$ for any isothermal process, the standard enthalpy change of the cell reaction can be seen to be:

$$\Delta H^{\theta}_{rxn} = n\mathscr{F}\left(T\frac{d\mathscr{E}^{\theta}}{dT} - \mathscr{E}^{\theta}\right) \tag{8.61}$$

Example: Calculating entropies with Eq. (8.60) presents us with a fine opportunity to practice numerical differentiation. Data for the Ag|AgCl cell (8.56) are given by Table 8.11 (also Figure 8.13). (Note that the value at 25°C differs slightly from that in Table 8.10.) The slope at 25°C can be estimated from the values at 20° and 30°C:

$$\frac{d\mathscr{E}^{\theta}}{dt} \cong \frac{0.21904 - 0.22557}{30 - 20} = -6.53 \times 10^{-4} \text{ V/K}$$

The more accurate formula of Appendix I can also be used [Eq. (I.21c)]; with an interval $h = 5°C$, and points at $t = 15°, 20°, 30°,$ and $35°C$ (Table 8.11), the derivative is:

$$\frac{d\mathscr{E}^{\theta}}{dT} \cong \frac{0.22857 - 8(.22557) + 8(0.21904) - (0.21565)}{(5)(12)}$$

$$= -6.553 \times 10^{-4} \text{ V/K}$$

Table 8.11   Standard cell potentials for Pt, $H_2(g)$ | HCl(aq) | AgCl(s) | Ag

t (°C)	$\mathscr{E}^{\theta}$ (volts)	t (°C)	$\mathscr{E}^{\theta}$ (volts)
0	0.23655	40	0.21208
5	0.23413	45	0.20835
10	0.23142	50	0.20449
15	0.22857	60	0.19649
20	0.22557	70	0.18782
25	0.22234	80	0.1787
30	0.21904	90	0.1695
35	0.21565		

Source: Reference 3, p. 457.

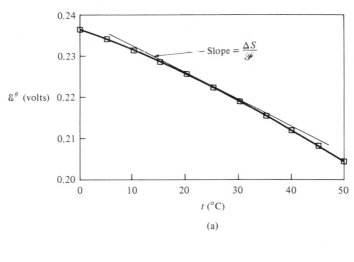

(a)

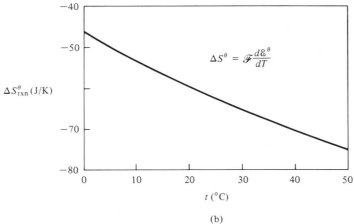

(b)

Figure 8.13 Standard emf vs. temperature. The standard emf vs. temperature for the cell Pt, $H_2(g) \mid HCl(Aq) \mid AgCl(s) \mid Ag$. The slope of this graph is related to the entropy of the cell reaction. Panel (b) shows the entropy of reaction for this cell vs. temperature, as calculated from the data in panel (a).

For the cell reaction (8.56b):

$$\Delta S_{rxn}^{\theta} = (96,487 \text{ C})(-6.553 \times 10^{-4} \text{ V/K}) = -63.23 \text{ J/K}$$

$$= S_m^{\theta}(\text{Ag}) + S_m^{\theta}(\text{H}^+) + S_m^{\theta}(\text{Cl}^-) - \tfrac{1}{2}S_m(\text{H}_2) - S_m^{\theta}(\text{AgCl})$$

The values of the standard entropy of Ag, AgCl, and H_2 at 25°C are given in Table 3.2, so the entropy of the ions can be calculated from this datum:

$$S_m^{\theta}(\text{H}^+, \text{aq}) + S_m^{\theta}(\text{Cl}^-, \text{aq}) = -63.23 - 42.68 + \tfrac{1}{2}(130.58) + 96.11$$

$$= 55.49 \text{ J/K}$$

It is marginally possible to determine absolute entropies of single ions, but sums and differences may be measured (as in this example) with much greater accuracy. Therefore, ionic entropies are usually listed *relative* to H^+; in effect $S_m^\theta(H^+)$ is arbitrarily set equal to zero. With this convention, $S_m^\theta(Cl^-) = 55.49$ J/K (relative to H^+); this answer should be compared to Table 8.4, where values for other ions are given.   ∎

Exercise: Calculate $d\mathscr{E}^\theta/dT$ for the cell of Table 8.11 using points at 5°, 15°, 35°, and 45°C (that is, $h = 10$) and, from that, $S_m^\theta(Cl^-, aq)$. (*Answer:* -6.465×10^{-4} V/K, 56.35 J/K.)   ∎

Example: For the simple electrodes in Table 8.10 (for example, Zn^{2+}/Zn as opposed to $Cl^-/AgCl/Ag$) the cell reaction — for example,

$$Zn^{2+} + 2\ e = Zn$$

— would be the reverse of the ion-formation reaction if we ignored the electrons. Now we remember that these emf's are for the cells *vs.* the hydrogen electrode, so the cell:

$$Pt, H_2 \,|\, H^+ \,\|\, Zn^{2+} \,|\, Zn$$

has $\mathscr{E}^\theta = -0.7628$ as listed. The cell reaction is (prove!):

$$Zn^{2+} + H_2 = 2\ H^+ + Zn$$

for which:

$$\Delta G_{rxn}^\theta = -2(96487)(-0.7628) = 147.2 \text{ kJ}$$
$$= 2\Delta G_f^\theta(H^+) - \Delta G_f^\theta(Zn^{2+})$$

Since H^+ is the standard, we get $\Delta G_f^\theta(Zn^{2+}) = 2(-0.7628)(96,487) = -147.2$ kJ (compare Table 8.4). Table 8.10 is, for those cases in which the electrode reaction is elemental, just a list of $-\Delta G_f^\theta$ of ions with units of volts.   ∎

Equilibrium Constants

We have frequently used the relationship:

$$\Delta G_{rxn}^\theta = -RT \ln K_a$$

to obtain thermodynamic equilibrium constants from free energies and vice versa. Now we see, from Eq. (8.59), that the standard emf is related to the equilibrium constant of the *cell reaction:*

$$\ln K_a = \frac{n\mathscr{F}\mathscr{E}^\theta}{RT} \tag{8.62}$$

The equilibrium constant for any redox pair for which the $\mathscr{E}^\theta$'s are known can be calculated by diagramming a cell, however hypothetical, with the reduction on the "right," and calculating:

$$\mathscr{E}^{\theta}_{\text{cell}} = \mathscr{E}^{\theta}_{R} - \mathscr{E}^{\theta}_{L}$$

for use in Eq. (8.62).

Example: It is of particular interest to oppose a compound electrode such as AgCl | Ag to the corresponding metal electrode — for example:

$$\text{Ag} \,|\, \text{Ag}^{+}(\text{aq}) \,\|\, \text{Cl}^{-}(\text{aq}) \,|\, \text{AgCl(s)} \,|\, \text{Ag}$$

This cell is somewhat hypothetical, since, if the chloride were in contact with the Ag^{+} electrolyte, AgCl would precipitate and there would be insufficient electrolyte to carry the current. A reasonable approximation to the cell could be made using a salt bridge (see below), but that is not necessary, since the $\frac{1}{2}$-cell potentials could be measured separately vs. any convenient standard.

The cell reaction is:

R (red.): $\text{AgCl(s)} + \text{e} = \text{Ag(s)} + \text{Cl}^{-}$ $\mathscr{E}^{\theta}_{R} = 0.2225 \text{ V}$

L (ox.): $\text{Ag(s)} = \text{Ag}^{+}(\text{aq}) + \text{e}$ $\mathscr{E}^{\theta}_{L} = 0.7991 \text{ V}$

Cell: $\text{AgCl(s)} = \text{Ag}^{+}(\text{aq}) + \text{Cl}^{-}(\text{aq})$ $\mathscr{E}^{\theta} = -0.5766 \text{ V}$

Thus, the cell reaction is just that of the solubility product. Therefore the thermodynamic solubility product of AgCl can be calculated:

$$\ln K_{a} = \frac{(-0.5766)(96{,}487)}{(8.3143)(298.15)}$$

$$K_{a} = 1.79 \times 10^{-10}$$

(Compare this to the earlier result, 1.60×10^{-10}, obtained from solubilities.)   ■

Exercise (optional): Show that the emf of any metal/metal salt compound electrode is related to that of the simple metal electrode by:

$$\mathscr{E}^{\theta}_{\text{cmpd}} = \mathscr{E}^{\theta}_{\text{M}^{+}|\text{M}} + \frac{RT}{n\mathscr{F}} \ln K_{a} \tag{8.63}$$

where K_{a} is the thermodynamic equilibrium constant of the solubility reaction. This equation applies to multiply charged ions with the appropriate value of n.   ■

Amalgam Electrodes

Constructing a metal ion/metal electrode seems easy enough; just put a bar of the metal into a solution of a salt of that metal. But if this were attempted with $\text{Na}^{+} \,|\, \text{Na}$ in water, the sodium would react instantly and violently with the water; a more subtle approach is needed in such cases. The chemical potential of a metal can be lowered by dissolving it in mercury to make an *amalgam* — for example, Na(Hg) — to make cells such as

$$\text{Ag} \,|\, \text{AgCl(s)} \,|\, \text{NaCl(aq}, m) \,|\, \text{Na(Hg)} \tag{8.64a}$$

Of course, the standard emf of the cell could be measured as in the earlier example [cell (8.56)], but the result would be:

$$\mathscr{E}_{cell}^{\theta} = \mathscr{E}_{amal} - \mathscr{E}_{Cl^-|AgCl|Ag}^{\theta} \tag{8.64b}$$

This is the standard emf vs. the amalgam electrode rather than vs. pure sodium. This latter quantity could subsequently be measured by opposing the same amalgam electrode to Na in a nonaqueous electrolyte; acetonitrile (CH_3CN) is useful for this purpose. The cell:

$$Na(Hg)\,|\,Na^+(\text{in acetonitrile})\,|\,Na \tag{8.65a}$$

will have an emf:

$$\mathscr{E}_{cell} = \mathscr{E}_{Na^+|Na}^{\theta} - \mathscr{E}_{amal} \tag{8.65b}$$

from which potentials vs. $Na^+|Na$ can be directly calculated. [The cell (8.65) is, strictly, a concentration cell. We shall analyze it in detail in Section 8.6.]

Example: Demonstrate that the cells (8.64) and (8.65), when added, give the same cell reaction as:

$$Ag\,|\,AgCl(s)\,|\,NaCl(aq)\,|\,Na(s)$$

from whose standard potential that of $Na^+(aq)\,|\,Na$ could be calculated.

The sum of the emf's of two cells is just the emf that would be measured if the cells were connected in series; the cell reactions are (acet = acetonitrile):

$$Ag\,|\,AgCl(s)\,|\,NaCl(aq)\,|\,Na(Hg)\,|\,NaCl(acet)\,|\,Na(s)$$

Left (cell 1):	$Ag + Cl^-(aq) = AgCl(s) + e$
Right (cell 1):	$Na^+(aq) + e = Na(Hg)$
Left (cell 2):	$Na(Hg) = Na^+(acet) + e$
Right (cell 2):	$Na^+(acet) + e = Na(s)$
Net:	$Ag + NaCl(aq) = AgCl(s) + Na(s)$

■

Measurement of Activity Coefficients

If you wished to measure the activity coefficient of an electrolyte, one way would be to construct an electrochemical cell such that the activity of that electrolyte appeared in the Nernst equation of the cell. Of course, the cell should not have a liquid junction, and its standard emf would have to be known or measurable. A single-electrolyte cell requires that one electrode be reversible to the cation while the other is reversible to the anion. Cations are rarely a problem (although some, like sodium, may have to be in an amalgam), but anions frequently are. If an insoluble salt of the anion is available, a compound electrode such as $Cl^-|AgCl|Ag$ is the best answer to this problem; for example, cell (8.64) could be used to measure activity coefficients of NaCl and cell (8.56) could be used for HCl. This leaves some problems; for example, there are no insoluble nitrate salts, so activity coefficients of nitrate salts must be measured by vapor pressure or some other colligative effect.

8.6   Concentration Cells

The cells that we have been discussing to this point have had emf's that arise from a chemical reaction. In *concentration cells* the potential is caused by the difference in chemical potential of a substance due to the difference in concentration. Two types of concentration cells can be distinguished: (1) electrode concentration cells and (2) electrolyte concentration cells. The latter class can be further divided into those with and without transference. We shall discuss these cases one at a time.

Electrode Concentration Cells

The simplest of this class is a pair of gas electrodes with the gas pressure different at each electrode — for example:

$$\text{Pt}, \text{H}_2(\text{g}, P_1) \,|\, \text{H}^+(\text{aq}, m) \,|\, \text{H}_2(\text{g}, P_2), \text{Pt} \tag{8.66a}$$

This cell can be analyzed in the usual manner:

L (ox.): $\text{H}_2(\text{g}, P_1) = 2\,\text{H}^+(\text{aq}, m) + 2\,\text{e}$

R (red.): $\underline{2\,\text{H}^+(\text{aq}, m) + 2\,\text{e} = \text{H}_2(\text{g}, P_2)}$

Cell rxn: $\text{H}_2(\text{g}, P_1) = \text{H}_2(\text{g}, P_2)$ $\qquad$ **(8.66b)**

Since the electrode reactions are the same, the standard emf of this cell is zero. Therefore, the Nernst equation is:

$$\mathcal{E} = \frac{RT}{2\mathcal{F}} \ln \frac{f_1}{f_2} \tag{8.66c}$$

where f_1 and f_2 are the fugacities of the gas at the left and right electrodes, respectively. At low pressures, the fugacity is approximately equal to the pressure of the gas (ideal gas approximation).

Another type of electrode concentration cell can be made with mercury solutions of metals — that is, amalgams. We can treat the activity of a metal in an amalgam as we did for other solutions of a nonelectrolytes (Chapter 7). For example:

$$\text{Tl}(\text{Hg}, a_1) \,|\, \text{Tl}^+(\text{aq}, m) \,|\, \text{Tl}(\text{Hg}, a_2) \tag{8.67a}$$

for which the cell reaction can be shown to be:

$$\text{Tl}(\text{Hg}, a_1) = \text{Tl}(\text{Hg}, a_2) \tag{8.67b}$$

The emf is:

$$\mathcal{E} = -\frac{RT}{\mathcal{F}} \ln \frac{a_2}{a_1} \tag{8.67c}$$

Example: A cell like (8.67) was constructed with thallium concentrations $X_1 = 0.01675$ and $X_2 = 0.003259$. Assuming ideal solution, $a = X$, the emf can be estimated as:

$$\mathcal{E} = \frac{(8.3143)(298.15)}{96{,}487} \ln \frac{0.01675}{0.003259} = 0.04206 \text{ V}$$

The measured value (ref. 1, p. 259) is 0.04555 V, so the ideal solution approximation is not exact, even at these low concentrations. ∎

Example: In an earlier example, cell (8.65), a quantity called "$\mathscr{E}_{amal}$" was introduced and related to the standard emf of the metal (sodium) via cell (8.65). It may now be apparent that cell (8.65) is just an electrode concentration cell, whose cell reaction can be shown to be:

$$Na(Hg) = Na(s, pure)$$

Since the standard emf of sodium refers to solid sodium, it is convenient to take pure solid sodium as the standard state to define an activity for the solute as:

$$a_2 = \frac{f(\text{Na in Hg})}{f(\text{pure Na})}$$

With this definition, the emf of cell (8.65) is:

$$\mathscr{E} = \frac{RT}{\mathscr{F}} \ln a_2$$

Then, using Eq. (8.65b), the amalgam emf can be seen to be:

$$\mathscr{E}_{amal} = \mathscr{E}^{\theta}_{Na^+/Na} - \frac{RT}{\mathscr{F}} \ln a_2 \tag{8.68}$$

For an ideal solution, a_2 can be approximated as the mole fraction of sodium in the amalgam, but this will not be very accurate. However, Eq. (8.68) is useful for reminding us that "$\mathscr{E}_{amal}$" is not a fixed quantity but depends on the concentration of sodium. ∎

Electrolyte Concentration Cells

A solute such as NaCl in water has a chemical potential that depends on concentration. This potential is reflected in cells such as:

$$Na(Hg) \,|\, NaCl(aq, m_1) \,|\, NaCl(aq, m_2) \,|\, Na(Hg) \tag{8.69a}$$

The electrodes are identical, but the cell will have a potential. This cell has a liquid junction and is called a cell *with transference* because its emf will depend on the transference or transport of material across this junction. Our earlier convention for cell reactions will require that Na^+ be produced at the left (oxidation) and used at the right (reduction), so that, within the cell, cations must move to the right while anions move to the left.

Before analyzing this cell, let's look at the simpler case of the same cell without the liquid junction:

$$Na(Hg) \,|\, NaCl(aq, m_1) \,\|\, NaCl(aq, m_2) \,|\, Na(Hg)$$

The liquid junction can be eliminated by use of a *bridge* that can transfer chloride ions from the right to the left as required by the cell reaction. One way of accomplishing this is to use a double electrode that is reversible to chloride — for example:

$$Na(Hg) \,|\, NaCl(aq, m_1) \,|\, AgCl(s) \,|\, Ag\text{-}Ag \,|\, AgCl(s) \,|\, NaCl(aq, m_2) \,|\, Na(Hg) \tag{8.70a}$$

This is really just two cells like (8.64) in series; the cells are identical except for the NaCl concentrations. The cells are analyzed as usual:

Left 1 (ox.): $Na(Hg) = Na^+(m_1) + e$

Right 1 (red.): $AgCl(s) + e = Ag(s) + Cl^-(m_1)$

Left 2 (ox.): $Ag(s) + Cl^-(m_2) = AgCl(s) + e$

Right 2 (red.): $Na^+(m_2) + e = Na(Hg)$

Cell rxn: $NaCl(m_2) = NaCl(m_1)$

The standard emf of these cells together is zero because of cancellation, so the potential from the Nernst equation is just:

$$\mathscr{E} = -\frac{2RT}{\mathscr{F}} \ln \frac{m_1 \gamma_{\pm 1}}{m_2 \gamma_{\pm 2}} \qquad \textbf{(8.70b)}$$

A cell such as (8.70) is a convenient way to measure relative activity coefficients; but it is, of course, no more than two cells in series, so its use is no different than using cell (8.64) at a series of different concentrations and comparing the results. More interesting things occur when the liquid junction is present.

Cells with Transference

The cell (8.69) will develop a potential at the liquid junction due to the transport of material across a change in chemical potential; this potential, $\mathscr{E}_J$, must be added to the Nernst potential [Eq. (8.70b)] that is developed by the electrode reactions. Remembering that, within the cell, cations will move right while anions move left, this cell will be analyzed in terms of the chemical processes that occur at the electrodes and at the liquid junction (LJ). If one faraday of electricity is transferred through the external circuit, the following will occur within the cell:

Left (ox.): $Na(Hg) = Na^+(m_1) + e$

LJ: $\begin{cases} t_+ Na^+(m_1) = t_+ Na^+(m_2) \\ t_- Cl^-(m_2) = t_- Cl^-(m_1) \end{cases}$

Right (red.): $Na^+(m_2) + e = Na(Hg)$

Cell rxn: $Na^+(m_2) + t_- Cl^-(m_2) + t_+ Na^+(m_1)$

$= Na^+(m_1) + t_+ Na^+(m_2) + t_- Cl^-(m_1)$

Next, the cell reaction is rearranged by collecting all m_2 terms on the left and all m_1 terms on the right:

$$(1 - t_+) Na^+(m_2) + t_- Cl^-(m_2) = (1 - t_+) Na^+(m_1) + t_- Cl^-(m_1)$$

By definition $t_+ + t_- = 1$, so the cell reaction is simply:

$$t_- NaCl(m_2) = t_- NaCl(m_1) \qquad \textbf{(8.69b)}$$

and the emf is readily shown to be:

$$\mathcal{E} = -\frac{RT}{\mathcal{F}} \ln \left(\frac{a_1}{a_2} \right)^{t_-}$$

or (with $a = m^2 \gamma_{\pm}^2$):

$$\mathcal{E} = -2t_- \frac{RT}{\mathcal{F}} \ln \frac{m_1 \gamma_{\pm 1}}{m_2 \gamma_{\pm 2}} \qquad \textbf{(8.69c)}$$

Such a cell has an obvious utility for measuring transference numbers; since these depend on concentration (see Table 8.5), the t_- of Eq. (8.69) is, roughly, the average of the values at m_1 and m_2.

Exercise: Cells such as (8.69) for which the cation is removed from or placed into solution at the electrodes are called *cation reversible*. Show that the emf of an *anion reversible* cell such as:

$$\text{Ag} \,|\, \text{AgCl(s)} \,|\, \text{NaCl}(m_1) \,|\, \text{NaCl}(m_2) \,|\, \text{AgCl(s)} \,|\, \text{Ag} \qquad \textbf{(8.71a)}$$

is

$$\mathcal{E} = 2t_+ \frac{RT}{\mathcal{F}} \ln \frac{m_1 \gamma_{\pm 1}}{m_2 \gamma_{\pm 2}} \qquad \textbf{(8.71b)}$$

∎

8.7   Liquid-Junction Potentials

Liquid-junction potentials have been mentioned several times, mostly as something to be avoided. Since they cannot always be avoided, we shall now present some useful results for estimating their size.

The potential of any cell can be written as a sum of the emf due to the electrode processes, $\mathcal{E}_N$, the *Nernst potential,* and that due to liquid junctions, $\mathcal{E}_J$:

$$\mathcal{E} = \mathcal{E}_N + \mathcal{E}_J \qquad \textbf{(8.72)}$$

The Nernst potential for the cell (8.69) can be derived by examining just the electrode processes:

$$\text{Left:} \quad \text{Na(Hg)} = \text{Na}^+(m_1) + \text{e}$$

$$\underline{\text{Right:} \quad \text{Na}^+(m_2) + \text{e} = \text{Na(Hg)}}$$

$$\text{Net:} \quad \text{Na}^+(m_2) = \text{Na}^+(m_1)$$

The potential for this reaction involves the ratio of the Na^+ activities, which we shall replace by the mean ionic activity $a_{\pm} = m\gamma_{\pm}$; the electrode potential is then:

$$\mathcal{E}_N = -\frac{RT}{\mathcal{F}} \ln \frac{m_1 \gamma_{\pm 1}}{m_2 \gamma_{\pm 2}}$$

The junction potential can be calculated by subtracting this result from Eq. (8.69c):

$$\mathcal{E}_J = (1 - 2t_-)\frac{RT}{\mathcal{F}} \ln \frac{m_1\gamma_{\pm 1}}{m_2\gamma_{\pm 2}} \tag{8.73a}$$

Using $1 = t_+ + t_-$, we can also write this as:

$$\mathcal{E}_J = (t_+ - t_-)\frac{RT}{\mathcal{F}} \ln \frac{m_1\gamma_{\pm 1}}{m_2\gamma_{\pm 2}} \tag{8.73b}$$

Equation (8.73) reveals the vital fact that junction potentials depend on the difference of the cation and anion mobilities and can be minimized if these quantities are of similar magnitude.

The general case of the junction potential between dissimilar electrolytes is much more complicated. The simplest case is if both are $1:1$ electrolytes:

$$MX(c_1)\,|\,M'X'(c_2) \tag{8.74a}$$

for which the results of Henderson (1907) give:

$$\mathcal{E}_J = \frac{RT}{\mathcal{F}}\left[\frac{c_1(\lambda_+ - \lambda_-) - c_2(\lambda_+' - \lambda_-')}{c_1\Lambda - c_2\Lambda'}\right]\ln\left[\frac{c_1\Lambda}{c_2\Lambda'}\right] \tag{8.74b}$$

where Λ is the equivalent conductivity of the solution at c_1, and Λ' is that for c_2. The derivation of this equation makes a number of nonthermodynamic assumptions, so it is not as exact as the thermodynamic equations used earlier. The reader should demonstrate that Eq. (8.74) reduces to Eq. (8.73) if the electrolytes are of the same type ($MX = M'X'$) and the activity is approximated by the concentration (c) — that is, ideal solution.

Exercise: Show [starting with Eq. (8.74)] that the liquid-junction potential between electrolytes at equal concentrations and with one common ion (either $M = M'$ or $X = X'$) is:

$$\mathcal{E}_J = \pm\frac{RT}{\mathcal{F}} \ln \frac{\Lambda}{\Lambda'} \tag{8.75}$$

(negative if the cations are identical, positive if the anions are identical). ∎

Example: Calculate the liquid-junction potential at 25°C for $HCl\,|\,KCl$ if $c = 0.01$ moles/dm^3 for both. From Eq. (8.75) with data from Table 8.6:

$$\mathcal{E}_J = \frac{RT}{\mathcal{F}} \ln \frac{412.00}{141.27} = 0.0275 \text{ V}$$

The observed value (ref. 2, p. 236) is 0.0268 V. This is representative of the sort of junction potentials that would be encountered with the Calomel electrode (cf. p. 400). ∎

Another simplification of Eq. (8.74) is possible if one of the salts is much more concentrated than the other. If $c_1 \gg c_2$, it is readily shown [do it!] that:

$$\mathcal{E}_J = \frac{RT}{\mathcal{F}}(t_+ - t_-) \ln \frac{c_1\Lambda}{c_2\Lambda'} \tag{8.76}$$

where t_+ and t_- are the transport numbers in the concentrated solution (c_1). This potential can be minimized by using a salt in the concentrated solution for which $t_+ \cong t_- \cong 0.5$ — that is, with equal cation and anion mobilities. A glance at Table 8.5 will show that NH_4Cl, KCl, KNO_3, and NH_4NO_3 are reasonably close to this requirement. Such salt solutions can be used as bridges to minimize the junction potential in cells such as (8.49); KCl is very popular for such a purpose; ammonium salts are to be avoided because of the ability of this ion to complex heavy metals. For example, the cell (8.49d) could be approximated with a KCl salt bridge as:

$$Zn \,|\, ZnCl_2(c_1) \,|\, KCl(sat.) \,|\, CuCl_2(c_2) \,|\, Cu$$

This arrangement has the added advantage that the two junction potentials have opposite signs and will tend to cancel.

Exercise (optional): Show that the double junction (all 1:1 electrolytes):

$$MX(c_1) \,|\, M''X''(c) \,|\, M'X'(c_2)$$

with $c \gg c_1$ or c_2 has a total junction potential:

$$\mathscr{E}_J = \frac{RT}{\mathscr{F}}(2t''_+ - 1) \ln \frac{c_1 \Lambda}{c_2 \Lambda'} \tag{8.77}$$

∎

Example: Silver ion solutions for the electrode $Ag^+ \,|\, Ag$ must use a soluble salt and must not contact a solution containing any of the many anions that form insoluble precipitates (including all halides). A cell such as:

$$Ag \,|\, AgNO_3(aq, c_1) \,\|\, KCl(aq, c_2) \,|\, Hg_2Cl_2(s) \,|\, Hg$$

would have to be bridged with something like a KNO_3 salt bridge, giving the liquid junctions:

$$AgNO_3(c_1) \,|\, KNO_3(c) \,|\, KCl(c_2)$$

The junction potential can be estimated from Eq. (8.77) if the concentration of KNO_3 is much greater than that of KCl or $AgNO_3$. Table 8.5 does not give the transference number of this salt at high concentrations, but the trend of the numbers suggests that $t_+ \cong 0.53$ might be reasonable. Assuming $c_1 = c_2 = 0.01$ mol/dm³, the equivalent conductances of Table 8.6 give [with Eq. (8.77)]:

$$\mathscr{E}_J = \frac{RT}{\mathscr{F}}[2(0.53) - 1] \ln \frac{141.28}{124.76} = 0.0002 \text{ V}$$

This value is clearly small but also significant, since it is within the accuracy of such measurements.

∎

Postscript

The principal emphasis of this chapter has been on the thermodynamics of ionic solutions, particularly on the relationship of thermodynamic properties to cell emf's. The practical side of electrochemical cells is, of course, batteries. The technology of batteries involves many practical issues besides those discussed in this chapter.

Batteries are very important because of their convenience and portability, but there is more to it than that. Of the chemical energy available from fuel combustion, only a portion can be made available to do mechanical work. Even in an ideal case, the amount of work that can be obtained in a heat engine (whether used directly, or used to generate electrical power) is limited by Carnot's formula [Eq. (3.2)]. In an ideal electrochemical cell (that is, a reversible cell), the work available is much greater (it is the ΔG of the reaction). Cells that can utilize fuel oxidation to provide electricity are called fuel cells. Fuel cells using hydrogen and oxygen have been developed and used in specialty applications such as space vehicles, but hydrocarbon fuel cells are a major current research effort and have not yet become widespread in commercial applications.

Another important application of batteries is for energy storage — an automobile battery is a good example. A large component of electrical generating costs depends on the peak load; forms of energy storage for "peak shaving" are eagerly sought. Also, solar power generation is inevitably spotty, and energy storage is needed. At present, batteries are not used for large-scale energy storage because of their cost; there is much room for improvement, and this too is an active research area.

This chapter brings to an end our coverage of thermodynamics; the assiduous student will, by now, have reached the threshold of the beginning of understanding. If it is any comfort, most persons who have been repeatedly exposed to this subject over a period of years agree that this repetition was necessary before they achieved a full appreciation of the subject. It can only be hoped that this exposure will have conveyed enough of the power, beauty, and utility of the subject so that you will, when necessary, be able to advance to the next level of understanding. "Now this is not the end. It is not even the beginning of the end. But it is, perhaps, the end of the beginning." (Churchill)

References

1. G. N. Lewis, M. Randall, K. S. Pitzer, and L. Brewer, *Thermodynamics,* 2d ed., 1961: New York, McGraw-Hill Book Co.
2. D. A. MacInnes, *The Principles of Electrochemistry,* 1939: New York, Van Nostrand Reinhold Company.
3. R. A. Robinson and R. H. Stokes, *Electrolyte Solutions,* 1955: London, Butterworth's Scientific Publications.

Problems

8.1   A 0.01-molal solution of a weak acid in water depresses the freezing point of water by 0.0208 K. Calculate the percent ionization of this acid (assume ideal solution for all species present).

8.2   Scatchard, Prentiss, and Jones [*J. Am. Chem. Soc.,* 56, 805 (1934)] measured freezing-point depressions of $KClO_4$ in H_2O. Use these data to calculate $\gamma_\pm$ at 0.001, 0.01, and 0.05 molal. (The concentrations given are low enough that it can be assumed that j/m vs. $1/\sqrt{m}$ is linear with reasonable accuracy.)

m	θ
0.003612	0.01316
0.006690	0.02421
0.009872	0.03509
0.016215	0.05712
0.030369	0.10541
0.048335	0.16359

8.3 Use data in Table 8.1 to calculate the mean ionic activities ($a_\pm$) for 0.5-molal solutions of (a) KCl, (b) HCl, (c) CuSO$_4$, and (d) Cr(NO$_3$)$_3$.

8.4 Calculate the ionic strength of a solution containing: $0.1m$ KNO$_3$, $0.15m$ K$_2$SO$_4$, and $0.023m$ La$_2$(SO$_4$)$_3$.

8.5 Use the Debye-Hückel theory to calculate the "distance of closest approach" from the experimental activity coefficients of KCl (Table 8.1). Note that the ionic radii from crystallography are 1.33 Å for K$^+$ and 1.81 Å for Cl$^-$.

8.6 Wood, Wicker, and Kreis [*J. Phys. Chem.*, 75, 2313 (1971)] determined activity coefficients of NaNO$_3$ in N-methylacetamide. Plot these results vs. $\sqrt{I}$ to see if they fit the Debye-Hückel limiting law. For this solvent $\alpha = 0.32531$ [calculated with Eq. (8.14a)].

m	$\gamma_\pm$
0.01	0.970
0.05	0.937
0.10	0.915
0.20	0.886
0.30	0.864

8.7 (a) Use data for $\gamma_\pm$ of HCl at 0.05 mol/kg (Table 8.1) to estimate the DHG β for this acid at 25°C.
(b) Use this β to calculate $\gamma_\pm$ from 0.01 to 0.5 mol/kg and compare the results to experimental.

8.8 Use the DHG equation to calculate the mean ionic activity coefficient at 25°C of CuSO$_4$ at 0.001, 0.1, and 1 mol/kg. Compare your answers to the experimental values given as Table 8.1.

8.9 (a) The van't Hoff osmotic factor gives, for the freezing-point depression, $\theta = iK_f m$. Use the Debye-Hückel limiting law for a 1:1 electrolyte, $\ln \gamma_\pm = -\alpha\sqrt{m}$, to derive (to the same approximation):

$$i = \nu(1 - 0.378\sqrt{m})$$

(b) Calculate the freezing point of a 0.5-molal KCl solution; the observed value is 271.49.

8.10 TlCl has a solubility of 1.42×10^{-2} mol/kg in water at 25°C.
(a) Calculate K_a.
(b) Estimate the solubility of TlCl in 0.1-molal NaNO$_3$.
(c) Estimate the solubility of TlCl in 0.1-molal NaCl.

8.11 The thermodynamic solubility product (K_a) of Ag$_2$CrO$_4$ is 2.0×10^{-7} at 25°C. Calculate the solubility of this salt in a solution with 0.1 moles AgNO$_3$/kg (a) assuming ideal solutions and (b) more accurately.

8.12 MgF$_2$ has $K_{sp} = 6.4 \times 10^{-9}$ (in water at 25°C).
(a) Calculate the solubility.
(b) Estimate the thermodynamic solubility product (K_a).

8.13 MgF_2(solid) has $\Delta G_f^\theta = -1049$ kJ/mole at 25°C. Calculate the thermodynamic solubility product (K_a) for this salt.

8.14   Calculate the molality of HCl in water for a solution for which $pH = 2.00$.

8.15   Dichloroacetic acid has an acid dissociation constant $K_a = 3.32 \times 10^{-2}$. Calculate the degree of dissociation and pH of a 0.01-mol/kg solution (a) for an ideal solution, (b) using the DHG formula for $\gamma_\pm$ ($\beta = 0$).

8.16   Calculate the degree of dissociation of chloroacetic acid ($K_a = 1.38 \times 10^{-3}$ at 25°C) when it is in solution at 0.1 mol/kg with 0.05 mol/kg HCl, (a) using ideal solution and (b) using DHG.

8.17   Show that Henry's law for a gas such as HCl that ionizes in solution should be written:

$$k_m = \lim_{m \to 0} \left(\frac{P}{m^2} \right)$$

8.18   In their famous book (1923), Lewis and Randall tell of a "distinguished chemist" who attempted to measure the hydrolysis of N_2 in water:

$$N_2(g) + 2 H_2O(liq) = NH_4^+(aq) + NO_2^-(aq)$$

They used $\Delta G_f^\theta = -56,560$ cal for H_2O, $-18,930$ cal for NH_4^+, -8500 cal for NO_2^-, to calculate the N_2 pressure required to create the ions at 10^{-3} mol/kg. Repeat this calculation.

8.19   Calculate the resistance (25°) of a conductivity cell with plates 1.2 cm apart and plate area 7.2 cm^2, if it is filled with a 0.1-mol/dm^3 solution of NaCl.

8.20   The cell of Problem 8.19 was used to measure the concentration of a $Ca(NO_3)_2$ solution. The measured resistance was 13 Ω. Assuming $\Lambda = \Lambda^\circ$, calculate the concentration of this salt.

8.21   Calculate the equivalent conductivity and cation transference number of the following at infinite dilution: (a) rubidium acetate, (b) ammonium sulphate, (c) $K_3Fe(CN)_6$.

8.22   Use data in this chapter to calculate the mobility of K^+ for (a) infinite dilution, (b) 0.1 mol/kg KCl, (c) 0.1 mol/kg KNO_3.

8.23   The equivalent conductances in ethanol (25°) are:

$$\Lambda^\circ(LiCl) = 39.2$$

$$\Lambda^\circ(NaCl) = 42.5$$

$$\Lambda^\circ(LiI) = 43.4$$

Calculate Λ° for NaI in this solvent.

8.24   The data below give values for the equivalent conductivity of NaBr in methanol at 25°C. Determine Λ° for this salt in methanol.

$10^4 c$	Λ
1	99.19
2	98.11
5	96.04
10	93.80
20	90.86

(Plot data; do not use linear regression.)

8.25   Use the data below for the equivalent conductivity of NaI in water (25°C) to calculate Λ° (use the Onsager equation).

c	Λ
0.005	121.25
0.01	119.24
0.02	116.70
0.05	112.79

8.26   The following are degrees of dissociation measured (by conductance) for picric acid in methanol (25°). Calculate the thermodynamic dissociation constant. (The DH constant $\alpha = 4.58$ for methanol at 25°C.)

c	α
0.001563	0.3131
0.003125	0.2408
0.00625	0.1820
0.0125	0.1379

8.27   Calculate the thermodynamic ionization constant of chloroacetic acid (in water) from the data:

$10^3 c$	$\Lambda \ (\Omega^{-1} cm^2)$
0.11010	362.10
0.30271	328.92
0.58987	295.58
1.3231	246.15

8.28   The cell:

$$\text{Pt, H}_2(1 \text{ atm}) \,|\, \text{HCl(aq}, m = 1.5346) \,|\, \text{AgCl(s)} \,|\, \text{Ag}$$

has $\mathscr{E} = 0.20534$ volts.
(a) Use $\mathscr{E}^\theta$ from Table 8.10 to calculate $\gamma_\pm$ for HCl.
(b) What is the pH of this solution?

8.29   The emf of the cell:

$$\text{Hg(liq)} \,|\, \text{Hg}_2\text{Cl}_2(s) \,|\, \text{HCl(aq}, 0.1m) \,|\, \text{Cl}_2(g, P), (\text{Pt-Ir})$$

was measured with the chlorine pressure reduced in nitrogen to prevent hydrolysis of Cl_2 in the water. When $P_{Cl_2} = 0.0124$ atm, $\mathscr{E} = 1.0330$ volts. Calculate the standard emf of this cell from this datum.

8.30   Use data in this chapter to calculate the emf of the cell:

$$\text{Zn(s)} \,|\, \text{ZnCl}_2(\text{aq}, m = 0.05) \,|\, \text{Cl}_2(g, P = 0.5 \text{ atm}), (\text{Pt-Ir})$$

8.31   Use data in this chapter to calculate the emf (25°C) of the cell:

$$\text{Cu(s)} \,|\, \text{CuSO}_4(\text{aq}, m = 0.2) \,|\, \text{PbSO}_4(s) \,|\, \text{Pb(s)}$$

8.32   The cell:

$$\text{Pb(Hg)} \,|\, \text{PbSO}_4(s) \,|\, \text{H}_2\text{SO}_4(\text{aq}, 0.001m) \,|\, \text{H}_2(1 \text{ atm}), \text{Pt}$$

has $\mathscr{E}^\theta = 0.3505$ and $\mathscr{E} = 0.09589$. Calculate $\gamma_\pm$ for the electrolyte.

8.33 Use the selected data (25°C) below [Harned and Ehlers, *J. Am. Chem. Soc.*, 54, 1350 (1932)] to calculate $\mathscr{E}^{\theta}$ for the cell:

$$\text{Pt, } H_2(g, 1 \text{ atm}) \,|\, HCl(aq, m) \,|\, AgCl(s) \,|\, Ag(s)$$

m	$\mathscr{E}$ (volts)
0.003215	0.52053
0.005619	0.49257
0.009138	0.46860
0.013407	0.44974
0.02563	0.41824

8.34 Use the emf data below [E. W. Canning and M. G. Bowman, *J. Am. Chem. Soc.*, 68, 2042 (1946)] for the cell (25°C):

$$\text{Pt, } H_2(g, 1 \text{ atm}) \,|\, H_2SO_4(m, CH_3OH) \,|\, Hg_2SO_4(s) \,|\, Hg(\text{liq})$$

to calculate the standard emf of this cell. Note that sulfuric acid in a dilute solution in methanol acts as a 1:1 electrolyte — that is, it ionizes to H^+ and HSO_4^-. Also, the Debye-Hückel constant for methanol at 25°C is $\alpha = 4.065$.

$10^3 m$	$\mathscr{E}$ (volts)	$10^3 m$	$\mathscr{E}$ (volts)
0.700	0.7289	8.111	0.6711
1.1184	0.7174	22.385	0.6509
2.412	0.6996	43.217	0.6388
5.475	0.6805	96.88	0.6249
6.778	0.6756		

8.35 Calculate the standard emf of the cell:

$$\text{Zn} \,|\, ZnSO_4(aq, m) \,|\, PbSO_4(s) \,|\, Pb(Hg)$$

from these data:

m	$\mathscr{E}$ (volts)
0.0005	0.61144
0.001	0.59714
0.002	0.58319
0.005	0.56598
0.01	0.55353

8.36 The standard potentials listed below are for:

$$\text{Pt, } H_2(g) \,|\, HBr(aq) \,|\, AgBr(s) \,|\, Ag(s)$$

(a) What is the cell reaction?
(b) Calculate ΔG^{θ}, ΔS^{θ}, and ΔH^{θ} for this reaction at 25°C.

t (°C)	$\mathscr{E}^{\theta}$ (volts)
5	0.07991
15	0.07595
25	0.07131
35	0.06597
45	0.05995

8.37   The standard potentials of the cell:

$$Pt, H_2(g) \,|\, HCl(solution) \,|\, AgCl(s) \,|\, Ag(s)$$

in 20% dioxane/water are $\mathscr{E}^\theta = 0.20674$ volts at 20°C, 0.20303 at 25°C, 0.19914 at 30°C. Calculate the standard entropy and free energy of H^+Cl^- in this solvent at 25°C.

8.38   (a) What is the cell reaction for:

$$Pb(s) \,|\, PbCl_2(s) \,|\, HCl(aq, m = 0.1) \,|\, Hg_2Cl_2(s) \,|\, Hg(liq) \;?$$

(b) Calculate the emf (25°C) of this cell from thermodynamic data in this text.
(c) Estimate the emf of this cell at 18°C.
(d) How much heat would be released if the cell were charged reversibly (at 25°C) with 2 faradays of electricity?
(e) How much heat would be released if the cell were discharged irreversibly (2 faradays, at 25°C) through a short circuit?

8.39   The cell:

$$Pb(s) \,|\, PbBr_2(s) \,|\, CuBr_2(aq, m = 0.01) \,|\, Cu(s)$$

has $\mathscr{E} = 0.442$ volts at 25°.
(a) Calculate $\mathscr{E}^\theta$. (Use the DHG formula to calculate the activity coefficient.)
(b) Calculate the thermodynamic solubility product of $PbBr_2$.

8.40   Use the standard emf (Table 8.10) to calculate the solubility product of AgBr.

8.41   The solubility product of AgSCN is $K_{sp} = 1.16 \times 10^{-12}$ (25°C). Calculate the standard emf of the electrode:

$$SCN^- \,|\, AgSCN(s) \,|\, Ag(s)^{\cdot}$$

(The actual value is 0.0859 V.)

8.42   The solubility product of $BaSO_4$ is 1.08×10^{-10} at 25°C. Calculate the standard emf of the cell:

$$Pt, H_2(g, 1 \text{ atm}) \,|\, H_2SO_4(aq) \,|\, BaSO_4(s) \,|\, Ba(s)$$

Is this a practical cell?

8.43   Calculate the ion product of water from the standard emf of $OH^- \,|\, H_2 \,|\, Pt$ (Table 8.10). (Devise a hypothetical cell and give the cell reaction.)

8.44   Iodine (I_2) has a solubility of 1.33×10^{-3} moles/dm³ in water (25°C). Use the standard emf's:

$$I^- \,|\, I_2(s) \,|\, Pt: \qquad \mathscr{E}^\theta = 0.5355 \text{ V}$$
$$I_3^-, I^- \,|\, Pt: \qquad \mathscr{E}^\theta = 0.5365 \text{ V}$$

to calculate the concentrations of I^- and I_3^- in a saturated solution.

8.45   (a) Use standard potentials (Table 8.10) to calculate the equilibrium constant of the reaction (25°C):

$$Cu^{2+} + Cu(s) = 2 \, Cu^+$$

(b) If a 1-molal solution of Cu^{2+} is equilibrated over copper, what will be the equilibrium concentration of Cu^+?

8.46 Calculate the fugacity coefficient (γ) for H_2(gas) at the pressures given, from the emf of the cell (at 25°C):

$$\text{Pt, } H_2(g, P) \,|\, HCl(aq, m = 0.1) \,|\, Hg_2Cl_2(s) \,|\, Hg(liq)$$

P (H_2, atm)	$\mathscr{E}$ (volts)
1.0	0.3990
110.2	0.4596
556.8	0.4844
1035.2	0.4975

8.47 Calculate the emf of the cell (using the ideal gas approximation):

$$\text{(Pt-Ir), } Cl_2(g, 1 \text{ atm}) \,|\, HCl(aq, m = 0.1) \,|\, Cl_2(g, 0.3 \text{ atm}), \text{(Pt-Ir)}$$

8.48 (a) Devise a cell that could be used to measure activity coefficients of $ZnCl_2$. (b) What is the Nernst equation for your cell?

8.49 The Ruben cell, which is frequently used in applications such as hearing aids, has a cell reaction:

$$Zn(s) + HgO(s) + H_2O + 2\,KOH = Hg(l) + K_2Zn(OH)_4$$

(a) Diagram this cell and calculate its standard emf.
(b) Assuming $\mathscr{E} = \mathscr{E}^\theta$, calculate the minimum weight of a cell that would last two weeks with a load of 5 mW.

8.50 Show that a concentration cell with a salt that has ν ions per "molecule" and transfers n equivalents of electricity per mole has an emf:

$$\mathscr{E} = \pm \frac{\nu t_i RT}{n\mathscr{F}} \ln \frac{a_{\pm 1}}{a_{\pm 2}}$$

where $t_i = t_+$ (+ sign) if the cell is anion reversible and $t_i = t_-$ (− sign) if the cell is cation reversible.

8.51 The emf of the cell:

$$Zn(Hg) \,|\, ZnI_2(m = 0.3, \gamma_\pm = 0.564) \,|\, ZnI_2(m = 0.1, \gamma_\pm = 0.581) \,|\, Zn(Hg)$$

is −0.02689 V (at 25°C). Calculate the cation transference number.

8.52 Calculate the emf of the cell:

$$\text{(Pt-Ir), } Cl_2(g, 1 \text{ atm}) \,|\, NaCl(m = 0.1) \,|\, NaCl(m = 0.02) \,|\, Cl_2(g, 1 \text{ atm}), \text{(Pt-Ir)}$$

8.53 Estimate the liquid-junction potential for:

$$KOH(c = 0.1) \,|\, KCl(c = 0.2)$$

(Use conductivities at infinite dilution, Λ°.)

8.54 Calculate the junction potential for:

$$NH_4Cl(c = 0.1) \,|\, LiCl(c = 0.1)$$

8.55 Devise a cell that could be used to measure the Nernst potential of:

$$Zn \,|\, ZnSO_4(aq) \,\|\, CuSO_4(aq) \,|\, Cu$$

(that is, a cell without a liquid junction).

The jaws of darkness do devour it up:
So quick bright things come to confusion.

– Shakespeare

9

Transport Properties

In the first eight chapters we have, for the most part, considered material properties which are uniform throughout the system. If there are no external forces, uniformity is what is to be expected at equilibrium. If some property such as temperature or concentration is not uniform, then there must be a flow (or transport) of that property in a direction which will, in time, make it uniform throughout. The properties related to such fluxes are heat conductivity and diffusion as well as viscosity.

The theory of transport properties is simplest for gases and, for such a case, the unifying phenomenon is the collisions of molecules. This topic, which we now explore, is also vital for understanding rates of chemical reactions (Chapter 10).

9.1 Molecular Collisions

The kinetic theory of gases (Chapter 1) provides the following results regarding the relationship between the molecular velocities and temperature: The distribution function for speeds (v) is:

$$F(v)\,dv = 4\pi\left(\frac{m}{2\pi kT}\right)^{3/2} e^{-mv^2/2kT} v^2\,dv \tag{9.1}$$

The average, rms average, and most probable speeds are:

$$\bar{v} = \sqrt{\frac{8RT}{\pi M}} \tag{9.2}$$

$$u = \sqrt{\frac{3RT}{M}} \tag{9.3}$$

$$v_p = \sqrt{\frac{2RT}{M}} \tag{9.4}$$

The number density of molecules is:

$$n^* = \frac{N}{V} = \frac{PL}{RT} \tag{9.5}$$

The simplest model for molecular collisions is to approximate the molecules as hard spheres with diameter σ. Two hard-sphere molecules with *diameter σ* (radius $\sigma/2$) will collide if they approach to the distance of σ. A molecule of diameter σ moving through space at a speed v will sweep out a cylindrical volume with *radius σ* (cross-sectional area $\pi\sigma^2$) and length (in time Δt) $v\Delta t$ and, hence:

$$\text{volume swept out in } \Delta t = \pi\sigma^2\, v\Delta t$$

Any molecule whose *center* is in this volume will be hit (Fig. 9.1); if the density of molecules is n^*, the collision frequency can be expected to be:

$$\text{collision frequency} = \pi\sigma^2 v n^*$$

This much should be obvious: the collision frequency is directly proportional to the molecular cross-sectional area, its speed, and the density of targets. What may be less

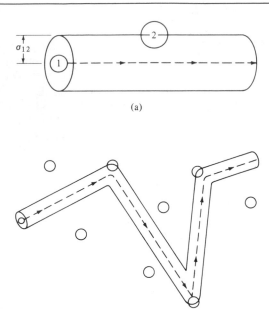

Figure 9.1 Molecular collisions. A molecule will collide with any other molecule whose center lies within a cylinder whose radius is the sum of its radius and that of the other molecule. (b) The mean free path is the average distance traveled by a molecule between collisions.

obvious is that, when this expression is averaged over all velocities, we do not want the average velocity but, rather, the average *relative* velocity; the reason is that the molecule is not hitting a fixed target (as for wall collisions, discussed in Section 1.5) but a moving target — another molecule.

Relative Velocity

The calculation of the average relative speed of a pair of hard spheres uses the center-of-mass (COM) coordinate system. This topic is discussed in Appendix III. Here, all we shall do is to briefly outline the results for the specific case of two identical spheres — that is, spheres with the same mass and diameter.

Two molecules with vector velocities $\mathbf{v}_1$ and $\mathbf{v}_2$ will have a relative velocity:

$$\mathbf{v}_{12} = \mathbf{v}_1 - \mathbf{v}_2$$

The relative speed (the magnitude of this vector):

$$v_{12} = |\mathbf{v}_1 - \mathbf{v}_2|$$

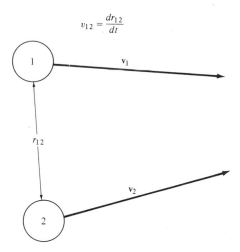

Figure 9.2 Relative velocity. The relative velocity is the rate at which two moving objects approach each other.

is the rate at which they are approaching each other (Fig. 9.2). At the same time, the COM of the pair is moving with a velocity:

$$\mathbf{V}_{12} = \frac{\mathbf{v}_1 + \mathbf{v}_2}{2}$$

The calculation of the average relative velocity requires the joint velocity distribution function $F(v_1, v_2)$; since the velocities of different molecules are independent variables, we can set this quantity equal to a product of the one-particle functions:

$$F(v_1, v_2) = F(v_1)F(v_2)$$

Then, the calculation of the relative speed is [using Eq. (9.1)]:

$$\langle v_{12} \rangle = \iint v_{12} F(v_1)F(v_2)\, dv_1\, dv_2$$

$$= (4\pi)^2 \left(\frac{m}{2\pi kT}\right)^3 \iint_0^\infty v_{12} e^{-b(v_1^2 + v_2^2)} v_1^2\, dv_1\, v_2^2\, dv_2$$

(with $b = m/2kT$). The transformation into the COM system allows the following substitutions:

$$v_1^2\, dv_1\, v_2^2\, dv_2 = V_{12}^2\, dV_{12} v_{12}^2\, dv_{12}$$

$$v_1^2 + v_2^2 = 2V_{12}^2 + \left(\frac{1}{2}\right) v_{12}^2$$

$$(b = m/2kT)$$

With these substitutions, the relative speed becomes:

$$\langle v_{12} \rangle = (4\pi)^2 \left(\frac{m}{2\pi kT} \right)^3 \int_0^\infty e^{-mv_{12}^2/4kT} v_{12}^3 \, dv_{12} \int_0^\infty e^{-mV_{12}^2/kT} V_{12}^2 \, dV_{12}$$

These integrals are similar to those we encountered earlier; the result is readily shown to be:

$$\langle v_{12} \rangle = \sqrt{\frac{16kT}{\pi m}} \tag{9.6}$$

Comparison to Eq. (9.2) shows that the average relative speed is related to the average speed as:

$$\langle v_{12} \rangle = \sqrt{2}\,\bar{v} \tag{9.7}$$

Collisions of Like Molecules: Mean Free Path

The collision frequency (z) is the number of collisions suffered by an individual molecule per unit time and, as discussed above, is proportional to the collision cross section ($\pi\sigma^2$), the density of targets (n^*), and the average relative velocity. For collisions of molecules with equal masses, this is:

$$z = \sqrt{2}\,\bar{v}\pi\sigma^2 n^* \tag{9.8}$$

Another very important parameter for the discussion of transport properties of gases is the *mean free path* (λ)—the average distance traveled by a molecule between collisions. An average molecule will travel a distance:

$$\Delta x = \bar{v}\Delta t$$

in a time interval Δt, and suffer a number of collisions:

$$z\,\Delta t$$

during that time. The mean free path is, therefore:

$$\lambda = \frac{\bar{v}}{z} \tag{9.9a}$$

Using Eq. (9.8), this becomes:

$$\lambda = \frac{1}{\sqrt{2}\,\pi\sigma^2 n^*} \tag{9.9b}$$

or, using the ideal gas law for n^* [Eq. (9.5)]:

$$\lambda = \frac{RT}{\sqrt{2}\,PL\pi\sigma^2} \tag{9.9c}$$

Example: Calculate the average velocity, collision frequency, and mean free path for Ar ($\sigma = 0.34$ nm) at STP. The atomic weight of Ar is 39.948 g = 0.039948 kg.

$$\bar{v} = \left[\frac{8(8.3143 \text{ J/K})(273.15 \text{ K})}{\pi(0.039948 \text{ kg})} \right]^{1/2}$$

$$= 380.48 \text{ m/s}$$

$$n^* = \frac{(101\,325 \text{ Pa})(6.02217 \times 10^{23})}{(8.3143 \text{ J/K})(273.15 \text{ K})} = 2.6868 \times 10^{25} \text{ m}^{-3}$$

$$z = \sqrt{2}\,\pi(0.34 \times 10^{-9} \text{ m})^2(380.48 \text{ m/s})(2.6868 \times 10^{25} \text{ m}^{-3})$$

$$= 5.25 \times 10^9 \text{ s}^{-1}$$

The mean free path would normally be calculated with either Eq. (9.9b) or (9.9c), but since the average velocity and collision frequency are calculated already, Eq. (9.9a) is easier in this case:

$$\lambda = \frac{380.48 \text{ ms}^{-1}}{5.25 \times 10^9 \text{ s}^{-1}} = 7.25 \times 10^{-8} \text{ m} = 72.5 \text{ nm}$$

In this case, the mean free path is about 200 molecular diameters. ∎

Collisions of Unlike Molecules

The mean relative speed of two molecules (A and B) with unequal masses is related to the reduced mass (Appendix III):

$$\mu = \frac{m_A m_B}{m_A + m_B} \tag{9.10a}$$

In this equation, m is the mass of an individual molecule (its molecular weight divided by Avogadro's number). It is somewhat more convenient to deal with the molar masses (M) and use:

$$L\mu = \frac{M_A M_B}{M_A + M_B} \tag{9.10b}$$

The average relative velocity can be shown to be:

$$\langle v_{AB} \rangle = \left[\frac{8kT}{\pi\mu} \right]^{1/2} = \left[\frac{8RT}{\pi(L\mu)} \right]^{1/2} \tag{9.11}$$

Exercise: Show that if $m_A = m_B$, Eq. (9.11) reduces to Eq. (9.6) (which was derived specifically for equal masses). ∎

If the molecules (A and B) have unequal diameters (σ_A and σ_B), the distance between their centers on collision will be:

$$\sigma_{AB} = \tfrac{1}{2}(\sigma_A + \sigma_B) \tag{9.12}$$

Then, using the same arguments as earlier, the frequency of collision by each A molecule with B molecules will be:

$$z_{A:B} = \pi\sigma_{AB}^2 \langle v_{AB} \rangle n_B^* \tag{9.13a}$$

and the frequency of collisions by each B molecule with A molecules will be:

$$z_{B:A} = \pi \sigma_{AB}^2 \langle v_{AB} \rangle n_A^* \qquad \textbf{(9.13b)}$$

Total Collisions

In chemical kinetics, the rate of a reaction is usually characterized in terms of changes in the concentration of a reacting species in time — that is, the number of molecules which react *per unit volume*. Therefore an important quantity for interpreting such reaction rates is the total collisions suffered by all molecules in a given volume (denoted Z).

For like molecules, if there are n_A^* molecules (type A) in a given volume, and each suffers z_A collisions per unit time, the total number of collisions is:

$$Z_{AA} = \frac{z_A n_A^*}{2}$$

(divided by 2 because each collision involves two molecules). With Eq. (9.8), this becomes:

$$Z_{AA} = \frac{\pi \bar{v}_A \sigma_A^2 (n_A^*)^2}{\sqrt{2}} \qquad \textbf{(9.14)}$$

The total collisions of A-type molecules with B-type molecules will be:

$$Z_{AB} = z_{A:B} n_A^* = z_{B:A} n_B^*$$

Using Eq. (9.13), this becomes:

$$Z_{AB} = \pi \sigma_{AB}^2 \langle v_{AB} \rangle n_A^* n_B^* \qquad \textbf{(9.15)}$$

9.2   Random Walks

It is clear from the various calculations of molecular speeds that molecules (at or around room temperature) are moving *very* fast; but are they getting anywhere? A molecule in the air at room temperature and atmospheric pressure is traveling ~ 500 m/s, but it also suffers $\sim 10^9$ collisions in each second. After each collision the direction of travel changes, and the net headway in one second is of the order of millimeters! With the large numbers of molecules that would be found in the smallest perceptible bit of matter, such a process is best treated as a random one. In this section we introduce a simple mathematical model — the random walk — which can be used to describe this and many other random molecular processes. For the moment we shall consider only the application of this model to movement (or *diffusion*) of molecules through a gas; but is has many other applications.

To keep the mathematics simple, we shall deal only with random walks in an unlimited, single dimension; Chandrasekhar (ref. 6) provides a thorough discussion of the entire problem. A very colorful analogy to a one-dimensional random walk has been used by George Gamow. A drunken sailor is leaning on a lamp post; he starts walking, up or down the street, randomly taking steps in either direction. After 100 such steps, how far is he likely to be from the post? (The answer, as we shall soon learn, is approximately 10 steps.)

There is a simpler model for this process. Place a marker on the center mark of a ruler; then toss a coin to decide whether to move the marker to the right or left. Let us label the starting position as $m = 0$:

$$-6 \quad -5 \quad -4 \quad -3 \quad -2 \quad -1 \quad \ 0 \quad \ 1 \quad \ 2 \quad \ 3 \quad \ 4 \quad \ 5 \quad \ 6$$

$$-m \ldots \bullet\!\!-\!\!\!-\!\!\bullet\!\!-\!\!\!-\!\!\bullet\!\!-\!\!\!-\!\!\bullet\!\!-\!\!\!-\!\!\bullet\!\!-\!\!\!-\!\!\bullet\!\!-\!\!\!-\!\!\bullet\!\!-\!\!\!-\!\!\bullet\!\!-\!\!\!-\!\!\bullet\!\!-\!\!\!-\!\!\bullet\!\!-\!\!\!-\!\!\bullet\!\!-\!\!\!-\!\!\bullet \ldots m$$

For each toss of the coin, we move one step positive for heads or one step negative for tails. After N tosses of the coin, there are p positive results (heads) and q negative results (tails). The current position is $m = p - q$; we also have $N = p + q$ and, therefore:

$$p = \frac{N + m}{2}, \qquad q = \frac{N - m}{2} \tag{9.16}$$

The probability (W) of being on step m after N tosses (steps) is given by the theory of permutations and combinations as:

$$W(m, N) = \left(\frac{1}{2}\right)^N \frac{N!}{p!q!} \tag{9.17}$$

This may also be written as:

$$W(m, N) = \left(\frac{1}{2}\right)^N C_p^N \tag{9.18}$$

with the **binomial coefficient:**

$$C_p^N = \frac{N!}{p!(N - p)!} \tag{9.19}$$

The function $W(m, N)$ is graphed in Fig. 9.3. For each trial, the chance of either heads or tails is $\frac{1}{2}$; so after N trials, the probability of any particular sequence of heads and tails is $(\frac{1}{2})^N$. The binomial coefficient simply gives the number of sequences which will land you on a particular step. For example, after 4 tosses there is only one sequence which lands you on step 4, namely *HHHH*; therefore $W(4, 4) = (\frac{1}{2})^4 = \frac{1}{16}$. However there are $4!/3!1!$ sequences which will land you on step 2, namely *HHHT*, *HHTH*, *HTHH*, *THHH*; each of these sequences has $(\frac{1}{2})^4 = \frac{1}{16}$ probability, and $W(2, 4) = \frac{4}{16}$.

To find out how far the coin will probably move, one might think to calculate:

$$\langle m \rangle = \sum_{m=0}^{N} m W(m, N) \tag{9.20}$$

But this quantity is going to be zero, perhaps that is obvious; all this tells us is that the marker is equally likely to go either way. It will be better to find the rms distance

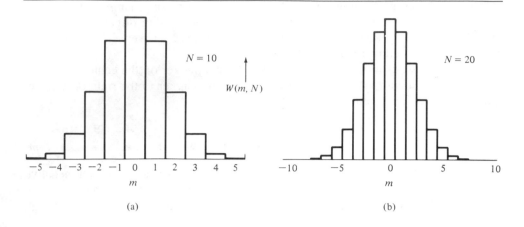

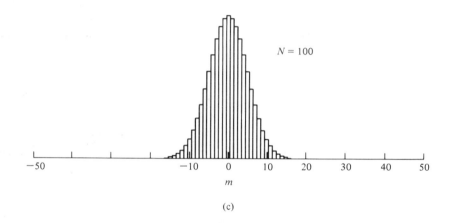

Figure 9.3 The random-walk probability distribution function. The function $W(m, N)$ [Eq. (9.17)] is shown for N = 10, 20, and 100. Since the number of steps (N) is even, the height of the bar for even values of m represents the probability that the walk will end on that step. Note that the width of this function is approximately $\sqrt{N}$.

to answer the "how far" question; to do this, we need:

$$\langle m^2 \rangle = \sum_{m=0}^{N} m^2 W(m, N) \tag{9.21}$$

To evaluate these sums, we use a very important property of the binomial coefficients, namely, that they are the coefficients in an expansion of $(1 + x)^N$:

$$(1 + x)^N = \sum_{p=0}^{N} x^p C_p^N \tag{9.22}$$

(The variable x is just a dummy variable, as we shall soon see.)

Exercise: The reader should confirm the expansions:

$$(1 + x)^2 = 1 + 2x + x^2$$
$$(1 + x)^3 = 1 + 3x + 3x^2 + x^3$$
$$(1 + x)^4 = 1 + 4x + 6x^2 + 4x^3 + x^4$$

These should be done directly (by polynomial multiplication) *and* using Eq. (9.22), unless the point is obvious.   ∎

Now, take the derivative of Eq. (9.22) with respect to x:

$$N(1 + x)^{N-1} = \sum_{p=1}^{N} px^{p-1}C_p^N$$

and let $x = 1$:

$$N \cdot 2^{N-1} = \sum_{p=1}^{N} pC_p^N$$

Divide by 2^N to get:

$$\frac{N}{2} = \sum_{p=1}^{N} p\left(\frac{1}{2}\right)^N C_p^N = \sum pW(m, N)$$

But the right-hand side of this expression is just $\langle p \rangle$, the average number of positive steps; from Eq. (9.16):

$$p = \frac{N + m}{2}, \qquad m = 2p - N$$

With $\langle p \rangle = N/2$, we see that $\langle m \rangle = 0$, as expected.

It gets more interesting if Eq. (9.22) is differentiated *twice* with respect to x:

$$N(N - 1)(1 + x)^{N-2} = \sum_{p=1}^{N} p(p - 1)x^{p-2}C_p^N$$

Again we let $x = 1$ and divide by 2^N; after a little algebra:

$$\frac{N(N - 1)}{4} = \sum_{p=1}^{N} p^2\left(\frac{1}{2}\right)^N C_p^N - \sum_{p=1}^{N} p\left(\frac{1}{2}\right)^N C_p^N$$

$$= \sum p^2 W(m, N) - \sum pW(m, N)$$

$$= \langle p^2 \rangle - \langle p \rangle$$

We have already seen that $\langle p \rangle = N/2$, so this becomes:

$$\frac{N^2}{4} - \frac{N}{4} = \langle p^2 \rangle - \frac{N}{2}, \qquad 4\langle p^2 \rangle = N^2 + N$$

Again from Eq. (9.16):

$$4p^2 = (N + m)^2 = N^2 + 2mN + m^2$$

Since $\langle m \rangle = 0$,

$$4\langle p^2 \rangle = N^2 + \langle m^2 \rangle$$

Then, using the result above for $4\langle p^2 \rangle$:

$$N^2 + N = N^2 + \langle m^2 \rangle$$

Finally:

$$\langle m^2 \rangle = N$$

$$(\Delta m)_{\text{rms}} = \sqrt{N} \qquad\qquad (9.23)$$

This is a very important result, which arises time and again in physical theory. For example, if you are trying to remove random noise from a repetitive signal, you can do so by adding the signals of successive measurements. The signal size will, in N measurements, be multiplied by N while the random noise will increase by only $\sqrt{N}$; therefore, after N measurements, the signal-to-noise ratio will improve by a factor of $\sqrt{N}$. Also, a polymer with N flexible links of length λ will have a mean end-to-end distance of $\sqrt{N}\,\lambda$ (cf. Problem 5.3).

Now let's go back to the original question: How far does a molecule in a gas move in a given time? The molecule takes a new "step" after each collision and, therefore, zt steps in time t. If we assume that each step has the mean length λ (the mean free path) the net distance traveled will be $\lambda \langle m \rangle$, or:

$$(\Delta x)_{\text{rms}} = \sqrt{N}\,\lambda = \sqrt{zt}\,\lambda \qquad\qquad (9.24)$$

Using the numbers calculated earlier (p. 429) for Ar at STP (x in meters, t in seconds):

$$(\Delta x)_{\text{rms}} = \sqrt{5.25 \times 10^9}\,(7.25 \times 10^{-8})\sqrt{t} = (5.25 \times 10^{-3})\sqrt{t}$$

In one second ($t = 1$) the net headway is 5.25 mm; in one minute ($t = 60$) it is 4.06 cm; in one hour ($t = 3600$) 31.5 cm. (We assume, of course, no convection currents or breezes.)

Small-Step Diffusion

Next we shall consider the case for which, in a given time interval, the molecule takes a very large number of small steps; the example above (Ar at STP) certainly fits this description. In such a case, the total number of steps (N) will be much greater than the net steps (m) for all the more probable values of m (see Fig. 9.3 for $N = 100$).

We begin by taking the logarithm of W [Eq. (9.17)]:

$$\ln W = \ln N! - \ln p! - \ln q! - N \ln 2$$

Next, Stirling's approximation is used for each of the factorials:

$$\ln N! = \tfrac{1}{2} \ln (2\pi) + (N + \tfrac{1}{2}) \ln N - N$$

$$\ln W = N \ln N + \tfrac{1}{2} \ln N - N + \tfrac{1}{2} \ln 2\pi - N \ln 2$$

$$- p \ln p - \tfrac{1}{2} \ln p + p - \tfrac{1}{2} \ln 2\pi$$

$$- q \ln q - \tfrac{1}{2} \ln q + q - \tfrac{1}{2} \ln 2\pi$$

Next, use the definitions:

$$p = \frac{N + m}{2}, \qquad q = \frac{N - m}{2}$$

(remembering that $p + q = N$) to derive (after some algebra):

$$\ln W = \frac{1}{2} \ln \left(\frac{N}{2\pi} \right) + N \ln \left(\frac{N}{2} \right)$$
$$- \frac{1}{2} (N + m + 1) \ln \left[\left(\frac{N}{2} \right) (1 + \varepsilon) \right]$$
$$- \frac{1}{2} (N - m + 1) \ln \left[\left(\frac{N}{2} \right) (1 - \varepsilon) \right]$$

with the definition:

$$\varepsilon = \frac{m}{N}$$

Since we will be considering the case of $N >> m$, then $\varepsilon << 1$.

Next, using:

$$\ln \left[\left(\frac{N}{2} \right) (1 + \varepsilon) \right] = \ln \left(\frac{N}{2} \right) + \ln (1 + \varepsilon)$$

(and similarly for the $1 - \varepsilon$ term), it can be shown (after a *lot* of algebra) that:

$$\ln W = \ln \left(\frac{2}{\sqrt{2\pi N}} \right) - \frac{1}{2} (N + 1) [\ln (1 + \varepsilon) + \ln (1 - \varepsilon)]$$
$$- \frac{1}{2} m [\ln (1 + \varepsilon) - \ln (1 - \varepsilon)]$$

Next, the logarithms of the second and third terms can be expanded in a Taylor series (for $\varepsilon << 1$):

$$\ln (1 + \varepsilon) = \varepsilon - \frac{\varepsilon^2}{2} + \cdots$$

$$\ln (1 - \varepsilon) = -\varepsilon - \frac{\varepsilon^2}{2} - \cdots$$

Neglecting higher terms, the equation above becomes:

$$\ln W = \ln \left(\frac{2}{\sqrt{2\pi N}} \right) - \frac{1}{2} (N + 1) \left(\varepsilon - \frac{\varepsilon^2}{2} - \varepsilon - \frac{\varepsilon^2}{2} \right)$$
$$- \frac{1}{2} m \left(\varepsilon - \frac{\varepsilon^2}{2} + \varepsilon + \frac{\varepsilon^2}{2} \right)$$

After a little more algebra, and using:

$$\varepsilon = \frac{m}{N}$$

this can be shown to give:

$$\ln W = \ln \left(\frac{2}{\sqrt{2\pi N}}\right) - \frac{m^2}{2N} + \frac{m^2}{2N^2}$$

The last term on the right-hand side can be neglected, since N is very large, and finally we obtain an equation for W:

$$W(m, N) = \frac{2}{\sqrt{2\pi N}} e^{-m^2/2N} \tag{9.25}$$

Figure 9.4 shows this function (called a Gaussian) for $N = 50$ together with the exact values as calculated by Eq. (9.18).

Equation 9.25 represents the probability as a continuous function, but this cannot be precisely correct: after an *even* number of steps (N), only *even* values of m are possible, no matter how large N is. Therefore Eq. (9.25) is correct only for steps for which m has the same *parity* (even/odd characteristic) as N; $W = 0$ for odd m when N is even and for even m when N is odd.

Now we wish to find the distribution function for distance (x) — that is, the probability $W(x, t)$ that the particle is in an interval x to $x + dx$ at some time t. We shall do this by assuming a constant step size (the mean free path, λ) and:

$$x = \lambda m$$

Because of the even/odd alternation discussed above, it will be necessary to consider intervals (dx) that span at least two steps — that is:

$$m \longrightarrow m + 2\, dm$$

$$\lambda m \longrightarrow \lambda m + 2\lambda\, dm$$

$$x \longrightarrow x + dx \qquad \text{with } dx = 2\lambda\, dm$$

Then Eq. (9.25) can be used to obtain the distribution function for x:

$$W(x, N)\, dx = W(m, N)\, dm$$

$$W(x, N)\, dx = \frac{1}{\sqrt{2\pi N}} e^{-m^2/2N}(2\, dm)$$

$$= \frac{1}{\sqrt{2\pi N}} e^{-x^2/2N\lambda^2} \left(\frac{dx}{\lambda}\right)$$

Next we assume that the molecule takes z steps per unit time, so:

$$N = zt$$

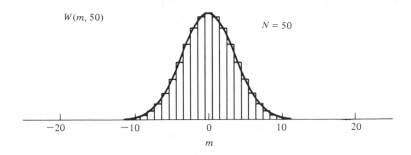

Figure 9.4 The random-walk probability distribution function, $W(m, N)$, for 50 steps. Superimposed on the actual distribution [Eq. (9.17)] is the Gaussian curve given by Eq. (9.25); it can be seen that the approximation [Eq. (9.25)] is reasonably good for 50 steps; it will be better for larger numbers of steps.

This gives:

$$W(x, t)\, dx = \frac{\exp\left(-\dfrac{x^2}{2z\lambda^2 t}\right)}{(2\pi\lambda^2 zt)^{1/2}}\, dx$$

This is generally written in terms of the definition:

$$D \equiv \tfrac{1}{2}z\lambda^2 = \tfrac{1}{2}\bar{v}\lambda \tag{9.26}$$

as

$$W(x, t)\, dx = \frac{1}{\sqrt{4\pi Dt}}\, e^{-x^2/4Dt}\, dx \tag{9.27}$$

Equation (9.27) gives the probability that a molecule which was at $x = 0$ when $t = 0$ will be at a position x at time t. In the next section, the constant D will be identified as the coefficient of diffusion (or *diffusion constant*). This formula is valid only for time intervals $t \gg 1/z$ and for the average probability over an interval of space $\Delta x \gg \lambda$. For a gas such as argon at STP (the example used earlier) it would be quite reasonable to use Eq. (9.27) for $t \sim \mu$s and $\Delta x \sim \mu$m; however, at low pressures its use could be very limited.

Combining the result calculated earlier for the rms distance traveled [Eq. (9.24)] with the definition of D [Eq. (9.26)] gives:

$$x_{\mathrm{rms}} = \langle x^2 \rangle^{1/2} = \sqrt{2Dt} \tag{9.28}$$

This equation was first derived by Einstein (1905, ref. 4) for the Brownian motion of small particles suspended in a stationary liquid. (This paper was an important landmark in the kinetic theory of heat and microscopic motion.)

Exercise: If the probability distribution function of Eq. (9.25) is correctly normalized, its integral over all space (representing the total probability that the molecule is *somewhere*) should be 1. Show that

$$\int_{-\infty}^{\infty} W(x, t)\, dx = 1$$

for any time t.   ■

Example: Calculate the probability that a particular molecule in Ar(gas) at STP will move 100 cm in one hour (3600 s). With the results calculated earlier:

$$D = 0.5\bar{v}\lambda = 0.5(380.48 \text{ m/s})(7.25 \times 10^{-8} \text{ m})$$

$$= 1.38 \times 10^{-5} \text{ m}^2 \text{ s}^{-1} = 0.138 \text{ cm}^2 \text{ s}^{-1}$$

$$4Dt = 1986 \text{ cm}^2$$

$$W(100, 3600) = \frac{1}{\sqrt{\pi(1986)}} \exp\left[-\frac{(100)^2}{1986}\right]$$

$$= 8.2 \times 10^{-5}$$ ■

Exercise (optional): Use Eq. (9.27) to calculate:

$$\langle x^2 \rangle = \int_{-\infty}^{\infty} x^2 W(x, t)\, dx$$

The result should be identical to Eq. (9.28).   ■

Random Walks in Three Dimensions

The probability function for a three-dimensional random walk can be derived by a procedure very similar to that used to derive the distribution function for speeds in Section 1.6 [Eq. (1.36)].

If the motions in the x, y, z directions are statistically independent, then:

$$W(x, y, z, t) = W(x, t)W(y, t)W(z, t)$$

Furthermore, Eq. (9.27) applies to any direction, x, y, or z, with only a change in symbol, so (using $r^2 = x^2 + y^2 + z^2$):

$$W(x, y, z, t)\, dx\, dy\, dz = (4\pi Dt)^{-3/2} e^{-r^2/4Dt}\, dx\, dy\, dz$$

This is the probability that the molecule will be found in a volume $dx\, dy\, dz$ at a point (x, y, z), a distance r from the origin. This must be summed over all directions using (Appendix III):

$$dx\, dy\, dz = r^2\, dr\, \sin\theta\, d\theta\, d\phi$$

with

$$\int_0^{2\pi} d\phi \int_0^{\pi} \sin\theta\, d\theta = 4\pi$$

Finally we get the probability that the molecule has moved a distance r in any direction:

$$W(r, t)\, dr = \frac{4\pi}{(4\pi Dt)^{3/2}} e^{-r^2/4Dt} r^2\, dr$$

$$= \frac{4}{\sqrt{\pi}} e^{-\rho^2} \rho^2\, d\rho \qquad (9.29)$$

(with $\rho = r/\sqrt{4Dt}$). [Compare Eq. (1.40b).]

Exercise: Calculate the rms distance traveled in a three-dimensional random walk.

$$\langle r^2 \rangle = \int_0^\infty r^2 W(r, t)\, dr$$

$$= \frac{4}{\sqrt{\pi}} (4Dt) \int_0^\infty \rho^4 e^{-\rho^2}\, d\rho$$

This integral is found on most standard tabulations.

$$\langle r^2 \rangle = \frac{4}{\sqrt{\pi}} (4Dt) \frac{3\sqrt{\pi}}{8} = 6Dt$$

$$\langle r^2 \rangle^{1/2} = \sqrt{6Dt} \qquad (9.30)$$

∎

9.3 Diffusion

Consider a solute whose concentration (c) is not uniform but varies in at least one direction (x). Since the concentration must be uniform at equilibrium, there will be a flow (flux) of material from regions with high concentration to regions with a lower concentration. Consider a plane (Fig. 9.5) with area A and width dx; assume that the concentration at x is c and at $x + dx$ is $c - dc$. The flow of material through the plane will be proportional to its area and the concentration difference per unit length,

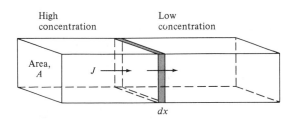

Figure 9.5 Diffusion. The flux J is the amount of material crossing a plane per unit area per unit time.

Albert Einstein (1879–1955)

Einstein is, of course, famed for his theory of relativity, which made him a household name and the most recognizable scientist of all time. He was the original "longhair" and the first "pop" scientist, but his fame among his fellow scientists was no less than among the hoi polloi.

In his early work, Einstein was interested in the kinetic-molecular theory, particularly in the physical significance of Boltzmann's constant. He showed that it was related to the degree of fluctuation of the energy and to the black-body radiation constants. His 1905 paper on the motion of suspended particles showed how kinetic theory could account for the Brownian motion (although he was apparently unaware that this phenomenon had been observed previously) and laid the theoretical basis for Perrin's work (described at the end of this chapter), which confirmed and removed all doubt as to the validity of the kinetic-molecular theory.

Einstein's religion made his relativity theory a target for anti-Semitic attacks (he was in Berlin at the time), and even his person was threatened with physical violence. However, he remained in Berlin until 1933, when Hitler's rise to power caused him to depart for America and the Institute for Advanced Study in Princeton. He was the forerunner of a flood of refugee scientists who made the United States, overnight, into the leading scientific nation on earth. His 1939 letter to President Roosevelt alerted the government to the potential of nuclear weapons and initiated their development; but this was not typical of Einstein, who, throughout his life, espoused pacifist and internationalist causes.

$-dc/dx$. The diffusional flux (J) is defined as:

$$J = \frac{1}{A}\frac{dn}{dt} = -D\frac{dc}{dx} \tag{9.31}$$

where the proportionality factor D is the *diffusion constant*. If c is in moles per cm^3 and t in seconds, unit analysis will show that the cgs units of D are cm^2/s (m^2/s in SI).

Equation (9.31) is called **Fick's first law of diffusion.** Since the net flow will be in the direction of decreasing concentration, the negative sign in Fick's law makes the diffusion constant a positive number.

A layer of the fluid (between x and $x + dx$) will have material entering from one side and leaving from the other. The net change of the concentration of this region in time will be:

$$\frac{\partial c}{\partial t} = \frac{1}{dx}[J(x) - J(x + dx)] = -\frac{\partial J}{\partial x}$$

Using Eq. (9.31), this becomes:

$$\frac{\partial c}{\partial t} = \frac{\partial}{\partial x}\left(D\frac{\partial c}{\partial x}\right) \tag{9.32}$$

This is **Fick's second law of diffusion.** Often it is assumed that D is independent of c (i.e., of x), and Eq. (9.32) can be written as:

$$\frac{\partial c}{\partial t} = D\frac{\partial^2 c}{\partial x^2} \tag{9.33}$$

If the fluid has a concentration which is nonuniform in more than a single direction, then Fick's law for $c(x, y, z, t)$ is:

$$\frac{\partial c}{\partial t} = D\,\nabla^2 c \tag{9.34}$$

where the **Laplacian** is:

$$\nabla^2 = \frac{\partial^2}{\partial x^2} + \frac{\partial^2}{\partial y^2} + \frac{\partial^2}{\partial z^2} \tag{9.35}$$

This and other operators are discussed in Chapter 11.

Equation (9.33) is an equation whose solution is discussed in many mathematics and engineering texts. In general, the nature of the solution depends on the *boundary condition;* in the present case, the boundary condition is the initial ($t = 0$) concentration, $c(x, 0)$, at various parts of the fluid. Solving this equation is beyond the scope of this book, but solutions will be presented and discussed for two particular cases. (In both cases it is assumed that D is independent of c.)

Instantaneous Point Source

Suppose that, at $t = 0$, all the material is concentrated in a single plane at $x = 0$. [Actually this is physically impossible if the mathematical definition of a plane — infinitesimal width — is used. But we require only that the material be concentrated in some small region of width Δx that is smaller than any region to be sampled later. Then $c(x)$ means the average concentration in a region of width Δx about the point x.] The solution of Eq. (9.33) is, for this case:

$$c(x, t) = \frac{c_0 \Delta x}{2\sqrt{\pi Dt}}e^{-x^2/4Dt} \tag{9.36}$$

where c_0 is the concentration in the initial region (width Δx about $x = 0$ at $t = 0$).

Figure 9.6 shows the function of Eq. (9.36) graphically for various values of Dt. Note that if $D \cong 10^{-5}$ cm^2 s^{-1} (a typical value for small molecules in a nonviscous liquid), and the scale for x is taken to be 1 cm to -1 cm, the times represented by the curves are (respectively) 10 s, 100 s, 10^3 s $= 16.7$ min, 10^4 s $= 2.78$ hours.

The reader may have noted the similarity between Eqs. (9.36) and (9.27). The diffusional process described above can be considered as a situation wherein each molecule that was at $x = 0$ at $t = 0$ undertakes an independent random walk.

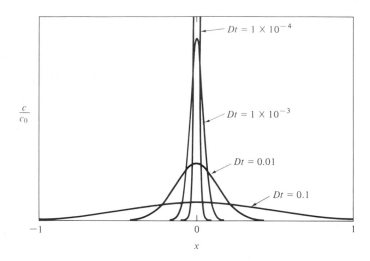

Figure 9.6 Concentration vs. distance for diffusion from a point source. At $t = 0$, the "curve" would be a point infinity at $x = 0$; the curve for $Dt = 1 \times 10^{-4}$ is off-scale at the center.

The average number to be found at some point x after time t will simply be the probability that any individual is there [as expressed by Eq. (9.27)] times the number of individuals that started. This justifies our identification of D [Eq. (9.28)] as the diffusion constant.

The earlier derivation made a number of assumptions (such as uniform steps) which are unnecessary in a rigorous discussion of diffusion in gases. The rigorous result for the diffusion coefficient of a hard-sphere gas is:

$$D = 0.599\lambda\bar{v} \tag{9.37}$$

[compare Eq. (9.28)].

Step-Function Source

From the point of view of measuring diffusion coefficients, a far more practical boundary condition is to have all the diffusing material (solute) on one side of a boundary at $x = 0$, with pure solvent on the other side:

$$c = c_0 \quad \text{for} \quad x < 0 \quad \text{at} \quad t = 0$$
$$c = 0 \quad \text{for} \quad x > 0 \quad \text{at} \quad t = 0$$

The solution of Eq. (9.33) for this initial condition is:

$$c(x, t) = c_0\left(\frac{1}{2} - \frac{1}{\sqrt{\pi}}\int_0^{\xi} e^{-\xi^2}\, d\xi\right) \tag{9.38}$$

with $\xi \equiv x/\sqrt{4Dt}$.

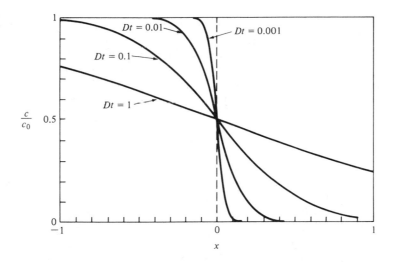

Figure 9.7 Concentration vs. distance for diffusion from an infinite plane source. At $t = 0$, the profile is a step function at $x = 0$ (dashed line), with height C_0.

Figure 9.7 shows $c(x)$ for various values of Dt. If the range of x on this graph were -1 cm $< x < 1$ cm, and $D = 10^{-5}$ cm^2 s^{-1}, then the times for each curve would be (respectively): 100 s, 10^3 s $= 16.7$ min, 10^4 s $= 2.78$ hr, 10^5 s $= 27.8$ hr.

Note that these equations presume that there is no stirring, convection, or any other driving force for mixing such as temperature gradients, density gradients, or pressure gradients (in the case of a gas). The dilution of a solution may involve some heat effect, and this could cause complications.

Self- and Interdiffusion

The preceding discussion generally assumed a situation which was nonuniform in a single material, which then diffused in an otherwise homogeneous medium; this is reasonable for the case of a solute in a dilute solution.

Suppose we had two miscible fluids (A and B) which, separated at $t = 0$, could diffuse into each other. In such a case, Eq. (9.31) (*et seq.*) could be used with the **mean interdiffusion constant:**

$$D_{AB} = X_A D_A + X_B D_B \qquad (9.39)$$

where D_A is the diffusion constant for A in B and D_B is the diffusion constant for B in A. (This statement is valid only to the extent that the diffusion constants are independent of concentration; the actual situation is much more complicated.)

The diffusion constant of a fluid in itself is called the *self-diffusion constant* — also called the *tracer diffusion constant,* since it can be measured by using isotopes. Nuclear magnetic resonance (NMR) can be used to measure self-diffusion constants for any material containing an abundant spin-$\frac{1}{2}$ nucleus (e.g., ^{1}H, ^{19}F, ^{31}P).

Diffusion in Solids

Ordinarily one thinks of diffusion as a phenomenon that occurs in fluids, but diffusion in solids is known.

Light gases such as hydrogen can diffuse through metals by moving interstitially. This is usually quite slow at room temperature but can be significant at elevated temperatures. For example, the diffusion coefficient of H_2 in γ iron is 1.5×10^{-5} cm^2 s^{-1} at 500°C, and 1.9×10^{-4} cm^2 s^{-1} at 1000°C. Comparison to Table 9.1 will show that this diffusion is as fast as for small molecules in ordinary liquids. The diffusion of carbon in γ iron is also interstitial, but much slower: $D = 1.8 \times 10^{-8}$ cm^2 s^{-1} at 500°C, 3×10^{-6} cm^2 s^{-1} at 1000°C.

Metals may also diffuse into metals; this process is generally by substitution of the diffusing metal for the host metal and depends a great deal on the presence of lattice defects. Such diffusion is *very* slow. For example Ni in γ iron has $D = 1 \times 10^{-19}$ cm^2 s^{-1} at 500°C, 2.5×10^{-12} cm^2 s^{-1} at 1000°C; Zn in GaAs has $D = 1.3 \times 10^{-24}$ cm^2 s^{-1} at 500°C, 1.5×10^{-18} cm^2 s^{-1} at 1000°C.

Example: Diffusion of Ni in γ iron at 500°C. If the unit for the x axis of Fig. 9.7 is taken to be 1 mm (that is, the diagram shows a concentration gradient over a length of ± 1 mm), then the Dt curves shown are appropriate for D in units of mm^2 s^{-1}:

$$D = 1 \times 10^{-19} \text{ cm}^2 \text{ s}^{-1} (10 \text{ mm/cm})^2 = 1 \times 10^{-17} \text{ mm}^2 \text{ s}^{-1}$$

The time required to reach the gradient shown for $Dt = 0.001$ would be:

$$t = \frac{0.001}{1 \times 10^{-17}} = 10^{14} \text{ s} = 3 \text{ million years} \qquad \blacksquare$$

Example: Diffusion of hydrogen through the side of an iron pipe at 500°C. Fick's first law gives:

$$J = \frac{1}{A} \frac{dn}{dt} = -D \frac{dc}{dx}$$

The concentration of hydrogen outside is effectively zero (it would rapidly diffuse away into the room) and the concentration inside can be assumed constant. Then, in the steady state, the average concentration can be assumed to be constant:

$$\frac{dc}{dx} \cong \frac{c}{\Delta x}$$

where Δx is the thickness of the wall. The rate of loss is then:

$$-\frac{dn}{dt} = \frac{Dc A}{\Delta x}$$

Assume a pipe with walls 1 cm thick, radius 50 cm, and length 100 m (10^4 cm). The area of the sides is:

$$A = 2\pi rh = 3.14 \times 10^6 \text{ cm}^2$$

Also assume a hydrogen concentration of 1.6×10^{-4} mol/cm^3 (this is a pressure of about

Table 9.1   Diffusion constants

Solute	Solvent	t (°C)	$10^5 D$ (cm^2 s^{-1})
Water	(Self-diffusion)	25	2.26
Iodine	Hexane	25	4.05
Iodine	CCl$_4$	25	2.13
Argon	CCl$_4$	25	3.63
CCl$_4$	n-Heptane	25	3.17
Glycine	Water	25	1.05
Glycine	Water	20	0.93
Dextrose	Water	25	0.67
Sucrose	Water	20	0.4586
Ribonuclease (a)*	Water	20	0.1068
Myosin (b)*	Water	20	0.0105
Rabbit papilloma virus (c)*	Water	20	0.0059

*Molecular weights: (a) 13,683 g, (b) 4.4×10^5 g, (c) 4.7×10^7 g.

10 atm). Then:

$$\frac{dn}{dt} = \frac{(1.5 \times 10^{-5}\ \text{cm}^2\ \text{s}^{-1})(1.6 \times 10^{-4}\ \text{mol/cm}^3)(3.14 \times 10^6\ \text{cm}^2)}{1\ \text{cm}}$$

$$= 7.5 \times 10^{-3}\ \text{mol/sec} = 27\ \text{moles per hour}$$

This is a small part of the total hydrogen, but hydrogen is light and tends to collect near the ceiling, so "leaks" of this type could have explosive consequences. Actually, a more serious consequence of hydrogen diffusion into iron is that the iron becomes brittle.   ■

9.4   Viscosity

Viscosity is a property of a fluid that resists flow; the greater ease of pouring milk than of pouring a thick syrup (molasses) is quantified by the coefficient of viscosity. By *flow* we mean a macroscopic motion of the fluid as a whole, as opposed to the random kinetic motions of its constituent molecules.

Actually, viscosity is not a resistance to fluid movement as such, but to a differential flow. If a bottle of syrup is lifted from the table, there is no resistance; but if the bottle is tipped and the syrup poured, the various portions of the fluid move with different velocities (the layer next to the wall of the container is stationary), and this type of flow is resisted by viscosity. Viscous forces are frictional forces which attempt to make all parts of a fluid move at the same velocity.

Newton's Law of Viscous Force

Consider two layers of a fluid (Fig. 9.8), at x and $x + dx$, moving with velocities v and $v + dv$, respectively; this defines a *velocity gradient, dv/dx*. According to

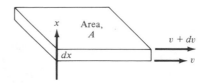

Figure 9.8 Velocity gradient. If two layers of a fluid, with area A and separated by a distance dx, are traveling at different velocities (difference $= dv$), the velocity gradient is dv/dx. Viscosity is related to the frictional force between the layers, which makes them tend toward equal velocities [Eq. (9.40)].

Newton, the frictional force between these layers will be proportional to the velocity gradient and the area of contact between the layers:

$$F = \eta A \frac{dv}{dx} \tag{9.40}$$

The proportionality constant (η) is called the coefficient of viscosity (or simply *the* viscosity).

Unit analysis of Eq. (9.40) will demonstrate that (in the cgs system) the unit of viscosity is:

$$g \, cm^{-1} \, s^{-1} \equiv poise \ (abbreviated \ ``p")$$

In the SI system the unit of viscosity is:

$$kg \, m^{-1} \, s^{-1} = 10 \ poise$$

Some viscosities are listed in Table 9.2.

Table 9.2   Viscosity of some fluids

(Units: poise $= g \, cm^{-1} \, s^{-1}$; temperature in parentheses)

Liquids	$10^3 \eta$				
Water	17.921(0°C)	10.050(20°C)	8.937(25°C)	8.007(30°C)	5.494(50°C)
Acetic acid	12.22(20°C)	10.396(30°C)	7.956(50°C)		
Acetone	4.013(0°C)	3.311(20°C)	2.561(50°C)		
Cyclohexane	10.3(17°C)	8.6(27°C)	7.5(35°C)		
Heptane	5.236(0°C)	4.163(20°C)	3.410(40°C)		
Ethylene glycol		173.3(25°C)			
Glycerol	42 200(2.8°C)	10 690(20°C)			

Gases	$10^6 \eta$			
Nitrogen	129.5(200 K)	178.6(300 K)	401.1(1000 K)	
Oxygen	147.6(200 K)	207.1(300 K)	472.0(1000 K)	
Argon	159.4(200 K)	227.0(300 K)	530.2(1000 K)	
CO_2(20°C)	148(1 atm)	156(20 atm)	166(40 atm)	

Poiseuille's Formula

The measurement of viscosity is generally based on the equations derived by J. L. Poiseuille (1844) for nonturbulent (laminar) flow through a tube. For such a flow, the layer nearest the wall is generally assumed to be stationary, and the flow (fastest at the center, cf. Fig. 9.9) will create a velocity gradient.

For an incompressible fluid (i.e., a liquid) the volume of flow (ΔV) through a tube (radius r, length l) in some time interval (Δt) is:

$$\frac{\Delta V}{\Delta t} = \frac{\pi r^4 \Delta P}{8 \eta l} \tag{9.41}$$

where ΔP is the pressure head driving the flow.

If the fluid is compressible (i.e., a gas), the rate of flow is:

$$\frac{\Delta V}{\Delta t} = \frac{\pi r^4}{16 \eta l} \left(\frac{P_i^2 - P_f^2}{P_0} \right) \tag{9.42}$$

where P_i is the inlet pressure, P_f is the outlet pressure, and P_0 is the pressure at which the volume of the gas was measured. Equation (9.42) applies in the case that the mean free path of the gas is small compared to the radius and length of the tube; at low pressures, Knudsen flow (Section 1.5) will occur. Also, the flow rate must be slow enough that there is no turbulence.

Stokes' Law

The motion of a solid object through a fluid is resisted by a frictional drag which (like all frictional forces) is proportional to its velocity. The force is:

$$F(\text{friction}) = -fv$$

where f is called the *friction constant*.

This friction is viscous in nature; the layer of fluid next to the object moves with it, while the fluid far away is stationary. Thus there is a velocity gradient created by the moving object.

G. G. Stokes (1851) showed that the friction constant of a sphere with radius r was:

$$f = 6\pi\eta r \tag{9.43}$$

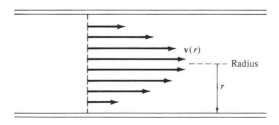

Figure 9.9 Velocity gradients for flow through a tube. The fluid at the center of the tube moves fastest; it is usually assumed in derivations that the fluid layer next to the wall is stationary ("stick" boundary conditions).

(This formula has been generalized by Perrin for ellipsoids and by others for other common shapes; a general proportionality to viscosity and "size" is always found.)

Stokes' law [Eq. (9.43)] finds application in the falling-ball viscometer. A ball with radius r and density ρ, falling through a fluid with density ρ_0, will be accelerated by the force of gravity:

$$F(\text{gravity}) = \left(\frac{4\pi r^3}{3}\right)(\rho - \rho_0)g$$

The acceleration is opposed by the frictional force, and the ball will reach a steady velocity of fall (v_s) for which the net force (i.e., acceleration) is zero:

$$m\frac{dv}{dt} = F(\text{gravity}) + F(\text{friction}) = 0$$

$$v_s = \frac{\left(\dfrac{4\pi r^3}{3}\right)(\rho - \rho_0)g}{f}$$

Using Eq. (9.43), this becomes:

$$v_s = \frac{2r^2(\rho - \rho_0)g}{9\eta} \qquad (9.44)$$

Stokes' law has also been used to relate molecular size to the diffusion coefficient of molecules in solution. This relationship is obtained from the equation derived by Einstein (1905) which relates the friction constant to the diffusion constant:

$$D = \frac{kT}{f} \qquad (9.45)$$

Using Stokes' law, the diffusion constant of a spherical molecule with radius r in a dilute solution in a solvent with viscosity η should be:

$$D = \frac{kT}{6\pi\eta r} \qquad (9.46)$$

Stokes' theory, like most hydrodynamic theories, treats the solvent as a continuous medium; therefore Eq. (9.46) (called the **Stokes-Einstein law**) should be restricted to molecules which are large compared to the solvent molecules (e.g., globular proteins in water). (The 1984 Hildebrand Award was given to Berni Adler, in part, for work which showed that the Stokes-Einstein relation with slip boundary conditions is applicable at the molecular level.) In its proper domain of application, the Stokes-Einstein law is a useful method for estimating particle sizes or diffusion constants from each other.

A somewhat more general form of Eq. (9.46) is:

$$\frac{D\eta}{T} = \text{constant} \qquad (9.47)$$

This is useful for finding the change of D with T from viscosity-vs.-temperature measurements on the solvent (an easy experiment compared to measuring diffusion constants).

Example: At 20°C, the diffusion constant of ribonuclease is $1.068 \times 10^{-6} \text{ cm}^2 \text{ s}^{-1}$. Estimate this quantity at 30°C. From Table 9.2, the viscosity of water is: $\eta(20°) = 10.05 \times 10^{-3}$ poise; $\eta(30°) = 8.007 \times 10^{-3}$. Equation (9.47) gives:

$$\frac{D(30°)(8.007 \times 10^{-3})}{(303 \text{ K})} = \frac{(1.068 \times 10^{-6})(10.05 \times 10^{-3})}{(293 \text{ K})}$$

$$D(30°) \cong 1.386 \times 10^{-6} \text{ cm}^2 \text{ s}^{-1} \qquad \blacksquare$$

Example: You were warned not to use the Stokes-Einstein law for small molecules — but, throwing caution to the winds, let us use data from Tables 9.1 and 9.2 to estimate the size of CCl_4.

$$D = 3.17 \times 10^{-5} \text{ cm}^2 \text{ s}^{-1} \qquad (CCl_4 \text{ in } n\text{-heptane at } 25°C)$$

$$\eta = 4.0 \times 10^{-3} \text{ g cm}^{-1} \text{ s}^{-1} \qquad (n\text{-heptane, interpolated at } 25°C)$$

Using Eq. (9.46) gives:

$$r = \frac{(1.38 \times 10^{-16} \text{ erg/K})(298 \text{ K})}{6\pi(4.0 \times 10^{-3} \text{ g cm}^{-1} \text{ s}^{-1})(3.17 \times 10^{-5} \text{ cm}^2 \text{ s}^{-1})} = 1.7 \times 10^{-8} \text{ cm}$$

This is certainly a correct order of magnitude, but a good bit too small. The van der Waals constant for CCl_4 ($b = 138.3 \text{ cm}^3$) gives $r = 4.8 \times 10^{-8}$ cm. $\qquad \blacksquare$

Viscosity of Gases

The viscosity of dense fluids (liquids) is strongly dominated by the attractive forces between molecules; yet even an ideal gas, for which there are no such forces, has a nonzero viscosity. (Strictly speaking, we shall be considering a hard-sphere gas at low densities for which the excluded volume effect is negligible; a true ideal gas is a gas of point masses with zero diameter and, hence, has no collisions.)

Suppose that two layers of a gas are moving at different velocities; the molecules within the layers are also moving randomly about in thermal motion (Fig. 9.10), and some of these molecules will diffuse from one layer to the other. The transfer of momentum between the layers will tend to make their velocities equal; viscosity is, thus, the transport of momentum.

Imagine that two trains are moving on parallel tracks at different speeds. The passengers (possibly an acrobatic troupe) amuse themselves by leaping from train to train. Each person jumping from the faster train to the slower train carries some of the momentum of the faster train to the slower and will cause the slower train to go faster. Persons jumping the other way will have the opposite effect, so the net effect will be to decrease the speed difference of the trains. That is, there is a viscous force between the trains caused by the jumping passengers. [This analogy, which has been widely used, implicitly assumes that the trains are in free motion (no other forces) on a level, frictionless track. It fails to consider the reaction of the engine or the engineer to the change in speed. The reaction of the engineer, one suspects, would be incredulity.]

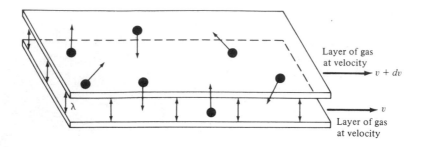

Figure 9.10 Viscosity of a gas. The flow of the layers (presumed to be separated by the mean free path λ) is superimposed on the random thermal motions; as a result of these random motions, molecules move between layers, creating a viscous force between the layers.

Newton's law of motion is:

$$F = m\frac{dv}{dt} = \frac{dp}{dt}$$

where $p = mv$ is the momentum. Combined with Eq. (9.40) for the viscous force, this becomes:

$$\frac{dp}{dt} = \eta A\left(\frac{dv}{dx}\right) \tag{9.48}$$

For simplicity, consider two layers of gas (Fig. 9.10) separated by the mean free path (λ); we assume that all molecules, when they move, jump with a step of constant size, the mean free path. The difference between the velocities of the two layers is:

$$\Delta v = \lambda\left(\frac{dv}{dx}\right)$$

Each molecule (mass m) which moves from one layer to the other transports a momentum change:

$$\Delta p(\text{per jump}) = m\lambda\frac{dv}{dx}$$

How many molecules will make such a jump? This question is answered by the equation derived in Section 1.5 for the collisions of a gas with a wall, per unit area, per unit time. In this case, the "wall" is the plane dividing the layers (area A), and the number of jumps (in *both* directions) will be:

$$2Z_{\text{wall}} A \Delta t$$

with [Eq. (1.33)]

$$Z_{\text{wall}} = n*\left(\frac{RT}{2\pi M}\right)^{1/2}$$

This equation is more frequently written in terms of the average velocity [$\bar{v}$, Eq. (9.2)] as:

$$Z_{\text{wall}} = \frac{n^* \bar{v}}{4} \tag{9.49}$$

The total momentum transfer per unit time is, therefore:

$$\frac{\Delta p}{\Delta t} = 2 Z_{\text{wall}} A m \lambda \left(\frac{dv}{dx}\right)$$

$$= \frac{m n^* \bar{v} \lambda}{2} A \left(\frac{dv}{dx}\right)$$

Comparison of this to Eq. (9.48) shows that the coefficient of viscosity is (using $mn^* = \rho$, the mass density):

$$\eta = 0.5 \rho \bar{v} \lambda \tag{9.50}$$

(The more rigorous derivation gives a nearly identical result, with a constant 0.499 in place of 0.5.)

Example: Calculate the viscosity of argon ($M = 39.918$ g/mol) at STP.

$$\rho = \frac{PM}{RT} = \frac{(1 \text{ atm}) (39.948 \text{ g})}{(82.06 \text{ cm}^3 \text{ atm K}^{-1}) (273.15 \text{ K})} = 1.778 \times 10^{-3} \text{ g/cm}^3$$

In an earlier example (p. 429) we calculated:

$$\lambda = 72.5 \text{ nm} = 72.5 \times 10^{-7} \text{ cm}$$

$$\bar{v} = 3.8048 \times 10^4 \text{ cm/s}$$

Therefore:

$$\eta = 0.5 (1.778 \times 10^{-3} \text{ g/cm}^3) (3.8048 \times 10^4 \text{ cm/s}) (72.5 \times 10^{-7} \text{ cm})$$

$$= 2.458 \times 10^{-4} \text{ g cm}^{-1} \text{ s}^{-1} = 245.8 \text{ } \mu p$$

The experimental value is 210 μp. ∎

Using the ideal gas density (PM/RT) and Eq. (9.9) for the mean free path, Eq. (9.50) becomes:

$$\eta = \frac{M \bar{v}}{2 \sqrt{2} \pi \sigma^2 L} \tag{9.51}$$

Written this way, this result demonstrates two surprising facts. First, the viscosity of a gas at a given temperature is predicted to be independent of pressure. This is due to a cancellation of effects: at higher pressures the number of molecules available (per unit volume) to transfer momentum is larger, but they can't go as far, because the mean free path is shorter. Of course, the limitations imposed by the assumptions made in deriving Eq. (9.50) limit its validity to the low pressure range; at high pressure, the viscosity will increase somewhat (see Table 9.2).

The second surprising prediction of Eq. (9.51) is that the viscosity of a gas will increase with increasing temperature. That this is true can be seen by inspection of Table 9.2. [However, the simple $T^{1/2}$ dependence of Eq. (9.51) is not exact, because attractive forces have been neglected.] This contradicts our ordinary experience ("slow as molasses in January"), which is based on our observations of liquids, and emphasizes the major difference in the mechanism of viscosities for gases and liquids. The viscosity of liquids is dominated by attractive forces and generally decreases with increasing temperature. The most viscous liquids are those with very strong forces such as hydrogen bonding; compare (Table 9.2) the viscosities of water, ethylene glycol, and glycerine to those of heptane or acetone.

Recalling the viscosity-between-trains discussion, the increase of viscosity of gases with temperature can be extended to the trains if, for some reason, the passengers jump more frequently as the temperature increases. If the passengers are barefooted, and the sun beating on the roof of the train makes it uncomfortably hot, an increase in the jump frequency and, hence, the viscosity, with temperature can be readily imagined.

The Effect of Intermolecular Forces

Even as a hard-sphere theory, Eq. (9.51) is deficient because it fails to take into account the excluded volume (van der Waals b for example). Enskog derived a more rigorous equation for the viscosity of a hard-sphere gas:

$$\eta = \eta_0 \left[1 + 0.175 \left(\frac{b_0}{V_m} \right) + 0.865 \left(\frac{b_0}{V_m} \right)^2 \right] \tag{9.52}$$

where $b_0 = 2\pi\sigma^3 L/3$ [compare Eq. (1.44)] and η_0 is the viscosity at low pressure [e.g., as given by Eq. (9.51)].

Accurate theories for the temperature dependence of gas viscosity involve the intermolecular potential (Section 1.7) and tend to be very complex. A simple, and fairly accurate, result was obtained by Sutherland (1893), who derived the equation:

$$\eta = \frac{k_s \sqrt{T}}{1 + \dfrac{S}{T}} \tag{9.53}$$

The *Sutherland constant* (S) is related to the attractive potential between the molecules. For the Sutherland potential:

$$U(r) = \infty \qquad (r < \sigma)$$

$$U(r) = -\varepsilon \left(\frac{\sigma}{r} \right)^n \qquad (r > \sigma)$$

(see Fig. 1.17), the Sutherland constant is related directly to the potential-well depth (ε):

$$S = C \left(\frac{\varepsilon}{k} \right) \tag{9.54}$$

In this equation, C is a constant that depends on the exponent n; for $2 < n < 8$, $C \cong 0.2$. (Reference 5, p. 550, gives the exact values.)

The other constant of Eq. (9.53) can be deduced by comparison of the earlier results [e.g., Eq. (9.50) at low pressure or Eq. (9.52)], which were derived neglecting attractive forces. These equations will be accurate at high temperatures for which the S/T term of Eq. (9.53) can be neglected. For low pressures, it is easily demonstrated [cf. Eq. (9.51)] that:

$$k_s = \frac{(RM)^{1/2}}{\pi^{3/2}\sigma^2 L} \tag{9.55}$$

Clearly, viscosity (and other transport properties) can be used to determine molecular potential constants (e.g., σ, ε). However, even with more accurate potential functions such as the Lennard-Jones potential (Section 1.7, Fig. 1.20), the parameters which work best for equilibrium properties such as second virial coefficients are often not the same as those which work best for transport properties. For that reason, two sets of parameters are generally tabulated. For example, reference 5 gives the Lennard-Jones constants for propane as: $\varepsilon/k = 254$ K, $\sigma = 5.061 \times 10^{-8}$ cm from viscosity and $\varepsilon/k = 242$ K, $\sigma = 5.637 \times 10^{-8}$ cm from second virial coefficients.

Viscosity of Polymer Solutions

Viscosity measurements on solutions have long been a popular method for studying polymers, both synthetic and biological. The basis for such studies is the relationship derived by Einstein (1906) for the viscosity of a dilute solution of suspended spheres:

$$\frac{\eta}{\eta_0} = 1 + 2.5\phi$$

where η_0 is the viscosity of the pure solvent and ϕ is the volume fraction of the suspended spheres. [This result has been extended to solutes of other shapes — ellipsoids, rods, and so on; only the numerical constant (2.5) is altered.]

The **specific viscosity** of a solution is defined as:

$$\eta_{sp} \equiv \frac{\eta}{\eta_0} - 1 = \frac{\eta - \eta_0}{\eta_0} \tag{9.56}$$

For a solution of volume V, containing N spheres of radius r, Einstein's result for the specific viscosity is:

$$\eta_{sp} = \frac{2.5N\left(\frac{4\pi r^3}{3}\right)}{V}$$

If each sphere has a mass m, this can be rewritten in terms of the mass concentration (c_m, mass of solute per unit volume) as:

$$\eta_{sp} = 2.5c_m\frac{\left(\frac{4\pi r^3}{3}\right)}{m}$$

The **intrinsic viscosity** is defined as:

$$[\eta] \equiv \lim_{c_m \to 0} \left(\frac{\eta_{sp}}{c_m} \right)$$

(9.57)

Einstein's result for spheres of radius r and mass m is:

$$[\eta] = \frac{2.5 \left(\dfrac{4\pi r^3}{3} \right)}{m}$$

(9.58)

This result was used by Einstein [together with diffusion constants and Eq. (9.45)] to estimate the molecular diameter of "sugar" and Avogadro's number; he calculated $r = 4.9 \times 10^{-8}$ cm, $L = 6.56 \times 10^{23}$.

Any macromolecular solute which is compact in solution (i.e., folded upon itself) should obey Eq. (9.58) (with perhaps a different constant if its shape is not spherical). But if one compared a series of solutes of different size and molecular weight which are made from the same constituents (e.g., H, C, O, and N for most organic molecules), one would expect the volume/mass ratio of the molecules to be very similar. That is, in such a case the intrinsic viscosity should be independent of molecular weight. This is generally true for biological macromolecules in their natural state. For example, the intrinsic viscosity (in water) of ribonuclease ($M = 13,683$) is 3.4 cm^3/g, that of serum albumin ($M = 67,500$) is 3.7, and that of bushy stunt virus ($M = 1.07 \times 10^8$) is 3.4.

Synthetic macromolecules (and denatured proteins) are not usually compact in solution but, rather, are strung out in what is called a *random coil*. Still, one might attempt to apply an equation such as (9.58) by assuming that the random coil moves as a unit, carrying the solvent which is within the coil along, and hence acts like a sphere with some effective radius (called the *hydrodynamic radius*).

How big is a random coil? One answer to this question is provided by the **radius of gyration** (R_G), the rms distance of the constituent masses of the polymer from the center of mass:

$$R_G = \left(\frac{\sum_i r_i^2}{N} \right)^{1/2}$$

(9.59)

(This is the simplest definition, as given by reference 2. Reference 3 defines R_G as a mass-weighted average:

$$R_G^2 = \frac{\sum_i m_i r_i^2}{\sum_i m_i}$$

For our present purposes, the distinction is not important.)

If this radius is used in place of the radius of a sphere in Eq. (9.58), then:

$$[\eta] = K' \frac{R_G^3}{M}$$

where M is the molecular weight and K' is a constant.

Now, if the polymer consists of N links (e.g., monomer units) attached by flexible joints, the random-walk model of Section 9.2 suggests that $R_G \propto \sqrt{N}$. Since the molecular weight is directly proportional to the number of monomer units, $R_G \propto M^{1/2}$ and, one would expect:

$$[\eta] = KM^{1/2}$$

A square-root dependence of intrinsic viscosity on molecular weight is, in fact, found for many cases — for example, polystyrene in cyclohexane, polyisobutylene in benzene, amylose in aqueous KCl. However, it is not universally true (see Table 9.3). Generally, random-coil polymers in solution are found to obey an equation:

$$[\eta] = KM^{\alpha} \tag{9.60}$$

The constant α usually has a value between 0.5 and 1, but K and α are totally empirical parameters which must be measured for each polymer/solvent pair. Despite the need for calibration (which can be done with a single polymer sample whose molecular weight has been determined by some other means such as osmotic pressure, sedimentation, or light scattering), viscosity measurement is a very popular method for determining polymer molecular weights because of its speed and simplicity. [Since α is usually irrational, it is simplest to regard M in Eq. (9.60) as unitless, so K has the same units as intrinsic viscosity, e.g., cm^3/g.]

The derivation above contains another interesting idea. If one thinks of the random coil as a kinetic phenomenon, maintained by the random thermal motions, then one would expect the radius of gyration to increase with temperature. Then, by the arguments presented above, one would expect the intrinsic viscosity (in effect, the polymer's contribution to the solution's viscosity) to increase with temperature. Since the solvent's viscosity decreases with temperature, this is a compensating effect, and the viscosity of a polymer solution will decrease less rapidly with temperature than that of the pure solvent. This effect (which is, in fact, observed) is utilized in making so-called "multigrade" motor oils. A motor oil must maintain a reasonable viscosity over a wide range of temperature; it must not be too viscous for cold starts, and it must not be too thin at the operating temperature of the engine. The use of polymers dissolved in oil gives a lubricant whose viscosity does not vary excessively over the necessary temperature range.

Table 9.3 Intrinsic viscosity parameters

Polymer	Solvent	$100K$ (cm^3/g)	α	t (°C)
Polystyrene	Cyclohexane	8.1	0.50	34°
Polystyrene	Benzene	0.95	0.74	25°
Polystyrene	Toluene	3.7	0.62	25°
Polyisobutylene	Benzene	8.3	0.50	24°
Polyisobutylene	Cyclohexane	2.6	0.70	30°

9.5   Sedimentation

In our discussion of diffusion, we implicitly assumed that a uniform distribution of solute in a solution was to be expected at equilibrium. This is strictly true only in the absence of external forces, and, because of gravity, one might expect there to be some variation of solute concentration with height. This effect is generally too small to be observed with small molecules, but for macromolecular solutes it is not only observable but a source of useful information.

The force exerted on a particle of mass m in a gravitational field (acceleration $g \cong 980$ cm s^{-2}) is:

$$F = (m - m_0)g$$

where m_0 is the mass of displaced fluid. If the solvent has a density ρ_0, and the solute has a partial specific volume $\bar{v}_s$, the force is:

$$F = mg(1 - \bar{v}_s \rho_0)$$

The *buoyancy correction* will be denoted b:

$$b = (1 - \bar{v}_s \rho_0)$$

The partial molar volume and its measurement was discussed in Chapter 7; the partial specific volume is the partial molar volume divided by the molecular weight:

$$\bar{v}_s = \frac{\bar{V}_m}{M}$$

Biological macromolecules typically have partial specific volumes of 0.75 cm^3/g, so (assuming the solvent is water, $\rho_0 \cong 1$ g/cm^3) the buoyancy correction is about 0.25. Note that any error in determining the partial specific volume is multiplied fourfold in the buoyancy correction (and, therefore, in the molecular weights determined from sedimentation); consequently it must be measured carefully.

The particle falling under the influence of gravity is resisted by a viscous frictional force, so:

$$F = m\frac{dv}{dt} = mgb - fv$$

The steady drift velocity (dx/dt) is then:

$$\frac{dx}{dt} = \frac{mgb}{f}$$

If the molecule is reasonably symmetrical or (as is usually the case) the sedimentation velocity is slow enough that there are no flow-orientation effects, the friction constant can be assumed to be that involved in diffusion, and Eq. (9.45) can be used:

$$f = \frac{kT}{D} = \frac{RT}{LD}$$

This gives (with $M = Lm$):

$$\frac{dx}{dt} = \frac{MDg}{RT}(1 - \bar{v}_s\rho_0) \tag{9.61}$$

Example: Estimate the gravity sedimentation rate for ribonuclease in water. From Table 9.1: $D = 1.068 \times 10^{-6}$ cm^2 s^{-1}, $M = 13{,}683$ g/mol. For an exact calculation we would need the partial specific volume, but for an estimate we will simply assume a buoyancy correction of 0.25:

$$\frac{dx}{dt} = \frac{(13\,683 \text{ g})(1.068 \times 10^{-6} \text{ cm}^2 \text{ s}^{-1})(980 \text{ cm s}^{-2})(0.25)}{(8.314 \times 10^7 \text{ erg/K})(300 \text{ K})}$$

$$= 1.4 \times 10^{-10} \text{ cm/sec} = 0.004 \text{ cm/year} \qquad \blacksquare$$

Ultracentrifuge Sedimentation

The example above illustrates that gravity sedimentation is apt to be very slow (although it may be practical for solutes as large as a virus — say $M > 10^7$). For that reason, sedimentation experiments are usually done in an *ultracentrifuge,* for which the acceleration may be about 10^7 times that of gravity.

The acceleration on a particle at a distance x from the axis of rotation in a centrifuge with angular velocity ω is $\omega^2 x$. [The velocity ω must have units of radians per second: 2π times the rate of rotation in cycles (or revolutions) per second.] Substitution of the centrifugal acceleration into Eq. (9.61) (in place of g) gives:

$$\frac{dx}{dt} = \frac{MD\omega^2 x (1 - \bar{v}_s\rho_0)}{RT} \tag{9.62}$$

Rates of sedimentation are generally measured optically, for example, through the variation of the index of refraction with solute concentration along the tube. In a typical arrangement (schlieren optics, ref. 2, p. 91) it is the index-of-refraction *gradient* which is measured, and this is directly proportional to the concentration gradient (dc/dx). A boundary between a solution and solvent (appearing as in Fig. 9.7 at various times) will appear as a peak for the index-of-refraction gradient (Fig. 9.11). The rate of progress of this peak through the tube during centrifugation is just the dx/dt required for Eq. (9.62). Also, over time the boundary will become more diffuse (Fig. 9.7) and the peak broader (Fig. 9.11); the broadening of this peak provides a direct measure of the diffusion coefficient.

Exercise: The concentration gradient is the derivative of Eq. (9.38):

$$\frac{\partial c}{\partial x} = \frac{c_0}{2(\pi D t)^{1/2}} e^{-x^2/4Dt} \tag{9.63}$$

(cf. Fig. 9.5.) The height of the $\partial c/\partial x$ curve (at $x = 0$) is then just:

$$H = \frac{c_0}{2(\pi D t)^{1/2}}$$

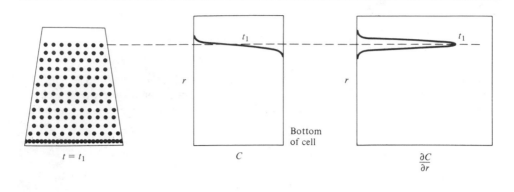

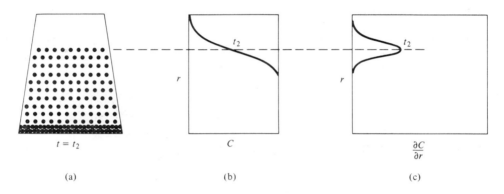

Figure 9.11 Sedimentation in an ultracentrifuge. Panel (a) shows schematically how the concentration changes as the solute molecules settle. Panel (b) shows the concentration profiles in the tube at two times. Panel (c) shows the concentration gradients, *dC/dx*, for the profiles of panel (b) as they could be measured in an experiment using schlierien optics.

Show that the area under the $\partial c/\partial x$ curve (*A*) is just c_0 and, therefore:

$$\frac{A}{H} = 2(\pi D t)^{1/2} \tag{9.64}$$

Show that the half-width of the $\partial c/\partial x$ curve (the value of *x*, measured from the center of the peak, for which the curve drops to half its maximum height) is:

$$\delta x_{1/2} = [4Dt \ \ln 2]^{1/2} \tag{9.65}$$

∎

The experimentally measurable quantities are used to calculate the **sedimentation coefficient:**

$$s \equiv \frac{(dx/dt)}{\omega^2 x} \tag{9.66}$$

which [from Eq. (9.62)] is related to the molecular parameters as:

$$s = \frac{MDb}{RT}$$

From this, the molecular weight can be calculated:

$$M = \frac{RTs}{D(1 - \rho_0 \bar{v}_s)} \tag{9.67}$$

Sedimentation coefficients typically have magnitudes $\sim 10^{-13}$ s; this quantity is called a *svedberg,* after The Svedberg (b. 1884), the inventor of the ultracentrifuge. Some values are given in Table 9.4.

Equilibrium Sedimentation

The flux of solute particles past some point x in the tube due to sedimentation is:

$$J_{\text{sed}} = c \frac{dx}{dt} = \frac{cMDb\omega^2 x}{RT}$$

The concentration gradient created by the sedimentation causes a diffusional flux in the opposite direction:

$$J_{\text{diff}} = -D \frac{dc}{dx}$$

After a long time (typically a day or more) a steady state will be reached along the tube and the next flux is zero:

$$J_{\text{sed}} + J_{\text{diff}} = 0$$

$$D \frac{dc}{dx} = \frac{cMDb\omega^2 x}{RT}$$

Table 9.4 Properties of biological macromolecules (20°C, in water)

Molecule	M (g/mol)	$[\eta]$ (cm^3/g)	$10^7 D$ (cm^2/s)	$\bar{v}_s$ (cm^3/g)	$10^{13} s$ (sec)
Ribonuclease	13 683	2.30	11.9	0.728	1.64
Lysozyme	14 400	—	11.2	0.703	1.91
Serum albumin	66 × 10^3	3.7	5.94	0.734	4.31
Hemoglobin	68 × 10^3	3.6	6.9	0.749	4.31
Myosin	570 × 10^3	217	1.0	0.728	6.4
Catalase	250 × 10^3	3.9	4.1	0.73	11.3
Bushy stunt virus	10.7 × 10^6	3.4	1.15	0.74	132
Tobacco mosaic virus	50 × 10^6	36.7	0.3	0.73	170

The diffusion constant will cancel (an important simplification), and:

$$\frac{dc}{c} = \frac{Mb\omega^2}{RT} x \, dx$$

The ratio of the concentrations at two points, c_1 at x_1 and c_2 at x_2, will thus be:

$$\frac{\ln\left(\dfrac{c_2}{c_1}\right)}{x_2^2 - x_1^2} = \frac{M(1 - \rho_0 \bar{v}_s)\omega^2}{2RT} \tag{9.68}$$

Although rather time consuming, this is considered to be the most accurate method for determining macromolecular weights in most cases.

Postscript

The research of Jean Perrin (ca. 1908) on colloidal suspensions was very important in establishing the kinetic-molecular theory. Actually there was abundant "proof" of the theory already, but, after all, molecules cannot be seen and the unimaginative remained skeptical. Perrin worked with particles of diameter ~0.1 μm which could, with the aid of a microscope, be seen directly.

In a microscopic colloidal suspension, the particles will sediment so that more will be near the bottom than the top; however, at equilibrium they are not all at the bottom. As discussed earlier, the force for sedimentation is:

$$F = mgb$$

where b is the buoyancy correction. A particle at a height h in the suspension, therefore, has a potential energy of $mgbh$, and the average of such potentials must also be the average kinetic energy of the particles. Perrin derived a formula for the concentration ratio (c_1 at x_1, c_2 at x_2):

$$\bar{\varepsilon} \ln \frac{c_2}{c_1} = \tfrac{3}{2} mgb(x_2 - x_1)$$

where $\bar{\varepsilon}$ is the average kinetic energy of the particles, the energy needed for them to remain suspended. Using the result from kinetic theory:

$$\bar{\varepsilon} = \tfrac{3}{2} kT$$

he obtained a value for Avogadro's number ($L = R/k$): 7.05×10^{23}.

In another study, Perrin observed the Brownian motion of the suspended particles (Fig. 9.12). Brownian motion refers to the continual and random motion of microscopic suspended particles which had been first reported by a botanist named Brown (1827). The theoretical results of Einstein [Eq. (9.28)] and Stokes [Eq. (9.46)] give, for the mean squared displacement of particles with radius r in time t:

$$\langle x^2 \rangle = \frac{RT}{L} \frac{t}{3\pi\eta r}$$

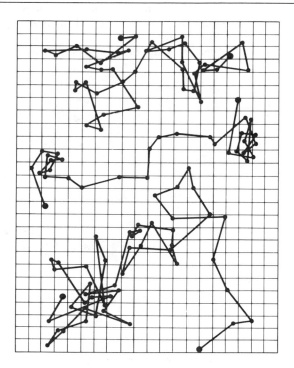

Figure 9.12   Brownian motion. The Brownian motion of particles as recorded by Perrin in his original work *(Oeuvres Scientifiques)*.

Measurements of this quantity gave yet another value for Avogadro's number: 7.15×10^{23}.

There are more accurate methods for determining this quantity; in fact, the first calculation by J. Loschmidt in 1865 gave 6.07×10^{23}, much closer to today's accepted value than Perrin's results. However, the agreement of these results with those obtained from entirely dissimilar physical phenomena made the kinetic-molecular theory unchallengeable.

Reference 7 gives a more thorough account of Perrin's work.

References

1. W. Kauzmann, *Kinetic Theory of Gases,* 1966: New York, W. A. Benjamin, Inc.
2. K. E. Van Holde, *Physical Biochemistry,* 1971: Englewood Cliffs, N.J., Prentice-Hall, Inc.
3. C. Tanford, *Physical Chemistry of Macromolecules,* 1967: New York, John Wiley & Sons, Inc.
4. A. Einstein, *Investigations on the Theory of the Brownian Movement* (edited, with notes, by R. Fürth), 1956: New York, Dover Publications, Inc.
5. J. O. Hirschfelder, C. F. Curtiss, and R. B. Bird, *Molecular Theory of Gases and Liquids,* 1964: New York, John Wiley & Sons, Inc.

6. S. Chandrasekhar, "Stochastic Problems in Physics and Astronomy," *Rev. Mod. Phys.*, *15*, 1 (1943). Reprinted in *Selected Papers on Noise and Stochastic Processes* (N. Wax, ed.) by Dover Publications, Inc.

7. G. L. Trigg, *Crucial Experiments in Modern Physics*, 1971: New York, Crane, Russak & Co., Inc.

Problems

9.1 Calculate the collision frequency for a molecule of neon ($\sigma = 0.2749 \times 10^{-9}$ m) in a gas at 1 atm and $T = 473$ and 673 K.

9.2 Calculate the mean free path of N_2 ($\sigma = 0.37$ nm) at 273 K, 1 atm.

9.3 Calculate the mean free path and the frequency of collisions of one molecule for He at 1 atm and $T = 300$ and 1000 K.

9.4 Treating air as a uniform gas with molecular weight $\overline{M} = 28.8$ g and $\sigma = 3.67 \times 10^{-8}$ cm, calculate the mean free path at 25°C for $P = 1$ atm, 1 torr, 10^{-3} torr.

9.5 Calculate the average velocities and average relative velocities (A-A, B-B, and A-B) for a mixture of CO and H_2 at STP.

9.6 For an equimolar mixture of H_2 and CO at STP, calculate the total H_2-H_2, CO-CO, and H_2-CO collisions per unit volume per second. (You may use the results of Problem 9.5.) Why is the last number so much larger than the first two?

9.7 Into a 1-m^3 container at 300 K are placed 20 moles of CH_4 and 5 moles of H_2. Calculate the number of H_2-CH_4 collisions in the container which occur in one second.

9.8 For N_2 gas at 298 K and 1 atm $\bar{v} = 454.2$ ms^{-1}, $\lambda = 6 \times 10^{-8}$ m, $z = 7.57 \times 10^9$ s^{-1}. Calculate the rms mean distance (x) traveled by an N_2 molecule in one minute under these conditions.

9.9 An indecisive professor sets out toward the dean's office (500 paces away) to ask for a raise. After each step he stops and flips a coin to decide whether to continue or return (this takes about 4 seconds). What is the probability that he will reach the dean's office in 8 hours?

9.10 The random-walk arguments which imply that the rms length of a polymer chain is $\sqrt{2N\lambda^2}$ presume that the links are universal. For N links with a fixed angle ϕ, the result is:

$$\langle r^2 \rangle \cong N\lambda^2 \frac{1 - \cos\phi}{1 + \cos\phi}$$

Estimate the rms average length of a polyethylene chain with molecular weight 10,000 assuming a C-C bond length of 1.54 Å, and the tetrahedral angle.

9.11 The diffusion constant of water (liquid) at 25°C is 2.26×10^{-5} cm^2 s^{-1}. How long would it take a molecule of water to travel across a container 1 cm wide? This can be estimated as the time required to have an rms displacement (Δx) of 1 cm.

9.12 Tobacco mosaic virus has a diffusion coefficient (water, 20°C) of 5.3×10^{-8} cm^2 s^{-1}. What time would be required for this material to have an rms displacement of 1 mm?

9.13 Calculate the diffusion coefficient for the molecules in air (assume $\overline{M} = 28.8$ g, $\sigma = 3.67 \times 10^{-8}$ cm) when $P = 1$ atm and $P = 1$ torr for $T = 300$ K and 1000 K.

9.14 The diffusion curves of Fig. 9.7 can be used to answer this question. If a solution of sucrose in water ($C_0 = 1$) is in contact with pure water, how long would it take for the concentration at a point 0.18 cm into the water to reach 0.1?

9.15 A solvent is layered over a solution ($C_0 = 1$). The diffusion coefficient of the solute is 2×10^{-5} cm^2 s^{-1}. After 24 hours, what will be the concentration 1.7 cm into the solution or solvent? (Numerical integration required.)

9.16 Hydrogen is in an iron container at 500°C. The diffusion coefficient for H_2 in iron is 1.5×10^{-5} cm^2 s^{-1}. If the container is a cylinder with radius 100 cm, height 1000 cm, and wall thickness 1 cm, how long would it take for half of the hydrogen to escape?

9.17 In constructing a flow viscometer, the tube diameter must be chosen to give reasonable flow rates, neither too fast nor too slow, for the range of viscosities to be measured. What would be the flow rate (cm^3/min) through a tube (length 5 cm, diameter 2 mm) for a liquid with a viscosity of 0.01 poise if the pressure head were 3 torr (3 torr is a water head of approximately 4 cm)?

9.18 A 1.50-cm radius spherical aluminum ball (density 2.70 g/cm^3) is dropped through a fluid of density 1.26 g/cm^3. Its terminal velocity was measured as 15.1 cm/s. What is the viscosity of the fluid?

9.19 The biological macromolecule lactalbumin has a diffusion coefficient in water (20°C) of 1.06×10^{-6} cm^2/s. Assuming the molecule to be spherical, calculate its radius using the Stokes-Einstein law.

9.20 Estimate the diffusion coefficient of sucrose in water at 50°C from its diffusion constant at 20°C. (Data: Tables 9.1 and 9.2.)

9.21 Chlorine and ethylene (C_2H_4) have similar diameters (to judge from their van der Waals b, Table 1.1) but greatly different masses. Which will have the higher viscosity, and by how much?

9.22 The viscosity of CO_2 at 1500 K is

$$\eta = 505.2 \times 10^{-6} \text{ poise}$$

Calculate the collision diameter and compare to the value given in Table 1.7 [from $B(T)$ data]. (Use the simple formula, which neglects intermolecular attractive forces.)

9.23 Estimate the molecular diameter (σ) of CH_4 from each of the following:

$$\text{van der Waals:} \quad b = 42.8 \text{ cm}^3$$

$$\text{viscosity:} \quad \eta = 1.116 \times 10^{-4} \text{ g cm}^{-1} \text{ s}^{-1} \text{ (at 300 K)}$$

$$\text{diffusion:} \quad D = 0.189 \text{ cm}^2 \text{ s}^{-1} \text{ (at 273 K, 1 atm)}$$

9.24 An experimental study of viscosities of ammonia (gas) in the range 300–400 K gave a Sutherland constant $k_s = 202.7 \times 10^{-7}$ pK$^{-1/2}$. Estimate the collision diameter of ammonia from this datum.

9.25 Use the data below for the viscosity of Ar to calculate the Sutherland constants. Then estimate the parameters for the Sutherland potential.

T	η (μp)
100	83.9
200	153.4
300	227.0
800	462.1

9.26   Estimate the viscosity of a solution of polystyrene ($\overline{M} = 1 \times 10^6$) in cyclohexane at 35°C, if the concentration is 20 mg/cm³.

9.27   Use the viscosity data below for a polystyrene sample in benzene (25°C) to calculate the average molecular weight of the sample.

c (mg/cm³)	η (mp)
0	6.04
2	6.41
5	6.98
10	8.02
20	10.38
50	19.69

9.28   Estimate the gravity sedimentation rate for rabbit papilloma virus in water (Table 9.1) assuming a buoyancy correction of 0.25.

9.29   In an ultracentrifuge with a speed of 60,000 rpm, the boundary of a solution (6 cm average distance from the axis) moved 5.1 mm in 3 hours. Calculate the sedimentation coefficient of the solute.

9.30   Serum globulin has a sedimentation coefficient in water ($\rho_0 = 0.9982$ g/cm³ at 20°C), $s = 7.1 \times 10^{-13}$ sec. Its diffusion constant (20°C) is 4.0×10^{-7} cm² s⁻¹ and its partial specific volume is 0.75 cm³/g. Calculate the molecular weight.

9.31   (a) Calculate the sedimentation coefficient for sucrose ($C_{12}H_{22}O_{11}$) (partial specific volume 0.630 cm³/g) in water at 20°C ($\rho_0 = 0.9982$ g/cm³). (Data: Table 9.1.)
(b) What angular velocity (rpm) would be required to obtain a sedimentation rate of 1 mm/hour if the radius is 5 cm?

9.32   For a sedimentation experiment to be useful, the sedimentation must be fast compared to diffusional mixing. For a macromolecule with $D = 10^{-7}$ cm² s⁻¹, $M = 10^5$ g, and a buoyancy correction $b = 0.25$, calculate the width of the $\partial c / \partial x$ curve after 10 hours. If the boundary started 5 cm from the axis, how far would it move in 10 hours if $\omega = 4000$ radians/sec?

9.33   An equilibrium sedimentation was carried out in an ultracentrifuge (80,000 rpm) in a 3-mm tube mounted with its bottom 1 cm from the rotation axis. The temperature was 23°C. The solute had a specific volume of 0.750 cm³/g and the solvent was water ($\rho_0 = 0.9951$ g/cm³ at 23°C). The solute concentration at the bottom of the tube was 4.2 times that at the top. Calculate the molecular weight.

9.34   (a) Show that, in a gravitational equilibrium sedimentation, the concentration ratio (c_1 at x_1, c_2 at x_2) is given by:

$$\frac{\ln\left(\frac{c_2}{c_1}\right)}{x_2 - x_1} = \frac{Mg(1 - \rho_0\bar{v}_s)}{RT}$$

[Note the similarity of this result to Eq. (4.5) for the variation of pressure with altitude.]
(b) Calculate the concentration ratio between the top and bottom of a 10-cm tube containing urease in water ($M = 4.9 \times 10^5$ g/mol, $\bar{v}_s = 0.73$ cm³/g, $\rho_0 = 0.9982$ g/cm³, $t = 20$°C).

Time is the devourer of all things.

–Ovid

10

Chemical Kinetics

Thermodynamics deals with the direction of spontaneous change for chemical reactions and other processes but has no concern with whether, in practice, the change will occur in a finite time. Thermodynamics is also conveniently independent of the details of the process, the often complex series of events and states that lead from the initial to the final states. Chemical kinetics deals with the rate at which chemical reactions proceed; as such it is necessarily concerned with the details of the process that thermodynamics could ignore. The practical interest in these details goes two ways. A reaction we want to happen and that is thermodynamically favorable is of little use if it does not occur at a finite rate. Undesirable reactions that are spontaneous may be prevented if we know how they occur. The deterioration, corrosion, and decay of natural or manufactured products are examples of such undesirable chemical processes; the thermodynamics cannot be reversed, but it may be possible to alter the rate so that the effect is not serious in some time interval (for example, during the useful lifetime of the product, or until the expiration of the guarantee). In effect, thermodynamics tells us what nature wants to do; chemical kinetics can tell us how to make nature do what we want it to do, provided the exact manner (mechanism) by which reactions occur can be discovered.

10.1   Rate Laws

The time dependence of a reaction is generally formulated in terms of the time rate of change of the concentration of one or the other of the reactants or products. For example, the rate of the reaction:

$$2\ NO + O_2 = 2\ NO_2$$

could be formulated in terms of the rate of disappearance of a reactant:

$$-\frac{d[NO]}{dt} \quad \text{or} \quad -\frac{d[O_2]}{dt}$$

or the rate of appearance of the product:

$$\frac{d[NO_2]}{dt}$$

This could lead to some ambiguity, however, since, because of the stoichiometry, the NO will react twice as fast as the oxygen. To avoid ambiguity, the **reaction velocity** (v) is defined as the rate of change of the concentration of any species (C_i) divided by its stoichiometric coefficient (ν_i):

$$v = \frac{1}{\nu_i}\frac{dC_i}{dt} \tag{10.1}$$

We use the convention that the stoichiometric coefficient of a *reactant* is negative, while that of a *product* is positive; thus, for the reaction above, $\nu = 2$ for NO_2, $\nu = -2$ for NO and $\nu = -1$ for oxygen. This convention has the effect that reaction velocities are always positive; for a reactant, dC/dt and ν are both negative while both are positive for products. The velocity of this reaction may be defined as:

$$v = \frac{1}{2}\frac{d[NO_2]}{dt}, \qquad v = -\frac{d[O_2]}{dt}, \qquad v = -\frac{1}{2}\frac{d[NO]}{dt}$$

Will all these velocities be the same? In most cases (including this one) they will be; but, as we shall see in due course, if the reaction involves intermediates with significant concentration and lifetimes, the various velocities that can be defined from the stoichiometric equation are not necessarily identical. Nonetheless, the rate of change of any reactant or product may be used to define the reaction velocity, and the practical choice may depend on which is more easily and/or more accurately measured.

A *rate law* is a mathematical statement of how the reaction velocity depends on concentration:

$$v = f\,(\text{concentration})$$

The reaction velocity must clearly depend on the concentration of the reactants in some manner, since, if any of them were missing, the reaction could not occur; however, we shall encounter cases where, for a significant portion of the reaction, the observed rate law is independent of the concentration of one or more of the reactants. The rate law may also depend on the concentration of some material that does not enter into the stoichiometric reaction; if this material increases the reaction velocity, it is called a *catalyst;* if it decreases the reaction velocity, it is called an *inhibitor*. The rate law may also involve the concentration of one or more of the products; if this increases the velocity, the reaction is called *autocatalytic;* if the products decrease the velocity, the reaction is called *self-inhibiting*.

In a large number of cases the rate law is found to depend on a simple power of the concentrations — for example:

$$v = k[A]^x[B]^y[C]^z \tag{10.2}$$

The proportionality constant (k) is called the *rate constant;* the manner by which the reaction velocity was defined ensures that k will always be a positive number. The exponents define the *order* of reaction; for example, if $x = 2$, the reaction is said to be second order in A. the sum of the exponents, $x + y + z$, is the overall order; if $x = 2$, $y = 0.5$, $z = 1$, the reaction is said to have an order of 3.5. Reaction orders are commonly integers; less commonly they may be ratios of integers (3/2, 4/3, and so on) or even irrational numbers. [The latter case is usually an indicator that the rate law is not really of the form of Eq. (10.2).] We shall also encounter examples of rate laws for which an order cannot be defined; that is, they do not have the mathematical form of Eq. (10.2).

Elementary Reactions

Before discussing the details of rate laws, we must clarify an important point that is the source of much of the confusion attendant upon the subject of chemical kinetics; this involves the various meanings that can be assigned to a chemical equation.

Primarily, a chemical equation is a stoichiometric statement; thus,

$$H_2 + I_2 = 2\ HI$$

means that 126.90 grams of iodine are required to react with each gram of hydrogen; this is more commonly expressed using the mole concept — one mole of hydrogen reacts with one mole of iodine to form two moles of HI. The stoichiometric equation also has meaning in thermodynamics in terms of an initial state (H_2 mixed with I_2) and a final state (HI).

Often, chemical equations are given an additional meaning in terms of what the *molecules* are doing. Thus, the equation above may be interpreted to mean that a molecule of hydrogen collides with a molecule of iodine, they exchange partners, and two molecules of HI depart. This is sometimes true, but in many cases (*including* this one) the sequence of events between reactants and products is more complex. In general the situation must be represented as:

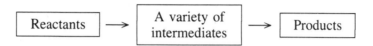

Figure 10.1 shows such a situation for the combustion of low-molecular-weight hydrocarbons in an abbreviated and simplified fashion.

We shall attempt to distinguish between these two interpretations by using the equals ($=$) for the stoichiometric reaction; reactions for which the molecular interpretation is valid are called *elementary reactions,* and for these we shall use an arrow ($\longrightarrow$). For example, the stoichiometric reaction:

$$2\ H_2 + O_2 = 2\ H_2O$$

is known to involve a dozen or so elementary steps; some of the more important ones are:

$$H_2 + O_2 \longrightarrow HO_2 + H$$

$$H_2 + HO_2 \longrightarrow OH + H_2O$$

$$OH + H_2 \longrightarrow H_2O + H$$

$$O_2 + H \longrightarrow OH + O$$

$$H_2 + O \longrightarrow OH + H$$

The series of elementary reactions that constitute the stoichiometric reaction is called the *mechanism.*

The reader is warned that many authors, indeed many workers in the field, do not make this distinction clearly or consistently. Furthermore, it is not always possible or desirable to be totally unambiguous; it is a common practice (which we shall also use) to summarize a series of reactions as elementary if they are *kinetically simple* — that is, if they behave kinetically just like a single-step, elementary reaction. This is often unavoidable, since few reactions have been studied in such detail that all the elementary steps are known.

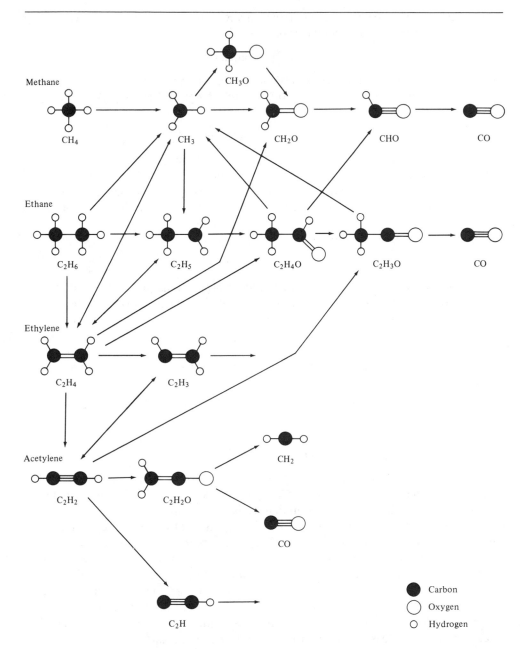

Figure 10.1 Hydrocarbon combustion. The burning of hydrocarbons is a complex process involving many intermediates; a few examples are shown here for C1 and C2 hydrocarbons. The intermediates O, H, OH, and H_2 are not shown on the diagram for simplicity's sake; likewise, the combustion products CO_2 and H_2O are not shown. (From W. C. Gardiner, Jr., *Scientific American*, February 1982)

Rate laws for elementary reactions are always just what one would expect. For example, a unimolecular decomposition:

$$A \longrightarrow \text{products}$$

will proceed at a rate that depends only on the concentration of eligible molecules in the reaction; that is, it is directly proportional to the concentration of A and hence is first-order:

$$v = k[A]$$

This type of kinetics will result if the species A is inherently unstable (perhaps as a result of some prior event such as a collision or reaction). Many first-order reactions are known, but most are thought to be more complex (see Section 10.6). The purest example of first-order kinetics is not chemical at all; it is the decay of radioactive nuclei.

The most common type of elementary reaction is the bimolecular reaction:

$$A + B \longrightarrow \text{products}$$

Here, we expect the frequency of reaction to be proportional to the concentration of eligible species, [A], times the concentration of eligible partners, [B], so that the rate law will be second-order:

$$v = k[A][B]$$

By the same process of reasoning, we expect a trimolecular (or termolecular) reaction:

$$A + B + C \longrightarrow \text{products}$$

to be third-order:

$$v = k[A][B][C]$$

At low concentrations, trimolecular collisions are several orders of magnitude less probable than bimolecular collisions. Generally speaking, a series of bimolecular steps will be faster than a trimolecular step that leads to the same result. A limited number of third-order gas-phase reactions are known; for most of these it is at least possible that they are not elementary. We shall discuss this point further in Section 10.6.

"Normal" chemical reactions, the sort of situation in which you can take chemicals off the shelf, mix them, and measure the rate of reaction, are generally not elementary. Elementary reaction rates can be and have been measured, but the procedures to do this are somewhat exotic. (See ref. 1 for some examples.) Although it may work in the majority of cases, it is generally dangerous to infer a rate law from the stoichiometric equation unless the reaction is known to be elementary.

Much of the confusion on this subject probably comes from the kinetic derivation of the mass-action equilibrium principle as it was formulated by Van't Hoff and as it is still taught in elementary courses. If a reaction:

$$A + B = C + D$$

is elementary, then the forward reaction velocity is $v_f = k_f[A][B]$, and the reverse reaction velocity is:

$$v_r = k_r[\text{C}][\text{D}]$$

At equilibrium, the forward and reverse velocities are equal, so:

$$v_f = v_r$$

$$k_f[\text{A}][\text{B}] = k_r[\text{C}][\text{D}]$$

The equilibrium constant is, therefore:

$$K_c = \frac{[\text{C}][\text{D}]}{[\text{A}][\text{B}]} = \frac{k_f}{k_r} \tag{10.3}$$

Equation (10.3) is clearly valid if the reaction is elementary — that is, single-step; but it is also required by thermodynamics and must therefore be true for complex, multi-step reactions as well. (The second equality, $K_c = k_f/k_r$, requires further justification; this will be discussed in Section 10.5.) However, it should not be presumed that this necessarily implies that the rate laws from which Eq. (10.3) was derived are correct. First, it is possible that there were factors that cancelled at equilibrium when the forward and reverse velocities were equated. Second, it is quite possible that the reaction mechanism and, hence, the rate law, may be different far from equilibrium than near equilibrium. All of this should not obscure the fact that simple, mass-action rate laws are correct in many cases.

Rate Laws: Examples

The reaction:

$$\text{H}_2 + \text{I}_2 = 2\,\text{HI}$$

was one of the earliest for which accurate kinetic data were available (M. Bodenstein, 1894). The equilibrium was approached from both directions, and the forward and reverse rate laws were found to be:

$$v_f = k_f[\text{H}_2][\text{I}_2]$$

$$v_r = k_r[\text{HI}]^2$$

This, of course, is exactly what one would expect for an elementary, bimolecular reaction and the law of mass action, and for years it was used as a textbook example of these ideas. However, seventy years later [J. H. Sullivan, *J. Chem. Phys.*, 46, 73 (1967)] it was demonstrated that this reaction is not, after all, a simple bimolecular reaction.

Those who think that a halogen is a halogen might expect the reaction:

$$\text{H}_2 + \text{Br}_2 = 2\,\text{HBr}$$

to be similar to the HI reaction. Kinetically, it clearly is not, since the rate law is:

$$v = \frac{k[\text{H}_2][\text{Br}_2]^{1/2}}{1 + \dfrac{k'[\text{HBr}]}{[\text{Br}_2]}}$$

This reaction is obviously not a simple bimolecular process. Also, this example illustrates that the concept of reaction order is not always applicable; that is, it is not of the form of Eq. (10.2). However, at the beginning of the reaction, the apparent order would be $\frac{3}{2}$, since the second term of the denominator would be small compared to 1. Changes in the apparent order of a reaction as the reaction progresses are generally indicative of a complicated rate law such as this.

The thermal decomposition of ethylene oxide:

$$\overset{\displaystyle O}{\overset{\diagup\ \diagdown}{CH_2=CH_2}} \ = \ CH_4 + CO$$

is first-order:

$$v = k[C_2H_4O]$$

This is a fairly common result for gas-phase decompositions. Such reactions are called "unimolecular"; their mechanism will be discussed in Section 10.6.

On the other hand, the thermal decomposition of acetaldehyde:

$$CH_3CHO = CH_4 + CO$$

has a rate law:

$$v = k[CH_3CHO]^{3/2}$$

The decomposition of ozone:

$$2\,O_3 = 3\,O_2$$

when catalyzed by N_2O_5 has a rate law:

$$v = k[O_3]^{2/3}[N_2O_5]^{2/3}$$

The preceding examples were all homogeneous reactions. Reactions that are catalyzed by a solid and that occur on the surface often give unusual rate laws — for example:

$$2\,SO_2 + O_2 = 2\,SO_3 \quad [\text{Pt catalysis}]$$

$$v = \frac{k[SO_2]}{[SO_3]^{1/2}}$$

$$2\,N_2O = 2\,N_2 + O_2 \quad [\text{Pt catalysis, } 741°C]$$

$$v = \frac{k[N_2O]}{1 + k'[N_2O]}$$

$$2\,NH_3 = N_2 + 3\,H_2 \quad [\text{on a tungsten filament, } 856°C]$$

$$v = k$$

The last is an example of a zeroth-order reaction. In view of the preceding discussion, you may find it difficult to imagine how the reaction order could be *zero;* a rate law independent of reactant concentration implies that the reaction would proceed even if there were *no* reactant! We shall return to this question in Section 10.9.

When one reactant is in excess — for example, if it is the solvent — the apparent order of the reaction may differ from the actual order. For example, the hydrolysis of an ester:

$$\text{ester} + H_2O = \text{alcohol} + \text{acid}$$

has a rate law:

$$v = k[H_2O][\text{ester}]$$

If the water is in great excess, its concentration will not change much in the course of the reaction and the rate law will be pseudo-first-order with an effective rate constant $k_e = k[H_2O]$.

These examples are perhaps more cautionary than typical, but they are intended to emphasize the point that the macroscopic rate law is an empirical expression whose form does not automatically follow from the stoichiometry of the reaction.

The Objectives of Chemical Kinetics

Now, having defined our terms, we are in a position to undertake an overview of the field of chemical kinetics and its objectives.

The first objective usually is to establish the empirical rate law; this involves determining both its mathematical form and the values of the rate constants as a function of temperature and other relevant conditions. Knowing the rate law and rate constants may be an end in itself; such data are important factors in process design.

A second objective may be to determine the mechanism of the reaction — that is, the sequence of elementary steps that constitute the overall reaction. This is an important step, because only through the study of elementary reactions can we obtain a store of basic data, theoretical understanding, and chemical intuition that can permit us to predict rate laws and anticipate how they may change with conditions. (Scale-up problems — the failure of reactions to proceed in the plant the way they did in the lab — are often related to reaction mechanism.) Understanding mechanisms is also useful when we wish to intervene and affect the course of the reaction; an everyday example of such intervention is the use of additives in gasoline to prevent engine knock, deterioration, and corrosion.

An important part of chemical kinetics is the empirical study of elementary reactions. These reactions often involve unstable species such as H, OH, and CH_3, which, while they are vital intermediates in many chemical reactions, cannot be isolated in large quantities. These experiments are clearly much more demanding than simple macroscopic kinetics.

The fourth part of this field is the theoretical calculation and prediction of reaction rates. The objective here is to use formalisms such as kinetic theory, statistical mechanics, and quantum mechanics to do for kinetics what they have done for thermodynamics (as illustrated, for example, in Chapter 5). The theoretical calculation of kinetic properties is substantially more difficult than the calculation of equilibrium properties, and this field is correspondingly less well developed. Despite the difficulties, considerable progress has been made, and theoretical kinetics is, in some cases, becoming a practical adjunct to empirical work.

A fifth aspect of chemical kinetics has developed rapidly in the last several decades. These studies, generally referred to as *chemical dynamics,* focus on the dynamics of collisions and reactions of individual molecules through studies of molecular collisions in high vacuum.

10.2   Determination of Rate Laws

The first fact with which one must deal in kinetics is that, ordinarily, experimental methods do not measure *rates* but, rather, concentrations as a function of time, $C(t)$. We must, therefore, first make a connection between the derivative, dC/dt, that enters into the rate law and $C(t)$. There are two ways to do this: integrate the rate law or differentiate the data. Each of these methods has its advantages and disadvantages. The techniques and inherent problems of numerical differentiation are discussed in Appendix I; briefly, the effect of limited precision and experimental error on a data set is magnified by differentiation. From the numerical point of view, it is preferable to deal with the integrated rate law; but this method is not without its problems. First, you must know what the rate law is before you integrate it; you cannot, strictly, prove that an integrated rate law is correct except by proving that all other possibilities are wrong. Also, integrated rate laws are usually more complex mathematically than the differential form. We shall discuss both of these approaches after we look briefly at how data may be generated.

Experimental Considerations

In most cases, the measurement of kinetic data is just a problem in analytical chemistry with an added dimension, time. Any analytical technique for measuring concentrations of the reactants and/or products can be used (and nearly all have been used), provided that the analysis can be completed in a time that is short compared to a time in which significant reaction occurs.

If a reaction is very slow (hours to days), the simplest procedure is possible: mix the reactants and analyze aliquots of the reacting mixture at various time intervals; furthermore, nearly any analytical technique is usable. If the time scale is of the order of minutes, it becomes necessary either to use a continuous method to monitor the extent of reaction or to stop the reaction for analysis. Reactions may be stopped by rapid cooling, by destroying one of the reactants, or by removal of a catalyst.

For continuous monitoring of the extent of the reaction, any property that varies with the extent of reaction may be used; for example: pressure (for gas reactions), conductivity (for ionic reactions in solution), solution volume (dilatometer), index of refraction, spectrophotometric absorption, optical rotation, viscosity. Also, spectroscopic techniques such as nuclear magnetic resonance (NMR) and infrared (IR) may be used to analyze the mixture as the reaction progresses. Note that if a gas-phase reaction is followed by pressure measurement (manometric), the data reflect the *total* pressure of the reaction, whereas the rate law involves the *partial* pressure. For example, a first-order reaction such as:

$$A \longrightarrow B + C$$

will have a pressure, $P = P_A + P_B + P_C$, whereas the rate law will be $v = kP_A$.

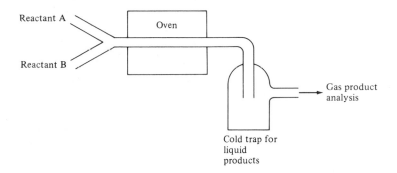

Figure 10.2 Flow reaction (schematic). The reactants are mixed and reacted in a tube in an oven (for the case of high-temperature reactions); the condensible products are collected in a cold trap for subsequent analysis, and the gaseous products are analyzed as evolved. The time of reaction is the ratio of the tube length to the flow rate.

For times shorter than a second, the principal problem becomes one of mixing the reactants thoroughly in the required time. Flow methods (Figure 10.2) are often used here, often with refinements such as stirred-flow or stopped-flow mixing.

For time scales of a millisecond or less, it is generally better to start with things already mixed. Reacting species may be created *in situ* by techniques such as flash photolysis (Figure 10.3) or pulse radiolysis; here, a flash of light or a pulse of high-energy electrons is used to dissociate molecules, and subsequent reactions of the fragments are followed by spectroscopy. For reactions that do not occur at ambient temperature, the reactants can be mixed and then heated rapidly by a supersonic shock wave. For very fast reactions, relaxation methods (see Section 10.12) start with an equilibrium mixture and, by altering some condition, change the equilibrium constant; the adjustment of the reactants to the new equilibrium conditions can be monitored spectroscopically.

Certain techniques are used in special situations. NMR can be used to study exchange-type reactions at equilibrium through their effect on line widths and relaxation times. Gas-phase ion-molecule reactions can be studied using ion cyclotron resonance (ICR) or chemical-ionization mass spectrometry. Reactions of individual molecules in the gas phase have been studied by observing reactive scattering in crossed molecular beams.

A full survey of all the techniques used in kinetics would require a book (or, at least, a chapter) of its own. Assuming that concentrations can be measured, let's now look at how the data are analyzed.

The Differential Method

The usual method for differentiating concentration data in chemical kinetics is to use a simple two-point derivative (Appendix I, Figure A.6):

$$\frac{dC}{dt} \cong \frac{C(t_2) - C(t_1)}{t_2 - t_1} \tag{10.4}$$

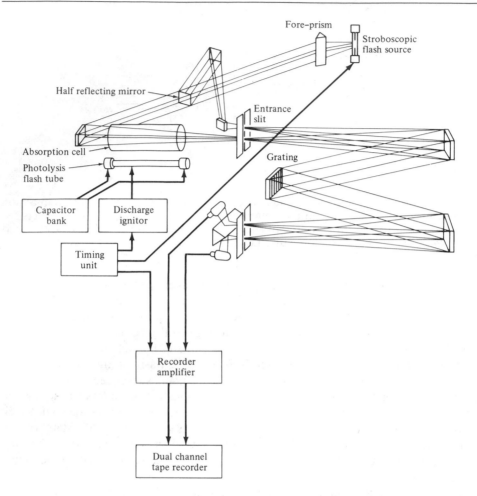

Figure 10.3 Flash photolysis. Schematic diagram of a high resolution flash photolysis spectrometer: H. H. Kramer, M. H. Hanes, and E. J. Bair, *J. Opt. Soc. Am.,* **51**, 755 (1961).

This is effectively the differential mean-value theorem (Appendix I), and it gives the approximate slope at the center of the interval C_1 to C_2. If C is the concentration of a reactant, the mean velocity is:

$$\bar{v} = -\frac{\Delta C}{\Delta t} \tag{10.5}$$

and the mean concentration is:

$$\overline{C} = \tfrac{1}{2}(C_1 + C_2) \tag{10.6}$$

If the rate law has a simple order—that is:

$$v = kC^n$$

the order can be calculated by linear analysis (graphical or least squares) of:

$$\ln \bar{v} = \ln k + n \ln \overline{C} \tag{10.7}$$

This may be useful even if the order is not simply defined, for more complex rate laws will show up as a change in the apparent order of reaction as the reaction progresses.

Practical use of this method reveals an inherent contradiction. Equation (10.5) will be accurate only for small Δt; small in this sense means with respect to the time for the reaction to progress significantly. However, if Δt is small, ΔC will be small and precision will be lost. For example, assume that C can be measured to three significant figures; if $C_1 = 0.521$ and $C_2 = 0.519$, then $\Delta C = 0.002$, which is of the order of the random error in C. This can be remedied by using a larger time interval — for example, with $C_1 = 0.521$, $C_2 = 0.479$ — but then Eq. (10.5) is less accurate. Thus, for this method to be reliable, measurements of very high precision at frequent time intervals are required.

There are more accurate methods for numerical differentiation. These are simplest if the concentrations are measured at equal time intervals. For example, if five measurements of concentration are made (C_0, C_1, C_2, C_3, C_4) at times (t_0, t_1, and so on) that are evenly spaced (Δt), then Eqs. (A-21) of Appendix I give the reaction velocity at the various times as:

$$v(t_0) = \frac{|-25C_0 + 48C_1 - 36C_2 + 16C_3 - 3C_4|}{12\,\Delta t} \tag{10.8a}$$

$$v(t_1) = \frac{|-3C_0 - 10C_1 + 18C_2 - 6C_3 + C_4|}{12\,\Delta t} \tag{10.8b}$$

$$v(t_2) = \frac{|C_0 - 8C_1 + 8C_3 - C_4|}{12\,\Delta t} \tag{10.8c}$$

$$v(t_3) = \frac{|-C_0 + 6C_1 - 18C_2 + 10C_3 + 3C_4|}{12\,\Delta t} \tag{10.8d}$$

$$v(t_4) = \frac{|3C_0 - 16C_1 + 36C_2 - 48C_3 + 25C_4|}{12\,\Delta t} \tag{10.8e}$$

(These equations have been written in terms of magnitudes, so we do not need to make separate provision for the cases when C is a reactant or product concentration; remember that the reaction velocity is always positive.) These equations are most accurate at the middle point [Eq. (10.8c)] and least accurate at the ends [Eqs. (10.8a) and (10.8e)]; in fact it is best not to use the first and last equations if it can be avoided. If the data set consists of N equally spaced points, the velocities at all but the first two and last two can be calculated with Eq. (10.8c), the most accurate formula. The second and penultimate velocities can be calculated with Eqs. (10.8b) and (10.8d). This gives $N - 2$ velocities; the first and last points would require us to use Eqs. (10.8a) and (10.8e), and this is probably inadvisable unless the $C(t)$ data are changing very smoothly and are measured very accurately.

If the data points cannot be obtained at equally spaced time intervals, there are a number of polynomial interpolation methods that can be used; a text on numerical analysis should be consulted.

Table 10.1   Decomposition of di-t-butyl peroxide

t (min)	C/C_0	$\bar{v}/C_0$	$\bar{C}/C_0$
0	1		
		0.0204	0.97960
2	0.9592		
		0.0180	0.95020
3	0.9412		
		0.01755	0.92365
5	0.9061		
		0.01770	0.89720
6	0.8884		
		0.01715	0.87120
8	0.8541		
		0.01680	0.8457
9	0.8373		
		0.01655	0.82075
11	0.8042		
		0.01360	0.79740
12	0.7906		
		0.01490	0.77570
14	0.7608		
		0.01390	0.75385
15	0.7469		
		0.01405	0.73285
17	0.7188		
		0.0153	0.71115
18	0.7035		
		0.01195	0.69155
20	0.6796		
		0.01420	0.67250
21	0.6654		

Example: The gas-phase decomposition of di-t-butyl peroxide (DTBP) was studied by Raley, Rust, and Vaughan [*J. Am. Chem. Soc.*, 70, 88 (1948)]:

$$(CH_3)_3COOC(CH_3)_3 = 2\ CH_3COCH_3 + C_2H_6$$

The data for concentration of DTBP vs. time are presented in Table 10.1 and Figure 10.4(a). (For convenience, the concentrations are given as ratios to the initial concentration, C/C_0; the original data are given in Problem 10.6.)

The data points at $t = 3$ and $t = 5$ can be used to calculate the average velocity in this interval by Eq. (10.5):

$$\bar{v} = -\frac{0.9412 - 0.9061}{3 - 5} = 0.01755$$

The average concentration is:

$$\bar{C} = \frac{0.9412 + 0.9061}{2} = 0.92365$$

The results of the other calculations are shown by Table 10.1.

Figure 10.4b shows a graph of $-\ln \bar{v}$ vs. $-\ln \bar{c}$; the slope of this graph is, by Eq. (10.7), the order of the reaction. Linear regression (see Appendix I, Section F) gives for the slope a value $n = 1.04$ ($r = 0.8810$). The standard deviation of n is calculated as described in Appendix I: $\sigma(n) = 0.16$. This means that the 90% confidence limits for the order (cf. Table A.1, $DF = 12$) are:

$$0.75 < n < 1.33$$

A conclusion that this is a first-order reaction is reasonably justified.

The rate constant can be calculated from the intercept of Figure 10.4b [cf. Eq. (10.7)]:

$$\ln k = -(3.93 \pm 0.07): \qquad k = (1.96 \pm 0.14) \times 10^{-2} \text{ min}^{-1}$$

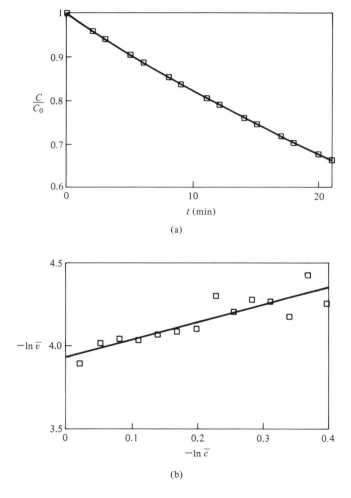

(a)

(b)

Figure 10.4 Reaction of di-*t*-butyl peroxide. (a) Concentration vs. time. (b) The determination of reaction order from reaction velocities (cf. Table 10.1). Note the large amount of scatter in this method, even when using relatively good data.

Because of the scatter in this method, rate constants so calculated cannot be relied upon. Once the order of reaction is established, the integral method (to be described shortly) should be used to calculate k. (This is done later; the result, $k = (1.93 \pm 0.01) \times 10^{-2} \text{ min}^{-1}$, is close to that calculated above, but, considering the wide range of uncertainty in the former method, the agreement can only be considered to be fortuitous.)   ■

Example: The reaction velocity can be calculated more accurately using Eqs. (10.8). For example, the points at $t = 3, 6, 12,$ and 15 can be used to calculate v at $t = 9$:

$$v = \frac{|0.9412 - 8(0.8884) + 8(0.7906) - 0.7469|}{12(3)} = 0.01634$$

Since we know that this reaction is first-order, the velocity can be calculated directly for purposes of comparison:

$$v = kC = (1.93 \times 10^{-2})(0.8373) = 0.01616$$

The velocity calculated with Eq. (10.8c) is within 1.2% of the correct velocity. This is clearly more accurate than the two-point method used above, but it requires a great deal more computation, and the points must be equally spaced in time.   ■

Method of Initial Velocities

A variant of the differential method is to determine the initial rate of the reaction for various initial concentrations. This method is convenient when the concentration of more than one species enters into the rate law. For example:

$$v = k[A]^x[B]^y$$

If the initial concentration of A is a, and of B is b, the initial velocity (v_0) is:

$$v_0 = ka^x b^y$$

$$\ln v_0 = \ln k + x \ln a + y \ln b \tag{10.9}$$

Since a and b can be varied independently, x and y can both be determined.

Some potential problems with this method include: (1) The concentration often drops sharply at $t = 0$, so v_0 may be difficult to calculate accurately; it is probably best to use Eq. (10.8a). (2) If the mechanism is multistep, only the rate law for the first step is discovered; this is not a problem if it is recognized. (3) Many complex reactions have initial transients or induction periods and follow a simple kinetic law only after this time; we shall discuss such a case in Section 10.5.

Method of Isolation

The reaction order of a mixed-order reaction such as:

$$v = k[A]^x[B]^y$$

can be analyzed as above by making the initial concentrations in stoichiometric ratio;

for example, if the stoichiometry is:

$$A + B = \ldots$$

and the initial concentration of A is equal to the initial concentration of B, then [A] = [B] for all times and the order will be $x + y$. If the stoichiometry is:

$$2A + B = \ldots$$

and the initial concentration of A is twice that of B, then [A] = 2[B] at all times and the empirical order is again $x + y$.

The individual order can be determined by *flooding*. If the initial concentration of A (denoted a) is much greater than that of B, it will be reasonably constant during the reaction, and the rate law will be:

$$v = ka^x[\text{B}]^y$$

The apparent order will be y, and the effective rate constant will be ka^x.

Integrated Rate Law: First-Order

If C is the concentration of a reactant, the first-order rate law is:

$$v = -\frac{dC}{dt} = kC \tag{10.10}$$

The variables C and t are easily separated and integrated from $t = 0$, $C = C_0$:

$$\int_{C_0}^{C} \frac{dC}{C} = -k \int_{0}^{t} dt$$

The result can be expressed in various ways:

$$\ln \frac{C_0}{C} = kt \tag{10.11a}$$

$$\ln C = \ln C_0 - kt \tag{10.11b}$$

$$C = C_0 e^{-kt} \tag{10.11c}$$

From Eqs. (10.10) and (10.11), it should be apparent that the first-order rate constant has units of $[\text{time}]^{-1}$ and is independent of the units used for C. Often, the rate law for gas reactions is formulated in terms of partial pressure rather than concentration; if the reactant is A:

$$\frac{-dP_A}{dt} = kP_A \tag{10.12}$$

For first-order, this change does not affect the value of k, and Eqs. (10.11) are still valid with the partial pressure in place of the concentration.

It is not usually necessary to measure actual concentrations; any property that is proportional to the concentration of the reactant or product will do as well. Suppose that λ is some property (such as index of refraction, absorption coefficient, conductivity) that varies linearly with the extent of reaction; we denote the initial ($t = 0$) value as λ_0 and the value at the end of the reaction ($t \rightarrow \infty$) as λ_∞. Then the concentration of reactant C is:

$$C(t) = p(\lambda - \lambda_\infty)$$

$$C_0 = p(\lambda_0 - \lambda_\infty) \tag{10.13}$$

where p is a proportionality constant. Using these relationships in Eq. (10.11a) gives:

$$\ln\left(\frac{\lambda_0 - \lambda_\infty}{\lambda - \lambda_\infty}\right) = kt \tag{10.14}$$

For first-order kinetics, it is not even necessary to know the value of the proportionality constant.

Methods of Data Analysis

Several methods may be used for calculating the rate constant from a data set. In discussing these, we shall presume a first-order reaction, but most comments apply equally to other orders.

Method 1.   Equation (10.11a) can be used to calculate k from each data point. This method gives a great deal of weight to the $t = 0$ point, C_0; this may be justified if C_0 is known more accurately than the other points, and there is no induction period or initial transient for the reaction (see discussion under method 4, below).

Method 2.   Pairs of points—for example, C_n at t_n and C_{n+1} at t_{n+1}—may be used to calculate k; with Eq. (10.11b) for each point, subtraction gives:

$$\ln \frac{C_n}{C_{n+1}} = k\,\Delta t$$

$$\Delta t = t_{n+1} - t_n \tag{10.15}$$

For a set of N data points, this method can be used for each adjacent pair of points to give $N - 1$ values of k. This is a good way to look for trends in the data that may indicate that the mechanism is not simple first-order for the entire reaction. However, the points must be far apart enough that $\Delta C = C_{n+1} - C_n$ is large compared to the uncertainty in C.

The $N - 1$ values of k calculated by this method should not be simply averaged, for this tends to give the greatest weight to the first and last points. In fact, if the time interval (Δt) is always the same, only the first and last point determine the "average" rate constant calculated this way. Suppose that there are seven equally spaced measurements, $C_0, C_1, \ldots, C_6$, and k is calculated using each successive pair:

$$k_1 = \frac{1}{\Delta t} \ln \frac{C_0}{C_1}, \quad k_2 = \frac{1}{\Delta t} \ln \frac{C_1}{C_2}, \quad \ldots, \quad k_6 = \frac{1}{\Delta t} \ln \frac{C_5}{C_6}$$

The average value of k will be:

$$\bar{k} = \frac{1}{6}(k_1 + k_2 + k_3 + k_4 + k_5 + k_6) = \frac{1}{6\,\Delta t}\left(\ln\frac{C_0}{C_1} + \ln\frac{C_1}{C_2} + \cdots + \ln\frac{C_5}{C_6}\right)$$

$$= \frac{1}{6\,\Delta t}\ln\frac{C_0}{C_6}$$

This is the same value that would have been calculated if the intermediate points had never been used! One way around this problem is to do a pyramid average, giving the center points extra weight. For the case of $N = 7$ (6 values of k), used in the preceding illustration, pyramid averaging gives:

$$\bar{k} = \frac{k_1 + 2k_2 + 3k_3 + 3k_4 + 2k_5 + k_6}{12}$$

A better way is to use method 4, below. (In this example, 12 in the denominator is just the sum of the weights, $1 + 2 + 3 + 3 + 2 + 1$; another value would be applicable if $N \neq 7$.)

Method 3.   Equation (10.11b) shows that a graph of $\ln C$ vs. t should be linear with a slope $-k$. An average line should be drawn through the points such that the data are scattered equally above and below the line. The plot should be inspected carefully to ensure that the deviations of the points from linearity are not systematic; this could indicate that the reaction is not really first-order. If the reaction is not elementary, deviations from linearity may be found near the beginning or end of the reaction.

Method 4.   Equation (10.11b) can be used with linear regression ($\ln C$ vs. t) to calculate the best average value of k. Linear regression is discussed in Appendix I; unless you have a calculator that does this automatically, or a computer, this procedure is rather tedious. Nonetheless, in serious work, it should always be used, for it gives an unbiased estimate of k with all points equally weighted; also, the confidence limits of k are easily estimated (see Appendix I). This method has two problems: (1) It does not distinguish between random and systematic errors; by systematic errors, we mean those due to the use of the wrong rate law. [Second-order data may give a reasonable fit to the first-order equations if the data span only a small portion of the reaction and the random error (δC) is large; see below.] For this reason, in serious work, one of the methods 1–3 should be used together with linear regression to look for trends in the data. (2) Taking the logarithm of C tends to give excessive weight to points at the end of the reaction where C is small and $\delta C/C$ comparatively large. This can be avoided by not using points for which $C/C_0 < 0.1$ or by using a weighted least-squares method. [Also, if a computer is available, Eq. (10.11c) can be used with a nonlinear least-squares method to calculate k.] An advantage of the regression method is that it treats the initial concentration, C_0, just like any other point; if the reaction is complex and has an induction period, the "C_0" calculated by linear regression may not be the same as the actual value.

Exercise: Analyze the data of Table 10.1 for the decomposition of DTBP using any of the methods above. Your results should be $k = (1.93 \pm 0.01) \times 10^{-2}$ min^{-1}.   ∎

Half-Life

Reaction kinetics are often characterized in terms of half-lives: the time required for the concentration of a reactant to decrease to one-half its initial value. From Eq. (10.11) it is easily shown that the half-life of a first-order reaction is:

$$t_{1/2} = \frac{\ln 2}{k} \tag{10.16}$$

A notable feature of first-order kinetics is that the half-life is independent of the initial concentration. Also, successive half-lives are equal; that is, if the concentration drops to $\frac{1}{2}$ in 10 seconds, it will drop $\frac{1}{2}$ again, to $\frac{1}{4}$, in 20 seconds; Figure 10.5 illustrates this point.

Exercise: The choice of one-half for characterizing a reaction is conventional but somewhat arbitrary. Show that the "63%" life, the time for C to go to 63% of its initial value, is, for a first-order reaction:

$$t_{63\%} = \frac{0.462}{k}$$

∎

Integrated Rate Law: Second-Order

The simplest case of second-order kinetics is:

$$A = \text{products}$$

$$v = k[A]^2 \tag{10.17}$$

If we denote [A] as C, the equation:

$$\frac{dC}{dt} = -kC^2$$

is easily integrated from $t = 0$, $C = C_0$, to give:

$$\frac{1}{C} = \frac{1}{C_0} + kt \tag{10.18}$$

From any of these equations we can see that the second-order rate constant must have units of $[\text{concentration}]^{-1} [\text{time}]^{-1}$; a typical choice is $dm^3\ mol^{-1}\ sec^{-1}$. The second-order rate law may also be written in terms of partial pressures:

$$-\frac{dP_A}{dt} = kP_A^2 \tag{10.19}$$

In this case, k has units $[\text{pressure}]^{-1} [\text{time}]^{-1}$; this may be, for example, $torr^{-1}\ min^{-1}$.

Exercise: The reaction $H_2 + I_2 = 2\ HI$ has a rate constant $k = 1.4 \times 10^{-8}\ cm^3\ mol^{-1}\ sec^{-1}$ at 400 K. Calculate k in $dm^3\ mol^{-1}\ sec^{-1}$ and $torr^{-1}\ min^{-1}$. [*Hint:* Use the ideal gas law.] (*Answer:* $1.4 \times 10^{-11}\ dm^3\ mol^{-1}\ sec^{-1}$, $3.4 \times 10^{-14}\ torr^{-1}\ min^{-1}$.) ∎

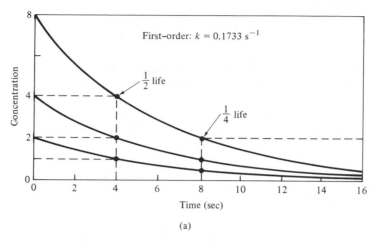

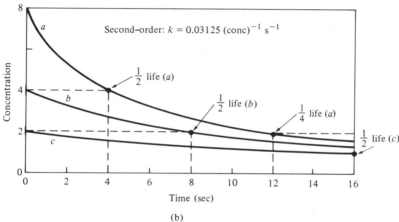

Figure 10.5 Reaction half-life. (a) The half-life of a first-order reaction is independent of the initial concentration; the fourth-life is twice the half-life. (b) The half-life of a second-order reaction is inversely proportional to the initial concentration; the fourth-life is three times the half-life. Note that the fourth-life is the half-life of what remains after the first half-life; the measurement of successive half-lives is one method for determining the reaction order.

Exercise: Show that the half-life of a second-order reaction [such as Eq. (10.17)] is:

$$t_{1/2} = \frac{1}{C_0 k} \tag{10.20}$$

Figure 10.5 contrasts the half-lives of first- and second-order reactions.

Second-order data can be analyzed with Eq. (10.18) using the same methods outlined under first-order: (1) use $(C_0/C - 1)$ for each point to calculate k; (2) use pairs of points to calculate:

$$k = \frac{1}{\Delta t}\left(\frac{1}{C_{n+1}} - \frac{1}{C_n}\right) \qquad (10.21)$$

(3) graph $1/C$ vs. t; the slope will be k; (4) linear regression of $1/C$ vs. t. The same comments and cautions apply here as for first-order.

Exercise: Show that if N data points are used by pairs to calculate $N - 1$ values of k with Eq. (10.21), if Δt is the same for all, the average k will be the same as the k calculated from the first and last points alone.   ■

Exercise: Show that a property (λ) that varies linearly with the extent of reaction [for example, Eqs. (10.13)] can be used to calculate a second-order rate constant for the rate-law equation, Eq. (10.17):

$$k = \frac{1}{C_0 t}\left(\frac{\lambda_0 - \lambda}{\lambda - \lambda_\infty}\right) \qquad (10.22)$$

■

The other type of second-order rate law is:

$$v = k[A][B] \qquad (10.23)$$

The integrated equation for this law depends on the stoichiometry of the reaction; some possibilities are:

$$A + B = \text{products}$$

$$A + 2\,B = \text{products}$$

$$A \longrightarrow \text{products} \qquad (B \text{ is a catalyst})$$

$$A \longrightarrow B + \ldots \qquad (\text{autocatalytic})$$

Each of these gives a different result, but we shall discuss only the most common case, $A + B = \ldots$. We shall denote the initial concentration of A as a, and that of B as b; the moles reacted per unit volume will be denoted x:

$$[A] = a - x, \qquad [B] = b - x$$

The reaction velocity is:

$$v = -\frac{d[A]}{dt} = -\frac{d[B]}{dt} = \frac{dx}{dt}$$

Thus, Eq. (10.23) becomes:

$$\frac{dx}{dt} = k(a - x)(b - x) \qquad (10.24)$$

Note that if $a = b$, then $[A] = [B]$ for all times, and this rate law reduces to Eq. (10.17) with Eq. (10.18) as the integrated version. If $a \neq b$, the differential equation:

$$\frac{dx}{(a - x)(b - x)} = k\,dt$$

can be simplified by use of partial fractions and integrated from $t = 0$, $x = 0$ to give:

$$\ln\left(\frac{[B]}{[A]}\right) = \ln\frac{b}{a} + (b - a)\,kt \qquad (10.25a)$$

$$k = \frac{1}{t(b - a)} \ln\left[\frac{a(b - x)}{b(a - x)}\right] \qquad (10.25b)$$

Equation (10.25a) is suitable for use with linear analysis (graphical or least squares); note that the difference of the initial concentrations $(a - b)$ must be known in order to calculate the rate constant; however, since the intercept gives a/b, this difference can be calculated if either a or b is known.

Exercise: Show that Eq. (10.25b) becomes indeterminate $(0/0)$ when $a = b$. (Optional) Find the limiting form of Eq. (10.25b) when $a = b$; either L'Hopital's rule or an expansion in powers of $\Delta = a - b$ can be used. The result should be Eq. (10.18) with $C = [A] = [B]$. ∎

Other Rate Laws

After first- and second-order, there are numerous other rate laws, and an attempt to list all of them would be lengthy and tedious. A small sampling of some of the more common ones will be given.

A rate law that is nth-order in a single-reactant concentration is easily solved in general:

$$v = -\frac{dC}{dt} = kC^n \qquad (10.26)$$

(Note that many other rate laws, such as $k[A]^x[B]^y$, can be put into this form if the initial concentrations are in stoichiometric ratio.) If $n \neq 1$, integration of Eq. (10.26) gives:

$$\frac{1}{C^{n-1}} = \frac{1}{C_0^{n-1}} + (n - 1)kt \qquad (10.27)$$

For example, if $n = \frac{3}{2}$, a plot of $1/\sqrt{C}$ vs. t should be linear.

Exercise: Show that the half-life for rate-law Eq. (10.26) is:

$$t_{1/2} = \frac{2^{n-1} - 1}{(n - 1)C_0^{n-1}k} \qquad (10.28)$$

∎

Equation (10.28) forms the basis for another method of determining reaction order.

If the half-lives of a reaction are measured for two different initial concentrations (C_0 and C_0') the reaction order can be calculated from:

$$\frac{t_{1/2}}{t_{1/2}'} = \left(\frac{C_0'}{C_0}\right)^{n-1}$$

$$\ln \frac{t_{1/2}}{t_{1/2}'} = (n - 1) \ln \frac{C_0'}{C_0} \tag{10.29}$$

Note that Eq. (10.29) also applies to the case $n = 1$, and that a similar relationship can be derived for any fractional life (for example, the 63% life mentioned in an earlier exercise).

Most known third-order reactions are of the type:

$$\text{(stoichiometry)} \qquad 2A + B = \text{products}$$

$$\text{(rate law)} \qquad v = k[A]^2[B] \tag{10.30}$$

This can be integrated by the method of partial fractions with $[A] = a - 2x$, $[B] = b - x$, to give:

$$\ln \frac{[A]}{[B]} + (a - 2b)\left(\frac{1}{a} - \frac{1}{[A]}\right) = \ln \frac{a}{b} + (a - 2b)^2 kt \tag{10.31}$$

Another case found occasionally is:

$$\text{(stoichiometry)} \qquad A + B = \text{products}$$

$$v = k[A]^{1/2}[B] \tag{10.32}$$

If $b > a$, the integrated rate law is:

$$\tan^{-1}\left(\frac{a - x}{b - a}\right)^{1/2} = \tan^{-1}\left(\frac{a}{b - a}\right)^{1/2} - \frac{\sqrt{b - a}}{2} kt \tag{10.33}$$

If $a > b$, the integrated rate law is:

$$\tanh^{-1}\left(\frac{a - b}{a - x}\right)^{1/2} = \tanh^{-1}\left(\frac{a - b}{a}\right)^{1/2} - \frac{\sqrt{b - a}}{2} kt \tag{10.34}$$

The following relationships will be helpful to anyone who wishes to attempt this derivation ($i = \sqrt{-1}$):

$$\tanh x = -i \tan ix$$

$$\tanh x = \frac{e^x - e^{-x}}{e^x + e^{-x}} = \frac{1}{2} \ln\left(\frac{1 + x}{1 - x}\right) \tag{10.35}$$

Rate laws of the form:

$$-\frac{dC}{dt} = \frac{kC}{K_m + C} \tag{10.36}$$

arise in some cases of enzyme catalysis and for certain surface reactions. The integrated form is:

$$K_m \ln \frac{C_0}{C} + (C_0 - C) = kt \qquad (10.37)$$

The reader will no doubt be relieved to learn that this is the last example.

Commentary

It is often suggested by texts that reaction order be discovered by use of the integrated equations — for example, plot $\ln C$ vs. t, then plot $1/C$ vs. t, until something "works." This is clearly something of a hit-or-miss procedure; it is impossible to define "works" absolutely, since deviations could be due to random or systematic errors, so the only criterion is that one works better than *all* the others. The differential method for determining the order combined with the use of the integrated equation to calculate the rate constants is, at least, objective. To be sure, if the data are not very accurate and do not cover a reasonable range of the reaction, answers such as order $= 2 \pm 1$ may be obtained; however, in such a case the "pick-and-plot" method is likely to be equally ambiguous. The proper conclusion is that more or better data are needed.

The data discussed earlier (Table 10.1) for the decomposition of DTBP are displayed in Figure 10.6 for three different presumed rate laws: $n = 1$ ($\ln C$ vs. t), $n = 1.5$ ($1/\sqrt{c}$ vs. t), and $n = 2$ ($1/C$ vs. t). At first glance all may appear linear; however, a close inspection of the second-order graph will clearly show a systematic trend in the deviations of the points from a straight line. The choice between $n = 1$ and $n = 1.5$ is by no means obvious; however, statistical analysis clearly favors first-order. (The correlation coefficients are $n = 1$, $r = 0.999968$; $n = 1.5$, $r = 0.999718$; $n = 2$, $r = 0.998621$.) It is relevant to note that the data used for this reaction span only about one-third of the reaction (until C drops to $\frac{2}{3}C_0$); data collected over two half-lives or longer would give a clearer indication of the order, and this is the recommended procedure.

10.3   Effect of Temperature on Rate Constants

Earlier [Eq. (10.3)] we saw that there was a relationship between rate constants and equilibrium constants. The temperature dependence of equilibrium constants is related to the energy change of the reaction by [compare Eq. (6.30)]:

$$\frac{d \ln K_c}{dt} = \frac{\Delta U_{rxn}}{RT^2} \qquad (10.38)$$

Svante Arrhenius (1889) proposed a similar equation for rate constants:

$$\frac{d \ln k}{dT} = \frac{E_a}{RT^2} \quad or \quad \frac{d \ln k}{d(1/T)} = -\frac{E_a}{R} \qquad (10.39)$$

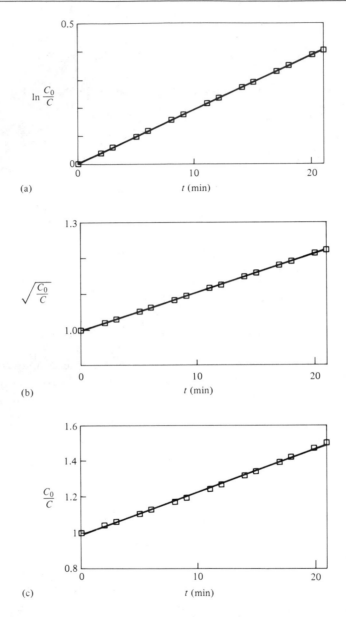

Figure 10.6 Decomposition of di-*t*-butyl peroxide. Plots shown will be linear for (a) first-order, (b) 1.5-order, and (c) second-order, respectively. Note that it is not easy to distinguish visually which reaction order is best. The data are for only about one-third of the reaction; in using this method to determine reaction order, it is generally preferred that data be collected over a half-life or more. Even so, considering that other possibilities (such as $\frac{4}{3}$-order) have been ignored, it can be seen that this method is somewhat subjective.

Table 10.2   Arrhenius factors for various reactions

First-Order	A (s^{-1})	E_a (kJ/mol)
$cis \longrightarrow trans\text{-}1,2\text{-dideuterocyclopropane}$	2.5×10^{16}	272
$CH_3NC \longrightarrow CH_3CN$	4×10^{13}	160
$C_2H_5I \longrightarrow C_2H_4 + HI$	2.5×10^{13}	209
$C_2H_6 \longrightarrow 2\ CH_3$	2.5×10^{17}	384
$N_2O_5 \longrightarrow NO_2 + NO_3$	6.3×10^{14}	88
$CH_3CO \longrightarrow CH_3 + CO$	1×10^{15}	43

Second-Order	A $(cm^3\ mol^{-1}\ s^{-1})$	E_a (kJ/mol)
$NO + O_3 \longrightarrow NO_2 + O_2$	7.9×10^{11}	10.5
$2\ NOCl \longrightarrow 2\ NO + Cl_2$	1×10^{13}	103.6
$NO + Cl_2 \longrightarrow NOCl + Cl$	4.0×10^{12}	84.9
$CH_3 + CH_3 \longrightarrow C_2H_6$	2×10^{13}	~0
$O + N_2 \longrightarrow NO + O$	10^{14}	315
$OH + H_2 \longrightarrow H_2O + H$	8×10^{13}	42
$Cl + H_2 \longrightarrow HCl + H$	8×10^{13}	23
$O_3 + C_3H_8 \longrightarrow C_3H_7O + HO_2$	10^9	51
$2\ NOBr \longrightarrow 2\ NO + Br_2$	4.15×10^{13}	58.1

Third-Order	A $(cm^6\ mol^{-2}\ s^{-1})$	E_a (kJ/mol)
$I + I + Ar \longrightarrow I_2 + Ar$	6.3×10^{15}	5.4
$H + H + H_2 \longrightarrow H_2 + H_2$	1×10^{16}	~0
$O + O + O_2 \longrightarrow O_2 + O_2$	1×10^{15}	~0
$O + O_2 + Ar \longrightarrow O_3 + Ar$	3.2×10^{12}	−9.6
$2\ NO + O_2 \longrightarrow 2\ NO_2$	1.05×10^9	−4.6
$2\ NO + Cl_2 \longrightarrow 2\ NOCl$	1.7×10^{10}	15
$2\ NO + Br_2 \longrightarrow 2\ NOBr$	3.2×10^9	~0

This equation is the definition of E_a, which is called the **Arrhenius activation energy.** If E_a can be regarded as a constant, Eq. (10.39) can be integrated to give:

$$\ln k = \ln A - \frac{E_a}{RT} \tag{10.40}$$

$$k = Ae^{-E_a/RT} \tag{10.41}$$

The factor A, which enters Eq. (10.40) as a constant of integration, is called the *preexponential* or *frequency* factor. Activation energies and frequency factors for some reactions are given in Table 10.2.

Experimentally it is found that plots of the logarithm of rate constants vs. $1/T$ are very close to linear over many orders of magnitude — see Figure 10.7 for an example. This implies that the Arrhenius factors A and E_a, if not exactly constant, vary quite slowly with temperature.

Arrhenius analysis has been used for a remarkably wide range of temperature-dependent phenomena including the chirp rate of crickets (*Oecanthus*), the creeping

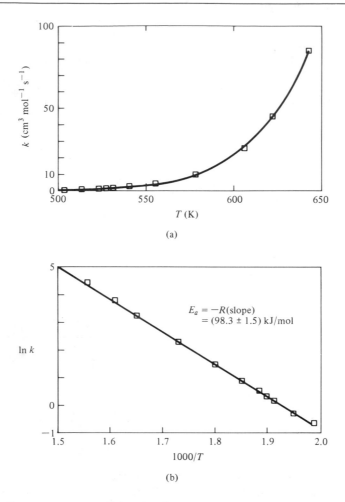

Figure 10.7 Rate constant vs. *T* for the dimerization of butadiene. (a) Many rate constants increase with temperature in this manner. (b) Graphs of ln *k* vs. 1/*T* are often close to linear (Arrhenius's law); the activation energy is −R times the slope of the line. [Data from Vaughan, *J. Am. Chem. Soc.*, 54, 3863 (1932)]

rate of ants (*Limetopum apiculetum*), and the flash rate of fireflies (Figure 10.8); Laidler (ref. 2) provides an interesting discussion of such applications. These are obviously very complex phenomena, yet the activation energies thus measured can provide valuable clues as to the nature of the underlying chemical processes.

The rate constants for the forward and reverse reactions of a reversible process are related to the equilibrium constant [Eq. (10.3)]:

$$K_c = \frac{k_f}{k_r}$$

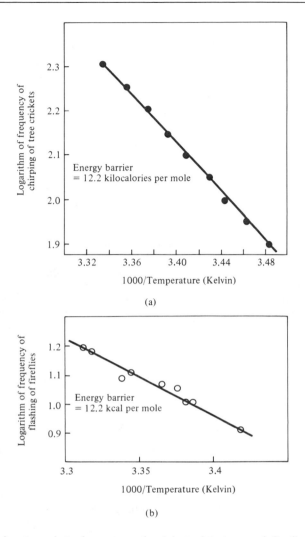

Figure 10.8 Arrhenius plots for rates of cricket chirping and firefly flashing. Many temperature-dependent processes are linear on such a plot; to some extent this is due to the "compressing effect" of logarithms. [From K. J. Laidler, *J. Chem. Ed.*, 49, 343 (1972)]

The temperature derivative of K_c, Eq. (10.38), gives:

$$\frac{d \ln K_c}{dT} = \frac{d \ln k_f}{dT} - \frac{d \ln k_r}{dT} = \frac{\Delta U_{rxn}}{RT^2}$$

Using Eq. (10.39) for the forward reaction (E_{af}) and the reverse reaction (E_{ar}) demonstrates that these activation energies are related to the ΔU of reaction:

$$\Delta U_{rxn} = E_{af} - E_{ar} \qquad \qquad \textbf{(10.42)}$$

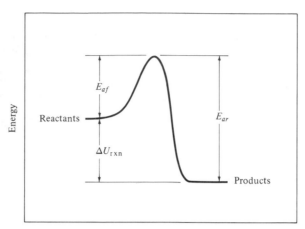

Reaction coordinate

Figure 10.9 Activation energy. This is a rough indication of energy changes during the course of a chemical reaction. The activation energy is the extra energy required before the reactants can cross over to the product state. This diagram also illustrates that the net (thermodynamic) energy of reaction must be the difference between the activation energies of the reverse and forward reactions.

[For solution reactions, ΔU is nearly the same as ΔH; for gases, the ideal gas relationship, $\Delta U = \Delta H + (\Delta n)RT$, can be used.]

The activation energy can be pictured as the extra energy required for the reactants state to pass over into the product state; this idea is diagrammed in Figure 10.9. For what is this extra energy required? Let's look at an example. The reaction:

$$CH_4 + Br \longrightarrow CH_3 + HBr$$

can be viewed as a process in which a CH bond is broken and an HBr bond is made. If this were a sequential process, the extra energy required would be the energy of the reaction:

$$CH_4 \longrightarrow CH_3 + H, \qquad \Delta H_{rxn} = 430.1 \text{ kJ/mol}$$

The net energy of the reaction is this plus the energy we get back when an HBr bond is formed:

$$H + Br \longrightarrow HBr, \qquad \Delta H_{rxn} = -364.0 \text{ kJ/mol}$$

By Hess's law, the ΔH of the total reaction is:

$$CH_4 + Br \longrightarrow CH_3 + HBr, \qquad \Delta H = 66.1 \text{ kJ/mol}$$

However, the activation energy of this reaction is only 74.5 kJ/mole, about 17% of the energy required to break the CH bond. Therefore, the reaction must begin to make the new bond before breaking the old one and must involve some intermediate state such as:

$$Br \cdots H \cdots C \overset{\displaystyle H}{\underset{\displaystyle H}{-}} H$$

This state, while it has more energy than the initial (reactant) state or the final (product) state, has substantially less energy than the separated state:

$$Br + H + CH_3$$

The reverse reaction:

$$HBr + CH_3 \longrightarrow CH_4 + Br$$

has $E_a = 8.4$ kJ/mol, which is about 2% of the energy of the bond (HBr) that must be broken. Note that from activation energies we can calculate ΔH (same as ΔU) for the reaction from Eq. (10.42):

$$\Delta H = 75.4 - 8.4 = 66.1 \text{ kJ/mol}$$

This is in agreement with the Hess's law value.

The simplest chemical reaction usually requires breaking at least one chemical bond, and there is a general correlation between bond energies and activation energies that permits observed activation energies to be used to speculate about the nature of the process. Hirschfelder [*J. Chem. Phys.*, 9, 645 (1941)] has provided a set of empirical rules; for example, a simple, exothermic atom or radical displacement reaction of the type:

$$A + BC \longrightarrow AB + C$$

has an E_a that is typically 5.5% of the energy of the bond that is broken. Evans and Polanyi [*Trans. Faraday Soc.*, 34, 11 (1938)] proposed a linear relationship between the ΔH of an exothermic reaction and the activation energy:

$$E_a = a\,\Delta H + c$$

where a and c are constant for a related series of reactions (cf. Figure 10.10). Such rules can be very helpful in deciding which steps in a complex reaction mechanism are the most likely to be important.

Arrhenius analysis is perfectly general, provided we do not insist that A and E_a be precisely constant. Various theories of reaction rates suggest that rate constants are of the form:

$$k = aT^m e^{-E'/RT} \tag{10.43}$$

where a and E' are constants and m has values such as $1, \frac{1}{2}, -\frac{1}{2}, -\frac{3}{2}, -2$. Since the Arrhenius parameters, A and E_a, are most commonly measured and tabulated, it will be useful to find the connection between the parameters of Eq. (10.43) and those of (10.41). At first sight these equations may seem to be contradictory; if (10.43) is correct, why use (10.41) at all? In practice, the exponential temperature dependence is so strong that it is very difficult to tell from the data whether the preexponential

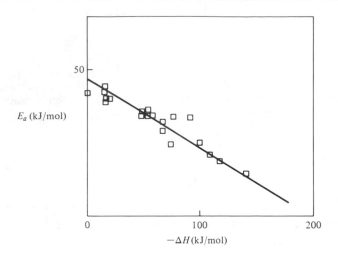

Figure 10.10 Example of an empirical relationship between activation energies and enthalpy of reaction. Such correlations can often be made for a homologous set of reactions. (From J. Nicholas, *Chemical Kinetics,* 1976: New York, John Wiley & Sons, Inc., Fig. 5.10)

depends on temperature or not; in most cases the preexponential cannot be measured accurately enough to determine what the value of m is, a priori.

Two theories of reaction (SCT and ACT, see below) suggest a preexponential of $\sqrt{T}$ [that is, $m = \frac{1}{2}$ in Eq. (10.43)] or T (that is, $m = 1$). The data for the dimerization of butadiene, which were presented on an Arrhenius plot by Figure 10.7, are shown on Figure 10.11 in forms that should be linear if $m = \frac{1}{2}$ (a) or $m = 1$ (b). Comparison of these plots, Figure 10.7 for $m = 0$ (Arrhenius), Figure 10.11(a) for $m = \frac{1}{2}$, and Figure 10.11(b) for $m = 1$, shows that all of them work equally well for these data. (Linear regression gives correlation factors $r = -0.999712$, -0.999714, and -0.999713, respectively.)

The logarithm of Eq. (10.43) gives:

$$\ln k = \ln a + m \ln T - \frac{E'}{RT}$$

Take the derivative with respect to T and equate to Eq. (10.39), the defining relationship for the activation energy:

$$\frac{d \ln k}{dT} = \frac{m}{T} + \frac{E'}{RT^2} = \frac{E_a}{RT^2}$$

Therefore:

$$E_a = E' + mRT \tag{10.44}$$

The preexponential is obtained by equating Eqs. (10.41) and (10.43):

$$k = Ae^{-E_a/RT} = aT^m e^{-E'/RT}$$

$$A = aT^m e^{(E_a - E')/RT}$$

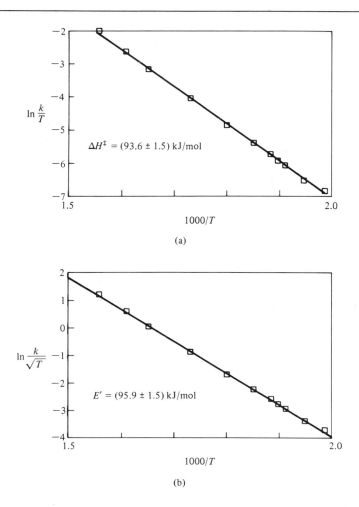

Figure 10.11 Temperature dependence of rate constant for dimerization of butadiene. The graphs shown use the preexponentials suggested by (a) activated complex theory (Eyring) and (b) simple collision theory. Comparison of these graphs to Figure 10.7 and to each other shows that it is very difficult to determine preexponential temperature dependences from experimental data.

Then, using Eq. (10.44) gives:

$$A = aT^m e^m \tag{10.45}$$

10.4 Theories of Reaction Rates

A gas at STP contains 10^{19} molecules in each cm^3 of volume, and each of these has 10^9 collisions per second, so there are, in all, $\sim 10^{28}$ bimolecular collisions per second in 1 cm^3; the collision velocities cover a wide range (cf. Figure 1.16) and, generally, not all collisions will result in a reaction. Thus, even for an elementary reaction, a macroscopic rate constant represents an average of a huge number of events.

If the molecules are not too complex, theoretical calculations are possible for the probability of reaction for any pair of molecules as a function of their relative velocity and the geometrical parameters that describe the possible modes of collision. This fundamental information can, using computer simulations, be projected to predict the average behavior of a large number of molecules. This approach is called *molecular dynamics*. This theory is clearly the most rigorous one for kinetics and the only one that can predict rate constants without egregious assumptions; it is, however, limited to relatively simple molecules. A full explanation of this theory would get us in rather deeper than seems desirable, so we shall discuss only two less rigorous theories: simple collision theory and activated complex theory. These will provide us with several useful ideas for understanding reaction rates. [Nicholas, ref. 6, chap. 8, provides a remarkably clear and simple discussion of molecular dynamics.]

Simple Collision Theory (SCT)

We start with the more or less obvious idea that molecules must be near each other to react. This leaves two questions: what does "near" mean, and how often will they react if they are near to each other?

In chapter 9, a collision of two hard spheres with diameter σ_A and σ_B was defined as an event when the distance between them was equal to:

$$\sigma_{AB} = \frac{\sigma_A + \sigma_B}{2}$$

The frequency of such events per unit volume is given by Eq. (9.15) as:

$$Z_{AB} = \pi \sigma_{AB}^2 \langle v_{AB} \rangle n_A^* n_B^*$$

The average relative velocity is:

$$\langle v_{AB} \rangle = \sqrt{\frac{8 k_b T}{\pi \mu}}$$

In this equation, k_b is Boltzmann's constant (called k in other chapters) and μ is the reduced mass: $\mu = m_A m_B / (m_A + m_B)$. It will be convenient to introduce the molecular weights, M_A, M_B, in place of the molecular masses, $m_A = M_A / L$; thus, the reduced mass is written as:

$$L\mu = \frac{M_A M_B}{M_A + M_B} \tag{10.46}$$

For chemical kinetics, cgs units are generally used, so $k_b = 1.38 \times 10^{-16}$ erg/K, M should be in grams and n^* is in molecules/cm^3. We introduce the **collision cross section:**

$$S_{AB} = \pi \sigma_{AB}^2 \tag{10.47}$$

(In many books the symbol σ is used for the cross section; in other books σ denotes the diameter, and it is this notation we used in Chapters 1 and 9; therefore we shall have to use S to denote the cross section.) If we also introduce the molar concentration $C = n*/L$ (moles per cm^3, in cgs units), the collision frequency becomes:

$$Z_{AB} = S_{AB} L^2 \sqrt{\frac{8RT}{\pi(L\mu)}} C_A C_B \qquad (10.48)$$

with $R = Lk_b = 8.314 \times 10^7$ erg K^{-1} mol^{-1}.

Equation (10.48) would reasonably represent the maximum reaction velocity if each collision resulted in a reaction; note that Z_{AB} is the number of collisions per second in a unit volume, whereas reaction velocities are generally given in terms of *moles* of reactive events per second in a unit volume. Therefore the maximum bimolecular reaction velocity will be:

$$v_{\max} = \frac{Z_{AB}}{L}$$

Actually, if S_{AB} is interpreted literally as the hard-sphere cross section, there are cases for which a reaction may occur faster than this expected maximum. This is generally interpreted as a *reaction cross section,* which is larger than that expected from hard-sphere theory; that is, the molecules can react at a distance larger than σ. Given that molecules have long-range attractive forces (Section 1.7), this should not be too surprising. The most extreme examples of this type occur with gas-phase ion-molecule reactions — for example:

$$H^- + HCN \longrightarrow H_2 + CN^-, \quad S_{AB} = 725 \text{ Å}^2, \quad \sigma_{AB} = 15 \text{ Å}$$
$$NH_4^+ + e \longrightarrow NH_3 + H, \quad S_{AB} = 1400 \text{ Å}^2, \quad \sigma_{AB} = 21 \text{ Å}$$

Why does not every collision result in a reaction? Two reasons may come to mind. First some specific orientation may be required; it is known from molecular beam experiments that the reaction $M + CH_3I \longrightarrow MI + CH_3$ is much more probable when the metal (M) approaches from the I end of methyl iodide [R. B. Bernstein and R. J. Beuhler, *J. Chem. Phys.*, 51, 5305 (1969)]. In simple collision theory this effect is accounted for by a *steric factor p.* Also, the probability of reaction can be expected to depend on the energy of the collision (E). The simplest approach is to define a minimum collisional energy ($E_{\min}$) required for reaction and to assume that the probability that a collision has this energy is exp $(-E_{\min}/RT)$. With this, the SCT reaction velocity becomes:

$$v = p\left(\frac{Z_{AB}}{L}\right) \exp\left(-\frac{E_{\min}}{RT}\right)$$

The specific rate constant is [using Eq. (10.48)]:

$$k = \frac{v}{C_A C_B}$$

and the SCT result for the rate constant is:

$$k = pLS_{AB} \sqrt{\frac{8RT}{\pi(L\mu)}} \exp\left(-\frac{E_{min}}{RT}\right) \qquad (10.49)$$

Unfortunately for the utility of SCT, there is no independent method for estimating the parameters p and E_{min}. Since Eq. (10.49) is of the form of Eq. (10.43) with $m = \frac{1}{2}$, we can obtain a relationship to the Arrhenius parameters:

$$E_{min} = E_a - \tfrac{1}{2}RT \qquad (10.50)$$

$$A = pLS_{AB}e^{1/2} \sqrt{\frac{8RT}{\pi(L\mu)}} \qquad (10.51)$$

Example: Calculate the Arrhenius preexponential A for the reaction:

$$NO + Cl_2 \longrightarrow NOCl + Cl$$

We shall assume that the steric factor $p = 1$ and $T = 300$. The hard-sphere cross section can be estimated from the van der Waals constant b, Table 1.1 and Eq. (1.4):

NO: $b = 27.9$ cm^3, $\sigma = 2.807 \times 10^{-8}$ cm $\Big\}$ $\sigma_{AB} = 3.18 \times 10^{-8}$ cm

Cl$_2$: $b = 56.2$ cm^3, $\sigma = 3.545 \times 10^{-8}$ cm $\Big\}$ $S_{AB} = 3.17 \times 10^{-15}$ cm^2

The molecular weights are $M(NO) = 30$, $M(Cl_2) = 71$; therefore:

$$L\mu = \frac{30(71)}{30 + 71} = 21 \text{ g/mol}$$

$$A = (6.02 \times 10^{23})(3.17 \times 10^{-15})(2.718)^{1/2}\left[\frac{8(8.314 \times 10^7)(300)}{\pi(21)}\right]^{1/2}$$

$$A = 1.7 \times 10^{14} \text{ cm}^3 \text{ mol}^{-1} \text{ sec}^{-1}$$

(The reader should check this calculation, particularly the units.) The observed value (Table 10.2) is 4.0×10^{12}.   ■

In the preceding example, SCT has, rather typically, overestimated the Arrhenius preexponential. We could ascribe this to a steric factor and calculate:

$$p = \frac{4.0 \times 10^{12}}{1.7 \times 10^{14}} = 0.024$$

but this may seem to be scarcely better than a "finagle factor." Actually, such calculation of "steric factors" is a useful exercise, since it permits one to compare reaction efficiencies, having removed such predictable factors as mass and molecular size (S_{AB}). However, since values of p as small as 10^{-6} have been observed, the interpretation of p as a "steric factor" would imply an unexpected and unlikely selectivity for orientation on the part of the reacting molecules.

What this discussion shows is that SCT is severely limited. In particular, it fails to include any ideas concerning the internal structure of the molecules. Nonetheless, it provides several helpful ideas and is useful for finding upper limits for preexponential factors and rate constants (with exceptions as noted above).

Potential-Energy Surface

Consider the interactions of three *atoms,* denoted A, B and C. Assume that these atoms may form diatomic molecules AB and BC but cannot form a stable triatomic molecule (ABC) nor the diatomic molecule AC. Thus, a collision between the atom A and the diatomic molecule BC may produce the bimolecular displacement reaction:

$$A + BC \longrightarrow AB + C$$

The potential energy of any three atoms can, in principle, be calculated from the laws of quantum mechanics as a function of their geometrical dispositions in space; this is conveniently written with two distances, R_{AB}, R_{BC}, and an angle θ:

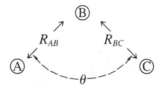

The energy of this system can be expressed parametrically as a function of these coordinates, $E(R_{AB}, R_{BC}, \theta)$. (Cf. Born-Oppenheimer approximation, Chapters 13 and 14.) This surface would require four dimensions to be pictured — a rather difficult situation. Normally, some particular angle θ is presumed; for example, the line-of-centers interaction with $\theta = 180°$:

$$A \ldots (R_{AB}) \ldots B \ldots (R_{BC}) \ldots C$$

Then the energy $E(R_{AB}, R_{AC})$ can be pictured as a surface in three-dimensional space; Figure 10.12 shows an example of such a surface both in perspective [(a) and (b)] and as an energy contour plot (c). These graphs are plotted "backward"; that is, large interatomic distances are at the lower left of the diagram and distances decrease toward the top and right.

In Figure 10.12(c), the plateau at the lower left represents the energy of the three atoms when they are widely separated — that is, the dissociated state A + B + C. The contour *a-a'* represents the interaction of the atoms B + C, with A being at a great distance. The contour along this line [Figure 10.12(d)] is then the potential energy of the BC diatomic molecule; the depth of the well is the bond energy, $D_e(BC)$, and the distance, $R_e(BC)$, is the bond length of the stable diatomic molecule. The contour *b-b'* represents the same situation for the diatomic molecule AB with C at a large distance.

The dashed line *c-d* in Figure 10.12(c) represents the energy as the atom A approaches the diatom BC along the line-of-centers. The maximum energy along this path [whose contour is shown by Figure 10.13(a)] is denoted ‡; this is called the *transition state*. The course of the reaction:

$$A + BC \longrightarrow (ABC)^{\ddagger} \longrightarrow AB + C$$

is traced by this path, which is called the *reaction coordinate*. Note that this transition state is a maximum along the reaction coordinate; more precisely, it is a *saddle point*

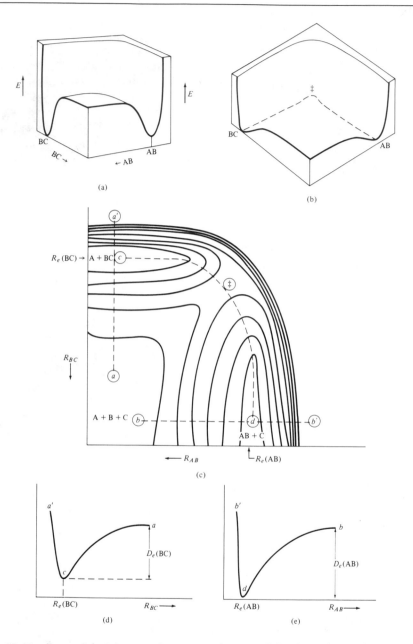

Figure 10.12 Potential-energy surfaces. Panels (a) and (b) show the surface in perspective; (c) is a contour plot of the surface. Panels (d) and (e) show cross sections of (c) for the lines *a'-a* and *b'-b*, respectively. These are graphs of potential energy vs. distance for the two-body interactions BC and AB. On panel (c), the dotted line shows the path of a reactive event, $A + BC \rightarrow AB + C$; this path is called the *reaction coordinate*. The activated complex (denoted ‡) is a saddle point on the surface.

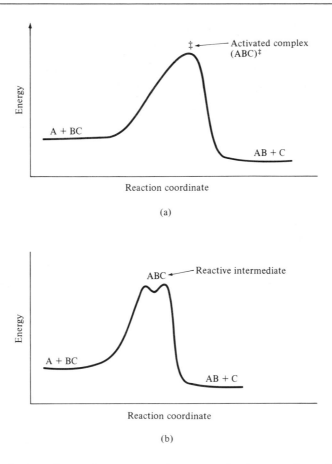

Figure 10.13 Activated complex compared to a reactive intermediate. Graph shows a contour along the dashed line of Figure 10.12(c). If there is a well at the intermediate state (with depth greater than kT), the intermediate is called a reactive intermediate rather than an activated complex.

on the energy surface. If there were a local minimum at this point [as shown by Figure 10.13(b)], the species ABC would be a true triatomic molecule and would be referred to as a *reactive intermediate*. The distinction between a transition state and a reactive intermediate can be a fine one, particularly if the depth of the local minimum is small; for the species to be considered a true molecule, a depth greater than $k_b T$ would be required.

Referring back to Figure 9.12(c), the energy difference between the floors of the valleys *BC* and *AB* must be related to the energy of the reaction. (Roughly, this difference is ΔH_0^θ as defined in Chapters 5 and 6; this point is discussed further in Chapter 13.) The height of the saddle point ($\ddagger$) above the valley floors must be related to the activation energy. (Compare Figures 10.13 and 10.9.)

Activated Complex Theory

Consider the bimolecular reaction:

$$A + B \longrightarrow products$$

The symbols A and B now may denote molecules, not necessarily atoms as in the preceding segment. Activated complex theory, suggested independently in 1935 by H. Eyring and M. G. Evans and M. Polanyi, treats this process as an equilibrium between the reactants and an activated complex (denoted ‡), with only the complex being able to proceed on to the product state:

$$A + B \rightleftharpoons (AB)^{\ddagger} \longrightarrow products$$

The activated complex is treated as an equilibrium species, and an equilibrium constant is defined:

$$K_c^{\ddagger} = \frac{[\ddagger]}{[A][B]} \tag{10.52}$$

Then, if we define f (the *transmission factor*) as the probability that the activated complex will, once formed, go over into the product state, the reaction velocity will be:

$$v = f[\ddagger]$$
$$= fK_c^{\ddagger}[A][B]$$

The specific bimolecular rate constant then is:

$$k = fK_c^{\ddagger} \tag{10.53}$$

From this point, two paths are possible. The more rigorous approach uses statistical mechanics to calculate $K_c^{\ddagger}$ as if it were a real equilibrium constant; the discussion in Chapters 5 and 6 will give some idea of how this could be done. This approach requires the evaluation of a partition function for the activated complex; since this is not a stable species subject to spectroscopic observation, there is some degree of speculation and adjustability in this step. Nonetheless, this approach has been used successfully to obtain reasonable answers for calculated rate constants. (It also has been used to calculate reasonable bimolecular rate constants for reactions that later were found not to be bimolecular.)

We shall discuss only the less rigorous, but more general, thermodynamic formulation of activated-complex theory. In this form the theory can be profitably applied to solution kinetics as well as to gas-phase kinetics. The only result we shall need from the statistical theory is that the transmission factor f is:

$$f = \frac{k_b T}{h} \tag{10.54}$$

where k_b is Boltzmann's constant and h is Planck's constant.

The thermodynamic formulation of activated-complex theory begins by defining the standard free-energy change for the formation of the activated complex:

$$\Delta G^{\ddagger} = \mu^{\theta}(\ddagger) - \mu^{\theta}(A) - \mu^{\theta}(B)$$

This is related to the equilibrium constant by:

$$\Delta G^{\ddagger} = -RT \ln (K_c^{\ddagger}C^{\theta}) \tag{10.55}$$

A very similar equation is used in Chapter 6, but there are important differences, which relate to differences in the conventions used in thermodynamics and kinetics. In thermodynamics, standard free energies generally refer to a standard state with $P^{\theta} = 1$ atm for gases or (in solution) unit concentration, either one mole per dm^3 (molarity) or one mole per kg solvent (molality). In kinetics, the standard state is always unit concentration; if cgs units are used, this is one mole per cm^3; if SI units are used, it is one mole per m^3. Also, $K_c^{\ddagger}$ was defined in Eq. (10.52) so that it has units of (concentration)$^{-1}$; since arguments of logarithms must be unitless, Eq. (10.55) contains a factor, C^{θ}, for the standard-state concentration. If cgs units are used, $C^{\theta} = 1$ mole/cm^3, but if SI units are to be used, $C^{\theta} = 1$ mole/m^3.

The standard free energy for the formation of the activated complex can be separated into enthalpic and entropic contributions with:

$$\Delta G^{\ddagger} = \Delta H^{\ddagger} - T \Delta S^{\ddagger}$$

Using this with Eqs. (10.53), (10.54), and (10.55) gives:

$$k = \frac{k_b T}{h} K_c^{\ddagger} \tag{10.56a}$$

$$k = \frac{k_b T}{hC^{\theta}} e^{-\Delta G^{\ddagger}/RT} \tag{10.56b}$$

$$k = \frac{k_b T}{hC^{\theta}} e^{\Delta S^{\ddagger}/R} e^{-\Delta H^{\ddagger}/RT} \tag{10.56c}$$

This is clearly rather similar to the Arrhenius formulation for the temperature dependence of the rate constant. To see the connection, we use the definition of the Arrhenius activation energy:

$$E_a = RT^2 \frac{d(\ln k)}{dT}$$

with Eq. (10.56a):

$$E_a = RT^2 \frac{d}{dT} \ln\left(\frac{k_b T}{h} K_c^{\ddagger}\right)$$

$$E_a = RT + RT^2 \frac{d \ln K_c^{\ddagger}}{dT}$$

With Eq. (10.38) for the temperature derivative of $\ln K_c^{\ddagger}$, this gives:

$$E_a = \Delta U^{\ddagger} + RT$$

The enthalpy is $H = U + PV$, so:

$$\Delta U^{\ddagger} = \Delta H^{\ddagger} - \Delta(PV)^{\ddagger}$$

with:

$$\Delta(PV)^{\ddagger} = (PV)^{\ddagger} - (PV)_A - (PV)_B$$

For solution reactions, P will be constant and the volume change for formation of the activated complex will be small, so:

$$\Delta U^{\ddagger} \cong \Delta H^{\ddagger}$$

$$\text{(solutions)} \quad E_a = \Delta H^{\ddagger} + RT \tag{10.57a}$$

The preexponential will then be:

$$A = \frac{e k_b T}{hC^{\theta}} e^{\Delta S^{\ddagger}/R} \tag{10.57b}$$

For gas-phase reactions, the ideal gas law gives:

$$\Delta U^{\ddagger} = \Delta H^{\ddagger} - \Delta n^{\ddagger} RT$$

We have discussed only bimolecular reactions for which $\Delta n^{\ddagger} = -1$; however, activated-complex theory can be used for unimolecular reactions ($\Delta n^{\ddagger} = 0$) and trimolecular reactions ($\Delta n^{\ddagger} = -2$). The Arrhenius parameters for these cases are:

(gas, uni-) $E_a = \Delta H^{\ddagger} + RT$ **(10.58a)**

$$A = \frac{ek_b T}{h} e^{\Delta S^{\ddagger}/R}$$ **(10.58b)**

(gas, bi-) $E_a = \Delta H^{\ddagger} + 2RT$ **(10.59a)**

$$A = \frac{e^2 k_b T}{hC^{\theta}} e^{\Delta S^{\ddagger}/R}$$ **(10.59b)**

(gas, tri-) $E_a = \Delta H^{\ddagger} + 3RT$ **(10.60a)**

$$A = \frac{e^3 k_b T}{h(C^{\theta})^2} e^{\Delta S^{\ddagger}/R}$$ **(10.60b)**

Example: Calculate $\Delta H^{\ddagger}$ and $\Delta S^{\ddagger}$ from the Arrhenius parameters (Table 10.2) for the reaction:

$$NO + Cl_2 \longrightarrow NOCl + Cl$$

Assume $T = 300$ K.

$$\Delta H^{\ddagger} = E_a - 2RT = 84.9 \times 10^3 \text{ J/mol} - 2(8.3)(300) = 79.9 \text{ kJ/mol}$$

$$e^{\Delta S^{\ddagger}/R} = \frac{AhC^{\theta}}{e^2 k_b T}$$

$$= \frac{(4.0 \times 10^{12} \text{ cm}^3 \text{ mol}^{-1} \text{ sec}^{-1})(6.63 \times 10^{-27} \text{ erg sec})(1 \text{ mole/cm}^3)}{(2.718)^2 (1.38 \times 10^{-16} \text{ erg/K})(300 \text{ K})}$$

$$= 8.67 \times 10^{-2}$$

$$\Delta S^{\ddagger} = -20.3 \text{ J K}^{-1} \text{ mol}^{-1}$$

If it is desired to compare such entropies to thermodynamic values, it must be remembered that gas-phase entropies (for example, Table 3.2) are for a standard pressure $P^{\theta} = 1$ atm. Since standard states refer to ideal gases, we can use the ideal gas law to calculate the pressure for even such a ludicrously high concentration as one mole per cm^3:

$$P = CRT = (1 \text{ mole/cm}^3)(82.06 \text{ cm}^3 \text{ atm/K})(300 \text{ K})$$

$$= 24{,}618 \text{ atm}$$

The entropy of an ideal gas at pressure P is given by Eq. (3.14) as:

$$S_m(P) = S_m^{\theta}(1 \text{ atm}) + R \ln P$$

The standard entropies per mole are therefore related by:

$$S_m^{\theta}(1 \text{ mole/cm}^3) = S_m^{\theta}(1 \text{ atm}) + 84.07 \text{ J K}^{-1} \text{ mol}^{-1}$$ **(10.61)**

Using data from Table 3.2, we get:

$$S_m^{\theta}(NO, 1 \text{ mole/cm}^3) = 210.62 + 84.07 = 294.69 \text{ J K}^{-1} \text{ mol}^{-1}$$

$$S_m^{\theta}(Cl_2, 1 \text{ mole/cm}^3) = 222.97 + 84.07 = 307.04 \text{ J K}^{-1} \text{ mol}^{-1}$$

Then, since:

$$\Delta S^{\ddagger} = S_m^{\theta}(\ddagger) - S_m^{\theta}(NO) - S_m^{\theta}(Cl_2)$$

we find the molar entropy of the activated complex to be:

$$S_m^{\theta}(\ddagger) = 581.4 \text{ J K}^{-1} \text{ mol}^{-1}$$

This is the standard molar entropy of the activated complex. ∎

The thermodynamic activated-complex theory is no better than collision theory for predicting rates *a priori*. However, discussion of rates in terms of the entropy of activation has proved to be generally more fruitful than discussion of "steric factors," and values of $\Delta H^{\ddagger}$ and $\Delta S^{\ddagger}$ have been measured for a great number of reactions. For example, the effect of catalysts can be clarified by separating the effect they have on the $\Delta H^{\ddagger}$ and $\Delta S^{\ddagger}$ "barriers" to the reaction; some enzymes appear to be effective through their ability to lower the entropy "barrier," perhaps by "preassembling" the reactants in the proper configuration for reaction. (Since entropy increases spontaneously, the "barrier" is more like a "trough"; it may be better to think about the *negentropy,* $-S$, and the "barrier" to reaction, $-\Delta S^{\ddagger}$.)

Activated-complex theory is also useful for discussing the effect of pressure on rate constants. Using (cf. Table 3.1):

$$\left(\frac{\partial G}{\partial P}\right)_T = V$$

$$\left(\frac{\partial \Delta G^{\ddagger}}{\partial P}\right)_T = \Delta V^{\ddagger} = V_m(\ddagger) - \sum V_m(\text{reactants})$$

where V_m is the molar volume (gases) or partial molar volume (solutions). This gives [taking the derivative of Eq. (10.56b) with respect to P]:

$$\left(\frac{\partial \ln k}{\partial P}\right)_T = -\frac{\Delta V^{\ddagger}}{RT} \tag{10.62}$$

10.5 Multistep Reactions

Most ordinary chemical reactions, the type likely to be encountered in a chemical laboratory or plant, involve multiple steps between the reactant and product states. In this section we shall examine the mathematics of multistep reactions; applications will be found for some of these results in subsequent sections.

Reversible Steps

It is certain that all chemical reactions are reversible to some extent, but if the equilibrium constant is greater than 100, this fact can usually be ignored for purposes of chemical kinetics.

The reversible first-order reaction:

$$A \underset{2}{\overset{1}{\rightleftharpoons}} B$$

has a net velocity:

$$v = v_1 - v_2 = k_1[A] - k_2[B]$$

If we assume initial concentrations, a for A, and b for B, and denote the amount reacted (per unit volume) as x, this equation becomes:

$$\frac{dx}{dt} = k_1(a - x) - k_2(b + x)$$

This equation can be rearranged as follows:

$$\frac{dx}{dt} = (k_1 a - k_2 b) - (k_1 + k_2)x$$

$$\frac{dx}{dt} = (k_1 + k_2)(r - x) \tag{10.63a}$$

$$r \equiv \frac{k_1 a - k_2 b}{k_1 + k_2} \tag{10.63b}$$

This equation is readily integrated to give a form very similar to that obtained earlier for the irreversible first-order reaction:

$$\ln \frac{r}{r - x} = (k_1 + k_2)t \tag{10.64}$$

Use of this equation requires a value for r; this can be found if the equilibrium constant:

$$K = \frac{k_1}{k_2}$$

is independently measured. Then, from Eq. (10.63b):

$$r = \frac{Ka - b}{K + 1} \tag{10.65}$$

Note that, if $K \gg 1$ ($k_1 \gg k_2$), $r = a$, and Eq. (10.64) becomes the equation used earlier for the irreversible first-order reaction.

Parallel Reactions

It is not uncommon for reactions to produce more than one product, and the reason is as often kinetic as thermodynamic. The sequence:

$$A \xrightarrow{1} B$$
$$A \xrightarrow{2} C$$

has a velocity:

$$v = -\frac{d[A]}{dt} = k_1[A] + k_2[A]$$

Clearly the loss of reactant is simple first-order, and:

$$[A] = ae^{-(k_1+k_2)t} \qquad\qquad (10.66a)$$

The rates of formation of the products are:

$$\frac{d[B]}{dt} = k_1[A], \qquad\qquad \frac{d[C]}{dt} = k_2[A]$$

$$\frac{d[B]}{dt} = k_1ae^{-(k_1+k_2)t}, \qquad\qquad \frac{d[C]}{dt} = k_2ae^{-(k_1+k_2)t}$$

These can be integrated to give:

$$[B] = \frac{k_1a}{k_1 + k_2}(1 - e^{-(k_1+k_2)t}) \qquad\qquad (10.66b)$$

$$[C] = \frac{k_2a}{k_1 + k_2}(1 - e^{-(k_1+k_2)t}) \qquad\qquad (10.66c)$$

The ratio of the products formed is therefore in proportion to their rate constants:

$$\frac{[B]}{[C]} = \frac{k_1}{k_2} \qquad\qquad (10.67)$$

Note that this is not necessarily the equilibrium ratio of these products. To reach equilibrium would require that either the direct reaction, $B \rightleftarrows C$, or the reverse reactions $B \rightarrow A$ and $C \rightarrow A$, occur at a finite rate.

Principle of Detailed Balance

The threefold reaction discussed above must, in general, be written:

$$A \underset{-1}{\overset{1}{\rightleftarrows}} B, \qquad A \underset{-2}{\overset{2}{\rightleftarrows}} C, \qquad C \underset{-3}{\overset{3}{\rightleftarrows}} B$$

The case discussed above assumed $k_1, k_2 \gg k_{-1}, k_{-2}, k_3, k_{-3}$. Drawing arrows so their length represents the magnitude of the rate constant, this would be:

On an activation energy diagram, we could picture A as a valley high in the mountains; B and C are both lower, but are separated from each other by a high mountain range; that is, the activation energy for direct conversion is very large and the rate correspondingly small.

It may seem at least possible that a dynamic equilibrium in such a system could be established by a cyclic reaction such as:

$$\begin{array}{c} B \\ \nearrow \quad \searrow \\ A \leftarrow C \end{array}$$

This would require that k_1, k_{-2}, $k_{-3} \gg k_{-1}$, k_2, k_3. Such a situation is strictly forbidden by the principle of *microscopic reversibility*. This principle can be derived rigorously from the fact that the mechanical laws governing nature (both quantum and classical) are invariant to time reversal. However, the activation-energy picture makes it seem fairly obvious; such a mechanism would require that the energy of A be higher than that of B which is higher than that of C which is, then, higher than that of A; this is a topological impossibility.

Microscopic reversibility requires that, in a multipath reaction, equilibrium be established along *each path individually*. This, the macroscopic manifestation of microscopic reversibility is called the *principle of detailed balance*. For the cyclic mechanism above, the principle of detailed balance requires, at equilibrium, the following:

Path 1: $\quad k_1[A] = k_{-1}[B]$, $\quad K_1 = \dfrac{[B]}{[A]} = \dfrac{k_1}{k_{-1}}$

Path 2: $\quad k_2[A] = k_{-2}[C]$, $\quad K_2 = \dfrac{[C]}{[A]} = \dfrac{k_2}{k_{-2}}$

Path 3: $\quad k_3[C] = k_{-3}[B]$, $\quad K_3 = \dfrac{[B]}{[C]} = \dfrac{k_3}{k_{-3}}$

The product of the second two equilibrium constants is:

$$K_2 K_3 = \frac{[C]}{[A]} \frac{[B]}{[C]} = \frac{[B]}{[A]}$$

But this ratio is K_1; therefore:

$$K_1 = K_2 K_3$$

$$\frac{k_1}{k_{-1}} = \frac{k_2}{k_{-2}} \frac{k_3}{k_{-3}}$$

Thus, the rate constants are not independent, and the condition for a cyclic mechanism cannot be met.

The principle of detailed balance is often very helpful in sorting out complex reaction schemes. In particular, it requires that the mechanism of the forward reaction be the same as that of the reverse reaction. Suppose that a chemical reaction A = B can be accomplished by two mechanisms: a direct conversion and a catalyzed conversion (C):

$$A + C \xrightarrow{1} B + C, \qquad v_1 = k_1[A][C]$$
$$A \xrightarrow{2} B, \qquad v_2 = k_2[A]$$

The reverse reactions are:

$$B + C \xrightarrow{-1} A + C, \qquad v_{-1} = k_{-1}[B][C]$$

$$B \xrightarrow{-2} A, \qquad v_{-2} = k_{-2}[B]$$

Dynamic equilibrium requires only that the net forward and reverse rates be equal:

$$v_1 + v_2 = v_{-1} + v_{-2}$$

$$k_1[A][C] + k_2[A] = k_{-1}[B][C] + k_{-2}[B]$$

The product to reactant ratio is, thus:

$$\frac{[B]}{[A]} = \frac{k_1[C] + k_2}{k_{-1}[C] + k_{-2}} \qquad (10.68a)$$

From this it appears that the equilibrium ratio may depend on the concentration of the catalyst, a situation forbidden by thermodynamics.

The principle of detailed balance requires that, at equilibrium, the forward and reverse rates be equal along *each path:*

$$v_1 = v_{-1}, \qquad \frac{[B]}{[A]} = \frac{k_1}{k_{-1}}$$

$$v_2 = v_{-2}, \qquad \frac{[B]}{[A]} = \frac{k_2}{k_{-2}}$$

This demonstrates that the ratio k_1/k_2 must be the same as the ratio k_{-1}/k_{-2}; the fraction of reaction by the two paths is the same for the forward and reverse reactions. The equilibrium ratios are, from Eq. (10.68a) above:

$$\frac{[B]}{[A]} = \frac{k_1}{k_{-1}} \left[\frac{[C] + \dfrac{k_2}{k_1}}{[C] + \dfrac{k_{-2}}{k_{-1}}} \right] = \frac{k_1}{k_{-1}} = \frac{k_2}{k_{-2}} \qquad (10.68b)$$

A note of caution: detailed balance, like mass action, should be applied only to elementary, single-step reactions.

Consecutive Reactions

Consider the consecutive first-order reactions:

$$A \xrightarrow{1} B \xrightarrow{2} C$$

The rate equations are:

$$-\frac{d[A]}{dt} = k_1[A] \qquad (10.69a)$$

$$\frac{d[B]}{dt} = k_1[A] - k_2[B] \qquad (10.69b)$$

$$\frac{d[C]}{dt} = k_2[B] \qquad (10.69c)$$

With initial ($t = 0$) concentrations, $[A] - a$, $[B] = 0$, $[C] = 0$, the first of these equations can be solved as indicated earlier:

$$[A] = ae^{-k_1 t} \tag{10.70a}$$

The second equation then becomes:

$$\frac{d[B]}{dt} = k_1 a e^{-k_1 t} - k_2[B]$$

$$\frac{d[B]}{dt} + k_2[B] = k_1 a e^{-k_1 t}$$

This equation can be integrated by a simple change in variable, $Y = [B]e^{k_2 t}$. The result is:

$$[B] = \frac{k_1 a}{k_2 - k_1}(e^{-k_1 t} - e^{-k_2 t}) \tag{10.70b}$$

The product concentration can be obtained from material balance:

$$[C] = a - [A] - [B]$$

$$[C] = a\left(1 - \frac{k_2 e^{-k_1 t} - k_1 e^{-k_2 t}}{k_2 - k_1}\right) \tag{10.70c}$$

Note that Eq. (10.70b) becomes indeterminate (0/0) if $k_1 = k_2$. It can be shown (Problem 10.33) that, if $k \equiv k_1 = k_2$:

$$[B] = kate^{-kt} \tag{10.70d}$$

Sample calculations are shown in Figure 10.14 for several combinations of the two rate constants; note that the rate of formation of the product (C) tends to be dominated by the smallest rate constant — that is, by the rate of the slowest step. [Compare Figure 10.14 (a) and (d) or (b) and (c).] This is the principle of the rate-determining step; in a series of consecutive reactions:

$$A \longrightarrow B \longrightarrow C \longrightarrow D \longrightarrow \cdots \longrightarrow P$$

the rate of product formation tends to be dominated by the slowest step. This fact often makes it difficult to distinguish between elementary and complex reactions.

Exercise: Show that Eq. (10.70c) in the limits $k_1 \gg k_2$ or $k_1 \ll k_2$ attains the form it would have if the reaction were simply $A \to C$ with a rate constant equal to the *smaller* of the two. ∎

Reaction Networks

A complex network of reactions — for example:

$$\begin{array}{ccc} A & \rightleftharpoons B & \rightleftharpoons C \\ \updownarrow & \searrow & \updownarrow \\ E & \rightleftharpoons & D \end{array}$$

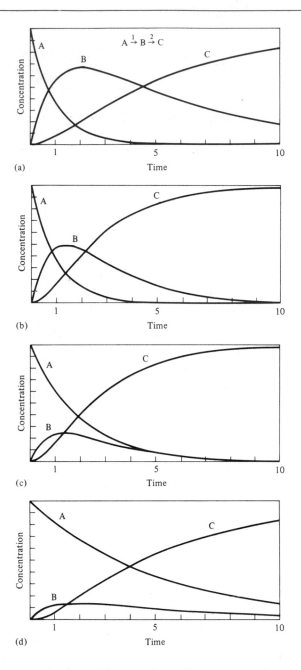

Figure 10.14   Consecutive first-order reactions. The concentrations of the reactant (A), intermediate (B), and product (C) are shown vs. time for several combinations of rate constants: (a) $k_1 = 1$, $k_2 = 0.2$, (b) $k_1 = 1$, $k_2 = 0.5$, (c) $k_1 = 0.5$, $k_2 = 1$, (d) $k_1 = 0.2$, $k_2 = 1$.

can be solved exactly if all steps are first-order. However, the mathematical form of the solution can become rather complex. In general, the solution is a sum of exponentials:

$$[\text{concentration}] = \sum_i C_i e^{-\lambda_i t}$$

where the C_i and λ_i are constants that can be related to the rate constants of the individual steps and the initial concentrations. A good example of such an application occurs for radioactive decay schemes. These solutions are of relatively limited interest in chemical kinetics because elementary reactions tend to be bimolecular and, hence, second-order.

The rate equations for a general reaction scheme with second-order steps cannot be solved exactly; in such a case, one usually must resort to numerical methods. A simple example of such a case is:

$$A + B \underset{2}{\overset{1}{\rightleftharpoons}} C$$

$$C \overset{3}{\longrightarrow} \text{products}$$

(Actually this reaction scheme can be solved in closed form, but the solutions are complicated; see ref. 5, p. 55, for a discussion of this and other second-order schemes that have been solved. We shall use it as an example for numerical solutions; this is the only method that works in all cases.) There is no problem in writing down the rate equations:

$$\frac{d[A]}{dt} = -k_1[A][B] + k_2[C] \tag{10.71a}$$

$$\frac{d[C]}{dt} = k_1[A][B] - k_2[C] - k_3[C] \tag{10.71b}$$

$$\frac{d[P]}{dt} = k_3[C] \tag{10.71c}$$

These three equations are not independent, as can be shown by introducing the material balance conditions (with only reactants present initially):

$$A \quad + \quad B \quad \rightleftharpoons \quad C \quad \longrightarrow \quad D$$

$$\text{concentration:} \quad (a - x) \quad (b - x) \quad (x - y) \quad y$$

Equations (10.71) then become:

$$\frac{dx}{dt} = k_1(a - x)(b - x) - k_2(x - y) \tag{10.72a}$$

$$\frac{dy}{dt} = k_3(x - y) \tag{10.72b}$$

The numerical solution of differential equations is a broad subject for which specialized texts should be consulted. However the simplest method, first-order Runge-Kutta, is easy to explain and quite illustrative of the general technique.

Equations (10.72) are of the form:

$$\frac{dx}{dt} = f(x, y), \qquad \frac{dy}{dt} = g(x, y)$$

The initial conditions, or boundary values, for x and y (in the case above, $x = 0$, $y = 0$) can be used to calculate the initial slopes:

$$\left(\frac{dx}{dt}\right)_0 = f(x_0 y_0), \qquad \left(\frac{dy}{dt}\right)_0 = g(x_0 y_0)$$

These slopes can be used to estimate the value of x and y after a short interval, $\Delta t = t_1 - t_0$:

$$x_1 = x_0 + \left(\frac{dx}{dt}\right)_0 \Delta t, \qquad y_1 = y_0 + \left(\frac{dy}{dt}\right)_0 \Delta t$$

These new values are then used to calculate the slopes at t_1:

$$\left(\frac{dx}{dt}\right)_1 = f(x_1 y_1), \qquad \left(\frac{dy}{dt}\right)_1 = g(x_1 y_1)$$

The general step in this iteration is:

$$x_{n+1} = x_n + f(x_n y_n) \Delta t, \qquad y_{n+1} = y_n + g(x_n y_n) \Delta t$$

Thus, a concentration-vs.-time curve can be generated in a point-by-point manner.

In principle, the solution obtained in this manner could be made arbitrarily accurate by making Δt sufficiently small. In practice, there are limitations due to the limited digital precision of the computational device and the cumulation of errors. The first problem indicates that, on a computer, double-precision arithmetic may be recommended. Both problems are relieved by using better methods; the fourth-order Runge-Kutta method is generally recommended. The numerical problem becomes particularly severe if the rate constants in a set of reactions [such as (10.72) above] differ greatly in magnitude; such equations are referred to as "stiff," and special techniques are needed. The reason for this problem is that the step size is set by the fastest reaction while the length of the calculation required is determined by the slowest step; therefore numerous iterations are required and round-off error becomes a more severe problem. Jordan (ref. 4) and Edelson (ref. 3) discuss this problem and provide references to good algorithms.

The Steady-State Approximation

Systems of equations such as that discussed above:

$$A + B \rightleftharpoons C \longrightarrow D$$

are most easily solved in the case when the intermediate (C) is unstable and short-lived. The rate equation (10.71b) for C has terms for both formation and destruction:

$$\frac{d[C]}{dt} = k_1[A][B] - (k_2 + k_3)[C]$$

Initially there will be a rapid build-up of C (and a corresponding rapid drop in A and B), but eventually the second term catches up and, at the maximum of [C], $d[C]/dt = 0$; thereafter the concentration of C declines. (Figure 10.14 illustrates this for consecutive first-order reactions.) If C is unstable (indicated by the fact that its maximum concentration is small), its concentration at this peak will be very small and its decline thereafter very slow (relative to the rate of reaction). In this region it may be appropriate to make a *steady-state* or *stationary-state* assumption regarding the intermediate — namely, that $d[C]/dt$ is very small. If, in Eq. 10.71b for example, we set $d[C]/dt = 0$, we can readily solve for the steady-state concentration:

$$[C]_{ss} = \frac{k_1[A][B]}{k_2 + k_3} \tag{10.73}$$

Exercise: Show that the steady-state reaction velocity as defined by either $-d[A]/dt$ [Eq. (10.71a)] or $d[P]/dt$ [Eq. (10.71c)] will be, assuming Eq. (10.73):

$$v = \frac{k_1 k_3}{k_2 + k_3}[A][B] \tag{10.74}$$

∎

Equation (10.74) demonstrates that this reaction in the steady-state region is second-order with an effective rate constant (denoted k_e):

$$k_e = \frac{k_1 k_3}{k_2 + k_3} \tag{10.75}$$

For all intents and purposes, it is as if there were a direct second-order reaction

$$A + B \longrightarrow P$$

with a rate constant k_e. The period of time before this steady-state is achieved is called the *induction period*.

Not all reactions involving intermediates will reach a steady state. There are a group of reactions involving several intermediates for which the concentrations of the intermediates *oscillate* in time; Jordan (ref. 4) has a good discussion of this interesting phenomenon. In Section 10.7 we shall encounter a class of reactions called branching-chain reactions for which the intermediates may increase without limit, resulting in an uncontrolled increase in the reaction rate — that is, explosions. Nevertheless, the steady-state hypothesis is widely useful for solving complex sequences of reactions.

Example: The reaction scheme:

$$A + B \underset{2}{\overset{1}{\rightleftharpoons}} C \overset{3}{\longrightarrow} P$$

was solved numerically using a fourth-order Runge-Kutta method, similar to that described above, in order to investigate the limitations of the steady-state equations just derived. For simplicity, it was assumed that the initial concentrations of A and B were both equal

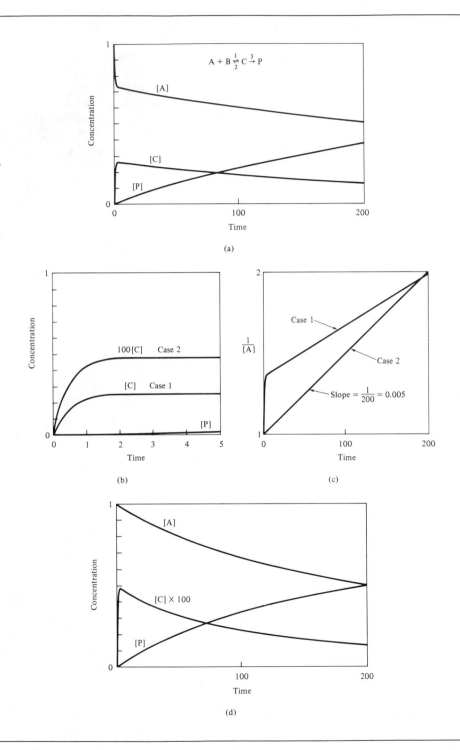

(a)

(b)

(c)

(d)

to 1 (arbitrary units); with $[A] = [B]$, the steady-state rate law is:

$$\frac{d[A]}{dt} = -k_e[A]^2$$

and plots of $1/[A]$ vs. t should be linear.

In case 1, the rate constants were:

$$k_1 = 0.5, \qquad k_2 = 1, \qquad k_3 = 0.010101$$

According to Eq. 10.75, $k_e = 0.005$ for this set of constants. The results are shown in Figure 10.15(a); note the initial transient due to the fast forward reaction (k_1) [shown in detail in Figure 10.15(b)]. In a real-data situation, it is quite possible that this initial transient would not be observed; for example, it could be over before the reactants were completely mixed and the first measurement made. It can be seen that the rate of decay of C is comparable to that of A, so the steady-state approximation is not appropriate in this case.

Figure 10.15(c) shows a graph of $1/[A]$ vs. t. For the constants of case 1, this graph is very nearly linear (after the initial transient) and could be mistaken for second-order kinetics; one might conclude, erroneously, that the steady-state equations were correct. However, the slope (about 0.0063) is not that predicted by steady-state (0.005), and the "intercept," the intersection of the linear portion at $t = 0$, is 1.74 rather than 1 (that is, the reciprocal of the initial concentration) as expected. This illustrates the wisdom of the procedure recommended earlier—namely, letting the intercept of linearized plots be a parameter to be determined rather than a fixed number.

Case 2 used:

$$k_1 = 0.01, \qquad k_2 = 1, \qquad k_3 = 1$$

Again, $k_e = 0.005$ [Eq. (10.75)]. In this case the intermediate is formed very slowly (k_1) but reacts very fast (k_2 and k_3). Figure 10.15(d) shows the results of this calculation. Note that the concentration of C is multiplied by 100, so its rate of decay appears to be comparable to that of A; actually $d[C]/dt$ is only about 1% of $d[A]/dt$ in this case.

The second-order plot, Figure 10.15(c), for case 2 is clearly linear with the exact slope predicted by the steady-state equations. (Actually, at the beginning there is a tiny transient, which is too small to be seen on this scale.) ∎

The Fast-Equilibrium Approximation

If the intermediate of a reaction is in rapid equilibrium with its progenitors, and decays slowly to products, the fast-equilibrium approximation may be useful. Using the same example as before:

$$A + B \underset{2}{\overset{1}{\rightleftharpoons}} C \overset{3}{\longrightarrow} P$$

Figure 10.15 Consecutive reactions. Concentrations vs. time for the reaction $A + B \rightleftharpoons C \rightarrow$ products: (a) case 1, (d) case 2 (rate constants in text). Panel (c) shows that the steady-state prediction is exactly correct for case 2; for case 1 the steady-state predicts the same result as for case 2, and this can be seen to be quite incorrect. Panel (b) shows the intermediate concentrations at early times; the steady-state approximation is valid when the intermediate concentration is very small, after the initial build-up. Note that in (b) and (d) the concentration is 100 times smaller than shown.

we assume:

$$[C]_{eq} = K[A][B], \qquad K = \frac{k_1}{k_2}$$

where K is the equilibrium constant of the first reaction. Like the steady-state hypothesis (of which this is a special case), it is required that [C] be small, therefore $K \ll 1$. The rate of formation of the product is then:

$$\frac{d[P]}{dt} = k_3[C] \cong k_3 K[A][B]$$

This result is clearly the limit of Eq. (10.74) for $k_2 \gg k_3$. The fast-equilibrium approximation is somewhat more useful in more complicated cases where the steady-state equations are not so simple. We shall encounter such a case shortly.

10.6   First- and Third-Order Reactions

The discussion of theories of reaction rates in Section 10.4 was directed primarily toward bimolecular reactions. The understanding of first-order decompositions (generally called "unimolecular") and third-order reactions (which may or may not be termolecular) in terms of collision-activation theory are topics that have occupied kineticists for the better part of this century.

Termolecular Reactions

There are two classes of gas-phase reactions for which the empirical rate law is third-order: atomic recombinations, and reactions of nitric oxide such as:

$$2\,NO + X_2 = 2\,NOX, \qquad v = k(NO)^2(X_2)$$

($X = O, Cl, Br,$ and I).

Rate constants for the nitric oxide reactions generally have preexponential factors $A \sim 10^9$ to 10^{10} cm^6 mol^{-2} sec^{-1} (10^3 to 10^4 dm^6 mol^{-2} sec^{-1}) and Arrhenius activation energies that are near zero or even negative. The preexponentials seem small; collision theory predicts $A \sim 10^{14}$ to 10^{16} cm^6 mol^{-2} sec^{-1}. Also, a little thought about what an "activation energy" means may convince you that a "negative activation energy" is a contradiction in terms.

Atomic recombinations in the gas phase must be termolecular—for example:

$$I + I + M = I_2 + M, \qquad v = k[I]^2[M]$$

where M is a third body (which could be I or I_2 or some foreign gas). The tendency to form a bond is effectively a very strong, long-range attraction for the dissociated atoms; qualitatively, this force is like the van der Waals forces discussed in Section 1.7, but about three orders of magnitude greater in strength. This strong attractive force will accelerate the atoms toward each other so that, when they collide, they have so much kinetic energy that they rebound without forming a bond. For the bond to be formed, a third body must be available while the atoms are within bonding distance to carry off the excess kinetic energy. More complex species do not need the third body because they have a sufficient number of internal degrees of freedom (vibrations) to dissipate the collisional energy. Simple free radicals such as OH and CH_3 are intermediate cases. Recombination reactions in solution are generally second-order, since solvent molecules are always nearby to take up the excess energy. The third body required to carry off the excess energy in gas-phase recombinations is referred to as the *chaperon,* "its function being to lead the partners into a stable union rather than leaving them in a highly excited state" (Jordan, ref. 4).

Atomic recombinations have near-zero activation energies; this is not unexpected. Nearly all radical recombinations have activation energies near zero, and, therefore, rate constants that are independent of temperature. Also, the preexponentials are $\sim 10^{15}$ cm^6 mol^{-2} sec^{-1}, as expected from collision theory. However, collision theory has difficulty in accounting for the widely varying efficiencies of various third bodies. For example, for the iodine recombination reaction:

$$M = Ar: \qquad A = 1.5 \times 10^{15}, \qquad E_a = -1.7 \text{ kJ/mol}$$

$$M = \text{benzene:} \qquad A = 8.6 \times 10^{15}, \qquad E_a = -7.1 \text{ kJ/mol}$$

$$M = I_2: \qquad A = 0.65 \times 10^{15}, \qquad E_a = -21 \text{ kJ/mol}$$

Many of these facts can be rationalized by assuming a preequilibrium mechanism such as (for a stoichiometry $2A + B = \ldots$):

$$A + B \underset{2}{\overset{1}{\rightleftharpoons}} C \qquad \text{fast}$$

$$C + A \overset{3}{\longrightarrow} \text{products} \qquad \text{slow} \qquad \qquad \textbf{(10.76)}$$

where C is a complex of A and B. The rate of product formation is:

$$\frac{d[P]}{dt} = k_2[C][A]$$

and, assuming the first reaction equilibrates rapidly:

$$[C] \cong K_c[A][B]$$

Therefore:

$$\frac{d[P]}{dt} = k_2 K_c[A]^2[B] \qquad \qquad \textbf{(10.77)}$$

The effective rate constant is, therefore, $k_3 K_c$, and:

$$\ln k_{\text{eff}} = \ln k_3 + \ln K_c$$

$$\frac{d}{dT}(\ln k_{\text{eff}}) = \frac{d \ln k_3}{dT} + \frac{d \ln K_c}{dT}$$

$$= \frac{E_{a3} + \Delta U_{\text{rxn}}}{RT^2}$$

The effective activation energy is, therefore:

$$E_a(\text{eff}) = \Delta U_{\text{rxn}} + E_{a3} \qquad \qquad \textbf{(10.78)}$$

where ΔU is the energy of the reaction $A + B = C$.

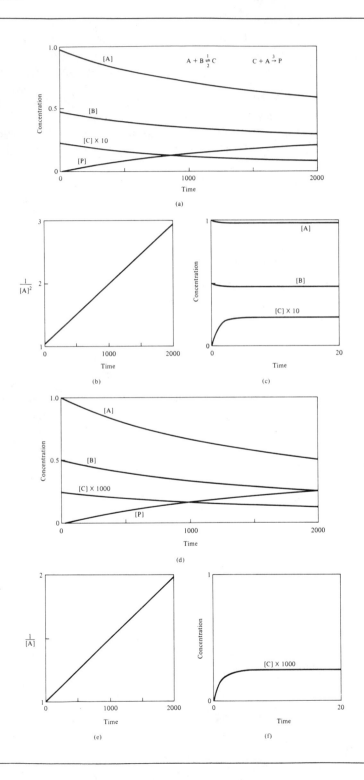

The complex (C) is surely a weak one, and the bond formed in step 2 is strong, so E_a for step 2 may be small. If the preequilibrium reaction is exothermic (ΔU negative), the effective activation energy could well be negative. Also, such a preequilibrium can account for the variation of third-body efficiencies in atomic recombinations; it would not be surprising if I + Ar had a much smaller K_c than I + benzene (benzene is well known for form charge-transfer complexes).

The preequilibrium mechanism can hardly be considered as acceptable for all termolecular reactions, since closer examination reveals a number of problems and inconsistencies. After decades of work, the mechanism of termolecular reactions is still something of an open question.

Example: The preceding kinetic scheme (10.76) can also be solved with a steady-state approximation. The reader should show that the result is:

$$[C]_{ss} = \frac{k_3[A][B]}{k_2 + k_3[A]}$$

$$\frac{d[A]}{dt} = \frac{-2k_1 k_3[A]^2[B]}{k_2 + k_3[A]}$$

Equation (10.77) can be seen to be a special case for $k_2 \gg k_3[A]$. ($K_c = k_1/k_2$.)

This kinetic scheme was also solved numerically in the manner described earlier; the results are shown in Figure 10.16. In both cases, the initial concentrations were taken to be $a = 1$, $b = 0.5$, so, because of the 2:1 stoichiometry, $[B] = 0.5[A]$ at all times. In such a case the rate law is:

$$\frac{d[A]}{dt} = \frac{-k_1 k_3[A]^3}{k_2 + k_3[A]}$$

Using the rate constants for case 1, this law is:

$$\frac{d[A]}{dt} = \frac{-0.0005[A]^3}{1 + 0.01[A]} \cong -0.0005[A]^3$$

The second term of the denominator is negligible, so this case should show third-order kinetics. Figure 10.16(b) shows a plot of $1/[A]^2$ that is very close to the linear plot expected for third-order reactions; note, however, that the apparent intercept is not 1, as expected, because of the initial transient. [This transient is shown more clearly on panel (c).]

For case 2, the rate constants give:

$$\frac{d[A]}{dt} = \frac{-0.0005[A]^3}{0.01 + (1)[A]} \cong -0.0005[A]^2$$

Figure 10.16 Consecutive reactions. The concentration vs. time curves for the reaction A + B $\rightleftharpoons$ C, C + A $\rightarrow$ products [Eq. (10.76)], are shown for two sets of rate constants. In the first case ($K_{eq} = 0.05$, $k_1 = 0.05$, $k_2 = 1.0$, $k_3 = 0.01$) [case 1, panel (a)] the reaction shows third-order kinetics as demonstrated by panel (b). Panel (c) shows the initial transients of the concentration vs. t curves for case 1. In the second case ($K_{eq} = 0.05$, $k_1 = 0.0005$, $k_2 = 0.01$, $k_3 = 1.0$) [case 2, panel (d)] the reaction shows second-order kinetics as demonstrated by panel (e). [Panel (f) shows the initial transient in the concentration of C.] Note that in panel (b) the "initial concentration," the intercept of the line, is not correct—it should equal 1 if the reaction were strictly third-order.

For the early part of the reaction (until the concentration of A becomes too small), the first term of the denominator will be negligible and second-order kinetics can be expected.

Figure 10.16(d) shows the results for this case. A graph of $1/[A]$ vs. t [panel (e)] is linear as expected. Note that there is a transient in this case [panel (f)], but it is two orders of magnitude smaller than for case 1.

There will, of course, be intermediate cases for which neither the second- nor third-order rate laws apply.   ∎

Unimolecular Reactions

Many gas-phase decompositions and isomerizations are first-order and, apparently, unimolecular. That is, they have a stoichiometry $A \longrightarrow \cdots$ and a rate law $v = k[A]$. However, the rate constants do change with pressure and are affected by the addition of foreign gases. A major breakthrough in the understanding of these reactions in terms of collision/activation ideas of chemical kinetics occurred with the mechanism proposed independently by J. A. Christiansen (1921) and F. A. Lindeman (1922); this mechanism is referred to as the Lindeman mechanism.

A reaction of a molecule (A) to form products is presumed to be preceded by an activating collision:

$$A + M \xrightarrow{\;1\;} A^* + M \tag{10.79a}$$

The molecule M is any molecule that can collide with A, including other molecules of A, products, or foreign gases. The activated molecule A* has an excess of vibrational energy; it can either lose this excess energy by a subsequent collision:

$$A^* + M \xrightarrow{\;2\;} A + M \tag{10.79b}$$

or decompose:

$$A^* \xrightarrow{\;3\;} products \tag{10.79c}$$

The reaction velocity as measured by the rate of appearance of products is:

$$v = \frac{d[P]}{dt} = k_3[A^*]$$

The steady-state assumption can be used to calculate the concentration of the activated molecules:

$$\frac{d[A^*]}{dt} = k_1[A][M] - k_2[A^*][M] - k_3[A^*] = 0$$

$$[A^*]_{ss} = \frac{k_1[M][A]}{k_3 + k_2[M]}$$

The reaction velocity will then be:

$$v = \frac{k_1 k_3[M][A]}{k_3 + k_2[M]}$$

This rate law is apparent-first-order with a rate constant (k_{uni}) that depends on pressure — that is, on [M] — which is proportional to the total pressure:

$$v = k_{uni}[A] \qquad \textbf{(10.80)}$$

$$k_{uni} = \frac{k_1 k_3 [M]}{k_3 + k_2[M]} \qquad \textbf{(10.81)}$$

At high pressure (when $k_2[M] \gg k_3$) the "unimolecular" rate constant, k_{uni}, will be independent of pressure; this value is called k_∞:

$$k_\infty = \frac{k_1 k_3}{k_2} \qquad \textbf{(10.82)}$$

Only in this region will the reaction be truly first-order.

The Lindeman mechanism accounts qualitatively for the observed decrease of the "unimolecular" rate constant with pressure but works rather poorly at low pressure. Improvements made by C. N. Hinshelwood (1927), O. K. Rice and H. C. Ramsperger (1927), L. S. Kassel (1928), and R. A. Marcus (1952) have brought the Lindeman mechanism into quantitative agreement with respect to both the pressure dependence and the values of the rate constant (Figure 10.17). These improvements deal primarily

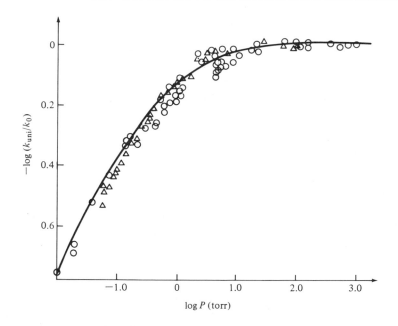

Figure 10.17 Unimolecular rate constants. Observed and predicted (by RRKM theory) rate constants for the decomposition of cyclobutane as a function of pressure. (From P. J. Robinson and K. A. Holbrook, *Unimolecular Reactions*, New York: Wiley-Interscience, 1972, p. 248)

with the mechanism of energy transfer within and between the molecules. The current theory, called RRKM (for Rice, Ramsperger, Kassel, Marcus), is considered to be one of the most outstanding successes of theoretical chemical kinetics.

10.7 Chain Reactions

A great number of important chemical reactions, both in solution and gas phase, proceed by a mechanism called a chain reaction. These reactions often have unusual rate laws, which may change unexpectedly with a change in reaction conditions. Some important processes that involve chain reactions include: hydrocarbon pyrolysis (petroleum cracking), combustions, explosions, photochemical smog production, the reactions of stratospheric ozone, organic chlorination, and many polymerizations. Most of these reactions involve free radicals, so we shall discuss this topic first. (Arrenhius factors for radical reactions are given in Table 10.3.)

Free Radicals

A free radical is a molecule (or atom) whose ground electronic state violates Lewis pairing; that is, they have unpaired electrons. Some free radicals are stable — for example, nitric oxide (NO), which has an odd number of electrons, and molecular oxygen (O_2), which has two unpaired electrons in its ground state (for reasons that will be explained in Chapter 13).

Most free radicals are either unstable or very reactive; often they will react upon their first collision with little or no activation energy required. As a result, when they appear in a chemical reaction, their lifetime will be very short and their concentration very small. Examples include atoms (H, Br, I, and so on), the hydroxyl radical (OH),

Table 10.3 Arrhenius factors for radical reactions

	$\log_{10} A$ (cm^3 mol^{-1} s^{-1})	E_a (kJ/mol)
$Cl + H_2 \longrightarrow HCl + H$	13.9	23
$Br + H_2 \longrightarrow HBr + H$	13.9	72
$I + H_2 \longrightarrow HI + H$	14.1	140
$H + HCl \longrightarrow H_2 + Cl$	13.6	19
$H + HBr \longrightarrow H_2 + Br$	13.2	4
$2\ CH_3 \longrightarrow C_2H_6$	13.3	~0
$2\ C_2H_5 \longrightarrow C_4H_{10}$	13.2	~0
$Cl + COCl \longrightarrow COCl_2$	14.6	3.3
Termolecular recombinations (A in cm^6 mol^{-2} s^{-1}):		
$H + H + H_2 \longrightarrow H_2 + H_2$	16.0	~0
$O + O + O_2 \longrightarrow O_2 + O_2$	15.0	~0
$O + O_2 + Ar \longrightarrow O_3 + Ar$	12.5	−9.6
$I + I + Ar \longrightarrow I_2 + Ar$	15.8	−5.4

Source: Ref. 6.

and the methyl radical (CH_3). We shall denote a general organic free radical as R· (with the corresponding hydrocarbon being RH). The dot will be used to indicate radicals when it seems required for clarity's sake.

Some organic radicals, particularly those in which the odd electron can be stabilized by a conjugated π system, may have reasonably long lifetimes — for example, the triphenyl methyl radical or diphenylpicrylhydrazine (DPPH). Even stable free radicals tend to be very reactive.

Free radicals play a part in many reactions; it is possible that all gas-phase reactions involve them in some way. But by no means are all reaction mechanisms involving free radicals chain reactions. Let us examine the essential features of a chain mechanism.

Components of Chain Reactions

Chain mechanisms have three distinct types of steps: (1) initiation, (2) propagation, (3) termination.

Propagation.   The propagation steps are a group of reactions in which unstable intermediates, called *chain carriers,* are both utilized and produced. Typically, a radical (R) may react with a molecule (M) to produce another molecule (M') and another radical (R'):

$$R + M \longrightarrow R' + M'$$

The new radical, R', undergoes subsequent reaction, eventually regenerating the original radical R; then the reaction sequence proceeds through another cycle. The simplest possibility is illustrated by the propagation steps of the chain reaction whose stoichiometry is:

$$H_2 + Cl_2 = 2\ HCl$$

The propagation steps are:

$$Cl + H_2 \longrightarrow HCl + H$$

$$H + Cl_2 \longrightarrow HCl + Cl$$

The radical chain carriers are H and Cl. If no other reactions occurred, this chain pair could go on forever producing product. Commonly, each chain carrier can produce thousands of product molecules; therefore, their concentration need not be very large for the rate of reaction to be fast.

There may be chain steps in which one carrier produces more than one new carrier; these are called *branched*-chain reactions. For example, the following sequence occurs in the combustion of hydrogen:

$$O + H_2 \longrightarrow OH + H$$
$$H + O_2 \longrightarrow OH + O$$
$$OH + H_2 \longrightarrow H_2O + H$$

In this sequence, one H radical may give birth to three; 6 cycles could produce

$3^6 = 729$ H radicals. Obviously, branched-chain reactions have the potential for accelerating out of control.

Initiation.   The initiation is a reaction, or a set of reactions, that provide carriers to get the chain started. For the HCl reaction above, this is probably a unimolecular thermal dissociation:

$$Cl_2 \longrightarrow 2\ Cl$$

(It will be understood from the discussion of the Lindeman mechanism, Section 10.6, that this is actually a more complex event than indicated.) This reaction could also be initiated by the thermal dissociation of $H_2 \longrightarrow 2\ H$, but this will be less common because the H-H bond is stronger and more stable than the Cl-Cl bond.

Initiation may be photochemical — for example:

$$Cl_2 + h\nu \longrightarrow 2\ Cl$$

We shall discuss this case later.

Initiation may be provided by a *sensitizer* — a substance that can react to provide radicals but cannot act as a chain carrier. An example is sodium vapor with chlorine:

$$Na + Cl_2 \longrightarrow NaCl + Cl$$

Unstable molecules such as organic peroxides may be added as *initiators;* for example di-*t*-butyl peroxide can be used to provide methyl radicals:

$$(CH_3)_3COOC(CH_3)_3 \longrightarrow 2\ (CH_3)_3CO$$

$$(CH_3)_3CO\cdot \longrightarrow CH_3COCH_3 + CH_3\cdot$$

Termination. These are reactions that remove carriers and, hence, terminate the chain. The most obvious type of termination is a second-order radical recombination:

$$2\ R\cdot \longrightarrow M, \qquad v_t = k_t[R\cdot]^2$$

A disproportionation of radicals to form stable molecules is also second-order. For example, ethyl radicals (C_2H_5) may combine to form butane:

$$2\ C_2H_5 \longrightarrow C_4H_{10}$$

or disproportionate to form ethane and ethylene:

$$2\ C_2H_5 \longrightarrow C_2H_4 + C_2H_6$$

Since both of these reactions are second-order, it matters little which type terminates the chain. (There is an effect on the product mixture, but since most products are formed by the propagation steps, this is usually minor.) As discussed earlier, gas-phase recombinations of atoms or simple radicals (OH, CH_3) must be third-order.

Chain carriers may also be terminated on the wall of the reaction vessel; this is usually first-order:

$$R\cdot \longrightarrow wall, \qquad v_t = k_{wall}[R\cdot]$$

The relative efficiencies of wall removal and bimolecular termination will depend on the surface/volume ratio of the vessel and on the mean free path of the gas. For a spherical or cylindrical vessel, the rate constant will depend on the diameter (d) and on the mean free path, which, in turn, depends on pressure (P) [Eq. (9.9)]:

$$k_{wall} = \frac{constant}{d^2 P} \tag{10.83}$$

Substances may be added to the reaction mixture for the expressed purpose of providing a termination; these substances may be called *inhibitors* or *scavengers*. For example, nitric oxide can terminate organic free radicals:

$$R\cdot + NO \longrightarrow RNO$$

Tetraethyl lead, which readily dissociates to form ethyl radicals (C_2H_5), has been added to gasoline to control the chain-reaction combustion and prevent engine knock:

$$C_2H_5\cdot + R\cdot \longrightarrow \text{stable molecules}$$

It is important for this purpose that the ethyl radical is a rather poor carrier and, therefore, will not participate in the propagation steps.

It should be understood that, except at the beginning of the reaction (during the induction period) these steps — initiation, propagation, termination — are not sequential; all are occurring simultaneously in the reaction mixture. However, in a given time, many thousands of propagation cycles may occur for each initiation or termination event. The **kinetic chain length** (ν) is defined as:

$$\nu = \frac{\text{velocity of propagation}}{\text{velocity of initiation}} \tag{10.84}$$

Chain lengths $\sim 10^5$ are not uncommon; this would mean that each initiation event will produce 10^5 reactant to product conversions.

Complex reactions such as hydrocarbon combustion may have hundreds of propagation-type steps; the examples we shall discuss will have only a few.

Example: It was mentioned earlier that the reaction

$$H_2 + Br_2 = 2\,HBr$$

has a rather unusual rate law. In fact, this was one of the earliest reactions for which a chain mechanism was proposed (1919). A mechanism that is consistent with the observed rate law is:

$Br_2 \xrightarrow{\ 1\ } 2\,Br$	(initiation),	$v_1 = k_1[Br_2]$
$Br + H_2 \xrightarrow{\ 2\ } HBr + H$	(propagation),	$v_2 = k_2[H_2][Br]$
$H + Br_2 \xrightarrow{\ 3\ } HBr + Br$	(propagation),	$v_3 = k_3[Br_2][H]$
$H + HBr \xrightarrow{\ 4\ } H_2 + Br$	(inhibition),	$v_4 = k_4[H][HBr]$
$Br + Br \xrightarrow{\ 5\ } Br_2$	(termination),	$v_5 = k_5[Br]^2$

(It should be understood that both steps 1 and 5 are more complex than indicated: this will not affect the rate law because of cancellation. Also, note that reaction 4 is the reverse of reaction 2.)

The overall reaction velocity can be written in terms of either the rate of appearance of HBr or the rate of disappearance of H_2 or Br_2; all of these will be equal in the steady-state.

$$v = -\frac{d[Br_2]}{dt} = v_1 + v_3 - v_5$$

$$= k_1[Br_2] + k_3[Br_2][H] - k_2[Br]^2$$

A macroscopic rate law is preferably written in terms of reactants and products only — the more so if the intermediate concentrations such as [Br] or [H] are very small and hard to measure, as in this case. The intermediate concentrations can be found using the steady-state approximation:

$$\frac{d[H]}{dt} = v_2 - v_3 - v_4 = 0$$

$$\frac{d[Br]}{dt} = 2v_1 - v_2 + v_3 + v_4 - 2v_5 = 0$$

Adding these equations, we get:

$$v_1 = v_5$$

rate of initiation = rate of termination

This is a fairly typical result for non-branched-chain reactions in the steady state. This means that the overall reaction velocity is:

$$v = v_1 + v_3 - v_5 = v_3 = k_3[Br_2][H]$$

The concentration of the radicals can be obtained from the conditions:

$$v_2 = v_3 + v_4, \qquad k_2[H_2][Br] = [H](k_3[Br_2] + k_4[HBr])$$

$$v_1 = v_5, \qquad k_1[Br_2] = k_5[Br]^2$$

$$[Br]_{ss} = \left(\frac{k_1}{k_5}\right)^{1/2}[Br_2]^{1/2}$$

$$[H]_{ss} = \frac{k_2\left(\frac{k_1}{k_5}\right)^{1/2}[Br_2]^{1/2}[H_2]}{k_3[Br_2] + k_4[HBr]}$$

The overall reaction velocity is thus:

$$v = \frac{k_2k_3\left(\frac{k_1}{k_5}\right)^{1/2}[H_2][Br_2]^{3/2}}{k_3[Br_2] + k_4[HBr]}$$

$$= \frac{k_2\left(\frac{k_1}{k_5}\right)^{1/2}[H_2][Br_2]^{1/2}}{1 + \frac{k_4}{k_3}\frac{[HBr]}{[Br_2]}} \qquad\qquad \textbf{(10.85)}$$

(compare p. 471). ∎

Photochemical Initiation

Photochemical reactions occur as a result of the absorption of a photon ($h\nu$) by a reactant molecule (A):

$$A + h\nu \longrightarrow \text{products}$$

For elementary reactions, a reactive event is generally associated with the absorption of a single photon; however, not all molecules that absorb a photon will react.

The easiest quantity to measure regarding photochemical reactions is the intensity of light absorbed (I_a). This will be related to the incident light intensity (I_0), the path length (l) of the light through the absorbing medium, the concentration (C) and molar absorptivity (ε) of the absorbing molecules by the law of Beer-Lambert-Bouguer:

$$I_a = I_0(1 - e^{-\varepsilon Cl}) \tag{10.86}$$

The simplest mechanism for photochemical reaction assumes that the absorbing species (A) is activated by the photon:

$$A + h\nu \xrightarrow{\ 1\ } A^*, \qquad v_1 = k_1 I_a$$

Then either it may become deactivated by collisions:

$$A^* + M \xrightarrow{\ 2\ } A + M, \qquad v_2 = k_2[A^*][M]$$

or it can react:

$$A^* \xrightarrow{\ 3\ } \text{products}, \qquad v_3 = k_3[A^*]$$

The reaction velocity is:

$$v = k_3[A^*]$$

The steady-state concentration of A* is easily shown to be:

$$[A^*]_{ss} = \frac{k_1 I_a}{k_3 + k_2[M]}$$

The reaction velocity is, therefore:

$$v = \left(\frac{k_1 k_3}{k_3 + k_2[M]}\right) I_a \tag{10.87}$$

The **quantum yield** (ϕ) is defined as the fraction of photons absorbed which result in reaction — in this case:

$$\phi = \frac{k_1 k_3}{k_3 + k_2[M]} \tag{10.88}$$

The reaction velocity can be written as:

$$v = \phi I_a \tag{10.89}$$

For an elementary process, the quantum yield must be less than or equal to 1. Sometimes quantum yields are defined in terms of the overall reaction, the rate of product formation; if the photochemical reaction is the initiation step of a long chain reaction, the quantum yield so defined could be a very large number. We shall use this term only to apply to elementary reactions — for example, the photochemical initiation of a chlorination reaction:

$$Cl_2 + h\nu \longrightarrow 2\ Cl, \qquad v_i = \phi I_a$$

Note that, according to Eq. 10.1, the velocity of this reaction is:

$$v_i = -\frac{d[Cl_2]}{dt} = \frac{1}{2}\frac{d[Cl]}{dt}$$

Therefore, the rate of production of Cl radicals is:

$$\frac{d[Cl]}{dt} = 2v_i = 2\phi I_a$$

Example: Many chlorination reactions can be photoinitiated, including the simplest one:

$$H_2 + Cl_2 = 2\ HCl$$

A reasonable mechanism for this reaction is:

$$Cl_2 + h\nu \xrightarrow{\ 1\ } 2\ Cl, \qquad\qquad v_1 = \phi I_a$$
$$Cl + H_2 \xrightarrow{\ 2\ } HCl + H, \qquad\quad v_2 = k_2[H_2][Cl]$$
$$H + Cl_2 \xrightarrow{\ 3\ } HCl + Cl, \qquad\quad v_3 = k_3[H][Cl_2]$$
$$Cl \longrightarrow (\text{wall removal}), \qquad v_4 = k_4[Cl]$$

(Note that the Cl on the wall will probably, in due course, show up again as Cl_2. This is of little concern; the kinetically significant fact is that the termination is first-order. This mechanism is similar to that presented above for HBr, but the inhibition reaction is missing; the reason for this difference is the greater bond energy of HCl as compared to HBr.)

The overall velocity will be:

$$v = \frac{1}{2}\frac{d[HCl]}{dt} = \frac{1}{2}(v_2 + v_3)$$

The steady-state equations are:

$$\frac{d[Cl]}{dt} = 2v_1 - v_2 + v_3 - v_4 = 0$$

$$\frac{d[H]}{dt} = \qquad v_2 - v_3 \qquad = 0$$

From this it should be obvious that $v_2 = v_3$ and $2v_1 = v_4$. The overall velocity is therefore:

$$v = \frac{1}{2}(v_2 + v_3) = v_2 = k_2[H_2][Cl]$$

Since $2v_1 = v_4$, $2\phi I_a = k_4[Cl]$, the steady-state concentration of chlorine atoms is:

$$[Cl]_{ss} = \frac{2\phi I_a}{k_4}$$

Finally, the rate law for this mechanism is:

$$v = \frac{2k_2\phi}{k_4}I_a[H_2] \tag{10.90}$$

The velocity of this reaction depends on $[Cl_2]$ only via I_a, cf. Eq. (10.86). ∎

Exercise: The nature of the termination step of a chain reaction has a major impact on the rate law. Show that if step 4 of the preceding mechanism is replaced by a second-order termination:

$$Cl + Cl \xrightarrow{5} Cl_2$$

the rate law will be $\frac{1}{2}$-order in I_a:

$$v = k_2\left(\frac{\phi}{k_5}\right)^{1/2}I_a^{1/2}[H_2] \tag{10.91}$$

∎

Explosions

Some reactions proceed so fast that the process becomes explosive. This is not a problem if intended (as for a propellant), but there are reactions that may proceed smoothly under some circumstances but become explosive when conditions are altered in a seemingly trivial way; such cases often are chain reactions.

Exothermic reactions will become hot if the rate of heat release due to reaction is faster than the rate of heat transfer out of the reaction zone. Most reaction rates will increase with temperature, so the rate of heat production will increase when the temperature increases. Also, the rate of cooling, which according to Newton's law of cooling is proportional to the temperature gradient, will increase, so the reaction may stabilize at some temperature. However, the increase of rate with temperature may cause the reaction to run away; this is called a *thermal explosion*.

Chain reactions with chain-branching steps may become explosive even without the thermal effect. Figure 10.18 shows this effect as it is found for the combustion of hydrogen. (There is also an effect from composition; hydrogen-air mixtures between 4% and 70% hydrogen by volume are explosive.) The reaction of H_2 and O_2 has several chain-branching steps; a probable (but simplified) mechanism is:

Initiation:	$H_2 + O_2 \xrightarrow{1} 2\ OH\ (wall)$
Propagation:	$OH + H_2 \xrightarrow{2} H_2O + H$
Branching:	$H + O_2 \xrightarrow{3} OH + O$
	$O + H_2 \xrightarrow{4} OH + H$
Termination:	$H \xrightarrow{5} wall$
	$H + O_2 + M \xrightarrow{6} HO_2 + M$

Below the first explosion limit (Figure 10.18), the reaction is kept under control by step 5, wall termination. As the pressure increases, wall termination becomes less

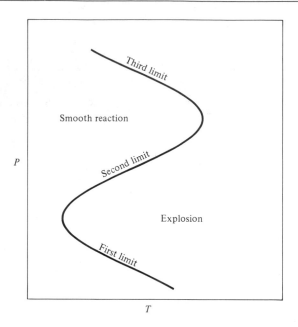

Figure 10.18 Explosion limits for a gas-phase chain reaction.

efficient because of the shortened mean free path—cf. Eq. (10.83). At the first explosion limit, the reaction is unable to reach a steady state, and the uncontrolled increase in the number of chain carriers results in an explosion. The location of this limit depends on the surface/volume ratio of the vessel and on the type of wall.

Above the second limit (Figure 10.18), the pressure is high enough for termolecular termination steps such as 6 to control the reaction. Note that HO_2, produced by step 6, is actually a radical, but it is relatively stable and probably ends up on the wall. The resumption of explosive behavior at the third limit is probably due to the participation of HO_2 (a "termination" product of reaction 6) in a propagation step:

$$HO_2 + H_2 \longrightarrow H_2O + OH$$

when, at higher pressures, it cannot reach the wall.

The Missing Steps

A thoughtful examination of the example mechanisms given above will show that numerous reactions that are possible are not included. Although—for brevity—reasons were not given, the choice was never arbitrary. For example, the reaction:

$$H_2 + X_2 = 2\,HX$$

when $X = Br$ included the reverse step:

$$H + HBr \longrightarrow H_2 + Br$$

When X — Cl, this step was not included; an examination of the empirical activation energies (Table 10.3) shows that this type of step will not be important for Cl until high temperatures. Also, this reaction is a chain reaction if X = Br or Cl; if X = I, the reaction is apparently a radical reaction, but the chain reaction is relatively unimportant below 800 K. Compare (Table 10.3) the very high E_a (140 kJ/mol) for the propagation step:

$$I + H_2 \longrightarrow HI + H$$

to the comparable steps for Cl and Br. Determining a mechanism involves more than simply picking steps that give the observed rate law; it is common to find that many mechanisms may "work" in this limited sense. To devise a plausible mechanism, the elementary steps must be understood and shown to be reasonable.

10.8   Vinyl Polymerization

An extraordinary fraction of the chemistry practiced in the world is related to the commercial production of polymers. These find application as elastomers, plastics, fibers, resins, and so on. Polymers fall into two groups with regard to their mode of formation. *Condensation polymers,* such as polyesters and polyamides, result from condensation of difunctional monomer units; these generally follow simple second-order kinetics. *Addition polymers* result from chain growth, with new monomers adding only at one end of the growing chain. Addition polymerizations are chain reactions, both literally and kinetically, and this is the mechanism we shall discuss.

Vinyl polymers are made from monomers such as:

$$\begin{array}{ccc} H & & H \\ \diagdown & & \diagup \\ & C{=}C & \\ \diagup & & \diagdown \\ H & & X \end{array}$$

These react by opening the double bond and joining end-to-end to form a chain:

$$\cdots CH_2{-}CHX{-}CH_2{-}CHX{-}CH_2{-}CHX{-}CH_2{-}CHX \cdots$$

Examples of some important commercial polymers are:

$$\begin{array}{ll} X = H, & \text{poly(ethylene)} \\ X = CH_3, & \text{poly(propylene)} \\ X = CN, & \text{poly(acrylonitrile)} \\ X = -OCCH_3, & \text{poly(vinylacetate)} \\ \quad\quad\; \underset{\displaystyle O}{\|} & \\ X = -\underset{\displaystyle O}{\overset{\displaystyle \|}{C}}-OCH_3, & \text{poly(methylacrylate) (PMA)} \end{array}$$

$$X = Cl, \qquad\qquad poly(vinyl\ chloride)\ (PVC)$$

$$X = phenyl, \qquad\qquad poly(styrene)$$

If the monomer is

$$CH_2=C \underset{\underset{O}{\overset{\|}{C}}-OCH_3}{\overset{CH_3}{}}$$

the polymer is poly(methylmethacrylate) (PMMA).

Initiation is usually by an added material called the initiator (In), which thermally decomposes to form a reactive radical:

$$In \longrightarrow R\cdot\ , \qquad v_i = fk_i[In] \qquad\qquad (10.92)$$

The efficiency factor (f) is included to account for the possibility that the radical formed may do something other than start a chain; for example, it could recombine or decompose further. The efficiency factor could be affected by factors that do not affect k_i (the rate constant for the decomposition of In), such as monomer concentration and solvent viscosity.

The reactive radical may react with a monomer:

$$R\cdot\ +\ CH_2{=}CHX \longrightarrow RCH_2{-}\underset{\underset{X}{|}}{\overset{\overset{H}{|}}{C}}\cdot$$

This is the new radical, which will react with another monomer (M) to extend the chain. Thus, there will be a series of propagation steps:

$$R\cdot + M \longrightarrow RM\cdot\ , \qquad v_0 = k_{p0}[M][R\cdot]$$

$$RM\cdot + M \longrightarrow RM_2\cdot\ , \qquad v_1 = k_{p1}[M][RM\cdot]$$

$$RM_2\cdot + M \longrightarrow RM_3\cdot\ , \qquad v_2 = k_{p2}[M][RM_2\cdot]$$

$$\vdots$$

$$RM_n\cdot + M \longrightarrow RM_{n+1}\ , \qquad v_n = k_{pn}[M][RM_n\cdot]$$

The rate of propagation (v_p) is measured by the rate of monomer loss:

$$v_p = -\frac{d[M]}{dt} = \sum_n v_{pn} = \sum_{n=0}^{\infty} k_{pn}[M][RM_n\cdot]$$

After the first few steps, the propagation rate constant is generally assumed to become independent of chain length, and the propagation rate can be approximated as:

$$v_p = k_p[M]\sum_{n=0}^{\infty}[RM_n\cdot]$$

The sum in this expression is simply the total concentration of actively growing chains, which we shall denote as $[R_x\cdot]$:

$$v_p = k_p[M][R_x\cdot] \tag{10.93}$$

Termination is typically second-order:

$$v_t = k_t[R_x\cdot]^2 \tag{10.94}$$

This may be by addition of two chains (length m and n) to form a completed polymer chain (P) of length $m + n$:

$$R_n\cdot + R_m\cdot \longrightarrow P_{n+m}$$

It may also be a disproportionation to form two shorter polymer chains:

$$R_n\cdot + R_m\cdot \longrightarrow P_n + P_m$$

This could happen by transfer of an H atom:

$$\left.\begin{array}{l} \cdots CH_2CHX\cdot \\ \cdots CH_2CHX\cdot \end{array}\right\} \longrightarrow \begin{array}{l} \cdots CH=CHX \\ \cdots CH_2CH_2X \end{array}$$

Termination may also be by chain transfer to the solvent (S):

$$R_x\cdot + S \longrightarrow P_x + S\cdot , \qquad v_t = k_t[R_x\cdot][S]$$

Since the solvent radical (S·) may begin a new chain, this is an effect less on the rate of reaction than on the mean chain length of the polymer molecules formed.

In the steady state, the rate of initiation will be equal to the rate of termination. For second-order termination:

$$k_i f[In] = k_t[R_x\cdot]^2$$

$$[R_x\cdot] = \sqrt{\frac{k_i f}{k_t}}[In]^{1/2}$$

Then, from Eq. (10.93), the rate of propagation will be:

$$v_p = k_p \sqrt{\frac{k_i f}{k_t}}[M][In]^{1/2} \tag{10.95}$$

A unique feature of polymerization kinetics is that the prime interest is less in how fast the reaction will proceed than in how long the resulting chains will be. A random distribution of chain lengths will be obtained, resulting in a distribution of molecular weights. The average molecular weight and the width of this distribution are important factors that affect the physical properties of the polymer product. The kinetic chain length is, from Eq. (10.84):

$$\nu = \frac{v_p}{v_i}$$

From Eqs. (10.92) and (10.95):

$$\nu = \frac{k_p[M]}{\sqrt{f k_i k_t}[In]^{1/2}} \tag{10.96}$$

If termination is by disproportionation, ν is the actual average polymer chain length; if termination is by addition, the average chain length will be twice ν.

10.9   Surface Catalysis

Reactions that occur in a single phase (usually gas or solution) are called homogeneous reactions. Many reactions occur on the surface of a solid. This may be intentional, as in the use of heterogeneous catalysts, or it may be unexpected, as when reactions occur on the surface of the reaction vessel. In either case, heterogeneous reactions have several distinctive features. To understand these features we must first discuss the adsorption of gases on solid surfaces.

Langmuir's Theory of Adsorption

First we must distinguish between adsorption and absorption. Paper towels absorb water; the material penetrates to all parts of the absorbing medium. Adsorption presumably occurs exclusively on the surface of the adsorbant. Adsorption is often categorized as either physical or chemical, based on some criterion for the strength of the gas-surface interaction. Physical adsorption is a weak interaction in which the adsorbed molecules apparently maintain their chemical identity. In chemical adsorption there is presumably a chemical bond formed between the adsorbent and the surface and a major alteration in the chemical nature of the adsorbent. Either type of adsorption could be involved in a surface reaction, but catalytic activity probably involves chemical adsorption.

Many substances will adsorb gases in substantial quantities at pressures below that required for the gas to condense (P less than the vapor pressure at that temperature); charcoal and silica are typical. Figure 10.19(a) shows, schematically, an apparatus for studying this phenomenon. The volume of gas adsorbed (corrected to STP) divided by the weight of adsorbent, is denoted V_{ads}. Plots of V_{ads} vs. P at constant temperature are called *adsorption isotherms;* a typical result is shown in Figure 10.19(b).

A major advance in the understanding of surface adsorption was provided by a theoretical insight of Irving Langmuir (1916). He proposed that the surface consisted of a finite number of *sites,* each of which could adsorb one molecule of gas. He further assumed that these sites were *identical* and *independent;* this implies that the probability of adsorption at a site will not depend on whether the adjacent sites are occupied. If we define θ as the fraction of sites occupied, the rate at which molecules are adsorbed on a surface will be proportional to the number of molecules in the gas phase—that is, the pressure—and the fraction of sites available, $1 - \theta$:

$$\text{rate of adsorption} = k_a P(1 - \theta)$$

Furthermore, the rate of desorption will depend only on the number already adsorbed:

$$\text{rate of desorption} = k_d \theta$$

At equilibrium, these two rates are equal:

$$k_d \theta = k_a P(1 - \theta)$$

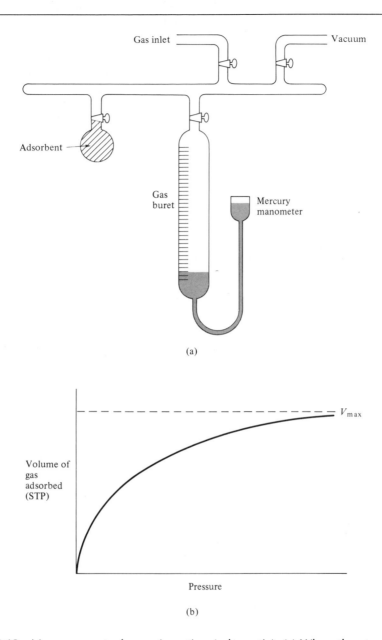

(a)

(b)

Figure 10.19 Measurement of gas adsorption (schematic). (a) When the stopcock between the vacuum manifold and the adsorbent is opened, gas is adsorbed, and the
volume adsorbed is measured with a gas buret. (b) A typical result for volume adsorbed
vs. pressure — this is called an *adsorption isotherm*.

This equation can be solved for θ as a function of the ratio:

$$b = \frac{k_a}{k_d}$$

$$\theta = \frac{bP}{1 + bP} \qquad (10.97)$$

The absorption constant (b) depends on the strength of the gas-surface interaction and on the nature of both.

Experimentally, θ can be related to V_{ads}, and its maximum value [V_{max}, cf. Figure 10.19(b)] by assuming that the maximum occurs when $\theta = 1$:

$$\theta = \frac{V_{ads}}{V_{max}} \qquad (10.98)$$

Equation (10.97) clearly is qualitatively similar to the isotherm of Figure 10.19(b); quantitative agreement is best tested by taking the reciprocal of Eq. (10.97); substituting Eq. (10.98), this gives:

$$\frac{1}{V_{ads}} = \frac{1}{V_{max}} + \left(\frac{1}{bV_{max}}\right)\frac{1}{P} \qquad (10.99a)$$

If Langmuir's equation is correct, plots of $1/V_{ads}$ vs. $1/P$ should be linear (cf. Figure 10.20). Another way by which Eq. (10.97) can be linearized is to graph P/V_{ads} vs. P:

$$\frac{P}{V_{ads}} = \frac{1}{bV_{max}} + \frac{1}{V_{max}}P \qquad (10.99b)$$

This form of Langmuir's isotherm is usually preferred for data analysis — Figure 10.20 contrasts these methods using data for the adsorption of CO on mica (90 K).

This theory is readily extended to gas mixtures if we assume that the only effect of one gas on the adsorption of the other is in making a certain number of sites unavailable; for two gases this gives:

$$k_{d1}\,\theta_1 = k_{a1}\,P_1(1 - \theta_1 - \theta_2)$$
$$k_{d2}\,\theta_2 = k_{a2}\,P_2(1 - \theta_1 - \theta_2)$$

With $b_1 = k_{a1}/k_{d1}$, $b_2 = k_{a2}/k_{d2}$, these equations can be solved for θ_1 and θ_2; with a bit of algebra, we obtain the result:

$$\theta_1 = \frac{b_1 P_1}{1 + b_1 P_1 + b_2 P_2}$$

$$\qquad (10.100)$$

$$\theta_2 = \frac{b_2 P_2}{1 + b_1 P_1 + b_2 P_2}$$

The generalization to any number of gases with the same assumptions is involved

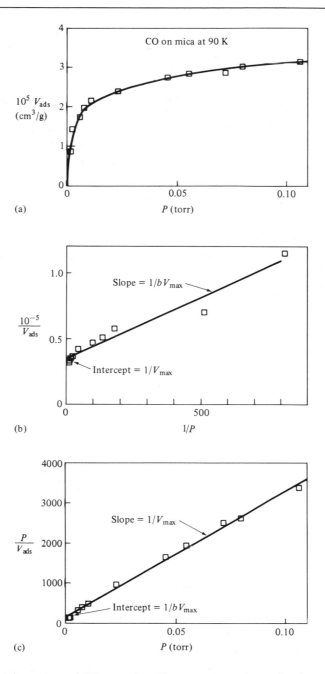

Figure 10.20 Adsorption of CO on mica. Three presentations of volume adsorbed vs. pressure are shown. Panels (b) and (c) should both be linear if Langmuir's isotherm were correct and there were no experimental error; for this set of data, the plot shown on panel (c) is more useful. [Data from Bawn, *J. Am. Chem. Soc.*, 54, 77 (1932)]

but straightforward; the result is:

$$\theta_i = \frac{b_i P_i}{1 + \sum_n b_n P_n} \tag{10.101}$$

Langmuir's theory unquestionably makes some rather broad assumptions. Like the ideal gas law, it is never completely correct but rarely totally wrong, and it provides a firm starting point for understanding a very complex phenomenon, surface adsorption. In subsequent discussions we shall use it without question, with the understanding that the results may not be exact for all situations.

Kinetics of Reactions on Surfaces

The rate of a surface reaction may be limited by several steps, including (1) the rate at which the reactant reaches the surface by diffusion, (2) the rate of adsorption, (3) the rate of reaction on the surface, (4) the rate of product desorption. While these steps may be studied separately, we shall summarize them by a single rate constant.

Consider first a unimolecular decomposition:

$$A \longrightarrow B + C$$

If this reaction occurs on a surface, the rate law (in pressure units) can be expected to be proportional to the surface area available (S_0) and the fraction of coverage (θ):

$$v = -\frac{dP_A}{dt} = kS_0 \theta_A$$

If no other gases are present, Eq. (10.101) gives:

$$v = \frac{kS_0 b_A P_A}{1 + b_A P_A + b_B P_B + b_C P_C} \tag{10.102}$$

We shall examine two limiting cases. First, if the products are not strongly adsorbed, b_B and $b_C \ll b_A$. The rate law will be:

$$\frac{-dP_A}{dt} = \frac{kS_0 b_A P_A}{1 + b_A P_A} \tag{10.103}$$

For high pressures and strong adsorption, it may happen that $b_A P_A \gg 1$, at least at the beginning of the reaction; the effective reaction order will then be *zero*.

$$-\frac{dP_A}{dt} = kS_0 \quad \text{(for } b_A P_A \ll 1\text{)}$$

As P_A decreases with time, the order will change; toward the end, the reaction will become first-order. This rate law was discussed earlier [Eqs. (10.36) and (10.37)] and is demonstrated in Figure 10.21 (which should be contrasted to Figure 10.5).

A bimolecular reaction:

$$A + B \longrightarrow \text{products}$$

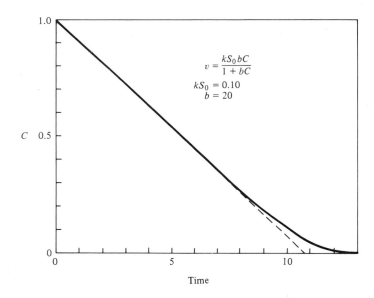

Figure 10.21 Apparent zeroth-order kinetics. A first-order decomposition that occurs on a surface, with the reactant adsorbing and obeying Langmuir's equation, and the products not adsorbed, will appear to be a zeroth-order reaction at early times. This is demonstrated on this graph by the fact that the concentration vs. time plot is linear up to about $t = 7$.

on a surface may happen two ways: (1) one reactant adsorbs on the surface and the other reacts with it on the surface without itself being adsorbed; (2) both reactants are adsorbed and subsequently meet and react on the surface. (Some adsorbents are quite mobile on the surface, and their wanderings can be treated as a two-dimensional diffusion process.)

In the first case, the reaction velocity will be:

$$v = kS_0 \, \theta_A P_B \tag{10.104}$$

If the products are not adsorbed, Eq. (10.97) gives:

$$v = \frac{kS_0 \, b_A P_A P_B}{1 + b_A P_A} \tag{10.105}$$

This reaction could vary from first to second order, depending on the pressure of A and the strength of its adsorption.

The second case would have a rate law:

$$v = kS_0 \, \theta_A \, \theta_B \tag{10.106}$$

This could take a variety of forms, depending on the relative strengths of product and reactant adsorption.

10.10   Enzyme Catalysis

Proteins are macromolecules that are polymeric chains of α-amino acids:

$$R-\overset{\overset{\displaystyle NH_3^+}{|}}{\underset{\underset{\displaystyle R'}{|}}{C}}-COO^-$$

which are linked together by peptide bonds:

$$\cdots N-\overset{\overset{\displaystyle R}{|}}{\underset{\underset{\displaystyle H}{|}}{C}}-\overset{\overset{\displaystyle}{\|}}{\underset{\underset{\displaystyle O}{}}{C}}-\overset{}{\underset{\underset{\displaystyle R'}{|}}{N}}-\overset{\overset{\displaystyle R}{|}}{\underset{\underset{\displaystyle H}{|}}{C}}-\overset{\overset{\displaystyle}{\|}}{\underset{\underset{\displaystyle O}{}}{C}}-\underset{\underset{\displaystyle R'}{|}}{N}-\overset{\overset{\displaystyle R}{|}}{\underset{\underset{\displaystyle H}{|}}{C}}-\overset{\overset{\displaystyle}{\|}}{\underset{\underset{\displaystyle O}{}}{C}}\cdots$$

Enzymes are proteins that act as catalysts for chemical reactions, particularly those which are involved in life processes. Although protein solutions are usually considered to be true molecular solutions, some proteins are so large (up to 100 nm diameter) that they may be considered to be colloidal suspensions. As we shall see, enzyme catalysis has much in common mechanistically with surface catalysis, so, in the end, whether the process is considered to be homogeneous or heterogeneous may be just a matter of point of view.

The Michaelis-Menten Mechanism

The simplest type of enzyme reaction is the conversion of a reactant, called the *substrate* (S), to a product. The rate law is often of the form of Eq. (10.36) with C representing the substrate concentration. The simplest mechanism that can account for this rate law is that proposed by L. Michaelis and M. L. Menten (1913). This mechanism postulates a reversible reaction of an enzyme (E) and substrate to form a complex (ES), with the product being formed from the complex and released from the enzyme:

$$E + S \underset{-1}{\overset{1}{\rightleftharpoons}} ES \overset{2}{\longrightarrow} \text{products} + E \tag{10.107}$$

Figure 10.22 shows the concentrations of S, ES, and product as a function of time for a fairly typical set of concentrations and rate constants. (For simplicity, all concentrations are given as ratios to the initial concentration of substrate, S_0.) Note the rapid transient and the small concentration of ES; this is clearly a case for which the steady-state approximation is appropriate.

The velocity of the overall reaction will be:

$$v = k_2[ES]$$

The enzyme-substrate complex concentration can be obtained using the usual steady-state assumption:

$$[ES]_{ss} = \frac{k_1[E][S]}{k_2 + k_{-1}} = \frac{[E][S]}{K_m} \tag{10.108}$$

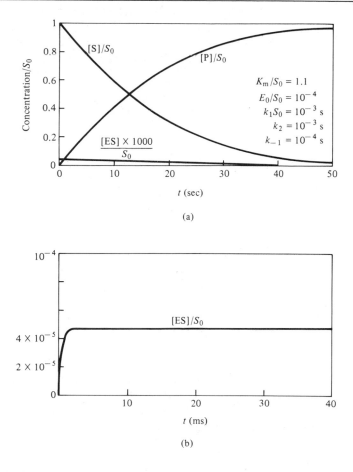

Figure 10.22 Enzyme kinetics. Concentration vs. time curves for substrate (S), product (P), and enzyme-substrate complex (ES) are shown for a typical case for the Michaelis-Menten mechanism. Panel (b) shows that the ES complex rises rapidly to its steady-state value and declines slowly thereafter; these are exactly the conditions required for the steady-state approximation to be valid. Note that on panel (a) the ES concentration is magnified by 1000; it would otherwise be indistinguishable from the baseline.

The ratio of rate constants:

$$K_m = \frac{k_2 + k_{-1}}{k_1}$$

is called the *Michaelis constant*. For the steady state, the rate law will be:

$$v = \frac{k_2[E][S]}{K_m}$$

This is rather similar to the result, Eq. (10.74), derived earlier for a similar mechanism; in this case, however, E is a catalyst rather than a reactant. For this reason, and because the concentration of the uncomplexed enzyme may be difficult to measure, it will be better to express the rate law in terms of the *total* enzyme concentration E_0; material balance gives:

$$E_0 = [E] + [ES]$$

Use of Eq. (10.108) for the steady state gives:

$$E_0 = [E] + \frac{[E][S]}{K_m}$$

$$[E] = \frac{E_0}{1 + \dfrac{[S]}{K_m}} = \frac{K_m E_0}{K_m + [S]}$$

Finally, the reaction velocity is:

$$v = \frac{k_2 E_0 [S]}{K_m + [S]} \tag{10.109}$$

Although Eq. (10.109) can be integrated [cf. Eq. (10.37) with $k = k_2 E_0$], enzyme data are usually analyzed by the differential method. There are two ways to do this: (1) measure v as a function of substrate concentration as the reaction proceeds; (2) measure initial velocity for a variety of substrate concentrations. The latter method has the advantage that it is affected less by complications such as the reversibility of the second step [Eq. (10.107)] or the formation of an enzyme-product complex.

A graph of Eq. (10.109) is shown in Figure 10.23; it can be seen that the reaction velocity approaches a maximum value:

$$v_{max} = k_2 E_0 \tag{10.110}$$

asymptotically; also $v = 0.5 v_{max}$ when $[S] = K_m$. However, determining v_{max} from such a plot would be difficult because of the slowness of the asymptotic approach. A better procedure is to take the reciprocal of Eq. (10.109); this can be shown to give:

$$\frac{1}{v} = \frac{1}{v_{max}} + \frac{K_m}{v_{max}} \frac{1}{[S]} \tag{10.111}$$

which demonstrates that a plot of $1/v$ vs. $1/[S]$ is linear (Figure 10.23). [A comparison to the Langmuir adsorption isotherm, Eq. (10.99), shows the similarity of the equations and method of analysis involved.]

Often, impure enzyme preparations are used for which E_0 is not known. The advantage of the analysis discussed above is that one can characterize enzyme activity through measurements of v_{max} and K_m without determining E_0 or the individual rate constants.

Enzyme Inhibition

Enzyme function can be affected by substances called inhibitors (I). Often a reversible association of the inhibitor with the enzyme prevents the formation of a product; that

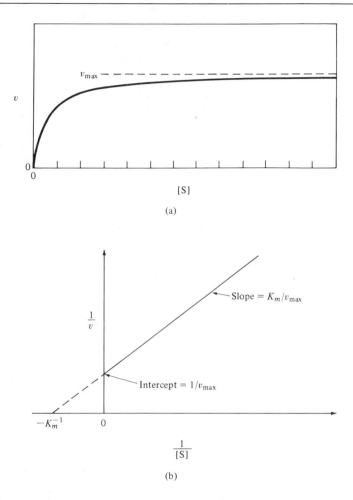

Figure 10.23 Enzyme kinetics: the Michaelis-Menten mechanism. (a) Reaction velocity vs. substrate (S) concentration. (b) A double-reciprocal plot will be linear if the Michaelis-Menten mechanism is correct.

is, either the enzyme-inhibitor complex is incapable of binding substrate or, if the substrate is bound, it cannot react to form product. There are two possible equilibria:

$$I + E \rightleftharpoons EI, \qquad K_I = \frac{[E][I]}{[EI]} \qquad (10.112)$$

$$I + ES \rightleftharpoons ESI, \qquad K_I' = \frac{[ES][I]}{[ESI]} \qquad (10.113)$$

Note that these equilibrium constants are the inverse of the usual definitions. Three extreme cases are usually distinguished.

1. *Competitive inhibition:* $K_I' \gg K_I$ [that is, only reaction (10.112)]. This is generally pictured as a case of the inhibitor occupying the same site on the enzyme that

the substrate needs; thus, the inhibitor prevents reaction by preventing the substrate from binding.

2. *Noncompetitive inhibition:* $K_I = K'_I$. Here, presumably, the inhibitor binds at a different site than the substrate, stopping the reaction by its effect on the enzyme itself.

3. *Uncompetitive inhibition:* $K'_I \ll K_I$ [that is, only reaction (10.113)]. The inhibitor in this case affects only the enzyme-substrate complex, possibly by binding to the already bound substrate and preventing product formation.

In any case, the principal effect of the inhibitor is on the enzyme material balance. Since only ES can react, we have as before:

$$v = k_2[ES]$$

Also the steady-state concentration of ES is as before [Eq. (10.109)]. [This is obviously true for competitive inhibition, where only E can react with I; the reader should demonstrate that it is also true when ES reacts with I, provided reaction (10.113) is at equilibrium.]

For competitive inhibition, the enzyme material balance is:

$$E_0 = [E] + [ES] + [EI]$$

$$E_0 = [E] + \frac{[E][S]}{K_m} + \frac{[E][I]}{K_I}$$

$$E_0 = [E]\left(1 + \frac{[S]}{K_m} + \frac{[I]}{K_I}\right)$$

The reader should demonstrate that the reaction velocity in this case is:

$$v = \frac{v_{max}[S]}{K_m + [S] + \frac{K_m}{K_I}[I]} \tag{10.114}$$

The reciprocal of this equation demonstrates that a $1/v$ vs. $1/[S]$ plot is still linear, but its slope depends on the inhibitor concentration:

$$\frac{1}{v} = \frac{1}{v_{max}} + \frac{K_{max}}{v_{max}}\left(1 + \frac{[I]}{K_I}\right)\frac{1}{[S]} \tag{10.115}$$

By a similar procedure, the rate law for noncompetitive inhibition can be shown to be:

$$v = \frac{v_{max}[S]}{K_m + \frac{K_m}{K_I}[I] + [S] + \frac{[I][S]}{K_I}} \tag{10.116}$$

The rate law for uncompetitive inhibition can be shown to be:

$$v = \frac{v_{max}[S]}{K_m + [S]\left(1 + \frac{[I]}{K'_I}\right)} \tag{10.117}$$

Exercise (optional): Show that plots of $1/v$ vs. $1/[S]$ for un- and noncompetitive inhibition are still linear, but with slopes and intercepts that change with [I] as shown in Figure 10.24. ∎

Most enzyme reactions have mechanisms that are considerably more complex than the Michaelis-Menten mechanism; like many other simple theories (such as the Langmuir theory of adsorption), it is only the starting point for understanding enzyme kinetics. The equations derived above often do work, but this is not a proof that the mechanism is, in fact, as simple as (10.107). Also, the plots illustrated by Figures 10.23 and 10.24 are generally used only for diagnostic purposes; for better methods of data analysis the reader should consult a specialized text such as that by Segel (ref. 9).

10.11 Solution Reactions

Much of what has been said so far applies as well to solution- or gas-phase reactions, but the theoretical explanations, when given, were generally for gas-phase reactions. In this section we shall examine briefly some distinctive features of solution kinetics.

The role of the solvent in a reaction may be passive or active. At the most passive, it may be no more than an energy-transfer agent, a "chaperon," and only its density makes its effect different than in the gas phase. At its most active it may be considered catalytic, but it is so ubiquitous that this role may not be explicitly recognized. The same solute-solvent interaction characteristics that affect thermodynamics (Chapters 7 and 8) will affect kinetics — for example, solvent polarity, specific interactions such as hydrogen bonding, and ionic strength.

Cage Effect

A major distinction between gas and solution phase is the limitation placed on the mobility of the reactants by the high density of solvent molecules. An example worked out in Section 9.2 demonstrated that a molecule in a gas at STP could diffuse a distance of 5 mm in one second; in a solution, this distance will be smaller by an order of magnitude or more. It is therefore somewhat surprising that some reactions, for which the gas- and solution-phase mechanism is the same, proceed at nearly the same rate in solution as in the gas; the reason for this is the *cage effect*. In solution, the reactants will meet less often; after they meet, however, they are held together for an extended period by a solvent *cage* and will undergo frequent collisions. Thus the *encounter* frequency is reduced, but the probability of reaction per encounter is greatly enhanced.

According to the "cage" explanation, one would expect a major difference in the maximum rates of reaction for reactions (such as radical recombinations) that occur on first encounter. In gases, as discussed earlier, the maximum rate constant of a bimolecular gas reaction is limited by the collision frequency to about 10^{16}–10^{17} cm^3 mol^{-1} sec^{-1}. In solution, such reactions will be limited by the rate of diffusion of the reactants and are therefore called *diffusion-controlled* reactions. In such

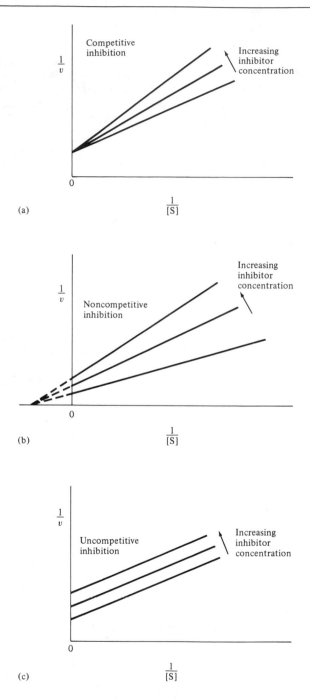

Figure 10.24 Enzyme inhibition. Double-reciprocal plots for three types of enzyme inhibition: (a) competitive, (b) noncompetitive, and (c) uncompetitive.

a case, the rate constant will be given by an equation derived by M. V. Smoluchowski (1917) and P. Debye (1942):

$$k = 4\pi\sigma_{12}L(D_1 + D_2) \tag{10.118}$$

The diffusion constants of the reactants (D_1 and D_2) in a nonviscous solvent are typically of the order of 10^{-5} cm^2 s^{-1}; if $\sigma_{12} \sim 4 \times 10^{-8}$ cm, the magnitude of the rate constant for a diffusion-controlled reaction can be expected to be $\sim 6 \times 10^{12}$ cm^3 mol^{-1} sec^{-1}, or about four orders of magnitude slower than in the gas phase. This expectation is generally borne out by experiment.

Ionic Reactions

Reactions of ions in solution (the solvent is usually water) can be treated using the Debye-Hückel theory, discussed in Chapter 8. The activated-complex theory (Section 10.3) must be modified to account for the nonideality of the ions. For a reaction A + B → with reactant charges z_A and z_B respectively, the activated complex will have a charge $z_A + z_B$; it will be less charged than the reactants if the reactant charges are of opposite sign and vice versa. Activated-complex theory predicts:

$$k = \frac{k_b T}{hC^\theta} \frac{\gamma_A \gamma_B}{\gamma_\ddagger} e^{\Delta S^\ddagger/R} e^{\Delta H^\ddagger/RT} \tag{10.119}$$

If the Debye-Hückel limiting law [Eq. (8.13)] is used for the activity coefficients, this formula will give the equation first derived by J. N. Brønsted (1922):

$$\ln k = \ln k^\circ + 2.34 z_A z_B \sqrt{I} \tag{10.120}$$

where I is the ionic strength (see Figure 10.25). For kinetics, concentration is generally preferred to molality, so:

$$I = \tfrac{1}{2} \sum_i z_i^2 C_i \tag{10.121}$$

The sum of Eq. (10.121) is over *all* ions in solution, including the reactants, products, and added salts. Thus, reaction rates of like ions (with the same sign of z) will be increased by an increase in the ionic strength and vice versa. This relationship also emphasizes that kinetic studies of ionic reactions, if they are to give consistent results, must control the ionic strength of the reacting mixture.

Equation (10.120) is also useful in working out reaction mechanisms; plots of $\ln k$ vs. $\sqrt{I}$ have a slope that gives the charge product ($z_A z_B$) of the reacting species involved in the rate-determining step.

10.12 Relaxation Methods

As mentioned earlier, the major problem in studying very fast reactions is not in measuring concentration, which could, for example, be done spectroscopically, but in getting the reactants totally mixed before the reaction is complete. For example, if you

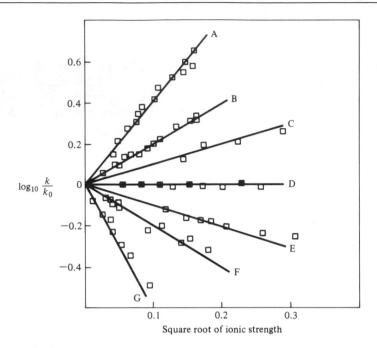

Figure 10.25   Effect of ionic strength on rate constants. A. ($z_A z_B = 4$) $Co(NH_3)_5Br^{2+}$ + Hg^{2+}. B. ($a_A z_B = 2$) $S_2O_8^{2-}$ + I^-. C. ($z_A z_B = 1$) $CO(OC_2H_5)N:NO_2^-$ + OH^-. D. ($z_A z_B = 0$) (open circles) $Cr(urea)_6^{3+}$ + H_2O (closed circles) $CH_3COOC_2H_5$ + OH^-. E. ($z_A z_B = -1$) $H^{++}Br^-$ + H_2O_2. F. ($z_A z_B = -2$) $Co(NH_3)_5Br^{2+}$ + OH^-. G. ($z_A z_B = -6$) Fe^{2+} + $Co(C_2O_4)_3^{3-}$. The slopes of the lines are equal to the charge product, $z_A z_B$, of the reacting species (From K. Laidler, *Chemical Kinetics*, 1965: New York, McGraw-Hill Book Co., Fig. 50)

attempted to measure the rate of the reaction:

$$H^+(aq) + OH^-(aq) = H_2O$$

by mixing acid and alkali, the reaction would occur as fast as the reactants were mixed. (The half-life of this reaction for initial concentrations of 0.001 mol/dm³ is 7×10^{-9} sec.)

Relaxation methods are techniques that begin with the reactants and products completely mixed and at equilibrium; then the equilibrium constant is changed by altering some condition, the "relaxation" of the concentrations to their new equilibrium values is measured, and, from this, the rate constants are measured. Manfred Eigen won the 1967 Nobel prize for his pioneering work in this field. A number of physical properties have been used for such experiments including:

T-jump. The sample is rapidly heated by an electrical discharge or by a high-power microwave pulse. Samples of volume ~1 cm³ can be heated in a few microseconds. The equilibrium constant will be changed provided ΔH_{rxn}^θ is not zero.

P-jump. The equilibrium constants of reactions for which ΔV_{rxn}^{θ} is not zero can be altered by pressure. This can be changed rapidly by pressurizing the solution with a gas, then releasing the gas via a rupture disk.

E-jump. The equilibrium constant of ionic reactions will be changed by the application of an electric field (the *Wien effect*). The sample is placed between the plates of a capacitor, which is charged and then rapidly discharged through an external circuit.

One of the earliest of the very fast reactions to be studied (Eigen, 1954) was the ionization of water mentioned above. Reactions of this type—

$$A + B \underset{1}{\overset{2}{\rightleftharpoons}} C$$

$$v = k_2[A][B] - k_1[C] \qquad (10.122)$$

—will be used to illustrate the general technique.

The concentrations can be related to their equilibrium values (A_{eq}, and so on) by a displacement parameter (x):

$$[A] = A_{eq} - x$$

$$[B] = B_{eq} - x$$

$$[C] = C_{eq} + x$$

(Note that x could be positive or negative, depending on the direction in which the equilibrium is shifted.)

The rate law is then:

$$\frac{dx}{dt} = k_2(A_{eq} - x)(B_{eq} - x) - k_1(C_{eq} + x)$$

$$= (k_2 A_{eq} B_{eq} - k_1 C_{eq}) - [k_1 + k_2(A_{eq} + B_{eq})]x + k_2 x^2$$

The equilibrium constant of this reaction is:

$$K = \frac{C_{eq}}{A_{eq} B_{eq}} = \frac{k_2}{k_1}$$

Therefore the first term of the rate law (above) is zero. The quadratic term of the rate law is negligible because x is small. Therefore:

$$\frac{dx}{dt} = -[k_1 + k_2(A_{eq} + B_{eq})]x \qquad (10.123)$$

The **relaxation time** (τ) of this reaction is defined as:

$$\tau = [k_1 + k_2(A_{eq} + B_{eq})]^{-1} \qquad (10.124)$$

Any property (λ) that varies linearly with the extent of the reaction—for example, spectroscopic absorption or solution conductance—is measured as a function of time following the application of the perturbation (*T*-jump and so on); from Eq. (10.123):

$$\lambda(t) = \lambda_0 e^{-t/\tau} \qquad (10.125)$$

where λ_0 is the value of the property at $t = 0$. (Since the change is not instantaneous,

Table 10.4   Rate constants in aqueous solution (25°)

Reaction	k(forward) $(dm^3\ mol^{-1}\ s^{-1})$	k(reverse) (s^{-1})
$H^+ + OH^- = H_2O$	1.4×10^{11}	2.5×10^{-5}
$D^+ + OD^- = D_2O$	8.4×10^{10}	2.5×10^{-6}
$H^+ + CH_3COO^- = CH_3COOH$	4.5×10^{10}	7.8×10^5
$H^+ + C_6H_5COO^- = C_6H_5COOH$	3.5×10^{10}	2.2×10^6
$H^+ + NH_3 = NH_4^+$	4.3×10^{10}	24.6
$OH^- + NH_4^+ = NH_3 + H_2O$	3.4×10^{10}	6×10^5

Source: F. Daniels and R. A. Alberty, *Physical Chemistry,* 3d ed., 1966: New York, John Wiley & Sons, Inc.

it may be convenient to choose the $t = 0$ point a brief time after the perturbation is applied.)

Example: Calculate the relaxation time for the neutralization of acetic acid for a concentration of 0.01 mol/dm³.

$$CH_3COO^- + H^+ \underset{1}{\overset{2}{\rightleftharpoons}} CH_3COOH$$

From Table 10.4, $k_1 = 7.8 \times 10^5\ s^{-1}$, $k_2 = 4.5 \times 10^{10}\ dm^3\ mol^{-1}\ s^{-1}$. The equilibrium constant is:

$$\frac{[CH_3COOH]}{[H^+][CH_3COO^-]} = \frac{k_2}{k_1} = 5.77 \times 10^4$$

The reader should solve this to get:

$$[H^+] = [CH_3COO^-] = 4.08 \times 10^{-4}\ mol/dm^3$$

Then from Eq. (10.124):

$$\tau^{-1} = (7.8 \times 10^5\ s^{-1}) + (4.5 \times 10^{10}\ dm^3\ mol^{-1}\ s^{-1})\,(2)\,(4.08 \times 10^{-4}\ mol/dm^3)$$

$$\tau = 2.7 \times 10^{-8}\ s = 27\ ns \qquad\blacksquare$$

Postscript

This chapter has introduced a number of concepts required for the understanding of rates of chemical reactions, including: reaction velocity, empirical rate laws, rate constants, and reaction mechanisms. For those who wish to learn more about this topic, the book by Gardiner (Ref. 8) has a somewhat more extensive coverage at about the same level as this chapter. There are a large number of books devoted to this subject, only a few of which are listed below.

At the beginning of this chapter, it was stated that an understanding of chemical kinetics was the key to controling chemical behavior. No where is this better illustrated than in the use of additives in gasoline. One of the early problems in the development of the internal combustion engines was the problem of engine knock. The internal combustion engine can be made more efficient by using higher compression ratios, but then the fuel in the cylinder tends to detonate rather than burn smoothly. This problem was first investigated thoroughly in 1916 by Thomas Midgley, Jr., working for Delco. Theoretical understanding was scant at the time, and Midgley systematically tested over 33,000 compounds in his search for an anti-knock additive; he found tetraethyl lead in 1921. As mentioned earlier, tetraethyl lead controls the chain reaction combustion by releasing ethyl radicals which provide a termination for the chains. "Ethyl" gasoline was introduced in 1923, and remained the standard fuel for the next fifty years. However, leaded gasoline was of concern on at least two counts. First, lead (like many other heavy metals) is very poisonous and its widespread dispersal through gasoline combustion was worrisome. Also, when catalytic converters were developed to remove harmful pollutants from automobile exhausts, it was found that lead in the exhaust poisoned the catalyst. Gasoline additives are also used to prevent autoxidation (cf. problem 10.39), corrosion of metal surfaces in pipelines and storage tanks, and to prevent icing and carbon deposits in the carburetor.

Another area where the control of chemical reactions is important is in the use of heterogeneous catalysts for hydrocarbon reforming. This field received a major impetus with the fuel shortages of the 1970s. Catalyst development has traditionally been totally empirical; it has been described as a "black art" and "the last stronghold of alchemy." In recent years, immense strides have been made in the understanding and planned development of catalysts. Modern spectroscopic techniques such as nuclear magnetic resonance (NMR), electron spin resonance (ESR), infrared (IR), and extended X-ray absorption fine structure (EXAFS) have been used to study both the catalytic surface and the adsorbed species. Progress in this field was reviewed by J. Haggin (Chemical & Engineering News, November 15, 1982, page 11).

References

1. W. C. Gardiner, Jr., "The Chemistry of Flames," *Scientific American,* February 1982.
2. K. J. Laidler, "Unconventional Applications of the Arrhenius Law," *J. Chem. Educ.,* 49, 343 (1972).
3. D. Edelson, "The New Look in Chemical Kinetics," *J. Chem. Educ.,* 52, 643 (1975).
4. P. C. Jordan, *Chemical Kinetics and Transport,* 1979: New York, Plenum Press.
5. S. W. Benson, *The Foundations of Chemical Kinetics,* 1960: New York, McGraw-Hill Book Co.
6. J. Nicholas, *Chemical Kinetics: A Modern Survey of Gas Reactions,* 1976: New York, John Wiley & Sons, Inc.
7. C. Walling, *Free Radicals in Solution,* 1975: New York, John Wiley & Sons, Inc.
8. W. C. Gardiner, Jr., *Rates and Mechanisms of Chemical Reactions,* New York, W. A. Benjamin, Inc.
9. Irwin H. Segel, *Enzyme Kinetics,* 1975: New York, Wiley-Interscience.
10. K. J. Laidler, *Chemical Kinetics,* 1965: New York, McGraw-Hill Book Co.

Problems

10.1   What would be the units for the rate constant of a $\frac{3}{2}$-order reaction? Use concentrations in mol/dm^3, time in seconds.

10.2   The decomposition of dimethyl ether:

$$CH_3OCH_3 = CH_4 + CO + H_2$$

was studied by C. N. Hinshelwood and P. J. Askey [*Proc. Roy. Soc.* (London), A115, 215 (1927)] by measuring the time required for the total pressure to double. Use the data below to determine the order of this reaction.

Initial P (torr)	t (sec)	Initial P (torr)	t (sec)
28	1980	321	625
58	1500	394	590
150	900	422	508
171	824	509	465
261	670	586	484

10.3   Use the data below for the decomposition of diacetylene at 1173 K [K. C. Hou and H. B. Palmer, *J. Phys. Chem.*, 69, 858 (1965)] to determine the reaction order. (Use the two-point differential method.)

10^7C (mol/cm^3)	t (sec)	10^7C (mol/cm^3)	t (sec)
0.532	0	0.298	0.150
0.454	0.030	0.267	0.200
0.420	0.050	0.237	0.250
0.364	0.100		

10.4   Use the data below for the partial pressure of N_2O_5 during a thermal decomposition at 45°C to determine the reaction order and rate constant.

t (min)	P (torr, N_2O_5)
0	348.4
20	185.2
40	105.4
60	58.6
80	33.1
100	18.6
160	2.8

10.5   T. Y. Chin and R. E. Connich [*J. Phys. Chem.*, 63, 1518 (1959)] measured the data below for the OH$^-$-catalyzed reaction:

$$OCl^- + I^- = OI^- + Cl^-$$

Determine the rate law and rate constant.

[OCl$^-$] (mol dm^{-3})	[I$^-$] (mol dm^{-3})	[OH$^-$] (mol dm^{-3})	Initial velocity (mol dm^{-3} s^{-1})
0.0017	0.0017	1.00	1.75×10^{-4}
0.0034	0.0017	1.00	3.50×10^{-4}
0.0017	0.0034	1.00	3.50×10^{-4}
0.0017	0.0017	0.50	3.50×10^{-4}

10.6 The reaction:

$$(CH_3)_3COOC(CH_3)_3 = 2\ CH_3COCH_3 + C_2H_6$$

was studied manometrically and found to be first-order. Use the data below to calculate the rate constant. (The measured pressures include 4.2 torr due to nitrogen that was present in the reaction container.)

t (min)	P (torr)	t (min)	P (torr)
0	173.5	12	244.4
2	187.3	14	254.4
3	193.4	15	259.2
5	205.3	17	268.7
6	211.3	18	273.9
8	222.9	20	282.0
9	228.6	21	286.8
11	239.8		

10.7 The first-order gas reaction:

$$SO_2Cl_2 = SO_2 + Cl_2$$

has $k = 2.20 \times 10^{-5}$ sec^{-1}. What percentage of the SO_2Cl_2 would be decomposed after 5 hours?

10.8 A first-order reaction has a half-life of 26.2 minutes. At what time will the reaction be 90% complete?

10.9 If a first-order reaction is 20% complete in 20 minutes, at what time will it be 90% complete?

10.10 A 0.0250-mol/dm^3 solution of oxalic acid in concentrated H_2SO_4 was studied by Lichty [*J. Phys. Chem.*, 11, 225 (1907)] by titration with $KMnO_4$ (volume V, below). Assume first-order kinetics and calculate the rate constant.

t (min)	V (KMnO$_4$, cm^3)	t (min)	V (KMnO$_4$, cm^3)
0	11.45	600	4.79
120	9.63	900	2.97
240	8.11	1440	1.44
420	6.22		

10.11 W. W. Heckert and E. Mack, Jr. [*J. Am. Chem. Sec.*, 51, 2706 (1929)] measured the following for the decomposition of ethylene oxide:

$$CH_2{=\!=\!}CH_2 = CH_4 + CO$$
$$\diagdown\ /$$
$$O$$

Assume that this reaction is first-order and calculate the rate constant. (The published total pressure P has been corrected for manometer dead space. If you use linear regression, omit the $t = 0$ point.)

t (min)	P (torr)	t (min)	P (torr)
0	115.30	10	129.10
6	122.91	11	130.57
7	124.51	12	132.02
8	126.18	13	133.49
9	127.53	18	140.16

10.12   F. G. Ciapetta and M. Kilpatrick [*J. Am. Chem. Soc.*, 70, 639 (1948)] studied the hydration of isobutene in perchloric acid; the reaction was found to be pseudo-first-order. Use the dilatometer readings (*h*) below to calculate the rate constant.

t (min)	*h*	*t* (min)	*h*
0	18.84	25	16.86
5	18.34	30	16.56
10	17.91	35	16.27
15	17.53	40	16.00
20	17.19	∞	12.16

10.13   The HCl-catalyzed isomerization of *N*-chloroacetanilide to *p*-chloroactanilide is pseudo-first-order. The reactant was destroyed with KI (to form I_2) followed by titration with sodium thiosulphate. Calculate the rate constant from the data below.

t (min)	Titer (ml)
0	24.5
15	18.1
30	13.2
45	9.7
60	7.1
75	5.2

10.14   A second-order reaction, $A + B \longrightarrow \cdots$, with rate law $v = k[A][B]$, has $k = 5.21 \, dm^3 \, mol^{-1} \, min^{-1}$. If a reaction mixture has initial concentrations $a = 0.1$, $b = 0.2 \, mol/dm^3$, what will be the concentrations of A and B after one minute?

10.15   A second-order reaction, $A + B \longrightarrow \cdots$, has a rate law $v = k[A][B]$ and $k = 1.23 \, dm^3 \, mol^{-1} \, sec^{-1}$. If A and B are mixed with equal initial concentrations of $0.365 \, mol/dm^3$, at what time will the reaction be 90% complete?

10.16   The second-order reaction:

$$OH^- + CH_3COOC_2H_5 = CH_3COO^- + C_2H_5OH$$

was investigated by measuring solution conductance (*L*). Use the results below for initial concentrations of $0.01 \, mol/dm^3$ (both reactants) to calculate the rate constant.

t (min)	$\dfrac{L}{L_0 - L_\infty}$	*t* (min)	$\dfrac{L}{L_0 - L_\infty}$
0	1.560	18	1.020
5	1.315	20	0.994
7	1.247	25	0.945
9	1.193	27	0.923
15	1.064	∞	0.560

10.17   The dimerization of 1,3-butadiene at 326°C was followed by measuring the total pressure. Assume second-order, and calculate the rate constant.

t (min)	P (torr)
0	632.0
3.25	618.5
12.18	584.2
24.55	546.8
42.50	509.3
68.05	474.6

10.18 Farkas, Lewin, and Bloch [*J. Am. Chem. Soc.*, 71, 1988 (1949)] studied the reaction:

$$Br^- + ClO^- = BrO^- + Cl^-$$

in water at 25°C. With initial concentrations $[ClO^-] = 3.230 \times 10^{-3}$ mol/dm³, $[Br^-] = 2.508 \times 10^{-3}$ mol/dm³, the results were:

t (min)	$[BrO^-]$ (mmol/dm³)
0	0
3.65	0.560
7.65	0.953
15.05	1.420
26.00	1.800
47.60	2.117
90.60	2.367

Assume a second-order rate law and calculate the rate constant.

10.19 F. M. Miller and M. L. Adams [*J. Am. Chem. Soc.*, 75, 4599 (1953)] have measured the kinetics for the alkaline hydrolysis of *p*-nitrosodiumdimethylaniline (NSA). This is a second-order reaction. Use the data below (concentrations, moles/dm³) to calculate the rate constant.

t (sec)	[NSA]	$[OH^-]$
0	0.0500	0.199
135	0.0413	0.190
380	0.0365	0.186
610	0.0325	0.182
945	0.0282	0.177
1880	0.0187	0.168

10.20 Derive the integrated rate law for the autocatalytic second-order reaction:

$$A \longrightarrow P$$

$$\frac{-d[A]}{dt} = k[A][P]$$

with initial $[A] = a$, $[P] = p$.

10.21 If f is defined as the fraction of the reactant remaining in time t (for example, $f = \frac{1}{2}$ for the half-life), derive an equation for the "f-life" of a general nth-order reaction.

10.22 A commonly used rule of thumb is that rate constants will double for a 10°C increase in temperature (at about room temperature). What does this imply about a typical Arrhenius activation energy?

10.23   Use the data in Table 10.2 to calculate the rate constant for the reaction:

$$C_2H_6 = 2\,CH_3$$

at 700 K.

10.24   Use the data (below) for the reaction:

$$N_2O_5 = N_2O_4 + \tfrac{1}{2}O_2$$

to calculate the Arrhenius parameters.

T (K)	k (s^{-1})	T (K)	k (s^{-1})
273.1	7.87×10^{-7}	313.1	2.47×10^{-4}
288.1	1.04×10^{-5}	318.1	4.98×10^{-4}
293.1	1.76×10^{-5}	323.1	7.59×10^{-4}
298.1	3.38×10^{-5}	328.1	1.50×10^{-3}
308.1	1.35×10^{-4}	338.1	4.87×10^{-3}

10.25   Calculate the Arrhenius parameters for the reaction:

$$2\,NO + O_2 = 2\,NO_3$$

from the data:

T (K)	$10^{-9}k$ (cm^6 mol^{-2} s^{-1})
270	9.12
370	4.67
470	3.28
570	2.75
670	2.49

10.26   Use the rate constants (below) for the reaction:

$$2\,HI = H_2 + I_2$$

to calculate the Arrhenius parameters.

T (K)	k (dm^3 mol^{-1} s^{-1})	T (K)	k (dm^3 mol^{-1} s^{-1})
556	3.52×10^{-7}	683	5.12×10^{-4}
575	1.22×10^{-6}	700	1.16×10^{-3}
629	3.02×10^{-5}	716	2.50×10^{-3}
647	8.59×10^{-5}	781	3.95×10^{-2}
666	2.19×10^{-4}		

10.27   Use simple collision theory to calculate the rate constant of the reaction:

$$CH_3 + CH_3 = C_2H_6$$

assuming $p = 1$ and $E_{min} = 0$. Assume $\sigma = 4 \times 10^{-8}$ cm, $T = 500$ K.

10.28   Use data from Table 10.2 to calculate $\Delta H^{\ddagger}$ and $\Delta S^{\ddagger}$ for the reaction:

$$CH_3NC = CH_3CN$$

(assume $T = 300$).

10.29   Use data from Table 10.2 to calculate $\Delta H^{\ddagger}$ and $\Delta S^{\ddagger}$ for the reaction:

$$OH + H_2 \rightarrow H_2O + H$$

(assume $T = 300$).

10.30 Use the Eyring equation to analyze the data of Problem 10.24 (for the N_2O_5 reaction) and calculate $\Delta H^{\ddagger}$ and $\Delta S^{\ddagger}$ of this reaction.

10.31 Analyze the data of Problem 10.26 (for the HI reaction) using the Eyring equations to determine $\Delta H^{\ddagger}$ and $\Delta S^{\ddagger}$.

10.32 The rate constant for the reaction:

$$\phi N(Me)_2 + EtI = \phi N(Me)_2 Et^+ + I^-$$

is 3.18×10^{-5} dm^3 mol^{-1} sec^{-1} at 52.5°C, 1 atm, and 12×10^{-5} at 1500 atm, same temperature. Calculate $\Delta V^{\ddagger}$ for this reaction.

10.33 The concentration of the intermediate (B) for consecutive first-order reactions is:

$$[B] = \frac{k_1 a}{k_2 - k_1}(e^{-k_1 t} - e^{-k_2 t})$$

Find the value of this expression if $k_1 = k_2$.

10.34 In consecutive first-order reactions with $k_1 = 0.25$ s^{-1} and $k_2 = 0.15$ s^{-1}, at what time will the intermediate reach its maximum concentration, and what percent of the total material present will it be at that time?

10.35 Use the data below for the gas-phase isomerization of cyclo-propane:

$$\begin{array}{c}CH_2\\ \diagup \quad \diagdown \\ CH_2 - CH_2 \end{array} \rightarrow CH_3CH{=}CH_2$$

to test the Lindeman mechanism.

P (torr)	$10^4 k$ (s^{-1})	P (torr)	$10^4 k$ (s^{-1})
84.1	2.98	1.37	1.30
34.0	2.82	0.569	0.857
11.0	2.23	0.170	0.486
6.07	2.00	0.120	0.392
2.89	1.54	0.067	0.303

10.36 The chlorination of vinyl chloride (VC):

$$C_2H_3Cl + Cl_2 = C_2H_3Cl_3 \ (P)$$

may be a chain reaction with an intermediate $R \cdot = C_2H_3Cl_2$. Derive the rate law for the following mechanism:

(1) $Cl_2 + h\nu \longrightarrow 2\,Cl$

(2) $Cl + VC \longrightarrow R\cdot$

(3) $R\cdot + Cl_2 \longrightarrow P + Cl$

(4) $R\cdot + R\cdot \longrightarrow$ stable molecules

10.37 Derive the steady-state rate law for the photoinitiated reaction:

$$H_2 + Br_2 = 2\,HBr$$

assuming the mechanism:

(1) $Br_2 + h\nu \longrightarrow 2\,Br$

(2) $Br + H_2 \longrightarrow HBr + H$

$(3) \quad H + Br_2 \longrightarrow HBr + Br$

$(4) \quad H + HBr \longrightarrow H_2 + Br$

$(5) \quad Br + Br \longrightarrow Br_2$

10.38 The mechanism for the nitrogen-pentoxide-catalyzed decomposition of ozone $(2\,O_3 = 3\,O_2)$ is given below. Derive a rate law by using the steady-state approximation for NO_2 and NO_3.

$(1) \quad N_2O_5 \longrightarrow NO_2 + NO_3$

$(2) \quad NO_2 + NO_3 \longrightarrow N_2O_5$

$(3) \quad NO_2 + O_3 \longrightarrow NO_3 + O_2$

$(4) \quad NO_3 + NO_3 \longrightarrow 2\,NO_2 + O_2$

10.39 The autoxidation of hydrocarbons to form peroxides:

$$RH + O_2 = ROOH$$

is an important reaction involved in the deterioration of oils, fats, and gasoline and the drying of oil-based paints. At high oxygen concentrations, the mechanism in solution appears to be:

$(1) \quad 2\,ROOH \longrightarrow ROO\cdot + RO\cdot + H_2O$

$(2) \quad ROO\cdot + RH \longrightarrow ROOH + R\cdot$

$(3) \quad R\cdot + O_2 \longrightarrow ROO\cdot$

$(4) \quad 2\,ROO\cdot + ROOR + O_2$

Derive the rate law for $-d[O_2]/dt$.

10.40 Sulfuryl chloride, SO_2Cl_2, is an effective chlorinating agent for hydrocarbons, provided some sort of initiator is present. The reaction is:

$$SO_2Cl_2 + RH = RCl + HCl + SO_2$$

A simplified mechanism for this reaction initiated by copper(I) chloride is given below. Derive the rate law.

$(1) \quad CuCl + SO_2Cl_2 \longrightarrow CuCl_2 + SO_2 + Cl$

$(2) \quad Cl + RH \longrightarrow R\cdot + HCl$

$(3) \quad R\cdot + SO_2Cl_2 \longrightarrow RCl + SO_2 + Cl$

$(4) \quad R\cdot + R\cdot \longrightarrow \text{stable molecules}$

10.41 Derive a relationship between the rate, kinetic chain length, and the light absorbed (I_a) for a photochemically initiated vinyl polymerization. Assume bimolecular termination.

10.42 Derive expressions for the rate of polymerization and kinetic chain length for a vinyl polymerization with a transfer termination:

$$R_x\cdot + S \longrightarrow P + S\cdot \qquad (k_{tr})$$

Assume that $S\cdot$ does not react further, and that initiation is via an added initiator (In).

10.43 The rate constants for the polymerization of vinyl acetate at 60°C are:

$$k_p = 2.3 \times 10^3 \text{ dm}^3 \text{ mol}^{-1} \text{ s}^{-1} \qquad (E_a = 26 \text{ kJ/mol})$$

$$k_t = 2.9 \times 10^2 \text{ dm}^3 \text{ mol}^{-1} \text{ s}^{-1} \qquad (E_a = 13 \text{ kJ/mol})$$

The initiator azobisisobutyronitrile has (at the same T):

$$k_i = 1.07 \times 10^{-5} \text{ s}^{-1} \qquad (E_a = 130 \text{ kJ/mol})$$

Assume $f = 1$ and calculate the rate of polymerization when $[\text{In}] = 0.001$, $[\text{M}] = 1 \text{ mol/dm}^3$. Calculate the activation energy for the rate of polymerization. Will (1) the rate, and (2) the average chain length, increase or decrease with temperature?

10.44 Use the data below for the adsorption of nitrogen on mica to determine the Langmuir adsorption parameters.

V_{ads} (cm^3/g)	P (torr)
0.494	2.1×10^{-3}
0.782	4.6×10^{-3}
1.16	1.3×10^{-2}

10.45 The data for the adsorption of krypton on charcoal at 193.5 K is given below. Find if these data fit Langmuir's isotherm and calculate the constants.

V_{ads} (cm^3/g)	P (torr)	V_{ads} (cm^3/g)	P (torr)
5.98	2.45	16.45	11.2
7.76	3.5	18.05	12.8
10.10	5.2	19.72	14.6
12.35	7.2	21.10	16.1

10.46 Use a graph of P/V vs. P to determine the Langmuir parameters for the adsorption of nitrous oxide on barium fluoride using the data below (-40°C).

P (torr)	V (cm^3)
35.9	3.70
64.5	5.09
120	6.70
232	8.48
357	9.92

10.47 The decomposition of ammonia:

$$NH_3 = \tfrac{1}{2}N_2 + \tfrac{3}{2}H_2$$

on a tungsten wire was investigated by Hinshelwood and Burk at 856°C. Use the results (below) to show that the reaction is approximately zero-order and determine the rate constant. Make a graph of $P(NH_3)$ vs. t and compare to Figure 10.21. ($P = 200$ torr at $t = 0$.)

t (sec)	P (torr)	t (sec)	P (torr)
100	214	800	292
200	227	1000	312
300	238	1200	332
400	248.5	1400	349
500	259	1800	378
600	270	2000	387

10.48   The Pt-catalyzed decomposition of NO (into $N_2 + O_2$) is found to have a rate law:

$$v = \frac{kP_{NO}}{P_{O_2}}$$

Show how this rate law could result from Langmuir adsorption and surface catalysis.

10.49   The adsorption of hydrogen on some surfaces has an adsorption isotherm of the form:

$$\theta = \frac{bP^{1/2}}{1 + bP^{1/2}}$$

Propose a mechanism and derive this isotherm; state assumptions clearly.

10.50   H. DeVoe and G. B. Kistiakowsky [*J. Am. Chem. Soc.*, 83, 274 (1961)] studied the kinetics of the reaction:

$$CO_2 + H_2O = HCO_3^- + H^+$$

as catalyzed by the enzyme bovine carbonic anhydrase at 0.5°C, $pH = 7.1$, $E_0 = 2.8 \times 10^{-9}$ mol dm^{-3}. Use the data below to calculate the kinetic parameters of this reaction.

$[CO_2]$ (mol/dm^3)	v (mol dm^{-3} s^{-1})
1.25×10^{-3}	2.8×10^{-5}
$2.5 \ \times 10^{-3}$	5.0×10^{-5}
$5 \ \ \times 10^{-3}$	8.3×10^{-5}
$20 \ \ \times 10^{-3}$	$17 \ \times 10^{-5}$

10.51   Use the data (below) for the hydrolysis of *N*-glutamyl-L-phenylalanine catalyzed by α-chymotrypsin to determine the kinetic constants for the Michaelis-Menten mechanism [*J. Chem. Ed.*, 50, 149 (1973)].

$[S]$ (mol/dm^3)	v (mol dm^{-3} min^{-1})
2.5×10^{-4}	2.2×10^{-6}
5.0×10^{-4}	3.8×10^{-6}
$1 \ \ \times 10^{-3}$	5.9×10^{-6}
1.5×10^{-3}	7.1×10^{-6}

10.52   Integrate the Michaelis-Menten rate law to obtain an equation for $[S]$ (initial value S_0) as a function of t and S_0.

10.53   In analyzing enzyme data, the Hanes plot, $[S]/v$ vs. $[S]$, is considered to be more reliable than the Lineweaver-Burk plot, $1/v$ vs. $1/[S]$, for obtaining kinetic constants from the data. How are the slope and intercepts of this plot related to v_{max} and K_m for the Michaelis-Menten mechanism.

10.54   Derive Eq. (10.116) for the velocity of an enzyme-catalyzed reaction with non-competitive inhibition.

10.55   Derive the rate law for the enzyme mechanism:

$$E + S \overset{1}{\rightleftharpoons} ES$$

$$ES \overset{2}{\longrightarrow} P$$

$$ES + I \rightleftharpoons ESI$$

10.56 A molecule that absorbs light (I_a) may reemit the light in a process called fluorescence or phosphorescence (depending on the time scale, cf. Chapter 13) or lose its energy by collision with another molecule (Q); this is called *quenching* of fluorescence. Show that the intensity of fluorescence (I_f) is related to the intensity of the light absorbed (I_a) by the Stern-Volmer equation:

$$\frac{1}{I_f} = \frac{1}{I_a}\left(1 + \frac{k_q[Q]}{k_f}\right)$$

where k_q is the rate constant for quenching and k_f is the rate constant for fluorescent emission.

10.57 The relaxation time for:

$$H^+ + OH^- = H_2O$$

with $pH = 7.00$ is 36 μs. Calculate the rate constants.

10.58 Calculate the relaxation time for:

$$H^+ + OH^- = H_2O$$

in water (25°C) for $pH = 5$.

10.59 Derive a formula for the relaxation time of reversible first-order reactions:

$$A \underset{-1}{\overset{1}{\rightleftharpoons}} B$$

10.60 Derive a formula for the relaxation time of

$$A + B \underset{-2}{\overset{2}{\rightleftharpoons}} C + D$$

"When I use a word," Humpty Dumpty said in a rather scornful tone, "it means just what I choose it to mean, neither more nor less."

−Lewis Carroll

I am not for imposing any sense on your words: you are at liberty to explain them as you please. Only I beseech you, make me understand something by them.

−Bishop Berkeley

11

Quantum Theory

In the last generation, no body of knowledge has had a greater effect on chemistry — what chemists do, how they approach and solve problems, the very language with which the subject is discussed — than the quantum theory. The language, images, and concepts of quantum theory permeate the teaching and practice of nearly all areas of chemistry. To be sure, the quantum theory taught in most chemistry courses is generally qualitative and pictorial — and this is valuable. But this fact makes it all the more important that students in chemistry and related subjects be exposed to the theoretical foundations of the subject, for only in this way can a proper and correct understanding of these words, concepts, and pictures be obtained. (A previous exposure to the ideas of quantum theory may seem to be an advantage; in fact, it can be somewhat of a hazard. Many of the concepts of quantum theory have become little more than catchwords in chemistry; when encountered in this and subsequent chapters, they may seem familiar — but this apparent familiarity can be misleading. The attitude of Humpty Dumpty, expressed at the beginning of this chapter, has much to recommend it here; the reader is advised to take words to be as defined and to reserve judgment until the last act.)

The role of theory in science is to unify and explain the experience and facts that characterize nature. By "unify" we mean that a theory will make it possible for us to discuss a wide range of otherwise disparate phenomena on a single conceptual basis, upon which new understanding will grow. Generally speaking, the wider the scope of a theory, the higher the level of abstraction. Thermodynamics, for example, unifies our understanding of a range of phenomena, from steam engines to chemical equilibria, by introducing somewhat abstract concepts such as enthalpy, entropy, and chemical potential. Quantum theory is more fundamental than thermodynamics and is almost unsurpassed in its scope. It achieves this, however, at the cost of a very high level of abstraction. Compared to thermodynamics, its concepts are far less tangible or subject to any intuitive understandings. Heat, mechanical work, temperature, pressure, surface tension, viscosity — these can all be understood and explained in terms of phenomena we can readily observe with our own eyes. Wave functions, orbitals, operators, eigenvalues, energy levels — these have no manifestation at all in the everyday, macroscopic level of experience. Quantum behavior can, of course, be observed experimentally, and some of these results are very graphic and convincing. However, these observations require sophisticated experiments, and, in any case, we cannot appreciate their significance without understanding the theory.

This, then, is the dilemma: we cannot appreciate the theory without knowing the facts, yet we cannot understand the facts without knowing the theory. Most books develop the theory first, with perhaps a brief mention of the experimental basis, and then develop the interpretation of the experiments; we shall follow this course. (A few authors, notably Gerhard Herzberg, do follow the opposite course — experiment to theory.) The approach that we intend does have some pedagogical problems. But forewarned is forearmed: the reader will, in this chapter, be required to learn material whose purpose will not be immediately apparent, and some patience is called for. Practical applications follow in subsequent chapters.

It may be helpful to begin with a list of areas and/or phenomena where the role of quantum theory has been decisive: chemical bonding, atomic and molecular spectra, black-body radiation (Chapter 15), heat capacity and other thermodynamic properties of gases (Chapter 5) and solids (Chapter 15), the PVT properties of light gases (second virial coefficient of hydrogen and helium), superconductivity, the superfluid properties of liquid helium (Chapter 4), the two types of hydrogen (*ortho* and *para*, Chapter 15), electron and neutron diffraction. And this list includes only things of importance in chemistry. Unfortunately we shall, in the end, gain only a very limited view of this gigantic landscape.

11.1 Particles and Waves

The motions of material bodies—objects possessing the property of mass—are governed by the laws of Newton. We shall call such objects *particles,* although the term may seem inappropriate for objects as big as the sun, moon, or Earth. Newton's theory did more than explain the observations of Galileo or Kepler; it placed on a common basis such phenomena as the motion of a pendulum, the trajectory of a cannonball, and the orbits of the planets. To do this, Newton invented the concepts of mass and force by which such phenomena are discussed. Newton's theory was remarkably resilient; new phenomena, unimagined by Newton, were included by simply stating a new force law (for example, Coulomb's law for the motion of charged particles in an electric field).

It was three centuries before it was discovered that the range of applicability of Newton's laws was limited; if the velocity is very high ($\sim 10^{10}$ cm/sec), the laws of relativity apply, and if the object is very small ($\sim 10^{-26}$ g), the laws of quantum mechanics are required. These twin pillars of twentieth-century physics—relativity and quantum theory—developed quite separately and have never coexisted peaceably. Einstein, on his part, never totally accepted the quantum theory, and quantum theory has had difficulty in assimilating relativity. The synthesis of these two theories is called quantum electrodynamics (Figure 11.1), but it is doubtful that the final word has been written on this subject.

Newton also groped with another natural phenomenon—light. Although it is with light that we observe the motion of material bodies, the nature of light is much more difficult to investigate and comprehend. (Would Marshall McLuhan think that the medium is less clear than the message?) Newton favored a model that pictured light as massless particles (called corpuscles). Huygens' wave model of light was eventually accepted because it could readily explain such phenomena as interference, refraction, and diffraction, but still the underlying nature of Huygens' waves was a mystery.

Maxwell's electromagnetic theory is an outstanding example of a successful theory. As intended, it unified electricity and magnetism and explained the experiments of Faraday, Ampere, and others. The theory also stimulated the experiments of Heinrich Hertz, which led, in due course, to technological developments from wireless telegraphy to television. It also clarified the nature of light, showing that Huygens' waves were electromagnetic in nature. The cost of all this was, of course, a higher level of abstraction. Huygens' waves couldn't be seen, but they could be pictured by analogy to waves in water; but how shall we "picture" *fields,*

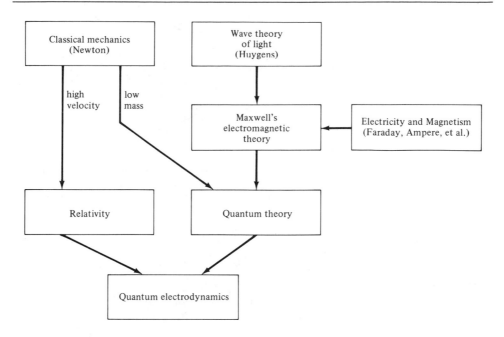

Figure 11.1   The genealogy of quantum mechanics

vector potentials, or *displacement?* With this theory, the picture of nature seemed complete — so long as one did not look too closely.

The situation prevailing in physics towards the end of the nineteenth century, with Newton ruling the particles while Maxwell ruled the waves, is referred to as *classical physics*. We shall not give a thorough historical development of the quantum theory; this has been done in many places. The writings of George Gamow (especially ref. 1) give a popularized version, while the two paperback volumes by A. D'Abro (ref. 2) provide a very thorough (but still relatively nonmathematical) account. A brief outline might be helpful.

This neat division of physical phenomena into waves (radiation) and particles (material bodies) ran into trouble with the discovery of the electron. Cathode rays, as electrons were called, were originally thought to be radiation, since (like X-rays, which were discovered slightly earlier) they cast shadows and could pass through a seemingly solid material without punching holes in it. The measurement of the charge-to-mass ratio by J. J. Thompson seemed to settle the question — how could waves have a mass? — but in fact it was still open. A number of phenomena, culminating in the discovery of electron diffraction by Davisson and Germer in 1927, showed that electrons had the characteristics of both particles and waves.

At the same time, Newton's corpuscular theory of light was making a comeback. The explanations by Planck of the black-body radiation distribution (Chapter 15) and by Einstein of the photoelectric effect suggested that light consisted of discrete entities

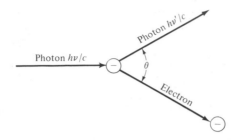

Figure 11.2 The Compton effect. The angle of scattering and change in frequency of the photon require the photon to have a momentum $h\nu/c$ in order for the law of conservation of momentum to be obeyed.

called *photons* having an energy:

$$\text{energy of a photon} = h\nu \qquad \qquad \textbf{(11.1)}$$

where h is Planck's constant and ν is the frequency of the electromagnetic wave.

The particulate nature of light is illustrated more vividly by the Compton effect — the elastic scattering of light by electrons (Figure 11.2). The interaction of a photon and an electron is rather like the collision of two billiard balls. The electron gains some energy, and the photon's loss of energy is reflected by a change in frequency in accord with Eq. (11.1). Furthermore, to explain the angles of scattering in accord with the law of conservation of momentum, it is required to assume that the photon, that massless object, must have *momentum:*

$$\text{momentum of a photon} = \frac{h\nu}{c} \qquad \qquad \textbf{(11.2)}$$

This is the crux: electromagnetic radiation has both wavelike and particlelike characteristics; particles often behave in a manner formerly thought to be characteristic only of waves. This is referred to as *wave-particle duality.* The wavelike characteristics of particles are observable only on the atomic scale of dimensions.

11.2 Bohr's Atomic Theory

First and foremost there was the problem of the structure of atoms. The experiments of Rutherford showed that the mass and positive charge of the atom was concentrated in a nucleus, apparently separate and distinct from the negatively charged electrons. For a model of how these could coexist one need only look at the sky; the moon circles the earth in a stable orbit in which the gravitational attraction is exactly counterbalanced by the centrifugal acceleration of the orbital motion. Electrons could similarly orbit about the nucleus where, in place of gravitation, the force is Coulomb's law.

More careful thought about this model reveals a formidable problem. If the negative electron is orbiting about the positive nucleus, this creates a dipole, which is oscillating at the frequency of rotation. Maxwell's theory and the experiments of Hertz showed that an oscillating dipole must emit radiation at the frequency of oscillation; you confirm this theory every time you listen to the radio or watch television. The continual radiative loss of energy by the electron would cause it to spiral inward; that is, the orbital model is not stable. Bohr resolved this difficulty with classic genius by stating the obvious — it doesn't happen — and deducing the consequences.

First, let's look at some of the clues that Bohr had available. Gaseous atoms do emit electromagnetic radiation when they are hot, but only at certain discrete wavelengths. (By way of contrast, hot solids emit radiation with a continuous spectrum — white light.) Spectroscopists had known for some time that the profusion of lines in an atomic spectrum could be derived from a smaller number of quantities called *terms*. Thus, each atom was characterized by a set of term values (P_i), and all observed lines in the spectrum had reciprocal wavelengths (or *wavenumbers*):

$$\tilde{\nu} = \frac{1}{\lambda}$$

which were related to differences of the term values:

$$\tilde{\nu} = P_i - P_j \tag{11.3}$$

This is called the **Ritz combination principle.** Bohr identified these terms as the *energy levels* or *stationary states* of the electron and showed that these would result if the angular momentum (p_ϕ) of the electron were quantized:

$$p_\phi = N\hbar \tag{11.4}$$

In Eq. (11.4), N is a positive integer called a *quantum number,* and:

$$\hbar = \frac{h}{2\pi} = 1.054 \times 10^{-27} \text{ erg sec}$$

The constant h is, of course, Planck's constant, which had been used by Planck to explain black-body radiation and by Einstein to explain the photoelectric effect. This restriction limited the permitted values of the radius of the electron's orbit to:

$$r = \frac{N^2\hbar^2}{Zme^2} \tag{11.5}$$

where Z is the atomic number of the nucleus. Bohr's theory, and the formulas of this section, apply to the hydrogen atom and any one-electron ion such as He^+, Li^{2+}, with an appropriate choice of Z. Also, the formulas of this section, as is common in this area, require cgs units; thus the electron charge (e) must be in esu (the cgs unit of charge) rather than coulombs. The collection of constants in Eq. (11.5) occurs frequently in quantum theory and is, therefore, given a special symbol and called **Bohr's radius:**

$$a_0 \equiv \frac{\hbar^2}{me^2} = 0.52918 \text{ Å} \tag{11.6}$$

An important consequence of Bohr's postulate is that the total energy (kinetic plus potential) of an electron is related to its quantum number (N) as

$$E_N = Z^2\left(\frac{e^2}{2a_0}\right)\frac{1}{N^2} \tag{11.7}$$

This formula is subject to two experimental comparisons. The first is the ionization potential (IP) of hydrogen atoms, which has an experimental value of 13.605 eV. In Bohr's theory, the energy required to remove an electron would be the difference between the energies [Eq. (11.7)] with $N = 1$ (the ground state) and with $N = \infty$ (the electron removed to an infinite distance):

$$E(\text{ionization}) = -\frac{Z^2 e^2}{2a_0}\left(\frac{1}{\infty} - \frac{1}{1}\right)$$

This quantity is generally measured in *volts* (an SI unit), so if e and a_0 are in cgs units, a unit conversion is required. The result is:

$$\text{IP} = \frac{299.7925 e Z^2}{2a_0} \qquad (11.8)$$

Example: Calculate the IP of the hydrogen atom ($Z = 1$).

$$\text{IP} = \frac{(299.7925)(4.80298 \times 10^{-10} \text{ esu})}{(2)(0.52918 \times 10^{-8} \text{ cm})} = 13.605 \text{ eV} \qquad \blacksquare$$

The second test of Bohr's theory comes from spectroscopy. It had been observed by Rydberg that the lines of the atomic spectrum of the hydrogen atom could be fitted by an empirical formula:

$$\tilde{\nu} = \mathcal{R}\left(\frac{1}{N_1^2} - \frac{1}{N_2^2}\right) \qquad (11.9)$$

In this formula, N_1 and N_2 are integers and the *Rydberg constant* is:

$$\mathcal{R} = 109{,}667.6 \text{ cm}^{-1}$$

The series of lines with $N_1 = 2$ ($N_2 = 3, 4, 5, \ldots, \infty$) is found in the visible region of the spectrum (Figure 11.3) and is called the Balmer series. Other series, corresponding to other values of N_1, are found in the infrared and ultraviolet (Figure 11.3).

According to Bohr, the electron in the atom may change its energy only among the quantum states as given by Eq. (11.7); the difference in energy must be the energy of the emitted photon [Eq. (11.1)], so the frequency of the photon is:

$$\nu = \Delta E / h \qquad (11.10a)$$

Since $\nu = c/\lambda = c\tilde{\nu}$, the wavenumbers are:

$$\tilde{\nu} = \frac{\Delta E}{hc} \qquad (11.10b)$$

With Eqs. (11.7) and (11.10) it is readily shown that:

$$\mathcal{R} = \frac{2\pi^2 m e^4 Z^2}{h^3 c} \qquad (11.11)$$

Exercise: Derive Eq. (11.11); remember that Bohr's radius is given by Eq. (11.6). Then calculate the value of the Rydberg constant implied by Eq. (11.11). Somewhat more accurate

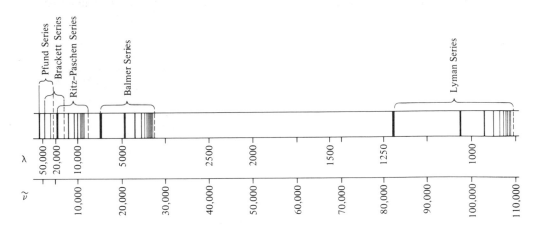

Figure 11.3 The spectrum of atomic hydrogen. Atomic spectra characteristically show such series—sharp lines with spacing and intensity decreasing towards shorter wavelengths. The series limit, shown as dashed lines, is the wavelength at which the series terminates (λ in Å, $\tilde{\nu}$ in cm^{-1}).

results will be obtained if you use, in place of the electronic mass (m), the reduced mass (Appendix III) of the proton (mass M) and electron:

$$\mu = \frac{Mm}{M + m} \qquad (11.12)$$

The reason for doing so is that, rather than the electron orbiting about the proton, the electron and proton are rotating about their center of mass. This correction is so small (since the mass of the nucleus, M, is 1836 times m) that we shall henceforth omit it. [*Answer:* 109,737 cm^{-1} (using m); 109,677.3 cm^{-1} (using μ).] ∎

It is hard to complain of a theory that gives six-significant-figure accuracy, but so we shall. The hydogen atom's spectrum when examined under very high resolution shows that the lines are split into several closely spaced lines; this is called *fine structure*. If the spectrum of the atoms in a magnetic field is examined, further splittings appear; this is called the *Zeeman effect*. Bohr's theory (even with refinements provided by Sommerfeld) could not comfortably explain all the facts concerning the fine structure and the Zeeman effect. But this may seem minor; a greater concern was that the theory worked only for the hydrogen atom (or one-electron ions). Since more than 100 atoms are known, that seems a rather poor score. Nevertheless, Bohr's theory was an enormous step, because it laid the basis for the next stage; it suggested that the electron in an atom was acting more like a wave than like a discrete particle. How is that?

De Broglie's Hypothesis

"Quantization"—that is, the occurrence of integers—arises naturally when wave motion is subject to *boundary conditions*. For example, standing waves on a string of length L, attached at each end to stationary walls, may have only wavelengths:

$$\lambda = \frac{2L}{N}$$

such that N is an integer; any other wavelength would require the attached ends of the string to move. Waves on a circular ring — for example, the compressional waves that occur if such a ring is struck — are subject to similar boundary conditions. The requirement is that the fluctuating quantity (for example, the density for a compressional wave) must be single-valued; this limits the wavelengths permitted to:

$$\text{wavelength} = \frac{\text{circumference}}{\text{integer}}$$

or

$$N\lambda = 2\pi r$$

This is illustrated in Figure 11.4.

De Broglie's hypothesis states that an object of mass m traveling with velocity v is associated with a wavelength:

$$\lambda = \frac{h}{mv} \tag{11.13}$$

Noting that $\lambda = c/\nu$ and $p = mv$, compare this to Eq. (11.2).

If this is combined with the conditions for standing waves on a ring, we get:

$$\frac{Nh}{mv} = 2\pi r$$

$$N\hbar = mvr = p_\phi$$

This is just Bohr's hypothesis for the quantization of angular momentum. The de Broglie hypothesis has also received abundant confirmation by electron- and neutron-diffraction experiments.

The message is clear: electrons sometimes behave like waves; photons sometimes behave like particles; evidently wave and particle behavior are manifestations of some common underlying nature.

After Bohr...

Bohr's theory was clearly only a stepping stone. It gave some correct results. It was correct in spirit; angular momentum is quantized but not quite in the way proposed [Eq. (11.4)]. Its symbolism is enduring — we still speak of the "Bohr radius" and "orbital" motions. The atom with its orbits still symbolizes "atomic" energy (which is really nuclear energy; the symbol would be inappropriate even if theoretically meaningful). But it was never more than a shadow of the correct picture.

The quantum theory, introduced independently by Schrödinger and Heisenberg in 1925, resolved all these problems and more, and, in effect, brought waves and particles under a single theoretical roof. As might be expected, it did so by introducing

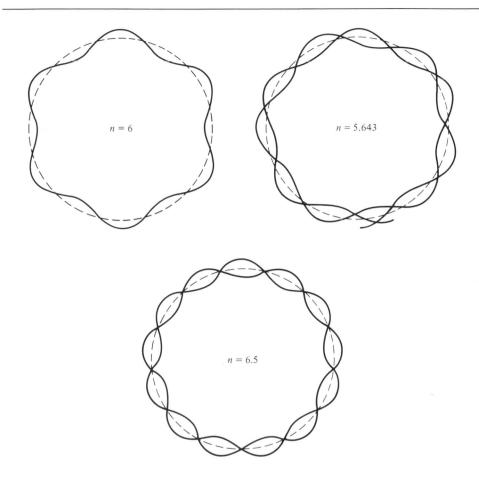

Figure 11.4 Standing waves on a ring. The requirement that the wave amplitude be single-valued — that is, that it have only one value at any point on the ring — means that only wavelengths that are integral divisors of the circumference are allowed (see the case for $n = 6$ above). Irrational numbers for the divisor ($n = 5.643$ above) would give multiple values for the amplitude at any given point. Half-integral divisors ($n = 6.5$ above) would give two amplitudes at each point — such a case is called double-valued.

a higher level of abstraction: the Bohr orbits had macroscopic counterparts that we could readily imagine; the wave functions that replace them are far more difficult to "picture."

The problem in understanding this theory lies not so much in the mathematics involved as in the absence of a "picture" of what is going on; the pictures we do have are of very limited validity. This is inevitable; quantum theory is necessary only at the submicroscopic level of nature, where, by definition, we cannot "see" what is going on. Since the words of language are inescapably linked to what we see and experience, we are bound to have difficulty in describing what cannot be seen. In the words of Jacob Bronowski (emphasis added):

When we step through the gateway of the atom, we are in a world which our senses cannot experience. There is a new architecture there, a way that things are put together which we cannot know: we only try to picture it by *analogy,* a new act of imagination. The architectural images come from the concrete world of our senses, because that is the only world that words describe. But all our ways of picturing the invisible are *metaphors,* likenesses that we snatch from the larger world of eye and ear and touch.*

Unfortunately that leaves us with only mathematics — although one could argue that mathematics itself is a metaphor. The words "analogy" and "metaphor" are very important, for many of the words we shall be using are no more than that.

In mathematical terms, quantum theory has three closely related formulations: the differential-equation (wave-mechanics) approach of Schrödinger, the matrix approach of Heisenberg, and the operator/linear vector-space approach of Dirac. We shall use a scaled-down version of Dirac's method; operator algebra is likely to be a new subject to the reader, but it is easier to develop at the level we require than the other two.

The theory itself will be presented in postulate form — a set of statements, assumed to be true, whose consequences are developed, and then compared to reality. First, we shall need some mathematical tools.

11.3 Operator Algebra

An algebra is a set of rules for manipulating symbols. Ordinary algebra, as taught in secondary school, defines procedures for the manipulation of real numbers with operations such as addition and multiplication. Extensions of the real-number concept, such as complex numbers, vectors, or matrices, require new rules — an algebra of their own (cf. Appendix IV). In this section we deal with symbols called *operators,* which will usually be denoted as letters with a circumflex above — for example, $\hat{\alpha}$, $\hat{d}$, $\hat{\imath}$. Each of these symbols represents a rule for transforming functions into new functions. If $\hat{\alpha}$ is an operator that transforms a function $f(x)$ into the function $g(x)$, the symbolic representation (called the **operator equation**) is:

$$\hat{\alpha}f(x) = g(x) \qquad \qquad \textbf{(11.14)}$$

Example: The operator $\hat{d} = d/dx$ means "take the derivative of the function with respect to x":

$$\hat{d}f(x) = \frac{df}{dx}$$

If $f(x) = e^{-x^2}$, then:

$$\hat{d}f(x) = -2xe^{-x^2}$$

This is a new function. We shall encounter other types of operators, but this one is the most important by far. ∎

Example: The operator $\hat{x}$ means "multiply the function by x." If $f(x) = e^{-x^2}$, then $\hat{x}f(x) = xe^{-x^2}$, a new function. Of course x could also be a function; the distinction between operators and functions can sometimes be a fine one. ∎

*J. Bronowski, *The Ascent of Man,* 1973: Boston, Little, Brown and Company.

Werner Heisenberg (1901–1976)

Erwin Schrödinger (1887–1961)

Paul Adrien Maurice Dirac (1902–1984)

Niels Bohr (1885–1962)

Example: The operator î (inversion) means "replace x by $-x$, y by $-y$, z by $-z$"; that is, invert the function through the origin. If

$$f(x) = x^2 - 3x + 5$$

the new function is:

$$\hat{\imath}f(x) = x^2 + 3x + 5$$

∎

Example: The operator $\hat{C}_4$ means "rotate the function by 90° about the z axis." This has the effect: $x \rightarrow y$, $y \rightarrow -x$, $z \rightarrow z$ (unchanged). If

$$f(x, y, z) = xy - xz + yz$$

then:

$$\hat{C}_4 f(x, y, z) = -yx - yz - xz$$ ■

You could define an operator for any type of mathematical manipulation — square root, logarithm, sine, and so on. Readers who have programmed computers will probably be familiar with this concept. Solutions to Rubik's Cube have been published in operator notation.*

Algebraic Rules for Operators

For these symbols we must carefully define operations such as equality, addition, and multiplication, because there are important differences from the usual meaning of these symbols.

1. Equality. For two operators $\hat{\alpha}$ and $\hat{\beta}$, the statement:

$$\hat{\alpha} = \hat{\beta}$$

means that the two operators have the same result when operated on any function; that is:

$$\left.\begin{array}{r}\hat{\alpha}f(x) = g(x)\\ \hat{\beta}f(x) = g(x)\end{array}\right\}\quad \text{the same new function}$$

Be careful! We showed earlier that the differential operator:

$$\hat{d}e^{-x^2} = -2xe^{-x^2}$$

but it is *incorrect* to say $\hat{d}$ is equal to $-2x$, for this result applies only to this particular function.

2. Addition. Given two operators such that, when operated on a function $f(x)$:

$$\hat{\alpha}f = g(x) \quad \text{and} \quad \hat{\beta}f = h(x)$$

where g and h are new functions, then:

$$(\hat{\alpha} + \hat{\beta})f = \hat{\alpha}f + \hat{\beta}f = g + h$$

That is, the distributive law is obeyed.

Linear operators are a class of operators that obey the distribution law with respect to addition of functions; that is:

$$\hat{\alpha}(f + g) = \hat{\alpha}f + \hat{\alpha}g$$

The operators, $\hat{d}$, $\hat{x}$, $\hat{\imath}$ are linear. The operator for square root is not linear, since:

$$\sqrt{f + g} \neq \sqrt{f} + \sqrt{g}$$

All the operators with which we shall be concerned are linear.

*Don Taylor, *Mastering Rubik's Cube*, 1980: New York, Holt, Rinehart and Winston.

3. Multiplication. The symbol $\hat{\alpha}\hat{\beta}$ means operate first with $\hat{\beta}$ and then (on the result) with $\hat{\alpha}$ — right to left. The operator $\hat{\alpha}^2$ means operate twice with $\hat{\alpha}$.

Example: Find $\hat{d}^2 f(x)$ for $f(x) = \sin 3x$.

$$\hat{d}(\sin 3x) = 3 \cos 3x$$

$$\hat{d}(3 \cos 3x) = -9 \sin 3x$$

$$\hat{d}^2(\sin 3x) = -9 \sin 3x \qquad \blacksquare$$

In many cases the order of operation matters; that is, operators do not necessarily commute:

$$\hat{\alpha}\hat{\beta} \neq \hat{\beta}\hat{\alpha}$$

What this inequality means is that the operations $\hat{\alpha}\hat{\beta}f(x)$ and $\hat{\beta}\hat{\alpha}f(x)$ do not necessarily give the same result.

Example: Consider the operation of $\hat{x}$ and $\hat{d}$ on the function $f(x) = e^{-x^2}$. If the order of operation is to use $\hat{d}$ first:

$$\hat{x}\hat{d}f(x) = x(-2xe^{-x^2}) = -2x^2 e^{-x^2}$$

If we reverse the order of operation, an entirely different function results:

$$\hat{d}\hat{x}f(x) = \frac{d}{dx}(xe^{-x^2}) = -2x^2 e^{-x^2} + e^{-x^2} = (1 - 2x^2)e^{-x^2} \qquad \blacksquare$$

4. Division. Division is not defined for operators, except in the following sense: if $\hat{\alpha}f(x) = g(x)$, there may exist some operator ($\hat{\beta}$) that reverses this transformation so that $\hat{\beta}g(x) = f(x)$. In such a case we may state:

$$\hat{\beta} = \hat{\alpha}^{-1}$$

meaning that $\hat{\beta}$ is the inverse operation to $\hat{\alpha}$.

Example: The inversion operator ($\hat{i}$, defined above) is its own inverse; since $\hat{i}f(x) = f(-x)$, then $\hat{i}f(-x) = f(x)$. Therefore $\hat{i}^2 f(x) = f(x)$; $\hat{i}^2 = 1$. The last is an operator equality meaning that $\hat{i}^2$ can be replaced by 1, the operator that means "multiply by one." $\qquad \blacksquare$

Commutators

In summary, the rules of "ordinary" algebra apply to linear operators except for division and the commutation rule for multiplication. The fact that certain operators do not commute has important consequences, which we shall encounter in due course. It will be necessary to define an operator:

$$\hat{C} = \hat{\alpha}\hat{\beta} - \hat{\beta}\hat{\alpha}$$

This operator is called the commutator and is usually symbolized as:

$$[\hat{\alpha}, \hat{\beta}] \equiv \hat{\alpha}\hat{\beta} - \hat{\beta}\hat{\alpha} \qquad \textbf{(11.15)}$$

What this operator equality means is that, for an arbitrary function $f(x)$, the operation:

$$(\hat{\alpha}\hat{\beta} - \hat{\beta}\hat{\alpha})f(x) = \hat{\alpha}(\hat{\beta}f(x)) - \hat{\beta}(\hat{\alpha}f(x)) = g(x)$$

can be replaced by an operation:

$$\hat{C}f(x) = g(x)$$

where $\hat{C}$ is called the commutator. Operators that commute have a commutator of zero: $\hat{C} = 0$.

Exercise: Prove the commutator:

$$[\hat{d}^2, x] = 2\hat{d}$$ ∎

Example: The commutator of $\hat{x}$ and $\hat{d}$ is the most important one for our purposes, and we will see later that it has an important physical significance. We need to find the result of $[\hat{d}, \hat{x}]$ operating on a general function, $f(x)$:

$$[\hat{d}, \hat{x}]f(x) = \hat{d}\hat{x}f(x) - \hat{x}\hat{d}f(x)$$

$$= \frac{d}{dx}(xf) - x\frac{df}{dx}$$

Using the rule from calculus for the differential of a product, we find:

$$\frac{d}{dx}(xf) = x\frac{df}{dx} + f\frac{dx}{dx} = x\frac{df}{dx} + f$$

Now we see:

$$[\hat{d}, \hat{x}]f(x) = x\frac{df}{dx} + f - x\frac{df}{dx} = f(x)$$

Therefore:

$$[\hat{d}, \hat{x}] = 1$$

$$\left[\frac{d}{dx}, x\right] = 1 \tag{11.16}$$

The right-hand side of this equality is the trivial operator "multiply by the number 1." Since the distinction between $\hat{x}$ and x is purely formal, we shall henceforth omit the circumflex on this operator. ∎

The Eigenvalue Equation

The general operator equation, $\hat{\alpha}f(x) = g(x)$, has an important special case in which the new function differs from the old one only by multiplication by a constant:

$$\hat{\alpha}f(x) = af(x) \tag{11.17}$$

This is called the eigenvalue equation, and the constant (a) is called the *eigenvalue*.

Remember that equations such as (11.17) cannot be replaced by the operator equation, "operator = constant," except for the trivial case when it is true for *all* functions.

For a given operator, there may exist a set of functions that obey Eq. (11.17); these are called the *eigenfunctions* of the operator. This set may have an infinity of members, but not all functions will be in the set (except in the trivial case that the operator is "multiply by a constant"). (For certain types of operators, the eigenfunctions may be a *complete set;* for our immediate purposes we can regard this term as a technicality into whose meaning we need not delve too deeply.)

Example: The function $f(x) = 7e^{-3x}$ is an eigenfunction of $\hat{d}$, since:

$$\hat{d}f(x) = \frac{d}{dx}(7e^{-3x}) = -3(7e^{-3x}) = -3f(x)$$

The eigenvalue is -3. On the other hand, the function $7e^{-3x^2}$ is not an eigenfunction of $\hat{d}$, since:

$$\frac{d}{dx}(7e^{-3x^2}) = -6x(7e^{-3x^2})$$

Multiplication by x produces a new function. Multiplication by a constant is not considered to produce a new function but merely a change in scale. For example, the reader should demonstrate that:

$$f(x) = 632e^{-3x}$$

is an eigenfunction of $\hat{d}$ with the eigenvalue of -3, as in the first example.

From this example it should be clear that all functions of the form:

$$f(x) = ce^{kx}$$

(with k and c constants) are eigenfunctions of $\hat{d}$ with eigenvalue $= k$.   ■

Compound Operators

Elements such as those illustrated above may be combined to form compound operators. One example for which we shall find use is:

$$\hat{h} = x^2 - \hat{d}^2$$

Also very important is the **Laplacian operator,** which is given the special symbol:

$$\nabla^2 = \frac{\partial^2}{\partial x^2} + \frac{\partial^2}{\partial y^2} + \frac{\partial^2}{\partial z^2} \tag{11.18}$$

in which x, y, and z are the Cartesian coordinates. The operation $\partial/\partial x$ means "take the derivative with respect to x, holding y and z constant."

We shall often need to use the Laplacian operator in spherical polar coordinates. (Readers unfamiliar with spherical polar coordinates should read Appendix III.) The derivation of this operator, unimportant for our purposes, can be found in refs. 3 and 5 and many other books. The result is:

$$\nabla^2 = \frac{1}{r^2}\frac{\partial}{\partial r}r^2\frac{\partial}{\partial r} + \frac{1}{r^2 \sin\theta}\frac{\partial}{\partial\theta}\sin\theta\frac{\partial}{\partial\theta} + \frac{1}{r^2 \sin^2\theta}\frac{\partial^2}{\partial\phi^2} \tag{11.19}$$

Example: When using compound operators of the type found in Eq. (11.19), the right-to-left rule must be followed strictly. The first term operating on the function e^{ar} gives:

$$\frac{1}{r^2}\left\{\frac{\partial}{\partial r}\left[r^2\left(\frac{\partial}{\partial r}e^{ar}\right)\right]\right\} = \frac{1}{r^2}\left[\frac{\partial}{\partial r}(ar^2e^{ar})\right]$$

$$= \frac{1}{r^2}(2are^{ar} + a^2r^2e^{ar})$$

$$= \left(\frac{2a}{r} + a^2\right)e^{ar}$$

Note that this is *not* an eigenvalue equation. Why?   ■

Exercise: Derive the following operator equalities using the product rule for differentiation:

$$\frac{1}{r^2}\frac{\partial}{\partial r}r^2\frac{\partial}{\partial r} = \frac{\partial^2}{\partial r^2} + \frac{2}{r}\frac{\partial}{\partial r} \qquad (11.20)$$

$$\frac{1}{\sin\theta}\frac{\partial}{\partial\theta}\sin\theta\frac{\partial}{\partial\theta} = \frac{\partial^2}{\partial\theta^2} + \cot\theta\frac{\partial}{\partial\theta} \qquad (11.21)$$

■

Example: The operator $\hat{h}$ (defined above) has an eigenfunction, $f(x) = e^{-x^2/2}$. What is the eigenvalue?

$$\hat{h}e^{-x^2/2} = (x^2 - \hat{d}^2)e^{-x^2/2} = x^2e^{-x^2/2} - \frac{d}{dx}\left(\frac{d}{dx}e^{-x^2/2}\right)$$

$$= x^2e^{-x^2/2} + \frac{d}{dx}(xe^{-x^2/2}) = x^2e^{-x^2/2} + e^{-x^2/2} - x^2e^{-x^2/2}$$

Therefore:

$$\hat{h}e^{-x^2/2} = e^{-x^2/2}$$

and the eigenvalue $= +1$.   ■

Exercise: Show that the function:

$$f(x) = xe^{-x^2/2}$$

is an eigenfunction of $\hat{h}$ with eigenvalue $= 3$. Show that e^{-x^2} is *not* an eigenfunction of $\hat{h}$.   ■

Simultaneous Eigenfunctions

Each operator has a set of eigenfunctions—is it possible that some functions could be eigenfunctions of two different operators?

THEOREM. Two operators that do not commute may not have a complete set of functions that are eigenfunctions of both. Conversely, if the operators do commute, such a set of simultaneous eigenfunctions will exist. Note that *all* possible eigen-functions of an operator need not be eigenfunctions of another, commuting operator; but some will be so.

This theorem is proved rigorously in many advanced texts (Ref. 4, for example). The following, which is not a rigorous proof, may suffice to give you the idea.

Suppose that two operators ($\hat{\alpha}$ and $\hat{\beta}$) both have some nontrivial eigenfunction (ψ). (By "nontrivial" we exclude such cases as ψ = constant; normally one would not wish to consider such sets as the set of real or imaginary numbers to be functions, but some types of functions, for example e^{kx} when $k = 0$, may reduce to a constant in a special case.) The eigenvalue equations are:

$$\hat{\alpha}\psi = a\psi, \qquad \hat{\beta}\psi = b\psi$$

Now we operate on the first equation with $\hat{\beta}$ and on the second with $\hat{\alpha}$:

$$\hat{\beta}\hat{\alpha}\psi = ab\psi, \qquad \hat{\alpha}\hat{\beta}\psi = ba\psi$$

These two equations may be subtracted to give:

$$\hat{\alpha}\hat{\beta}\psi - \hat{\beta}\hat{\alpha}\psi = (ba - ab)\psi$$

Since a and b are numbers, $ab - ba = 0$, and:

$$[\hat{\alpha}, \hat{\beta}]\psi = 0 \qquad\qquad (11.22)$$

There are three ways in which this equation could be true: (1) $\psi = 0$; this can occur only as a special case—for example, $\sin kx$ with $k = 0$. (2) ψ is an eigenfunction of the commutator with eigenvalue = 0; this again can occur, but only in special cases. (3) Clearly this relationship will be true if the operators commute:

$$[\hat{\alpha}, \hat{\beta}] = 0$$

This is the only general case for which Eq. (11.22) will be true—for a nontrivial set of such simultaneous eigenfunctions to exist, the operators must commute.

Ladder Operators

Among the various types of commutators, some are of the form:

$$[\hat{\alpha}, \hat{\beta}] = k\hat{\beta} \qquad\qquad (11.23)$$

where k is a constant. Such operators have the property that the operator $\hat{\beta}$ is a *generator* for the eigenfunctions of $\hat{\alpha}$. We prove this statement as follows: let ψ be an eigenfunction of $\hat{\alpha}$ so that:

$$\hat{\alpha}\psi = a\psi$$

Since $\hat{\alpha}$ and $\hat{\beta}$ do not commute, operation by $\hat{\beta}$ on ψ will generally produce a new function:

$$\hat{\beta}\psi = \Phi$$

Here, Φ is some new function; is Φ an eigenfunction of $\hat{\alpha}$? To find out, we operate with $\hat{\alpha}$ on Φ:

$$\hat{\alpha}\Phi = \hat{\alpha}\hat{\beta}\psi$$

The commutator [Eq. (11.23)] permits us to replace the operation:

$$\hat{\alpha}\hat{\beta} = \hat{\beta}\hat{\alpha} + k\hat{\beta}$$

$$\hat{\alpha}(\hat{\beta}\psi) = \hat{\beta}(\hat{\alpha}\psi) + k(\hat{\beta}\psi)$$

but $\hat{\alpha}\psi = a\psi$ by the definition above, so:

$$\hat{\alpha}(\hat{\beta}\psi) = a(\hat{\beta}\psi) + k(\hat{\beta}\psi)$$

$$\hat{\alpha}\Phi = (a + k)\Phi$$

We discover that Φ is also an eigenfunction of $\hat{\alpha}$ with eigenvalue $(a + k)$. Repeating the procedure, $\hat{\beta}\Phi = \hat{\beta}^2\psi$ will be seen to be an eigenfunction with eigenvalue $= a + 2k$, and so on; thus $\hat{\beta}$ generates a *ladder* of eigenfunctions for $\hat{\alpha}$, and the relationship of commutators to "quantization" may be becoming apparent.

Example: The operator $\hat{h} = x^2 - \hat{d}^2$ is associated with an operator:

$$\hat{A} = x - \hat{d}$$

The commutator of this operator with $\hat{h}$ is:

$$[\hat{h}, \hat{A}] = 2\hat{A}$$

This is calculated as follows:

$$[\hat{h}, \hat{A}] = (x^2 - \hat{d}^2)(x - \hat{d}) - (x - \hat{d})(x^2 - \hat{d}^2)$$

$$= (x^3 - \hat{d}^2x - x^2\hat{d} + \hat{d}^3) - (x^3 - \hat{d}x^2 - x\hat{d}^2 + \hat{d}^3)$$

$$= -\hat{d}^2x - x^2\hat{d} + \hat{d}x^2 + x\hat{d}^2$$

The usual procedure for simplifying such expressions is to use the basic commutator [Eq. (11.16)] to exchange some or all of the $x\hat{d}$, $\hat{d}x$ pairs; Eq. (11.16) can be written as:

$$\hat{d}x = x\hat{d} + 1 \quad \text{or} \quad x\hat{d} = \hat{d}x - 1$$

$$\text{Commutator} = \begin{cases} \text{1st term:} & -\hat{d}\hat{d}x = -\hat{d}(x\hat{d} + 1) = -\hat{d}x\hat{d} - \hat{d} \\ \text{2nd term:} & -xx\hat{d} = -x(\hat{d}x - 1) = -x\hat{d}x + x \\ \text{3rd term:} & \hat{d}xx = (x\hat{d} + 1)x = x\hat{d}x + x \\ \text{4th term:} & x\hat{d}\hat{d} = (\hat{d}x - 1)\hat{d} = \hat{d}x\hat{d} - \hat{d} \end{cases}$$

Summing these, we get:

$$[\hat{h}, \hat{A}] = -2\hat{d} + 2x = 2\hat{A} \qquad \blacksquare$$

Exercise: Given:

$$\hat{B} = x + \hat{d}$$

Prove:

$$[\hat{h}, \hat{B}] = -2\hat{B} \qquad \blacksquare$$

Exercise: We showed earlier that $f(x) = e^{-x^2/2}$ was an eigenfunction of $\hat{h}$ with eigenvalue $= 1$.

1. Show that $\hat{A}f(x) = (x - (d/dx))e^{-x^2/2}$ is equal to $2xe^{-x^2/2}$.
2. Show by direct operation that the new function, $2xe^{-x^2/2}$, is an eigenfunction of $\hat{h}$ with eigenvalue $= 2 + 1 = 3$.
3. Show that $B(2xe^{-x^2/2})$ reproduces the original function (within a constant). That is, the operators $\hat{A}$ and $\hat{B}$ are *raising* and *lowering* operators for the eigenfunctions of $\hat{h}$. ▨

In due course it will become apparent that the concepts of this section — in particular, eigenfunctions, commutators and ladder operators — have physical significance. All we have done in this section is to forge a set of tools; in the next section we shall present the plan; then we will be ready to build.

We shall need two more mathematical tools, specifically *complex numbers* and the concept of *vectors*. The reader who is not familiar with these subjects should read Appendix IV before proceeding. Appendix III will help review coordinate systems — the spherical polar coordinates are especially important.

11.4   Postulates of Quantum Theory

In a postulate approach we start with a series of statements that are presumed to be true and then investigate the consequences. The "proof" of the postulates is their ability to explain the appropriate experimental observations. The postulate approach was first used by Euclid to introduce geometry. Euclid regarded his postulates as "self-evident." Several millenia later it was shown by Lobachevski, Einstein, and others that several of the postulates were neither self-evident nor necessary. No one is likely to regard the postulates of quantum theory as obvious or self-evident, so some patience will be required until we get around to examining the experimental results. The postulates as presented here make no claim to those happy mathematical adjectives such as "rigorous," "complete," or "irreducible"; they are chosen only to provide us with a practical starting point.

POSTULATE I. The state of a system is defined by a function (usually denoted ψ and called the *wave function* or *state function*) that contains all the information that can be known about the system.

By "system" we shall usually mean a particle (for example, an electron) or a group of particles (for example, an atom or molecule). Generally, the state function is a function of the coordinates and time; we shall be dealing here only with time-independent functions representing *stationary states*.

The analogous situation in classical (Newtonian) mechanics is to define a state by giving the coordinates and velocities of a particle at some time t; then, given a force law for the particle, the behavior (that is, trajectory) of the particle can be calculated for all future times. The implicit assumption is that the initial coordinates and velocity *can* be specified to any degree of accuracy required. By way of contrast, the wave function does not usually specify exact values of the position or velocity, and the concept of a trajectory (a series of positions occupied successively in time) is of limited meaning.

What are the coordinates for the state function? These are simply the spatial variables that in classical mechanics would indicate the position of a particle in space; they may be Cartesian coordinates (x, y, z), spherical polar coordinates (r, θ, ϕ), or any other orthogonal set of coordinates that is convenient for the problem at hand. Classically, a system consisting of N particles requires $3N$ coordinates to specify its state at any given time. [For most of our applications we shall be able to use the center-of-mass (COM) coordinate system, so only $3N$-3 coordinates are required.] The same will be true in quantum mechanics — the state function for N particles will be a function of the $3N$ spatial coordinates ($3N$-3 in the COM system). For the time being, we shall assume a one-particle system with $\psi(x, y, z)$; for simplicity we may at times assume a one-dimensional system with $\psi(x)$.

The physical significance of the state function is that $\psi^*\psi$ is a probability distribution function. (The state function is often a complex function, and the asterisk denotes the complex conjugate — see Appendix IV.) The probability that the particle will have coordinates x to $x + dx$, y to $y + dy$, z to $z + dz$, is:

$$\psi^*\psi \, dx \, dy \, dz = |\psi|^2 \, d\tau \tag{11.24}$$

In Eq. (11.24) we introduce the symbol $d\tau$ for the volume element. For a single-particle system the volume element is, in Cartesian coordinates, $d\tau = dx \, dy \, dz$. In spherical polar coordinates, it is shown in Appendix III that:

$$d\tau = r^2 \, dr \sin \theta \, d\theta \, d\phi$$

By way of unlearning misconceptions that may have been picked up in other classes, the reader is warned that a wave function or state function does not necessarily refer to the coordinates of an electron, in an atom or anywhere else; of course, at times it may do so.

As a consequence of its physical significance, the state function must meet several criteria:

1. Like all functions, the state function must be single-valued. This means that at any point in space (x, y, z), $|\psi|^2$ may have only one possible numerical value.
2. The state function must be finite and continuous at all points in space. Likewise, its first and second derivatives must be finite and continuous.
3. The function $|\psi|^2$ must have a finite integral over all space:

$$\int_{\substack{\text{all} \\ \text{space}}} \psi^*\psi \, d\tau = \text{(a finite number)} \tag{11.25}$$

For a single particle, "integral over all space" means, in Cartesian coordinates:

$$\int_{-\infty}^{\infty} dx \int_{-\infty}^{\infty} dy \int_{-\infty}^{\infty} |\psi|^2 \, dz$$

or, in spherical polar coordinates:

$$\int_{0}^{2\pi} d\phi \int_{0}^{\pi} \sin \theta \, d\theta \int_{0}^{\infty} |\psi|^2 r^2 \, dr$$

Since a function is not altered in its essential nature when multiplied by a constant, it is always possible to scale the state function so that the "finite number" of Eq. (11.25) is equal to 1. This can be accomplished by multiplying the state function by a number (called the *normalization constant*); if this is done, the function is said to be *normalized,* and Eq. (11.25) becomes:

$$\int_{\substack{\text{all} \\ \text{space}}} \psi^* \psi \, d\tau = 1 \tag{11.26}$$

Given the meaning of $|\psi|^2$ as a probability distribution function, normalization is a sensible thing to do, but it is not always necessary or desirable. (This and other topics relevant to probability distribution functions are discussed in more detail in Chapter 1, particularly Section 1.6.)

The probability of finding the particle in a finite portion of space can be found by summing the infinitesimals:

$$\text{probability} \ (x_1 \leq x \leq x_2) = \frac{\displaystyle\int_{x_1}^{x_2} |\psi|^2 \, dx}{\displaystyle\int_{-\infty}^{\infty} |\psi|^2 \, dx} \tag{11.27}$$

If the state function is normalized, the integral in the denominator of Eq. (11.27) is equal to 1.

POSTULATE II. Every physical observable is represented by a linear operator.

In classical mechanics, the principal variables are the position (x, y, z) and the velocity $(\mathbf{v})$ or momentum $(\mathbf{p} = m\mathbf{v})$ vectors; the same is true for quantum mechanical "observables," except that momentum is strongly preferred over velocity. The basic operators are:

Position: The operators are x, y, z (multiplication by the appropriate coordinate).
Linear momentum:

$$\hat{p}_x = -i\hbar \frac{\partial}{\partial x}, \qquad \hat{p}_y = -i\hbar \frac{\partial}{\partial y}, \qquad \hat{p}_z = -i\hbar \frac{\partial}{\partial z} \tag{11.28}$$

The imaginary $i = \sqrt{-1}$ is discussed in Appendix IV; note that $1/i = -i$, so

Eqs. (11.28) are often written:

$$\hat{p}_x = \frac{\hbar}{i}\frac{\partial}{\partial x} \qquad \text{and so on}$$

There are other important physical observables, such as energy, angular momentum, dipole moment, and magnetic moment; the operators for these are constructed from the basic operators (above) in a manner that we shall detail as each is encountered. (Operators that represent physical observables must be of a particular type called *Hermitian operators;* but that is a technicality with which we need not concern ourselves.)

POSTULATE III. The measurement of a physical observable will give a result that is one of the eigenvalues of the corresponding operator for that observable.

This immediately raises the possibility of quantization, since only certain values (eigenvalues) are possible; however, it is possible that the set of eigenvalues for a specific case may be infinite or continuous or both — that is, any real number could be a possible result of a measurement.

The foremost observable is the total energy, E. Its operator is called the *Hamiltonian* and is given the special symbol $\hat{H}$. The eigenvalue equation for this operator is called the **Schrödinger equation:**

$$\hat{H}\psi = E\psi \qquad \textbf{(11.29)}$$

The eigenvalues of the Hamiltonian operator are the allowed energies of the system.

At this point we have a sufficient number of postulates to get started. However we shall, in due course, introduce several additional postulates which, in a more rigorous course, could be derived.

The Hamiltonian

Commonly, a quantum-mechanical problem will begin with the definition of the total-energy operator of the system, the Hamiltonian; next comes the mathematical problem of finding the eigenfunctions and eigenvalues of the system.

The first step is relatively simple for the case of velocity-independent forces (by this restriction, we exclude, for example, the case of a charged particle in a magnetic field). First we write down an expression for the classical total energy of the system as a function of momentum and position; there will be two terms: the kinetic energy, T, which will depend on the momenta, and a potential-energy function, V, which will depend on the coordinates:

$$E_{cl} = T + V \qquad \textbf{(11.30)}$$

For a single particle of mass m the kinetic energy is:

$$T = \frac{p^2}{2m} = \frac{1}{2m}(p_x^2 + p_y^2 + p_z^2)$$

This can be made into an operator with the prescription provided by Postulate II, Eq. (11.28):

$$\hat{p}_x^2 = \left(\frac{\hbar}{i}\frac{\partial}{\partial x}\right)^2 = -\hbar^2\frac{\partial^2}{\partial x^2} \qquad \text{(and so on)}$$

$$\hat{T} = \frac{-\hbar^2}{2m}\left(\frac{\partial^2}{\partial x^2} + \frac{\partial^2}{\partial y^2} + \frac{\partial^2}{\partial z^2}\right) = \frac{-\hbar^2}{2m}\nabla^2 \qquad (11.31)$$

Since the potential energy depends only on coordinates (for the cases we shall consider), nothing need be done to V to make it an operator. The Hamiltonian operator for a single particle results from using Eq. (11.31) in Eq. (11.30) with the result:

$$\hat{H} = \frac{-\hbar^2}{2m}\nabla^2 + V(x, y, z) \qquad (11.32)$$

Equation (11.32) is, then, the starting point for all single-particle problems, and it is only the potential function that differentiates one problem from another. However, the potential function makes the difference between a simple, easy-to-solve problem and one that is very difficult.

The Average-Value Theorem

If the state function is an eigenfunction of a particular operator, then a measurement of the corresponding physical property will give, as an answer, the eigenvalue. If the state function (ψ) is not an eigenfunction of a particular operator ($\hat{\alpha}$), measurement will give *one* of the eigenvalues of that operator, but we cannot predict which one. The *average value* that will be obtained from repeated measurements can be calculated; this quantity is denoted $\langle\hat{\alpha}\rangle$, also called the *expectation value*, and is calculated by:

$$\langle\hat{\alpha}\rangle = \frac{\displaystyle\int \psi^*(\hat{\alpha}\psi)\,d\tau}{\displaystyle\int \psi^*\psi\,d\tau} \qquad (11.33)$$

These integrals are over all space; the integral in the denominator will be equal to 1 if the state function is normalized.

In Eq. (11.33) the symbol ($\hat{\alpha}\psi$) means the result (generally some new function) of $\hat{\alpha}$ operating upon the state function ψ. There is an interesting special case when the operator represents a coordinate, say $\hat{\alpha} = \hat{x}$ (the multiply-by-x operation). Then Eq. (11.33) becomes:

$$\langle x\rangle = \frac{\displaystyle\int x(\psi^*\psi)\,d\tau}{\displaystyle\int (\psi^*\psi)\,d\tau} \qquad (11.34)$$

Comparison to the discussion of average values given in Chapter 1 will demonstrate that ($\psi^*\psi$) is just a distribution function for the coordinate, as stated earlier.

Orthogonality and Completeness

The set of eigenfunctions of the type of operators that may represent physical variables has two important properties: orthogonality and completeness. For the beginning treatment of quantum mechanics this book gives, these properties are not totally essential; however, in a rigorous course they would be crucial. The ensuing material (through p. 593) is rather difficult, and a complete understanding is not essential for subsequent developments. However, the conclusions are very important and will shed light on the meaning of a physical measurement in quantum theory.

Two functions, $F(x)$ and $G(x)$, are said to be orthogonal if the integral of their product over all space is zero:

$$\int_{-\infty}^{\infty} F(x)^*G(x)\, dx = 0$$

The eigenfunctions of a Hermitian operator (the type that may represent physical variables) are mutually orthogonal (as long as they have different eigenvalues). If we assume also that these functions are *normalized,* they are called an *orthonormal* set:

$$\int_{\substack{\text{all} \\ \text{space}}} \psi_i^* \psi_j\, d\tau = \begin{cases} 1 & \text{if} \quad i = j \\ 0 & \text{if} \quad i \neq j \end{cases} \tag{11.35}$$

A complete set of functions, $\{\Phi_i\}$, has the property that an arbitrary function (ψ) can be expanded exactly as a linear conbination of this set:

$$\psi = \sum_i c_i \Phi_i \tag{11.36}$$

This is called the *superposition principle*. This idea has wide application outside of quantum mechanics, and the reader may have encountered it before, possibly in a different guise. For example, the set $\{x^N\}$ for $N = 0, 1, 2, \ldots, \infty$ is a complete set; in this case, Eq. (11.36) becomes the *power series* (Appendix I). This set is, however, not an orthogonal set. The sets of functions $\{\cos Nkx\}$ or $\{\sin Nkx\}$ for $N = 0, 1, 2, \ldots$, are both complete and orthogonal; with these, Eq. (11.36) becomes the *Fourier series*. A number of other such sets are generally useful in mathematics, including the Legendre, Laguerre, Hermite, and Chebyshev functions.

Measurement and Uncertainty

Earlier we raised two possibilities for the result of a physical measurement: (1) The state function is an eigenfunction of the operator; in this case a measurement will always give the same result — the eigenvalue. (2) The state function is not an eigenfunction of the operator; in this case a variety of answers may be obtained, with the average value given by Eq. (11.33). In the latter case, there is some *uncertainty* in the value of that physical quantity; to see how this can be calculated, let's review some ideas from statistics.

Suppose we measure a quantity (x) N times with a set of results $\{x_i\}$. The average value is just:

$$\langle x \rangle = \frac{\sum x_i}{N}$$

The variance (σ^2) of this set of measurements is:

$$\sigma^2 = \frac{1}{N} \sum (x_i - \langle x \rangle)^2$$

$$= \frac{1}{N} \left\{ \sum x_i^2 - 2\langle x \rangle \sum_i x_i + \langle x \rangle^2 \right\}$$

Using $\langle x \rangle = \sum x_i / N$:

$$\sigma^2 = \frac{1}{N} \left\{ \sum x_i^2 \right\} - \langle x \rangle \langle x \rangle$$

The first term is just the average value of x^2:

$$\langle x^2 \rangle = \frac{\sum x_i^2}{N}$$

Then:

$$\sigma^2 = \langle x^2 \rangle - \langle x \rangle^2 \tag{11.37}$$

This is translated into the language of quantum theory as follows: Suppose we have some physical variable (denoted A) represented by an operator $\hat{\alpha}$, with eigenvalues a_i. The uncertainty in the measurement of A (denoted here as δA) can be calculated with the average-value theorem [Eq. (11.33)] as:

$$\delta A = [\langle \hat{\sigma}^2 \rangle - \langle \hat{\sigma} \rangle^2]^{1/2} \tag{11.38}$$

In the event that the state function is an eigenfunction of the operator $\hat{\alpha}$, this uncertainty is zero, since the same value (the eigenvalue) will be obtained for each measurement.

Incompatible Variables: Heisenberg's Uncertainty Principle

Suppose we have two physical quantities (A and B) represented by operators:

A: operator $\hat{\alpha}$ with eigenfunctions $\{\psi_i\}$ (eigenvalues a_i)

B: operator $\hat{\beta}$ with eigenfunctions $\{\Phi_i\}$ (eigenvalues b_i)

As stated earlier, these sets of eigenfunctions can be the same only if the operators commute. If the system is in a state represented by the function ψ_i, a measurement of A will give the result a_i; then a measurement of B will give as a result *one* of the allowed values $\{b_j\}$. According to the superposition principle, the state function can be expanded in terms of the eigenfunctions of $\hat{\beta}$:

$$\psi_i = \sum_j c_{ij} \Phi_j$$

The average-value theorem will give (assuming the eigenfunctions are all normalized):

$$\langle \hat{\beta} \rangle = \int \psi_i^*(\hat{\beta}\psi_i)\, d\tau$$

$$= \int \left(\sum_j c_{ij}\Phi_j \right)^* \hat{\beta}\left(\sum_k c_{ik}\Phi_k \right) d\tau$$

Since the Φ functions are eigenfunctions of $\hat{\beta}$, this becomes:

$$\langle \hat{\beta} \rangle = \int \left(\sum_j c_{ij}^*\Phi_j^* \right)\left(\sum_k b_k c_{ik}\Phi_k \right) d\tau$$

$$= \sum_j \sum_k c_{ij}^* c_{ik} b_k \int \Phi_j^*\Phi_k\, d\tau$$

But the set $\{\Phi_j\}$ is orthogonal, so only the terms with $j = k$ will give a nonzero result; for normalized functions this is:

$$\langle \hat{\beta} \rangle = \sum_j |c_{ij}|^2 b_j$$

The coefficient $|c_{ij}|^2 = c_{ij}^* c_{ij}$ is, therefore, the *probability* that a particular value (b_j) will be obtained for measurement of B when the system is in a state for which the measured value of A is a_i.

Suppose now that we measure B and get a result b_i; we now know that the system is in the state represented by Φ_i. The superposition principle can be used in reverse to show:

$$\Phi_i = \sum_j c'_{ij}\psi_j$$

The measurement of A will now be subject to the statistical uncertainties that previously applied to B.

There is a fundamental incompatibility in the measurement of physical variables that are represented by noncommuting operators; a measurement of one causes an uncertainty in the other. This idea was new and unique to quantum mechanics. Before the advent of quantum theory, it had been obvious that physical measurements contained some degree of uncertainty, but it had been assumed that this was merely a limitation imposed by the experimental method or apparatus or both. According to the quantum theory, this limitation is, in some cases, fundamental and unavoidable. The product of the uncertainties of two physical quantities (A and B) can be shown to be:

$$\delta A\, \delta B \geq \tfrac{1}{2}|\langle [\hat{\alpha}, \hat{\beta}]\rangle| \tag{11.39}$$

For a normalized state function (ψ) we have, from the average-value theorem:

$$\langle [\hat{\alpha}, \hat{\beta}]\rangle = \int \psi^*(\hat{\alpha}\hat{\beta} - \hat{\beta}\hat{\alpha})\psi\, d\tau$$

This is the general statement of the **Heisenberg uncertainty principle.**

The uncertainty principle is commonly stated for two particular incompatible variables, the position (x) and the corresponding momentum (p_x). Using the operators as defined in Postulate II, we can readily calculate the commutator:

$$[\hat{p}_x, x] = \hat{p}_x x - x \hat{p}_x = \frac{\hbar}{i}\left[\frac{\partial}{\partial x}x - x\frac{\partial}{\partial x}\right]$$

This commutator was calculated earlier [Eq. (11.16)]; therefore:

$$[\hat{p}_x, x] = \frac{\hbar}{i} \tag{11.40}$$

The magnitude of this quantity is:

$$|[\hat{p}_x, x]| = \hbar$$

so Eq. (11.39) (with A representing p_x and B representing x) becomes:

$$\delta p_x\, \delta x \geq \frac{\hbar}{2} \tag{11.41}$$

The specification of a position and momentum of a particle is fundamental to classical mechanics. If the position and momentum of a particle are known at any instant in time, and the particle is subject to known forces, Newton's law:

$$\text{force} = ma = m\frac{dv}{dt} = \frac{dp}{dt}$$

can be solved to give the particle's position and momentum at any future time; this specifies the "state" of the particle in classical mechanics. If there is any uncertainty in the initial position or momentum, there will be an uncertainty in its path (trajectory) such that, at some future time, its "state" will be unpredictable. However, classical mechanics never questioned that the position and momentum *could* be measured with whatever degree of accuracy was required; it is this presumption that quantum mechanics denies.

Let us look at how position and momentum are measured. Suppose we have an object in force-free space (and, therefore, traveling in a straight line). To determine its velocity, we photograph it at two successive times using light with wavelength λ. If we examine the photograph closely to precisely determine the position, we will see that the edges of the image are fuzzy. This fuzziness is probably due to the grain size of the film, but even if we had perfect film, there would be some scatter due to the diffraction of light at the edge of the object; this phenomenon has been known for centuries and is readily explained with Huygens' wave theory. The uncertainty in the position will be $\sim\lambda$, and we can reduce it only by using light of shorter wavelength. However, the Compton effect (Figure 11.2) demonstrated that a photon has a momentum:

$$p = \frac{h\nu}{c} = \frac{h}{\lambda}$$

so that the act of measurement will cause an uncertainty in the momentum of the body of this magnitude:

$$\delta x\, \delta p \sim \lambda\left(\frac{h\nu}{c}\right) \sim h$$

Thus, qualitatively, the Heisenberg uncertainty principle can be seen to be a consequence of wave-particle duality.

Another experiment to illustrate these ideas is shown by Figure 11.5. Suppose we have a beam of particles traveling in the x direction with a momentum p_x (therefore $p_y = 0$). We try to determine a particle's y position by interposing a slit with a width δy. However, the particles will act like waves with a de Broglie wavelength $\lambda = h/p_x$ [Eq. (11.13)], so they will be deflected by diffraction with an angle given by:

$$\sin \theta \cong \frac{\lambda}{\delta y}$$

The momentum in the y direction will now be changed to $\delta p_y = p_x \sin \theta = (h/\lambda) \sin \theta$; therefore:

$$\delta p_y \, \delta y \cong \left(\frac{\lambda}{\sin \theta}\right)\left(\frac{h \sin \theta}{\lambda}\right) \cong h$$

The smallness of Planck's constant ensures that this uncertainty will not be a practical problem for macroscopic-sized objects.

The Plan

Now that we have stated the principles, it remains to illustrate them by solving some representative problems. Only a handful of quantum-mechanical problems can be solved exactly, and only a few of these can be solved without a lengthy digression on differential equations. In the next three sections we shall discuss three problems that have three major virtues: (1) they are relatively simple, (2) they are exactly solvable,

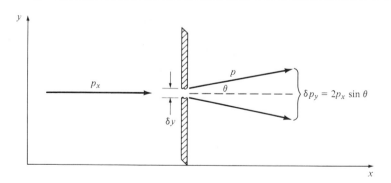

Figure 11.5 The Heisenberg uncertainty principle. A particle is traveling in the x direction, having, therefore, zero momentum in the y direction, and an attempt is made to locate its y coordinate by interposing a slit. In passing through the slit, the particle is deflected by diffraction, introducing an uncertainty in its momentum in the y direction; the smaller the slit is made, the better to define the particle's y position, the greater the maximum deflection and, hence, the uncertainty in the y momentum. Particles, as shown here, behave like waves—though this is evident only at the atomic scale of masses and dimensions—and waves have characteristics of particles (cf. Figure 11.2).

(3) they have physical applications that will be detailed in subsequent chapters. The applications and the experimental evidence will be discussed in due course; the present objective is to illustrate the principles already stated and to introduce several new ones.

11.5   The Particle in a Box

The mathematical difficulty of a quantum-mechanical problem depends primarily on the potential function; the case of $V = 0$, the free particle, is therefore the simplest. This problem is most straightforward (and, for our purposes, most interesting) if the particle is free only in some finite region of space—the "box." One important application of this problem is the calculation of the energies (and related quantities such as heat capacity and entropy) for the translational motions of a gas in a rigid-walled container (a "box"); this is done in Chapter 15. However, this problem is far more important as an illustration of the quantum principles and as a vehicle to introduce some new and valuable ideas.

One-Dimensional Box: Derivation

Although it is not physically realistic, the problem of particle motion in one dimension is a good place to start because it is very simple and because it forms the basis for the more realistic three-dimensional solution. Suppose (Figure 11.6) that the potential energy of a particle is zero for $0 < x < a$, and infinitely large elsewhere. This is somewhat unrealistic, since it requires that the force $(-dV/dx)$ be infinite at the edges of the box ($x = 0$ and $x = a$); nonetheless we can learn a lot from it.

In the regions where $V = \infty$, the state function will be $\psi = 0$; that is, the particle is never there at all. We now seek solutions for the region $0 < x < a$ where $V = 0$. The Hamiltonian in this region is:

$$\hat{H} = \frac{\hat{p}^2}{2m} = \frac{-\hbar^2}{2m} \frac{d^2}{dx^2}$$

The eigenvalue equation for the energy (the Schrödinger equation) is $\hat{H}\psi = E\psi$, which is, thus:

$$\frac{-\hbar^2}{2m} \frac{d^2\psi}{dx^2} = E\psi$$

$$\frac{d^2\psi}{dx^2} = -\frac{2mE}{\hbar^2} \psi \qquad (11.42)$$

What types of functions will, when differentiated twice, reproduce themselves? There are three: $\sin kx$, $\cos kx$, and e^{kx}. For the first of these:

$$\frac{d}{dx}(\sin kx) = k \cos kx$$

$$\frac{d^2}{dx^2}(\sin kx) = \frac{d}{dx}(k \cos kx) = -k^2 \sin kx$$

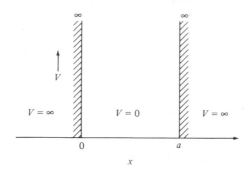

Figure 11.6 Potential energy for the particle in a box. The potential (V) is zero for some finite region ($0 < x < a$) and infinite elsewhere.

This function can therefore satisfy Eq. (11.42) with:

$$k^2 = \frac{2mE}{\hbar^2}$$

The reader should demonstrate that the functions cos kx and e^{kx} give similar results.

But not all these functions are allowed eigenfunctions. The wave function must be zero for $x = a$; if k is a real number, e^{kx} cannot fit this boundary condition. (If k is imaginary, this is not an independent solution; cf. the Euler relationships, Appendix IV. It will be simplest if we assume that k is real.) The boundary conditions also require that $\psi = 0$ when $x = 0$; this cannot be true for the cosine function, since cos $0 = 1$. Therefore, the allowed eigenfunctions must be of the form:

$$\psi = A \sin kx$$

(A is a constant, which can be chosen to normalize this function.) This function fits the boundary condition at $x = 0$, since sin $0 = 0$. To fit the boundary condition $\psi = 0$ at $x = a$, we must require:

$$\sin ka = 0$$

This will be true only for specific values of k:

$$ka = \pi, 2\pi, 3\pi, \ldots, n\pi$$

Therefore:

$$k = \frac{n\pi}{a}$$

for $n = 1, 2, 3, \ldots, \infty$. (The boundary conditions are also met when $n = 0$, but then $\psi = 0$ *everywhere* in the box; that is, the particle doesn't exist!)

Therefore, we get an allowed eigenfunction for each nonzero integer value of n:

$$\psi_n = A \sin \frac{n\pi x}{a} \tag{11.43}$$

These wave functions are shown by Figure 11.7. The energies can be calculated from our earlier result:

$$k^2 = \frac{2mE}{\hbar^2}, \qquad k = \frac{n\pi}{a}$$

Remembering that $\hbar = h/2\pi$, this can be shown to give:

$$E_n = \frac{n^2 h^2}{8ma^2} \tag{11.44}$$

Example: Show that the functions of Eq. (11.43) are orthogonal, and determine the value that the constant A must have to make them normalized. Since these functions are valid only for $0 < x < a$, the integrals over "all space" mean, in this instance, $x = 0$ to a; elsewhere $\psi = 0$ and the integral is zero. We wish to calculate the integral (I):

$$I = \int_0^a \psi_n \psi_m \, dx = A^2 \int_0^a \sin \frac{n\pi x}{a} \sin \frac{m\pi x}{a} \, dx$$

This can be simplified with the change in variable $\theta = \pi x/a$:

$$I = A^2 \frac{a}{\pi} \int_0^\pi \sin n\theta \sin m\theta \, d\theta$$

This integral can be found in many tables as:

$$\int \sin n\theta \sin m\theta \, d\theta = \frac{\sin[(n-m)\theta]}{2(n-m)} - \frac{\sin[(n+m)\theta]}{2(n+m)} \tag{11.45}$$

The reader should demonstrate that this integral (with limits of 0 and π) is zero if $n \neq m$; therefore the eigenfunctions are orthogonal. If $n = m$, Eq. (11.45) is indeterminate, and the limit must be found using L'Hopital's rule. More directly, we could evaluate:

$$I = \int_0^a \psi_n^2 \, dx = A^2 \frac{a}{\pi} \int_0^\pi \sin^2 n\theta \, d\theta$$

The reader should do this (either way) and show that the functions will be normalized ($I = 1$) if:

$$A = \sqrt{\frac{2}{a}} \tag{11.46}$$

■

One-Dimensional Box: Discussion

The quantum-mechanical solution is complete at this point—but what have we learned? Since the wave function presumably "knows" everything that can be known about the system, let's ask it some questions.

Where is the particle? Classically we would expect it to be anywhere in the box with equal probability, so:

$$\text{classical probability } (x \text{ to } x + dx) = \frac{dx}{a}$$

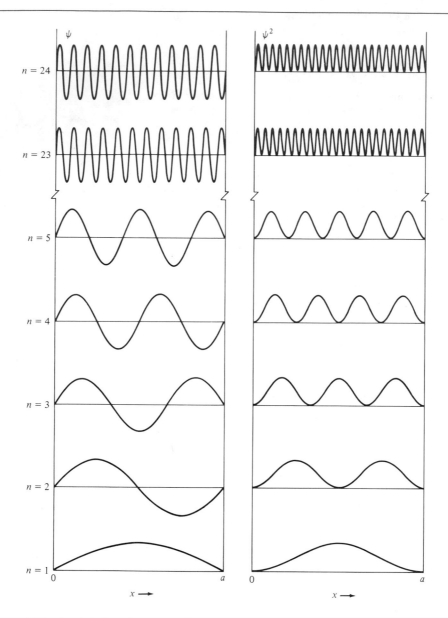

Figure 11.7   Particle-in-a-box wave functions. Wave functions (on the left) and probabilities (the square of the wave functions, on the right) are shown for several values of the quantum number (*n*). Note the similarity of the wave functions to the amplitudes of standing waves on a string. Also note that as the quantum number becomes large, the macroscopic expectation that the particle should, with equal probability, be in any portion of the box is approached more closely.

But from quantum mechanics we find:

$$\text{probability } (x \text{ to } x + dx) = \psi^2 \, dx$$

which is by no means a uniform distribution (Figure 11.7). However, as the quantum number becomes large, the distribution becomes more uniform and like the classical limit. This is an example of the *Bohr correspondence principle: quantum mechanics approaches classical mechanics in the limit of large quantum numbers.* For macroscopic masses and box sizes, the quantity $h^2/8ma^2$ is so small that very large quantum numbers would be required to get a significant amount of energy.

How fast is the particle moving? If you calculate the expectation value of the momentum by the average-value theorem [Eq. (11.33), with $\hat{p}_x = (\hbar/i)[\partial/\partial x)]$, you will find:

$$\langle \hat{p}_x \rangle = \frac{\displaystyle\int_0^a \psi_n(\hat{p}_x \psi_n) \, dx}{\displaystyle\int_0^a \psi_n^2 \, dx} = 0$$

This may mean only that the particle is equally likely to be traveling in each direction.

From the average-value theorem, or more directly from the classical formula ($E = p^2/2m$), one can show [using Eq. (11.44) for E]:

$$\langle \hat{p}_x^2 \rangle = 2mE = \frac{n^2 h^2}{4a^2}, \qquad |\hat{p}_x| = \frac{nh}{2a} \tag{11.47}$$

This result could occur if the speed (or magnitude of the velocity) were:

$$v = \frac{nh}{2ma}$$

We note that there is a minimum or zero-point energy corresponding to $n = 1$, which suggests that the particle cannot be sitting still. A minimum velocity is required by Heisenberg's uncertainty principle, since, if the particle were stationary, the uncertainty in its momentum would be zero and its position would be totally uncertain; however the particle must be in the box ($0 < x < a$), so its position is never totally uncertain.

Example: Calculate the minimum speed of a 1-g mass in a 1-cm box.

$$v = \frac{(1)(6.6256 \times 10^{-27} \text{ g cm}^2 \text{ s}^{-1})}{2(1 \text{ g})(1 \text{ cm})} = 3.3 \times 10^{-27} \text{ cm s}^{-1}$$

This amounts to about 10^{-17} centimeters displacement (less than the width of an atomic nucleus) *per century!* ∎

Exercise: Calculate the minimum speed of an electron in a 20-Å box. (*Answer:* 1.8×10^7 cm s^{-1}) ∎

We must use these ideas carefully! We cannot picture a submicroscopic particle as "moving" in the sense that if we measured its position at $t = 0$, and knew it to be moving to the right at a certain velocity, we could predict the result of a measurement at a later time with any accuracy. Look at the $n = 2$ state function; it has a node (zero) at $x = a/2$. The average value of x can be calculated easily from the average-value theorem; however, the fact that ψ^2 is symmetric about the center of the box should make it obvious that the result should be:

$$\langle x \rangle = \frac{a}{2}$$

But, if after a given measurement we know that the particle is in the left half of the box, where will it be the next time we measure it? If we measure x numerous times we must find $\langle x \rangle = a/2$—equal occurrences of $x < a/2$ and $x > a/2$. But how can it get between these two regions if it can never be at $x = a/2$ (where $\psi = 0$)? There is no answer except that the notion of a trajectory is not supported by the quantum theory.

De Broglie's Hypothesis Revisited

The state functions of this problem [Eq. (11.43)] are just those obtained for the standing-wave problem in classical physics. If we are discussing the vibrations of a string attached at $x = 0$ and a, then ψ represents the amplitude of the string vibrations for the standing waves with wavelengths:

$$\lambda = \frac{2a}{n}$$

(For a compressional standing wave in a metal bar, ψ would represent the density fluctuations.) If this wavelength is combined with Eq. (11.47), we get:

$$p = \frac{nh}{2a} = \frac{h}{\lambda}$$

This is just de Broglie's hypothesis [Eq. (11.13)].

The conclusion is that the particle is behaving like a wave, particularly if it is a "light" particle in a "small" box. For "large" boxes and "heavy" masses, it will behave in the classical manner. This wave-like behavior shows why the question, discussed above—how can the particle cross a node?—is meaningless. We would never ask how the excitation of a standing wave "crosses" a node; it is on both sides of the node simultaneously.

Note that the "quantization" goes away as the box gets larger; unrestricted particles, even electrons, have no restrictions on their kinetic energy.

Three-Dimensional Box: Separation of Variables

The more realistic problem of a particle in a three-dimensional box brings new lessons on the meaning of quantum theory. For simplicity we assume a cubical box with the origin at one corner, so (Figure 11.8):

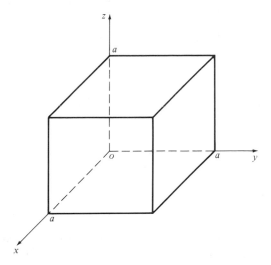

Figure 11.8   The cubic box. For the three-dimensional particle in a box, the potential is zero inside a cube and infinite elsewhere. This could represent the situation of a particle inside a container with perfectly rigid, inpenetrable walls.

$$V = 0 \qquad \text{for } \begin{cases} 0 < x < a \\ 0 < y < a \\ 0 < z < a \end{cases}$$

and $V = \infty$ elsewhere. The momentum is:

$$\mathbf{p} \cdot \mathbf{p} = p_x^2 + p_y^2 + p_z^2$$

and the Hamiltonian (in the box where $V = 0$) is:

$$\hat{H} = -\frac{\hbar^2}{2m} \nabla^2 \tag{11.48}$$

If we call the state functions $\Psi(x, y, z)$, the Schrödinger equation becomes:

$$-\frac{\hbar^2}{2m} \left(\frac{\partial^2 \Psi}{\partial x^2} + \frac{\partial^2 \Psi}{\partial y^2} + \frac{\partial^2 \Psi}{\partial z^2} \right) = E\Psi \tag{11.49a}$$

The operator is, in this case, a sum of three independent terms, each containing derivatives with respect to only *one* of the Cartesian coordinates. Whenever this occurs, it is *generally true* that the state function can be written as a product:

$$\Psi(x, y, z) = \psi_x(x)\psi_y(y)\psi_z(z)$$

We shall run into this again: *whenever the operator is a sum of independent terms, the eigenfunctions can be written as a product.* Let's try it and see what happens.

Since the operation $\partial/\partial x$ is a derivative holding y and z constant:

$$\frac{\partial^2}{\partial x^2}(\psi_x\psi_y\psi_z) = \psi_y\psi_z\frac{\partial^2\psi_x}{\partial x^2} = \psi_x''\psi_y\psi_z$$

Using this in Eq. (11.49a), we get:

$$-\frac{\hbar^2}{2m}[\psi_x''\psi_y\psi_z + \psi_x\psi_y''\psi_z + \psi_x\psi_y\psi_z''] = E\psi_x\psi_y\psi_z$$

and dividing by ψ we get:

$$-\frac{\hbar^2}{2m}\left[\frac{\psi_x''}{\psi_x} + \frac{\psi_y''}{\psi_y} + \frac{\psi_z''}{\psi_z}\right] = E \qquad (11.49b)$$

This equation is of the form:

$$f(x) + g(y) + h(z) = \text{constant}$$

Because x, y, and z are independent variables, we shall argue that each term of this equation must be constant. Suppose we vary x to $x + dx$, holding y and z constant; this equation now becomes:

$$f(x + dx) + g(y) + h(z) = \text{constant}$$

Subtracting previous equation from this gives:

$$f(x + dx) - f(x) = 0 \quad \text{or} \quad \frac{df}{dx} = 0$$

The solution of this equation is $f(x) = \text{constant}$.

Now, going back to Eq. (11.49b), we can write the right-hand side as a sum of three constants:

$$E = E_x + E_y + E_z$$

Since each term on the left-hand side must be constant, Eq. (11.49b) becomes three separate equations; one of these is:

$$-\frac{\hbar^2}{2m}\frac{\psi_x''}{\psi_x} = E_x, \quad \text{or} \quad \frac{d^2\psi_x}{dx^2} = -k_x^2\psi_x$$

with $k_x^2 = 2mE_x/\hbar^2$. There are two similar equations for y and z. These equations have been solved earlier [cf. Eq. (11.42)], so the results should be fairly obvious. For each direction (x, y, and z) the boundary conditions produce a quantum number:

$$k_x = \frac{n_x\pi}{a}, \qquad E_x = \frac{n_x^2 h^2}{8ma^2}, \qquad \psi_x = A\sin\frac{n_x\pi x}{a}$$

$$k_y = \frac{n_y\pi}{a}, \qquad E_y = \frac{n_y^2 h^2}{8ma^2}, \qquad \psi_y = A\sin\frac{n_y\pi y}{a}$$

$$k_z = \frac{n_z\pi}{a}, \qquad E_z = \frac{n_z^2 h^2}{8ma^2}, \qquad \psi_z = A\sin\frac{n_z\pi z}{a}$$

In each case the eigenfunction will be normalized if $A = \sqrt{2/a}$.

The total energy will be the sum of these three:

$$E = (n_x^2 + n_y^2 + n_z^2)\frac{h^2}{8ma^2} \tag{11.50}$$

The state function will be a product:

$$\Psi(x, y, z) = \left(\frac{2}{a}\right)^{3/2} \sin\frac{n_x \pi x}{a} \sin\frac{n_y \pi y}{a} \sin\frac{n_z \pi z}{a} \tag{11.51}$$

Degeneracy

Each quantum number may take a value $1, 2, 3, \ldots, \infty$. Each set of quantum numbers is an independent state, which we shall denote (n_x, n_y, n_z). The first several of these are [with energies calculated with Eq. (11.50)]:

$$(1, 1, 1) \text{ with } E = 3\frac{h^2}{8ma^2}$$

$$(1, 1, 2) \text{ with } E = 6\frac{h^2}{8ma^2}$$

$$(1, 2, 1) \text{ with } E = 6\frac{h^2}{8ma^2}$$

$$(2, 1, 1) \text{ with } E = 6\frac{h^2}{8ma^2}$$

The occurrence of states with the same energy is called a *degeneracy* (denoted with the symbol g) — the *energy level* $E = 6(h^2/8ma^2)$ is said to be 3-fold degenerate; $g = 3$. Table 11.1 lists the states with energies up to $56(h^2/8ma^2)$. Numerous examples of degeneracies can be noted; 163 states are listed, but only 34 energy levels. (Also see Figure 11.9.)

Careful study of Table 11.1 or Figure 11.9 may convince you that the density of energy levels (that is, the number of states in a fixed interval ΔE) is increasing. From 3 to 33 units of energy there are 66 states; from 33 to 56 units there are 97. (With patience and persistence you can discover a total of 410 states and 53 levels up to energy = 100 units.) This fact is important physically, because macroscopic objects are not observed to have quantized energies — all velocities seem to be possible. In accord with the Bohr correspondence principle, the state density for large quantum numbers is so large that, effectively, any velocity is possible. [Note that the quantum numbers n_x, n_y, and n_z are, effectively, direction cosines; therefore, quantization affects not only speed but direction of travel. The Maxwell-Boltzmann distribution for velocities (Figure 1.16) was derived in Section 1.6 with the assumption that all velocities were possible for molecules; for this purpose molecules can be assumed with little or no error to be "macroscopic" objects with classical behavior.]

Table 11.1   Energy levels and degeneracies for the particle in a box

(The energy is given in units of $h^2/8ma^2$)

Energy	g	States[a]	Energy	g	States[a]
3	1	$(1, 1, 1)$	34	3	$(4, 3, 3)$
6	3	$(2, 1, 1)$, etc.	35	6	$(5, 3, 1)$
9	3	$(2, 2, 1)$	36	3	$(4, 4, 2)$
11	3	$(3, 1, 1)$	38	9	$(5, 3, 2)\,(6, 1, 1)$
12	1	$(2, 2, 2)$	41	9	$(4, 4, 3)\,(6, 2, 1)$
14	6	$(3, 2, 1)$	42	6	$(5, 4, 1)$
17	3	$(3, 2, 2)$	43	3	$(5, 3, 3)$
18	3	$(4, 1, 1)$	44	3	$(6, 2, 2)$
19	3	$(3, 3, 1)$	45	6	$(5, 4, 2)$
21	6	$(4, 2, 1)$	46	6	$(6, 3, 1)$
22	3	$(3, 3, 2)$	48	1	$(4, 4, 4)$
24	3	$(4, 2, 2)$	49	6	$(6, 3, 2)$
26	6	$(4, 3, 1)$	50	6	$(5, 4, 3)$
27	4	$(3, 3, 3)\,(5, 1, 1)$	51	6	$(5, 5, 1)\,(7, 1, 1)$
29	6	$(4, 3, 2)$	53	6	$(6, 4, 1)$
30	6	$(5, 2, 1)$	54	12	$(5, 5, 2)\,(6, 3, 3)\,(7, 2, 1)$
33	6	$(4, 4, 1)\,(5, 2, 2)$	56	6	$(6, 4, 2)$

[a]And permutations.

Example: What is the average quantum number for a He atom in a 1-cm^3 box at 1 K? The average translational energy must be:

$$E = \tfrac{3}{2}kT = 2 \times 10^{-16}\ \text{ergs} = (n_x^2 + n_y^2 + n_z^2)\frac{h^2}{8ma^2}$$

Also:

$$\frac{h^2}{8ma^2} = \frac{(6.626 \times 10^{-27})^2}{8(6.6 \times 10^{-24}\ \text{g})\,(1\ \text{cm})^2} = 8.32 \times 10^{-31}\ \text{ergs}$$

Therefore:

$$n_x^2 + n_y^2 + n_z^2 = \frac{2 \times 10^{-16}}{8.32 \times 10^{-31}} = 2.4 \times 10^{14}$$

In other words, the average quantum number would have to be of the order of 10 million! For heavier molecules and higher temperatures the quantum numbers will be even larger; we conclude that molecular translations can be considered to be in the classical limit.   ∎

The equations derived in this section are useful for a variety of applications, but for our purposes they are less important than others to come. The important points illustrated, for which we shall find applications time and again, are: separation of variables, degeneracy, and the Bohr correspondence principle.

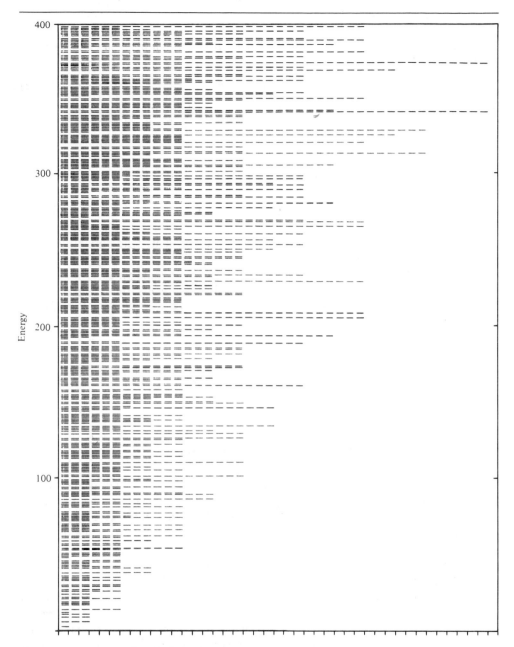

Figure 11.9 Energy levels for a particle in a box. Note the increase in the number of states (degeneracy) with increasing energy. (There are 3719 states shown.) Particles as massive as even an atom, in any container large enough to be seen with the naked eye, have such a huge number of states available (with energy $< kT$) that their energy is effectively a continuum, and the translational motions can be treated using classical mechanics without significant error. (On the vertical scale, the energy is given in units of $h^2/8ma^2$.)

11.6   The Harmonic Oscillator

Next we consider the case of a particle of mass μ, in one dimension (x), subject to a potential [Figure 11.10(a)]:

$$V = \tfrac{1}{2}kx^2 \tag{11.52}$$

This is called a *harmonic* potential. The total classical energy will then be:

$$E_{cl} = \frac{p_x^2}{2\mu} + V(x)$$

The usual prescription from Postulate III gives the Hamiltonian:

$$\hat{H} = -\frac{\hbar^2}{2\mu}\frac{d^2}{dx^2} + \frac{1}{2}kx^2 \tag{11.53}$$

What sort of physical system might this problem represent? One possibility is the case of a mass (μ) attached to a rigid wall by a spring [Figure 11.10(b)]. If we denote the distance of the mass from the wall as R, there will be some value of R (denoted R_e, the *equilibrium* distance) at which no forces will act — that is, the spring is neither compressed nor stretched. The *displacement* of the mass from this equilibrium distance defines x:

$$x = R - R_e$$

For small displacements, the restoring force will be given by Hooke's law:

$$\text{force} = F = -kx$$

The potential is related to the force as $F = -dV/dx$, so:

$$V = -\int F\,dx = \tfrac{1}{2}kx^2 + \text{constant}$$

The constant of integration merely defines the zero of our scale for energy, and for the time being we can conveniently let it equal zero. Thus, the harmonic potential will be the result for the displacement of any system from equilibrium if the restoring force can be approximated by Hooke's law; this works in many situations if the displacement is small.

The harmonic oscillator [any physical system with a Hamiltonian like Eq. (11.53)] is useful for a wide variety of applications. Some discussed in this book are vibrations of diatomic molecules (Chapters 13 and 5), vibrations of polyatomic molecules (Chapter 14), heat capacity of crystalline solids (Chapter 15), and black-body radiation (Chapter 15). A diatomic molecule can be approximated as two masses (the nuclear masses m_1 and m_2) attached by a rigid spring [Figure 11.10(c)]; if the bond distance is R, and the bonding energy has a minimum at $R = R_e$, the vibrations can be described in terms of a displacement coordinate, $x = R - R_e$. For small displacements, Hooke's law can be used and the potential of Eq. (11.52) will apply. In the COM coordinate system (Appendix III) it can be shown that this problem is equivalent to the one with a single mass (μ) attached to a wall [Figure 11.10(b)], provided we

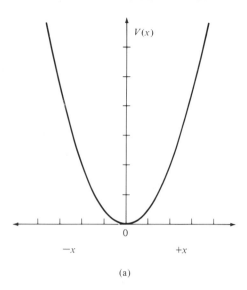

(a)

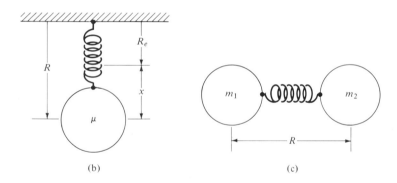

(b)

(c)

Figure 11.10 The harmonic oscillator. (a) The parabolic potential function is also called the harmonic potential. (b) A mass on a spring that obeys Hooke's law and is attached to a wall will vibrate in a harmonic potential such as that shown in panel (a); the displacement coordinate x is the difference between the length of the spring and the length (R_e) at which there is no force (the equilibrium length). (c) A diatomic molecule can be modeled as two masses connected by a stiff, weightless spring. If the spring obeys Hooke's law, the vibrations of this diatom are harmonic and the potential is as given in panel (a) (with $x = R - R_e$), where R is the distance between the masses and R_e is the equilibrium bond length. The center-of-mass (COM) coordinate transformation shows that the diatomic oscillator is identical to the weight on a spring [panel (b)], provided the mass on the spring is the reduced mass ($\mu = m_1 m_2 / (m_1 + m_2)$) of the diatom.

identify μ as the *reduced mass:*

$$\mu = \frac{m_1 m_2}{m_1 + m_2}$$

Then, the Hamiltonian for the vibrational motion can be shown to be given by Eq. (11.53).

For the next several chapters the identification of the harmonic oscillator with molecular vibrations is by far the most important application. For the moment, however, let's just view the operator of Eq. (11.53) as another quantum-mechanical problem to be solved, for it has that most important attribute — it can be solved.

The Quantum Harmonic Oscillator

The Schrödinger equation:

$$\hat{H}\psi = E\psi$$

for the harmonic oscillator is, with Eq. (11.53):

$$\frac{-\hbar^2}{2\mu}\frac{d^2\psi}{dx^2} + \frac{1}{2}kx^2\psi = E\psi \tag{11.54}$$

It will make the notation a great deal simpler if we change Eq. (11.54) into a unit-less form; this can be done by replacing x with a scaled variable given by:

$$x = \alpha y$$

[Be careful! In other sections x and y denote orthogonal Cartesian coordinates; in this section x is the displacement coordinate $(R - R_e)$ and y is just x in different units: $y \equiv (R - R_e)/\alpha$. Also, other texts use variants such as $y = \alpha x$ or $x^2 = \alpha y^2$, and so forth — be careful when making comparisons!]

Substitution of αy for x in Eq. (11.54) gives:

$$\frac{-\hbar^2}{2\mu\alpha^2}\frac{d^2\psi}{dy^2} + \frac{1}{2}k\alpha^2 y^2\psi = E\psi$$

This can be multiplied by $2\mu\alpha^2/\hbar^2$ to give:

$$-\frac{d^2\psi}{dy^2} + \left(\frac{\mu k \alpha^4}{\hbar^2}\right)y^2\psi = \left(\frac{2\mu\alpha^2 E}{\hbar^2}\right)\psi$$

The simplest thing we can do is to choose the constant α as:

$$\alpha = \left(\frac{\hbar^2}{k\mu}\right)^{1/4} \tag{11.55}$$

This will make the set of constants on the left-hand side equal to 1. On the right-hand side we define a unitless energy as:

$$\varepsilon = \frac{2\mu\alpha^2 E}{\hbar^2} = \frac{2\mu E}{\hbar\sqrt{k\mu}} = \frac{4\pi}{h}\sqrt{\frac{\mu}{k}}\,E$$

The Schrödinger equation now looks like this:

$$-\frac{d^2\psi}{dy^2} + y^2\psi = \varepsilon\psi$$

The operator of this equation is the one denoted earlier (p. 581) as:

$$\hat{h} = y^2 - \frac{d^2}{dy^2} \tag{11.56}$$

The eigenvalue equation of this operator, a unitless Schrödinger equation, is:

$$\hat{h}\psi = \varepsilon\psi \tag{11.57}$$

The eigenvalues of this operator are unitless energies — related to the real energy as above; this relationship can be rewritten conveniently as:

$$E = \frac{\hbar}{2}\sqrt{\frac{k}{\mu}}\,\varepsilon \tag{11.58}$$

Ladder Operators

Next we have the problem of finding the allowed eigenvalues and eigenfunctions of Eq. (11.57). This is most conveniently done using the ladder operations mentioned earlier:

$$\hat{A} = y - \frac{d}{dy}, \qquad \hat{B} = y + \frac{d}{dy} \tag{11.59}$$

The example of p. 584 demonstrated that the commutators are:

$$[\hat{h}, \hat{A}] = 2\hat{A}, \qquad [\hat{h}, \hat{B}] = -2\hat{B}$$

For present purposes these will be more conveniently written as:

$$\hat{A}\hat{h} = \hat{h}\hat{A} - 2\hat{A} \tag{11.60a}$$

$$\hat{B}\hat{h} = \hat{h}\hat{B} + 2\hat{B} \tag{11.60b}$$

Only one more idea is required — that there is a minimum eigenvalue (denoted E_0 or ε_0) for this system. The total energy is a sum of the kinetic and potential energies of the particle; if the particle is motionless, the kinetic energy is $T = 0$; if it is located at $x = 0$ [the bottom of the potential curve — Figure 11.10(a)], then $V = 0$. Thus we might expect the minimum energy to be zero. Actually it cannot be that small; if the oscillator is motionless, $p_x = 0$, and if it is at $x = 0$, its momentum and position are exactly fixed — this violates the Heisenberg uncertainty principle. But we don't need that fact; it is built into Eqs. (11.56) to (11.60) (as we shall see). All we need to know is that there *is* a minimum energy (ε_0).

Let us denote the eigenfunction for ε_0 as ψ_0. The eigenvalue equation for this state is, then:

$$\hat{h}\psi_0 = \varepsilon_0\psi_0$$

Now we operate on both sides of this equation with $\hat{A}$:

$$\hat{A}\hat{h}\psi_0 = \varepsilon_0 \hat{A}\psi_0$$

The product operator on the left-hand side can be replaced with Eq. (11.60a) (which is a result of the commutation relationship) to give:

$$(\hat{h}\hat{A} - 2\hat{A})\psi_0 = \varepsilon_0 \hat{A}\psi_0$$

With a little algebra, this becomes:

$$\hat{h}(\hat{A}\psi_0) = (\varepsilon_0 + 2)(\hat{A}\psi_0)$$

This demonstrates that the function $\hat{A}\psi_0$ (which we shall call ψ_1) is an eigenfunction of $\hat{h}$ with an eigenvalue two units larger than the minimum:

$$\hat{h}\psi_1 = (\varepsilon_0 + 2)\psi_1, \qquad \text{with } \psi_1 = \hat{A}\psi_0$$

If this eigenvalue equation is operated upon with $\hat{A}$, the same procedure will show that:

$$\hat{h}\psi_2 = (\varepsilon_0 + 4)\psi_2, \qquad \text{with } \psi_2 = \hat{A}\psi_1 = \hat{A}^2\psi_0$$

If this procedure is repeated, a whole series of eigenfunctions is generated; the general term is:

$$\hat{h}\psi_n = (\varepsilon_0 + 2n)\psi_n, \qquad \text{with } \psi_n = \hat{A}^n\psi_0$$

Next, let's operate on the eigenvalue equation for ψ_n with $\hat{B}$:

$$\hat{B}\hat{h}\psi_n = (\varepsilon_0 + 2n)\hat{B}\psi_n$$

The commutator, Eq. (11.60b), can be used to give:

$$(\hat{h}\hat{B} + 2\hat{B})\psi_n = (\varepsilon_0 + 2n)\hat{B}\psi_n$$

$$\hat{h}(\hat{B}\psi_n) = (\varepsilon_0 + 2n - 2)(\hat{B}\psi_n)$$

But the eigenfunction with this eigenvalue, $\varepsilon_0 + 2n - 2 = \varepsilon_0 + 2(n - 1)$, must be ψ_{n-1}; therefore:

$$\hat{B}\psi_n = (\text{constant})\psi_{n-1}$$

(The value of this constant is worked out in Problem 11.30, but it need not concern us now; in the end, all the eigenfunctions will be multiplied by some constant to normalize them.)

Thus, the operator $\hat{A}$ takes us up the ladder, while $\hat{B}$ takes us down the ladder; they are called *raising* and *lowering* operators, respectively. Imagine Jacob's ladder, with an infinite number of steps ascending to heaven. But (and this is a dirty trick) the bottom rung is 100 meters off the ground. If you are on the ladder, you can step up or down freely. But the bottom rung is unique—if you step down from there, you will fall off and break your neck. Similarly with the ladder of eigenfunctions: the operator $\hat{B}$ upon any function ψ_n will give the next lowest function (ψ_{n-1}) unless $n = 0$, for then no lower function may exist. Put another way, the operation $\hat{B}\psi_n = \psi_{n-1}$ (within a constant) can always be reversed by $\hat{A}$:

$$\hat{A}\hat{B}\psi_n = \hat{A}\psi_{n-1} = (\text{constant})\psi_n$$

unless $n - 0$. We can meet this requirement if:

$$\hat{B}\psi_0 = 0$$

for then $\hat{A}(\text{zero}) = (\text{zero})$ does not restore the function ψ_0. Using the definition of $\hat{B}$ [Eq. (11.59)], this requirement gives:

$$\hat{B}\psi_0 = 0, \qquad \left(y + \frac{d}{dy}\right)\psi_0 = 0$$

$$\frac{d\psi_0}{dy} = -y\psi_0$$

This equation can be integrated directly to give:

$$\psi_0 = A_0 e^{-y^2/2} \tag{11.61}$$

(The constant A_0 can be chosen to normalize this eigenfunction.)

Exercise: Earlier (p. 585) it was shown that:

$$\left(y - \frac{d}{dy}\right)e^{-y^2/2} = 2ye^{-y^2/2}$$

We conclude therefore that $\psi_1 = A_1(2y)e^{-y^2/2}$. Now show that $\hat{A}\psi_1$ gives:

$$\left(y - \frac{d}{dy}\right)2ye^{-y^2/2} = (4y^2 - 2)e^{-y^2/2}$$

Therefore $\psi_2 = A_2(4y^2 - 2)e^{-y^2/2}$. The general result is given below — Eq. (11.62). ∎

Exercise: Now that we know what ψ_0 is, the minimum eigenvalue follows from Eq. (11.57):

$$\hat{h}\psi_0 = \varepsilon_0\psi_0, \qquad \left(y^2 - \frac{d^2}{dy^2}\right)e^{-y^2/2} = \varepsilon_0 e^{-y^2/2}$$

Carry out this operation to show that $\varepsilon_0 = 1$. (As predicted, it is not zero, so Heisenberg's principle is obeyed.) ∎

Wave Functions and Energies

In the preceding, the number n is clearly an integer (it's the number of steps up the ladder). For the harmonic oscillator, the symbol v is generally used in lieu of n. With this change in symbol, the preceding results may be summarized:

Eigenfunctions:

$$\psi_v = A_v H_v(y)e^{-y^2/2} \tag{11.62}$$

(A_v is the normalization constant, H_v is the Hermite polynomial — Table 11.2.)

Eigenvalues: $\varepsilon_v = 1 + 2v$. Using Eq. (11.58) for the energy:

$$E_v = (v + \tfrac{1}{2})\hbar\sqrt{\frac{k}{\mu}}$$

Table 11.2   Hermite polynomials[a]

v	$H_v(y)$
0	1
1	$2y$
2	$4y^2 - 2$
3	$8y^3 - 12y$
4	$16y^4 - 48y^2 + 12$
5	$32y^5 - 160y^3 + 120y$
6	$64y^6 - 480y^4 + 720y^2 - 120$
12	$4096y^{12} - 135{,}168y^{10} + 1{,}520{,}640y^8 - 7{,}096{,}320y^6$ $+ 13{,}305{,}600y^4 - 7{,}983{,}360y^2 + 665{,}280$

[a]Recursion formula: $H_{v+1} = 2yH_v - 2vH_{v-1}$

It is usual to define a quantity called the *vibrational constant:*

$$\nu_0 = \frac{1}{2\pi} \sqrt{\frac{k}{\mu}} \tag{11.63}$$

so that the energy is:

$$E_v = (v + \tfrac{1}{2}) h\nu_0 \tag{11.64}$$

The normalization condition for these functions is:

$$\int_{-\infty}^{\infty} \psi_v^2 \, dx = 1 \tag{11.65}$$

(Table 11.3 lists some integrals that are useful for the harmonic oscillator problem.)

Table 11.3   Some definite integrals of use for the harmonic-oscillator problem[a]

$$I_n = \int_{-\infty}^{\infty} y^{2n} e^{-y^2} \, dy$$

n	I_n	n	I_n
0	$\sqrt{\pi}$	4	$105\sqrt{\pi}/16$
1	$\sqrt{\pi}/2$	5	$945\sqrt{\pi}/32$
2	$3\sqrt{\pi}/4$	6	$10{,}395\sqrt{\pi}/64$
3	$15\sqrt{\pi}/8$	7	$135{,}135\sqrt{\pi}/128$

[a]Recursion formula: $I_{n+1} = \dfrac{I_n(2n + 1)}{2}$

Comparison to the Classical Oscillator

The vibrational constant, Eq. (11.63), is just the classical vibrational frequency; that is, a classical oscillator with this potential is found to have a trajectory:

$$x = x_0 \sin(2\pi\nu_0 t + \phi) \tag{11.66}$$

where x_0 is the maximum amplitude of the oscillation and ϕ is a phase factor. Does this mean the quantum harmonic oscillator is actually oscillating? The quantum theory postulates say nothing about a trajectory, but we can compare the probability function for x:

$$\psi_v^2 \, dx$$

to the classical expectation.

The classical situation is that, during the oscillation, the total energy:

$$E_{cl} = \tfrac{1}{2}mv^2 + V(x)$$

remains constant. This has the effect that, when $x = 0$ and V is a minimum, the particle is moving with maximum velocity. At the extreme of the vibration, $V = E$ and the velocity becomes zero before the oscillator turns back—this is the *classical turning point*. Thus we would expect to find the oscillator most often near the turning point, where it moves slowest, and least often near $x = 0$, where it moves fastest. Figure 11.11 shows that the classical expectation is diametrically opposed to the quantum-theory prediction for the $v = 0$ state; at larger v, the quantum prediction more closely approximates the classical expectation as required by the Bohr correspondence principle. Also note that the quantum probabilities do not go sharply to zero at the classical turning point; this suggests that we will sometimes find the coordinate x such that $V(x) > E$, a situation strictly forbidden by classical mechanics. As for the particle in a box, the existence of nodes in the probability for $v > 1$ also contradicts the notion of a trajectory.

As mentioned in the beginning, the harmonic oscillator is the basis for treating vibrations of diatomic molecules and, by extension, polyatomic molecules. The quantum-theory results are amply confirmed by the observed heat capacities (Chapter 5) and by spectroscopic evidence, which will be discussed later (Chapters 13 and 14).

11.7 Angular Momentum

The quantum theory of angular momentum is an essential feature for many areas of chemistry and physics, including atomic structure and spectroscopy, molecular dynamics, theory of nuclear structure, rotational (microwave) spectroscopy, magnetic-resonance spectroscopy (NMR, ESR), and Mössbauer spectroscopy. We shall use the results of this section time and again throughout the remainder of the book; applications we shall encounter include atomic structure and spectroscopy (Chapter 12), rotations of diatomic molecules (Chapter 13), rotations of polyatomic

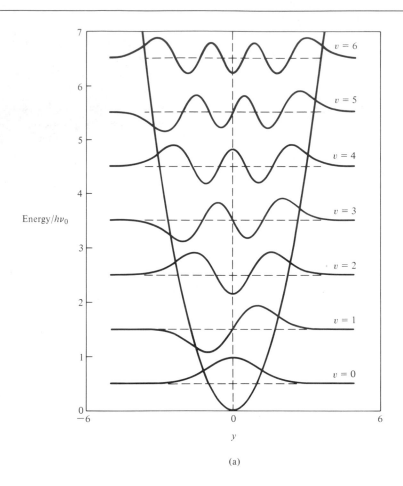

(a)

Figure 11.11 The quantum harmonic oscillator. (a) The potential energy (*V*) and total energy (*E*, horizontal dashed lines) are shown, together with the wave functions super-imposed for illustrative purposes. Note that there is a finite probability for the displacement coordinate (*y*) to have a value that would make the potential energy larger than the total energy—that is, the oscillator is capable of extending past its classical turning point. (b) (See following page.) The probabilities for the quantum oscillator [solid lines,

molecules (Chapter 14), and the effect of nuclear spin on thermodynamic properties (Chapter 15).

The theory for all types of angular momentum is exactly the same, but different symbols are used in different applications. At one place or another the symbols may be *L*, *S*, *J*, *N*, *I*, or *T*, but the results of this section (which will utilize the symbol *L*) are applicable in each case. (Some conventional choices for symbols are indicated in Table 11.4.) For this reason, the ensuing discussions will be phrased rather abstractly in terms of the rotation of a body with a moment of inertia (*I*). (It will be assumed

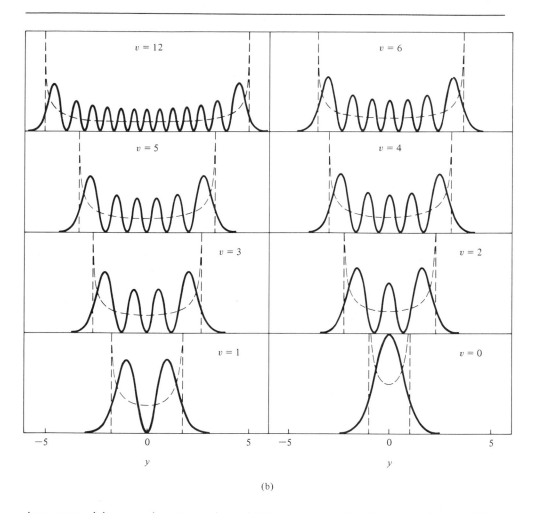

(b)

the square of the wave functions of panel (a)] are compared to the classical probabilities (dashed lines) for the displacement (y) for various quantum states. Note that as the quantum number (v) becomes large (as illustrated by the $v = 12$ case shown), the quantum prediction more closely matches the classical one, and the fraction of the probability outside the classical turning points diminishes.

that only one moment of inertia is required—the more general case is discussed in Chapter 14.)

Classical Angular Momentum

Classically, angular momentum is a vector quantity that, in the absence of external torques, is conserved in both magnitude and direction. (Table 11.5 compares angular motion to the more familiar linear motion. A review of the material on vectors in

Table 11.4   Symbols for angular momentum

General	Molecular rotation[a]	Electron orbital	Electron spin	Nuclear spin
$\mathbf{L}$	$\mathbf{J}$	$\mathbf{L}$	$\mathbf{S}$	$\mathbf{I}$
$\hat{L}^2$	$\hat{J}^2$	$\hat{L}^2$	$\hat{S}^2$	$\hat{I}^2$
l	J	l or L	s or S	I
m	M, M_J	m, M_L	m_s, M_S	M_I

[a]These symbols are also used for the total angular momentum (spin + orbital) of an atom.

Table 11.5   Comparison of linear and circular motions

	Linear	Circular
Velocity	$v = \dfrac{dx}{dt}$	$\omega = \dfrac{d\phi}{dt}$
Momentum	$p = mv$	$L = I\omega$
Kinetic energy	$T = \frac{1}{2}mv^2$	$T = \frac{1}{2}I\omega^2$
	$= \dfrac{p^2}{2m}$	$= \dfrac{L^2}{2I}$
Classical dynamics	force $= \dfrac{dp}{dt}$	torque $= \dfrac{dL}{dt}$

Appendix IV would be appropriate at this point.) Three coordinates are required to specify the rotation of a rigid body. One choice is the magnitude of the angular velocity (ω, radians per second) and the direction of the axis of rotation in spherical polar coordinates (θ, ϕ). Another choice, which we shall prefer, is to discuss the angular momentum vector ($\mathbf{L}$); this can be done either with the magnitude ($|\mathbf{L}| = I\omega$) and direction (θ, ϕ) or with the Cartesian components (L_x, L_y, L_z—see Figure 11.12).

The angular momentum of a particle of mass m, momentum $\mathbf{p}$, and a position vector $\mathbf{r}$ is given by

$$\mathbf{L} = \mathbf{r} \times \mathbf{p} \tag{11.67}$$

This vector is perpendicular to both $\mathbf{r}$ and $\mathbf{p}$ and has a magnitude $|\mathbf{L}| = |\mathbf{p}||\mathbf{r}| = mvr$. The rules for the vector cross product (Appendix IV) give, for the Cartesian components:

$$L_x = yp_z - zp_y, \qquad L_y = zp_x - xp_z, \qquad L_z = xp_y - yp_x \tag{11.68}$$

A rotating body with a moment of inertia I will have a kinetic energy:

$$T = \frac{1}{2}I\omega^2 = \frac{|L|^2}{2I} \tag{11.69}$$

The application in which we shall be most interested is the case of two point masses (m_1 and m_2) connected by a rigid, weightless rod—the rigid rotor. This will be found (in Chapter 13) to be a very good approximation for the rotational energy of a diatomic

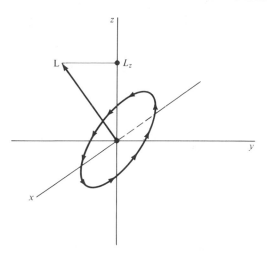

Figure 11.12 Angular momentum. An orbiting or rotational motion in a plane has an angular momentum (**L**) that is a vector perpendicular to that plane. The z component of the angular momentum (L_z) is the projection of the angular momentum vector on the z axis.

molecule. If the masses are separated by R, and the reduced mass is:

$$\mu = \frac{m_1 m_2}{m_1 + m_2}$$

then the moment of inertia will be:

$$I = \mu R^2 \tag{11.70}$$

This formula is also correct for a mass μ connected to a fixed point in space by a rigid massless rod of length R.

Example: Calculate the moment of inertia of the molecule $^{12}C^{16}O$ assuming an internuclear distance $R = 1.11284 \times 10^{-8}$ cm.

$$\mu = \frac{(12)(16)}{(12 + 16)(6.022 \times 10^{23})} = 1.14 \times 10^{-23} \text{ g}$$

$$I = (1.14 \times 10^{-23} \text{ g})(1.11284 \times 10^{-8} \text{ cm})^2$$

$$= 1.41 \times 10^{-39} \text{ g cm}^2 \qquad \blacksquare$$

Operators for Angular Momentum

For a general, unspecified, angular momentum we shall use the following symbols:

1. Classical quantity: $\mathbf{L}, L_x, L_y, L_z, L^2 = L_x^2 + L_y^2 + L_z^2$
2. Operators: $\hat{L}_x, \hat{L}_y, \hat{L}_z, \hat{L}^2$
3. Quantum numbers: l, m

The Cartesian components of the angular-momentum operators can be derived from the classical quantities [Eq. (11.68)] with the prescription given by Postulate III; the results are:

$$\hat{L}_x = \frac{\hbar}{i}\left(y\frac{\partial}{\partial z} - z\frac{\partial}{\partial y}\right) \tag{11.71a}$$

$$\hat{L}_y = \frac{\hbar}{i}\left(z\frac{\partial}{\partial x} - x\frac{\partial}{\partial z}\right) \tag{11.71b}$$

$$\hat{L}_z = \frac{\hbar}{i}\left(x\frac{\partial}{\partial y} - y\frac{\partial}{\partial x}\right) \tag{11.71c}$$

The commutators for these operators can be derived in a straightforward fashion from the basic commutators, $[\partial/\partial x, x] = 1$, and so on. The results are:

$$[\hat{L}_x, \hat{L}_y] = i\hbar\hat{L}_z, \qquad [\hat{L}_y, \hat{L}_z] = i\hbar\hat{L}_x, \qquad [\hat{L}_z, \hat{L}_x] = i\hbar\hat{L}_y \tag{11.72}$$

Example: Derive the first of these commutators.

$$\hat{L}_x\hat{L}_y - \hat{L}_y\hat{L}_x = \hat{C}$$

$$\hat{C} = (y\hat{p}_z - z\hat{p}_y)(z\hat{p}_x - x\hat{p}_z) - (z\hat{p}_x - x\hat{p}_z)(y\hat{p}_z - z\hat{p}_y)$$

These multiplications must be carried out with due regard to the *order* of multiplication; for example, $\hat{p}_z$ and z cannot be exchanged.

$$\hat{C} = (y\hat{p}_z z\hat{p}_x - y\hat{p}_z x\hat{p}_z - z\hat{p}_y z\hat{p}_x + z\hat{p}_y x\hat{p}_z$$
$$- z\hat{p}_x y\hat{p}_z + z\hat{p}_x z\hat{p}_y + x\hat{p}_z y\hat{p}_z - x\hat{p}_z z\hat{p}_y)$$

We note, for example, that $\hat{p}_z$ and x, z and $\hat{p}_y$, and so on, can be exchanged, since they do commute. Close examination will reveal that some cancellation occurs — specifically the second and seventh terms and the third and sixth terms. With some simplification, the remaining terms are:

$$\hat{C} = [y\hat{p}_x(\hat{p}_z z - z\hat{p}_z) - x\hat{p}_y(\hat{p}_z z - z\hat{p}_z)]$$
$$= [(\hat{p}_z z - z\hat{p}_z)(y\hat{p}_x - x\hat{p}_y)]$$
$$= (-i\hbar)(-\hat{L}_z) = i\hbar\hat{L}_z \qquad \blacksquare$$

Exercise: Derive the commutator $[\hat{L}_y, \hat{L}_z]$. $\qquad \blacksquare$

It will prove useful to place the angular-momentum operators into spherical polar coordinates; the derivation is lengthy (but see, for example, refs. 3 and 5), so we will just present the results.

$$\hat{L}_x = i\hbar\left(\sin\phi\frac{\partial}{\partial\theta} + \cot\theta\cos\phi\frac{\partial}{\partial\phi}\right) \tag{11.73a}$$

$$\hat{L}_y = -i\hbar\left(\cos\phi\frac{\partial}{\partial\theta} - \cot\theta\sin\phi\frac{\partial}{\partial\phi}\right) \tag{11.73b}$$

$$\hat{L}_z = -i\hbar\frac{\partial}{\partial\phi} \tag{11.73c}$$

Another important operator is that which represents the total squared angular momentum; in various forms, it is:

$$\hat{L}^2 = \hat{L}_x^2 + \hat{L}_y^2 + \hat{L}_z^2 \tag{11.74a}$$

$$\hat{L}^2 = -\hbar^2 \left[\frac{1}{\sin\theta} \left(\frac{\partial}{\partial\theta} \left(\sin\theta \frac{\partial}{\partial\theta} \right) \right) + \frac{1}{\sin^2\theta} \frac{\partial^2}{\partial\phi^2} \right] \tag{11.74b}$$

$$\hat{L}^2 = -\hbar^2 \left(\frac{\partial^2}{\partial\theta^2} + \cot\theta \frac{\partial}{\partial\theta} + \frac{1}{\sin^2\theta} \frac{\partial^2}{\partial\phi^2} \right) \tag{11.74c}$$

Notice that $\hat{L}^2$ is very similar to the angular part of ∇^2 in polar coordinates [Eq. (11.19)]; the derivation of these operators is part of the general problem of writing ∇^2 in other coordinate systems.

Example: Let us prove the operator identity for $\hat{L}_z$ implied by Eqs. (11.71c) and (11.73c):

$$\frac{\partial}{\partial\phi} = x\frac{\partial}{\partial y} - y\frac{\partial}{\partial x}$$

For a general function $f(x,y,z)$ the chain rule gives:

$$\frac{\partial f}{\partial\phi} = \frac{\partial f}{\partial x}\frac{\partial x}{\partial\phi} + \frac{\partial f}{\partial y}\frac{\partial y}{\partial\phi} + \frac{\partial f}{\partial z}\frac{\partial z}{\partial\phi}$$

With:

$$x = r\sin\theta\cos\phi$$
$$y = r\sin\theta\sin\phi$$
$$z = r\cos\theta$$

we see:

$$\frac{\partial x}{\partial\phi} = r\sin\theta(-\sin\phi) = -y$$

$$\frac{\partial y}{\partial\phi} = x \quad \text{and} \quad \frac{\partial z}{\partial\phi} = 0$$

Therefore:

$$\frac{\partial f}{\partial\phi} = -\left(\frac{\partial f}{\partial x}\right)y + \left(\frac{\partial f}{\partial y}\right)x + (\text{zero}) = \left(x\frac{\partial}{\partial y} - y\frac{\partial}{\partial x}\right)f$$

which was to be proven. ∎

The fact that $\hat{L}_x$, $\hat{L}_y$, and $\hat{L}_z$ do not commute with each other means that no set of functions can be eigenfunctions of more than one of these operators. However, they all can be shown to commute with $\hat{L}^2$:

$$[\hat{L}^2, \hat{L}_x] = 0, \qquad [\hat{L}^2, \hat{L}_y] = 0, \qquad [\hat{L}^2, \hat{L}_z] = 0 \tag{11.75}$$

It is, therefore, possible to find simultaneous eigenfunctions of $\hat{L}^2$ and *any one* of the operators $\hat{L}_x$, $\hat{L}_y$, or $\hat{L}_z$. Usually $\hat{L}_z$ is chosen because of its mathematical simplicity in spherical polar coordinates [Eq. (11.73c)], but there is nothing fundamental about this choice.

The Particle on a Ring

Before doing the general case, we shall discuss a mathematically simpler problem that shows many of the important features of angular momentum in general.

A particle of mass μ is confined to travel on a circle (radius R) by an infinite potential. For convenience, we center the circle on the origin and place it in the xy plane (Figure 11.13):

$$\begin{cases} V = 0 & \text{when } r = R, \ \theta = 90° \\ V = \infty & \text{otherwise} \end{cases}$$

This is, effectively, a circular "box." Because of the location of the circle, $L_x = L_y = 0$ and the kinetic energy is:

$$T = \frac{L^2}{2I} = \frac{L_z^2}{2I}$$

with $I = \mu R^2$. When $V = 0$, the Hamiltonian operator becomes [from Eq. (11.73c)]:

$$\hat{H} = -\frac{\hbar^2}{2I} \frac{\partial^2}{\partial \phi^2}$$

and the eigenvalue equation is:

$$\hat{H}\Psi = E\Psi, \qquad -\frac{\hbar^2}{2I} \frac{\partial^2 \Psi}{\partial \phi^2} = E\Psi \tag{11.76}$$

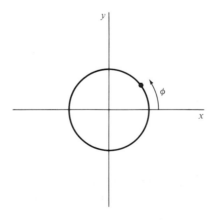

Figure 11.13 The particle on a ring. A particle is confined to a ring in the *xy* plane, centered on the origin, by an infinite potential; the potential is zero on the ring.

We also have an eigenvalue equation for the operator $\hat{L}_z$:

$$\hat{L}_z\Phi = k\Phi, \qquad -i\hbar\frac{\partial\Phi}{\partial\phi} = k\Phi \qquad (11.77)$$

where k is a constant. The functions Φ and Ψ may be the same function, since $\hat{L}_z$ and $\hat{H}$ commute, but this is not required.

The energy eigenvalue equation is just like Eq. (11.42), discussed earlier, and has possible solutions:

$$\cos m\phi, \qquad \sin m\phi, \qquad e^{im\phi}$$

where m is a real constant.

Exercise: Find the results of operating $\hat{H}$ and $\hat{L}_z$ on the functions $\cos m\phi$, $\sin m\phi$, $e^{im\phi}$. ∎

The third of these functions is the only one that is an eigenfunction of both $\hat{H}$ and L_z, so let's use it for the moment; that is, we assume:

$$\Psi = Ae^{im\phi}$$

This function is required to be *single-valued,* so:

$$\Psi(\phi) = \Psi(\phi + 2\pi)$$

$$e^{im\phi} = e^{im(\phi+2\pi)}$$

$$e^{im2\pi} = 1$$

We now use Euler's relationship (Appendix IV):

$$e^{im2\pi} = \cos 2\pi m + i\sin 2\pi m = 1$$

This will be true only if $m = 0, \pm1, \pm2, \ldots$.

The solution to the problem is therefore [using Eqs. (11.76) and (11.77)]:

$$\Psi_m = Ae^{im\phi} \qquad (11.78)$$

$$E_m = \frac{m^2\hbar^2}{2I} \qquad (11.79)$$

$$k = m\hbar \quad \text{(eigenvalue of } \hat{L}_z) \qquad (11.80)$$

Thus the angular momentum and the energy are quantized (cf. Bohr's results, Section 11.2).

Example: Calculate the value of A [Eq. (11.78)] that will normalize the function of Eq. (11.78):

$$1 = A^2\int_0^{2\pi}(e^{-im\phi})(e^{im\phi})\,d\phi = A^2\int_0^{2\pi}d\phi = 2\pi A^2$$

hence:

$$A = \frac{1}{\sqrt{2\pi}} \qquad\qquad ∎$$

These functions (except for $m = 0$) are doubly degenerate; for example, $m = 2$ and $m = -2$ have the same energy, $E = 4\hbar^2/2I$. However, the angular momentum of these states is $+2\hbar$ and $-2\hbar$, respectively, so these degenerate states must represent the particle as rotating in opposite directions.

But what are we to make of the other two functions ($\cos m\phi$ and $\sin m\phi$) that are also eigenfunctions of the Hamiltonian? The functions:

$$\Psi_{ma} = A \cos m\phi = \left(\frac{A}{2}\right)(e^{im\phi} + e^{-im\phi})$$

$$\Psi_{mb} = A \sin m\phi = \left(\frac{A}{2i}\right)(e^{im\phi} - e^{-im\phi})$$

are degenerate eigenfunctions of $\hat{H}$ with the same energy [as given by Eq. (11.79)]. If we were to measure the energy and find, for example, $E = 4\hbar^2/2I$ (which means $m = \pm 2$), the state function could be either $e^{2i\phi}$, $e^{-2i\phi}$, $\cos 2\phi$, or $\sin 2\phi$. However, if the angular momentum were measured for the same state, the answer would be either $+2\hbar$ or $-2\hbar$. This is a general occurrence for degenerate eigenfunctions, so the theorem below is relevant.

THEOREM. Linear combinations of degenerate eigenfunctions of some operator are also eigenfunctions of that operator with the same eigenvalue.

If an operator ($\hat{\alpha}$) has two eigenfunctions (f_1 and f_2) with the same eigenvalue:

$$\hat{\alpha} f_1 = a f_1, \qquad \hat{\alpha} f_2 = a f_2$$

then the function:

$$\psi = c_1 f_1 + c_2 f_2$$

(in which c_1 and c_2 are constants) has the same eigenvalue:

$$\hat{\alpha}\psi = c_1\hat{\alpha}f_1 + c_2\hat{\alpha}f_2 = a(c_1 f_1 + c_2 f_2) = a\psi$$

The General Theory

In this section we shall derive the properties of a general angular momentum as required by the postulates of quantum theory. This derivation is rather complicated, and one need not follow all the details in order to understand subsequent material; the conclusions are, however, essential.

The method we shall use utilizes ladder operators and is rather similar to that used for the harmonic oscillator (Section 11.6). First, we define two operators:

$$\hat{L}_+ = \hat{L}_x + i\hat{L}_y, \qquad \hat{L}_- = \hat{L}_x - i\hat{L}_y \qquad \textbf{(11.81)}$$

These operators have the useful property that they are ladder operators for the eigenfunctions of $\hat{L}_z$. We can demonstrate this, using their commutators.

Example: Find the commutator $[\hat{L}_z, \hat{L}_+]$.

$$\hat{C} = \hat{L}_z\hat{L}_+ - \hat{L}_+\hat{L}_z = \hat{L}_z(\hat{L}_x + i\hat{L}_y) - (\hat{L}_x + i\hat{L}_y)\hat{L}_z$$

This can be multiplied out (with careful regard to the order of multiplication, since these operators do not commute); the reader should demonstrate that the result is:

$$\hat{C} = i(\hat{L}_z\hat{L}_y - \hat{L}_y\hat{L}_z) + (\hat{L}_z\hat{L}_x - \hat{L}_x\hat{L}_z)$$

The commutators of Eq. (11.71) can now be used:

$$\hat{C} = i(-i\hbar\hat{L}_x) + (i\hbar\hat{L}_y) = \hbar(\hat{L}_x + i\hat{L}_y)$$

This result is just $\hat{L}_+$. Therefore:

$$[\hat{L}_z, \hat{L}_+] = \hbar\hat{L}_+, \quad \text{or} \quad \hat{L}_+\hat{L}_z = \hat{L}_z\hat{L}_+ - \hbar\hat{L}_+ \qquad \textbf{(11.82)}$$

■

Exercise: Prove:

$$[\hat{L}_z, \hat{L}_-] = -\hbar\hat{L}_-, \quad \text{or} \quad \hat{L}_-\hat{L}_z = \hat{L}_z\hat{L}_- + \hbar\hat{L}_- \qquad \textbf{(11.83)}$$

■

These commutators are of the type discussed earlier [Eqs. (11.23) *et seq.*], which were seen to be ladder operators.

The operators $\hat{L}^2$ (representing the squared length of the angular-momentum vector) and $\hat{L}_z$ (representing the z component of that vector) commute [Eq. (11.75)]; therefore there will exist a set of eigenfunctions $\{\psi_i\}$ that are simultaneous eigenfunctions of both operators:

$$\hat{L}^2\psi_i = a\psi_i, \qquad \hat{L}_z\psi_i = b\psi_i$$

Since we know that these eigenvalues must have units of $\hbar^2$ and $\hbar$, respectively, it will be convenient to define dimensionless constants k and m such that these eigenvalue equations become:

$$\hat{L}^2\psi_i = k^2\hbar^2\psi_i, \qquad \hat{L}_z\psi_i = m\hbar\psi_i$$

The physical meaning of these constants is that the length of the angular-momentum vector is $k\hbar$, while the z projection is $m\hbar$. For this reason, we must require:

$$|m| \le k$$

This implies that the constant m has, for a given value of k, both an upper and lower limit.

For a given value of k, m will have some maximum value, which we shall call l:

$$l \equiv m(\text{max}), \qquad l \le k$$

The eigenfunction corresponding to this eigenvalue will be denoted as ψ_0:

$$\hat{L}_z\psi_0 = l\hbar\psi_0$$

We now operate upon this equation with $\hat{L}_-$ and use the commutator [Eq. (11.83)]:

$$\hat{L}_-\hat{L}_z\psi_0 = l\hbar\hat{L}_-\psi_0$$

$$(\hat{L}_z\hat{L}_- + \hbar\hat{L}_-)\psi_0 = l\hbar\hat{L}_-\psi_0$$

$$\hat{L}_z(\hat{L}_-\psi_0) = (l - 1)\hbar(\hat{L}_-\psi_0)$$

We have generated a new function, $\psi_1 = \hat{L}_-\psi_0$, whose eigenvalue for $\hat{L}_z$ is reduced by one unit. The eigenvalue for $\hat{L}^2$ has not been changed: since $\hat{L}^2$ commutes with $\hat{L}_x$ and $\hat{L}_y$ [Eq. (11.75)], it also commutes with $\hat{L}_-$. The eigenvalue equation for $\hat{L}^2$ operating on ψ_0 was:

$$\hat{L}^2\psi_0 = k^2\hbar^2\psi_0$$

Operating on this equation with $\hat{L}_-$:

$$\hat{L}_-\hat{L}^2\psi_0 = k^2\hbar^2(\hat{L}_-\psi_0)$$

$$\hat{L}^2\hat{L}_-\psi_0 = k^2\hbar^2(\hat{L}_-\psi_0)$$

$$\hat{L}^2\psi_1 = k^2\hbar^2\psi_1$$

We have demonstrated that operation with $\hat{L}_-$ alters only the z component of the angular momentum — that is, its direction in space — and leaves the length unchanged.

If this procedure is repeated, a "ladder" of eigenfunctions (Table 11.6) will be generated with the general term:

$$\psi_n = \hat{L}_-^n\psi_0, \qquad \hat{L}^2\psi_n = k^2\hbar^2\psi_n, \qquad \hat{L}_z\psi_n = (l - n)\hbar\psi_n$$

The effect of the operator $\hat{L}_+$ can be demonstrated as follows:

$$\hat{L}_z\psi_n = (l - n)\hbar\psi_n$$

$$\hat{L}_+\hat{L}_z\psi_n = (l - n)\hbar(\hat{L}_+\psi_n)$$

Using the commutator, Eq. (11.82):

$$(\hat{L}_z\hat{L}_+ - \hbar\hat{L}_+)\psi_n = (l - n)\hbar(\hat{L}_+\psi_n)$$

$$\hat{L}_z(\hat{L}_+\psi_n) = (l - n + 1)\hbar(\hat{L}_+\psi_n)$$

Thus, $\hat{L}_+$ is a raising operator that increases the eigenvalue of $\hat{L}_z$ by one unit. But if this is done upon ψ_0, the eigenvalue cannot increase, since it was already the maximum by definition. Therefore we must require that:

$$\hat{L}_+\psi_0 = 0$$

Since $\hat{L}_-(\text{zero}) = (\text{zero})$, this can also be written as:

$$\hat{L}_-\hat{L}_+\psi_0 = 0$$

Table 11.6   The angular-momentum ladder

Eigenfunction	Eigenvalue of $\hat{L}_z$ (units of $\hbar$)	
ψ_0	$m = l$	(maximum)
$\psi_1 = \hat{L}_-\psi_0$	$m = l - 1$	
$\psi_2 = \hat{L}_-\psi_1 = \hat{L}_-^2\psi_0$	$m = l - 2$	
$\vdots$		
$\psi_n = \hat{L}_-^n\psi_0$	$m = l - n$	(general term)
$\vdots$		
$\psi_N = \hat{L}_-^N\psi_0$	$m = l - N$	(minimum)

From this we will derive an important consequence. [Note that for any other function on the ladder, this operation regenerates the original function: $\hat{L}_-\hat{L}_+\psi_n =$ (constant)ψ_n.]

Example: We can express the operator $\hat{L}_-\hat{L}_+$ in terms of the Cartesian operators using the definitions of Eq. (11.81):

$$\hat{L}_-\hat{L}_+ = (\hat{L}_x - i\hat{L}_y)(\hat{L}_x + i\hat{L}_y)$$
$$= \hat{L}_x^2 + \hat{L}_y^2 + i(\hat{L}_x\hat{L}_y - \hat{L}_y\hat{L}_x)$$

The last term is a commutator [Eq. (11.72)], and the first two are part of $\hat{L}^2$ [Eq. (11.74a)]; using $\hat{L}_x^2 + \hat{L}_y^2 = \hat{L}^2 - \hat{L}_z^2$, we get:

$$\hat{L}_-\hat{L}_+ = \hat{L}^2 - \hat{L}_z^2 - \hbar\hat{L}_z \qquad \textbf{(11.84)}$$

∎

Exercise: Prove:

$$\hat{L}_+\hat{L}_- = \hat{L}^2 - \hat{L}_z^2 + \hbar\hat{L}_z \qquad \textbf{(11.85)}$$

∎

The necessary condition for ψ_0 to be the top rung on the ladder, $\hat{L}_-\hat{L}_+\psi_0 = 0$, can be restated using Eq. (11.84):

$$(\hat{L}^2 - \hat{L}_z^2 - \hbar\hat{L}_z)\psi_0 = 0$$

But this is an eigenvalue equation, since ψ_0 was defined as an eigenfunction of $\hat{L}^2$ (eigenvalue $= k^2\hbar^2$) and of $\hat{L}_z$ (with eigenvalue $= l\hbar$); therefore:

$$(k^2 - l^2 - l)\hbar^2 = 0$$

We have demonstrated that the maximum value of m (that is, l) is related to the length of the angular-momentum vector ($k\hbar$) as:

$$k^2 = l(l + 1) \qquad \textbf{(11.86)}$$

The ladder that we have generated must also have a lower rung, since the discussion above indicated that $|m| \le k$. We shall denote the lowest eigenfunction as ψ_N. Since it is the lowest rung, we must require $\hat{L}_-\psi_N = 0$, so:

$$\hat{L}_+\hat{L}_-\psi_N = 0$$

This operation can be replaced by Eq. (11.85) to give:

$$(\hat{L}^2 - \hat{L}_z^2 + \hbar\hat{L}_z)\psi_N = 0$$

This is an eigenvalue equation. Using:

$$\hat{L}^2\psi_N = k^2\hbar^2\psi_N \qquad \text{[and } k^2 = l(l + 1)]$$
$$\hat{L}_z\psi_N = (l - N)\hbar\psi_N$$

We get:

$$[l(l + 1) - (l - N)^2 + (l - N)]\hbar^2\psi_N = 0$$

This equation can be solved to give:

$$l = \frac{N}{2} \tag{11.87}$$

Since N is obviously an integer (the number of operations between ψ_0 and ψ_N), we conclude that l must be either an integer (if N is even) or a half-integer (if N is odd).

How many rungs are on the ladder? Remembering that we started counting with zero (ψ_0), and N repetitions of the operations brought us to the last step, $\psi_N = \hat{L}_-^N \psi_0$, we see [using Eq. (11.87)]:

$$(\# \text{ rungs}) = N + 1 = 2l + 1 \tag{11.88}$$

We started with the definition of a quantity called m (the eigenvalue of $\hat{L}_z$ was $m\hbar$); the maximum value of m was called l. We now see that m is permitted $2l + 1$ values:

$$m = l, \, l - 1, \, l - 2, \, \ldots, \, -l \tag{11.89}$$

Summary of the Quantum Properties of Angular Momentum

Angular momentum is represented by a set of functions that are entirely specified by two quantum numbers l and m: $\psi_{l,m}$. These functions are orthogonal and may be normalized.

$$\int \psi_{l,m}^* \psi_{l',m'} \, d\tau \quad \begin{cases} = 1 & \text{if } l = l' \quad and \quad m = m' \\ = 0 & \text{if } l \neq l' \quad or \quad m \neq m' \end{cases} \tag{11.90}$$

The eigenvalues are:

$$\hat{L}^2 \psi_{l,m} = l(l + 1)\hbar^2 \psi_{l,m}, \qquad \hat{L}_z \psi_{l,m} = m\hbar \psi_{l,m} \tag{11.91}$$

The quantum number l is restricted to either integer $(0, 1, 2, \ldots)$ or half-integer $(\frac{1}{2}, \frac{3}{2}, \frac{5}{2}, \ldots)$ values. For each value of l, $2l + 1$ values of m are possible, as given by Eq. (11.89). These functions are not eigenfunctions of $\hat{L}_x$ or of $\hat{L}_y$ (and cannot be, since these operators do not commute with $\hat{L}_z$), so the x and y components of the angular momentum are not specified by these functions.

The physical significance of these results is (Figure 11.14):

1. The magnitude of the angular momentum is quantized, but not as an integer multiple of $\hbar$ as Bohr postulated; rather, the lengths are $\sqrt{l(l + 1)}\,\hbar$ with l restricted as stated above. (See Table 11.7 for some examples.)
2. The orientation of the angular-momentum vector in space is quantized; the only allowed directions are those which permit a z projection of $m\hbar$. The polar angle (θ) must therefore be:

$$\theta = \cos^{-1}\left[\frac{m}{\sqrt{l(l + 1)}}\right] \tag{11.92}$$

The azimuthal angle (ϕ) is not quantized, and any direction is permitted. In fact, if the angular momentum had definite values of both θ and ϕ, all the Cartesian components (L_x, L_y, L_z) would be specified; this would violate Heisenberg's uncertainty

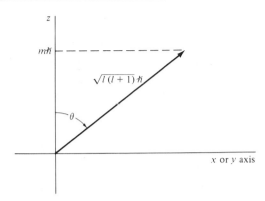

Figure 11.14 Quantum angular momentum. According to the quantum theory of angular momentum, the angular-momentum vector is quantized in length and in direction (with respect to the z axis); there is no restriction on the x or y components of **L**, nor on the azimuthal angle (ϕ).

Table 11.7 Quantum mechanics of angular momentum

$$\hat{L}^2\psi_{l,m} = l(l+1)\hbar^2\psi_{l,m}, \qquad \hat{L}_z\psi_{l,m} = m\hbar\psi_{l,m}$$

$$\hat{L}_+\psi_{l,m} = (const.)\psi_{l,m+1} \qquad (\text{zero if } m = l)$$

$$\hat{L}_-\psi_{l,m} = (const.)\psi_{l,m-1} \qquad (\text{zero if } m = -l)$$

Allowed values for l	Number of orientations	Length of vector (units of $\hbar$)	Allowed values for m
$\frac{1}{2}$	2	$\sqrt{\frac{3}{4}} = 0.866$	$\frac{1}{2}, -\frac{1}{2}$
1	3	$\sqrt{2} = 1.414$	$1, 0, -1$
$\frac{3}{2}$	4	$\sqrt{\frac{15}{4}} = 1.936$	$\frac{3}{2}, \frac{1}{2}, -\frac{1}{2}, -\frac{3}{2}$
2	5	$\sqrt{6} = 2.449$	$2, 1, 0, -1, -2$
$\frac{5}{2}$	6	$\sqrt{\frac{35}{4}} = 2.958$	$\frac{5}{2}, \frac{3}{2}, \frac{1}{2}, -\frac{1}{2}, -\frac{3}{2}, -\frac{5}{2}$
3	7	$\sqrt{12} = 3.464$	$3, 2, 1, 0, -1, -2, -3$
$\vdots$			
l	$2l + 1$	$\sqrt{l(l+1)}$	$l, l-1, \ldots, -l$

principle, since these quantities are represented by operators that do not commute with each other [cf. Eq. (11.72)].

Eigenfunctions for Angular Momentum

Perhaps the most remarkable thing about the derivation above is that we learned so much without ever actually solving for the eigenfunctions, $\psi_{l,m}(\theta, \phi)$. It turns out that this is often a superfluous step; Eqs. (11.90) and (11.91) [plus a few others, which can be derived from Eqs. (11.84) and (11.85)] tell us all we need to know about the angular momentum.

However, it is often useful or necessary to know what these functions are. Equation (11.91), with the operator as defined by Eq. (11.74), can be written as:

$$\frac{\partial^2 \psi}{\partial \theta^2} + \cot \theta \frac{\partial \psi}{\partial \theta} + \frac{1}{\sin^2 \theta} \frac{\partial^2 \psi}{\partial \phi^2} + l(l + 1)\psi = 0 \qquad (11.93)$$

This is Legendre's differential equation, which is solved in many texts (for example, Ref. 5). We shall not concern ourselves with the manner of the solution but only with the results.

In the solution of Legendre's equation, the quantum numbers l and m appear as a result of the boundary condition that $\psi(\theta, \phi)$ be single-valued (some idea of how this occurs can be obtained from the earlier discussion of the particle on a ring). However, there is one major difference between this and the result obtained earlier: the requirement that the eigenfunctions be single-valued permits only *integer* values for the quantum numbers l and m.

When l is an integer, the eigenfunctions of Eq. (11.91) are the *spherical harmonics,* generally denoted with the symbol Y. (These spherical harmonics arise also in many other physical problems aside from quantum mechanics.) The solutions are:

$$\psi_{l,m} = Y_{l,m}(\theta, \phi) = AP_l^{|m|}(\theta)e^{im\phi} \qquad (11.94)$$

The constants (A) that normalize these functions and the associated Legendre polynomials (P) are listed on Table 11.8. Normalization in this case requires:

$$\int_0^{2\pi} d\phi \int_0^{\pi} Y_{l,m}^* Y_{l,m} \sin \theta \, d\theta = 1 \qquad (11.95)$$

The shapes of the Legendre polynomials are shown by Figure 11.15. (The alert reader may note a resemblance between these shapes and those of the atomic orbitals — that this is no coincidence will be developed in the next chapter.)

Example: Show that:

$$Y_{1,0} = \sqrt{\frac{3}{4\pi}} \cos \theta$$

is a normalized solution to Legendre's equation [Eq. (11.93)]. Equation (11.92) with $l = 1$ $[l(l + 1) = 2]$ becomes:

$$\frac{\partial^2 (\cos \theta)}{\partial \theta^2} + \cot \theta \frac{\partial (\cos \theta)}{\partial \theta} + \frac{1}{\sin^2 \theta} \frac{\partial^2 (\cos \theta)}{\partial \phi^2} + 2 \cos \theta = 0$$

$$-\cos \theta + \frac{\cos \theta}{\sin \theta}(-\sin \theta) + 0 + 2 \cos \theta = 0$$

$$-\cos \theta - \cos \theta + 2 \cos \theta = 0$$

Table 11.8   Spherical harmonics and associated Legendre polynomials

$$Y_{l,m} = AP_l^{|m|}(\theta)e^{im\phi}$$

l	$|m|$	A	$P_l^{|m|}(\theta)$
0	0	$\sqrt{\dfrac{1}{4\pi}}$	1
1	0	$\sqrt{\dfrac{3}{4\pi}}$	$\cos\theta$
1	1	$\sqrt{\dfrac{3}{8\pi}}$	$\sin\theta$
2	0	$\sqrt{\dfrac{5}{6\pi}}$	$3\cos^2\theta - 1$
2	1	$\sqrt{\dfrac{15}{8\pi}}$	$\sin\theta\cos\theta$
2	2	$\sqrt{\dfrac{15}{32\pi}}$	$\sin^2\theta$
3	0	$\sqrt{\dfrac{7}{16\pi}}$	$\cos\theta\,(5\cos^2\theta - 3)$
3	1	$\sqrt{\dfrac{21}{64\pi}}$	$\sin\theta\,(5\cos^2\theta - 1)$
3	2	$\sqrt{\dfrac{105}{32\pi}}$	$\sin^2\theta\cos\theta$
3	3	$\sqrt{\dfrac{35}{64\pi}}$	$\sin^3\theta$

Normalization requires [from Eq. (11.95)]:

$$\frac{3}{4\pi}\int_0^\pi \cos^2\theta\,\sin\theta\,d\theta\int_0^{2\pi} d\phi = 1$$

$$\int_0^\pi \cos^2\theta\,\sin\theta\,d\theta = -\int_0^\pi \cos^2\theta\,d(\cos\theta) = -\frac{\cos^3\theta}{3}\Big]_0^\pi = \frac{2}{3}$$

The other integral is obviously equal to 2π, so:

$$\left(\frac{3}{4\pi}\right)\left(\frac{2}{3}\right)(2\pi) = 1 \qquad\blacksquare$$

Spin Angular Momentum

In view of the preceding, what are we to make of half-integer quantum numbers? Clearly they are allowed by quantum theory—they simply represent ladders with an

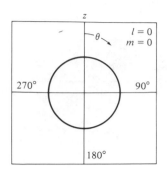

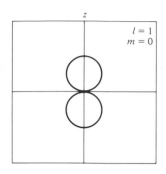

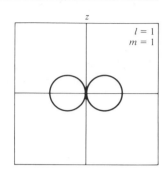

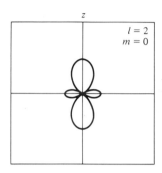

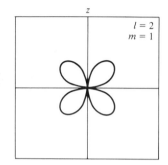

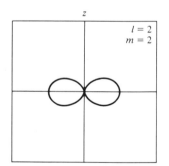

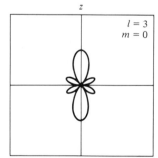

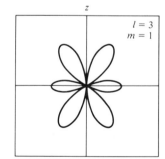

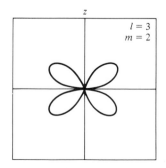

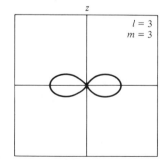

even number of rungs, and nothing in the postulates forbids that. However, the solution to Legendre's equation demonstrates that *half-integer quantum numbers may not represent an angular momentum that results from the rotation of a physical body.* One might be tempted to dismiss them as an artifact—except that they have been observed experimentally. Electrons, protons, and neutrons all have an intrinsic angular momentum that has an "*l*-type" quantum number of $\frac{1}{2}$; atomic nuclei (collections of protons and neutrons) also have half-integer "*l*-type" quantum numbers (some, however, are integers). This property, which cannot be a result of a physical rotation, is referred to as *spin*.

The word *spin* is based on an analogy. The planet Earth has two types of angular momentum—"orbital" due to its rotation about the sun, and "spin" due to its rotation upon its own axis. By analogy, particles are said to have orbital angular momentum with integer quantum numbers and spin angular momentum with half-integer quantum numbers. But this is only a metaphor (in Bronowski's phrase); if the electron were really spinning, it would have an angular momentum, but an integer quantum number would be required.

The existence of "spin" has been shown to be a result of relativity. The original quantum theory of Schrödinger and Heisenberg did not take into account the laws of relativity; a few years later, P.A.M. Dirac modified the quantum theory to account for relativity and explained thereby the known experimental fact that electrons have a "spin" of $\frac{1}{2}$ (that is, the angular momentum of an electron corresponded to a quantum number $l = \frac{1}{2}$). The full integration of relativity and quantum theory, called quantum electrodynamics, occurred much later; the 1956 Nobel Prize in physics was awarded to Richard Feynmann, Julian Schwinger, and Shin-Ichio Tomonaga for this feat. For our purposes we will just take the existence of "spin" and half-integer angular momentum as established facts.

Rotational Energy

The rotational kinetic energy of an object with a moment of inertia I is $T = L^2/2I$. In the absence of external fields, there is no potential energy, so the Hamiltonian operator is derived by simply using the operator $\hat{L}^2$ in the expression for the kinetic energy:

$$\hat{H} = \hat{L}^2/2I \qquad \textbf{(11.96)}$$

The eigenfunctions of this operator are just the eigenfunctions of $\hat{L}^2$ discussed above, so:

$$\hat{H}\psi_{l,m} = \frac{1}{2I}\,\hat{L}^2\psi_{l,m} = \frac{l(l+1)\hbar^2}{2I}\,\psi_{l,m}$$

Figure 11.15 Polar plots of the associated Legendre polynomials. These are the angular-momentum eigenfunctions for integer values of the quantum numbers (l, m). [In such polar graphs, the magnitude of the function $P(\theta)$ at a given angle θ is given by the length of the radius vector from the origin to the curve at that angle.]

The permitted energies are, therefore, given by (using $\hbar = h/2\pi$):

$$E_l = \frac{l(l + 1)h^2}{8\pi^2 I} \tag{11.97}$$

In this case the quantum number l must be restricted to integers.

11.8 Spectroscopy

The absorption or emission of electromagnetic radiation by matter is the principal experimental technique by which the quantized energy levels of atoms and molecules are measured. A full discussion of the interaction of radiation with matter would involve us in time-dependent quantum mechanics, a subject we would prefer to avoid. Instead, in this section we shall look at some of the important conclusions of this theory and discuss applications in general. Chapters 12 through 14 will each have one or more sections on spectroscopy in which more explicit applications will be discussed.

Transitions between two stationary states (ψ_i, ψ_j) of an atom or molecule may occur when it absorbs or emits a photon of energy $h\nu$, with the frequency of the photon given by:

$$\nu = \frac{\Delta E}{h} \tag{11.98a}$$

where ΔE is the difference between the energies of the particle before and after photon absorption:

$$\Delta E = E(\text{final}) - E(\text{initial})$$

This may also be written in terms of the wavenumber frequency $\tilde{\nu} = 1/\lambda$:

$$\tilde{\nu} = \frac{\Delta E}{hc} \tag{11.98b}$$

(This was one of Bohr's postulates.)

Selection Rules

Transitions between all pairs of stationary states are not permitted by photon absorption, even if Eq. (11.98) is obeyed. The rules that state which combinations are allowed are called *selection rules;* these are usually stated in terms of the permitted changes in the quantum numbers when the atom or molecule interacts with electromagnetic radiation.

According to Maxwell's theory of electromagnetic radiation, the emission of such radiation is associated with an oscillation of an *electric dipole moment*. A charge distribution consisting of point charges (q_i) with coordinates x_i, y_i, z_i can be described in terms of the dipole-moment vector with components:

$$\mu_x = \sum_i q_i x_i, \qquad \mu_y = \sum_i q_i y_i, \qquad \mu_z = \sum_i q_i z_i$$

Of particular interest is the case of two point charges, $+q$ and $-q$, separated by a distance r; the COM coordinates can be either polar (r, θ, ϕ) or Cartesian (x, y, z). Then the dipole moment will be:

$$\mu_x = qx = qr \sin \theta \cos \phi$$

$$\mu_y = qy = qr \sin \theta \sin \phi$$

$$\mu_z = qz = qr \cos \theta \qquad \textbf{(11.99)}$$

The quantum theory associates the absorption and emission of electromagnetic radiation with *dipole-moment operators*, which (since $\hat{x} = x$) are given by Eq. (11.99). A transition $\psi_i \longrightarrow \psi_j$ is electric-dipole allowed if any of the integrals $\int \psi_i^* \mu \psi_j \, d\tau$ is nonzero. We shall define the following integrals:

$$I_x \equiv \int \psi_i^* x \psi_j \, d\tau$$

$$I_y \equiv \int \psi_i^* y \psi_j \, d\tau$$

$$I_z \equiv \int \psi_i^* z \psi_j \, d\tau \qquad \textbf{(11.100)}$$

The intensity of the absorption or emission is given by:

$$\text{intensity} \propto I_x^2 + I_y^2 + I_z^2 \qquad \textbf{(11.101)}$$

If all the integrals of Eq. (11.100) are zero, the transition $\psi_i \longleftrightarrow \psi_j$ is said to be *forbidden;* what this means is that the frequency $\nu = (E_i - E_j)/h$ will be either absent from the spectrum or very weak.

Example: Determine the selection rules for the one-dimensional particle in a box. For a transition between states with quantum number $n \longrightarrow m$:

$$I_x = \int_0^a x \psi_n \psi_m \, dx$$

With the wave functions, Eq. (11.43):

$$I_x = \frac{2}{a} \int_0^a x \, \sin\left(\frac{n\pi x}{a}\right) \sin\left(\frac{m\pi x}{a}\right) dx$$

This can be simplified with the definition $\theta = \pi x / a$:

$$I_x = \frac{2a}{\pi^2} \int_0^\pi \theta \, \sin(n\theta) \, \sin(m\theta) \, d\theta$$

This equation can be integrated by parts if we define:

$$u = \theta, \qquad dv = \sin(n\theta) \sin(m\theta) \, d\theta:$$

$$\int_a^b u \, dv = \left[uv \right]_a^b - \int_a^b v \, du$$

Integrating dv with Eq. (11.45):

$$v = \frac{\sin(n - m)\theta}{2(n - m)} - \frac{\sin(n + m)\theta}{2(n + m)}$$

The reader should fill in the remaining steps to get:

$$I_x = \frac{2a}{\pi^2}\left[\frac{\cos(n-m)\theta}{2(n-m)^2} - \frac{\cos(n+m)\theta}{2(n+m)^2}\right]_0^\pi$$

To evaluate this expression at its limits, we must find

$$\cos N\theta\Big]_0^\pi = \cos N\pi - 1$$

for $N = n + m$ and $N = n - m$; this will be zero if N is even and nonzero if N is odd.

Conclusion: A transition $n \longrightarrow m$ is forbidden ($I_x = 0$) if $n + m$ is even. If $n + m$ is odd, the relative intensity is:

$$I_x^2 = \left[\frac{1}{(n-m)^2} - \frac{1}{(n+m)^2}\right]^2 \tag{11.102}$$

(We have dropped the constant $-2a/\pi^2$, since we can calculate only *relative intensities* in any case.)

Transitions originating in the ground ($n = 1$) state will have frequency:

$$\nu = \frac{E_n - E_1}{h} = \frac{h}{8ma^2}(n^2 - 1^2)$$

The frequencies and intensities [calculated with Eq. (11.102)] are as follows:

Transition	Relative intensity (I_x^2)	Frequency
$1 \longrightarrow 2$	1.2346	$3\,(h/8ma^2)$
$1 \longrightarrow 3$	0	
$1 \longrightarrow 4$	0.0228	$15\,(h/8ma^2)$
$1 \longrightarrow 5$	0	
$1 \longrightarrow 6$	0.0036	$35\,(h/8ma^2)$
$1 \longrightarrow 7$	0	
$1 \longrightarrow 8$	0.0011	$63\,(h/8ma^2)$

We see that this system would have only one absorption frequency with any significant intensity.   ■

Energy Units

The symbol E as used in this chapter implies units of *ergs,* but in fact molecular energies are rarely reported in that unit. Molecular energies are nearly always reported in the units that are measured, and this, in turn, is related to the technology involved. A brief discussion of the various regions of spectroscopy (Table 11.9) may make this clearer.

The technology involved in a spectrometer depends on the wavelength (λ) of the radiation relative to a typical dimension of the sample being studied ($l \sim 1$ cm).

In the *radiofrequency region,* the wavelength of the radiation is large, $\lambda \gg l$. The technology is the same as that of radio and television—transistors, capacitors, inductors, and so on. The quantity measured directly is frequency (ν) in units of cycles per second (cps, also called hertz, abbreviated Hz), and it is in this unit that energies are reported.

Table 11.9   Regions of spectroscopy

Frequency range	Wavelengths	Region	Energies investigated
1 MHz–1 GHz	300 m to 30 cm	Radiofrequency	Nuclear spin
1–300 GHz	30 to 0.1 cm	Microwave	Electron spin, molecular rotation
10–13,333 cm^{-1}	0.1 cm to 750 μm	Optical–infrared	Molecular vibrations
13,333–25,000 cm^{-1}	750 to 400 μm	Optical–visible	Electrons
25,000 cm^{-1} and up	400 μm and down	Optical–ultraviolet	Electrons
	~0.1 nm	X-ray	Inner electrons

In the *microwave region,* $\lambda \sim l$. The technology is that of radar, satellite communications, and microwave ovens and involves such devices as klystrons and wave guides. Again it is frequency that is directly measured, and energies are reported in hertz.

In the *optical region,* $\lambda \ll l$; here devices such as prisms, diffraction gratings, slits, and lenses are used. The details vary, but in all cases it is the wavelength of the radiation that is measured and energies are reported as reciprocal wavelengths (or wavenumbers); we shall use the symbol $\bar{\nu}$ for this quantity; $\bar{\nu} = 1/\lambda$.

All of these are related, of course, by the velocity of light (c):

$$\nu = c/\lambda = c\bar{\nu}$$

Note that the velocity of light and wavelength depend on the index of refraction of the medium (frequency does not).

The optical region is subdivided into infrared (IR), visible, and ultraviolet (uv), based on the sensitivity of the human eye to light. At high energies (far uv and X-ray), energies are commonly reported in electron volts (eV).

If E is an energy in ergs, the other common units are:

$$\text{energy in Hz} = \frac{E}{h}$$

$$\text{energy in cm}^{-1} = \frac{E}{hc}$$

$$\text{energy in eV} = 300\,\frac{E}{e} \qquad\qquad \textbf{(11.103)}$$

(The constants h, c, e are in cgs units.)

Exercise: Calculate the conversion from cm^{-1} to eV. (*Answer:* 8065 cm^{-1}/eV) ∎

Exercise: For purposes of knowing when Boltzmann population factors are important, it is useful to know kT in the practical units. Calculate this quantity at 300 K. (*Answer:* 4×10^{-21} J, 4×10^{-14} ergs, 6×10^{12} Hz, 200 cm^{-1}, 0.03 eV. This is the energy of a photon in the far-IR region.) ∎

Postscript

In this chapter we have developed a number of tools and fragments that we will use as, in the next three chapters, we discuss some of the experimental manifestations of quantum theory.

More material on operator algebra can be found in Ref. 3. Reference 7 gives a brief account of these topics (and more) from the "wave equation" point of view; this might be helpful. References 1 and 2 are highly recommended, but the others will probably not be helpful until later.

References

1. G. Gamow, *Biography of Physics,* 1961: New York, Harper & Row; *Mr. Tompkins in Paperback,* 1967: New York, Cambridge University Press.
2. A. D'Abro, *The Rise of the New Physics,* 2 vols., 1939: New York, Dover Publications, Inc.
3. James R. Barrante, *Applied Mathematics for Physical Chemistry,* 1974: Englewood Cliffs, N. J., Prentice-Hall, Inc.
4. I. N. Levine, *Quantum Chemistry,* 2d ed., 1974: Boston, Allyn & Bacon, Inc.
5. H. Margenau and G. M. Murphy, *The Mathematics of Physics and Chemistry,* 1957: New York, John Wiley & Sons, Inc.
6. M. E. Rose, *Elementary Theory of Angular Momentum,* 1957: New York, John Wiley & Sons, Inc.
7. J. W. Linnett, *Wave Mechanics and Valency,* 1960: New York, John Wiley & Sons, Inc.

Problems

11.1   Calculate the Bohr radius as defined [Eq. (11.6)] and confirm the numerical value given.

11.2   Coulomb's law for the potential between a charge $+Ze$ and a charge $-e$ is:

$$V = -\frac{Ze^2}{r}$$

For the first Bohr orbit ($N = 1$) of the hydrogen atom ($Z = 1$), calculate the total (E), potential (V), and kinetic (T) energies of the electron. You will note that $E = -T = \frac{1}{2}V$ — an example of the *virial theorem*.

11.3   (a) Calculate the wavelength of the line of the hydrogen atom spectrum corresponding to $N_1 = 2$, $N_2 = 4$.
(b) Repeat the calculation for $N_2 = 9$, 10, 11, . . ., until it is obvious that you are approaching a *series limit*. What is that limit for $N_1 = 2$?

11.4   Calculate the velocity of the electron in the first Bohr orbit of a hydrogen atom. (Remember that angular momentum $= mvr$.) Repeat the calculation for Hg ($Z = 80$).

11.5   Find the result of operating with $\hat{A} = y - (d/dy)$ and $\hat{B} = y + (d/dy)$ on the function $f(y) = e^{-y^2/2}$.

11.6   Find the value of the commutator $[d/dx, \sin x]$.

11.7   Calculate the commutators:

(a) $\left[\dfrac{d}{dx}, x^2\right]$ (b) $\left[\dfrac{d^2}{dx^2}, x\right]$

11.8   Find the result of operating with:

$$\hat{O} = i\frac{d}{d\phi} \quad \text{on} \quad f(\phi) = 3e^{i\phi}$$

Is it an eigenfunction? What is the eigenvalue?

11.9   Find the result of operating with:

$$\hat{O} = \frac{d^2}{dx^2} - 4x^2 \quad \text{on the function} \quad \psi = e^{-ax^2}$$

What must be the value of a if ψ is to be an eigenfunction of this operator?

11.10   Find the result of operating with the operator:

$$\hat{O} = \frac{1}{r^2}\frac{d}{dr}r^2\frac{d}{dr} + \frac{2}{r}$$

on the function $\psi = Ae^{-br}$. What values must the constants have for this to be an eigenfunction?

11.11   Show that the Legendre polynomials $P_1 = \cos\theta$ and $P_2 = 3\cos^2\theta - 1$ are eigenfunctions of:

$$\hat{O} = \frac{1}{\sin\theta}\frac{d}{d\theta}\left(\sin\theta\frac{d}{d\theta}\right)$$

What are the eigenvalues?

11.12   Find the result of ∇^2 on the function $(x^2 + y^2 + z^2)$. Is it an eigenfunction? Try the same problem in polar coordinates; you should get the same answer.

11.13   Operate with ∇^2 (in Cartesian coordinates) on the function $\psi = \exp(ax + by + cz)$. Is it an eigenfunction?

11.14   Show that the momentum operator $\hat{p}_x$ has eigenfunctions of the form $\psi = e^{ikx}$. What is the physical significance of the constant k?

11.15   Calculate $\langle x \rangle$ and $\langle x^2 \rangle$ for the one-dimensional particle in a box.

11.16   Find the probability that $(0 < x < 0.25a)$ for the one-dimensional particle in the box. Show that this probability approaches the classical limits as $n \longrightarrow \infty$.

11.17   Prove that the functions $\psi_1 = A\sin(\pi x/a)$ and $\psi_2 = A\sin(2\pi x/a)$ are orthogonal.

11.18   Calculate the value of A so that $\psi_n = A\sin(n\pi x/a)$ is normalized in the region $0 < x < a$.

11.19   Show that the one-dimensional particle in a box obeys Heisenberg's uncertainty principle for the $n = 1$ state.

11.20   The particle in a square box has energy:

$$E_{n_x n_y} = \frac{h^2}{8ma^2}(n_x^2 + n_y^2)$$

Calculate the energies and degeneracies of the ten lowest energy levels.

11.21 Calculate the zero-point translational energy of (a) an H_2 molecule in a 1-mm cubic box, (b) an electron in a 1-Å cubic box.

11.22 Calculate the average translational quantum number for H_2 molecules in a box 1 mm on a side when $T = 10$ K.

11.23 If an electron is in a 10^{-8}-cm cubic box, what would be the wavelength of the photon that would raise it from the lowest to the second energy level?

11.24 Derive a formula for the energy of a particle in a box with nonequal sides — that is, $0 < x < a, 0 < y < b, 0 < z < c$.

11.25 Calculate the value for the constant A_v required to normalize the $v = 2$ harmonic-oscillator wave function.

11.26 In the harmonic-oscillator problem, what are the units of the following quantities (cgs system)? k, μ, α, ψ.

11.27 Prove that the harmonic-oscillator wave function for $v = 0$ is orthogonal to those for $v = 1$ and $v = 2$.

11.28 Calculate the probability that a harmonic oscillator in the $v = 0$ state will be found outside its classical turning points.

11.29 (a) For the harmonic oscillator, calculate:

$$\langle V \rangle = \tfrac{1}{2} k \langle x^2 \rangle$$

for the $v = 0$ state.
(b) Using $T = p^2/2\mu = E - V$, calculate $\langle p^2 \rangle$ for the harmonic oscillator.
(c) Show that $\delta p\, \delta x = \hbar/2$ for the harmonic-oscillator $v = 0$ state.

11.30 (a) Prove the following for the harmonic-oscillator operators: $\hat{B}\hat{A} = \hat{h} + 1$; $\hat{A}\hat{B} = \hat{h} - 1$; $[\hat{B}, \hat{A}] = 2$.
(b) Use the results of part (a) to prove that the eigenvalue of ψ_0 is $\varepsilon_0 = 1$, independent of any specific functional form for ψ_0.
(c) With the definition $\psi_{n+1} = \hat{A}\psi_n$, and the eigenvalue equation $\hat{h}\psi_n = (2n + 1)\psi_n$, show [using the results of parts (a) and (b)] that $\hat{B}\psi_{n+1} = 2(n + 1)\psi_n$. (This constant was ignored earlier.)

11.31 Show that the functions:

$$\psi_m = \frac{1}{\sqrt{2\pi}} e^{im\phi}$$

are orthonormal — that is:

$$\int \psi_m^* \psi_{m'}\, d\phi \begin{cases} = 0 & \text{if } m \neq m' \\ = 1 & \text{if } m = m' \end{cases}$$

11.32 Use the commutators $[\hat{L}_x, \hat{L}_y]$ and so on to prove $[\hat{L}^2, \hat{L}_x] = 0$.

11.33 (a) Prove that, in spherical polar coordinates:

$$\hat{L}_+ = \hbar \left(e^{i\phi} \frac{\partial}{\partial \theta} + i \cot \theta\, e^{i\phi} \frac{\partial}{\partial \phi} \right)$$

(b) Find the result $\hat{L}_+ \cos \theta$ and $\hat{L}_+^2 \cos \theta$. Explain these results.

11.34 Prove that a state function that is radially symmetric, such as $\psi(r) = A e^{-cr}$, has no angular momentum.

11.35   (a) Calculate the moment of inertia of HCl^{35}; use $R = 0.1275$ nm.
(b) Assuming R to be the same, calculate I for the other isotopes: HCl^{37}, DCl^{35}, DCl^{37}.

11.36   (a) Calculate the average l of a small gyroscope that has $I = 1$ g cm^2 and that is rotating with an angular velocity $\omega = 159$ rad/sec. What is the minimum amount by which the velocity may change for this gyroscope? Is the orientation "quantized"?
(b) Calculate the average l of CO molecules ($I = 14.48 \times 10^{-40}$ g cm^2) with an average rotational energy of kT at $T = 300$ K.

11.37   Calculate the commutator: $[\hat{L}_+, \hat{L}_-]$.

11.38   Prove, by direct operation, that the function $(3 \cos^2 \theta - 1)$ is an eigenfunction of $\hat{L}^2$. What must the quantum numbers (l, m) be for this function?

11.39   (a) Assuming that a harmonic oscillator is capable of absorbing radiation, show that the transition $v = 0 \longrightarrow 1$ is allowed.
(b) Show that the transition $v = 0 \longrightarrow 2$ is not allowed.

11.40   Determine the electric dipole selection rules for a particle on a ring.

11.41   Calculate the wavelength of light that an electron in a (1×10^{-7})-cm box (one dimension) must absorb to change its quantum number from $n = 1 \longrightarrow 2$. In what region of the spectrum will this be found? What will be the wavelength of the $n = 1 \longrightarrow 3$ transition? Will it be a strong absorption?

11.42   Assuming that a diatomic rotor can absorb radiation and has $I = 10^{-38}$ g cm^2, calculate the frequency of the radiation that will cause a transition from the $l = 1$ to the $l = 2$ state. In what region of the spectrum is this?

11.43   If a harmonic oscillator is capable of absorbing energy, the selection rule is $\Delta v = \pm 1$. Show that the frequency absorbed is identical to the *vibrational constant*, v_0.

Twinkle, twinkle, little star,
I don't wonder what you are
For by the spectroscopic ken
I know that you are hydrogen.

—Author unknown

12
Atoms

As quantum theory is applied to chemistry, atomic theory, especially that of the hydrogen atom, forms the basis for all subsequent theories and applications. For that reason, the theory of the hydrogen atom, or more generally a system consisting of one negatively charged electron interacting with a massive and positively charged nucleus, must be studied very carefully and be thoroughly understood before one proceeds to more chemically interesting systems. Also, the student is cautioned that preconceptions garnered from earlier courses are best left behind at this point. Approach this chapter as a totally new subject, and leave any attempted correlations with what is taught in qualitative courses until the end of the chapter.

The potential energy between a nucleus of charge $+Ze$ (Z is the atomic number) and an electron of charge $-e$ is given by Coulomb's law (Figure 12.1); in cgs units:

$$V(r) = \frac{-Ze^2}{r}$$

By analogy to the particle in a box, this could be called the particle in a bottomless pit. The question of why the particle doesn't fall limitlessly into the pit is the heart of the problem; how can oppositely charged particles in such close proximity be stable? We begin with a standard opening—write down a Hamiltonian.

12.1 The Hydrogen Atom

Assuming a nucleus that is in a fixed position in space, the total classical energy of an electron is:

$$E_{cl} = \frac{p^2}{2m} - \frac{Ze^2}{r}$$

Note that there is only *one* electron—if $Z = 2$, then we mean the one-electron ion, He^+. By the usual prescription, we get the Hamiltonian operator:

$$\hat{H} = \frac{-\hbar^2}{2m}\nabla^2 - \frac{Ze^2}{r} \tag{12.1}$$

whose eigenvalue equation:

$$\hat{H}\psi = E\psi$$

is to be solved.

Angular Momentum: Separation of Variables

Because the potential is a function of r only, spherical polar coordinates will be more useful than Cartesian coordinates in this problem. (Note that $r = \sqrt{x^2 + y^2 + z^2}$, so the potential is rather complicated in Cartesian coordinates.) As stated in Chapter 11, the Laplacian operator in spherical polar coordinates is:

$$\nabla^2 = \frac{1}{r^2}\frac{\partial}{\partial r}r^2\frac{\partial}{\partial r} + \frac{1}{r^2 \sin\theta}\frac{\partial}{\partial\theta}\sin\theta\frac{\partial}{\partial\theta} + \frac{1}{r^2 \sin^2\theta}\frac{\partial^2}{\partial\phi^2}$$

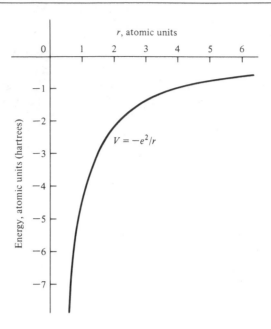

Figure 12.1   The Coulombic potential. The radius is given in multiples of Bohr's radius, $a_0 = 0.5292$ Å; the energy is given in atomic units (1 a.u. $= e^2/a_0 = 27.21$ eV). The atomic unit of energy is also called the *hartree*.

The angular part of this operator is, except for a factor of $\hbar^2$, simply $\hat{L}^2$ as defined by Eq. (11.74), so the Laplacian can also be written as:

$$-\hbar^2\nabla^2 = \frac{-\hbar^2}{r^2}\frac{\partial}{\partial r}r^2\frac{\partial}{\partial r} + \frac{1}{r^2}\hat{L}^2$$

Therefore, the one-electron Hamiltonian is:

$$\hat{H} = \left(\frac{-\hbar^2}{2mr^2}\frac{\partial}{\partial r}r^2\frac{\partial}{\partial r} - \frac{Ze^2}{r}\right) + \frac{1}{2mr^2}\hat{L}^2 \tag{12.2}$$

When discussing the three-dimensional particle in a box (Chapter 11), we noted that if an operator was a sum of independent terms, the eigenfunction would be a product of terms. The operator of Eq. (12.2) meets this criterion, since the first part (in parentheses) takes derivatives only with respect to r, while the $\hat{L}^2$ term takes derivatives only with respect to θ and ϕ. Therefore, the eigenfunctions of this Hamiltonian can be written as:

$$\psi(r, \theta, \phi) = R(r)S(\theta, \phi)$$

Since $\hat{L}^2$ commutes with the Hamiltonian, it is not unreasonable to require the angular functions (S) to be eigenfunctions of $\hat{L}^2$. In Chapter 11 we saw that the spherical harmonics were eigenfunctions of the $\hat{L}^2$ operator; therefore, we can use for the angular functions:

$$S(\theta, \phi) - Y_{lm}(\theta, \phi)$$

with the eigenvalue relationship:

$$\hat{L}^2 Y_{lm}(\theta, \phi) = l(l + 1)\hbar^2 Y_{lm}(\theta, \phi) \qquad (12.3)$$

(The spherical harmonics are given by Table 11.8; also, see Figure 11.15.) With these substitutions, the Schrödinger equation becomes:

$$\left[\hat{h}(r) + \frac{1}{2mr^2}\hat{L}^2\right]R(r)Y_{lm}(\theta, \phi) = ER(r)Y_{lm}(\theta, \phi)$$

where $\hat{h}(r)$ is the radial operator [the part in the parentheses of Eq. (12.2)].

The Radial Equation

Using Eq. (12.3), the eigenvalue equation of $\hat{L}^2$, and performing the indicated operations in the equation above, we obtain:

$$\left[\hat{h}(r) + \frac{l(l + 1)\hbar^2}{2mr^2}\right]R(r)Y(\theta, \phi) = ER(r)Y(\theta, \phi)$$

Since there are no other operators for θ and ϕ in this equation, we can divide out the angular function (Y) from both sides to get the **radial equation:**

$$\left[\frac{-\hbar^2}{2mr^2}\frac{\partial}{\partial r}r^2\frac{\partial}{\partial r} - \frac{Ze^2}{r} + \frac{l(l + 1)\hbar^2}{2mr^2}\right]R(r) = ER(r) \qquad (12.4)$$

The third term in the brackets is called the centrifugal potential. The radial equation can be simplified by using the Bohr radius:

$$a_0 = \frac{\hbar^2}{me^2}$$

to define a unitless variable:

$$\sigma = \frac{Zr}{a_0}$$

Using Bohr's radius, we can write Eq. (12.4) as:

$$\left[\frac{-a_0 e^2}{2r^2}\frac{\partial}{\partial r}r^2\frac{\partial}{\partial r} - \frac{Ze^2}{r} + \frac{l(l + 1)a_0 e^2}{2r^2}\right]R(r) = ER(r)$$

This equation is multiplied by $-2a_0/Z^2 e^2$ to give:

$$\left[\frac{a_0^2}{Z^2 r^2}\frac{\partial}{\partial r}r^2\frac{\partial}{\partial r} + \frac{2a_0}{Zr} - \frac{l(l + 1)a_0^2}{Z^2 r^2}\right]R(r) = -\frac{2a_0 E}{Z^2 e^2}R(r)$$

Now, introducing σ, we get:

$$\left[\frac{1}{\sigma^2}\frac{\partial}{\partial \sigma}\sigma^2\frac{\partial}{\partial \sigma} + \frac{2}{\sigma} - \frac{l(l + 1)}{\sigma^2}\right]R(r) = \varepsilon R(r) \qquad (12.5)$$

with:

$$E = \frac{Z^2 e^2}{2a_0} \varepsilon$$

The reader should fill in the missing steps in the derivation of Eq. (12.5).

Equation (12.5) is **Laguerre's differential equation,** which we shall not solve; the solution can be found in many mathematics, physics, or engineering texts.

Example: A solution to Eq. (12.5) when $l = 2$ is $R = \sigma^2 e^{-\sigma/3}$. Find ε.

$$\frac{\partial R}{\partial \sigma} = 2\sigma e^{-\sigma/3} - \frac{\sigma^2}{3} e^{-\sigma/3}$$

$$\sigma^2 \frac{\partial R}{\partial \sigma} = 2\sigma^3 e^{-\sigma/3} - \frac{1}{3} \sigma^4 e^{-\sigma/3}$$

$$\frac{1}{\sigma^2} \frac{\partial}{\partial \sigma} \left(\sigma^2 \frac{\partial R}{\partial \sigma} \right) = 6e^{-\sigma/3} - 2\sigma e^{-\sigma/3} + \frac{1}{9} \sigma^2 e^{-\sigma/3}$$

This is the first term of the eigenvalue equation. The others are:

$$\frac{2}{\sigma} R = 2\sigma e^{-\sigma/3} \qquad \text{(second term)}$$

$$\frac{-l(l + 1)R}{\sigma^2} = -6e^{-\sigma/3} \qquad \text{(third term)}$$

Added together, these give:

$$\frac{1}{9} \sigma^2 e^{-\sigma/3}$$

Therefore, the eigenvalue must be $\varepsilon = \frac{1}{9}$.   ■

It is found that well-behaved solutions to Eq. (12.5) — that is, those which do not become infinite as $r \to \infty$ — will exist for only certain values of ε, namely:

if $l = 0$: $\varepsilon = 1, \frac{1}{4}, \frac{1}{9}, \frac{1}{16}, \ldots$

if $l = 1$: $\varepsilon = \frac{1}{4}, \frac{1}{9}, \frac{1}{16}, \ldots$

if $l = 2$: $\varepsilon = \frac{1}{9}, \frac{1}{16}, \ldots$

In general it is required that $\varepsilon = 1/n^2$, where n is an integer greater than l. (We have seen already in Chapter 11 that l must be an integer.)

Eigenvalues and Eigenfunctions

The general solution to Eq. (12.5) is:

$$R_{nl} = L_{nl}(\sigma)e^{-\sigma/n}$$

where L_{nl} are the associated Laguerre polynomials (Table 12.1). Finally, the eigen-

Table 12.1   Radial wave functions for one-electron atoms or ions

$$\sigma = \frac{Zr}{a_0}, \qquad a_0 = \frac{\hbar^2}{me^2}, \qquad R_{nl} = L_{nl}e^{-\sigma/n}$$

n	l	$L_{nl}(\sigma)$
1	0	1
2	0	$1 - \dfrac{\sigma}{2}$
2	1	σ
3	0	$1 - \dfrac{2\sigma}{3} + \dfrac{2\sigma^2}{27}$
3	1	$\sigma\left(1 - \dfrac{\sigma}{6}\right)$
3	2	σ^2
4	0	$\left(1 - \dfrac{3\sigma}{4} + \dfrac{\sigma^2}{8} - \dfrac{\sigma^3}{192}\right)$
4	1	$\sigma\left(1 - \dfrac{\sigma}{4} + \dfrac{\sigma^2}{80}\right)$
4	2	$\sigma^2\left(1 - \dfrac{\sigma}{12}\right)$
4	3	σ^3

functions of the Hamiltonian [Eq. (12.1)] are:

$$\psi_{nlm}(r, \theta, \phi) = R_{nl}(r)Y_{lm}(\theta, \phi) \tag{12.6}$$

for $l = 0, 1, 2, \ldots, n - 1$, and $m = l, l - 1, \ldots, -l$. (Be careful: m is the quantum number for $\hat{L}_z$, not the electron mass.) The energy eigenvalues are (Figure 12.2):

$$E_n = -\frac{Z^2 e^2}{2a_0} \frac{1}{n^2} \tag{12.7}$$

The energy, Eq. (12.7), is exactly Bohr's result—a good thing, since Bohr's theory was correct in this instance—but the physical picture is quite different. A very important difference is that the lowest energy, $n = 1$, must have a quantum number $l = 0$, which means *there is no orbital angular momentum;* this kills any idea that the electron is "orbiting" or that a centripetal force is responsible for the atom's stability. The "picture" is in the wave functions—so let's look at them.

Degeneracy and Normalization

Since the energy depends only on n, when more than one value of l is possible, a degeneracy of $(2l + 1)$ for each permitted value of l will result. For example:

$$n = 1, \quad l = 0, \qquad g = 1 \qquad \text{(not degenerate)}$$

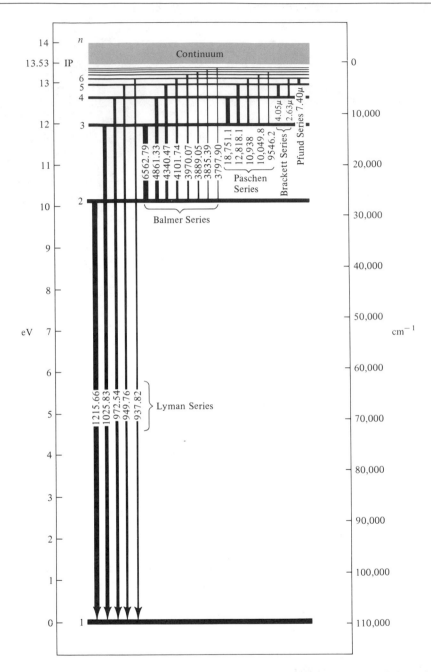

Figure 12.2 Energy levels of the hydrogen atom. Energy is given in eV (left scale with the *n* = 0 state taken as zero) and cm⁻¹ (right scale, with the ionization limit taken as zero). The vertical lines are the transitions observed in the spectrum; the numbers give wavelengths (in Å unless otherwise specified). The spectrum is shown by Figure 11.3.

$$n = 2, \quad l = 0, 1, \qquad g = 1 + 3 \qquad \text{(4 states have the same energy)}$$

$$n = 3, \quad l = 0, 1, 2, \quad g = 1 + 3 + 5 \quad \text{(9 states have the same energy)}$$

and so on.

If it is desired to have normalized wave functions, the requirement is:

$$\int_0^{2\pi} \int_0^{\pi} \int_0^{\infty} \psi_{nlm}^* \psi_{nlm} r^2 \, dr \, \sin\theta \, d\theta \, d\phi = 1$$

It is usual to normalize the radial and angular parts separately. The spherical harmonics as given in Table 11.8 are normalized for integration over θ and ϕ. The radial functions in Table 12.1 are not normalized; to make them so, we must multiply them by a constant so that:

$$\int_0^{\infty} R_{nl}^2 r^2 \, dr = 1 \tag{12.8}$$

Example: Find the proper constant to normalize the hydrogen atom ($Z = 1$) $n = 1$ wave function. From Table 12.1, $R(r) = Ae^{-\sigma}$ with $\sigma = r/a_0$:

$$\int_0^{\infty} R^2 r^2 \, dr = a_0^3 A^2 \int_0^{\infty} e^{-2\sigma} \sigma^2 \, d\sigma = 1$$

From a table of integrals we get:

$$\int_0^{\infty} x^n e^{-ax} \, dx = \frac{n!}{a^{n+1}} \tag{12.9}$$

Using this, the normalization condition becomes:

$$1 = \frac{a_0^3 A^2 2}{2^3} = \frac{a_0^3 A^2}{4}$$

The constant:

$$A = 2a_0^{-3/2}$$

will normalize R. To get the total wave functions we need the properly normalized spherical harmonic from Table 11.8: $Y_{00} = 1/\sqrt{4\pi}$. Therefore:

$$\psi_{100} = R_{10} Y_{00} = \frac{2a_0^{3/2}}{\sqrt{4\pi}} e^{-r/a_0} \qquad \blacksquare$$

The Wave Functions

Now let's look at several of the wave functions individually. (Constants will be omitted, since they do not affect the fundamental nature of the functions.) These functions are pictured in Figure 12.3.

The 1s function: $n = 1, l = 0, m = 0$:

$$(1s) \equiv \psi_{100} = e^{-\sigma}$$

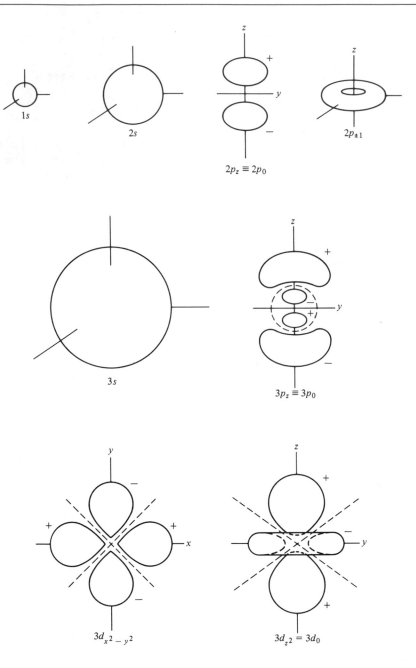

Figure 12.3 Shapes of some of the hydrogen-atom wave functions. The positive and negative signs give the sign of the wave function in that region. Nodes are shown by dashed lines.

This function is radially symmetric; note that $\psi \neq 0$ when $r = 0$; there is a finite probability that the electron will be found at the site of the nucleus.

The 2s function: $n = 2$, $l = 0$, $m = 0$:

$$(2s) \equiv \psi_{200} = \left(1 - \frac{\sigma}{2}\right)e^{-\sigma/2}$$

This is again a radially symmetric function. It has a node at $r = 2a_0$.

The 2p functions: $n = 2$, $l = 1$, $m = 0, \pm 1$. When $m = 0$:

$$(2p_0) \equiv \psi_{210} = \sigma e^{-\sigma/2} \cos \theta$$

This is also called the $2p_z$ function, since $z = r \cos \theta$ and, therefore:

$$(2p_0) = (2p_z) = \frac{z}{a_0} e^{-\sigma/2}$$

This function is not radially symmetric but has a maximum (for a given r) when $\theta = 0°$ and is zero in the xy plane ($\theta = 90°$). Note that the sign of the function is positive for $0 < \theta < 90°$ but negative for $90° < \theta < 180°$.

When $m = \pm 1$:

$$(2p_1) \equiv \psi_{211} = \sigma e^{-\sigma/2} \sin \theta \, e^{i\phi}$$

$$(2p_{-1}) \equiv \psi_{21-1} = \sigma e^{-\sigma/2} \sin \theta \, e^{-i\phi}$$

The spatial shapes of these functions are the same, since:

$$|\psi|^2 = \sigma^2 e^{-\sigma} \sin^2 \theta$$

in each case. The angular part, $\sin^2 \theta$, is a toroid of revolution about the z axis. The density is zero on the z axis. Operation upon the $2p$ functions with:

$$\hat{L}_z = \frac{\hbar}{i} \frac{\partial}{\partial \phi}$$

will demonstrate that the functions p_1 and p_{-1} have z projections of angular momentum of $\hbar$ and $-\hbar$, respectively. This would be the case if the electron in the p_{-1} state were circulating clockwise about the z axis; the electron in the p_1 state is presumably circulating counterclockwise.

In the absence of the magnetic field, the functions $2p_1$ and $2p_{-1}$ are degenerate (as are $2p_0$ and $2s$, for that matter), so we could equally well use any linear combination of them as eigenfunctions. Two combinations that are often used are called:

$$(2p_x) = \frac{1}{2}[(2p_1) + (2p_{-1})] = \sigma e^{-\sigma/2} \sin \theta \cos \phi$$

$$(2p_y) = \frac{1}{2i}[(2p_1) - (2p_{-1})] = \sigma e^{-\sigma/2} \sin \theta \sin \phi$$

(Cf. Euler's formula, Appendix IV.) Since $x = r \sin \theta \cos \phi$ and $y = r \sin \theta \sin \phi$, these can also be written:

$$(2p_x) = \frac{x}{a_0} e^{-\sigma/2}$$

$$(2p_y) = \frac{y}{a_0} e^{-\sigma/2}$$

The 3d functions: When $n = 3$, there are three sets of functions, $3s\,(l = 0)$, $3p\,(l = 1)$, and $3d\,(l = 2)$. For $n = 3$, $l = 2$, $m = 0$:

$$(3d_0) = \psi_{320} = \sigma^2 e^{-\sigma/3}(3 \cos^2 \theta - 1) = \frac{3z^2 - r^2}{a_0^2} e^{-\sigma/3}$$

This function is also called d_{z^2}. The functions for $m = \pm 1$, $3d_1$ and $3d_{-1}$, may also be written as linear combinations called d_{xz} and d_{yz}. The functions for $m = \pm 2$, $3d_2$ and $3d_{-2}$, can be combined to form d_{xy} and $d_{x^2-y^2}$.

Exercise: Use:

$$\hat{L}_z = \frac{\hbar}{i} \frac{\partial}{\partial \phi}$$

to prove:

$$\hat{L}_z(2p_1) = +\hbar(2p_1)$$

$$\hat{L}_z(2p_{-1}) = -\hbar(2p_{-1})$$

$$\hat{L}_z(2p_x) = \frac{\hbar}{i} (2p_y)$$

$$\hat{L}_z(2p_z) = 0$$

Note that $(2p_x)$ and $(2p_y)$ are eigenfunctions of $\hat{L}^2$ but not of $\hat{L}_z$. ■

Electron Density Distribution Functions

If the electron is not on an orbit, where is it? The distribution function:

$$\psi^* \psi \, d\tau$$

tells us the relative probability that an electron will be found in a volume element:

$$d\tau = dx\,dy\,dz = r^2\,dr \, \sin \theta\,d\theta\,d\phi$$

that is located at coordinates (x, y, z) [(r, θ, ϕ) if expressed in polar coordinates].

The polar and radial distributions are usually discussed separately. In spherical polar coordinates:

$$\psi^* \psi \, d\tau = R^2 r^2 \, dr \, |Y(\theta, \phi)|^2 \sin \theta\,d\theta\,d\phi$$

is the probability of finding the electron between r and $r + dr$ in a solid angle

$d\Omega = \sin\theta\,d\theta\,d\phi$. If we sum this formula over all angles — that is, integrate $0 < \theta < \pi$ and $0 < \phi < 2\pi$ — we get the radial distribution function (Figure 12.4):

$$f(r)\,dr = R^2 r^2\,dr \tag{12.10}$$

This is the probability of finding the electron in an *annulus* between spheres of radius r and $r + dr$, regardless of direction. The factor r^2 accounts for the fact that the volume of an annulus of a given width dr will increase as r^2. Thus, for an s-type function, $R(0)$ is not zero but $f(0)$ is equal to zero because the size of the annular volume becomes zero when $r = 0$; this does not mean that an s electron cannot be at the nucleus, but only that that part of its distribution function makes a negligible contribution to any average value.

Unless normalized, $f(r)$ represents only a relative probability, and the probability of finding $r_1 < r < r_2$ must be written as:

$$\frac{\displaystyle\int_{r_1}^{r_2} f(r)\,dr}{\displaystyle\int_{0}^{\infty} f(r)\,dr} \tag{12.11}$$

Example: Calculate $\langle r \rangle$ for the ground state of atomic hydrogen.

$$\langle r \rangle = \frac{\displaystyle\int_{0}^{\infty} r f(r)\,dr}{\displaystyle\int_{0}^{\infty} f(r)\,dr} = \frac{\displaystyle\int_{0}^{\infty} r^3 e^{-2r/a_0}\,dr}{\displaystyle\int_{0}^{\infty} r^2 e^{-2r/a_0}\,dr}$$

From Eq. (12.9) we get:

$$\langle r \rangle = \frac{3a_0}{2} = 0.79375 \times 10^{-8}\ \text{cm} \qquad \blacksquare$$

Exercise: Prove that $f(r)$ for the hydrogen $1s$ state has a maximum at the Bohr radius, $r = a_0$. $\qquad \blacksquare$

12.2 Electron Spin

Certain inexplicable features of atomic spectra caused Uhlenbeck and Goudsmit (1925) to postulate that the electron had an intrinsic angular momentum, or "spin"; Dirac later showed that this fact was a consequence of relativistic effects.

If any distribution of mass is spinning about an axis, it will certainly have an angular momentum. In such a case, however, the boundary-value conditions on the eigenvalue equation (Chapter 11) will require the quantum numbers to be integers. In fact, the electron has a fixed "l-type" quantum number (called s) of $\frac{1}{2}$. Evidently the term "spin," like the term "orbital," as they apply to angular momentum, is another metaphor. We shall take the existence of the electron's intrinsic angular momentum with quantum number $s = \frac{1}{2}$ as an additional postulate — one that would be unnecessary in a rigorous treatment.

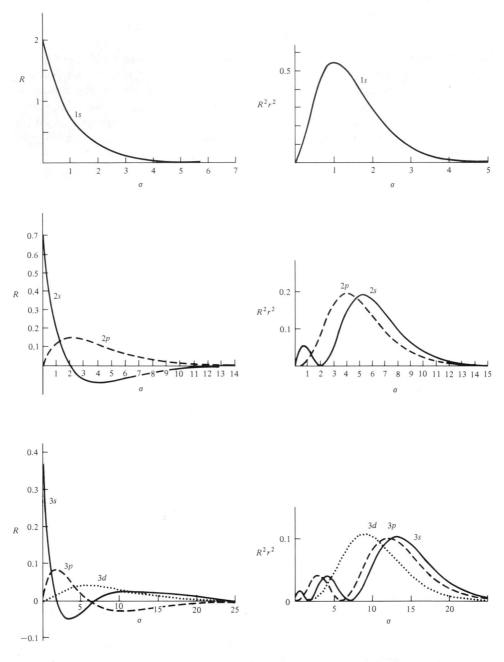

Figure 12.4 Hydrogen-atom radial wave functions. The radial function (R) and the radial distribution function (R^2r^2) are shown for several hydrogen-atom states.

Operator Equations for Spin

The general theory of angular momentum, as developed in Chapter 11, provides the following requirements for a "spin" angular momentum **S** with quantum number $s = \frac{1}{2}$:

1. There will be two eigenfunctions with projection quantum numbers $m_s = +\frac{1}{2}$ and $-\frac{1}{2}$, respectively; these functions are called α and β. (α and β are eigenfunctions, not operators.)

2. The length of **S** is $\sqrt{s(s+1)}\,\hbar = \sqrt{\frac{3}{4}}\,\hbar$, since $s = \frac{1}{2}$. The eigenvalue equation for $\hat{S}^2$ is:

$$\hat{S}^2\alpha = s(s+1)\hbar^2\alpha = \tfrac{3}{4}\hbar^2\alpha \tag{12.12}$$

   and the same for the β function.

3. The eigenvalue equations for $\hat{S}_z$ are:

$$\hat{S}_z\alpha = \tfrac{1}{2}\hbar\alpha, \qquad \hat{S}_z\beta = -\tfrac{1}{2}\hbar\beta \tag{12.13}$$

4. The functions α and β are connected by the raising and lowering operators (as in Chapter 11); it can be shown that:

$$\hat{S}_+\alpha = 0, \qquad \hat{S}_-\alpha = \hbar\beta$$
$$\hat{S}_+\beta = \hbar\alpha, \qquad \hat{S}_-\beta = 0 \tag{12.14}$$

   From Eqs. (12.14) it can be shown (see the example below) that the operators for the x and y components of the angular momentum have the effect:

$$\hat{S}_x\alpha = \frac{\hbar}{2}\beta, \qquad \hat{S}_x\beta = \frac{\hbar}{2}\alpha$$

$$\hat{S}_y\alpha = \frac{i\hbar}{2}\beta, \qquad \hat{S}_y\beta = -\frac{i\hbar}{2}\alpha \tag{12.15}$$

5. Finally, these functions are orthogonal and normalized:

$$\int \alpha\alpha\,d\tau_e = 1, \qquad \int \beta\beta\,d\tau_e = 1, \qquad \int \alpha\beta\,d\tau_e = 0 \tag{12.16}$$

These integrals are over the internal electron coordinates — that is, the orientation of the electron spin axis in space (θ_e, ϕ_e). Actually it is never necessary to know the actual functional forms of $\alpha(\theta_e\phi_e)$ or $\beta(\theta_e\phi_e)$ to do any integral involving these functions; Eqs. (12.2) to (12.16) tell us all we shall ever need to know about spin angular momentum.

Example: Find the result $\hat{S}_x\alpha$ from Eq. (12.14). [Raising and lowering operators are defined by Eq. (11.81).] Since $\hat{S}_+ = \hat{S}_x + i\hat{S}_y$ and $\hat{S}_- = \hat{S}_x - i\hat{S}_y$, then $\hat{S}_x = (\hat{S}_+ + \hat{S}_-)/2$:

$$\hat{S}_x\alpha = \tfrac{1}{2}(\hat{S}_+\alpha + \hat{S}_-\alpha) = \tfrac{1}{2}(0 + \hbar\beta) = \frac{\hbar}{2}\beta \qquad \blacksquare$$

Exercise: Find the result $\hat{S}_y\beta$ using Eq. (12.14). $\qquad\qquad\qquad\blacksquare$

Spin Eigenfunctions

A complete wave function for a one-electron problem (such as H, He$^+$) must include both a spatial part, $\psi_{nlm}(r\theta\phi)$, and a spin part, $\chi = \alpha$ or β:

$$\text{wave function} = (\text{spatial part})(\text{spin part})$$

$$\Psi = \psi_{nlm}\chi_{m_s} \tag{12.17}$$

with:

$$\chi_{1/2} = \alpha, \qquad \chi_{-1/2} = \beta$$

As a result, the $1s$ ground state is doubly degenerate, with eigenfunctions that we shall call $(1s\alpha)$ and $(1s\beta)$. Therefore, the ground state of the hydrogen atom does have angular momentum — not what Bohr's theory said ($\hbar$), but $\sqrt{\frac{3}{4}}\hbar$.

12.3   The Helium Atom

In any two-electron system, H$^-$ ($Z = 1$), He ($Z = 2$), Li$^+$ ($Z = 3$), and so on, with the position of the nucleus fixed, there are six spatial coordinates, three for each electron. These are:

$$\text{(Cartesian)} \qquad x_1, y_1, z_1, x_2, y_2, z_2$$

$$\text{(polar)} \qquad r_1, \theta_1, \phi_1, r_2, \theta_2, \phi_2$$

Each electron has its own operators — for example:

$$\hat{p}_{x1} = -i\hbar\frac{\partial}{\partial x_1}, \qquad \hat{p}_{x2} = -i\hbar\frac{\partial}{\partial x_2}$$

$$\hat{L}_{z1} = -i\hbar\frac{\partial}{\partial \phi_1}, \qquad \hat{L}_{z2} = -i\hbar\frac{\partial}{\partial \phi_2}$$

which operate only on their own coordinates. That is:

$\dfrac{\partial}{\partial x_1}$ is a derivative holding y_1, z_1, x_2, y_2, z_2 constant

$\dfrac{\partial}{\partial \phi_1}$ is a derivative holding $\theta_1, r_1, r_2, \theta_2, \phi_2$ constant

Example: Find the result of operating with $\hat{L}_{z1}$ on the function $F(\phi_1\phi_2) = e^{ia\phi_1}e^{ib\phi_2}$.

$$\hat{L}_{z1}F(\phi_1\phi_2) = -i\hbar\frac{\partial}{\partial \phi_1}e^{ia\phi_1}e^{ib\phi_2} = a\hbar F(\phi_1\phi_2) \qquad ■$$

Exercise: Prove $\hat{L}_{z2}F(\phi_1\phi_2) = b\hbar F(\phi_1\phi_2)$.   ■

The Hamiltonian

The Hamiltonian for a two-electron atom is (see Figure 12.5 for definitions of the coordinates):

$$\hat{H} = -\frac{\hbar^2}{2m}\nabla_1^2 - \frac{\hbar^2}{2m}\nabla_2^2 - \frac{Ze^2}{r_1} - \frac{Ze^2}{r_2} + \frac{e^2}{r_{12}} + \left(\begin{matrix}\text{spin}\\\text{orbit}\end{matrix}\right) \qquad \textbf{(12.18)}$$

These terms represent, in order: (1) kinetic energy of electron 1, (2) kinetic energy of electron 2, (3) Coulombic attraction of nucleus to electron 1, (4) Coulombic attraction of nucleus to electron 2, (5) the electron-electron (ee) Coulombic repulsion, (6) the spin-orbit coupling. Except for the last term, all of these follow from our earlier discussion. The spin-orbit coupling term is a result of a magnetic interaction between the spin and orbital motions of the electron; it is quite small for light atoms and will be ignored for now.

The major difficulty in finding eigenfunctions of this Hamiltonian is term 5, the ee repulsion. Since:

$$r_{12} = |\mathbf{r}_1 - \mathbf{r}_2|$$

this term is very complicated mathematically. Because of this interaction we shall not be able to find exact, closed-form solutions for two-electron atoms, let alone many-electron atoms, so everything from this point onward will be approximate.

A cynic may remember that Bohr's theory was criticized because it worked only for hydrogen; now we seem to be saying that quantum theory works only for a one-electron atom while all others are only approximations. This is not quite fair; while we can never get simple formulas for the wave functions or energies of many-electron atoms, numerical solutions of high accuracy have been calculated for many of the lighter atoms, and good approximations can be calculated for the heavier ones. This difficulty is not unique to quantum mechanics; the classical three-body problem (for example, the orbit of a spaceship traveling between the earth and moon) is similarly difficult, while the classical many-body problem (for example, the entire solar system) is equally "unsolvable". The fact that Voyager II traveled 1.24×10^9 miles to Saturn with an error of only 41 miles demonstrates that the classical many-body problem can be solved as accurately as needed, using numerical methods and computers. The same

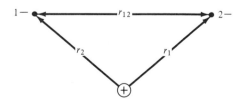

Figure 12.5 Coordinates used for the helium atom. The + represents the position of the nucleus, and 1- and 2- the positions of the two electrons.

can be said for atoms; in fact, many of the methods used — for example, perturbation theory — are the same for the quantum and classical many-body problems.

Zero-order Solution for Two Electrons

We define a one-electron operator for electron i:

$$\hat{h}_0(i) = -\frac{\hbar^2}{2m}\nabla_i^2 - \frac{Ze^2}{r_i} \tag{12.19}$$

and write the two-electron Hamiltonian [Eq. (12.18)] as:

$$\hat{H} = \hat{h}_0(1) + \hat{h}_0(2) + \frac{e^2}{r_{12}}$$

If the last term is (temporarily) neglected, the *zeroth-order* Hamiltonian:

$$\hat{H}_0 = \hat{h}_0(1) + \hat{h}_0(2)$$

is a sum of two independent terms. Once again we have a situation where an operator is a sum of two terms, and once again we may expect the eigenfunctions to be a product. The eigenfunctions of $\hat{H}_0(\Psi_0)$ will therefore be products of one-electron functions:

$$\Psi_0 = \psi(1)\psi(2)$$

$$\hat{H}_0\Psi_0 = E^{(0)}\Psi_0$$

($E^{(0)}$ is the zeroth-order approximate energy.) The one-electron functions are called *orbitals*. The energy is calculated as follows:

$$\hat{H}_0\Psi_0 = [\hat{h}_0(1) + \hat{h}_0(2)]\psi(1)\psi(2)$$
$$= [\hat{h}_0(1)\psi(1)]\psi(2) + [\hat{h}_0(2)\psi(2)]\psi(1)$$

The one-electron eigenvalue problem — for example:

$$\hat{h}_0(1)\psi(1) = E\psi(1)$$

is just that which was solved earlier — see Eqs. (12.6) and (12.7). We conclude that the zeroth-order energy is just the sum of terms like Eq. (12.7), one for each electron:

$$E^{(0)} = -Z^2 E_H\left(\frac{1}{n_1^2} + \frac{1}{n_2^2}\right) \tag{12.20}$$

where n_1 and n_2 are the principal quantum numbers of electron 1 and 2, respectively, and:

$$E_H \equiv \frac{e^2}{2a_0} \tag{12.21}$$

Note that E_H is the ionization energy of the H atom (13.6 eV or 109,679 cm^{-1}). Also:

$$\Psi_0 = \psi_{n_1 l_1 m_1}\psi_{n_2 l_2 m_2} \tag{12.22}$$

is a product of two one-electron functions as defined by Eq. (12.6). Thus, we can approximate two-electron wave functions as products of the one-electron functions. The lowest energy, the configuration $(1s)^2$ of He ($Z = 2$), has a zeroth-order state function:

$$\Psi_0(1s^2) = e^{-\sigma_1}e^{-\sigma_2} = e^{-2r_1/a_0}e^{-2r_2/a_0} \tag{12.23}$$

and an energy given by Eq. (12.20) with $n_1 = 1$ and $n_2 = 1$:

$$E^{(0)}(1s^2) = -8E_H \tag{12.24}$$

Example: A good way to test the accuracy of these formulas is to calculate the first IP for He — that is, the energy for He $\longrightarrow$ He$^+$ + e. $E(\text{He}^+)$ can be calculated from Eq. 12.7 with $Z = 2$ (since this species has only one electron); the energy of He$^+$ in the $n = 1$ state is easily shown to be $-4E_H$. Then, the ionization potential is:

$$\text{IP} = E(\text{He}^+) - E(\text{He}) \cong -4E_H + 8E_H = 4E_H$$

Using $E_H = 13.6$ eV, we get:

$$\text{IP} = 4(13.6) = 54.4 \text{ eV}$$

The experimental value is 24.6 eV! ∎

The preceding has described an independent electron model. We see that this model gives a 100% error in calculating the IP of helium. Obviously some improvements are needed.

Perturbation Theory

One improvement would be to restore the neglected term of the Hamiltonian, the ee repulsion, to the calculations. In that case, Ψ_0 is no longer an eigenfunction of the Hamiltonian; but it could be used with the average-value theorem to estimate the *first-order energy:*

$$E^{(1)} = \langle \hat{H} \rangle = E^{(0)} + \left\langle \frac{e^2}{r_{12}} \right\rangle$$

For the $1s^2$ configuration with Ψ_0 as in Eq. (12.23):

$$\left\langle \frac{e^2}{r_{12}} \right\rangle = \frac{\displaystyle\iint \frac{e^2}{r_{12}}(e^{-2r_1/a_0}e^{-2r_2/a_0})^2 \, d\tau_1 \, d\tau_2}{\displaystyle\iint (e^{-2r_1/a_0}e^{-2r_2/a_0})^2 \, d\tau_1 \, d\tau_2}$$

These integrals can be evaluated with modest difficulty (see, for example, ref. 1) to give:

$$\left\langle \frac{e^2}{r_{12}} \right\rangle = 2.5E_H$$

The first-order approximate energy is, therefore:

$$E^{(1)} = -8E_H + 2.5E_H = -5.5E_H \tag{12.25}$$

This formula gives for the IP of He:

$$IP = E(\mathrm{He}^+) - E(\mathrm{He}) = -4E_H + 5.5E_H$$

$$= 1.5(13.6) = 20.4 \text{ eV}$$

Equation (12.25) is still in error by 20% (or 4.2 eV); this is because Ψ_0 is not the correct wave function. This is no small error; 4.2 eV per molecule is equal to 404 kJ/mol, a number that is very large by chemical energy standards.

Variation Theory

More accurate energies can be obtained using the *variation theorem*. In this method a trial function (χ) is used for the ground state and an approximate energy is calculated using the average-value theorem and the true Hamiltonian [for example, Eq. (12.18)]:

$$E \cong E_{var} = \langle \hat{H} \rangle$$

$$E_{var} = \frac{\displaystyle\iint \chi^*(\hat{H}\chi)\, d\tau_1\, d\tau_2}{\displaystyle\iint \chi^*\chi\, d\tau_1\, d\tau_2} \tag{12.26}$$

A common method for two-electron atoms is to modify Ψ_0 [Eq. (12.22)] by using an effective nuclear charge Z' to account for the screening effect of one electron on the other, and use the trial function [for the configuration $(1s)^2$]:

$$\chi = e^{-Z'r_1/a_0} e^{-Z'r_2/a_0} \tag{12.27}$$

The *variation theorem* states that the approximate energy calculated by Eq. (12.26) is always greater than the lowest eigenvalue of the Hamiltonian. So long as we seek the ground-state energy, the best value of Z' can be determined by minimizing E_{var}. [That is, find the value of Z' for which the integral, Eq. (12.26), is a minimum.] The evaluation of Eq. (12.26) with the trial function, Eq. (12.27), is lengthy but straightforward (ref. 1). The result is:

$$Z' = Z - \tfrac{5}{16} \tag{12.28}$$

$$E_{var} = -2(Z')^2 E_H \tag{12.29}$$

For helium, of course, $Z = 2$ and $Z' = \tfrac{27}{16}$, but Eqs. (12.28) and (12.29) are the same for any two-electron atom or ion, for example, H^- or Li^+.

Example: Estimate the first IP of He using the variation-theory results.

$$Z' = 2 - \tfrac{5}{16} = \tfrac{27}{16}$$

$$E_{var} = -2(\tfrac{27}{16})^2 E_H = -5.6953 E_H \qquad \text{for He}$$

Table 12.2 Ionization potential of helium

Model	Value (eV)
Independent electron	54.4
First-order perturbation	20.4
Variation theory, effective Z	23.1
Taylor and Parr (configuration interaction)	24.4
Observed	24.6

For He^+, $E = -4E_H$, so:

$$IP = -4E_H + 5.6953E_H = 23.1 \text{ eV}$$

Table 12.2 summarizes the results for calculating the He IP by various methods. ■

Approximate Wave Functions and the Aufbau Principle

The trick to variation theory is in choosing realistic trial functions. In Chapter 11 we saw that commuting operators can have simultaneous eigenfunctions; it is a useful exercise to look for operators that commute with the Hamiltonian and to use trial functions that are eigenfunctions of those operators. The most important operators of this type are those representing angular momentum and symmetry. Most of our effort from now on will be to seek such operators and to classify the states of atoms and molecules according to their eigenvalues. Even when energy calculations are impractical or impossible, this classification effort will pay off by permitting us to interpret the atomic spectra by which the energies are *measured*.

The picture of the He atom as two electrons, each in a "$1s$"-type orbital, is the beginning of a useful scheme for the building up (German: Aufbau) of the entire periodic table using the hydrogen wave functions as "orbitals." We have seen that, for He, this independent electron model—the picture of electrons as individuals occupying distinct "orbitals"—is not very accurate. Computational methods that are beyond the scope of this book can produce very accurate energy calculations; but the more accurate the calculation, the less clear the "picture"; this is a common difficulty when dealing with atoms and molecules. The independent electron model is very useful; it is widely used, and we shall use it, but its limitations should be appreciated.

12.4 The Pauli Exclusion Principle

In the preceding section we discussed only the spatial wave functions for the He atom $(1s)^2$ configuration. If the spin of the electron is included, as in eq. (12.17), there will be four possible product functions for the two electrons:

$$\Psi_a = \psi_{1s}(1)\alpha(1)\psi_{1s}(2)\alpha(2) \equiv 1s(1)1s(2)\alpha(1)\alpha(2)$$

$$\Psi_b = \psi_{1s}(1)\alpha(1)\psi_{1s}(2)\beta(2) \equiv 1s(1)1s(2)\alpha(1)\beta(2)$$

$$\Psi_c = \psi_{1s}(1)\beta(1)\psi_{1s}(2)\alpha(2) \equiv 1s(1)1s(2)\beta(1)\alpha(2)$$

$$\Psi_d = \psi_{1s}(1)\beta(1)\psi_{1s}(2)\beta(2) \equiv 1s(1)1s(2)\beta(1)\beta(2)$$

As before, α and β are the spin $\frac{1}{2}$ eigenfunctions [Eq. (12.13)], and $1s(1)$ or $1s(2)$ denotes the spatial function [cf. Eq. (12.23)]:

$$1s(1)1s(2) = e^{-Zr_1/a_0}e^{-Zr_2/a_0}$$

It appears from this that the He ground state is fourfold degenerate; in fact it is not degenerate — there is only one wave function for the lowest energy state of He. The way to avoid this embarrassment of functions is to use the *Pauli exclusion principle*. The simple form of this principle, "electrons in the same orbital must have opposite spins," can eliminate two of these functions (*a* and *d*), but we still have one too many. Let's look at the situation more closely.

Indistinguishability of Electrons

The electrons of an atom or molecule are indistinguishable. When we write operators [such as Eq. (12.18)] or wave functions [Eqs. (12.23) or (12.27)] we label the electrons as a matter of convenience, but no physical result (eigenvalues, average values, electron densities) may depend on this labeling. Even in an approximate wave function or operator we must require that the electrons be indistinguishable in the sense that, if the labels are exchanged, the physical results must be unchanged. The two-electron Hamiltonian [Eq. (12.18)] clearly meets this requirement — but what about the wave functions? Since physical results such as:

$$\text{electron density} = \Psi^*\Psi$$

$$\text{average values} = \int \Psi^*(\text{Operator})\Psi \, d\tau$$

and so on always involve two wave functions, it is possible for Ψ to be either *even* (symmetric) or *odd* (antisymmetric) with respect to exchange of electrons — that is, either:

$$\Psi \longrightarrow \Psi \qquad \text{when labels are exchanged}$$

or:

$$\Psi \longrightarrow -\Psi \qquad \text{when labels are exchanged}$$

The Pauli exclusion principle states: *wave functions of electrons must be antisymmetric with respect to exchange of indistinguishable electrons.* Since, in general:

$$\Psi = (\text{spatial part})(\text{spin part})$$

we shall also require that both spatial and spin parts be either symmetric or antisymmetric with respect to exchange. The Pauli principle requires that the total function be antisymmetric (odd); this can be satisfied in two ways:

$$\Psi_{\text{odd}} = (\text{odd spatial})(\text{even spin})$$

Wolfgang Pauli (1900–1958)

   Shortly after receiving his doctorate in 1922, Pauli became interested in the yet-unexplained anomalous Zeeman effect in atomic spectra. He was the first to realize that the spectra of atoms were due entirely to the outer electrons and that the inner core had no angular momentum. He explained the core structure by generalizing Bohr's aufbauprinzip with his famous exclusion principle: "There can never be two equivalent electrons in an atom for which, in a strong field, the values of all the quantum numbers are the same." He had to specify a strong magnetic field, for in such cases the electrons are uncoupled from each other (the Paschen-Back effect) and the quantum numbers of individual electrons become "good." This principle alone, without any computations or further speculation, can explain the periodicity of the elements, which had been known since Mendeleev, and gave rise to the modern form of the periodic table.

   According to George Gamow, Pauli was, like many theoreticians, clumsy with experimental equipment and among experimentalists was known for the "Pauli effect": expensive and sophisticated equipment would break down if he even walked into the laboratory. On one occasion, the explosion of a piece of equipment in Göttingen was said to have taken place just as a train carrying Pauli from Zurich to Copenhagen made a brief stop at the railway station in that town.

or

$$\Psi_{\text{odd}} = (\text{even spatial})(\text{odd spin})$$

   Back to the He atom. The spatial wave function for a $(1s)^2$ configuration [Eqs. (12.23) and (12.27)] is necessarily even — that is, there is no change when the electron coordinates r_1 and r_2 are exchanged. Therefore, if the Pauli principle is to be obeyed, the spin part must be odd. The spin functions $\alpha(1)\alpha(2)$ and $\beta(1)\beta(2)$ are clearly even — this excludes functions a and d, as anticipated. However, the functions b and c are neither even nor odd; in fact:

$$\Psi_c \xleftrightarrow{\text{exchange}} \Psi_b$$

$$\alpha(1)\beta(2) \xleftrightarrow{\text{exchange}} \alpha(2)\beta(1)$$

We can "symmetrize" these functions by taking linear combinations:

$$\Psi_{c'} = \Psi_c + \Psi_b = 1s(1)1s(2)[\alpha(1)\beta(2) + \beta(1)\alpha(2)]$$

$$\Psi_{b'} = \Psi_c - \Psi_b = 1s(1)1s(2)[\alpha(1)\beta(2) - \beta(1)\alpha(2)]$$

Only the function b' is antisymmetric when the electron labels are exchanged; hence this is the proper wave function for the ground $(1s)^2$ configuration of the He atom.

To simplify the notation, product functions are usually written without the electron labels — for example:

$$\Psi_{b'} = (1s)(1s)(\alpha\beta - \beta\alpha)$$

with the convention that the first function of a product is for electron 1, the second is for electron 2, and so on. [That $(\alpha\beta - \beta\alpha)$ is not equal to zero has nothing to do with commutators; these are functions, not operators. It is just a convention that $\alpha\beta$ *means* α(electron 1)β(electron 2) while $\beta\alpha$ *means* β(electron 1)α(electron 2).]

All of the results of the Section 12.3 were achieved without worrying about electron spins — why bring them up now? In the first place, we have been ignoring the only part of the Hamiltonian that explicitly involves electron spin operators — the spin-orbit coupling term; this term is small for helium but is much larger for heavier atoms. Also, the Pauli principle will be of critical importance when we consider excited states of He (next section) and when we start discussing many-electron atoms.

12.5 Excited States of Helium

As mentioned earlier (Section 11.2), the Ritz combination principle permits the analysis of atomic spectra in which the observed wavelengths are differences of term values (P):

$$\tilde{\nu} = \frac{1}{\lambda} = P_i - P_j$$

As Bohr anticipated, these terms are simply the energy levels of the atom (with units of cm^{-1}). Figure 12.6 shows the measured term diagram of He, as deduced directly from the He spectrum. The classification of the terms into groups called ^1S, ^1P, ^1D, ^1F (singlets) and ^3S, ^3P, ^3D, ^3F (triplets) is due to the spectroscopists, who used it to show that only certain terms could be combined to give an observed line of the spectrum — that is, not *all* possible term differences are observed. (In general only differences of adjacent columns, $S \leftrightarrow P \leftrightarrow D \leftrightarrow F$, are observed, and singlets do not combine with triplets.)

The nature of these terms can be deduced by extending the orbital model to include excited states by placing one electron into a higher orbital. [Terms corresponding to two excited electrons are observed but are usually very high in energy. For example, the He$(2s)^2$ term is above the ionization potential; that is, it is unstable with respect to ionization to He$^+(1s)$ + e.]

Singlet and Triplet States

For the $1s\,2s$ configuration of He, two spatial functions can be constructed from the products: $1s(1)2s(2)$ and $2s(1)1s(2)$ (abbreviated $1s\,2s$, $2s\,1s$, respectively).

$$1s\,2s \equiv \psi_{1s}(1)\psi_{2s}(2) = e^{-2r_1/a_0}\left(1 - \frac{r_2}{a_0}\right)e^{-r_2/a_0}$$

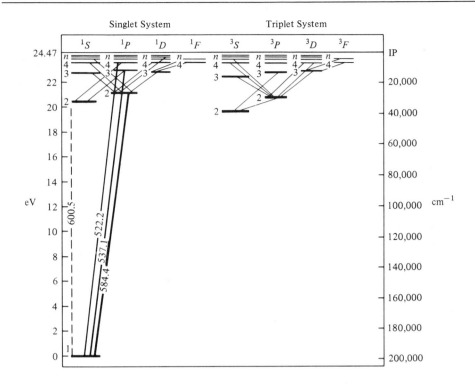

Figure 12.6 Term diagram for helium. Energy levels are shown as horizontal lines, with the principle quantum number (n) of the excited electron to the left. The vertical lines show some allowed transitions found in the helium spectrum; the numbers are the wavelengths in angstroms. Note that no singlet-to-triplet transitions are observed.

$$2s\ 1s \equiv \psi_{2s}(1)\psi_{1s}(2) = e^{-2r_2/a_0}\left(1 - \frac{r_1}{a_0}\right)e^{-r_1/a_0}$$

To account for the Pauli principle, these must be arranged into symmetric and anti-symmetric combinations:

$$\psi_s = (1s\ 2s) + (2s\ 1s)$$

$$\psi_a = (1s\ 2s) - (2s\ 1s) \tag{12.30}$$

The spin functions can likewise be formed into three symmetric functions:

$$\chi_s = \begin{cases} \alpha\alpha \\ \alpha\beta + \beta\alpha \\ \beta\beta \end{cases} \tag{12.31}$$

and one antisymmetric function:

$$\chi_a = \alpha\beta - \beta\alpha \tag{12.32}$$

The total wave function, obeying Pauli's principle, can be either:

$$\psi_s \chi_a \quad \text{or} \quad \psi_a \chi_s$$

Thus the (1s)(2s) configuration gives rise to four states—the state called ^1S ("singlet S"):

$$\Psi(^1S) = [(1s\,2s) + (2s\,1s)](\alpha\beta - \beta\alpha) \qquad (12.33)$$

and three called ^3S ("triplet S"):

$$\Psi(^3S) = [(1s\,2s) - (2s\,1s)] \begin{Bmatrix} \alpha\alpha \\ \alpha\beta + \beta\alpha \\ \beta\beta \end{Bmatrix} \qquad (12.34)$$

These are, of course, not exact wave functions but "zeroth-order" or "trial functions," as discussed earlier. Nonetheless, the symmetrization required by the Pauli principle produces a startling result—the four functions [Eqs. (12.33) and (12.34)] are not degenerate, but constitute two separate energy levels.

The zeroth-order energy (neglecting the ee repulsion) for both the ^1S and ^3S functions is the same; it is just the energy of Eq. (12.20) for quantum numbers $n = 1$ and 2:

$$E^{(0)} = -4E_H \left(\frac{1}{1} + \frac{1}{4} \right) = -5E_H$$

To get the first-order energy we must evaluate the e^2/r_{12} term. For the ^1S function:

$$\left\langle \frac{e^2}{r_{12}} \right\rangle = \iint \frac{e^2}{r_{12}} [(1s)(2s) + (2s)(1s)]^2 \, d\tau_1 \, d\tau_2$$

$$= \iint \frac{e^2}{r_{12}} \psi_{1s}(1)^2 \psi_{2s}(2)^2 \, d\tau_1 \, d\tau_2 + \iint \frac{e^2}{r_{12}} \psi_{1s}(2)^2 \psi_{2s}(1)^2 \, d\tau_1 \, d\tau_2$$

$$+ 2 \iint \frac{e^2}{r_{12}} \psi_{1s}(1) \psi_{2s}(1) \psi_{1s}(2) \psi_{2s}(2) \, d\tau_1 \, d\tau_2$$

The first two terms are called the "Coulomb integral" (symbol J); the evaluation of these integrals gives:

$$J = 0.839 E_H$$

The last term is called the "exchange integral" (symbol K), and the calculation yields:

$$K = 0.088 E_H$$

The first-order energy calculation gives:

$$E^{(1)}(^1S) = E^{(0)} + J + K = -4.073 E_H$$

The experimental value is $-4.293 E_H$, but the accuracy, or lack thereof, is not the most important result. If we do the same calculation for the ^3S state:

$$\left\langle \frac{e^2}{r_{12}} \right\rangle = \iint \frac{e^2}{r_{12}} [(1s)(2s) - (2s)(1s)]^2 \, d\tau_1 \, d\tau_2$$

the result is the same except for the *sign* of the cross term—that is, the exchange integral. Therefore:

$$E^{(1)}(^3S) = E^{(0)} + J - K = -4.249 E_{\mathrm{H}} \qquad (\text{observed}, -4.351 E_{\mathrm{H}})$$

The energy of the triplet is *lower* than the singlet by $2K = 0.176 E_{\mathrm{H}}$. (The experimental difference is 6360 cm^{-1}, less than half this value.)

The accuracy of the calculations above was not great, but it established that the energy difference of singlet and triplet states can be accounted for by the Pauli principle; this result followed directly from the form of the symmetrized wave functions of Eq. (12.30). This is an important point: merely specifying a configuration, in this case $(1s)(2s)$, does not tell you the energy states. In this example there are two distinct energy states—singlet and triplet—with distinctly different energies. The triplet state is usually the lower in energy in such cases.

Other Excited States

The next excited configuration is $(1s)(2p)$. There are three p functions, either (p_1, p_0, p_{-1}) or (p_x, p_y, p_z); it will not matter which set we use. There will therefore be three spatial functions for each symmetry:

$$\text{(symmetric spatial)} \qquad \psi_{s1} = (1s)(2p_1) + (2p_1)(1s)$$
$$\psi_{s2} = (1s)(2p_0) + (2p_0)(1s)$$
$$\psi_{s3} = (1s)(2p_{-1}) + (2p_{-1})(1s)$$
$$\text{(antisymmetric spatial)} \qquad \psi_{a1} = (1s)(2p_1) - (2p_1)(1s)$$
$$\psi_{a2} = (1s)(2p_0) - (2p_0)(1s)$$
$$\psi_{a3} = (1s)(2p_{-1}) - (2p_{-1})(1s)$$

As before, there is one antisymmetric spin function, $\chi_a = (\alpha\beta - \beta\alpha)$, and three symmetric spin functions [χ_s of Eq. (12.31)]. The products of spatial and spin functions that obey the Pauli principle fall into two groups: the three "singlet P" functions:

$$\Psi(^1P) = \psi_{s1}\chi_a, \quad \psi_{s2}\chi_a, \quad \psi_{s3}\chi_a \tag{12.35}$$

and the nine "triplet P" functions:

$$\Psi(^3P) = \psi_{a1}\chi_s, \quad \psi_{a2}\chi_s, \quad \psi_{a3}\chi_s \tag{12.36}$$

with:

$$\chi_s = (\alpha\alpha) \quad \text{or} \quad (\alpha\beta + \beta\alpha) \quad \text{or} \quad (\beta\beta)$$

The zeroth-order energy [Eq. (12.20)] of these states is the same as that of the S states, since the H-atom energy depends only on the principal quantum number (n) and not on l:

$$E^{(0)}(1s\,2p) = -5E_H$$

However, the Coulomb and exchange integrals used to evaluate $\langle e^2/r_{12}\rangle$ are different for the p functions than they were for the s functions, so:

$$E^{(1)}(^1P) = -5E_H + J_{2p} + K_{2p}$$
$$E^{(1)}(^3P) = -5E_H + J_{2p} - K_{2p}$$

In the helium atom the P states are higher than the corresponding S states, owing to differences in the Coulomb integrals. From the term diagram (Figure 12.6) the differences can be calculated as:

$$E(2^1P) - E(2^1S) = 4859 \text{ cm}^{-1}$$
$$E(2^3P) - E(2^3S) = 9235 \text{ cm}^{-1}$$

A misconception that students often hold (presumably due to a misunderstanding obtained in some previous course) is that s and p electrons with the same n (such as $2s$, $2p$ or $3s$, $3p$) have different energies. A brief glance at Eq. (12.7) will show that this is false for one-electron systems — only the principal quantum number (n) is in the formula, not l. However, the $(1s)(2s)$ and $(1s)(2p)$ configurations of He do differ in energy, because the electron-electron repulsion energy is different — that is, the Coulomb and exchange integrals (J and K).

Other excited states of He result from other orbitals. Examining Figure 12.6: the states under the ^1S column result from the configurations $(1s)(2s)$, $(1s)(3s)$, $(1s)(4s)$, and so on; the states under ^3S result from the same configurations. The ^1P and ^3P states are from configurations $(1s)(2p)$, $(1s)(3p)$, $(1s)(4p)$, and so on. The ^1D and ^3D states have configurations $(1s)(3d)$, $(1s)(4d)$, ..., while the ^1F and ^3F states have configurations $(1s)(4f)$, $(1s)(5f)$, ...

But what do these names mean? To answer that, and to be able to handle more complicated cases, we must first discuss the addition of angular momentum.

Example: How many states are there for the $(1s)(3d)\ ^3D$ state of He? The spin parts of a triplet function [Eqs. (12.31)] are symmetric, so the spatial parts must be antisymmetric; they are:

$$\psi_{a1} = (1s)(3d_2) - (3d_2)(1s)$$
$$\psi_{a2} = (1s)(3d_1) - (3d_1)(1s)$$
$$\psi_{a3} = (1s)(3d_0) - (3d_0)(1s)$$
$$\psi_{a4} = (1s)(3d_{-1}) - (3d_{-1})(1s)$$
$$\psi_{a5} = (1s)(3d_{-2}) - (3d_{-2})(1s)$$

Each of these is multiplied by each of the three spin functions — $\alpha\alpha$, $(\alpha\beta + \beta\alpha)$ and $\beta\beta$ — to give 15 functions in all. ∎

12.6 The Vector Model of the Atom

The calculation of the energy levels for many-electron atoms is a rather complicated business — already for the two-electron cases we have encountered integrals that we preferred to avoid (although they *can* be solved). Of course, these energies are readily measured from the atomic spectrum and its associated term diagram, but to understand these diagrams we must understand how atomic states are classified — and that is the subject of this section.

Earlier, the point was made that commuting operators may have simultaneous eigenfunctions — that is, the eigenfunctions of these operators may be simultaneously specified for any particular stationary state. Without a computer, and a great deal more theory, we cannot calculate the eigenvalues of the many-electron Hamiltonian, but we can easily classify those energy states according to the eigenvalues of any operators that commute with the Hamiltonian; this, in turn, will permit us to interpret the spectra. For atoms, the most useful operators for this purpose are the angular-momentum operators.

Coupling of Electrons

Although we may discuss atoms in terms of individual electrons, each with its spin and orbital quantum numbers (s, m_s, l, m), the only legitimate observable is the total angular momentum of the atom; we shall denote this quantity as **J**. There are two systems for resolving the total angular momentum into contributions from individual electrons:

1. *LS coupling* (also called Russell-Saunders coupling). We define a total spin angular momentum $(\mathbf{S}_T)$, an operator with quantum numbers S and M_S (cf. Table 12.3), as the vector sum of the spins of each individual electron:

$$\mathbf{S}_T = \sum_i \mathbf{S}_i \tag{12.37}$$

Also, we define an operator $(\mathbf{L}_T)$ for the vector sum of the orbital angular momenta:

$$\mathbf{L}_T = \sum_i \mathbf{L}_i \tag{12.38}$$

This operator will have quantum numbers L and M_L.

Table 12.3 Symbols for angular momentum in atoms

	One electron		Total for many electrons		
	Spin	Orbital	Spin	Orbital	Total
Operator	$\hat{S}$	$\hat{L}$	$\hat{S}_T$	$\hat{L}_T$	$\hat{J}$
Quantum number	s	l	S	L	J
Projection	m_s	m	M_S	M_L	M_J

The total angular momentum of the atom is then the sum of spin plus orbital:

$$\mathbf{J} = \mathbf{L}_T + \mathbf{S}_T \tag{12.39}$$

The quantum numbers for this operator are J and M_J.

2. *jj coupling*. For each individual electron we calculate a total angular momentum:

$$\mathbf{j}_i = \mathbf{L}_i + \mathbf{S}_i$$

The atom's angular momentum is then the vector sum of these:

$$\mathbf{J} = \sum_i \mathbf{j}_i$$

This model is required for heavier atoms for which the spin-orbit coupling is comparable to or greater than the ee repulsion. Otherwise, the LS model is preferred, and we shall discuss in detail only this model.

Addition of Angular Momenta

To carry out the additions required by Eqs. (12.37), (12.38), and (12.39), we need a few simple rules for angular-momentum addition. These rules will be derived for two general vectors called $\mathbf{J}_1$ and $\mathbf{J}_2$, which will be added to give the resultant vector, $\mathbf{J}_T$. The quantum rules derived in Section 11.7 apply to each of these vectors: each angular momentum has an "l-type" quantum number, which must be an integer or half-integer; the length of the vector (in units of $\hbar$) is $\sqrt{l(l + 1)}$; the orientation of each vector is quantized such that there will be $(2l + 1)$ "m-type" quantum numbers for each possible value of the "l-type" quantum number.

The vector addition:

$$\mathbf{J}_T = \mathbf{J}_1 + \mathbf{J}_2$$

is diagrammed in Figure 12.7; we use the following definitions:

$\mathbf{J}_1$ has quantum numbers j_1 and m_1 and length $\sqrt{j_1(j_1 + 1)}$

$\mathbf{J}_2$ has quantum numbers j_2 and m_2 and length $\sqrt{j_2(j_2 + 1)}$

$\mathbf{J}_T$ has quantum numbers J and M and length $\sqrt{J(J + 1)}$

For the sake of argument, assume that the $\mathbf{J}_2$ vector is longer than $\mathbf{J}_1$ (as illustrated in Figure 12.7). For any type of vector, the resultant cannot be longer than the sum of the lengths, $|\mathbf{J}_T| \le |\mathbf{J}_2| + |\mathbf{J}_1|$, nor shorter than the difference, $|\mathbf{J}_T| \ge |\mathbf{J}_2| - |\mathbf{J}_1|$. The quantum rules for angular momentum place additional restrictions on these lengths. From the quantum rules we know that J may have integer or half-integer values; but it cannot have both.

The z-components of the vector addition of the operators are:

$$\hat{\mathbf{J}}_{zT} = \hat{\mathbf{J}}_{z1} + \hat{\mathbf{J}}_{z2}$$

From this we get a simple relationship for the projection quantum numbers:

$$M = m_1 + m_2 \tag{12.40}$$

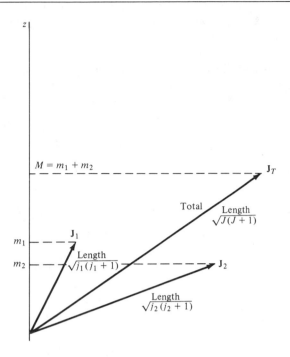

Figure 12.7 Addition of angular-momentum vectors. Quantum-theory rules as to the length and orientation of angular-momentum vectors (Figure 11.14) apply to the component vectors and to the resultant of their addition. This places restrictions on the allowed values of the quantum number (J) of the resultant vector.

The "l-type" quantum number was originally defined as the maximum projection: $l \equiv m(\text{max})$. Thus the "m-type" quantum number must be integer or half-integer according to which its "l-type" quantum number is. Equation (12.40) therefore tells us that J must be an integer if j_1 and j_2 are both integers *or* both half-integers; J will be half-integer if one of the quantum numbers j_1 or j_2 is half-integer, while the other is integer.

These restrictions on the length of the vector and the permitted values of J permit the following values for the quantum number J:

$$J = (j_2 + j_1), (j_2 + j_1 - 1), \ldots, |j_2 - j_1| \tag{12.41}$$

The quantum number (J) for the resultant vector may have values between the sum of the component quantum numbers ($j_1 + j_2$) and the smallest positive difference ($|j_1 - j_2|$), in integer steps. This derivation may seem complicated, but the rule, Eq. (12.41), is simple and easy to apply — cf. examples below.

We can generalize these rules for the addition of more than two angular momenta:

$$\mathbf{J_T} = \mathbf{j_1} + \mathbf{j_2} + \mathbf{j_3}$$

by simply adding two, then adding the third to this sum.

These rules apply to any type of angular momentum with no more than a change in the symbol; for example, we shall use L, S, or I in place of J for specific applications (recall Table 11.4).

Example: What states (J, M) will be the result of the addition of angular momenta with $j_1 = 1$ and $j_2 = 2$? The first vector can have $(2j_1 + 1) = 3$ orientations, while the second can have $(2j_2 + 1) = 5$ orientations; there will therefore be $3 \times 5 = 15$ states (all possible combinations) for the resultant vector.

According to Eq. (12.41), the quantum number J may have values between $2 + 1 = 3$, and $2 - 1 = 1$; that is, $J = 3, 2, 1$. The possible resultant states are, therefore:

$$J = 3, \quad M = 3, 2, 1, 0, -1, -2, -3 \quad \text{(7 states)}$$
$$J = 2, \quad M = 2, 1, 0, -1, -2 \quad\quad \text{(5 states)}$$
$$J = 1, \quad M = 1, 0, -1 \quad\quad\quad\quad \text{(3 states)}$$

The total number of states is $7 + 5 + 3 = 15$, as expected.   ■

Example: What states (J, M) result from the addition of $j_1 = \frac{5}{2}$ and $j_2 = 1$? The total number of states is:

$$(2j_1 + 1)(2j_2 + 1) = 6 \times 3 = 18$$

The permitted values of J range from $(\frac{5}{2} + \frac{2}{2}) = \frac{7}{2}$ to $(\frac{5}{2} - \frac{2}{2}) = \frac{3}{2}$, in integer steps:

$$J = \tfrac{7}{2}, \quad M = \tfrac{7}{2}, \tfrac{5}{2}, \ldots, -\tfrac{5}{2}, -\tfrac{7}{2} \quad \text{(8 states)}$$
$$J = \tfrac{5}{2}, \quad M = \tfrac{5}{2}, \ldots, -\tfrac{5}{2} \quad\quad\quad \text{(6 states)}$$
$$J = \tfrac{3}{2}, \quad M = \tfrac{3}{2}, \tfrac{1}{2}, -\tfrac{1}{2}, -\tfrac{3}{2} \quad\quad \text{(4 states)}$$

The total number of states is $8 + 6 + 4 = 18$, as expected.   ■

Electron Spins

The total spin quantum number (S) of the electrons in an atom is a very important characteristic of the atom's quantum states. Each electron has a spin quantum number $s = \frac{1}{2}$; the rules given above tell us that the maximum value of the total spin quantum number (S) for the resultant of N spins $\frac{1}{2}$ will be $N(\frac{1}{2}) = N/2$; that is, all spins add. If N is even, the minimum value of S is zero (all spins cancel); if N is odd, the minimum value of S must be $\frac{1}{2}$ (all spins but one cancel). This idea can be explained with the simple arrow ($\uparrow$) model of the electron used in elementary chemistry courses in terms of spin pairing ($\uparrow\downarrow$), but this model has severe limitations. These limitations may be appreciated if we examine in detail the case of two spins (Figure 12.8). (The states of two electrons were discussed from another point of view in Sections 12.4 and 12.5.)

Two spins $\frac{1}{2}$ may, according to Eq. (12.41), have a resultant quantum number $S = 1$ or 0. The $S = 0$ state has only $M_S = 0$ for the projection quantum number and is called a *singlet;* the $S = 1$ state has three projections, $M_S = 1, 0, -1$, and is called a *triplet.*

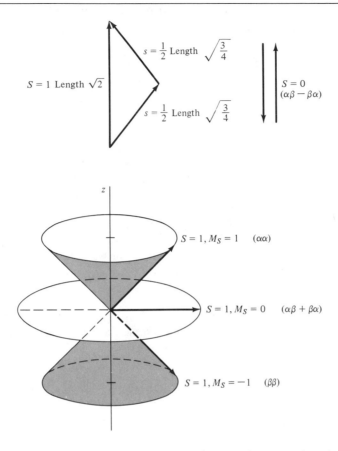

Figure 12.8 Singlet and triplet states. Two spins $\frac{1}{2}$ may either cancel each other's angular momentum (giving the singlet state, $S = 0$) or add to give a resultant state with $S = 1$ (the triplet state). The triplet state (like all angular momenta with quantum number 1) can have three orientations with respect to the z axis, corresponding to projection quantum numbers $M_S = 1$, 0, or -1.

Each individual electron has two orientations, which we earlier called α and β. Two spins can have four states, which may be denoted as $\alpha\alpha$, $\alpha\beta$, $\beta\alpha$, and $\beta\beta$; however, the Pauli exclusion principle requires us to use the states $\alpha\alpha$, $\alpha\beta + \beta\alpha$, $\beta\beta$ (called triplet) and $\alpha\beta - \beta\alpha$ (called singlet). From Eq. (12.40) we can see that the $\alpha\alpha$ and $\beta\beta$ states have $M_S = \frac{1}{2} + \frac{1}{2} = 1$ and $M_S = -\frac{1}{2} - \frac{1}{2} = -1$, respectively, and are therefore two of the three states with $S = 1$. Similarly the states $\alpha\beta + \beta\alpha$ and $\alpha\beta - \beta\alpha$ have $M_S = 0$; there are indeed two states with $M_S = 0$, one each for $S = 1$ and 0. With some difficulty it can be proven that the $\alpha\beta + \beta\alpha$ function is the state $S = 1$, $M_S = 0$, while $\alpha\beta - \beta\alpha$ is the state $S = 0$, $M_S = 0$. Thus the words "singlet" and "triplet" as used earlier actually are specifications of the resultant quantum number,

S. In conclusion, the spin functions (χ) for two spins $\frac{1}{2}$ (illustrated in Figure 12.8) are:

$$S = 0, \qquad M_S = 0, \qquad \chi = \alpha\beta - \beta\alpha \qquad \text{(singlet)}$$

$$\left.\begin{array}{lll} S = 1, & M_S = 1, & \chi = \alpha\alpha \\ S = 1, & M_S = 0, & \chi = \alpha\beta + \beta\alpha \\ S = 1, & M_S = -1, & \chi = \beta\beta \end{array}\right\} \quad \text{(triplet)}$$

[Compare above, Eqs. (12.31) through (12.36)]

What about the arrow model? Symbols such as ↑↑ or ↓↓ are readily understood to be triplet states; however, the symbol for "spins paired," ↑↓, is ambiguous — it could denote the singlet state or the $M_S = 0$ orientation of the triplet state. Thus the arrow model tells only the projection quantum numbers and gives no information about the total angular momentum. It has only limited utility and must be used with caution.

Since the operator $\hat{S}_T^2$ commutes with the Hamiltonian, each energy state of the atom can be labeled by the quantum number *S*. For historical reasons, atomic states are labeled by the *spin multiplicity* (g_S); this is simply the number of orientations (degeneracy) of the total spin vector:

$$g_S = 2S + 1$$

Examples for more than two electrons are shown by Table 12.4; the reader should study these results carefully in order to understand the general pattern.

Orbital Angular Momentum

In the orbital model of the atom, each electron is characterized by quantum numbers (*l*) for orbital angular momentum:

$$s \text{ electrons:} \qquad l = 0 \quad (m = 0)$$

$$p \text{ electrons:} \qquad l = 1 \quad (m = 1, 0, -1)$$

Table 12.4   States for spins $\frac{1}{2}$

Number of spins	S	M_S	Multiplicity, g_s	Name	Arrow symbol
1	$\frac{1}{2}$	$\frac{1}{2}, -\frac{1}{2}$	2	Doublet	↑
2	$\begin{cases} 1 \\ 0 \end{cases}$	$\begin{array}{l} 1, 0, -1 \\ 0 \end{array}$	$\begin{array}{l} 3 \\ 1 \end{array}$	Triplet / Singlet	↑↑ / ↑↓
3	$\begin{cases} \frac{3}{2} \\ \frac{1}{2} \end{cases}$	$\begin{array}{l} \frac{3}{2}, \frac{1}{2}, -\frac{1}{2}, -\frac{3}{2} \\ \frac{1}{2}, -\frac{1}{2} \end{array}$	$\begin{array}{l} 4 \\ 2 \end{array}$	Quartet / Doublet	↑↑↑ / ↑↑↓
4	$\begin{cases} 2 \\ 1 \\ 0 \end{cases}$	$\begin{array}{l} 2, 1, 0, -1, -2 \\ 1, 0, -1 \\ 0 \end{array}$	$\begin{array}{l} 5 \\ 3 \\ 1 \end{array}$	Quintet / Triplet / Singlet	↑↑↑↑ / ↑↑↑↓ / ↑↑↓↓
5	$\begin{cases} \frac{5}{2} \\ \frac{3}{2} \\ \frac{1}{2} \end{cases}$	$\begin{array}{l} \frac{5}{2}, \ldots, -\frac{5}{2} \\ \frac{3}{2}, \ldots, -\frac{3}{2} \\ \frac{1}{2}, -\frac{1}{2} \end{array}$	$\begin{array}{l} 6 \\ 4 \\ 2 \end{array}$	Sextet / Quartet / Doublet	↑↑↑↑↑ / ↑↑↑↑↓ / ↑↑↑↓↓

d electrons: $l = 2$ $(m = 2, 1, 0, -1, -2)$

f electrons: $l = 3$ $(m = 3, 2, 1, 0, -1, -2, -3)$

The total resultant orbital angular momentum has quantum numbers L and M_L with $M_L = L, L - 1, \ldots, -L (2L + 1$ values). The quantum number L may have values between the sum of the l values of the individual electrons and the smallest positive difference of these numbers. For example, two p electrons ($l = 1$) may have L quantum numbers from $1 + 1$ to $1 - 1$: $L = 2, 1, 0$. It should be obvious from this discussion that L must always be an integer.

Example: Find the total angular-momentum states (L, M) for two electrons, one p-type and one d-type. The quantum number L may have values between $2 + 1 = 3$ and $2 - 1 = 1$. There are five orientations for a d electron and three for a p electron and, therefore, $3 \times 5 = 15$ combinations. These are:

$$L = 3, \quad M_L = 3, 2, 1, 0, -1, -2, -3 \quad \text{(7 states)}$$

$$L = 2, \quad M_L = 2, 1, 0, -1, -2 \quad \text{(5 states)}$$

$$L = 1, \quad M_L = 1, 0, -1 \quad \text{(3 states)}$$

The total number of states is $7 + 5 + 3 = 15$, as expected.   ■

Term Symbols

It can be shown that the operator for the total squared orbital angular momentum $(\hat{L}_T^2)$ commutes with the atomic Hamiltonian; therefore its quantum number (L) can be used to label the energy states (terms) of the atoms. For historical reasons, the states of the atom are given symbols that specify the quantum number L:

$$L = 0, 1, 2, 3, 4, 5, 6, 7, 8, 9 \ldots$$

$$\text{symbol:} \quad S, P, D, F, G, H, I, K, L, M \ldots$$

(Sober Physicists Don't Find Giraffes Hiding In Kitchens Like Mine)

The *term symbol* of an atomic state is a symbol (as given above) with an upper left superscript that gives the spin multiplicity, $g_S = 2S + 1$. For example, the symbol ^3P means a state with $L = 1$ and $S = 1$ [cf. Eq. (12.36)], while the symbol ^1P means $L = 1$ and $S = 0$ [cf. Eq. (12.35)]. Such symbols are really just "codes" denoting the quantum numbers L and S that characterize a particular atomic state.

A term symbol really denotes a group of states [for example, ^3P denotes the nine states of Eq. (12.36)]. The spin angular momentum has $2S + 1$ orientations, and the orbital angular momentum has $2L + 1$ orientations; the total possible combinations are:

$$\text{total states in a term} = (2L + 1)(2S + 1)$$

If spin-orbit coupling can be neglected, these states all have the same energy, so we shall call this a *degeneracy* (g); the effect of spin-orbit coupling will be discussed in Section 12.8.

Example: Give the symbols and degeneracies (g) for the following states:

$$L = 3, S = 1: \qquad \text{symbol } {}^{3}F, \quad g = 7 \times 3 = 21$$

$$L = 1, S = \tfrac{3}{2}: \qquad \text{symbol } {}^{4}P, \quad g = 3 \times 4 = 12$$

$$L = 6, S = \tfrac{9}{2}: \qquad \text{symbol } {}^{10}I, \quad g = 13 \times 10 = 130 \qquad \blacksquare$$

[In the scientific literature, the symbols s, p, d, f and S, P, D, F are often used interchangeably. If there is only one electron with orbital angular momentum, then $l = L$, and for all practical purposes the notations are identical. Also, as we shall see in the next section, if there is only one electron outside a closed shell (or one "hole" in a closed shell), then L is the same as the l of that electron and, for example, the symbols P and p are interchangeable. Until you understand it better, you should make the distinction carefully.]

12.7   Many-Electron Atoms

The Hamiltonian of an atom with N electrons (neglecting spin-orbit coupling) is:

$$\hat{H} = \sum_{i=1}^{N} \hat{h}_0(i) + \sum_{i>j} \frac{e^2}{r_{ij}} \tag{12.42}$$

In this equation, $\hat{h}_0$ is the one-electron operator of Eq. (12.19). The zeroth-order wave functions, eigenfunctions of the Hamiltonian when the ee repulsion term is omitted, will be products of the one-electron wave functions — that is, the H-atom orbitals. The building-up of many electron states from one-electron orbitals is called the *aufbau principle*. We can avoid violating the Pauli principle by using each one-electron function (including spin — for example, $1s\alpha$ and $1s\beta$) only once. One such product for the carbon atom ($Z = 6$) would be:

$$\psi_{1s}(1)\alpha(1)\psi_{1s}(2)\beta(2)\psi_{2s}(3)\alpha(3)\psi_{2s}(4)\beta(4)\psi_{2p_1}(5)\beta(5)\psi_{2p_0}(6)\beta(6)$$

We can write a great number of other products by exchanging the labels of the identical electrons or by using other combinations of the $2p$ orbitals. Linear combinations of these products must be formed so as to obey the Pauli principle for exchange of identical electrons and, preferably, so that these functions will correspond to definite values of the "good" quantum numbers, L, M_L, S, M_S (that is, quantum numbers of operators that commute with the Hamiltonian). We cannot presume that all of these states that correspond to a given configuration will have the same energy. [Recall that the He$(1s)(2s)$ configuration gave two energy levels, ${}^{1}S$ and ${}^{3}S$.]

Rules for Term Symbols

The rules for addition of angular momentum can be used to classify these states and to identify them with the measureable spectroscopic terms. With a dozen or more electrons, this could be a complicated task; it is made easier by two rules:

1. (Unsölds theorem.) All filled ("closed") shells have zero spin and orbital angular momentum: $L = 0$, $S = 0$. The configurations s^2, p^6, d^{10}, f^{14} therefore correspond

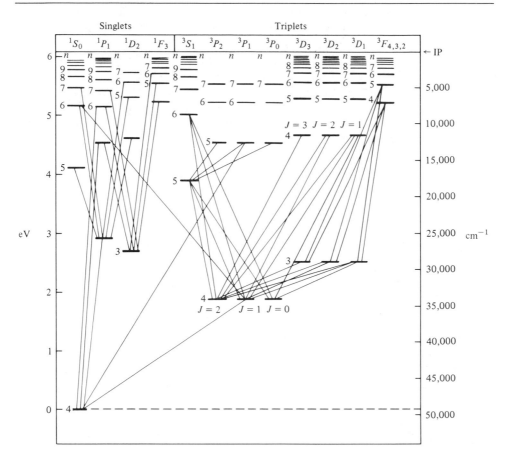

Figure 12.9 Term diagram of calcium atoms. The numbers to the left of the terms are the principal quantum numbers of the outermost electron. This should be compared to that of helium, Figure 12.6; note that there are a number of single-triplet transitions for calcium. The helium terms are similarly split by spin-orbit coupling, but this is not shown in Figure 12.6 because it is too small.

to ^1S states. The spectroscopic terms (also called "Rydberg terms"), those observed by absorption or emission spectroscopy, generally result from the excitation of one outer electron; therefore this rule permits us to ignore the inner shells and warns us to expect similar spectra from $He(1s)^2$, $Mg[Ne](3s)^2$, $Ca[Ar](4s)^2$, and so on. (Compare Figures 12.6 and 12.9.) (The symbol [Ne] indicates the neon configuration $1s^2 2s^2 2p^6$, and so on.) One way to see this is to realize that the distribution function of, for example, six p electrons, two in each orbital p_x, p_y, and p_z, is:

$$F(r) = p_x^2 + p_y^2 + p_z^2$$

This function is radially symmetric (see Problem 12.13); if there is no angular dependence, $\hat{L}^2 F(r) = 0$ necessarily [remember Eq. (11.74) for this operator]. Another point of view is that for this configuration: $M_l = \Sigma_i m_i = 0$ and that this is the *only* (and therefore the maximum) value of M_L; therefore, $L \equiv M_L(\text{max})$ is equal to zero. A similar argument can be applied to $S \equiv M_S(\text{max})$.

2. The state of the "hole" is the state of the electron. A partially filled shell, for example p^5, can be viewed as a closed shell minus an electron (a "hole" is left), and the resultant states of the holes are the same as the states of the removed electron(s). By conservation of angular momentum, the vector sum of the angular momentum of the electron plus that of the hole must be the angular momentum of the closed shell — which is zero. Therefore, the states of p^5 are identical to those of p^1, those of d^8 are identical to those of d^2, and so on.

Examples of Configurations and Terms

It may be useful to give some examples of the possible term types which can result from several configurations. The reader should derive these results.

(a) s^1; examples: $H(1s)$, $Li(1s)^2(2s)$, $Na[Ne](3s)$. Term ^2S.

(b) s^2; examples: $He(1s)^2$, $Sr[Kr](5s)^2$. Term ^1S.

(c) ss; example: $He(1s)(2s)$ (nonequivalent electrons). Terms ^1S, ^3S.

(d) sp; example: $He(1s)(2p)$. Terms ^1P, ^3P.

(e) s^2p; example: $B(1s)^2(2s)^2(2p)$. Term ^2P.

(f) pp; example: $(2p)(3p)$ (nonequivalent electrons). For the first time we have two electrons with orbital momentum. If $l_1 = 1$ and $l_2 = 1$, the possible resultants are $L = 2, 1, 0$ and the terms are (spins paired) ^1D, ^1P, ^1S and (spins parallel) ^3D, ^3P, ^3S.

Example: Find the total possible states for a $(2p)(3p)$ configuration. There are six "boxes" into which we can place two electrons. The first electron can go into either $2p_1$, $2p_0$, or $2p_{-1}$ either spin up (α) or spin down (β) — six ways. There are six ways to put the other electron into the $3p$ orbitals, so there are $6 \times 6 = 36$ possible combinations. The term degeneracy sum for the terms of this configuration given above is $^1D(g = 5) + ^1P(g = 3) + ^1S(g = 1) + ^3D(g = 15) + ^3P(g = 9) + ^3S(g = 3)$ for a total of 36 states, as expected. ∎

(g) p^2; example: $C(1s)^2(2s)^2(2p)^2$ (equivalent electrons). The same arguments can be used as for pp above to predict $L = 2, 1, 0$ and $S = 1, 0$, but not all combinations can be used because of the Pauli principle. For example, a ^3D state ($L = 2$, $S = 1$) would have to have, as one of its 15 substates, the quantum numbers $M_L = 2$ and $M_S = 1$. This could happen only if both electrons had $m = 1$ and $m_s = \frac{1}{2}$; this state:

$$(2p_1)(2p_1)\alpha\alpha$$

is totally symmetric and violates the Pauli principle. We conclude that the p^2 states that are ^3D (all 15) are impossible. [The same argument does not apply to nonequivalent electrons. The $(2p)(3p)^3D$ substate with $M_L = 2$, $M_S = 1$, could be:

$$(2p_13p_1 - 3p_12p_1)\alpha\alpha$$

which *is* properly antisymmetric.]

Table 12.5 Wave functions for the p^2 configuration

Wave functions[a] (antisymmetric)	Quantum numbers		Allowed states[b]		
	M_L	M_S	^1D	^1S	^3P
1. $p_1p_1(\alpha\beta - \beta\alpha)$	2	0	✓		
2. $(p_1p_0 - p_0p_1)\alpha\alpha$	1	1			✓
3. $(p_1p_0 - p_0p_1)(\alpha\beta + \beta\alpha)$	1	0			✓
4. $(p_1p_0 - p_0p_1)\beta\beta$	1	−1			✓
5. $(p_1p_0 + p_0p_1)(\alpha\beta - \beta\alpha)$	1	0	✓		
6. $(p_1p_{-1} - p_{-1}p_1)\alpha\alpha$	0	1			✓
7. $(p_1p_{-1} - p_{-1}p_1)(\alpha\beta + \beta\alpha)$	0	0			✓
8. $(p_1p_{-1} - p_{-1}p_1)\beta\beta$	0	−1			✓
9. $(p_1p_{-1} + p_{-1}p_1)(\alpha\beta - \beta\alpha)$	0	0	✓		
10. $p_0p_0(\alpha\beta - \beta\alpha)$	0	0		✓	
11. $(p_0p_{-1} - p_{-1}p_0)\alpha\alpha$	−1	1			✓
12. $(p_0p_{-1} - p_{-1}p_0)(\alpha\beta + \beta\alpha)$	−1	0			✓
13. $(p_0p_{-1} - p_{-1}p_0)\beta\beta$	−1	−1			✓
14. $(p_0p_{-1} + p_{-1}p_0)(\alpha\beta - \beta\alpha)$	−1	0	✓		
15. $p_{-1}p_{-1}(\alpha\beta - \beta\alpha)$	−2	0	✓		

[a]The symbols indicate one-electron functions for $p(l = 1)$ electrons with $m = 1$ (p_1), $m = 0$ (p_0), and $m = 1$ (p_{-1}), respectively.
[b]The assignment of wave functions to states is somewhat arbitrary, since states with the same set of quantum numbers (for example, states 9 and 10) cannot always be distinguished.

In fact, there are only 15 ways to put two electrons into the three p orbitals that do not violate the Pauli principle (shown in Table 12.5); these correspond to states ^3P, ^1D, and ^1S. In zeroth order (neglecting ee repulsion) these 15 states all have the same energy. When the ee repulsion is included, however, they are no longer degenerate. Experimentally (Figure 12.10) it is found for the carbon atom that the energy (relative to the ^3P ground state) of the ^1S state is 21,648 cm^{-1} and that of the five ^1D states is 10,193 cm^{-1}. [The nine ^3P states do not all have the same energy because of spin-orbit coupling. This small energy (43 cm^{-1}) will be ignored for now, and we shall continue to refer to a "degeneracy" of 9 for the ^3P ground state.]

The fact that a single configuration can give rise to such a large range of energies demonstrates the limitations of the orbital model of the atom. The "orbital energy" of a $2p$ electron in a carbon atom is defined as the energy of the process:

$$C(2p)^2 \longrightarrow C^+(2p) + e$$

But which $(2p)^2$ state do we mean? The "uncertainty" of 20,000 cm^{-1} (2.7 eV or 260 kJ/mole) is not insignificant. We shall return to this point later, but for now we must conclude that the configuration is an insufficient specification of the state of an open-shell atom.

Determining the allowed terms for open shells is straightforward (if tedious), but we shall not pursue it further. Table 12.6 lists the results for many common cases. Fig. 12.11 shows the terms for the p^3 configuration of the nitrogen atom.

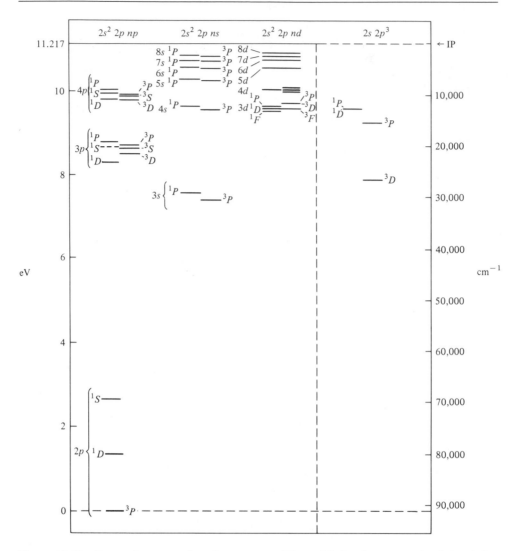

Figure 12.10 Term diagram of carbon atoms. The orbitals shown to the left of the states indicate the orbital of the excited electron. Note that each electron configuration gives rise to several energy levels. The states to the right have a configuration $(1s)^2(2s)^1(2p)^3$.

Hund's Rules

There is no simple rule that will predict the order of the term energies, but there are simple rules — Hund's rules — to find the ground state:

1. The ground state of the lowest electronic configuration will be the state of maximum spin multiplicity (maximum S).

Table 12.6 Allowed terms of equivalent electrons

Configurations	Terms
p^2, p^4	^1S, ^1D, ^3P
p^3	^2P, ^2D, ^4S
p, p^5	^2P
d, d^9	^2D
d^2, d^8	^1S, ^1D, ^1G, ^3P, ^3F
d^3, d^7	^2P, ^2D (twice), ^2F, ^2G, ^2H, ^4P, ^4F
d^4, d^6	^1S (twice), ^1D (twice), ^1F, ^1G (twice), ^1I, ^3P (twice), ^3D, ^3F (twice), ^3G, ^5D
d^5	^2S, ^2P, ^2D (thrice), ^2F (twice), ^2G (twice), ^2H, ^2I, ^4P, ^4D, ^4F, ^4G, ^6S

2. If several terms have the same maximum multiplicity, the ground state will be that of largest L.

Example: The ground state of d^2 (Table 12.6) is, by rule 1, either ^3P or ^3F. By rule 2 we choose the largest L, so ^3F is the ground state. ∎

Determining the Lowest Term of a Configuration

There is a quick way to find the term of the ground state for any configuration without the bother of finding them all and then applying Hund's rules. From the definition of the L- or S-type quantum numbers:

$$S = M_S(\text{max}) = \left[\sum_i m_s(i) \right]_{\text{max}}$$

$$L = M_L(\text{max}) = \left[\sum_i m(i) \right]_{\text{max}}$$

We shall represent the electron with arrows up (↑) for $m_s = +\frac{1}{2}$ or down (↓) for $m_s = -\frac{1}{2}$. We can get $M_S(\text{max}) = S$ by putting the electrons into the orbitals with a maximum of up states consistent with the Pauli principle. If we do this starting with the maximum value of m (for example, p_1 or d_2), we will get the maximum M_L, and this will be L. For p^2 this is:

orbital ↑ ↑ ☐

$$m = \quad 1 \quad 0 \quad -1$$

which gives $M_L = 1$, $M_S = 1$ for a ^3P state. Note that this picture gives only *one* of the nine ^3P substates (number 2 in Table 12.5); it does not "represent" the ^3P state but is only an indicator.

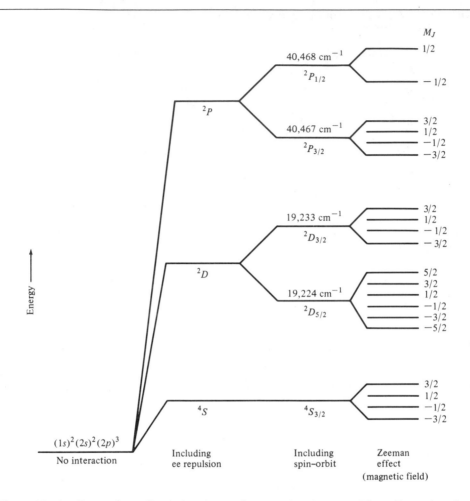

Figure 12.11 Ground configuration terms for atomic nitrogen. The effect of various terms of the atomic Hamiltonian on the energy levels of nitrogen atoms. The Zeeman effect applies only when the atoms are in a magnetic field — see Section 12.10. Energies are not shown to scale, but the observed energies of the various terms are given relative to the lowest state.

Example: Find the ground state of the d^6 configuration.

orbital | ⇅ | ↑ | ↑ | ↑ | ↑ |

$m = \quad 2 \quad\quad 1 \quad\quad 0 \quad -1 \quad -2$

$S = M_S(\text{max}) = 2$, $L = M_L(\text{max}) = (2 \times 2 + 1 + 0 - 1 - 2) = 2$. The term is therefore ^5D. ∎

12.8   Spin-Orbit Coupling

The analysis of this section assumes that the spin-orbit energy is small compared to the *ee* repulsion energy. If this is not true, as will happen in some heavy atoms, the *jj* coupling model must be used.

The total angular momentum of an atom is a sum of spin plus orbital angular momenta:

$$\mathbf{J} = \mathbf{L} + \mathbf{S} \tag{12.43}$$

As before, the quantum number of the resultant angular momentum (J) may have values:

$$J = (L + S), \quad (L + S - 1), \ldots, |L - S| \tag{12.44}$$

Example: What are the allowed J values for a ^3P state? $L = 1$ and $S = 1$, so $J = 1 + 1, \ldots, |1 - 1|$ or $J = 2, 1, 0$. This means that the nine ^3P substates [Eq. (11.36)] can be divided into three groups:

$$J = 2: \quad g_J = 2J + 1 = 5 \quad (M_J = 2, 1, 0, -1, -2)$$

$$J = 1: \quad g_J = 2J + 1 = 3 \quad (M_J = 1, 0, -1)$$

$$J = 0: \quad g_J = 2J + 1 = \underline{1} \quad (M_J = 0)$$

$$\text{9 states total} \qquad \blacksquare$$

The independent electron model gives a simple formula for the spin-orbit energy of a state with quantum numbers J, L, S:

$$E_{SO} = \tfrac{1}{2}hcA[J(J + 1) - L(L + 1) - S(S + 1)] \tag{12.45}$$

[A is called the spin-orbit coupling constant and has units cm^{-1}. This constant will have a different value for each (n, L, S) state.] This energy is added to the other electronic energies, with the result that the states of a given configuration with a given L and S — for example, ^3P — are no longer degenerate. The term symbol is modified by adding the J value as a subscript; for example, ^3P becomes $^3P_2$, $^3P_1$, $^3P_0$. When either $L = 0$ or $S = 0$, there is only one J possible; the J subscript may be included, but it is redundant — for example, $^1S_0$, $^2S_{1/2}$, $^1P_1$.

The sign of the spin-orbit coupling constant (A) determines which J state will be lower in energy. For the ground state, Hund's (third) rule is:

3a. If the shell is less than half-filled, the smallest J is lowest (A is positive).
3b. If the shell is more than half-filled, the largest J is lowest (A is negative).

Do we need a rule for half-filled shells? Hund's rule strictly applies only to the *ground state*, which always has $L = 0$ for half-filled shells; therefore $J = S$ and there is only one J state and, hence, nothing to order.

Example: For p^2, the energy order of states is $^3P_2 > {}^3P_1 > {}^3P_0$; $^3P_0$ is the ground state. For p^4 the order is reversed and $^3P_2$ is lowest. This can be illustrated by C($2p^2$) and O($2p^4$). The experimental energies (relative to the ground state = 0) are:

	Oxygen	Carbon
$^3P_2$	0	43 cm^{-1}
$^3P_1$	158 cm^{-1}	16 cm^{-1}
$^3P_0$	227 cm^{-1}	0

[An alert reader may note that these energies cannot be calculated from Eq. (12.45) with a single value of A; this indicates that this equation is approximate.] ■

12.9   Atomic Spectroscopy

The absorption and emission of radiation by atoms is a prime method for measuring their energy levels. Figure 12.12 shows, schematically, an arrangement for experimentally studying the emissions of hot, gaseous atoms.

A Rydberg series, a group of lines spaced closer as they go to longer wavelengths and converging to a series limit, is a characteristic of all atomic spectra (cf. Figure 11.3). However, only the series found in atomic hydrogen can be explained by a simple formula such as Eq. (11.9).

The Na atom is superficially similar to the hydrogen atom, having one electron outside the Ne core. The spectrum of Na in the visible to near ultraviolet (uv) region (Figure 12.13) can be resolved into several overlapping Rydberg-type series, each converging to a series limit. These are called the principal, sharp, and diffuse series; the principal series is the only one observed in absorption. The dominating feature of the spectrum, the one that gives sodium lamps their obvious color, is an intense, yellow line called the Na-D line. Examined with a spectroscope, the Na-D line is clearly a doublet; that is, it consists of two lines of nearly the same wavelength (589.593 nm and 588.996 nm). Under high resolution, all the lines of the sodium atom will be seen to be doublets (although the diffuse series may appear as compound doublets — see the example below). These small splittings are called *fine structure*.

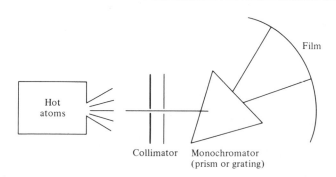

Figure 12.12   Spectrometer (schematic). Light emitted by hot atoms is diffracted by a prism or grating and recorded on a strip of film.

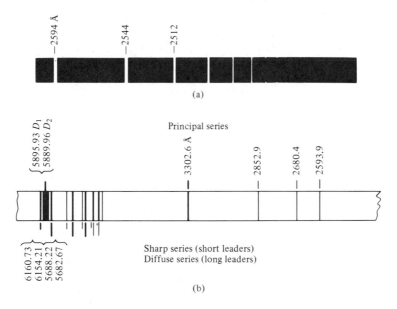

Figure 12.13 Atomic spectra. (a) The absorption spectrum of the sodium atom is obtained by shining white light through a sodium vapor and analyzing the transmitted light with a spectrometer; the absence of the absorbed light is indicated by white bands on the film. The portion shown here includes only the higher-frequency portion of the principal series, starting with the fifth line. (b) The emission spectrum of the sodium atom consists of several Rydberg-type series called *sharp, principal,* and *diffuse.* The wavelengths of the principal series are identical to those found in the absorption spectrum. (From G. Herzberg, *Atomic Spectra and Atomic Structure,* 1944: New York, Dover Publications, Inc., Figs. 2, 3, and 31.)

Selection Rules

The selection rules for atomic spectra are:

1. $\Delta S = 0$. No change in multiplicity. This rule is frequently violated, especially when spin-orbit coupling is large.
2. $\Delta L = \pm 1$. That is, $S \leftrightarrow P \leftrightarrow D \leftrightarrow F$, and so on, but $S \not\leftrightarrow S$, $S \not\leftrightarrow D$, ...
3. $\Delta J = 0, \pm 1$ except $(J = 0) \not\leftrightarrow (J = 0)$.
4. The Laporte rule. Transitions within a configuration are forbidden; in effect, one quantum number of one electron (n or l) must change. Another point of view is that, in an allowed transition, the electron must change its orbital. Earlier (Figure 12.10) we saw that the $C(1s)^2(2s)^2(2p)$ configuration gave rise to three states ($^3P, {}^1D, {}^1S$); transitions among these states are forbidden by the Laporte rule (also by one or both of rules 1 and 2). Another example: the d^3 configuration (Table 12.6) has a number of states, such as ^2P and ^2D, that are allowed transitions according to rules 1 and 2; such a transition is forbidden by the Laporte rule because the configuration is not changed.

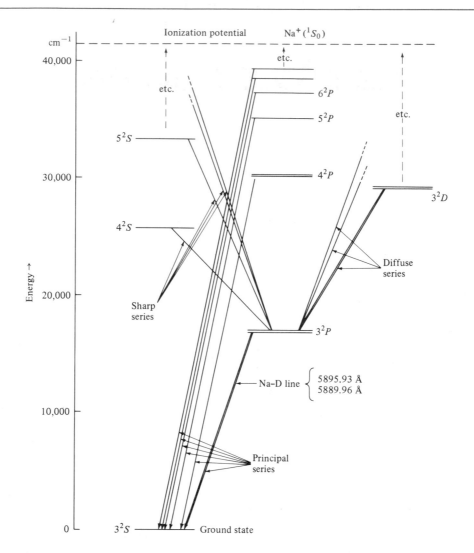

Figure 12.14 Term diagram of sodium atoms. All lines of the sharp and principal series'
are doublets; those of the diffuse series are compound doublets.

The Sodium Spectrum

Figure 12.14 explains the source of the series of the Na spectrum. The fine structure
is due to spin-orbit coupling, which splits ^2P into $^2P_{3/2}$ and $^2P_{1/2}$ levels.

A careful study of this diagram may convince you that not all possible transitions
are shown; what about transitions such as $5^2S \rightarrow 4^2P$ or $6^2P \rightarrow 3^2D$? Such transitions
are allowed, and may, in fact, be observed weakly. The theory of absorption and
emission of radiation due to Einstein shows that the transition probability for spontaneous emission is proportional to the *cube* of the frequency of the emitted photon. For

that reason, an atom in an excited state will, other things being equal, prefer to emit the highest-frequency photon possible — that is, to take the largest jump permitted by the selection rules. The size of the integrals such as Eq. (11.100) will also affect the intensities and, in fact, is responsible for the weakening of the lines of a series as they approach the series limit.

Example: What is the fine structure of a $^2D \rightarrow {}^2P$ transition? With the rule $\Delta J = 0, \pm 1$ we get:

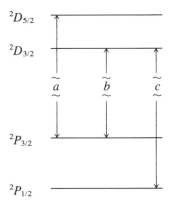

There are three lines, but they are called (for historical reasons) a "compound doublet" rather than a "triplet." If the spin-orbit coupling of the D state is small compared to that of the P state, the lines a and b will be very close in frequency and, perhaps, not resolved, so this fine structure may appear to be a doublet. ∎

Forbidden Transitions: Ca and Hg

The selection rules listed above can all be violated at one time or another. Comparison of the He term diagram (Figure 12.6) to that of Ca (Figure 12.9) shows many more "forbidden" transitions, especially singlet-triplet ("intersystem crossing"), for the heavier Ca atom, in which the spin-orbit coupling is much stronger.

As a final example, we shall discuss the spectrum of mercury (Figure 12.15) which, like that of He and Ca, shows singlet and triplet spectra from the excitation of its $(6s)^2 \, {}^1S_0$ ground configuration. The strongest emission line is $6^1P_1 \rightarrow 6^1S_0$ at 54,000 cm^{-1}; this and many of the other mercury lines are ultraviolet. Mercury lamps enclosed in quartz are used in the laboratory as sources of uv radiation for photochemistry; such lamps can be a danger to the eye. (Mercury lamps used for street lighting are enclosed in glass so that the dangerous uv radiation cannot escape. Although fluorescent materials can be used to convert uv to visible light, the greater efficiency of Na lamps over Hg lamps is due largely to the fact that their principal emission, the Na-D line, is in the visible region.) Transitions between the 6^3P states and the ground state are allowed only weakly; such states are metastable and may have relatively long lifetimes and a chemistry of their own; for example, the hydrogen will react with metastable ^3P mercury atoms but not with ground-state (^1S) atoms:

$$\mathrm{Hg}(^3P) + \mathrm{H}_2 \longrightarrow \mathrm{HgH} + \mathrm{H}$$

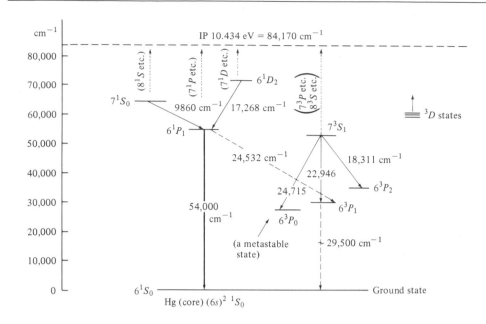

Figure 12.15   Term diagram of mercury atoms. Only selected energy levels and transitions are shown. The single-triplet transition shown (dashed line) is weak; the 54,000-cm^{-1} line is the most intense in the spectrum. The 6^3P_0 state is referred to as metastable because transitions from this state to the ground state (the only one that is lower) are strongly forbidden; consequently this state will have a long lifetime.

12.10   Atomic Magnetism

It was mentioned in Section 11.7 that the quantization condition on the eigenvalues of $\hat{L}_z$ placed a restriction on the directions an angular momentum could have with respect to the z axis. Can this be observed experimentally? Since an isolated atom has a radially symmetric potential, in field-free space all directions are equivalent. Therefore, any direction could be called "z" and any orientation will be possible. The degeneracy of the p orbitals showed this; each linear combination of the three degenerate functions represents a different direction for **L**, and all are possible state functions by the theorem proved in Chapter 11 (p. 622). If an external field (with which the atom can interact) is applied, the M_L degeneracy is raised and only certain orientations of **L** with respect to that field will be allowed—space, as seen by the atom, is no longer symmetric but has a unique direction, which will be called the z axis. For atoms, this field could be electric (the Stark effect) or magnetic (the Zeeman effect). The magnetic interactions of an atom are the direct experimental manifestation of its angular-momentum quantization.

Magnetic Moments

An orbiting electron can be viewed as a circulating electrical current, which will, by Faraday's law, produce a magnetic moment perpendicular to the plane of the orbit and parallel to the angular-momentum vector. According to classical mechanics, a rotating charge e with a mass m will have a magnetic dipole moment:

$$\boldsymbol{\mu}_L = \frac{e}{2mc}\,\mathbf{L} \tag{12.46}$$

If we replace this angular momentum with its quantum-mechanical eigenvalue [Eq. (11.91)], we get the quantum-mechanical formulas for the magnetic moment:

$$\mu_{L_z} = \beta M_L \tag{12.47}$$

$$|\boldsymbol{\mu}_L| = \beta\sqrt{L(L+1)} \tag{12.48}$$

with

$$\beta = \frac{e\hbar}{2mc} = 9.2732 \times 10^{-21}\ \mathrm{erg/gauss} \tag{12.49}$$

The quantity β is called the Bohr magneton.

The electron spin is similarly associated with a magnetism, but it is related to the spin angular momentum in an anomalous fashion, which reinforces our feeling that the term "spin" is a metaphor:

$$|\boldsymbol{\mu}_S| = 2\beta\sqrt{S(S+1)} \tag{12.51}$$

The factor of 2 in Eq. (12.51) is called the "g factor"; actually, $g = 2.002322$ for a free electron—but $g = 2$ will be close enough for us.

The total magnetic moment of an atom will be a sum of the orbital plus spin moments:

$$\boldsymbol{\mu}_J = \boldsymbol{\mu}_L + \boldsymbol{\mu}_S$$

$$|\boldsymbol{\mu}_J| = g_L\beta\sqrt{J(J+1)} \tag{12.52}$$

$$\mu_z = g_L\beta M_J \tag{12.53}$$

The quantity g_L, called the *Landé g factor,* is a geometrical factor resulting from the vector addition of $\boldsymbol{\mu}_L$ and $\boldsymbol{\mu}_S$ and can be shown to be:

$$g_L = 1 + \frac{J(J+1) + S(S+1) - L(L+1)}{2J(J+1)} \tag{12.54}$$

The Zeeman Effect

The classical energy of a magnetic moment in a magnetic field $\mathbf{B}$ (the *Zeeman energy*) is:

$$E_Z = -\boldsymbol{\mu}\cdot\mathbf{B}$$

If the field is uniform, and its direction is defined as the z axis (unit vector, $\hat{\mathbf{k}}$), then:

$$\mathbf{B} = B_0\hat{\mathbf{k}}, \qquad E_Z = -\mu_z B_0$$

The quantum-mechanical Zeeman energy will then be:

$$E_Z = -M_J g_L \beta_0 \tag{12.55}$$

This quantity must be added to the atom's energy when the atom is in a magnetic field. The result of this additional energy is that atomic transitions, when observed in a magnetic field, will spit into several lines; this is called the *Zeeman effect*.

The selection rules for changes in the projection quantum number, M_J, for electric dipole transitions are:

$\Delta M_J = 0$ with light polarized parallel to magnetic field

$\Delta M_J = \pm 1$ with light polarized perpendicular to magnetic field

Figure 12.16 shows a typical experimental arrangement. Light emitted by the sample in the field that comes out of the sides of the pole gap has a direction of propagation perpendicular to the magnetic field. Therefore the light traveling in this direction could be polarized either parallel or perpendicular to the field, and all transitions, $\Delta M_J = \pm 1$ *and* $\Delta M_J = 0$, will be seen. (Furthermore, they could be distinguished using a polarizer.) However, if there is a hole in the pole gap, we can observe radiation that is propagating parallel to the field; such light cannot have its plane of polarization parallel to the field, so photons corresponding to transitions with $\Delta M_J = 0$ will not be observed in this direction. Note that this is a restriction, not on the transitions the atoms can make, but on the direction in which the photons can travel. Photons for $\Delta M_J = 0$ transitions have a maximum probability for coming out of the side of the magnet and a minimum (*zero*) probability for coming out parallel to the field.

Example: Calculate the Zeeman effect on the $^2S_{1/2} \longrightarrow {}^2P_{1/2}$ transition of Na (one of the Na-D doublet lines). We will assume a field B_0 such that $\beta B_0/hc = 3$ cm^{-1}. The unperturbed frequency (without field) is $\nu_0 = 16,961$ cm^{-1}. (See Figure 12.17.)

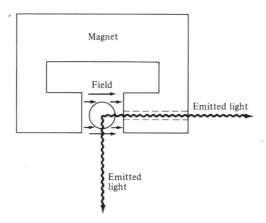

Figure 12.16 Zeeman effect (schematic). Light emitted to the side of the pole gap is propagated in a direction perpendicular to the field; it therefore could be polarized either parallel or perpendicular to the field. Light emitted toward the pole faces (which can be observed via a hole in the pole face) is propagated parallel to the magnetic field and, therefore, can only be polarized perpendicular to the field.

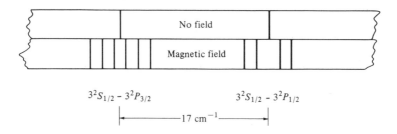

Figure 12.17 Zeeman effect on the Na-D doublet. The splitting of the lines will be proportional to the strength of the magnetic field (B). The 17-cm^{-1} splitting is due to spin-orbit coupling.

From Eq. (12.54) we calculate $g_L = 2$ for $^2S_{1/2}$ and $g_L = \frac{2}{3}$ for $^2P_{1/2}$.

$$^2S_{1/2} \begin{cases} M = \tfrac{1}{2}: & \dfrac{E}{hc} = -\left(\tfrac{1}{2}\right)(2)(3) = -3 \text{ cm}^{-1} & \text{(level } a) \\[2em] M = -\tfrac{1}{2}: & \dfrac{E}{hc} = +3 \text{ cm}^{-1} & \text{(level } b) \end{cases}$$

$$^2P_{1/2} \begin{cases} M = \tfrac{1}{2}: & \dfrac{E}{hc} = \tilde{\nu}_0 - \left(\tfrac{1}{2}\right)\left(\tfrac{2}{3}\right)(3) = \tilde{\nu}_0 - 1 \text{ cm}^{-1} & \text{(level } c) \\[2em] M = -\tfrac{1}{2}: & \dfrac{E}{hc} = \tilde{\nu}_0 + 1 \text{ cm}^{-1} & \text{(level } d) \end{cases}$$

The transition frequencies are:

$$\begin{aligned} c - a: & \quad \tilde{\nu} = \tilde{\nu}_0 + 2 \text{ cm}^{-1} \\ c - b: & \quad \tilde{\nu} = \tilde{\nu}_0 - 4 \text{ cm}^{-1} \\ d - a: & \quad \tilde{\nu} = \tilde{\nu}_0 + 4 \text{ cm}^{-1} \\ d - b: & \quad \tilde{\nu} = \tilde{\nu}_0 - 2 \text{ cm}^{-1} \end{aligned}$$

This transition will therefore be split into four lines by the magnetic field (Figure 12.17). ■

Exercise: What was the field strength (B_0) for the preceding example? (*Answer:* 64,266 gauss) ■

12.11 Photoelectron Spectroscopy

Chemists generally view atoms as a collection of individual electrons, each in its own "box," each with its own energy. This is certainly a useful point of view — it accounts neatly for the most fundamental fact of chemistry, the periodicity of the elements. We have seen that this picture has its limitations, both quantitatively (as in the calculation of the He IP) and conceptually (what *is* the energy of a 2p electron in atomic carbon?). For these reasons we have strongly emphasized the states of the atom as a whole, for these are the only legitimate observables. Nonetheless, the orbital picture is very useful, nowhere more than in the interpretation of X-ray photoelectron spectroscopy (XPES — the technique is also called ESCA, Electron Spectroscopy for Chemical Analysis).

The photoelectric effect was of great historical importance in physics; it was predicted by Einstein, who postulated that a photon of frequency ν had an energy $h\nu$ and that each electron emitted was a result of an interaction with a single photon. This fact is utilized in photoelectron spectroscopy. A sample is irradiated by high-energy photons (X-ray or ultraviolet), and the kinetic energy of the ejected electron is analyzed. The binding energy (BE) of the electron can be calculated by conservation of energy:

$$BE = h\nu - \text{(kinetic energy of electron)} \qquad (12.56)$$

A plot of the number of electrons reaching the detector (the detector current) vs. the electron kinetic energy (that is, the detector discriminator voltage) gives a spectrum of the energy levels of the ions that are produced when the electron is ejected by the photon. There are two ways to interpret these spectra — the atomic-state model, and the atomic-orbital model.

The simplest interpretation is for isolated atoms in a closed-shell ground state; the process by which a photon is absorbed by an atom M and a $3d$ electron is ejected can be represented as:

$$M(^1S_0) + h\nu \longrightarrow M^+(\overline{3}{}^2D) + e^-$$

(Since we earlier used the symbol 3^2D to represent the state of a $3d$ electron, we now introduce $\overline{3}$ to indicate the state of the *hole* left in the $3d$ shell.) The ejection of other electrons will give different ion states and different electron energies; for example, the ejection of a $3p$ electron gives:

$$M(^1S_0) + h\nu \longrightarrow M^+(\overline{3}{}^2P) + e^-$$

Thus the electron kinetic-energy spectrum traces the excited states of the *ion*. Since we are dealing in this case with single-electron states (actually single-hole states), the quantum numbers $l = L$ and lower-case letters (s, p, d, f) can be, and are, used to label the spectrum. Spin-orbit coupling ($j = l \pm \frac{1}{2}$) is also observed and indicated as $3p_{1/2}$, $3p_{3/2}$, $3d_{3/2}$, $3d_{5/2}$, and so on.

These spectra are also consistent with an orbital picture of an atom in which the electrons are arranged in shells, each shell labeled by quantum numbers n, l, j; each shell has a specific binding energy, which is that revealed by its XPES.

The power of the orbital model can be illustrated by the XPES of gold (Figure 12.18). Isolated Au atoms have a $d^{10}s^1$ configuration and a ^2S ground state; from this we would expect the spectrum of Au^+ to show singlets or triplets. In contradiction to this picture, the observed spectrum shows doublets. The reason for this seeming contradiction is that the XPES of Figure 12.18 is not of isolated Au atoms but, rather, of solid, metallic gold, and the atomic-state picture is irrelevant. In metals, the outer-shell (valence) electrons are not associated with a specific atom but are in a conduction band that is shared by all the atoms in the lattice. However, the inner-shell electrons are localized on a specific nucleus and have a shell structure very much

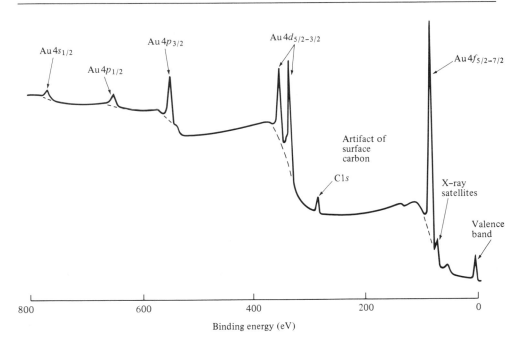

Figure 12.18 XPES spectrum of Au metal. The sharp peaks are evidence that the inner electrons are arranged in shells or orbitals. Spin-orbit splittings are clearly observed in this spectrum. (Spectrum by courtesy of Charles Ginnard)

like that of the isolated atom. Thus the holes produced by the X-ray can be analyzed in exactly the same manner as for an isolated ^1S atom. In this case the orbital picture is more revealing than the state picture, since, even in metals or compounds, it is still meaningful for the inner electrons.

The insensitivity of the inner electrons to environment (for example, isolated atom, metal lattice, compound) is responsible for the great value of XPES as an analytical technique. The ability to study a solid sample without vaporizing it in a carbon arc (as for atomic spectroscopy) is also a great advantage. XPES is specifically sensitive to the surface layer of the sample — a possible disadvantage, but a major benefit if the surface is what you *wish* to study.

An ion with a missing electron in an inner shell is in a highly excited state; when the ion decays to its ground state (a missing electron in its outermost shell), a great deal of energy must be dissipated. Much of this can be emitted radiatively, but, if the energy of the ion is greater than the atom's second IP, it may emit additional (secondary) electrons called *Auger electrons:*

$$M^+(\text{excited}) \longrightarrow M^{2+} + e^-$$

A related technique in which uv photons are used to study the valence electrons (VPES) will be discussed in Chapter 13. Also, Figure 13.23 shows a schematic diagram of the experiment.

12.12   Clocks, Nebulae, etc.

The study of radio and microwave emissions from outer space is a valuable tool for the astronomer. An important advance was made in 1956 by Ewen and Purcell, who discovered a ubiquitous microwave emission with a 21-cm wavelength coming from interstellar space. This emission turns out to be due to the hyperfine coupling between the electron and the proton in the ground (^2S) state of atomic hydrogen.

Hyperfine Coupling

The proton, like the electron, has an intrinsic angular momentum (spin) with a quantum number $I = \frac{1}{2}$. An electron in an s state has a finite probability of being at the site of the nucleus; this causes the Fermi contact interaction, a magnetic interaction between the electron and the nucleus (also called the hyperfine coupling). This interaction is of the form:

$$\hat{H}_{hf} = hA\hat{\mathbf{I}} \cdot \hat{\mathbf{S}} \tag{12.57}$$

(In this equation, A is the *hyperfine coupling constant*.) The hyperfine energy is best analyzed by introducing an operator for the total electron plus nuclear spin angular momentum:

$$\hat{\mathbf{T}} = \hat{\mathbf{I}} + \hat{\mathbf{S}} \tag{12.58}$$

As we have seen, the angular momentum is usually analyzed through squared operators such as $\hat{T}^2 = \hat{\mathbf{T}} \cdot \hat{\mathbf{T}}$ (an operator like $\hat{L}^2$ in Chapter 11). If Eq. (12.58) is squared, the result:

$$\hat{T}^2 = \hat{I}^2 + \hat{S}^2 + 2\hat{\mathbf{I}} \cdot \hat{\mathbf{S}}$$

can be used to eliminate $\hat{\mathbf{I}} \cdot \hat{\mathbf{S}}$ from Eq. (12.57), giving:

$$\hat{H}_{hf} = \tfrac{1}{2}hA(\hat{T}^2 - \hat{I}^2 - \hat{S}^2) \tag{12.59}$$

The eigenvalues of this Hamiltonian follow directly from the results of Section 11.7:

$$E_{hf} = \tfrac{1}{2}hA[T(T + 1) - I(I + 1) - S(S + 1)] \tag{12.60}$$

The quantum number T, using the usual rules for angular-momentum addition (Section 12.6), has the values:

$$T = (I + S), \ldots, (I - S)$$

In the hydrogen atom, $I = \frac{1}{2}$, $S = \frac{1}{2}$, and the levels are $T = 1$ and 0. This hyperfine coupling causes the ground ^2S state of H to be split into two levels with:

$$\frac{\Delta E}{h} = \frac{3 \times 10^{10} \text{ cm s}^{-1}}{21 \text{ cm}} = 1.4 \text{ GHz}$$

(We ignored this effect earlier because it is so small.)

Since this energy is in the microwave region, its frequency can be measured very accurately; the absorption frequency of the atomic hydrogen $^2S(T = 0 \rightarrow 1)$ transition is, precisely:

$$\nu = 1,420,405,751.786 \text{ cycles per second}$$

The hyperfine coupling constant of atomic hydrogen is one of nature's most accurately measured physical constants.

Atomic Clocks

When a frequency is reported to 13 significant figures, it is well to ask whether it is possible to measure the length of the second to such accuracy. In fact, the old definition of the second, based on the rotation period of the earth, was not that reproducible. The accuracy with which hyperfine frequencies can be measured suggested that they could provide a better time standard. Cesium atoms, like hydrogen, have a ^2S ground state, but a nuclear spin $I = \frac{7}{2}$. The hyperfine levels (for $T = 4$ and 3) have a transition with frequency (approximate, by the old time standard):

$$\nu = 9,192,631,770 \text{ cycles per second}$$

A second is now *defined* as the time required for the Cs hyperfine emission to make just that many cycles. This system, the *atomic clock,* provides a convenient and reproducible method for time measurement. (The shift of this frequency when a Cs clock is traveling at high speed in an airplane has been used recently to check Einstein's theory of relativity; the theory was correct; this would not have surprised Einstein.)

Nebular Spectra

Optical spectroscopy, especially visible and uv, has long been used as a tool in astronomy. Because of its characteristic spectrum, helium was discovered on the sun before it was found on earth; scientists concluded it was a new element because the spectrum matched no spectrum observed, to that time, from a terrestrial source. Some years ago, spectroscopic analysis of light from nebulae revealed a number of lines never observed in terrestrial sources; these were initially thought to be due to a new element — *nebulinium.* Later it was realized that these emissions were from forbidden transitions of the *inner* states of O^+, O^{2+}, and N^+. By *inner* states we mean those due to the ground electronic configuration — specifically:

$O^+(1s)^2(2s)^2(2p)^3$ with states $^4S, \, ^2D, \, ^2P$ (Figure 12.19)
$O^{2+}(1s)^2(2s)^2(2p)^2$ with states $^3P, \, ^1D, \, ^1S$ (cf. carbon, Table 12.5)
$N^+(1s)^2(2s)^2(2p)^2$ with states $^3P, \, ^1D, \, ^1S$

All the transitions among these states are forbidden by several selection rules including the Laporte rule — but forbidden is not forever. Normally, an excited atom or ion that cannot lose its energy by emission will lose it through collisions with the walls or with other atoms or molecules. Excited-state lifetimes of greater than 1 ms are considered long. The nebular density is so small that it is estimated that the average time between collisions is between 1 and 100 seconds — by that time, even "forbidden" transitions can occur.

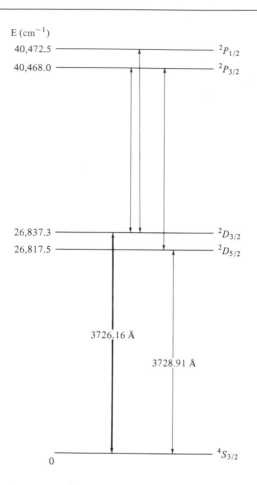

Figure 12.19 Nebular lines of O^+. Shown is a portion of the term diagram of oxygen atomic ion that accounts for some emission lines observed in nebulae. These are the "inner terms" — that is, they are all due to the ground-state configuration: $O^+(1s)^2 \cdot (2s)^2(2p)^3$. Compare this to Figure 12.11 for the isoelectronic nitrogen atom.

In a similar vein, the red and green auroral lines have been explained as $^1S \rightarrow {}^1D$ and $^1D \rightarrow {}^3P$ transitions (respectively) of neutral oxygen atoms. (See ref. 3.)

Postscript

In situations that are complicated enough to be called chemistry, the quantum-mechanical equations cannot be solved without numerical methods. Even where numerical solutions are possible, the accuracy is often poor compared to chemical-scale energies. We have seen in this chapter that an understanding of the principles of quantum theory, combined with experimental results, can provide a consistent and useful picture of even very complicated atoms. An understanding of angular momentum was a key to this process. This theme will be continued as we go on to more complicated, and much more interesting, systems — molecules.

A major objective of this chapter has been to put into perspective that widely used and abused concept — the orbital. An orbital is a one-electron wave function. The orbital approximation is to write a zero-order wave function of a many-electron atom as a product of these one-electron functions — the orbitals. (Thus, we can answer that archetypal freshman question, "What is an empty orbital?" Since any number or function raised to the zeroth power is equal to 1, an "empty orbital" means that the atom's wave function is multiplied by the number 1.)

The orbital picture of the atom, with electrons arranged in shells characterized by quantum numbers n, l, j, is useful but limited; it cannot, for example, explain the lower electronic states of carbon (Figure 12.10), nitrogen (Figure 12.11), or O^+ (Figure 12.19). The state model of the atom (with quantum numbers n, L, S, J) is mandatory if atomic spectra are to be understood.

Another theme in this chapter has been the role of angular momentum in the classification of atomic states and the interpretation of atomic spectra. This will continue, and the reader should be certain to understand terms such as "singlet," "doublet," and "triplet" before proceeding.

References

1. M. Karplus and R. N. Porter, *Atoms and Molecules*, 1971: New York, W. A. Benjamin, Inc.
2. J. W. Linnett, *Wave Mechanics and Valency*, 1960: London, Mctheuen & Co.
3. G. Herzberg, *Atomic Spectra and Atomic Structure*, 1944: New York: Dover Publications, Inc.
4. I. N. Levine, *Quantum Chemistry*, 1974: Boston: Allyn & Bacon.
5. I. N. Levine, *Molecular Spectroscopy*, 1975: New York, Wiley-Interscience.
6. C. J. Ballhausen, *Ligand Field Theory*, 1962: New York, McGraw-Hill Book Co.
7. C. E. Moore, *Atomic Energy Levels*, Nat. Bur. Standards Circ. 467, vol. 1, 1949; vol. 2, 1952; vol. 3, 1958.

Problems

12.1 Prove, by direct substitution, that the function

$$R = \sigma e^{-\sigma/2}$$

is an eigenfunction of the hydrogen-atom radial equation if $l = 1$. What is the eigenvalue?

12.2 Find the number and location of the radial nodes of the hydrogen 3s wave function.

12.3 Find the constant that will normalize $\Psi_{2p_z} = A\sigma e^{-\sigma/2} \cos \theta$.

12.4 Calculate the probability that a 1s electron is outside the Bohr radius.

12.5 Calculate $\langle r^2 \rangle$ for the 1s state.

12.6 Find the maximum of the radial distribution function $[f(r) = R^2 r^2]$ for a 2s function.

12.7   Calculate $\langle 1/r \rangle$ for the $1s$ hydrogen state.

12.8   In Problem 11.2 it was shown for the Bohr atom that $E = \frac{1}{2}V$ (the virial theorem). Calculate the quantum-mechanical $\langle V \rangle$ for the hydrogen atom $1s$ state and show that this theorem still holds.

12.9   What are the units of $R(r)$ (assumed normalized)? What are the units of R^2 and $R^2 r^2$? (Use the cgs system.)

12.10   (a) Show that the function whose general form is:

$$3d_{xz} = xzF(r)$$

can be written as a linear combination of the $3d_1$ and $3d_{-1}$ functions. [$F(r)$ is some function of r.]

(b) Show that the $d_{xy} = xyF(r)$ wave function is a linear combination of the d_2 and d_{-2} functions. The other linear combination should give $x^2 - y^2$.

12.11   Prove that:

$$\hat{L}_x(2p_x) = 0$$

where:

$$(2p_x) = \left(\frac{x}{a_0}\right) e^{-r/2a_0}$$

(This may be easier to do in Cartesian coordinates.)

12.12   Calculate the wavelength of light absorbed for a hydrogen atom to go from the $n = 1 \rightarrow 2$ states. Do the same for $n = 2 \rightarrow 3$.

12.13   A $2p$ electron with three degenerate states (p_x, p_y, p_z or p_1, p_0, p_{-1}) available is likely to spend its time in all of them, so its distribution will be given by (for example):

$$\psi = p_x + p_y + p_z$$

Because of orthogonality, the electron density is:

$$\psi^*\psi = p_x^2 + p_y^2 + p_z^2$$
$$= p_1 p_1^* + p_0 p_0 + p_{-1} p_{-1}^*$$

Show that this distribution is radially symmetric. This can also be interpreted to mean that a closed shell (for example, p^6) is radially symmetric and, hence, has no angular momentum.

12.14   Show that the one-electron spin function:

$$\chi = \alpha + \beta$$

is an eigenfunction of $\hat{S}_x$ and not an eigenfunction of $\hat{S}_z$. Can you find another such function?

12.15   Find linear combinations of the electron spin functions (α and β) that will be eigenfunctions of $\hat{S}_y$.

12.16   Calculate the second and third IP's of the lithium atom using formulas in this chapter. The observed values are 75.6 and 122.4 eV, respectively.

12.17   Use the variation-theory formula for the energy of a two-electron atom or ion to calculate the electron affinity of atomic hydrogen — that is, the energy of the process:

$$H(1s) + e = H^-(1s^2)$$

The experimental value is 0.754 eV.

12.18   (a) Prove that the function:

$$\chi_s = \frac{1}{\sqrt{2}}[\alpha(1)\beta(2) + \beta(1)\alpha(2)]$$

is normalized if the functions α and β are normalized and orthogonal.
(b) Prove that:

$$\chi_a = \frac{1}{\sqrt{2}}[\alpha(1)\beta(2) - \beta(1)\alpha(2)]$$

is normalized and is orthogonal to χ_s.

12.19   The first ionization potential of He is 24.6 eV (from the ground state). The transition $1^1S \rightarrow 2^1P$ absorbs energy at $\lambda = 584.4$ Å. Calculate the IP of He(2^1P) atoms.

12.20   In the He spectrum, the wavelengths of the transitions are:

$$1^1S - 2^1P: \qquad 584.4 \text{ Å}$$
$$2^1S - 2^1P: \qquad 20{,}582 \text{ Å}$$

Calculate $E(2^1S) - E(1^1S)$ from these numbers.

12.21   If two angular momenta, $j_1 = \frac{3}{2}$, $j_2 = \frac{5}{2}$, are added, what are the quantum numbers (J) of the total angular-momentum states? What are the degeneracies? How many states are there in all?

12.22   What are the quantum numbers L and S of a state ^4G? What is the total degeneracy of this state?

12.23   Derive the ground-state term symbols of the following atoms or ions: H, F, F$^-$, Na, Na$^+$, P, Sc, Sc^{2+}[Ar]$(3d)^1$.

12.24   Given that the operator form of the spin-orbit Hamiltonian is:

$$\hat{H}_{SO} = hcA\hat{\mathbf{L}}\cdot\hat{\mathbf{S}}$$

derive Eq. (12.45).

12.25   (a) Show that the spacing of a spin-orbit multiplet should be [by Eq. (12.45)]:

$$\frac{E_J - E_{J-1}}{hc} = AJ$$

(b) Test this formula on the data for oxygen 2^3P ground state:

$$J = 2: \qquad E/hc = 0 \text{ (ground state)}$$
$$J = 1: \qquad E/hc = 158.265 \text{ cm}^{-1}$$
$$J = 0: \qquad E/hc = 226.997 \text{ cm}^{-1}$$

12.26   The energies of the Ca 4^3P excited state are:

$$J = 0: \qquad E/hc = 15{,}157.9 \text{ cm}^{-1}$$
$$J = 1: \qquad E/hc = 15{,}210.0 \text{ cm}^{-1}$$
$$J = 2: \qquad E/hc = 15{,}315.9 \text{ cm}^{-1}$$

Calculate the spin orbit coupling constant of this state.

12.27 (a) What is the ground-state term symbol for the f^9 configuration? (b) What J values are possible, and which is lowest in energy? (c) What is the term symbol including J of the ground state and what is its degeneracy?

12.28 What are the possible J values for the following terms?
(a) ^6S (b) ^1F (c) ^2H (d) ^4P (e) ^3D

12.29 In the Na spectrum, the following wavelengths are absorbed:

$$4^2P - 3^2S: \qquad \lambda = 330.26 \text{ nm}$$

$$3^2P - 3^2S: \qquad \lambda = 589.593 \text{ nm}, \ 588.996 \text{ nm}$$

$$5^2S - 3^2P: \qquad \lambda = 616.073 \text{ nm}, \ 615.421 \text{ nm}$$

Calculate the energies of the 4^2P and 5^2S states with respect to the gound (3^2S) state.

12.30 The principal line of the potassium atomic spectrum ($4^2S \longleftrightarrow 4^2P$) is a doublet with wavelengths 3933.66 Å and 3968.47 Å. Calculate the spin-orbit coupling constant.

12.31 The transition

$$\text{Al[Ne]}(3s)^2(3p)^1 \longleftrightarrow \text{Al[Ne]}(3s)^2(4s)^1$$

has two lines, $\bar{\nu}_1 = 25{,}354.8 \text{ cm}^{-1}$, $\bar{\nu}_2 = 25{,}242.7 \text{ cm}^{-1}$. The transition

$$(\text{ground}) \longleftrightarrow \text{Al[Ne]}(3s)^2(3d)^1$$

has three lines, $\bar{\nu}_3 = 32{,}444.8 \text{ cm}^{-1}$, $\bar{\nu}_4 = 32{,}334.0 \text{ cm}^{-1}$, $\bar{\nu}_5 = 32{,}332.7 \text{ cm}^{-1}$. Sketch an energy-level diagram of the states involved and explain the source of all lines. Calculate spin-orbit coupling constants when possible.

12.32 What would be the multiplicity of a $^3P_2 \longrightarrow {}^3S_1$ transition in a magnetic field?

12.33 Diagram the Zeeman effect of a $^3S_1$ to $^3P_2$ transition. How many lines are there?

12.34 How many Zeeman lines will be found when the Na $^2S_{1/2} \longrightarrow {}^2P_{3/2}$ transition is examined in a magnetic field? What is g_L for $^2P_{3/2}$?

12.35 Although the Landé formula is strictly valid only for gaseous atoms or ions, it is often used to interpret the magnetic moments of ions in solution. Calculate the magnetic moment (in Bohr magnetons) of $Co^{2+}(3d)^7$ (4.1 to 5.2β observed) and $Er^{3+}(4f)^{11}$ (9.4β observed).

12.36 The application of an electric field ($\mathscr{E}$) to the H atom adds a potential:

$$V_S = e\mathscr{E}z = e\mathscr{E}r \cos \theta$$

to the Hamiltonian. This added potential could alter both the wave functions and the energy levels of the H atom. Show that the first-order Stark effect (that is, $\langle V_s \rangle$ as calculated with the unperturbed wave functions) is zero. (Do the integrals for the $1s$ and $2p$ functions as an example.)

12.37 The second-order Stark effect in atoms gives energy-level perturbation that depends on the *square* of M_J; $E_s = KM^2$, where K is a collection of constant factors. Diagram this effect on a spectrum line for a $^1S_0 \longrightarrow {}^1P_1$ transition.

12.38 In the XPES spectrum of Au (Figure 12.18), what is the theoretical intensity ratio of the $p_{3/2}$ to the $p_{1/2}$ line? Show that the effect will be less pronounced for the $4d$ and $4f$ lines. Assign the lines of the $4d$ transition. (In this context, "intensity" refers to height above background.)

12.39 Calculate the spin-orbit coupling constant of a $4d$ electron in Au from its XPES spectrum (Figure 12.18).

12.40 The magnetic moment of a nucleus is given by $|\mu| = g_N \beta_N \sqrt{I(I + 1)}$, where $\beta_N = e\hbar/2Mc$ is the nuclear magneton. The mass of a proton is 1836 times the mass of electron, and $g_N = 5.585$. Calculate the ratio of the electron/proton magnetic moments.

12.41 The ^{23}Na nucleus has spin $I = \frac{3}{2}$. What are the total nuclear spin quantum numbers (T) for Na^{2}S atoms?

12.42 The selection rule for hyperfine transitions is $\Delta T = \pm 1$. Derive a formula relating the transition frequency to the coupling constant and the spin I (assuming $S = \frac{1}{2}$). Calculate the hyperfine coupling constant of cesium from its frequency (given in text).

12.43 The theory of angular momentum shows that the integral of a triple product of Legendre polynomials:

$$\iint P_{l_1}^{m_1} P_{l_2}^{m_2} P_{l_3}^{m_3} \, d\Omega$$

is zero unless $m_1 + m_2 + m_3 = 0$ and the numbers l_1, l_2, and l_3 can form a triangle. Show that this requires a selection rule $\Delta L = \pm 1$ for atomic spectra.

I am a little world made cunningly of elements.

–John Donne

13

Diatomic Molecules

Over the last century, chemists have increasingly viewed their domain from the point of view of atomic-molecular theory: "Compounds are created from elements," the macroscopic view, has been gradually supplanted by the microscopic view, "Molecules are created from atoms." For this reason the understanding of molecular properties and structure has become a vital area of chemistry. In this chapter we shall make a beginning by examining the simplest of chemical compounds, the diatomic molecule.

13.1 The Molecular Hamiltonian

In the most complicated problem, we can always write the exact (nonrelativistic) Hamiltonian in a straightforward manner using the prescriptions developed in the previous chapters. The only new development in going from atoms to molecules is that the nuclear motions become much more important. When we were discussing atoms, we assumed that the nucleus was at the center of mass (COM) and that its kinetic energy, the translational motion of the COM, had no effect on the electronic states; these assumptions are *very* accurate. Many of the complications that arise when we deal with molecules result from the fact that the nuclei, which are the Coulombic force centers, do not lie at or near the COM and, in fact, are moving with respect to it.

To specify the positions of two nuclei and n electrons requires $3n + 6$ coordinates. By a COM transformation (Appendix III), three of these become the coordinates that specify the position of the COM in space. Three coordinates will give the location of the nuclei with respect to the COM; these are conveniently chosen to be R — the distance between the nuclei, and (θ, ϕ) — the orientation of the internuclear vector in space. These will be abbreviated as $\mathbf{R} = (R, \theta, \phi)$. Specification of the positions of n electrons requires $3n$ coordinates, which will be abbreviated:

$$\mathbf{r} = (x_1, y_1, z_1, x_2, y_2, z_2, \ldots, x_n, y_n, z_n)$$

In the COM coordinate system the Hamiltonian is:

$$\hat{H} = \hat{T}_N + V_{NN} + \hat{T}_e + V_{Ne} + V_{ee} \tag{13.1}$$

Using the distances as specified in Figure 13.1, the individual terms are:

1. The nuclear kinetic energy:

$$\hat{T}_N = -\frac{\hbar^2}{2\mu} \nabla_N^2 \tag{13.2}$$

where μ is the reduced mass:

$$\mu = \frac{m_A m_B}{m_A + m_B} \tag{13.3}$$

This is a kinetic-energy term for a single particle of mass μ; the fact that the motions of the two nuclei can be so represented is demonstrated by the COM transformation of Appendix III.

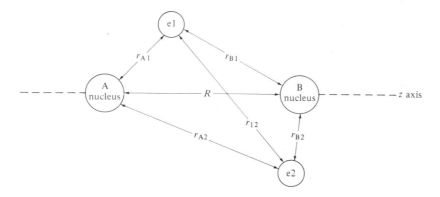

Figure 13.1   Diatomic coordinates. The nuclei are denoted A and B, and the electrons e1 and e2.

The Laplacian in polar coordinates is:

$$\nabla_N^2 = \frac{1}{R^2}\frac{\partial}{\partial R}R^2\frac{\partial}{\partial R} + \frac{1}{R^2 \sin\theta}\frac{\partial}{\partial\theta}\sin\theta\frac{\partial}{\partial\theta} + \frac{1}{R^2 \sin^2\theta}\frac{\partial^2}{\partial\phi^2} \qquad (13.4)$$

The angular part of this term is related to the angular-momentum operator for the rotation of the nuclear masses — this is generally given the symbol **J**; from Eq. (11.74) (note change in symbol) the operator for the total angular momentum is:

$$\hat{J}^2 = -\hbar^2\left(\frac{1}{\sin\theta}\frac{\partial}{\partial\theta}\sin\theta\frac{\partial}{\partial\theta} + \frac{1}{\sin^2\theta}\frac{\partial^2}{\partial\phi^2}\right)$$

Using this in Eq. (13.4) gives the nuclear kinetic-energy term of the diatomic molecule as:

$$\hat{T}_N = \frac{-\hbar^2}{2\mu R^2}\frac{\partial}{\partial R}R^2\frac{\partial}{\partial R} + \frac{1}{2\mu R^2}\hat{J}^2 \qquad (13.5)$$

2. The nuclear-nuclear Coulombic repulsion [which occurs because the nuclei each have a positive charge, $Z_A e$ and $Z_B e$, respectively]:

$$V_{NN} = \frac{Z_A Z_B e^2}{R} \qquad (13.6)$$

3. The electronic kinetic energy:

$$\hat{T}_e = \sum_{i=1}^{n}\frac{-\hbar^2}{2m}\nabla_i^2 \qquad (13.7)$$

In this case, the angular parts of the Laplacian are operators for the orbital angular momentum of the electrons.

4. The nuclear-electron Coulombic attraction [each electron (i) is attracted to each nucleus (A and B)]:

$$V_{Ne} = -\sum_{i=1}^{n} \left(\frac{Z_A e^2}{r_{Ai}} + \frac{Z_B e^2}{r_{Bi}} \right) \qquad \textbf{(13.8)}$$

5. The electron-electron Coulombic repulsion:

$$V_{ee} = \sum_{i>j} \frac{e^2}{r_{ij}} \qquad \textbf{(13.9)}$$

Note that the quantities r_{Ai}, r_{Bi}, and r_{ij} (Figure 13.1) are relative distances; Eqs. (13.8) and (13.9) would be very complicated if written using the COM coordinates. More about that later. (Figure 13.2 shows the Coulombic potential experienced by an electron due to the nuclei.)

Writing down the Hamiltonian is not solving the problem, but a proper understanding of the meaning and relative sizes of the terms can tell us a great deal about molecular energy levels without an excessive amount of mathematics; the manner of accomplishing this is called the Born-Oppenheimer approximation.

13.2 The Born-Oppenheimer Approximation

The Hamiltonian we have just described involves, for n electrons, $3n + 3$ coordinates. The three-dimensional particle in a box and the hydrogen atom illustrated how such problems could be simplified by separation of variables; the principle used was: if the Hamiltonian is a sum of independent terms, the wave function will be a product of single coordinate functions, and the energy will be a sum of terms, one for each coordinate motion. This separation was rigorous for the particle in a box and for the hydrogen atom, but when we dealt with helium and many-electron atoms, the rigorous separation of variables was prevented by the ee repulsion terms of the Hamiltonian. We resolved such problems approximately by solving the one-electron problem and constructing many-electron wave functions as products of the one-electron functions (the orbitals).

The same approximation can be used for diatomic molecules, but the one-electron problem still has six coordinates — three for the electron and the three COM nuclear coordinates for the nuclei (R, θ, ϕ). What is needed is a separation of variables for the nuclear and electronic coordinates; if this is possible, we can then discuss molecular energies in terms of a sum of electronic energies and the energy due to nuclear motion (rotation and vibration). This separation of variables cannot be done rigorously because the molecular Hamiltonian, even for one electron, is not a sum of independent terms; for example, the nuclear-electronic potential for one electron:

$$V_{Ne} = -\frac{Z_A e^2}{r_A} - \frac{Z_B e^2}{r_B}$$

involves distances (r_A and r_B) that depend on the electron's coordinates (**r**) and the internuclear distance (R) in a rather complicated fashion. (This potential is shown in Figure 13.2.) The separation of variables for molecules can be done using a method introduced by M. Born and J. R. Oppenheimer.

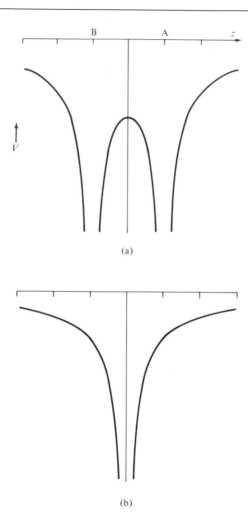

Figure 13.2 Diatomic nuclear potential. (a) The potential along the line connecting the two nuclei. (b) The potential along a line perpendicular to the internuclear line and passing through one of the nuclei.

The physical basis for the Born-Oppenheimer approximation is that the heavy nuclei move slowly compared to the electrons. The picture is that, as the nuclei plod slowly back and forth, the electrons can, at each distance R, adjust to a state with an energy that they would have if the nuclei were stationary at that position. Thus, the electronic energy is a parametric function of $R: E_e(R)$. The coordinate R is treated as a parameter (a constant whose value can be changed) rather than as a variable.

The first step is to define an effective Hamiltonian for a diatomic molecule with stationary nuclei; this will be Eq. (13.1) with the nuclear kinetic-energy term (T_N) set

J. Robert Oppenheimer (1904–1967)

The Born-Oppenheimer approximation, a development absolutely fundamental to the quantum theory of molecules and the interpretation of molecular spectra, was published while Oppenheimer was a postdoctoral student with Max Born in 1927. Upon his return to the United States as a professor of physics at Berkeley, Oppenheimer's interest turned to nuclear physics. He gained fame as the director of the Los Alamos laboratory during World War II and has been called the father of the "atomic bomb."

After the war, in 1954, Oppenheimer was accused of being disloyal and traitorous and, in a "hearing" by the AEC (really more of a star-chamber trial), had his security clearance revoked. There were spies and traitors at Los Alamos, of course, but if Oppenheimer was one of them, it is hard to comprehend why criminal charges were never brought against him. As it was, the penalty was meaningless (Oppenheimer no longer worked for the government anyway) and ultimately self-defeating, since it alienated many scientists without improving security. (Indeed, it appears that the F.B.I. was so busy watching Oppenheimer that they didn't notice when Klaus Fuchs sold the whole program to the Soviet Union.) There was a strong suspicion that the charges were motivated by Oppenheimer's lack of enthusiasm for the hydrogen bomb and that his rival in science and defense policy, Edward Teller, was responsible. In the long run, it was Teller who was the big loser; Oppenheimer became a martyr while Teller was ostracized by many scientists. Whatever the motivation, Oppenheimer was but one of the victims of the witch-hunt of the 50s, which, in the end, did far more harm than good to America's security.

to zero:

$$\hat{H}_{\text{eff}} = \hat{T}_e + V_{NN} + V_{Ne} + V_{ee} \tag{13.10}$$

Of course the distance R is in this operator [most obviously in V_{NN}, Eq. (13.6), but in V_{Ne} as well], but it is as a parameter rather than as a variable. The eigenfunctions of this operator will be functions of the electronic coordinates, $\mathbf{r}$, and R: $\psi_e(R, \mathbf{r})$. The eigenvalues are the electronic energies, $E_e(R)$, mentioned above:

$$\hat{H}_{\text{eff}}\psi_e(R, \mathbf{r}) = E_e(R)\psi_e(R, \mathbf{r}) \tag{13.11}$$

This equation, solved for each value of R, will give the electronic energies as a function of the internuclear distance R. Here is a major difference between atomic and molecular theory: the electronic energy levels of atoms were numbers, the eigenvalues of the Hamiltonian; the electronic energy "levels" of a molecule are sets of numbers that constitute the functions, $E_e(R)$. [In earlier chapters the point was strongly made that eigenvalues were constants — that is, numbers. Saying that the "eigenvalues" of Eq. (13.11) are parametric functions is a fine distinction indeed. This is in fact just an artifact of the approximation; when nuclear motion is included, the molecular energy levels are a set of constants.]

The full Schrödinger equation of the diatomic molecule can be obtained by restoring the nuclear kinetic energy, T_N [Eq. (13.5)]:

$$(\hat{T}_N + \hat{H}_{eff})\psi(\mathbf{r}, \mathbf{R}) = E\psi(\mathbf{r}, \mathbf{R}) \tag{13.12}$$

This is a true eigenvalue equation and its eigenvalues (E) are the energy levels of the molecule. We now approximate the wave function of Eq. (13.12) as a product of a nuclear function, $\psi_N(\mathbf{R})$, and the electronic functions that are solutions to Eq. (13.11):

$$\psi(\mathbf{r}, \mathbf{R}) = \psi_N(R, \theta, \phi)\psi_e(R, \mathbf{r})$$

Equation (13.12) then becomes:

$$(\hat{T}_N + \hat{H}_{eff})\psi_e\psi_N = E\psi_e\psi_N$$

Using Eq. (13.11), this becomes:

$$[\hat{T}_N + E_e(\mathbf{R})]\psi_e\psi_N = E\psi_e\psi_N$$

The electronic function (ψ_e) can now be divided out to give an eigenvalue equation that involves only nuclear coordinates:

$$[\hat{T}_N + E_e(R)]\psi_N(R, \theta, \phi) = E\psi_N(R, \theta, \phi) \tag{13.13}$$

The reader may recall that one of the first statements about Hamiltonians (in Section 11.4) was that the Hamiltonian of a single-particle system had a kinetic-energy and a potential-energy term, $\hat{H} = \hat{T} + V$. That is exactly what Eq. 13.13 represents, provided we identify the electronic energy, $E_e(R)$, as a potential energy for the nuclear motion. The Born-Oppenheimer approximation thus gives us the picture of the nuclei moving in a potential field due to the electrons.

In principle, the calculation of the energy levels of diatomic molecules could proceed in two steps: (1) Solve Eq. (13.11) to find the function $E_e(R)$. (2) For each of these functions, solve Eq. (13.13) for the energy levels; each curve is a separate *electronic state* with its own set of energy levels. The first step is by far the more difficult. When we discuss it (Section 13.4), we shall discover, as we did for atoms, that the simple, pictorial, easy-to-understand theories do not give quantitative results; the theoretical methods that can (and do) give accurate results are not easy to understand and cannot be carried out (except in the simplest cases) without very sophisticated computers.

On the other hand, the molecular energies are empirically observable — most directly by spectroscopy, but just as well from the thermodynamic functions of diatomic gases. The effect of the molecular energy levels on thermodynamics is discussed in Chapters 5 and 15; in this chapter we shall discuss only the spectroscopic aspects of molecular energy quantization. The next topic then is to discuss the solution of the nuclear equation (13.13), treating the electronic potential $E_e(R)$ as an empirical quantity whose nature can be deduced from the spectroscopic results.

13.3 Molecular Vibrations and Rotations

Figure 13.3 shows a typical potential curve (E_e) for the lowest (ground) electronic state of a stable diatomic molecule. This figure introduces two eminently measurable parameters. R_e, the distance between the nuclei when E_e is a minimum, is called the equilibrium *bond length*; D_e, the depth of the potential well, is closely related to the bond strength. That D_e is not precisely the thermodynamic dissociation energy of a diatomic molecule will be developed shortly.

Separation of Variables

Using Eq. (13.5) for the nuclear kinetic energy, Eq. (13.13) becomes:

$$\left[\frac{-\hbar^2}{2\mu R^2} \frac{\partial}{\partial R} R^2 \frac{\partial}{\partial R} + \frac{1}{2\mu R^2} \hat{J}^2 + E_e(R) \right] \psi_N(R, \theta, \phi) = E\psi_N(R, \theta, \phi) \quad \textbf{(13.14)}$$

This equation is analogous to the Schrödinger equation that we encountered for the hydrogen atom [Eq. (12.2)], but the magnitudes of the quantities in this equation are quite different [for example, the nuclear reduced mass of Eq. (13.14), which is three orders of magnitude greater than the electronic mass]; also the potential function is quite different — compare Figure 13.3 to Figure 12.1. Nonetheless, the procedure for the separation of variables in Eq. (13.14) is exactly like that we used for the hydrogen atom. The function of Eq. (13.14) is written as a product of a radial function, $\chi(R)$, and an angular function $Y(\theta, \phi)$:

$$\psi_N(R, \theta, \phi) = \chi(R)Y(\theta, \phi)$$

The angular part should, of course, be an eigenfunction of $\hat{J}^2$, and, as we have seen before, these are the spherical harmonics of Table 11.8 — however, we shall denote

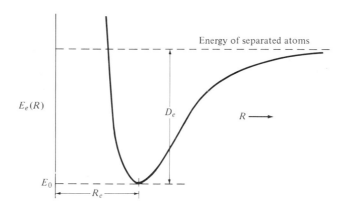

Figure 13.3 Diatomic electronic energy. The electronic energy of a stable diatomic molecule as a function of the internuclear distance (R). Two important parameters are introduced here: the well depth, D_e, and the equilibrium internuclear distance, R_e.

the quantum number as J rather than l. The eigenvalue equation for angular momentum [cf. Eq. (11.94) and Table 11.8] is:

$$\hat{J}^2 Y_{JM}(\theta, \phi) = J(J + 1)\hbar^2 Y_{JM}(\theta, \phi) \qquad (13.15)$$

Now, we substitute:

$$\psi_N = \chi(R) Y_{JM}(\theta, \phi) \qquad (13.16)$$

into Eq. (13.14), carry out the $\hat{J}^2$ operation [Eq. (13.15)], and, just as we did for the H atom, divide out the angular function to get a radial equation:

$$\left[\frac{-\hbar^2}{2\mu R^2} \frac{\partial}{\partial R} R^2 \frac{\partial}{\partial R} + \frac{J(J + 1)\hbar^2}{2\mu R^2} + E_e(R) \right] \chi(R) = E\chi(R) \qquad (13.17)$$

Equation (13.17) is analogous to the radial equation of the hydrogen atom [Eq. (12.4)], but both the form of the potential energy and the procedure from this point onward are quite different.

The procedure used for the solution of Eq. (13.17) is based on two physical facts that apply to the ground state of most stable diatomic molecules: (1) The vibrational energy (that which results from changes in R) is small compared to the depth of the potential well. (2) The rotational energy [the "J" term of Eq. (13.17)] is small compared to the vibrational energy. This permits the second stage of the Born-Oppenheimer approximation, which is to write the molecular energy of Eq. (13.17) as a sum of terms for the electronic energy at the bottom of the well (E_0, Figure 13.3), the vibrational (E_v) and the rotational (E_J) energy:

$$E = E_0 + E_v + E_J \qquad (13.18)$$

As always we have some choice as to the location of zero on the energy scale. There are several common choices: (1) If zero energy is defined as the separated-atom state, then $E_0 = -D_e$. (2) The zero can be chosen to make $E_0 = 0$; this is the most convenient one for present purposes. (3) A third choice, which is used in Chapter 5, is to choose the zero at the lowest vibrational state; this one will not be used in this chapter.

There are now two steps for calculating the energy: (1) Solve Eq. (13.17) with $J = 0$ — that is, for a molecule that is not rotating; the resulting energy is then entirely vibrational. (2) Next, assume that the vibrational energy is the same when the molecule is rotating, and calculate the rotational energy by the average-value theorem:

$$E_J = \frac{J(J + 1)\hbar^2}{2\mu} \langle R^{-2} \rangle \qquad (13.19)$$

using the vibrational wave functions of part 1 to calculate the average value.

Vibrational Energy

In this section we shall discuss Eq. (13.17) for the case $J = 0$:

$$\left[\frac{-\hbar^2}{2\mu R^2} \frac{\partial}{\partial R} R^2 \frac{\partial}{\partial R} + E_e(R) \right] \chi(R) = (E_0 + E_{\text{vib}})\chi(R)$$

We shall assume that the vibrational energy is small compared to the well depth and expand $E_e(R)$ in a Taylor series about $R = R_e$:

$$E_e(R) = E_0 + \left(\frac{\partial E_e}{\partial R}\right)_{R_e}(R - R_e) + \frac{1}{2}\left(\frac{\partial^2 E_e}{\partial R^2}\right)_{R_e}(R - R_e)^2$$

$$+ \frac{1}{3!}\left(\frac{\partial^3 E_e}{\partial R^3}\right)_{R_e}(R - R_e)^3 + \frac{1}{4!}\left(\frac{\partial^4 E_e}{\partial R^4}\right)_{R_e}(R - R_e)^4 + \cdots \qquad \textbf{(13.20)}$$

The first term of Eq. (13.20) is the energy at the minimum of the potential, which we have called E_0. The second term is zero because, at the minimum, the slope $(\partial E_e/\partial R) = 0$. If the third term is compared to Eq. (11.52), it will be apparent that this is a *harmonic potential* such as was discussed in Section 11.6 [cf. Eq. (11.52)]:

$$V(\text{harmonic}) = \frac{1}{2}kx^2 \qquad \text{with } x = R - R_e$$

For that reason, we shall define a *force constant* as the *curvature* of the potential curve $E_e(R)$ at the minimum ($R = R_e$):

$$k = \left(\frac{\partial^2 E_e}{\partial R^2}\right)_{R_e} \qquad \textbf{(13.21)}$$

The remainder of the terms in this expansion are lumped together as the *anharmonic potential*, V_{anh}. With these definitions, Eq. (13.20) becomes:

$$E_e(R) = E_0 + \frac{1}{2}k(R - R_e)^2 + V_{\text{anh}}$$

When this is substituted into Eq. (13.17) (with $J = 0$) we get:

$$\left[\frac{-\hbar^2}{2\mu R^2}\frac{\partial}{\partial R}R^2\frac{\partial}{\partial R} + \frac{1}{2}k(R - R_e)^2 + V_{\text{anh}}\right]\chi(R) = E_v\chi(R) \qquad \textbf{(13.22)}$$

where the vibrational energy has been defined as $E_v = E - E_0$. Since the leading term in the anharmonic potential is proportional to $(R - R_e)^3$, we will be able to neglect this term for small oscillations of R about R_e. Without the anharmonic term, Eq. (13.22) looks a good bit like the harmonic-oscillator problem [section 11.6, Eq. (11.54)], but the kinetic-energy operator is somewhat different. The operator:

$$\frac{1}{R^2}\frac{\partial}{\partial R}R^2\frac{\partial}{\partial R}$$

can be replaced by the simpler operator $\partial^2/\partial R^2$ if we replace the function $\chi(R)$ in Eq. (13.22) with another function (ψ_v) defined by:

$$\psi_v(R) = R\chi(R) \qquad \textbf{(13.23)}$$

Exercise: If $f(R)$ is a general function of R, show that:

$$\frac{1}{R^2}\frac{\partial}{\partial R}R^2\frac{\partial}{\partial R}\left(\frac{f}{R}\right) = \frac{1}{R}\frac{\partial^2 f}{\partial R^2} \qquad \blacksquare$$

If Eq. (13.23) is substituted into Eq. (13.22), the eigenvalue equation for the vibrational energy becomes:

$$\left[\frac{-\hbar^2}{2\mu} \frac{\partial^2}{\partial R^2} + \frac{1}{2} k(R - R_e)^2 + V_{anh} \right] \psi_v = E_v \psi_v \qquad (13.24)$$

Now, neglecting the anharmonic term and substituting

$$y \equiv \frac{R - R_e}{\alpha} \qquad \text{with } \alpha = \left(\frac{\hbar^2}{k\mu} \right)^{1/4}$$

Eq. (13.24) becomes:

$$\left(\frac{-\partial^2}{\partial y^2} + y^2 \right) \psi_v = \varepsilon_v \psi_v$$

with ε defined such that:

$$E_v = \varepsilon_v \frac{\hbar}{2} \sqrt{\frac{k}{\mu}}$$

This should be recognized as the harmonic-oscillator problem that was solved in Section 11.6; the reader should review that derivation at this point. The results will be briefly summarized:

The eigenfunctions are written in terms of the Hermite polynomials $H_v(y)$ as given in Table 11.2:

$$\psi_v = H_v(y)e^{-y^2/2} \qquad (13.25)$$

The vibrational energy is written in terms of the **vibrational constant:**

$$\omega_e = \frac{1}{2\pi c} \sqrt{\frac{k}{\mu}} \qquad (13.26)$$

(Note the change in notation as compared to Chapter 11; this notation is consistent with that normally used in molecular spectroscopy.)

The vibrational energy is given by:

$$E_v \cong (v + \tfrac{1}{2})hc\omega_e \qquad (13.27)$$

This equation is approximate because it is derived from Eq. (13.24) by neglecting the anharmonic potential.

Bond Energy

We are now in a position to discuss the relationship of the well-depth parameter, D_e, to the bond energy. The thermodynamic dissociation energy will be ΔU^θ for the process that breaks a diatom (AB) into atoms:

$$AB(gas) = A(gas) + B(gas)$$

The energy required to dissociate a molecule that is in a vibrational state with quantum number v will be:

$$D_e - (v + \tfrac{1}{2})hc\omega_e$$

The number of molecules in any vibrational state (N_v) will be given by the Boltzmann distribution; the population relative to the ground-state population is:

$$\frac{N_v}{N_0} = e^{-(E_v - E_0)/k_b T} = \exp\left(\frac{-vhc\omega_e}{k_b T}\right) \tag{13.28}$$

where h is Planck's constant and k_b is Boltzmann's constant. It will be helpful to recall that the collection of constants in Eq. (13.28) has the value:

$$\frac{hc}{k_b} = 1.4388 \text{ cm K}$$

and, at room temperature:

$$\frac{k_b T}{hc} \sim 200 \text{ cm}^{-1}$$

From Table 13.1 it can be seen that vibrational constants for stable diatomic molecules tend to be much larger than 200 cm^{-1}; this means that, at room temperature, most molecules will be in the lowest ($v = 0$) vibrational state.

Example: The vibrational constant of N_2 is 2360 cm^{-1} (Table 13.1). The relative populations of the $v = 1$ and $v = 0$ states of this molecule at 300 K are:

$$\frac{N_1}{N_0} = \exp\left[-\frac{1.4388(2360)}{300}\right] = 1.21 \times 10^{-5} \qquad \blacksquare$$

The energy required to dissociate a diatomic molecule, strictly an average over all vibrational states, is approximately the energy required to dissociate a molecule in the

Table 13.1   Ground-state molecular constants for some diatomic molecules

Molecule	Term symbol	ω_e (cm^{-1})	$\omega_e x_e$ (cm^{-1})	$\bar{B}_e$ (cm^{-1})	α_e (cm^{-1})	R_e (Å)	D_0 (eV)
^{1}H^{1}H	$^1\Sigma_g^+$	4395.2	117.91	60.81	2.993	0.7417	4.4763
^{12}C^{16}O	$^1\Sigma^+$	2170.21	13.461	1.9314	0.01748	1.1282	11.108
^{1}H^{35}Cl	$^1\Sigma^+$	2989.74	52.05	10.5909	0.3019	1.27460	4.430
^{35}Cl^{35}Cl	$^1\Sigma_g^+$	564.9	4.0	0.2438	0.0017	1.988	2.475
^{14}N^{14}N	$^1\Sigma_g^+$	2359.61	14.456	2.010	0.0187	1.094	9.756
^{1}H^{79}Br	$^1\Sigma^+$	2649.67	45.21	8.473	0.226	1.414	3.60
^{1}H^{127}I	$^1\Sigma^+$	2309.5	39.73	6.551	0.183	1.604	—
^{127}I^{35}Cl	$^1\Sigma^+$	384.18	1.465	0.1141619	0.00053	2.32069	2.152
^{127}I^{79}Br	$^1\Sigma^+$	268.4	0.78	—	—	—	1.817
^{1}H^{19}F	$^1\Sigma^+$	4138.52	90.069	20.939	0.770	0.9171	—
^{2}D^{19}F	$^1\Sigma^+$	2998.25	45.71	11.007	0.293	0.9170	—

$v = 0$ state; this quantity is called D_0 and differs from the well depth by the zero-point vibrational energy:

$$D_0 = D_e - \tfrac{1}{2}h\omega_e \qquad (13.29)$$

At zero kelvin, all molecules would be in the $v = 0$ state, so the energy of dissociation — ΔH_0^θ (as introduced in Chapter 5) — is just:

$$\Delta H_0^\theta = LD_0 \qquad (13.30)$$

where L is Avogadro's number.

The preceding discussion assumed that the energies D_e or D_0 had units of either ergs or joules; in fact they are generally tabulated as either cm^{-1} or eV. We shall treat this as a simple unit conversion; for placing bond energies in eV into perspective, it will be helpful to remember that:

$$1 \text{ eV/molecule} = 96,487 \text{ J/mole}$$

Example: Table 13.1 lists D_0 of H_2 as 4.4763 eV. For the reaction:

$$\tfrac{1}{2}H_2 = H$$

the zero-kelvin energy of reaction is:

$$\Delta H_0^\theta = \tfrac{1}{2}96,487(4.4763) = 215.95 \text{ kJ/mol}$$

This is the formation enthalpy per mole for H atoms; Table 6.4 gives a value of 216.0 kJ/mol for this quantity. ∎

Anharmonicity Corrections

More accurate vibrational energies are obtained by treating the anharmonic potential by perturbation theory. The first step (remembering what was done for the He atom, Section 12.3) is to use the average-value theorem to calculate:

$$\langle V_{\text{anh}} \rangle$$

using the harmonic-oscillator wave function [Eq. (13.25)]; this quantity is then added to the approximate energy of Eq. (13.27).

The complete procedure, which we shall not attempt to describe, results in an equation for the vibrational energy as a power series in the vibrational quantum number v:

$$E_v = hc\left[(v + \tfrac{1}{2})\omega_e - (v + \tfrac{1}{2})^2\omega_e x_e + (v + \tfrac{1}{2})^3\omega_e y_e + \cdots\right]$$

The higher-order terms are not usually necessary. The *first anharmonicity constant* ($\omega_e x_e$) is always a positive number; the *second anharmonicity constant* ($\omega_e y_e$) may be positive or negative and is usually a lot smaller than the first anharmonicity constant. These constants could be calculated if the potential $E_e(R)$ were known, but more often they are treated as empirical; some values of $\omega_e x_e$ are tabulated in Table 13.1. The second anharmonicity constants have been measured for only a few molecules and will not be needed for the accuracy we shall require; therefore, the vibrational energy in units of cm^{-1} (E_v/hc) will be taken to be:

$$\frac{E_v}{hc} = (v + \tfrac{1}{2})\omega_e - (v + \tfrac{1}{2})^2\omega_e x_e \qquad (13.31)$$

The Morse Potential

The approach we have been using did not require us to know what the potential function $E_e(R)$ was; only certain characteristics such as D_e, R_e, k (that is, ω_e), and $\omega_e x_e$ were needed. It is often useful to have a function that describes the potential, and for these purposes the empirical potential proposed by P. M. Morse is most commonly used (Figure 13.4):

$$E_e(R) = D_e(1 - e^{-\beta x})^2$$

$$x = R - R_e \qquad (13.32)$$

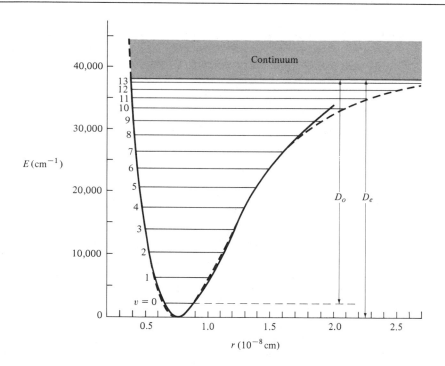

Figure 13.4 Potential curve for H_2. The solid curve is the experimental energy and the dashed curve is calculated with the Morse potential. Vibrational states are shown as horizontal straight lines with the quantum number (v) to the left. (From G. Herzberg, *Molecular Spectra and Molecular Structure: I. Spectra of Diatomic Molecules*, 1950: New York, Van Nostrand Reinhold Company, Fig. 50)

Exercise: Construct a graph of the Morse potential [Eq. (13.32)]. For purposes of graphing, a convenient choice of constants is $D_e = 10$, $\beta = 1$; for this value of β, you should plot the values of x: $-1 < x < 6$. It will also be instructive to repeat the plot with a different value of β such as 2.

From this graph you should deduce that D_e is the well depth (as in Figure 13.3) and the zero of energy is such that $E_e(R_e) = 0$. Also, larger values of β will give a narrower well. ∎

The Morse constant β can be related to the empirical parameters defined earlier by expanding Eq. (13.32) as a power series in x:

$$e^{-\beta x} = 1 - \beta x + \frac{\beta^2 x^2}{2} - \frac{\beta^3 x^3}{3!} + \cdots$$

$$E_e(R) = D_e(\beta^2 x^2 - \beta^3 x^3 + \tfrac{7}{12}\beta^4 x^4 - \cdots)$$

From this, by comparison to Eq. (13.20), we see that the Morse constant (β) is related to the harmonic force constant k [Eq. (13.21)] as:

$$\beta = \left(\frac{k}{2D_e}\right)^{1/2} \tag{13.33}$$

Calculation of vibrational energies using the Morse potential is not easy, but it can be done, and the solution is given by some advanced texts. A useful (but approximate) result that can be used to estimate anharmonicity constants is:

$$\omega_e x_e = \frac{\omega_e^2}{4(D_e/hc)} \tag{13.34}$$

Note that D_e/hc is the well depth D_e in units of cm^{-1}.

Vibrational Spectroscopy

The vibrational constants of heteronuclear diatomic molecules can be measured by direct vibrational spectroscopy. [The constants of homonuclear diatomics must be measured by other techniques such as Raman spectroscopy (Section 14.10) or electronic spectroscopy (Section 13.6)]. These absorptions commonly fall in the infrared (IR) region, so the technique is usually called IR spectroscopy. The harmonic-oscillator selection rule (Problem 11.40) requires the vibrational quantum number to change by 1; but because of the anharmonic potential, transitions for $\Delta v = 2, 3$, and so on may be observed weakly.

When $hc\omega_e \gg k_b T$ (a common case when T is near room temperature), most molecules will be in the $v = 0$ state and the principal absorption will be from the $v = 0$ to the $v = 1$ state; this is called the *fundamental*. From Eq. (13.31) we find:

$$v = 0: \quad \frac{E_0}{hc} = 0.5\omega_e - (0.5)^2\omega_e x_e$$

$$v = 1: \quad \frac{E_1}{hc} = 1.5\omega_e - (1.5)^2\omega_e x_e$$

The fundamental absorption frequency is, therefore:

$$\tilde{\nu}_0 = \frac{E_1 - E_0}{hc} = \omega_e - 2\omega_e x_e \tag{13.35}$$

A weaker absorption will be observed for the $v = 0 \longrightarrow 2$ transition; this is called the *first overtone,* because its frequency is approximately twice that of the fundamental. The reader should show that the frequency of the first overtone is:

$$\tilde{\nu}_1 = \frac{E_2 - E_0}{hc} = 2\omega_e - 6\omega_e x_e \tag{13.36}$$

The transition $v = 1 \rightarrow 2$ is an allowed transition like the fundamental, but it will be weak because fewer molecules will be in the $v = 1$ state than in the $v = 0$ state; since the strength of this absorption will increase with temperature, it is called the *hot band.* The reader should show that the frequency of the hot band is:

$$\tilde{\nu}_h = \frac{E_2 - E_1}{hc} = \omega_e - 4\omega_e x_e \tag{13.37}$$

Now is probably a good time to remember that we have been ignoring the molecular rotation energy. When this is included, we shall have to refine the term "vibrational absorption frequency."

Example: The observed vibrational absorption frequencies for $^1\text{H}^{35}\text{Cl}$ are $\tilde{\nu}_0 = 2885.9 \text{ cm}^{-1}$ and $\tilde{\nu}_1 = 5668.0 \text{ cm}^{-1}$. Calculate the vibrational constants. Also calculate the frequency and intensity at $T = 300 \text{ K}$ of the hot band relative to the fundamental. From Eqs. (13.35) and (13.36) (solved as simultaneous equations in two unknowns):

$$2\tilde{\nu}_0 - \tilde{\nu}_1 = 2\omega_e x_e$$

$$\omega_e x_e = \tfrac{1}{2}[2(2885.9) - 5668.0]$$

$$= 51.9 \text{ cm}^{-1}$$

Then, using Eq. (13.35), calculate the vibrational constant:

$$\omega_e = 2885.9 + 2(51.9) = 2989.7 \text{ cm}^{-1}$$

Now use Eq. (13.37) to calculate $\tilde{\nu}_h = 2782.1 \text{ cm}^{-1}$. The intensity of the hot band relative to the fundamental is approximately proportional to the relative population of the $v = 1$ and $v = 0$ states. From Eq. (13.28), using:

$$\frac{hc}{k_b} = 1.4388 \text{ cm K}$$

the population ratio at 300 K is:

$$\frac{N_1}{N_0} = e^{-1.4388(2989.7)/300} = 6 \times 10^{-7}$$

At this temperature, the hot band would be unobservable. ∎

Isotopes

In infrared and other types of spectroscopy, individual molecules are observed. Therefore, for example, molecules containing chlorine must be treated as a mixture of isotopes, ^{35}Cl (about 75%) and ^{37}Cl (about 25%), which will have different spectra. When calculating reduced masses, we must use the isotopic nuclear masses rather than average molecular weights.

Most of the problems and exercises of this chapter will assume that the nuclear mass is the mass number divided by Avogadro's number; for example, for ^{35}Cl:

$$m = \frac{35}{6.02217 \times 10^{23}} = 5.81186 \times 10^{-23} \text{ g}$$

This is not quite correct; the mass of this nucleus is actually 5.80647×10^{-23} g (error 0.1%). For accurate work, the exact nuclear masses should be looked up in a reference book. (When doing so, check whether the new atomic mass scale is being used; you can do this by checking ^{12}C, which should have a mass per mole of 12.0000 g exactly if the new scale is being used.)

Molecules that differ only by isotopic substitution, such as $^{1}H^{35}Cl$, $^{2}D^{35}Cl$, $^{1}H^{37}Cl$, will have different vibrational constants. Generally it is sufficiently accurate to assume that a change of only nuclear mass will not affect the electronic energy $E_e(R)$; therefore the force constant (k) and well depth (D_e) will be unchanged and the only difference will be the reduced mass. The change in the vibrational constant can then be deduced from Eq. (13.26). Suppose we know this constant for some diatomic molecule AX and wish to calculate it for an isotopically substituted molecule AX'. Using Eq. (13.26) and assuming k is unchanged, we get:

$$\frac{\omega_e(AX')}{\omega_e(AX)} = \left[\frac{\mu(AX)}{\mu(AX')}\right]^{1/2} \tag{13.38}$$

Similarly, Eq. (13.34) (derived from the Morse potential) demonstrates that the ratio of the anharmonicity constants will be (assuming that D_e is the same for AX and AX'):

$$\frac{\omega_e x_e(AX')}{\omega_e x_e(AX)} = \frac{\mu(AX)}{\mu(AX')} \tag{13.39}$$

The change in the vibrational constant upon isotopic substitution will affect the bond energy (D_0) as defined by Eq. (13.29). This is the principal cause of the "isotope effect" on chemical reactivity, and is most pronounced when deuterium is substituted for hydrogen.

Rotational Energy

The rotational energy of a diatomic molecule, as given by Eq. (13.19), is:

$$E_J = \frac{J(J+1)\hbar^2}{2\mu}\langle R^{-2}\rangle$$

As a first approximation we can assume that $R \cong R_e$; this is called the rigid-rotor approximation, and the rotational energy will be:

$$E_J \cong \frac{J(J + 1)\hbar^2}{2\mu R_e^2}$$

Using $\hbar = h/2\pi$ and defining the **moment of inertia:**

$$I = \mu R_e^2 \tag{13.40}$$

we get:

$$\text{(rigid rotor)} \qquad E_J = \frac{J(J + 1)h^2}{8\pi^2 I} \tag{13.41}$$

Since energies are usually measured spectroscopically in units of Hz (E/h) or cm^{-1} (E/hc), it is usual to define **rotational constants** (B) as follows:

$$B_e = \frac{h}{8\pi^2 I} \qquad \text{(units: Hz)}$$

$$\bar{B}_e = \frac{h}{8\pi^2 I c} \qquad \text{(units: cm}^{-1}) \tag{13.42}$$

With these definitions the rigid-rotor approximation to the rotational energy becomes:

$$\frac{E_J}{h} \cong J(J + 1)B_e \qquad \text{(units: Hz)}$$

$$\frac{E_J}{hc} \cong J(J + 1)\bar{B}_e \qquad \text{(units: cm}^{-1}) \tag{13.43}$$

[The distinction between B_e and $\bar{B}_e$ may not seem worthwhile. As we shall soon see, the rotational energy can be measured by either microwave or IR spectroscopy. In IR, the wavelength of the photon is measured and a value for $\bar{B}_e$ can be deduced; in the microwave region frequencies (Hz) are measured and hence B_e. The ratio of $B_e/\bar{B}_e$, measured on the same molecule by these techniques, is one of the most accurate methods for measuring the velocity of light (see Problem 13.16). For our purposes, $c = 3 \times 10^{10}$ cm/s will usually be accurate enough to convert one to the other.]

Equation (13.43) is approximate, partly because the average R^{-2} in Eq. (13.19) is not precisely R_e^{-2}; in fact, this average and, hence, the rotational constant will be different for each vibrational state. This correction is made by introducing a *vibration-rotation interaction* constant (α_e, see Table 13.1) and B_v, the rotational constant for the vibrational state with quantum number v:

$$\bar{B}_v = \bar{B}_e - (v + \tfrac{1}{2})\alpha_e$$

$$B_v = B_e - (v + \tfrac{1}{2})(\alpha_e c) \tag{13.44}$$

[Equation (13.44) presumes that the value of α_e has units of cm^{-1}; this is how it is often tabulated.]

Another approximation in the rigid-rotor energy [Eq. (13.43)] is the neglect of *centrifugal stretching;* a rotating molecule, especially one in a high J state, will be subject to centrifugal forces, which will tend to stretch the bond and, therefore, decrease $\langle R^{-2} \rangle$ and the rotational energy. This is accounted for by subtracting a term:

$$J^2(J + 1)^2 D_c$$

from the rotational energy; D_c is called the *centrifugal distortion constant.* (This quantity is more commonly denoted as D_e; the present notation is adopted to avoid confusion with the well-depth parameter of Figure 13.3.) It can be shown (ref. 1) that this constant is related to the other molecular constants as:

$$D_c = \frac{4B_e^3}{\omega_e^2 c^2} \qquad \text{(units: Hz)} \qquad \textbf{(13.45)}$$

Centrifugal distortion corrections usually are small and often are omitted.

Putting it all together, we get the following for the rotational energy:

$$\frac{E_J}{h} = J(J + 1)B_v - J^2(J + 1)^2 D_c$$

$$B_v = B_e - (v + \tfrac{1}{2}) \alpha_e c \qquad \textbf{(13.46)}$$

Example: Calculate the rotational energy of $^1H^{79}Br$ in the $J = 1$ state. The centrifugal stretching constant is calculated using Eq. (13.45) with data from Table 13.1:

$$B_e = \tilde{B}_e c = (8.473 \text{ cm}^{-1})(3 \times 10^{10}) = 2.542 \times 10^{11} \text{ Hz}$$

$$D_c = \frac{4(2.542 \times 10^{11})^3}{(2649.67)^2 (3 \times 10^{10})^2} = 1.04 \times 10^7 \text{ Hz}$$

From Eq. (13.44):

$$B_v = B_e - \frac{1}{2}(0.226)(3 \times 10^{10}) = 2.508 \times 10^{11} \text{ Hz}$$

From Eq. (13.46):

$$\frac{E_J}{h} = 2(2.508 \times 10^{11}) - 4(1.04 \times 10^7) = 5.016 \times 10^{11} \text{ Hz}$$

Note that the centrifugal stretching correction is negligible to this number of significant figures! Also, if the rotation energy is expressed in wavenumbers:

$$\frac{E_J}{hc} = \frac{5.016 \times 10^{11}}{3 \times 10^{10}} = 16.72 \text{ cm}^{-1}$$

it can be seen that the rotational energy is, as expected, very small compared to the vibrational energy (1325 cm^{-1} in the $v = 0$ state).   ∎

Rotational Spectroscopy

Pure rotational spectra, in which a molecule absorbs a photon and changes its J quantum number (leaving v unchanged), are usually found in the microwave region of the spectrum. (A few molecules, those with the smallest moments of inertia, will have rotational spectra in the far IR.) The selection rules will be discussed in Section 13.9; for now we shall simply state the results: (1) The molecule must have a permanent dipole moment; this means that heteronuclear diatomics (for example, HCl) will show microwave spectra but homonuclear diatomics (for example, H_2, Cl_2) will not. (2) The quantum number J must change by 1.

Example: We shall derive a formula for the frequencies of a rotational spectrum. Assume the molecule is in state J (with $v = 0$) and absorbs a photon to go to a new state J' (also $v = 0$) with $J' = J + 1$. The energies are, according to Eq. (13.46):

$$\text{final:} \qquad \frac{E_{J'}}{h} = J'(J' + 1)B_0 - [J'(J' + 1)]^2 D_c$$

$$\text{initial:} \qquad \frac{E_J}{h} = J(J + 1)B_0 - [J(J + 1)]^2 D_c$$

The frequency of absorption will be, therefore:

$$\nu = [J'(J' + 1) - J(J + 1)]B_0 - [(J')^2(J' + 1)^2 - J^2(J + 1)^2]D_c$$

Using $J' = J + 1$, the first term becomes:

$$(J + 1)(J + 2) - J(J + 1) = J^2 + 3J + 2 - J^2 - J = 2(J + 1)$$

The second term becomes:

$$(J + 1)^2(J + 2)^2 - J^2(J + 1)^2 = (J + 1)^2(J^2 + 4J + 4 - J^2)$$

$$= 4(J + 1)^3$$

Finally, the frequency of a $J \rightarrow J + 1$ transition is:

$$\nu = 2B_0(J + 1) - 4D_c(J + 1)^3 \qquad (13.47)$$

The rotational constant in the $v = 0$ state is, from Eq. (13.46):

$$B_0 = B_e - \tfrac{1}{2}\alpha_e \qquad\qquad\blacksquare$$

It was pointed out earlier that the vibrational-state spacing was large compared to kT, so observed absorption spectra will mostly be due to molecules in the $v = 0$ state. At room temperature, the rotational energy is usually small compared to $k_b T$, so many rotational states will be occupied. The ratio of the population of a J state (N_J) to the population of the $J = 0$ state (N_0) is given by Boltzmann's law as:

$$\frac{N_J}{N_0} = (2J + 1)e^{-J(J+1)hB_e/k_b T} \qquad (13.48)$$

Figure 13.5 shows an example of these populations for CO at 300 K.

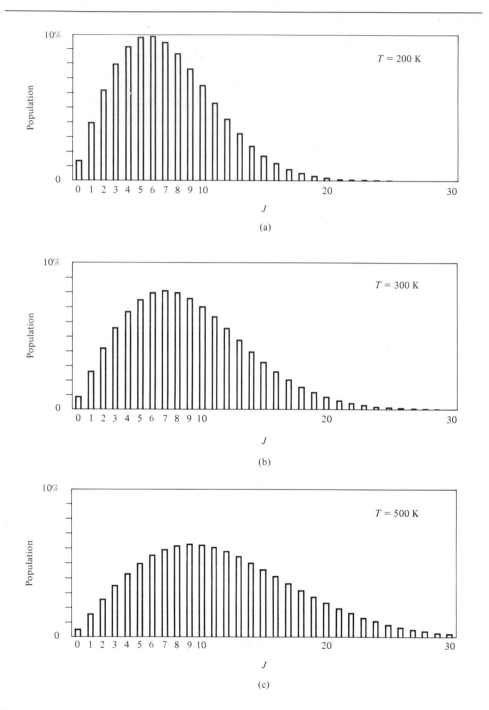

Figure 13.5 Rotational populations of carbon monoxide. The percent population in each quantum rotational state (J) is shown for T = 200 K, 300 K, and 500 K.

Example: Calculate the relative population of the $J = 2$ state of CO at 300 K using $B_e = 57.636$ GHz.

$$\frac{hB_e}{k_bT} = \frac{(6.6256 \times 10^{-27} \text{ erg sec})(57.636 \times 10^9 \text{ sec}^{-1})}{(1.38054 \times 10^{-16} \text{ erg/K})(300 \text{ K})}$$

$$= 9.2204 \times 10^{-3}$$

$$\frac{N_2}{N_0} = (5)e^{-6(9.2204 \times 10^{-3})} = 4.73$$

(Compare Figure 13.5.) ∎

Vibration-Rotation Spectroscopy

The earlier discussion of vibrational spectroscopy, and the formulas such as (13.35), (13.36), and (13.37) for the absorption frequencies, neglected the molecular rotation. Because of this rotation, IR spectra of gases consist of a band of lines centered on the frequencies of Eqs. (13.35) to (13.37); Figure 13.6 shows several examples of this phenomenon. (However, this structure will be resolved only if the sample is a gas at low pressure and the molecule has a small moment of inertia; also, the spectrometer must have better resolution than the typical analytical IR spectrometer found in chemistry laboratories.)

These spectra can be interpreted by combining Eq. (13.31) for the vibrational energy and Eq. (13.46) for the rotational energy to get the total energy of a state with quantum numbers v and J:

$$\frac{E_{v,J}}{hc} = (v + \tfrac{1}{2})\omega_e - (v + \tfrac{1}{2})^2\omega_e x_e + J(J + 1)\bar{B}_e - J(J + 1)(v + \tfrac{1}{2})\alpha_e \quad \textbf{(13.49)}$$

(The centrifugal distortion term has been omitted because it is usually too small to be measured by IR spectroscopy.)

The IR transitions will now occur between a ground state (v'', J'') and an excited state (v', J') with frequencies:

$$\tilde{\nu} = \frac{E_{v'J'} - E_{v''J''}}{hc}$$

The selection rules for such transitions will be discussed later (Section 13.9); for now we shall simply state them: (a) the molecule must have a dipole moment and (b) the rotational quantum number must increase or decrease by 1:

$$J' = J'' \pm 1$$

There are two distinct cases (diagrammed in Figure 13.7):

1. The R branch—$J' = J'' + 1$ ($J'' = 0, 1, 2, \dots$). In this case the molecule gains rotational energy and the absorption frequencies will be higher than the "pure vibration" frequencies of Eqs. (13.35) to (13.37).
2. The P branch—$J' = J'' - 1$ ($J'' = 1, 2, 3, \dots$). In this case the molecule loses rotational energy and the absorption frequencies will be lower than those of Eqs. (13.35) to (13.37).

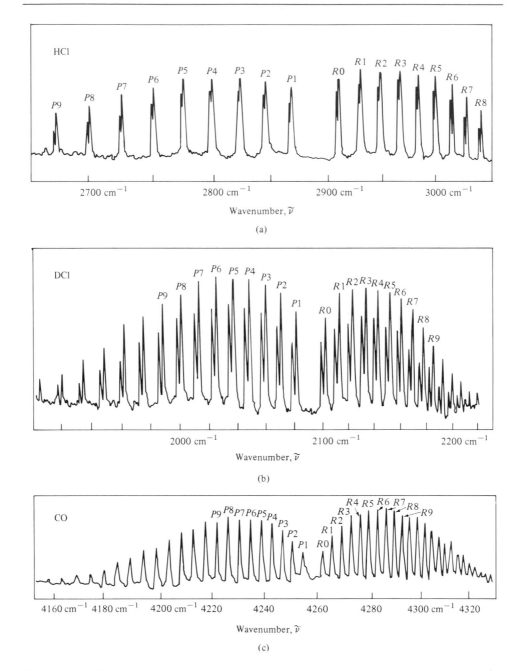

Figure 13.6   IR spectra of diatomic molecules. Compare the intensity pattern to Figure 13.5. The closely spaced lines in the spectra of HCl and DCl are due to the mass 35 and 37 chlorine isotopes.

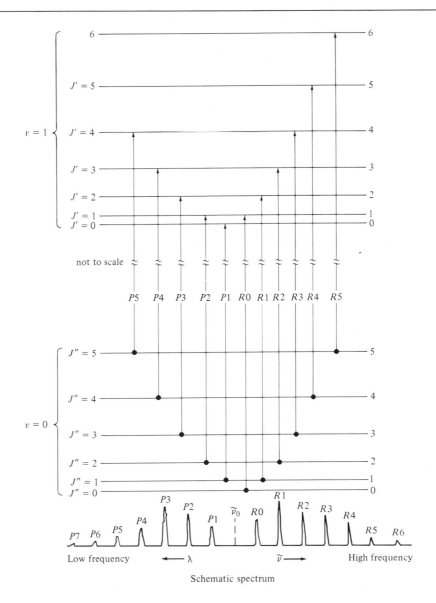

Figure 13.7 The rotational-vibrational energy levels of a diatomic molecule. The rotation-vibration states involved in the fundamental IR band are shown; vertical lines indicate the source of the lines of the spectrum, which is shown schematically at the bottom.

Another interesting fact will emerge if you compare the Boltzmann distribution of Figure 13.5 to the intensity pattern of the spectra (Figure 13.6). Each distinct line in the IR band is due to absorption by molecules in a particular J state. The intensity will therefore reflect the population of that state. The intensity of the R branch, $J'' = 0$ line (called $R\,0$) will be proportional to the population N_0; the R branch, $J'' = 1$ ($R\,1$) and P branch, $J'' = 1$ ($P\,1$) transitions will have intensities that are proportional to N_1, and so on. Thus the intensity pattern of the spectrum directly reflects the Boltzmann distribution and, hence, the temperature of the sample.

Exercise (optional): Show that the frequencies of the fundamental band ($v'' = 0$ to $v' = 1$) of an IR spectrum are:

$$(R \text{ branch}) \qquad \bar{\nu}_R = \bar{\nu}_0 + 2(J'' + 1)\bar{B}_e - (J'' + 1)(J'' + 3)\alpha_e \qquad \textbf{(13.50)}$$

$$(P \text{ branch}) \qquad \bar{\nu}_P = \bar{\nu}_0 - 2J''\bar{B}_e - J''(J'' - 2)\alpha_e \qquad \textbf{(13.51)}$$

The *band origin:*

$$\bar{\nu}_0 = \omega_e - 2\omega_e x_e$$

is just what was earlier called the fundamental vibration frequency [Eq. (13.35)]. ∎

The vibrational and rotational spectroscopy of diatomic molecules shows graphically the quantization of energy. However, we began our discussion with the electronic energy $E_e(R)$, and these spectroscopic results tell us very little about that. The most valuable result of the Born-Oppenheimer approximation has been to permit us to discuss rotational and vibrational energy independently of the more elusive electronic energy; now we must go back and see what can be learned about the electronic states of the molecule.

13.4   Molecular Orbital Theory

A number of theoretical methods have been used in the attempt to understand the nature of the chemical bond. By far the most widely used, general, and informative approach is the molecular orbital theory. This approach is quite similar to the atomic orbital method used in Chapter 12: the one-electron problem is solved and these eigenfunctions, called orbitals, are used to build up wave functions for the many-electron systems. For atoms the one-electron prototype was the hydrogen atom; for diatomic molecules the prototype is the one-electron system, the hydrogen molecule ion, H_2^+.

The Hamiltonian for H_2^+

If our system contains two hydrogen nuclei (labeled A and B) separated by a distance R, and a single electron whose separation from nucleus A is r_A and from nucleus B, r_B (see Figure 13.8), the effective Hamiltonian will be [from Eq. (13.10)]:

$$\hat{H}_{eff} = \frac{-\hbar^2}{2m}\nabla^2 - \frac{e^2}{r_A} - \frac{e^2}{r_B} + \frac{e^2}{R} \qquad \textbf{(13.52)}$$

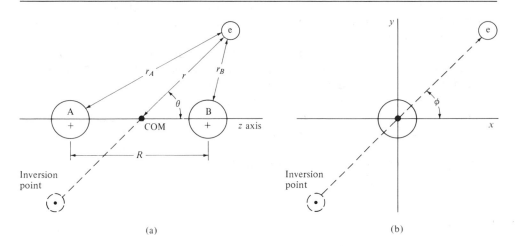

Figure 13.8   Coordinates used for the H_2^+ problem. (a) The COM coordinates shown are r and the polar angle θ; the azimuthal angle ϕ is the angle of rotation of the electron's vector about the z axis. [Panel (b), end view] The inversion point shown is the position that results upon inversion of the electron's COM coordinates.

The eigenfunctions of this operator, the state functions of H_2^+, are called molecular orbitals (MO's) and can be used to construct wave functions for more complex diatomic molecules. This is very similar to what was done for atoms, but there is an important difference. The one-electron atom, being a two-body problem, could be solved exactly. For diatomic molecules, however, the simplest case is already a three-body problem and, from the mathematical point of view, as complex as the helium atom. This means that our building blocks, the MO's, are already either very complicated or very approximate. The situation is, as it was for atoms, one where the simple, pictorial theories that so potently shape the chemist's picture of atoms and molecules are very inaccurate; the theories that can and do predict accurately such parameters as the bond strength and length are very complicated and do not provide simple pictures upon which an intuitive understanding can easily be based. In this book we shall never go beyond some very simple and crude approximations for MO's and molecular wave functions; for that reason, we shall start by looking at some very simple but *exact* properties that the diatomic molecular orbitals and wave functions must possess.

Symmetry and Angular Momentum

In our discussion of atoms in the last chapter we got a lot of information about the states of many-electron atoms by classifying those states according to their angular-momentum eigenvalues. A similar thing can be done for molecular orbitals and wave functions, not only with angular momentum but with regard to the *symmetry* that the orbitals and wave functions must possess.

Picture the diatomic molecule as two positive charges (motionless in the Born-Oppenheimer approximation) lying on the z axis of a molecule-fixed coordinate system (Figure 13.8) and surrounded by a fluid cloud of electrons. The electron density $(\psi^*\psi)$ must conform in shape to the shape of the electrostatic field of the nuclei; from an end view (along the z axis) it must be cylindrically symmetric and it must be symmetric end to end. [This statement is true, at least, for the ground state, in the absence of external electric or magnetic fields. Once the mysticism with which quantum theory is often endowed is swept away, this should be intuitively obvious. The same situation in atoms argues that the electron density must be spherically symmetric — that is, $\psi^*\psi = F(r)$. It might be objected that, for example, the $2p$ functions of the H atom are not spherically symmetric, but remember that these three functions are *degenerate,* so an electron in the $2p$ state would be shared among all of them; it was shown in Chapter 12 (problem 12.13) that $p_x^2 + p_y^2 + p_z^2 = F(r)$ was spherically symmetric.]

Inversion Symmetry

In a homonuclear diatomic molecule, because of the symmetry of the potential field of the nuclei, the electron density at some point (x, y, z) in the COM coordinate system must be the same as the density at the opposite point, $(-x, -y, -z)$. This can be expressed mathematically by introducing an operator called the *inversion* operator $(\hat{\imath})$ whose effect on a function is to change the sign of the Cartesian coordinates:

$$\hat{\imath}f(x, y, z) = f(-x, -y, -z)$$

There are three possibilities. The inversion operation may produce a new function; for example, if $f = (x + y - z)e^{-x}$, then:

$$\hat{\imath}(x + y - z)e^{-x} = (-x - y + z)e^{x}$$

Also, some functions will be unchanged by this operation; for example, if $f = (x^2 + y^2 + z^2)e^{-x^2}$, then:

$$\hat{\imath}(x^2 + y^2 + z^2)e^{-x^2} = (x^2 + y^2 + z^2)e^{-x^2}$$

Such functions are said to be *even* with respect to inversion. The German word *gerade* and the symbol "g" are used to denote such functions.

Functions may also be *odd* with respect to inversion; for example, if $f = (x + y + z)e^{-x^2}$, then:

$$\hat{\imath}(x + y + z)e^{-x^2} = -(x + y + z)e^{-x^2}$$

For such functions the German word *ungerade* and the symbol "u" are used. (Note that functions such as xe^{-x} are not ungerade, since the inverted function, $-xe^{x}$, is a new function, not just the old function multiplied by -1.)

The electron density of a homonuclear diatomic molecule must be an even function with respect to inversion:

$$\hat{\imath}(\psi^*\psi) = (\psi^*\psi) \tag{13.53}$$

Two types of state functions will meet this requirement, gerade and ungerade:

$$\hat{\imath}\psi_g = \psi_g, \qquad \hat{\imath}\psi_u = -\psi_u \tag{13.54}$$

We conclude that wave functions of homonuclear diatomic molecules, including the MO's that are wave functions of a one-electron molecule, can be classified as either gerade or ungerade. Wave functions of heteronuclear diatomics need not have this symmetry, since, if the nuclear charges are unequal, the nuclear potential does not have this symmetry. Therefore the g/u classification cannot be used for heteronuclear diatomic molecules.

Orbital Angular Momentum

The nuclear force field of a diatomic molecule is also cylindrically symmetric about the internuclear axis — the z axis. The wave function in COM polar coordinates will be a function of ϕ, θ, and R — $\psi(r, \theta, \phi)$ (Figure 13.8). However, the electron density, $\psi^*\psi$, may not depend on the azimuthal angle ϕ. This requirement could be met if the wave functions were of the form:

$$\psi(r, \theta, \phi) = F(r, \theta)e^{i\lambda\phi} \tag{13.55}$$

since then:

$$\psi^*\psi = [F(r, \theta)]^2$$

does not depend on the azimuthal angle ϕ.

It should not escape your attention that the function of Eq. (13.55) is an eigenfunction of the angular-momentum operator:

$$\hat{L}_z = \frac{\hbar}{i}\frac{\partial}{\partial\phi} \tag{13.56}$$

In fact:

$$\hat{L}_z\psi = \frac{\hbar}{i}\frac{\partial}{\partial\phi}[F(r, \theta)e^{i\lambda\phi}] = \lambda\hbar F(r, \theta)e^{i\lambda\phi}$$

$$\hat{L}_z\psi = \lambda\hbar\psi \tag{13.57}$$

so λ is simply the eigenvalue of this operator, and the quantity $\lambda\hbar$ is the value of the projection of the electron's angular momentum on the internuclear (z) axis. Recalling our discussion of the particle on a ring in Section 11.6, it may be evident that the boundary conditions on Eq. (13.57) will require that the quantum number λ be an integer:

$$\lambda = 0, \pm 1, \pm 2, \pm 3, \ldots$$

(In Chapters 11 and 12 we used the letter m for this quantum number. In molecular theory, the symbol m is reserved for the projection of the angular momentum on a space-fixed z axis as defined by an external electric or magnetic field, while λ denotes the projection on the molecule-fixed z axis in the COM coordinate system — cf. Figure 13.9.)

The implication of Eqs. (13.55) through (13.57) is that the operator $\hat{L}_z$, because of the symmetry of the nuclear electric field, commutes with the Hamiltonian, and, therefore, the eigenfunctions of a diatomic molecule can be classified according to

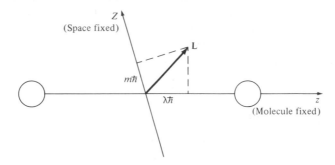

Figure 13.9 Projections of the angular-momentum vector. The quantum number m is for the projection on the space-fixed Z axis; the quantum number λ is for the projection of L on the diatomic axis, which is the z axis of a molecule-fixed coordinate system. Both projections must obey the quantum rules for angular momentum derived in Chapter 11.

their eigenvalues for this operator. The reader should recall the analogous situation for atoms (Chapter 12), when eigenfunctions of the Hamiltonian were classified according to their eigenvalues for the operator $\hat{L}^2$.

It is not essential that all eigenfunctions of the diatomic Hamiltonian be eigenfunctions of $\hat{L}_z$; it is possible to have state functions such as:

$$\psi_x = F(r, \theta)\sin(\lambda\phi), \qquad \psi_y = F(r, \theta)\cos(\lambda\phi)$$

However, such functions must always occur as degenerate pairs whose linear combinations do produce eigenfunctions of $\hat{L}_z$, such as Eq. (13.55) — for example:

$$F(r, \theta)\cos(\lambda\phi) + iF(r, \theta)\sin(\lambda\phi) = F(r, \theta)e^{i\lambda\phi}$$

(Remember Euler's equations, Appendix IV. This is exactly the situation we encountered for atoms where we had a choice of using the p_1 and p_{-1} functions or the p_x and p_y functions; both pairs of functions are eigenfunctions of the Hamiltonian, but only the first pair are eigenfunctions of $\hat{L}_z$.)

The fact that $\hat{L}_z$ commutes with the diatomic Hamiltonian is a consequence of the cylindrical symmetry of the potential; this means that the potential energy cannot depend on the azimuthal angle ϕ. However, the potential in which the electron moves does depend on the polar angle θ, a fact that may be obvious from an inspection of Figure 13.8 (cf. Figure 13.2). An important consequence of this is that the $\hat{L}^2$ operator [which contains the operator $\partial/\partial\theta$, cf. Eq. (11.74)] does *not* commute with the Hamiltonian, so its quantum number cannot be used to classify molecular states.

The projection quantum number λ is, however, very useful. Molecular orbitals are given symbols, which are the Greek equivalents of the letters s, p, d used for atomic orbitals (AO) (Table 13.2):

$$\lambda = 0, \ \pm1, \ \pm2$$

$$\text{orbital symbol:} \quad \sigma, \quad \pi, \quad \delta$$

Table 13.2 Orbital types

MO type	λ	Correlates to AO's	m
σ	0	s, $p_0(p_z)$, $d_0(d_{z^2})$	0
π	± 1	$p_{\pm 1}(p_x, p_y)$, $d_{\pm 1}(d_{xz}, d_{yz})$	± 1
δ	± 2	$d_{\pm 2}(d_{xy}, d_{x^2-y^2})$	± 2

Note, however, that there is a fundamental difference in the meaning of the AO and MO symbols. The AO symbols specify an "l-type" quantum number with a $2l + 1$ degeneracy, while the MO symbols specify a projection quantum number (non-degenerate if $\lambda = 0$, doubly degenerate otherwise). In accord with Eq. (13.55), σ-type functions do not depend on ϕ:

$$\psi_\sigma = F(r, \theta)$$

while π- and δ-type functions are the form:

$$\psi_\pi = F(r, \theta)e^{\pm i\phi} \qquad \psi_\delta = F(r, \theta)e^{\pm 2i\phi}$$

(The function F is not the same for all, however.)

Mirror Planes

Another type of symmetry that diatomic molecules may have is the mirror planes, denoted $\hat{\sigma}$. Homonuclear diatomic molecules will have a plane of symmetry through the COM and perpendicular to the bond axis (the xy plane). This is called a *horizontal mirror plane,* denoted $\hat{\sigma}_h$. The homonuclear molecular orbitals must be eigenfunctions of this operator:

$$\hat{\sigma}_h \psi = \pm \psi$$

If the orbital is odd with respect to this operation (eigenvalue $= -1$), there must be a *node* in the center. Such orbitals are called *antibonding,* because MO's with such a node will contribute little to the bond stability—this point will be elaborated upon shortly. Antibonding orbitals are denoted with an asterisk: σ^*, π^*, and so on. Bonding orbitals (eigenvalue for $\hat{\sigma}_h = +1$, no node) are denoted by symbols without the asterisk.

Heteronuclear diatomic molecules need not have this symmetry, but their orbitals are still classified as bonding or antibonding based on the absence or presence of a node in the electron density between the nuclei.

Because of the cylindrical symmetry of the electron density about the bond axis (the z axis in the COM system), both homo- and heteronuclear diatomic molecules will have an infinity of *vertical mirror planes* (denoted $\hat{\sigma}_v$) that contain the z axis (for example, the xz and yz planes). This symmetry is somewhat less important than the others and is, by far, the most difficult to understand; we shall therefore wait a while before discussing it.

From this point onward, we shall be dealing with fairly crude approximations to the diatomic molecular orbitals and wave functions. It cannot be overemphasized that the true functions (whatever they may be) must have the same characteristics as will the approximate functions; that is, they must be eigenfunctions of the operators $\hat{i}$, $\hat{L}_z$, $\hat{\sigma}_h$, and $\hat{\sigma}_v$. This fact will permit us to get a great deal of exact information from approximate theories.

Approximate Molecular Orbitals for H_2^+: LCAO

The most commonly used and easily pictured method for constructing MO functions is by use of linear combinations of the atomic orbitals (LCAO). This method has the advantage that the MO functions automatically have the correct limiting behavior — that is, when the bond distance (R) of the molecule is increased to infinity, the molecule becomes two atoms; therefore the molecular orbital functions should become atomic orbital functions when $R \rightarrow \infty$.

For the ground state of H_2^+, it seems reasonable to use the $1s$ atomic orbitals; there will be two such orbitals, $1s_A$ centered on nucleus A, and $1s_B$ centered on B:

$$(1s_A) = Ke^{r_A/a_0}, \qquad (1s_B) = Ke^{-r_B/a_0} \qquad \textbf{(13.58)}$$

Example: The implications of the term "centered on" can be illustrated by examining a simple function — the circle. The equation of a circle of radius a whose center is at the origin is, in Cartesian coordinates:

$$x^2 + y^2 = a^2$$

In polar coordinates, the equation is even simpler: $r = a$. However, if the center of the circle is at the point $r = b$, $\phi = \phi_0$ (in polar coordinates), the formula is much more complicated:

$$r^2 - 2rb \cos(\phi - \phi_0) = a^2 - b^2$$

The functions $(1s_A)$ and $(1s_B)$ are mathematically simple *only* in a coordinate system centered on nuclei A and B, respectively. This fact causes complications in the so-called two-center integrals — those involving products of $(1s_A)$ and $(1s_B)$.   ∎

Using the two functions of Eq. (13.58), we can construct two MO's:

$$\Phi_+ = C_+(1s_A + 1s_B), \qquad \Phi_- = C_-(1s_A - 1s_B) \qquad \textbf{(13.59)}$$

In Eq. (13.59), the quantities C_+ and C_- are the normalization "constants" — they are not really constant, except in the Born-Oppenheimer sense, but are parametric functions of the internuclear distance R.

We can normalize Φ_+ by requiring:

$$\int \Phi_+^* \Phi_+ \, d\tau = 1$$

Using Eq. (13.59), we get:

$$C_+^2 \left[\int (1s_A)^2 \, d\tau + \int (1s_B)^2 \, d\tau + 2 \int (1s_A)(1s_B) \, d\tau \right] = 1$$

Assuming that we are using normalized AO's, the first two integrals will be equal to 1, so:

$$C_+^2\left[1 + 1 + 2\int (1s_A)(1s_B)\,d\tau\right] = 1$$

The third integral is called the overlap integral:

$$S \equiv \int (1s_A)(1s_B)\,d\tau \qquad (13.60a)$$

Little benefit would derive from grinding through this or any of the other integrals in this section, so we shall simply state the results. In this case, if we define:

$$\rho \equiv \frac{R}{a_0}$$

as the internuclear distance in units of Bohr's radius (such units are called *atomic units* or a.u.), the overlap integral can be shown to be:

$$S = \left(\frac{\rho^2}{3} + \rho + 1\right)e^{-\rho} \qquad (13.60b)$$

Now, the normalization condition for this MO becomes:

$$2C_+^2(1 + S) = 1$$

A similar procedure for the Φ_- function will give:

$$2C_-^2(1 - S) = 1$$

Therefore, the properly normalized MO's are:

$$\Phi_+ = \frac{(1s_A) + (1s_B)}{\sqrt{2}(1 + S)^{1/2}}, \qquad \Phi_- = \frac{(1s_A) - (1s_B)}{\sqrt{2}(1 - S)^{1/2}} \qquad (13.61)$$

It will be instructive to look at the electron density for these functions. The squares of the functions of Eq. (13.61) are:

$$\Phi_+^2 = \frac{\frac{1}{2}[(1s_A)^2 + (1s_B)^2 + 2(1s_A)(1s_B)]}{1 + S}$$

$$\Phi_-^2 = \frac{\frac{1}{2}[(1s_A)^2 + (1s_B)^2 - 2(1s_A)(1s_B)]}{1 - S} \qquad (13.62)$$

The principal difference between these two functions is the product term, $(1s_A)(1s_B)$; this is a function whose value exists primarily in the region between the nuclei. Figure 13.10 shows each of these terms, as well as the sums and differences that occur in Eq. (13.62).

The first thing that can be seen in Figure 13.10 is that the Φ_- function has a mode between the nuclei while Φ_+ does not; therefore Φ_- is the type of orbital that we earlier referred to as antibonding. In all of the problems we have solved exactly — the particle in a box, the harmonic oscillator, and the hydrogen atom — the state functions with the larger number of nodes had higher energy. Therefore we expect Φ_- to have a higher energy than Φ_+.

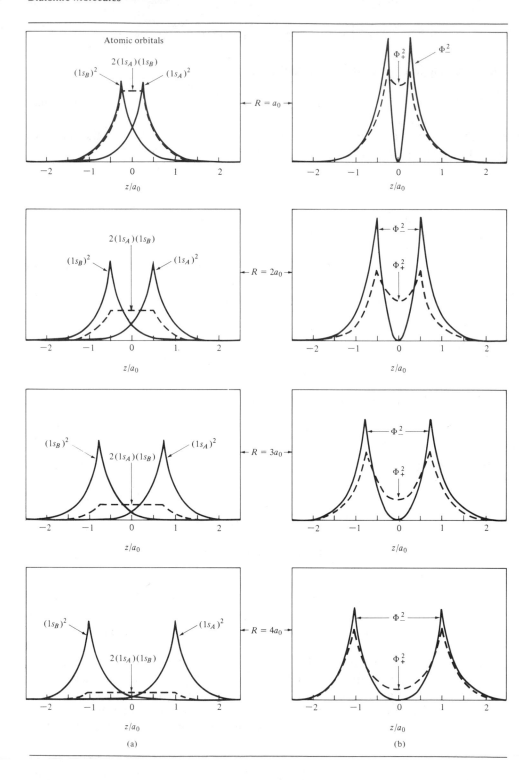

(a)

(b)

But will Φ_+ be stable? That is, will a bond form? The nuclei of the diatomic molecule are both positively charged, so they will repel each other strongly as they are brought together. Now, if we had an electron to "glue" these antipathetic particles together, we would probably wish to place it between the nuclei in order to "neutralize" their repulsive forces. Again, from Figure 13.10 it can be seen that the Φ_+ function does just that—it concentrates the electron density in the region between the nuclei—and this concentration can be expected to provide a stable bond.

The same question can be answered quantitatively by calculating the bond energy. These functions are not, of course, eigenfunctions of the effective Hamiltonian, but the average-value theorem can be used to calculate a first-order energy:

$$E_+^{(1)} = \langle \hat{H}_{\text{eff}} \rangle$$

$$\langle \hat{H}_{\text{eff}} \rangle = \int \Phi_\pm^* (\hat{H}_{\text{eff}} \Phi_\pm) \, d\tau \qquad (13.63)$$

With the functions of Eq. (13.61), Eq. (13.63) will be seen to involve the following types of integrals:

$$H_{AA} = \int (1s_A) \hat{H}_{\text{eff}} (1s_A) \, d\tau$$

$$H_{BB} = \int (1s_B) \hat{H}_{\text{eff}} (1s_B) \, d\tau$$

$$H_{AB} = \int (1s_A) \hat{H}_{\text{eff}} (1s_B) \, d\tau$$

$$H_{BA} = \int (1s_B) \hat{H}_{\text{eff}} (1s_A) \, d\tau \qquad (13.64)$$

Because of symmetry (A and B refer to the same type of nucleus) the following equalities must hold: $H_{AA} = H_{BB}$; $H_{AB} = H_{BA}$. The values of these integrals can be shown to be:

$$H_{AA} = H_{BB} = \frac{-e^2}{2a_0} + e^2 \left(\frac{1}{a_0} + \frac{1}{R} \right) e^{-2\rho}$$

$$H_{AB} = H_{BA} = S \left(\frac{e^2}{R} - \frac{e^2}{2a_0} \right) - \frac{e^2}{a_0} (1 + \rho) e^{-\rho} \qquad (13.65)$$

As before, S is the overlap integral [Eq. (13.60)], $\rho = R/a_0$, and a_0 is Bohr's radius. The reader should be able to show that, in terms of the symbols defined by

---

Figure 13.10 Electron densities for the hydrogen molecule ion. Figures show a cross section of electron density along the internuclear vector for several values of the internuclear distance (R). The left panels show the individual terms on the right-hand side of Eq. (13.62); the right-hand panels show the sums, which give the electron density for the bonding (Φ_+) and antibonding (Φ_-) states, respectively. Note that the bonding combination concentrates more of the electron density between the nuclei.

Eq. (13.64), the first-order energy of Eq. (13.63) will be:

$$E_+^{(1)} = \frac{H_{AA} + H_{AB}}{1 + S} \tag{13.66a}$$

By a similar procedure, the energy of the antibonding orbital Φ_- can be shown to be:

$$E_- = \frac{H_{AA} - H_{AB}}{1 - S} \tag{13.66b}$$

The energies of Eq. (13.66), as calculated using Eqs. (13.65) and (13.60), are plotted in Figure 13.11. [*Note:* The energies of Eq. (13.66) are referred to the ionized state ($H^+ + H^+ + e$) as zero; hence $E_+ \rightarrow -E_H$ as $R \rightarrow \infty$ ($E_H = e^2/2a_0 = 13.6$ eV). It is customary to add E_H to these energies so that the zero of energy is the dissociated state ($H + H^+$); this has been done in Figure 13.11.] The curve for the bonding state, E_+, can be seen to be qualitatively similar to what we expected for a stable molecule (cf. Figure 13.3), so this theory predicts the most fundamental fact about H_2^+ —that a stable bond will form. This is major victory for LCAO-MO theory, and it gives us the picture of bond formation as a consequence of constructive interference between the wave functions of the bonding atoms. Quantitatively the calculation is not too accurate; for the equilibrium bond length, it predicts $R_e = 2.5a_0 = 1.32$ Å (the experimental value is 1.06 Å); for the well depth, this calculation predicts $D_e = 1.76$ eV (experimental: 2.79 eV). This is, of course, just a first-order estimate, and better methods give more accurate answers. Despite the computational problems, the LCAO picture of molecular bonding, because it is easily pictured and used, has become a major part of the chemist's understanding of bond formation.

The energy curve for the Φ_- state (Figure 13.11) has no minimum and therefore does not represent a bonding state; the name we have been using for it, antibonding, is thus justified.

Classification of MO's

According to the classification scheme outlined earlier, we must determine the effect of the symmetry operations on the MO's of Eq. (13.61). The effect of the inversion operator $\hat{\imath}$ will be to exchange the atomic orbitals centered on A and B:

$$\hat{\imath}(1s_A) = (1s_B), \qquad \hat{\imath}(1s_B) = (1s_A)$$

The effect on the MO's of Eq. (13.61) is, therefore:

$$\hat{\imath}\Phi_+ = +\Phi_+, \qquad \hat{\imath}\Phi_- = -\Phi_-$$

The bonding orbital is therefore *gerade,* while the antibonding orbital is *ungerade.* Also, these orbitals contain atomic orbitals that have an angular-momentum quantum number $m = 0$; therefore they will have $\lambda = 0$—that is, they are σ-type orbitals. The proper names for the MO's of Eq. (13.61) are therefore:

$$\Phi_+ = (\sigma_g 1s), \qquad \Phi_- = (\sigma_u^* 1s)$$

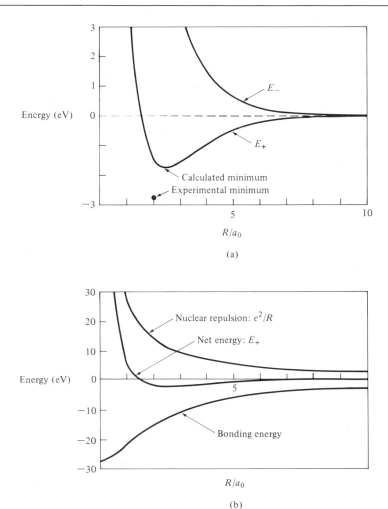

Figure 13.11 Calculated electronic energy of the hydrogen molecule ion. The theoretical energies given by Eq. (13.66) [using Eq. (13.65)] are plotted vs. internuclear distance (R). Panel (b) shows how the bonding energy breaks down into nuclear repulsion and electronic bonding; the fact that two large terms add to give a small result (the net energy) means that small errors in calculating the electronic energy are magnified. In panel (a) it can be seen that the position of the calculated minimum is quite a bit different than the experimental value.

The other MO's excited states of the one-electron molecule, are constructed from the atomic orbitals in a similar manner. All atomic orbitals with $m = 0$ will form σ-type MO's ($\lambda = 0$); some of these are (Figure 13.12):

$$(\sigma_g 2s) = c[(2s_A) + (2s_B)], \qquad (\sigma_u^* 2s) = c[(2s_A) - (2s_B)]$$

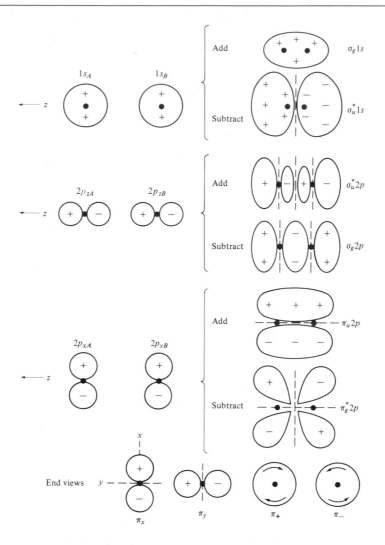

Figure 13.12 Diatomic molecular orbitals from LCAO. Nodes are shown as dashed lines. The plus and minus signs denote the phase of the orbital. The end views for all σ MO's are cylindrically symmetric. Two end views are shown for the π orbitals corresponding to the two pictures developed in the text for this degenerate pair. The arrows denote a rotational angular momentum for the two orbitals π_+ and π_-.

and similarly for $3s$, $4s$, and so on. The p_0 (that is, p_z) orbitals also have $m = 0$, so the following MO's result:

$$(\sigma_g 2p) = c[(2p_{0A}) - (2p_{0B})], \qquad (\sigma_u^* 2p) = c[(2p_{0A}) + (2p_{0B})]$$

and similarly for $3p_0$ and so on. Likewise the $d_0(d_z2)$ atomic orbitals give σ-type MO's (see Table 13.2).

Atomic orbitals with $m = \pm 1$ will form π MO's with $\lambda = \pm 1$; these will occur as degenerate pairs. Such orbitals could be visualized either as linear combinations of the p_1 and p_{-1} orbitals or (more easily) of the p_x and p_y orbitals; in either case, a degenerate pair results:

$$\left. \begin{array}{l} \pi_+ = (2p_{1A}) + (2p_{1B}) \\ \pi_- = (2p_{-1A}) + (2p_{-1B}) \end{array} \right\} \quad \text{or} \quad \left\{ \begin{array}{l} \pi_x = (2p_{xA}) + (2p_{xB}) \\ \pi_y = (2p_{yA}) + (2p_{yB}) \end{array} \right.$$

The reader may recall that the degenerate p atomic orbitals were related by Euler's equation (Appendix IV) as:

$$p_1 = p_x + i p_y, \qquad p_{-1} = p_x - i p_y$$

Similarly these two representations of the π degenerate pair are related:

$$\pi_+ = \pi_x + i\pi_y, \qquad \pi_- = \pi_x - i\pi_y$$

Note that the symbol "π" (unadorned) means the *pair* of degenerate orbitals, which may be referred to individually as either π_+, π_-, or π_x, π_y. The distinction is that π_+ and π_- are eigenfunctions of $\hat{L}_z$ (with eigenvalues $+\hbar$ and $-\hbar$ respectively), whereas π_x and π_y are not. These orbitals are pictured in Figure 13.12. The π_+ and π_- orbitals have a cylindrically symmetrical electron density, but in the π_- orbital the electron is rotating in the opposite sense to that of one in the π_+ orbital.

The atomic orbitals d_{+1} and d_{-1} will also form π molecular orbitals, while the d_{+2}, d_{-2} atomic orbitals form δ MO's, a degenerate pair with $\lambda = \pm 2$. Table 13.2 summarizes the types of MO's and the AO's with which they correlate — that is, the types of AO's that they must become as the internuclear distance $R \to \infty$.

The Hydrogen Molecule

The procedure used to construct approximate wave functions for many-electron molecules is precisely analogous to that used for atoms. In particular the two-electron hydrogen molecule is much like the helium atom discussed in Sections 12.3, 12.4, and 12.5. A major difference is that the atomic orbitals, the one-electron wave functions, were exact solutions of the one-electron problem, whereas the MO's already contain approximations.

We can obtain the effective Hamiltonian for the hydrogen molecule from Eq. (13.1) and subsequent definitions by dropping the nuclear kinetic energy ($\hat{T}_{NN}$) and letting $Z_A = Z_B = 1$;

$$\hat{H}_{\text{eff}} = \frac{e^2}{R} - \left(\frac{\hbar^2}{2m} \nabla_1^2 + \frac{e^2}{r_{A1}} + \frac{e^2}{r_{B1}} \right) - \left(\frac{\hbar^2}{2m} \nabla_2^2 + \frac{e^2}{r_{A2}} + \frac{e^2}{r_{B2}} \right) + \frac{e^2}{r_{12}} \quad \textbf{(13.67)}$$

The first term in this operator is just a constant (in the Born-Oppenheimer approximation) and will cause no problems. The two terms in parentheses are one-electron operators whose eigenfunctions will be those of H_2^+ — the molecular orbitals. As with the helium atom, it is the last term, the ee repulsion, that causes the problems. If this term is neglected, the eigenfunctions of this operator will be products of the MO's (which must, however, be antisymmetrized with respect to electron exchange in accord with the Pauli principle).

Again by analogy with the procedure used in Section 12.3, the ground state of H_2 will be the $(\sigma_g 1s)^2$ configuration with the approximate wave function:

$$\psi \cong (\sigma_g 1s)(\sigma_g 1s)(\alpha\beta - \beta\alpha) \tag{13.68}$$

(As before, α and β represent the spin states of the electron, and the convention is that the first term in a product represents electron 1, the second represents electron 2, and so on.)

The approximate wave function of Eq. (13.68) can be used with the operator of Eq. (13.67) to calculate the first-order approximate energy:

$$E^{(1)} = \langle \hat{H}_{eff} \rangle$$

The results of this calculation are qualitatively correct, but, as we found for the helium atom, the error is rather large: R_e(calc.) = 0.850 Å (obs. 0.740 Å), D_e(calc.) = 2.68 eV (obs. 4.75 eV). Variational calculations using an effective nuclear charge— that is, with atomic orbitals $(1s) = \exp(-Z'r/a_0)$—give somewhat better results: R_e(calc.) = 0.732 Å, D_e(calc.) = 3.49 eV.

In chemical terms, the error in this calculated bond energy is very large. The error is partly because of the rather crude approximations we used in deriving the MO's, but in large part it is a fundamental limitation of orbital models. Because of the e^2/r_{12} term in the Hamiltonian, the exact wave function cannot be a simple product of one-electron functions, no matter how complex, but must depend on the instantaneous distance between the electrons—r_{12}; this is called the *correlation* problem. However, the fundamental value and correctness of MO theory is demonstrated more by its ability to explain and classify the observed molecular states than by its accuracy in calculating bond energies. This most valuable classification scheme that MO theory provides is the topic of the next section. First, however, we shall take a brief look at how better calculations can be obtained.

Configuration Interaction

It is very instructive to look at the spatial part of the approximate wave function of Eq. (13.68) with the explicit MO of Eq. (13.59); leaving out the normalization factors, the spatial wave function is:

$$\psi(spat.) = [1s_A(1) + 1s_B(1)][1s_A(2) + 1s_B(2)]$$

$$= [1s_A(1)1s_A(2) + 1s_B(1)1s_B(2)] + [1s_A(1)1s_B(2) + 1s_B(1)1s_A(2)]$$

The first two terms of this function represent a state in which both electrons are on one atom; as $R \to \infty$ they would represent an ionized state, $H^- + H^+$. These are usually called the "ionic" terms. The last two terms of this function represent states in which the electrons are shared equally, one to each atom; as $R \to \infty$, these functions would represent two neutral atoms, $H + H$. These are called the "covalent" terms.

$$\psi_{ionic} = 1s_A(1)1s_A(2) + 1s_B(1)1s_B(2)$$

$$\psi_{covalent} = 1s_A(1)1s_B(2) + 1s_B(1)1s_A(2) \tag{13.69}$$

The LCAO-MO wave function assumes that the molecular state is an equal mixture of the so-called ionic and covalent functions of Eq. (13.69) at all distances (R):

$$\psi(\text{LCAO-MO}) = \psi_{\text{ionic}} + \psi_{\text{covalent}}$$

But this is not realistic; it is known experimentally that H_2 in its ground state will dissociate into neutral atoms — not ions:

$$H_2 \longrightarrow H + H$$

$$\text{not} \qquad H_2 \longrightarrow H^+ + H^-$$

That is, as $R \to \infty$, the correct wave function should approach ψ_{covalent}. Yet, at finite R, use of only the covalent function is less correct than the LCAO-MO approximation.

The resolution of this problem is to use a wave function that is a linear combination of the functions of Eq. (13.69), but with coefficients that vary with R:

$$\psi_{\text{CI}} = a(R)\psi_{\text{covalent}} + b(R)\psi_{\text{ionic}} \qquad \textbf{(13.70)}$$

This general procedure is called *configuration interaction* (CI). With the restriction that $a \to 1$ and $b \to 0$ as $R \to \infty$, these functions can be used with the variation principle (Section 12.3) to minimize the energy. The best calculations of this sort (which include many other configurations, including excited states) give the bond energy of H_2 to within 0.0001 eV (~ 10 J/mol) of the observed value.

13.5 Molecular State Symbols

The wave functions of homonuclear diatomic molecules are classified according to the following characteristics:

1. *Orbital angular momentum.* The quantum number for the projection of **L** on the molecular axis will be a sum of the λ quantum numbers for each individual electron (λ_i) and is denoted Λ:

$$\Lambda = \sum_i \lambda_i \qquad \textbf{(13.71)}$$

The states are symbolized as follows:

$$\Lambda = 0, \ \pm 1, \ \pm 2, \ \pm 3, \ \ldots$$

$$\text{symbol:} \ \ \Sigma, \ \ \Pi, \ \ \Delta, \ \ \Phi$$

This is similar to, but actually simpler than, the situation in atoms discussed in Section 12.7; note that Eq. (13.71) is a scalar addition, as contrasted to the vector addition required for atomic states. The ground state of H_2, whose configuration was given as $(\sigma_g 1s)^2$, must be Σ, since there is no net angular momentum.

2. *Spin angular momentum.* Molecular states are classified as to their electron spin angular momentum exactly as were the atomic states — see Table 12.4. As before, the spin degeneracy:

$$g_s = 2S + 1$$

is given as an upper left superscript. The two electrons in the ground state of H_2 must have $S = 0$ to conform to the Pauli exclusion principle [See Eq. (13.68)], so the state symbol is $^1\Sigma$. In states for which $\Lambda \neq 0$ and $S \neq 0$ (for example, a $^2\Pi$ state) there will be spin-orbit coupling as there was for atoms, but we shall not go into this complication.

3. *Inversion symmetry.* The wave functions are products of orbitals, and the orbitals must be either even (gerade) or odd (ungerade) with respect to inversion. If the orbitals are gerade, their product is likewise gerade. If two occupied orbitals are ungerade, their product is also gerade — the product of two odd functions is even; the same will apply to any case with an *even* number of ungerade occupied orbitals. However, a state such as:

$$(\sigma_g)(\sigma_g)(\sigma_u)$$

will be odd (Σ_u in this case). The ground state of H_2 should now be seen to be $^1\Sigma_g$.

4. *Bond order.* Although it is not reflected in the state symbol, the bonding-antibonding characteristics of the individual electrons are an important indicator of the stability of molecular states with respect to dissociation. The **bond order** (BO) is defined as:

$$\text{BO} = \tfrac{1}{2}[(\# \text{ bonding electrons}) - (\# \text{ antibonding electrons})] \quad \textbf{(13.72)}$$

The bond order of the H_2 ground state is equal to 1; the physical significance of this quantity will become clearer as we encounter more examples.

5. *The Vertical Mirror plane* $(\hat{\sigma}_v)$. Σ-type states are labeled either Σ^+ or Σ^- according to their symmetry with respect to a reflection in a vertical mirror plane (a plane of symmetry that contains the bond axis):

$$\hat{\sigma}_v \psi(\Sigma^+) = \psi(\Sigma^+), \qquad \hat{\sigma}_v \psi(\Sigma^-) = -\psi(\Sigma^-)$$

The σ MO's have a perfect cylindrical symmetry and are always symmetric with respect to this operation: $\hat{\sigma}_v(\sigma) = +(\sigma)$. Therefore, any molecular wave function that is a product of σ-type MO's will be Σ^+, including the ground state of H_2 (whose complete symbol can now be seen to be $^1\Sigma_g^+$). It is much more difficult to see how a Σ^- state could arise, and we shall defer this discussion until a specific example is at hand. Π and Δ states are not so classified; these types always occur as degenerate pairs, which can be made into linear combinations that are *either* symmetrical or unsymmetrical with respect to $\hat{\sigma}_v$. Table 13.3 summarizes the symmetry properties of the orbitals and states.

Deriving State Symbols

In determining the state symbols of many-electron diatomics, the following rules are useful:

1. All filled shells (for example, $\sigma^2, \pi^4, \delta^4$) are $^1\Sigma_g^+$. (The analogous situation for atoms, Chapter 12, had all filled shells being ^1S.) This has the effect that we need examine only outer electrons to determine the state.

2. The state of the "hole" is the state of the electron. As for atoms, the state can be determined from the holes that must be filled to complete a shell.

Table 13.3   Symmetry and angular momentum of diatomic orbitals and states

State symbol	Orbital symbol	Λ	$\hat{\imath}$	$\hat{\sigma}_v$	Degeneracy, g
Σ_g^+	σ_g	0	+	+	1
Σ_g^-	(a)	0	+	−	1
Σ_u^+	σ_u	0	−	+	1
Σ_u^-	(a)	0	−	−	1
Π_g	π_g	±1	+	(b)	2
Π_u	π_u	±1	−	(b)	2
Δ_g	δ_g	±2	+	(b)	2
Δ_u	δ_u	±2	−	(b)	2

(a) No orbital has this symmetry.
(b) The degenerate pair may be combined to give either symmetry.

Example: The following examples will illustrate the general procedure.

$(\sigma_g)^1$ — The single unpaired electron gives a doublet state, so the symbol must be $^2\Sigma_g^+$.

$(\sigma_u)^1$ — The state is $^2\Sigma_u^+$.

$(\sigma_u)^2$ — The electrons must be paired; that is, $S = 0$ — hence the state is a singlet. The product of two ungerade functions is gerade. Therefore the state symbol is $^1\Sigma_g^+$.

$(\pi_u)^1$ — The electron could be in either the π_+ ($\lambda = +1$) state or the π_- ($\lambda = -1$) state. Therefore there is a degenerate pair of states with $\Lambda = \pm 1$, and the state is $^2\Pi_u$.

$(\pi_u)^3$ — There are two possible configurations:

$$(\pi_+)^2(\pi_-)^1: \quad \Lambda = 1 + 1 - 1 = +1$$

$$(\pi_+)^1(\pi_-)^2: \quad \Lambda = 1 - 1 - 1 = -1$$

Therefore $\Lambda = \pm 1$ and the state is Π. Also, the state function is a product of three ungerade functions — hence ungerade. The state symbol is therefore $^2\Pi_u$; note that this is identical to what we obtained for the $(\pi_u)^1$ configuration.

The π^2 configuration is rather more complicated, and we shall look at it in detail later. Briefly, we would find the following configurations:

$$(\pi_+)^2(\pi_-)^0: \quad \Lambda = +2, \left.\right\}$$
$$(\pi_+)^0(\pi_-)^2: \quad \Lambda = -2, \left.\right\} \quad \text{a } \Delta \text{ state}$$

$$(\pi_+)^1(\pi_-)^1: \quad \Lambda = 0, \quad \text{a } \Sigma \text{ state}$$

The electrons in the Δ state must be paired, since they are in the same spatial orbital, so it is $^1\Delta$. The electrons in the Σ state could be either paired ($S = 0$) or parallel ($S = 1$), so we expect $^1\Sigma$ and $^3\Sigma$. The occurrence of multiple states when degenerate orbitals are partly filled should remind the reader of the analogous situation for atoms — for example, the p^2 configuration dicussed in detail in Chapter 12. We shall return to this point. ■

Excited States of H_2

As for atoms, the states observed by absorption or emission spectroscopy, called Rydberg states, are derived from the ground-state configuration by the excitation of a single electron from the outer shell. Doubly excited states are usually above the

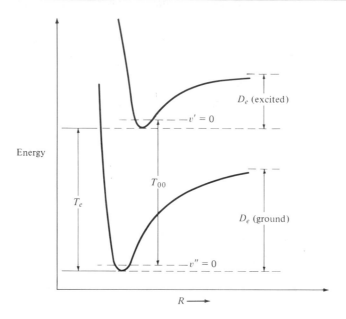

Figure 13.13   Parameters used to discuss excited states of diatomic molecules. Each state is characterized by its own parameters—R_e, D_e, and so on. The energy difference between the minima of the excited and ground electronic states is called T_e. The energy difference between the lowest vibrational level of the excited state ($v' = 0$) and the lowest vibrational level of the ground state ($v'' = 0$) is called $T_{0,0}$.

ionization potential. Each state is characterized by an electronic energy curve, $E_e(R)$, with its parameters D_e, R_e, ω_e, and so on. In addition, the energy difference between the minimum of the energy curve and that of the ground state is called T_e—see Figure 13.13, which illustrates this parameter. The ground state is denoted by the symbol X, while excited states are denoted a, b, c, and so on for singlet or doublet states, and A, B, C, and so on for triplets and quartets.

Some of the observed excited states of H_2 are shown in Figure 13.14; examples of the parameters that characterize these states are given in Table 13.4. The reader should confirm that the state symbols are as shown.

The first excited state of H_2 would be a product of the $(\sigma_g 1s)$ and $(\sigma_u^* 1s)$ orbitals; but because the electrons are identical, symmetrized product functions, analogous to Eq. (12.30) for the He atom, must be used:

$$\psi_s = (\sigma_g 1s)_1(\sigma_u^* 1s)_2 + (\sigma_u^* 1s)_1(\sigma_g 1s)_2$$

$$\psi_a = (\sigma_g 1s)_1(\sigma_u^* 1s)_2 - (\sigma_u^* 1s)_1(\sigma_g 1s)_2$$

To account for the Pauli exclusion principle, the symmetric spatial function must be multiplied by the antisymmetric spin function, $\alpha\beta - \beta\alpha$, which is, of course, the "singlet" spin state (cf. Figure 12.8):

$$\Psi(^1\Sigma_u^+) = \psi_s(\alpha\beta - \beta\alpha)$$

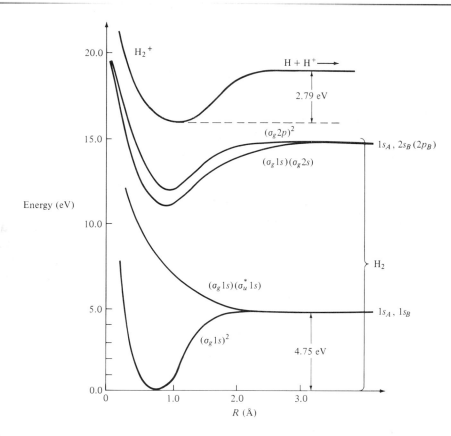

Figure 13.14 Energy states for the hydrogen molecule. Only singlets are shown. Electron configurations for the molecular states and the separated atoms (to the right) are given. (From W. H. Flygare, *Molecular Structure and Bonding*, 1978: John Wiley & Sons, Inc., Fig. 6.2)

Table 13.4 Electronic states of molecular hydrogen*

(Only spectroscopically observed singlet states are listed)

State	Configuration	T_e	ω_e	$\omega_e x_e$	$\bar{B}_e$	α_e	R_e
$X^1\Sigma_g^+$	$(\sigma_g 1s)^2$	0	4395.2	117.9	60.8	2.993	0.7416
$a^1\Sigma_u^+$	$(\sigma_g 1s)(\sigma_u^* 1s)$	Unstable					
$b^1\Sigma_u^+$	$(\sigma_g 1s)(\sigma_u^* 2p)$?	91,689	1356.90	19.932	20.016	1.1933	1.29270
$c^1\Pi_u$	$(\sigma_g 1s)(\sigma_u 2p)$	100,043.0	2442.72	(67.03)?	31.340	1.626	1.0331
$e^1\Sigma_g^+$	$(\sigma_g 1s)(\sigma_g 2s)$	100,062.8	2588.9	(130.5)?	32.68	(1.818)?	1.011

*All units cm^{-1} except R_e(Å).

The antisymmetric spatial function must be multiplied by the symmetric "triplet" spin functions:

$$\Psi(^3\Sigma_u^+) = \begin{cases} \psi_a \alpha\alpha \\ \psi_a(\alpha\beta + \beta\alpha) \\ \psi_a \beta\beta \end{cases}$$

Both of these states have a bond order of zero and are unstable (cf. Figure 13.4). In this manner, the other excited states of H_2 may be derived.

Exercise: What excited-state terms result for H_2 if an electron is promoted to the $(\pi_u 2p)$ orbital? (Answer: $^1\Pi_u$ and $^3\Pi_u$) ∎

Ground States of First-Row Diatomics

The ground-state configurations of many-electron diatomics can be derived by "filling" the MO's with the appropriate number of electrons. The energy order of the orbitals varies with the nuclear charge. The emprical filling order for neutral diatomics is:

$$\sigma_g 1s, \sigma_u^* 1s, \sigma_g 2s, \sigma_u^* 2s, \pi_u 2p, \sigma_g 2p, \pi_g^* 2p, \sigma_u^* 2p$$

Next we shall look at several first-row diatomic molecules to illustrate how their ground electronic states are constructed from these orbitals.

The diatomic He_2 does not form a stable molecule. The ground-state configuration would be:

$$He_2(\sigma_g 1s)^2(\sigma_u^* 1s)^2$$

which has a bond order, BO $= 0$.

The diatomic Li_2 has 6 electrons and a ground state:

$$Li_2(\sigma_g 1s)^2(\sigma_u^* 1s)^2(\sigma_g 2s)^2 \, {}^1\Sigma_g^+$$

This molecule has a bond order of 1 and is stable. It has been observed spectroscopically in the gas phase at high temperatures. (At room temperature, elemental lithium is a solid.)

The diatomic N_2 is a stable molecule. Its configuration (14 electrons) is:

$$N_2(\sigma_g 1s)^2(\sigma_u^* 1s)^2(\sigma_g 2s)^2(\sigma_u^* 2s)^2(\pi_u 2p)^4(\sigma_g 2p)^2 \, {}^1\Sigma_g^+$$

with BO $= 3$. The bond order corresponds to the "triple bond" predicted by the Lewis structure:

$$:N:::N: \quad \text{or} \quad :N\equiv N:$$

and the bond is very stable. Some of the excited states of N_2 are shown in Figure 13.15.

Skipping oxygen for the moment, the configuration of F_2 with 4 more electrons than N_2 is:

$$F_2[N_2](\pi_g^* 2p)^4 \, {}^1\Sigma_g^+$$

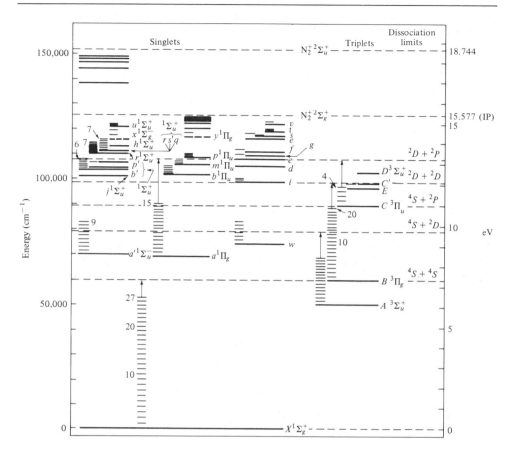

Figure 13.15 Energy levels of the nitrogen molecule. This is an alternative presentation of diatomic energy levels. Heavy horizontal lines indicate the energies in the lowest ($v = 0$) vibrational state; the short horizontal lines indicate the location of the vibrational states. Horizontal dashed lines show the dissociation limits for each molecular state (the atomic term symbols to the right of these lines give the states of the dissociated atoms). The horizontal dashed line at 15.577 eV indicates the ionization potential to a ground-state ion; the line at 16.744 eV is the first excited state of the molecular ion.

The addition of 4 antibonding electrons reduces the bond order to 1; this corresponds to the single bond predicted by the Lewis structure:

$$: \ddot{F} : \ddot{F} :$$

Of course the next element, the inert Ne, would add two more electrons, which must go into the antibonding ($\sigma_u^* 2p$) orbital to give Ne_2 with BO = 0. Thus the simple LCAO-MO theory predicts the known facts about these elements — Ne and He do not bond, N forms a triple bond, F forms a single bond. The elements B, and C are ordinarily solids, but their diatomic molecules can be observed in a high-temperature gas phase, and their characteristics are in accord with those predicted by LCAO-MO.

Exercise: Beryllium does not form a stable diatomic molecule. Write the configuration of the molecule Be_2 and show that the bond order is zero. ∎

The remaining first-row element, O_2, presents some interesting and instructive complications.

The States of the π^2 Configuration

Oxygen presents us with our first example of a diatomic open-shell configuration:

$$O_2[N_2](\pi_g^*2p)^2$$

The complications that this causes are very similar to those we encountered for the p^2 configuration; a review of Table 12.5 and related material would be useful at this point.

The two degenerate π spatial orbitals can be combined into four products for two electrons, namely $\pi_+\pi_+$, $\pi_+\pi_-$, $\pi_-\pi_+$, and $\pi_-\pi_-$. To account for the Pauli principle regarding the exchange of identical electrons, these functions must be arranged into combinations that are either symmetric or antisymmetric with respect to electron exchange:

$$\pi_+\pi_+, \quad \pi_-\pi_-, \quad (\pi_+\pi_- + \pi_-\pi_+), \quad (\pi_+\pi_- - \pi_-\pi_+)$$

These are multiplied by the electron spin functions:

$$\text{(singlet)} \quad \alpha\beta - \beta\alpha$$

$$\text{(triplet)} \quad \alpha\alpha, \alpha\beta + \beta\alpha, \beta\beta$$

such that the total wave function, spatial times spin, is antisymmetric with respect to exchange of identical electrons. There will be six such functions which obey the Pauli principle:

$$\left. \begin{aligned} \psi_1 &= \pi_+\pi_+(\alpha\beta - \beta\alpha) \\ \psi_2 &= \pi_-\pi_-(\alpha\beta - \beta\alpha) \end{aligned} \right\} \quad {}^1\Delta_g$$

$$\psi_3 = (\pi_+\pi_- + \pi_-\pi_+)(\alpha\beta - \beta\alpha) \quad {}^1\Sigma_g^+$$

$$\left. \begin{aligned} \psi_{4a} &= (\pi_+\pi_- - \pi_-\pi_+)\alpha\alpha \\ \psi_{4b} &= (\pi_+\pi_- - \pi_-\pi_+)(\alpha\beta + \beta\alpha) \\ \psi_{4c} &= (\pi_+\pi_- - \pi_-\pi_+)\beta\beta \end{aligned} \right\} \quad {}^3\Sigma_g^- \tag{13.73}$$

The fact that the term symbols are as indicated should be fairly obvious from the earlier discussion, *except* for the distinction between Σ^+ and Σ^-. This distinction is one of the most subtle ones in the theory of diatomic molecules, and understanding it is not essential to subsequent material. It will be discussed later in this chapter.

The ground configuration of oxygen, π^2, therefore gives rise to three states. These are illustrated in Figure 13.16 and Table 13.5 along with some of the other excited states of oxygen. The ground state of O_2 is the ${}^3\Sigma_g^-$ state, which has a spin quantum

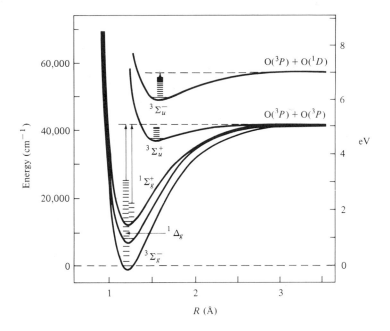

Figure 13.16 Electronic states of molecular oxygen. Not all observed states are shown. The source of some of these states can be seen by comparison to Table 13.5. (From Herzberg, *Spectra of Diatomic Molecules*, Fig. 195)

number $S = 1$, corresponding to two unpaired electrons. The explanation of the empirical fact that oxygen has unpaired electrons in its ground state was one of the greatest victories for the LCAO-MO theory. The singlet states are also very important; these will be discussed later.

Table 13.5 Electronic states of molecular oxygen

State	T_e (cm^{-1})	ω_e (cm^{-1})	$\omega_e x_e$ (cm^{-1})	$\bar{B}_e$ (cm^{-1})	R_e (Å)
Configuration	$(\pi_u 2p)^4 (\pi_g^* 2p)^2$				
$X\,^3\Sigma_g^-$	0	1580.361	12.07	1.4456	1.20739
$a\,^1\Delta_g$	7,918.1	1509	12	1.4260	1.2155
$b\,^1\Sigma_g^+$	13,195.2	1432.7	13.95	1.4004	1.22675
Configuration	$(\pi_u 2p)^3 (\pi_g^* 2p)^3$				
$c\,^1\Sigma_u^-$	~32,700				
$C\,^3\Delta_u$	~34,300				
$A\,^3\Sigma_u^+$	36,096	819	22.5		(1.42)?
$B\,^3\Sigma_u^-$	49,802.1	700.36	8.002	0.819	1.60

The Vertical Mirror Plane

This section will attempt to explain the distinction between the Σ^+ and Σ^- states that arise from the π^2 configuration. It is a rather difficult section, and could be omitted with little harm; however, we will need the conclusions, summarized at the end.

One of the difficulties in explaining the effect of the vertical mirror planes is that there are an infinity of them, and it may appear that not all have the same effect. However, once it is realized that our system has cylindrical symmetry, it may be apparent that the "x" axis could be chosen in any direction. Therefore, if we determine the effect of a reflection in the xz plane, $\sigma_v(xz)$, the conclusions will be generally true. As discussed earlier, the degenerate pair of π orbitals can be written in several ways:

$$\pi_+ = f(r, \theta)e^{i\phi}, \qquad \pi_- = f(r, \theta)e^{-i\phi}$$

$$\pi_x = \frac{\pi_+ + \pi_-}{2} = \frac{1}{2} f(r, \theta) \cos \phi = \frac{1}{2} f(r, \theta) \frac{x}{r}$$

$$\pi_y = \frac{\pi_+ - \pi_-}{2i} = \frac{1}{2i} f(r, \theta) \sin \phi = \frac{1}{2i} f(r, \theta) \frac{y}{r}$$

A reflection in the xz plane will change y to $-y$ and leave x unchanged:

$$\hat{\sigma}_v \pi_x = \pi_x, \qquad \hat{\sigma}_v \pi_y = -\pi_y$$

Since:

$$\pi_+ = \pi_x + i\pi_y$$

the effect on this function will be:

$$\hat{\sigma}_v \pi_+ = \hat{\sigma}_v \pi_x + i\hat{\sigma}_v \pi_y = \pi_x - i\pi_y$$

But this function is just π_-, so:

$$\hat{\sigma}_v \pi_+ = \pi_-, \qquad \hat{\sigma}_v \pi_- = \pi_+ \qquad \textbf{(13.74)}$$

A more graphic proof of Eq. (13.74) can be seen if we view the molecule along the z axis, and represent the π_+ state with an arrow indicating a clockwise rotation and the π_- state by an arrow in the opposite direction. Reflection of the figure through a mirror plane will reverse the direction of the arrow:

$$\pi_- \qquad\qquad\qquad \pi_+$$

The effect of $\hat{\sigma}_v$ on the product functions of Eqs. (13.73) is, using Eq. (13.74):

$$\hat{\sigma}_v(\pi_+\pi_- + \pi_-\pi_+) = (\pi_+\pi_- + \pi_-\pi_+)$$

This state is symmetric and is called Σ^+. For the other function:

$$\hat{\sigma}_v[\pi_+\pi_- - \pi_-\pi_+] = [\pi_-\pi_+ - \pi_+\pi_-] = -[\pi_+\pi_- - \pi_-\pi_+]$$

This state is antisymmetric and is called Σ^-.

The situation is even more complicated if the electrons are in different π orbitals — for example, the configuration $(\pi_u 2p)^1 (\pi_g^* 2p)^1$. We shall call the four spatial orbitals π_+, π_-, π'_+, π'_-. The possible product functions are:

$$\psi_1 = \pi_+ \pi'_+, \quad \psi_2 = \pi'_+ \pi_+, \quad \psi_3 = \pi_- \pi'_-, \quad \psi_4 = \pi'_- \pi_-$$

$$\psi_5 = \pi_+ \pi'_-, \quad \psi_6 = \pi_- \pi'_+, \quad \psi_7 = \pi'_- \pi_+, \quad \psi_8 = \pi'_+ \pi_-$$

The first two have $\Lambda = +2$ while the second two have $\Lambda = -2$; these are clearly part of a Δ state. To obey the Pauli principle they must be combined as follows:

$$\left. \begin{array}{l} (\psi_1 + \psi_2)(\alpha\beta - \beta\alpha) \\ (\psi_3 + \psi_4)(\alpha\beta - \beta\alpha) \end{array} \right\} \quad {}^1\Delta$$

$$\left. \begin{array}{l} (\psi_1 - \psi_2)\alpha\alpha \\ (\psi_1 - \psi_2)(\alpha\beta + \beta\alpha) \\ (\psi_1 - \psi_2)\beta\beta \\ (\psi_3 - \psi_4)\alpha\alpha \\ (\psi_3 - \psi_4)(\alpha\beta + \beta\alpha) \\ (\psi_3 - \psi_4)\beta\beta \end{array} \right\} \quad {}^3\Delta$$

The spatial functions 5–8 have $\Lambda = 0$ and must belong to Σ states. They must be arranged into linear combinations that are even (Σ^+) or odd (Σ^-) with respect to $\hat{\sigma}_v$ and even (singlet) or odd (triplet) with respect to electron exchange. A little thought may convince you that $\hat{\sigma}_v$ [with Eq. (13.74)] and electron exchange have the following effects:

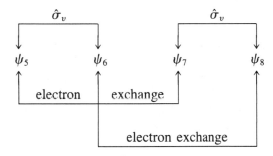

The proper combinations are, therefore:

$$(\psi_5 + \psi_6 + \psi_7 + \psi_8)(\alpha\beta - \beta\alpha) \quad {}^1\Sigma^+$$

$$(\psi_5 - \psi_6 + \psi_7 - \psi_8)(\alpha\beta - \beta\alpha) \quad {}^1\Sigma^-$$

$$(\psi_5 + \psi_6 - \psi_7 - \psi_8) \begin{cases} \alpha\alpha \\ \alpha\beta + \beta\alpha \\ \beta\beta \end{cases} \quad {}^3\Sigma^+$$

$$(\psi_5 - \psi_6 - \psi_7 + \psi_8) \begin{cases} \alpha\alpha \\ \alpha\beta + \beta\alpha \\ \beta\beta \end{cases} \qquad {}^3\Sigma^-$$

In conclusion: Two π electrons in different spatial orbitals may have the following states:

$$ {}^1\Delta, \ {}^3\Delta, \ {}^1\Sigma^+, \ {}^1\Sigma^-, \ {}^3\Sigma^+, \ {}^3\Sigma^- $$

Two π electrons in the same spatial orbital (π^2) may have only the states:

$$ {}^1\Delta, \ {}^1\Sigma^+, \ {}^3\Sigma^- $$

The Inner Orbitals

For second-row diatomics—for example, N_2, whose configuration was given on p. 744—writing the configuration of the inner electrons as:

$$(\sigma_g 1s)^2(\sigma_u^* 1s)^2$$

implies that these electrons are delocalized over the entire molecule and, perhaps, actually participate in the bond formation. In fact, these four electrons contribute little to the bond; this could be viewed as a consequence of a cancellation of the bonding and antibonding contributions. But there is more than that; calculations (for example, Figure 13.17) show that the electron density in these orbitals consists of two nearly separate lobes, one centered on each atom; in effect, they look very much like the $1s$ atomic orbitals from which the molecular orbitals were constructed in the first place. Considering these electrons to be in localized AO's, such as:

$$(1s_A)^2(1s_B)^2$$

rather than delocalized MO's, such as:

$$(\sigma_g 1s)^2(\sigma_u^* 1s)^2$$

may be just a different point of view, but it is a very useful one; it means that in more complex molecules the MO's responsible for bonding can be constructed using only outer-shell AO's, and the inner shells can be left as "nonbonding" orbitals, localized on the atoms and relatively unaffected by the bond formation. For heavier atoms this idea has strong experimental support from X-ray photoelectron spectroscopy (XPES). As discussed in Chapter 12, XPES explores the energies of the inner electrons; it will show features distinctive for the atoms involved, even when those atoms are part of a molecule.

Reflecting the fact that inner electrons are strongly localized, the $1s$ inner shells are often abbreviated with the symbol K (the old nomenclature for this shell); for example, the configuration of Li_2 may be written as:

$$Li_2 \, KK \, (\sigma_g 2s)^2 \, {}^1\Sigma_g^+$$

[The second shell ($2s, 2p$) is abbreviated as L when filled, the third as M.]

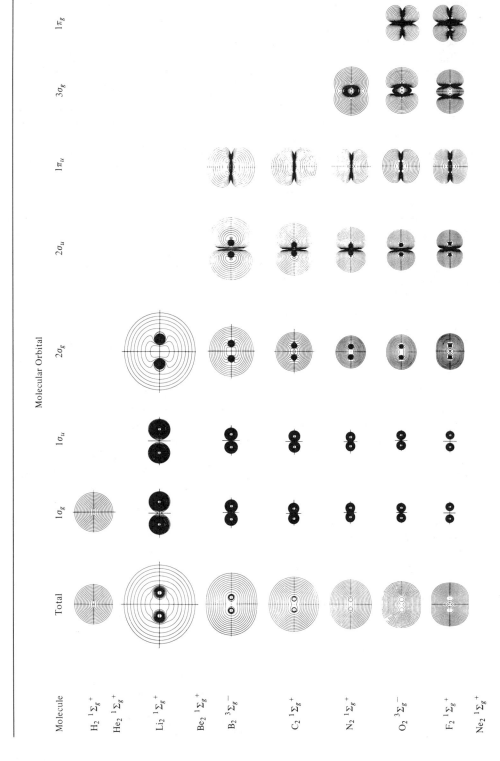

Figure 13.17 Electronic densities for diatomic molecules. Lines are contours of constant electron density; shown for the individual orbitals and the net state (on the left). [From A. C. Wahl, *Scientific American*, 322, 4 (April 1970)]

Heteronuclear Diatomics

The decision to construct LCAO-MO's from conbinations such as $1s + 1s$, $2s + 2s$, rather than $1s + 2s$, for example, was based on the proximity of the orbital energies in the separated atoms. In the heteronuclear case, $Z_A \neq Z_B$, the choice is not always so straightforward. If the atomic numbers are very close — for example, CO, NO, CN — then the MO's will be very similar to the homonuclear MO's described above; of course, the g/u designation must be omitted, since the electron density will not be invariant to the inversion operation.

We take as an example the molecule NO with 15 electrons. The configuration will be:

$$NO(\sigma 1s)^2(\sigma^* 1s)^2(\sigma 2s)^2(\sigma^* 2s)^2(\pi 2p)^4(\sigma 2p)^2(\pi^* 2p)^1$$

For this configuration, $\Lambda = \pm 1$ depending on whether the electron is in the π_+ or π_- orbital and the state symbol is $^2\Pi$. Figure 13.18 shows some of the states for NO, CN, and CO.

Figure 13.18(c) for CO shows most of the known states up to the ionization limit. Note the distinction between *dissociation:*

$$CO \longrightarrow C + O$$

and *ionization:*

$$CO \longrightarrow CO^+ + e$$

As an example of a heteronuclear diatomic for which the atomic numbers are very different, consider the case of HF. The electron on the hydrogen atom is in the $1s$ orbital, and this is the only orbital we need to consider for the ground state; the other hydrogen AO's are very much higher in energy. The fluorine atom, on the other hand, has a configuration $(1s)^2(2s)^2(2p)^5$ and a number of orbitals that could participate in bonding; however our discussion above regarding inner electrons will lead us to conclude, correctly, that the $1s$ electrons will remain on the fluorine atoms and, hence, be nonbonding. (Note the distinction between *nonbonding* electrons, which remain localized on one atom and do not participate in bonding, and *antibonding* electrons, which are in delocalized orbitals but destabilize the bond because of destructive interference.)

The discussion of H_2^+ above suggests that the reason a bond is formed is that the overlap of the AO's builds up electron density in the region between the atoms, thereby neutralizing the nuclear-nuclear repulsion. Therefore, one might expect the most useful LCAO-MO's for HF to be those which can overlap the hydrogen $1s$ orbital to the greatest extent. The axis between the atoms is, by definition, the z axis, so the $2p_z$ orbital of fluorine will be in the proper position to overlap with the hydrogen $1s$ orbital; it would therefore be reasonable to construct LCAO-MO's as:

$$\sigma = c_1(1s_H) + c_2(2p_{zF})$$

$$\sigma^* = c_1'(1s_H) - c_2'(p_{zF})$$

The p_x and p_y orbitals (or p_1, p_{-1} if you prefer) cannot overlap usefully with the H-atom orbitals (the $2p$ orbitals of H, with which they could overlap, are too high in

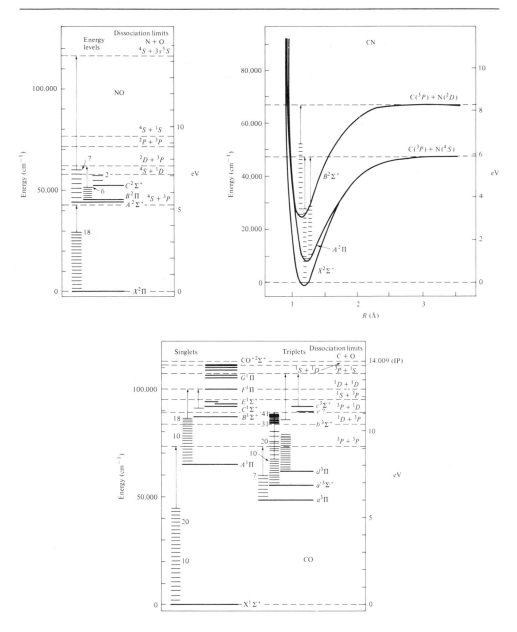

Figure 13.18 Energy states for CO, CN, and NO.

energy) and so become nonbonding orbitals; however, since the nuclear potential field has cylindrical symmetry, they are properly called "π" rather than "p".

What about the fluorine $2s$ orbital? The LCAO method for approximating molecular orbitals is not limited to using only "pure" atomic orbitals. In this case there may be some advantage to using a mixture of the $2p_z$ and $2s$ orbitals:

$$a(2p_z) + b(2s)$$

In such a case, the atomic orbital is called a *hybrid* orbital; the concept of hybrid orbitals is most useful for polyatomic molecules, so we shall defer further discussion until Chapter 14. In the case of HF, computations show that the $2s$ orbital does not participate in bonding to any great extent, so it is probably best viewed as nonbonding. However there is experimental evidence (for example, NMR spin-spin coupling constants) to indicate that this bond must have some degree of "s-character."

The concepts discussed in the preceding paragraph have a wide currency in chemistry and are undoubtedly useful. However, one must bear in mind that the LCAO concept is itself just a first approximation, so such ideas are bound to have their limitations.

As mentioned in the discussion of H_2, accurate calculations of bond energies are possible by considering configuration interaction (CI) with other structures. The ionic configurations for HF would include:

$$H^+F^- \quad \text{and} \quad H^-F^+$$

However, these structures will not make equal contributions; in fact chemical intuition would suggest that the former will be more important. (In aqueous solution, the fluoride anion F^- is well known, the cation is unknown.) Section 13.8 discusses this point further.

13.6   Electronic Spectroscopy

Spectroscopic transitions in which electrons are promoted from one orbital to another will usually fall in the visible or ultraviolet region of the spectrum. A typical uv spectrum of a diatomic molecule, Figure 13.19, for example, is distinctively different from an atomic spectrum (for example, Figure 12.13). Instead of sharp individual lines, there are a series of *bands,* which, under high resolution, are seen to consist of a large number of closely spaced lines; these lines are clumped together at one end, called the *band head.*

Absorption Spectra

Absorption spectra are a great deal simpler than emission spectra, so we shall deal with them first. Figure 13.20 shows the situation diagrammatically. Leaving out the rotational energy and anharmonicity for the moment, a molecule in the ground electronic state with a vibrational quantum number $v'' = 0$ has energy:

$$\frac{E_g}{hc} = \tfrac{1}{2}\omega_e''$$

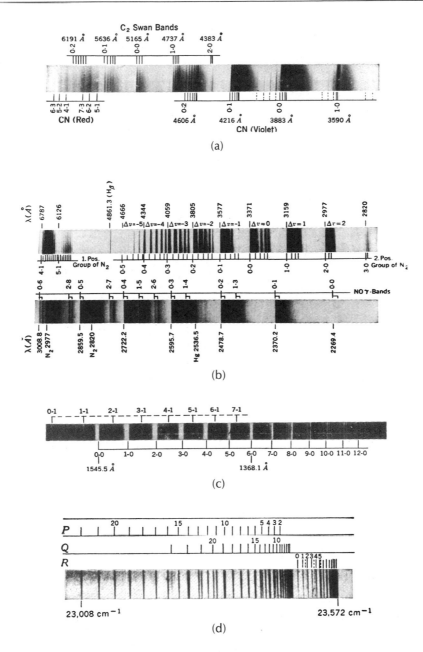

Figure 13.19 Band spectra of diatomic molecules. (a) Spectrum from a carbon arc, showing lines from CN and C_2. (b) A spectrum from an air-filled Geissler tube, showing lines from molecular nitrogen. (c) Absorption spectrum of CO. (d) The fine structure of one of the bands of AlH. Note the greatly different scale between this spectrum and those above ($23\,008\ \text{cm}^{-1}$ corresponds to a wavelength of 4346 Å, $23{,}572\ \text{cm}^{-1}$ to 4242 Å). (From Herzberg, *Spectra of Diatomic Molecules*, Figs. 7, 8, 14, and 20)

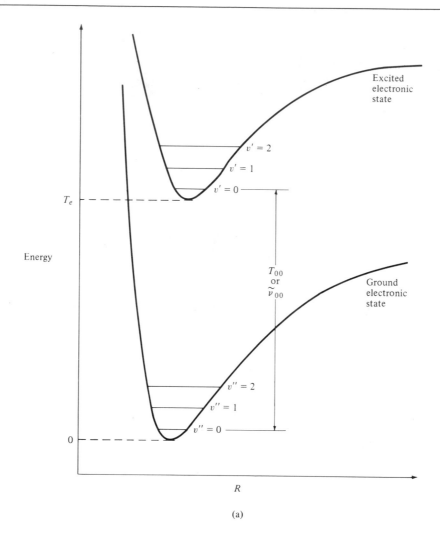

(a)

Figure 13.20 Diatomic energy levels. (a) A portion of the energy levels (electronic-vibrational). (b) A detail to explain the source of the bands observed in molecular spectra (cf. Figure 13.19). Note that double primes denote ground electronic state vibrational states, single primes refer to the excited electronic state. The conventional

where ω_e'' is the vibrational constant in the ground electronic state. An excited electronic state with a vibrational quantum number $v' = 0$ has energy:

$$\frac{E_x}{hc} = T_e + \tfrac{1}{2}\omega_e'$$

where ω_e' is the vibrational constant of the excited state, and T_e is defined in Figure 13.20. The frequency of the 0-0 band (also called T_{00}) will then be,

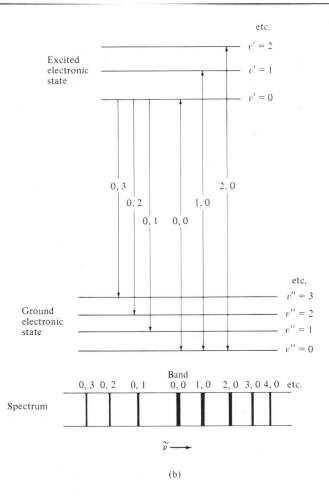

(b)

naming of bands gives the upper state quantum number first—for example, the 0-1 band is a transition between the $v' = 0$ and $v'' = 1$ states, and the 1-0 band is between the $v' = 1$ and $v'' = 0$ states.

approximately

$$\tilde{\nu}_{00} = \frac{E_x - E_g}{hc} \cong T_e + \tfrac{1}{2}(\omega'_e - \omega''_e)$$

When anharmonicity is included, this frequency is exactly:

$$T_{00} = \tilde{\nu}_{00} = T_e + \tfrac{1}{2}\omega'_e - \tfrac{1}{4}\omega'_e x'_e - \tfrac{1}{2}\omega''_e + \tfrac{1}{4}\omega''_e x''_e \qquad \textbf{(13.75a)}$$

If the final state is $v' = 1$, the excited-state energy is, approximately:

$$\frac{E_x}{hc} = T_e + \tfrac{3}{2}\omega'_e$$

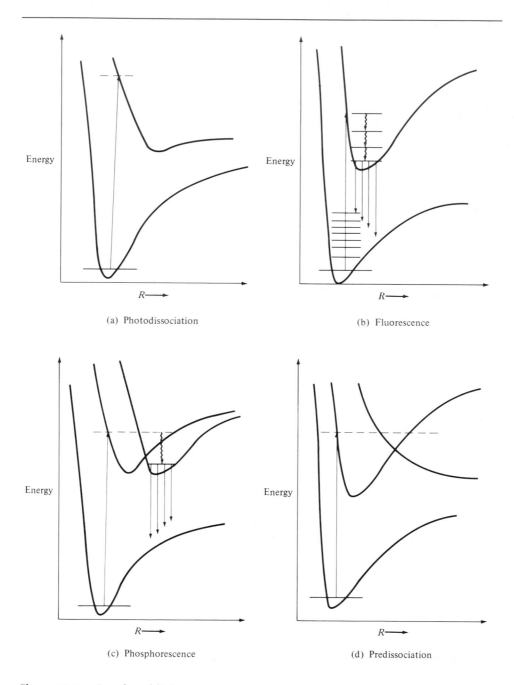

(a) Photodissociation

(b) Fluorescence

(c) Phosphorescence

(d) Predissociation

Figure 13.21   Results of light absorption. (a) If light is absorbed into an excited state above its dissociation limit, the molecule will immediately dissociate into atoms. Such absorption is a continuum. (b) If light if absorbed into an excited state below its dissociation limit, it will be reemitted when the molecule returns to the ground state. The

Then the 1-0 absorption band frequency is:

$$\tilde{\nu}_{10} \cong T_e + \tfrac{3}{2}\omega_e' - \tfrac{1}{2}\omega_e''$$
$$= \tilde{\nu}_{00} + \omega_e'$$

In other words, its frequency exceeds that of the 0-0 band by ω_e' — a fact that may be evident upon close inspection of Figure 13.20.

It can be shown that a general absorption frequency ν'-0 is

$$\tilde{\nu}_{\nu'-0} = \tilde{\nu}_{00} + \nu'(\omega_e' - \omega_e'x_e') - (\nu')^2\omega_e'x_e' \qquad \textbf{(13.75b)}$$

The rotational energy terms, neglected in Eq. (13.75), are responsible for the band structure of these transitions. They will add to Eq. (13.75):

$$\tilde{B}_e' J'(J' + 1) - \tilde{B}_e'' J''(J'' + 1)$$

Analysis of the band structure is complicated; suffice it to say that this analysis permits the measurement of the rotational constants, hence the R_e's, for both states. In the absence of a more detailed analysis, we shall interpret Eq. (13.75) as the position of the band head.

Emission spectra could arise from any of the ν' vibrational levels of the excited electronic state and go to any of the vibrational levels of the ground state (called ν''). There are obviously a great number of these, and their analysis is lengthy but straightforward — paralleling the derivation of Eq. (13.75).

Fluorescence Spectra, etc.

When a molecule absorbs uv radiation, it ends up in a state that is electronically excited and, usually, vibrationally excited. Afterward a number of things can happen. A few possibilities include (1) dissociation, (2) reemission, (3) relaxation (loss of energy through collision), and (4) crossing into another excited state. Figure 13.21 diagrams several possibilities: (a) If the energy of the molecule after absorption is above the dissociation limit of the excited state, photo-dissociation will result. In the case shown, one of the resulting atoms (at least) is in an excited state. (b) Absorption is followed by vibrational relaxation and then reemission at a longer wavelength — this is called *fluorescence*. (Figure 13.22 shows how fluorescent emission is measured.) (c) Absorption followed by crossing into another excited state, vibrational

---

light reemitted at longer wavelengths is called *fluorescence*. (c) Light is absorbed for an allowed transition into an excited state, but the molecule "crosses over" to a forbidden state — a state whose transition to the ground state is forbidden. Often the states involved are singlet and triplets. Because of the selection rule, the excited state has a longer lifetime before the energy is reemitted. In such a case, the reemitted light is called *phosphorescence* — the operational distinction between fluorescence and phosphorescence is the lifetime of the excited state rather than (as is often stated) whether singlet and triplets are involved. (d) If a bonding excited state crosses a dissociative state, absorption into that state may be followed by dissociation. However, in contrast to case (a), the absorption spectrum does have bands rather than a continuum, but the bands are broad. This is called *predissociation*.

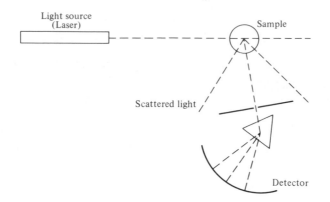

Figure 13.22   Schematic of a fluorescence spectrometer. In contrast to an absorption spectrometer, the analyzer is oriented perpendicular to the light source, so only scattered light will be detected.

relaxation, and reemission at longer wavelengths. If the second state is a forbidden transition to the ground state, the lifetime of the excited state may be long — the emission is then usually called *phosphorescence*. (d) Absorption followed by crossing into another state and dissociation from that state. This is called *predissociation*.

Selection Rules

The selection rules for the absorption and emission of photons by diatomic molecules are: (1) $\Delta\Lambda = 0, \pm 1$, (2) $\Delta S = 0$, (3) $g \leftrightarrow u$ (in homonuclear diatomics) (4) $\Sigma^+ \not\leftrightarrow \Sigma^-$. (These will be discussed in more detail in Section 13.9.)

Electronic States of O_2

These rules will be illustrated with the example of oxygen (Figure 13.16 and Table 13.5). When ground-state oxygen atoms combine, the molecule formed could be in a number of excited states. Since all of these states are forbidden to radiate into the ground state, the excited molecules will likely end up in the $^1\Delta_g$ state. The transition $a^1\Delta_g \rightarrow X^3\Sigma_g^-$ is forbidden by three selection rules (which?), so this state can live for a very long time — approximately 45 minutes. [This is the radiative lifetime; the molecule can also relax (lose energy) by collisions, so it may not really last this long.] Such a long-lived state is almost a separate chemical species and the study of the chemistry of "singlet oxygen" is an active field [J. Brand, "Biochemical Effects of Excited State Molecular Oxygen," *J. Chem. Ed.*, 53, 274 (1976)].

Singlet oxygen can be produced by some chemical reactions — for example:

$$H_2O_2 + OCl^- \longrightarrow O_2(^1\Delta_g) + H_2O + Cl^-$$

Subsequently:

$$O_2(^1\Delta_g) + O_2(^1\Delta_g) \longrightarrow 2\,O_2(^3\Sigma_g^-) + h\nu$$

with the emission of a photon:

$$\bar{\nu} \sim 2 \, T_e(^1\Delta_g) \sim 15,800 \quad cm^{-1}$$

This photon is in the red region of the spectrum. The production of light directly from a chemical reaction, of which this is the simplest example, is called *chemiluminescence*.

The first strongly allowed absorptions of O_2 are the transitions $X^3\Sigma_g^- \rightarrow B^3\Sigma_u^-$, called the Schumann-Runge bands. This absorption begins at 49,363 cm^{-1} and goes to a continuum shortly thereafter. Because of this strong absorption, it is impossible to do uv spectroscopy in air at wavelengths shorter than 200 nm, since the oxygen in the air will absorb all the radiation. The portion of the uv below 200 nm is called the *vacuum uv,* because a special spectrometer from which air can be removed is required. The strong $X \rightarrow B$ absorption by atmospheric oxygen causes the dissociation of oxygen in the upper layers of the atmosphere and aids in shielding the earth from high-energy radiation.

The transition $X \rightarrow A^3\Sigma_u^+$ is only weakly forbidden and can be observed; these are called the Herzberg bands.

Exercise: Air is more N_2 than O_2. Examine the term diagram of N_2 (Figure 13.15) and determine the lowest allowed transition. Will this be a problem when doing uv spectroscopy? ∎

13.7 Photoelectron Spectroscopy

Photoelectron spectroscopy, utilizing X-rays to eject inner electrons (XPES), was discussed in Chapter 12. This technique is also useful for molecules, but here we shall discuss a related technique, in which an ultraviolet photon is used to eject an electron from the valence shell of molecules — valence-electron photoelectron spectroscopy (VPES). (See Figure 13.23.)

Figure 13.24a contrasts the spectra obtained from atoms (H and He) and a molecule. The removal of an electron from H_2 produces H_2^+ in one of several vibrationally excited states:

$$H_2[(\sigma_g 1s)^2, v = 0] + h\nu = H_2^+[(\sigma_g 1s), v'] + e$$

The energy balance gives:

$$KE \text{ of electrons} = h\nu - (IP \text{ of } H_2) - (vib. \text{ energy of } H_2^+) \qquad \textbf{(13.76)}$$

Thus the electron energy spectrum will show a structure that corresponds to the vibrational energies of the H_2^+ ion. These transitions are illustrated by Figure 13.24(b).

Figure 13.25 shows the states involved in the VPES of nitrogen. The configuration for the outer electrons of neutral N_2 is:

$$N_2[\text{inner e}] \, (\sigma_u^* 2s)^2 (\pi_u 2p)^4 (\sigma_g 2p)^2 \; {}^1\Sigma_g^+$$

If an electron is removed from the $\sigma_g 2p$ shell, the resulting state is:

$$N_2^+[\ldots] \, (\sigma_u^* 2s)^2 (\pi_u 2p)^4 (\sigma_g 2p)^1 \; {}^2\Sigma_g^+$$

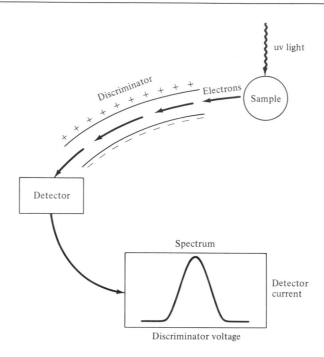

Figure 13.23 Schematic of a photoelectron spectrometer. The sample is irradiated with uv radiation (shown here as a helium source), which causes electrons to be ejected. The electrons are passed through an electrostatic analyzer, which will pass only those with a specific kinetic energy (the others will strike the sides of the analyzer). A detector provides a voltage proportional to the number of electrons reaching it, as the analyzer voltage is swept.

(The reader should derive these state symbols as an exercise.) In this manner the energies of the ion's states are measured, and, from this, the orbital energies of the neutral are inferred. Reference 4 (from which Figures 13.23 through 13.25 were obtained) has a good discussion of the use of VPES to study molecular energies.

13.8   Ionic Bonding and Dipole Moments

The type of bonding we discussed earlier (Section 13.4) is called covalent bonding; the bonding forces that hold the atoms together are a result of the sharing of electrons by open-shell atoms. The interactions of closed-shell atoms — for example, He + He, Ne + Ne, or Ar + Ar — are very weak and do not result in the formation of a stable bond (except at very low temperatures). (These weak interactions are the van der Waals forces, which were discussed in Chapter 1.) Open-shell heteronuclear diatoms may bond covalently, but they also have another type of interaction available, called the *ionic bond*.

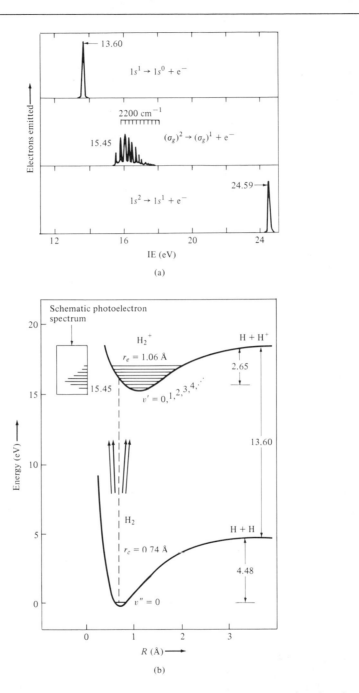

Figure 13.24 Photoelectron spectra of hydrogen. (a) The spectrum of molecular hydrogen, together with those of atomic hydrogen and helium. (b) The molecular and ion states involved and the source of the vibrational structure of the VPES spectrum. (From DeKock and Gray, *Chemical Structure and Bonding*, Figs. 4.22 and 4.21)

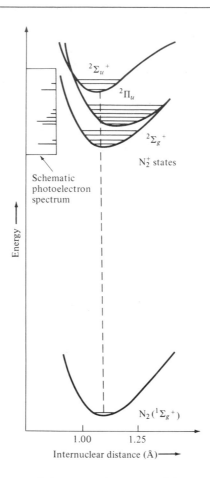

Figure 13.25 VPES spectrum of nitrogen. The states of the nitrogen ion observed in the VPES spectrum (shown in box to the left). (From DeKock and Gray, *Chemical Structure and Bonding*, Fig. 4.32)

Ionic bonding is well illustrated by the alkali halides, such as KCl. At ordinary temperatures, alkali halides (such as KCl) are ionic crystals with alternate lattice sites occupied by K^+ and Cl^- ions, but they are not diatomic molecules — each ion interacts equally with a number of neighboring ions. In the gas phase, however, observable spectroscopically at high temperatures, diatomic molecules such as KCl can be observed and studied like any other diatomic molecule. Table 13.6 shows some typical data, which should be compared to the data of Table 13.1; generally, the bond energies are similar, but the bond lengths of the alkali halides are longer.

The neutral-atom interaction, K + Cl, is of course an open-shell interaction, which could form a covalent bond. However, it takes very little energy to exchange an electron and form K^+ and Cl^-. This energy is calculated as follows.

Table 13.6 Properties of alkali halide diatomics

	ω_e (cm^{-1})	R_e (Å)	D_e (eV)	μ (debyes)
NaCl	364.6	2.3609	4.22	9.00
KCl	279.8	2.6668	4.37	10.48
KBr	219.17	2.8208	3.92	10.41
RbI	119.20	3.3152	3.57	12.1
LiF	910.34	1.5639	5.99	6.32

The ionization potential (IP) of K is (Table 13.7):

$$\text{K} \longrightarrow \text{K}^+ + \text{e}: \qquad \Delta E = \text{IP} = 4.339 \text{ eV}$$

The electron affinity (EA) of Cl is:

$$\text{Cl} + \text{e} \longrightarrow \text{Cl}^-: \qquad \Delta E = -\text{EA} = -3.613 \text{ eV}$$

Adding these, we get the net energy required to ionize this pair:

$$\text{K} + \text{Cl} \longrightarrow \text{K}^+ + \text{Cl}^-: \qquad \Delta E = \text{IP} - \text{EA} = 0.726 \text{ eV}$$

This ionization energy is fairly small, as we can see by comparing it to a similar calculation for H + Cl:

$$\text{H} + \text{Cl} \longrightarrow \text{H}^+ + \text{Cl}^-: \qquad \Delta E = 13.6 - 3.6 = 10.0 \text{ eV}$$

But still, energy is required to ionize this pair, so at large distances the neutrals will be more stable. However, the ions have a very strong Coulombic attraction, so that inside some distance the K$^+$ + Cl$^-$ pair will be more stable than K + Cl.

The ions, K$^+$ and Cl$^-$, are closed-shell; in fact they are both isoelectronic with Ar. The covalent interactions of K$^+$ + Cl$^-$ will therefore be similar to those of Ar + Ar—that is, very weak. However, the Coulombic attraction of the ions provides a very strong force for holding them together, and this interaction is primarily responsible for the formation of a bond in such molecules.

Table 13.7 Ionization potentials and electron affinities

Atom	Z	IP (eV)	EA (eV)
H	1	13.595	0.754
Li	3	5.390	0.62
C	6	11.264	1.25
O	8	13.614	1.47
F	9	17.42	3.4
Na	11	5.138	0.54
Cl	17	13.01	3.613
K	19	4.339	0.726
Br	35	11.84	3.363
Rb	37	4.176	0.42

The picture we have developed is this: two closed-shell ions, which will repel each other at short distances because of the repulsive interactions of their electrons, are held together by the Coulombic attraction of their net charges. How accurate is this picture? All heteronuclear diatomics have both ionic bonding *and* covalent bonding. The degree to which ionic bonding will contribute will depend on such things as the energy required to exchange an electron; the calculations above, which showed that H + Cl required 10.0 eV to ionize, would lead us to expect that HCl will be less ionic and more covalent than KCl. Also, there are several ways to check the degree of ionic bonding with experimental results — which we shall now do.

Ionic-Bond Energy

One way to compare the ionic-bond theory to experiment is to calculate the electronic energy curve, $E_e(R)$. In contrast to the situation outlined in Section 13.4, an accurate electronic energy curve for ionic interactions can be obtained from an amazingly simple model; we need only add together the energy contributions for attraction, repulsion, and ionization.

Two ions with net charges $+e$ and $-e$, respectively, which are separated by a distance R, will have a Coulombic energy of $-e^2/R$. In cgs units, $e = 4.803 \times 10^{-10}$ esu and R is in cm, so e^2/R will have units of *ergs*. It will be better to change these units to eV in the manner indicated earlier [Eq. (11.103)]; in units of eV (rounding to three significant figures):

$$E_e(\text{Coulombic}) = \frac{-300e}{R}$$

The repulsive interactions of the closed electronic shells are usually approximated by an exponential:

$$E_e(\text{repulsive}) = Ae^{-BR/R_e}$$

The constants A and B can, as we shall soon see, be evaluated with experimental data.

Both of these terms approach zero as $R \to \infty$. It is conventional to choose the zero of energy as the neutral atoms at $R = \infty$, rather than the ions — for example, K + Cl rather than K^+ + Cl^-. For this reason we must add to the energy an amount equal to the energy required to exchange an electron at an infinite distance; as discussed above, this will be the difference between the ionization potential (IP) of the atom that will lose the electron and the electron affinity (EA) of the atom that will gain the electron:

$$E_e(\text{at } R = \infty) = \text{IP} - \text{EA}$$

Putting these together, we get (units: eV):

$$E_e(R) = Ae^{-BR/R_e} - \frac{300e}{R} + (\text{IP} - \text{EA}) \qquad \textbf{(13.77)}$$

This curve is plotted in Figure 13.26, using parameters derived in the example below.

This can be compared to experiment, and the constants A and B evaluated, via three measurable quantities. (1) The bond energy (D_e) is the depth of the potential well:

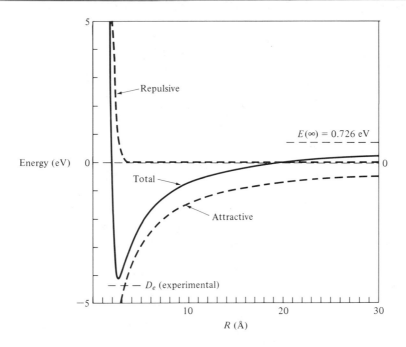

Figure 13.26   Ionic bonding in KCl. The energy curve for ionic bonding in KCl is shown as calculated by Eq. (13.78). For this calculation, an exponential repulsive potential was assumed; the dashed lines on this diagram show the repulsive and attractive parts of the binding energy separately.

$E_e(R_e) = -D_e$. (2) The bond length (R_e) is the value of R for which $E_e(R)$ has a minimum — that is, $(dE_e/dR) = 0$ when $R = R_e$. (3) The vibrational force constant (k) is related to d^2E_e/dR^2 as indicated by Eq. (13.21) — this is in turn related to the vibrational constant [ω_e, Eq. (13.26)]. The parameters R_e and ω_e can be measured accurately by the spectroscopic methods discussed in Section 13.3.

Example: Calculate the bond energy (D_e) of K^{35}Cl using Eq. (13.77) and:

$$\omega_e = 279.8 \text{ cm}^{-1}, \quad R_e = 2.6668 \times 10^{-8} \text{ cm}, \quad \text{IP} - \text{EA} = 0.726 \text{ eV}$$

$$L\mu = \frac{35 \times 39}{35 + 39} = 18.45 \text{ g}$$

First we shall calculate the force constant with Eq. (13.26):

$$k = \mu(2\pi c\omega_e)^2$$

$$= \frac{18.45 \text{ g}}{6.022 \times 10^{23}}[2\pi(3 \times 10^{10} \text{ cm s}^{-1})(279.8 \text{ cm}^{-1})]^2$$

$$= 8.522 \times 10^4 \text{ g/s}^2$$

(Note that $g/s^2 = erg/cm^2$.) Since Eq. (13.77) is written with units of eV, we need the force constant in eV (k'):

$$k' = \frac{300k}{e} = 5.323 \times 10^{16} \text{ eV/cm}^2$$

Next we take the derivative of Eq. (13.77) with respect to R:

$$\frac{dE_e}{dR} = -\frac{BA}{R_e} e^{-BR/R_e} + \frac{300e}{R^2}$$

This slope is zero when $R = R_e$, so we get:

$$BAe^{-B} = \frac{300e}{R_e} = \frac{300(4.803 \times 10^{-10})}{2.6668 \times 10^{-8}} = 5.403 \text{ eV}$$

The second derivative of Eq. (13.77) gives:

$$\frac{d^2E_e}{dR^2} = \frac{B^2A}{R_e^2} e^{-BR/R_e} - \frac{2(300e)}{R^3}$$

This will be equal to k' (the force constant with units eV/cm^2) when $R = R_e$; therefore (multiplying by R_e^2):

$$B^2Ae^{-B} - 2\left(\frac{300e}{R_e}\right) = k'R_e^2$$

Using the results above (obtained from the first derivative), we get:

$$(5.403)B - 2(5.403) = k'R_e^2$$

$$B = \frac{2(5.403) + (5.323 \times 10^{16})(2.6668 \times 10^{-8})^2}{5.403} = 9.01$$

Then, from the first derivative:

$$A = \frac{5.403}{B} e^B = 4.91 \times 10^3 \text{ eV}$$

The electronic energy of KCl is then, using Eq. (13.77) (units: eV):

$$E_e(R) = (4.91 \times 10^3)e^{-9.01R/R_e} - \frac{5.403R_e}{R} + 0.726 \qquad \textbf{(13.78)}$$

When $R = R_e$, this gives:

$$-D_e = 4.91 \times 10^3 e^{-9.01} - 5.403 + 0.726 = -4.08 \text{ eV}$$

Table 13.6 gives $D_e = 4.37$ eV for KCl, so this simple model has accounted for 93% of the observed bond energy. [Also, see Figure 13.26 for a graph of Eq. (13.78).]   ■

Dipole Moments and Ionic Character

Another way to test the accuracy of the ionic bonding theory is by comparison of the theoretical dipole moment to the experimental value.

A distribution of electrical charges q_i will have an electric dipole moment $\boldsymbol{\mu}$ with Cartesian components:

$$\mu_x = \sum_i q_i x_i, \qquad \mu_y = \sum_i q_i y_i, \qquad \mu_z = \sum_i q_i z_i \qquad \textbf{(13.79)}$$

In a diatomic molecule, the only nonzero component of $\boldsymbol{\mu}$ in the molecule-fixed COM coordinate system will be along the bond axis z. We shall call this μ_0, the permanent dipole moment of the molecule:

$$\mu_0 = \sum_i q_i z_i \qquad \textbf{(13.80)}$$

If KCl is purely ionic, then there would be two charges, $q_+ = e$, $q_- = -e$, separated by $R_e = z_K - z_{Cl}$, and the dipole moment would be:

$$\mu_0 = e(z_K - z_{Cl}) = eR_e$$
$$= (4.803 \times 10^{-10} \text{ esu})(2.6668 \times 10^{-8} \text{ cm})$$
$$= 12.81 \times 10^{-18} \text{ esu-cm} \qquad \textbf{(13.81)}$$

The unit 10^{-18} esu-cm is called the debye (symbol: D). The experimentally observed value of the dipole moment of KCl is 10.48 D. We can contrast this to HCl, for which Eq. (13.81) predicts $\mu_0 = 6.12$ D while the measured value is 1.03 D. The ionic-bond model works well for KCl, poorly for HCl. These calculations are used to define:

$$\% \text{ ionic character} = \frac{100\mu_0}{eR_e} \qquad \textbf{(13.82)}$$

The examples above give KCl, 82% ionic, and HCl, 17% ionic. This concept can be used to combine ionic and covalent bonding by defining a configuration interaction function including ionic terms:

$$\psi = A\psi(\text{ionic}) + B\psi(\text{covalent})$$

and choosing the coefficients such that they represent the fractional ionic character as defined by the observed dipole moment:

$$\frac{A^2}{A^2 + B^2} = \frac{\mu_0}{eR_e}$$

Ionization potentials and electron affinities can also be used to access the relative importance of various ionic configurations in bonding. In an earlier discussion of HF, it was suggested that the ionic configurations H^+F^- and H^-F^+ would need to be included in CI calculations. Their relative importance can be estimated by considering the energy required to create these configurations when the atoms are at an infinite distance. From the data in Table 13.7 we get the following:

$$H + e \longrightarrow H^-: \qquad EA = 0.754 \text{ eV}$$

$$F \longrightarrow F^+ + e: \qquad IP = 17.42 \text{ eV}$$

Therefore, ionization of the neutral atoms at $R = \infty$ requires:

$$H + F \longrightarrow H^- + F^+ : \quad \Delta E = IP - EA = 16.67 \text{ eV}$$

On the other hand:

$$H \longrightarrow H^+ + e : \quad IP = 13.595 \text{ eV}$$

$$F + e \longrightarrow F^- : \quad EA = 3.4 \text{ eV}$$

Therefore:

$$H + F \longrightarrow H^+ + F^- : \quad \Delta E = IP - EA = 10.195 \text{ eV}$$

This supports our intuition that H^+F^- will be the more important ionic configuration; however, comparison of these numbers to that calculated for KCl ($\Delta E = 0.726$ eV) tells us that HF is far more covalent than KCl.

Measurement of Dipole Moments

From the preceding discussion it can be seen that dipole moments are an important quantity in bonding theory, but they are equally important in many other areas of chemistry, including the theory of ionic solutions, gas laws, intermolecular forces of polar molecules, and chemical kinetics. It therefore seems worthwhile to take a brief look at how dipole moments are measured.

The classical method, due largely to Debye, measures the electrical capacitance of two parallel metal plates. The ratio of the capacitance (C) when a material fills the capacitor to the capacitance in a vacuum (C_0) defines the dielectric constant:

$$\varepsilon = \frac{C}{C_0} \tag{13.83}$$

This quantity is related to the polarizability of a material (α) and the dielectric constant by the formula (cgs units):

$$\frac{\varepsilon - 1}{\varepsilon + 2} \frac{M}{\rho} = \frac{4\pi L}{3}\left(\alpha + \frac{\mu_0^2}{3k_b T}\right) \tag{13.84}$$

(M = molecular weight, ρ = density, k_b = Boltzmann's constant. This is called the Clausius-Mosotti equation.)

The polarizability relates the size of an induced dipole moment to the strength of the electrical field ($\mathcal{E}$). Consider an atom; the negative charge has the same center as the positive charge, so there is no dipole moment. However, if the atom is placed between plates with opposite electrical charge — that is, in an electric field — the atom will be *polarized* as the electrons are attracted toward the positive plate and the nucleus is attracted toward the negative plate. The size of the resulting dipole defines the polarizability constant α:

$$\mu_{\text{ind}} = \alpha\mathcal{E} \tag{13.85}$$

Measurements of the dielectric constant as a function of T can, with Eq. (13.84), determine both μ_0 and α.

A more modern and accurate method for measuring dipole moments utilizes the effect of an electrical field on the microwave rotational spectra — the *Stark effect.*

A rotating molecule in field-free space with a quantum number J has a rotational degeneracy corresponding to the $(2J + 1)$ allowed values of the quantum numbers $M = J, \ldots, -J$. When an electric field is applied, an additional energy:

$$E_{\text{Stark}} = -\frac{I\mu_0^2 \mathscr{E}^2}{\hbar^2} \left[\frac{J(J + 1) - 3M^2}{J(J + 1)(2J - 1)(2J + 3)} \right] \tag{13.86}$$

lifts this degeneracy. (I = moment of inertia.)

The displacement and splitting of a microwave transition $J'' \rightarrow J'$ by the Stark effect is useful for measuring the dipole moment and also as an assignment aid — that is, as an aid in determining which J states give rise to the observed line. The selection rules are $\Delta M = 0$ for microwave radiation polarized parallel to the electric field and $\Delta M = \pm 1$ for radiation polarized perpendicular to the field.

Example: A $J = 1$ state in an electric field will split into two levels, $M = 0$ and $M = \pm 1$. A $J = 2$ state will split into $M = 0, \pm 1, \pm 2$. For radiation polarized perpendicular to the field, the $J = 1 \rightarrow 2$ transition will show three lines:

$$J = 1 \begin{cases} M = 0 \\ M = \pm 1 \end{cases} \qquad \begin{matrix} M = 0 \\ M = \pm 1 \\ M = \pm 2 \end{matrix} \Bigg\} J = 2$$

■

13.9 Selection Rules

The manner by which selection rules could be derived was discussed briefly in Chapter 11; we can now be a little more specific. Atoms and molecules can interact with either the electric or magnetic fields of an electromagnetic (em) wave. Interaction with the electric field gives rise to *electric dipole* transitions whose intensity depends on integrals of the type:

$$\int \psi_1 \boldsymbol{\mu} \psi_2 \, d\tau \tag{13.87}$$

The definition of the dipole, Eq. (13.79), should make it evident that the integrals of Eq. (13.87) are just the integrals $(I_x, I_y, \text{and } I_z)$ as defined by Eq. (11.100). These "transition dipoles" should not be confused with the permanent dipole moment, which is defined by the average-value theorem for the ground state (ψ_0) as:

$$\mu_0 = \frac{\int \psi_0^* \mu \psi_0 \, d\tau}{\int \psi_0^* \psi_0 \, d\tau} \tag{13.88}$$

Transitions resulting from interactions with the magnetic part of the em field are called *magnetic dipole* transitions. Such transitions are generally weaker than the electric dipole transitions and, in general, have different selection rules. Weak, "forbidden," transitions can also occur owing to electric quadrupole interactions.

All the selection rules mentioned in this chapter are electric dipole rules and could have been derived by calculating the transition integrals I_x, I_y, or I_z. We have omitted most of these integrations because they tend to be rather involved. However, many selection rules can be derived without doing any integrals at all; we shall illustrate this by looking at the homonuclear diatomic selection rules.

Consider the function $F(x)$ in one dimension. The integral:

$$I = \int_{-\infty}^{\infty} F(x)\,dx$$

can be broken into two parts:

$$I = \int_{-\infty}^{0} F(x)\,dx + \int_{0}^{\infty} F(x)\,dx$$

If we change the sign of x and the limits of the first integral (letting dx become $-dx$):

$$I = \int_{0}^{\infty} F(-x)\,dx + \int_{0}^{\infty} F(x)\,dx = \int_{0}^{\infty} [F(x) + F(-x)]\,dx$$

If $F(x)$ is odd (ungerade), then $F(x) = -F(-x)$ and the integrand is zero. Therefore, for any ungerade function (F_u):

$$\int_{-\infty}^{\infty} F_u(x)\,dx = 0$$

The functions x, y, z that enter into the transition integral are clearly odd. Therefore the products of x, y, and z with functions that are either even or odd will be as follows:

$$\psi_g x \psi_g \quad \text{is odd}$$
$$\psi_g x \psi_u \quad \text{is even}$$
$$\psi_u x \psi_u \quad \text{is odd}$$

The only type of transition integral that will *not* be zero is:

$$\int_{-\infty}^{\infty} \psi_g x \psi_u\,dx$$

Therefore the selection rules must be:

$$g \longleftrightarrow u, \qquad g \longleftrightarrow\!\!\!/ \, g, \qquad u \longleftrightarrow\!\!\!/ \, u$$

This is admittedly the simplest example, but the lesson should be clear: in deriving selection rules, and in many other types of applications, a great deal of effort can be avoided by using symmetry to determine what integrals must be zero.

Harmonic-Oscillator Selection Rule

A diatomic molecule with a partial separation of charge, $+q$ on one atom and $-q$ on the other, will have a dipole moment:

$$\mu = qR$$

where R is the internuclear separation. The distance R is changing during the vibration, so μ is not a constant; however, we can define a dipole moment:

$$\mu_e = qR_e$$

for a molecule at a fixed distance (R_e), which will be approximately equal to the permanent dipole moment, μ_0:

$$\mu_e \cong \mu_0 = \langle qR \rangle$$

The selection rule for a transition $v \rightarrow v'$ will depend on the integral:

$$I = (\text{const.}) \int \psi_v \mu \psi_{v'} \, dy$$

with $y = (R - R_e)/\alpha$ as before. This is evaluated by expanding μ as a Taylor series about R_e:

$$\mu = \mu_e + \left(\frac{\partial \mu}{\partial R}\right)(R - R_e)$$

Therefore the transition integral becomes:

$$I = (\text{const.}) \left[\mu_e \int \psi_v \psi_{v'} \, dy + \frac{\partial \mu}{\partial R} \int \psi_v \psi_{v'} (R - R_e) \, dy \right]$$

The first integral must be zero, because the harmonic oscillator wave functions are orthogonal; therefore, using $y = (R - R_e)/\alpha$:

$$I = (\text{const.}) \frac{\partial \mu}{\partial R} \int \psi_v \psi_{v'} \, y \, dy$$

The evaluation of this integral is not entirely straightforward. If done, it will show that, for the integral to be nonzero, v' must be equal to either $v + 1$ or $v - 1$ (see the example below). But even without doing an integral, we can see that if $\partial \mu / \partial R$ is zero, as it is for homonuclear diatomics, the transition is always forbidden. (This argument applies strictly only to electric dipole transitions. There are some exceptions; for example, oxygen does have an IR spectrum due to magnetic dipole transitions associated with its unpaired electrons.)

Example (optional): The evaluation of the transition integral for the harmonic oscillator proceeds as follows:

The wave functions are given by Eq. (11.62) and Table 11.2 as:

$$\psi_v = A_v H_v(y) e^{-y^2/2}$$

Since we only wish to determine whether or not the integral is zero, all constants such as A_v will be dropped. The transition integral for $v \rightarrow v'$ will be:

$$I = \int y H_v H_{v'} e^{-y^2} \, dy$$

The theory of Hermite polynomials gives the recursion formula:

$$H_{v+1} = 2y H_v - 2v H_{v-1}$$

(This formula was given in Table 11.2; the reader should verify it by deriving H_3 from H_2 and H_1 given in that table.) This formula can be solved for yH_v:

$$yH_v = \tfrac{1}{2}H_{v+1} + vH_{v-1}$$

Using this in the transition integral, we get:

$$I = \int (\tfrac{1}{2}H_{v+1} + vH_{v-1})H_{v'}e^{-y^2}\,dy$$

$$= \tfrac{1}{2}\int H_{v+1}H_{v'}e^{-y^2}\,dy + v\int H_{v-1}H_{v'}e^{-y^2}\,dy$$

The orthogonality of these functions now demonstrates that the first integral is zero unless $v' = v + 1$, and the second is zero unless $v' = v - 1$. Therefore, the transition is allowed ($I \neq 0$) if either $v' = v + 1$ or $v' = v - 1$. ∎

Rotational Selection Rules

The rigid-rotor rotational transition J, M to J', M' will have the transition integral:

$$I = \int \psi_{JM}^* \mu \psi_{JM}\,d\tau$$

with (in polar coordinates) $d\tau = \sin\theta\,d\theta\,d\phi$. The dipole moment will, in this case, be written most conveniently in polar coordinates; for example:

$$\mu_x = qx = qR\,\sin\theta\,\cos\phi$$

Assuming R to be constant, we can use $\mu_0 = qR$, giving:

$$\mu_x = \mu_0\,\sin\theta\,\cos\phi, \quad \mu_y = \mu_0\,\sin\theta\,\sin\phi; \quad \mu_z = \mu_0\,\cos\theta$$

The transition integral for z will therefore be:

$$I_z = \mu_0 \int \psi_{JM}^*\,\cos\theta\,\psi_{J'M'}\,\sin\theta\,d\theta$$

These integrals are a bit involved (but see, for example, Problem 12.43), but without doing them we reach an important conclusion: direct rotational (microwave) spectroscopy will be impossible unless the molecule has a permanent dipole moment—that is, $\mu_0 \neq 0$. Therefore, homonuclear diatomic molecules, which must have $\mu_0 = 0$, cannot be observed by microwave spectroscopy.

Postscript

Several lessons of this chapter have great generality—in particular: (1) the Born-Oppenheimer approximation, which permits us to discuss separately the electronic, vibrational, and rotational energies of a molecule, (2) the use of symmetry to describe, classify, and correlate the states of a molecule, and (3) molecular orbital theory. We shall continue these themes as, in the next chapter, we investigate the nature of polyatomic molecules.

Reference 4 gives a relatively simple discussion of chemical bonding, with many examples from photoelectron spectroscopy. Reference 1 contains more detailed accounts of many of the topics we have covered in this chapter.

References

1. M. Karplus and R. N. Porter, *Atoms and Molecules,* 1971: New York, W. A. Benjamin, Inc.
2. G. Herzberg, *Molecular Spectra and Molecular Structure: I. Spectra of Diatomic Molecules,* 1950: New York: Van Nostrand Reinhold Company.
3. I. N. Levine, *Molecular Spectroscopy,* 1975: New York, Wiley-Interscience.
4. R. L. DeKock and H. B. Gray, *Chemical Structure and Bonding,* 1980: New York, Benjamin/Cummings.
5. W. H. Flygare, *Molecular Structure and Dynamics,* 1978: Englewood Cliffs, N.J., Prentice-Hall, Inc.
6. K. P. Huber and G. Herzberg, *Molecular Spectra and Molecular Structure: IV. Constants of Diatomic Molecules,* 1979: New York: Van Nostrand Reinhold Company.

Problems

13.1 A particle confined to a sphere of radius a by an infinite potential has $V = 0$ for $r \le a$ and $V = \infty$ for $r > a$. Solve the Schrödinger equation for the eigenstates and energies of such a system for the case of no angular momentum — that is, for wave functions $\psi(r)$ independent of θ and ϕ. (You need not normalize the eigenfunctions.)

13.2 The anharmonic potential can be written $V_{\text{anh}} = a(R - R_e)^3 + b(R - R_e)^4$.
(a) Identify a and b as they are related to $E_e(R)$.
(b) Calculate the first-order correction to the harmonic oscillator energy for the ground ($v = 0$) state.

13.3 Calculate the force constant for H_2 from its vibrational constant.

13.4 Calculate ΔH_0^θ for the reaction $\frac{1}{2} H_2(g) + \frac{1}{2} Cl_2(g) = HCl(g)$ from spectroscopic data.

13.5 Prove that the Morse potential approaches the harmonic potential ($\frac{1}{2} kx^2$) for small values of $x = R - R_e$.

13.6 (a) What is the value of β in the Morse approximation to the potential of $^1H\,^{35}Cl$?
(b) Plot the Morse potential for HCl.

13.7 Derive a formula for the vibrational frequency (band origin) of the second overtone.

13.8 Analyze the vibrational frequencies (band origins) for $^1H\,^{35}Cl$ (below) to determine ω_e and $\omega_e x_e$.

    Fundamental: 2885.9 cm^{-1}
    1st overtone: 5668.0
    2nd overtone: 8346.9

13.9 Calculate the vibrational constant ω_e and $\omega_e x_e$ for DF from those of HF (Table 13.1) assuming that only the reduced mass will change on isotopic substitution.

13.10 If it is assumed that molecules that differ only by isotopic substitution, such as H_2 and D_2, have identical electronic energies, then D_e will be the same. However, the bond energy D_0 will differ because of the change in the zero-point vibrational energy upon isotopic substitution. Calculate the bond energy of D_2 from that of H_2 (Table 13.1).

13.11 Calculate the classical rotation frequency, the number of rotations performed in one second, for a CO molecule in the $J = 10$ state. Do this by equating the classical kinetic energy, $\frac{1}{2}I\omega^2$, to the quantum mechanical energy.

13.12 The rotational constant of $^{133}Cs\,^{35}Cl$ measured by microwave spectroscopy is 2163.8 MHz. Calculate R_e.

13.13 Calculate the bond lengths R_e from the rotational constants of HF and DF as listed on Table 13.1.

13.14 Calculate the populations of the $J = 1$ through 5 rotational state of HBr at 300 K (relative to the $J = 0$ state population).

13.15 What J state of $^{127}I\,^{35}Cl$ will have the maximum population at 300 K?

13.16 The $J = 0 \rightarrow 1$ microwave frequency for $^2D\,^{35}Cl$ was measured [Cowan and Gordy, *Phys. Rev.*, 111, 209 (1958)] as 323 295.8 MHz. Estimate the centrifugal distortion constant from data and equations in this chapter and calculate B_0 from this number (keep six significant figures).

From the IR of the same compound it was determined that $B_0 = 5.39226$ cm^{-1}. Calculate the velocity of light from these data to six significant figures.

13.17 Microwave rotational frequencies have been measured for $^{12}C\,^{16}O$ as:

$$\nu(0 \rightarrow 1) = 115\,271.20 \text{ MHz}$$

$$\nu(1 \rightarrow 2) = 230\,537.97 \text{ MHz}$$

Calculate B_0 and D_c.

13.18 Derive the formulas for the P and R branches of the first-overtone vibrational band.

13.19 Analyze the vibrational frequencies (below) for the fundamental IR band of $^{12}C^{16}O$ to find the band origin, $\tilde{B}_e$ and α_e. (units: cm^{-1}):

J''	R	P
0	2147.084	—
1	2150.858	2139.427
2	2154.599	2135.548
3	2158.301	2131.633
4	2161.971	2127.684
5	2165.602	2123.700
6	2169.200	2119.681
7	2172.759	2115.632

13.20 In an IR spectrum of CO, the maximum-intensity line is associated with $J'' = 8$. What is the temperature of the sample? For CO, $\tilde{B}_0 = 1.9227$ cm^{-1}.

13.21 A diatomic molecule has a series of absorption bands in the far IR with frequencies 83.03, 104.1, 124.30, 145.03, 165.51, 185.86, 206.38, and 226.50 cm^{-1}. Identify the source of these lines and analyze their frequencies for the appropriate molecular constants.

13.22 What are the degeneracies of the following diatomic states? (Neglect spin-orbit coupling.) (a) $^1\Sigma^+$ (b) $^1\Sigma_u^-$ (c) $^2\Pi$ (d) $^3\Delta_g$ (e) $^3\Sigma_g^-$

13.23 Determine the configuration and state symbol and bond order for the ground states of B_2^+, B_2, B_2^-, C_2.

13.24 Use MO theory to explain the trends in bond lengths observed in the O_2 ions below:

	R_e (Å)
O_2^+	1.1227
O_2	1.20741
O_2^-	1.26

13.25 Determine the state symbol and bond order for F_2, F_2^+, and F_2^-. Which molecule should be the most stable?

13.26 There is some inconsistency in the literature as to the ground state of C_2, involving the order of the $\pi_u 2p$ and $\sigma_g 2p$ orbitals. According to JANAF tables, the order is:

$$X^1\Sigma_g^+, \quad {}^3\Pi_u (610 \text{ cm}^{-1}), \quad {}^3\Sigma_g^- (6243.5 \text{ cm}^{-1})$$

(Some sources give ${}^3\Pi_u$ as the ground state). What are the likely configurations for these states?

13.27 Calculate the relative equilibrium populations of the $X^3\Sigma_g^-$ and the $a^1\Delta_g$ states of O_2 at 1000 K.

13.28 Write out the configuration and determine the ground-state symbol for CN.

13.29 Determine the probable configuration and ground-state symbol of the diatomic molecule BO. What is the bond order?

13.30 The ground state of CN is $X^2\Sigma^+$. Spectroscopically observed excited states include $A^2\Pi$ ($T_e = 9242 \text{ cm}^{-1}$) and $B^2\Sigma^+$ ($T_e = 25,752 \text{ cm}^{-1}$). What are the probable electron configurations for these states?

13.31 Some observed states of NH are: $X^3\Sigma^-$, $a^1\Delta$, $b^1\Sigma^+$, $A^3\Pi$, $c^1\Sigma^+$. List all electric-dipole allowed transitions among these states.

13.32 (a) Use data in Table 13.4 to calculate the wavelength of the 0-0 band of the H_2 $X \rightarrow b$ transition. Is this an allowed transition?
(b) From this, calculate the wavelengths of the 1-0 and 2-0 absorption bands. [The first number is the vibrational quantum number (v') for the excited state.]

13.33 The uv spectrum shows the following frequencies for the $v'' = 0 \rightarrow v'$ bands of PN:

v'	$\tilde{\nu}$ (cm^{-1})
0	39 699.1
1	40 786.8
2	41 858.9
3	42 919.0
4	43 962.0
5	44 991.3

Calculate the vibrational constants ω_e and $\omega_e x_e$ for the excited state.

13.34 Use the wavelengths of the C_2 Swan bands (Figure 13.19) to calculate the vibrational constants of this molecule.

13.35 The absorption frequencies for the Schumann-Runge bands of O_2 include (near the dissociation limit):

v	$\tilde{\nu}$	v	$\tilde{\nu}$
16	56 719.50	19	57 030.18
17	56 852.41	20	57 082.83
18	56 954.54	21	57 114.77

(a) Plot the differences $\delta = \bar{\nu}_{i+1} - \bar{\nu}_i$ vs. $\bar{\nu}$ and determine the dissociation energy by extrapolating to $\delta \to 0$.

(b) From this and the atomic oxygen energies, $O(^1D) - O(^3P) = 15{,}867.862$ cm^{-1}, calculate D_0 for ground-state O_2.

13.36 Calculate the percent ionic character of LiF from its dipole moment (Table 13.6).

13.37 Calculate the energy of the gas-phase ionization at $R = \infty$:

$$\text{Na} + \text{Cl} \longrightarrow \text{Na}^+\text{Cl}^-$$

from the data in Table 13.7.

13.38 Use the potential function of Eq. (13.77) and data for NaCl to calculate the constants (A, B) and bond energy of this diatomic molecule.

13.39 The potential energy between two ions such as $K^+ + Cl^-$ is: $V = F(R) - e^2/R + \Delta E_\infty$, where $F(R)$ is the repulsive potential and $\Delta E_\infty = \text{IP} - \text{EA}$ is the ionization energy at $R = \infty$.

(a) Prove that if $F(R) = a/R^{12}$, then $a = e^2 R_e^{11}/12$ and $D_e = 11e^2/12R_e - \Delta E_\infty$ (units: ergs).

(b) Use $\Delta E_\infty = 0.726$ eV and $R_e = 2.67 \times 10^{-8}$ cm to estimate D_e for KCl using this model.

13.40 If a microwave $J = 2 \to 3$ transition is observed in an electric field, how many lines will appear? Assume the electric field is perpendicular to the electric vector of the microwaves.

13.41 Calculate the Stark splitting (in Hz) of the $J = 1$ state of KCl in an electric field of 3000 volts/cm. (The cgs unit of voltage is 1 esu = 300 volts, so, in cgs units, $\mathscr{E} = 10$.)

13.42 If a linear rotating molecule interacts with radiation polarized with its electric vector along the z axis, the transition dipole is:

$$I = \iint \psi_{JM} \mu_z \psi_{J'M'}^* \, d\Omega$$

where z is a direction in space, not the molecular axis.

(a) Prove $I = 0$ unless $M = M'$.

(b) Do the integral for $\psi_{0,0} = 1/\sqrt{4\pi}$, $\psi_{1,0} = \sqrt{3/4\pi} \cos \theta$.

13.43 The molar polarization:

$$P = \frac{\varepsilon - 1}{\varepsilon + 2} \frac{M}{\rho}$$

has been measured for CHCl$_3$ (below). Use these data to calculate the polarizability (α) and dipole moment (μ_0) of this molecule. (These values are for solutions in hexane.)

T (K)	P (cm^3)	T (K)	P (cm^3)
193	62.1	273	51.1
213	58.9	293	49.7
233	56.0	313	48.3
253	53.1	323	47.5

13.44 In Figure 13.24(b), what is the "meaning of" (what words describe) the numbers 4.48, 13.60, 2.65 (all eV)?

Symmetry is not merely a descriptive nicety . . .
it penetrates to the harmony in nature.

–Jacob Bronowski

14

Polyatomic Molecules

The lessons we learned from diatomic molecules tell us much of what we need to know about bonding and spectroscopy; it is, after all, sufficient for many purposes to consider a molecule to be just a sum of its two-center bonds. However, the theory of polyatomic molecules brings us one final lesson of great importance and generality — the use of symmetry in analyzing nature.

Polyatomic molecules fall into two classes, linear and nonlinear. Linear polyatomics have rather more in common with diatomics than with nonlinear molecules, and the theory applicable to these is nearly identical to that discussed in the last chapter. For nonlinear molecules, orbital angular momentum is of little or no use for classifying states or interpreting spectroscopy. Spin angular momentum is still useful, so we will still have singlets, doublets, and so on, but the overriding theme will be symmetry.

The principles of symmetry will be discussed first. Subsequent sections on the theory of molecular structure and spectroscopy will utilize these ideas and methods.

14.1   Symmetry Operations

A symmetry operator can be thought of as operating on a mathematical function or on a physical object such as a molecule. Symmetry operations that leave the molecule, more specifically the nuclei of the molecule, in a position indistinguishable from its initial position are the *symmetry elements* of that molecule. Since such an operation leaves the potential field and the molecular Hamiltonian invariant, the wave functions, orbitals, and so on of the molecule must be eigenfunctions of that operator. For example, if all the nuclei of a molecule lie in a plane, the electron density above and below that plane must be the same. Denoting an operator for reflection in that plane as $\hat{\sigma}$, the electron density $|\psi|^2$ must obey the equation:

$$\hat{\sigma}|\psi|^2 = |\psi|^2 \tag{14.1}$$

Generally this will be true for two types of functions, symmetric and antisymmetric:

$$\hat{\sigma}\psi = +\psi, \qquad \hat{\sigma}\psi = -\psi \tag{14.2}$$

(An exception must be made for the case of degenerate functions; this topic will be discussed in Section 14.4.) This, of course, is just what was said for diatomic molecules in Chapter 13 — the inversion operator, for example [Eq. (13.54)] — but for polyatomic molecules there is a greater variety of symmetry elements and an equation such as (14.2) must apply for each symmetry element of the molecule. Therefore, we shall discuss the symmetry elements first.

Example: The effect of the inversion operation on ammonia:

does not leave the molecule in an indistinguishable position, so $\hat{\imath}$ *is not* a symmetry element of NH_3. The same operation on benzene:

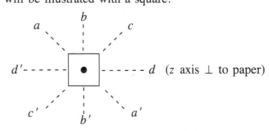

leaves it in a position indistinguishable from its initial position, so î *is* a symmetry element of benzene. The wave functions of benzene must be either gerade or ungerade, as we found for homonuclear diatomics; however, the wave functions of ammonia cannot be so classified. ∎

Elements of Symmetry

The following symmetry operations may be symmetry elements of a molecule. All these operations will leave one point in the molecule, the center of mass (COM), invariant and therefore must pass through that point.

1. $\hat{E}$: The *identity* operation. This does nothing at all; the need to define such an operation will become apparent in due course.
2. $\hat{C}_n$: The *proper rotation* operator. This operator rotates the object or function about an axis through the COM by an angle $2\pi/n$. For example, $\hat{C}_3$ rotates by $120°$, $\hat{C}_3^2$ by $240°$. $\hat{C}_3^{-1}$ rotates by $-120°$. Note the operator equalities $\hat{C}_3^2 = \hat{C}_3^{-1}$ and $\hat{C}_3^3 = \hat{E}$. The operator $\hat{C}_4$ rotates by $90°$ and $\hat{C}_2$ by $180°$, therefore $\hat{C}_4^2 = \hat{C}_2$. If the object has several rotation axes, that of highest order (largest n) is called the *principal axis*. In the case of a tie (several axes of the same order), all may be considered to be principal axes.
3. î: The *inversion* operation. Also called a *center of symmetry*.
4. $\hat{\sigma}_h$: The *horizontal mirror plane* reflection. This type of mirror plane is perpendicular to a principal axis.
5. $\hat{\sigma}_v$: *Vertical mirror plane* reflection. If we picture the object with its principal axis up and down, these mirror planes are also up and down; that is, they include the principal axis.
6. $\hat{\sigma}_d$: *Dihedral mirror plane* reflection. These mirror planes are also vertical. The distinction between $\hat{\sigma}_v$ and $\hat{\sigma}_d$ is often somewhat arbitrary; for as far as we will get into the subject, it will not be an important distinction.

Example: These operations will be illustrated with a square.

(a) There is a $\hat{C}_4$ element for rotation about the z axis (perpendicular to the center of the square). This also implies the operations $\hat{C}_4^2 = \hat{C}_2$ and $\hat{C}_4^3$. This is the principal axis.
(b) There are four $\hat{C}_2$ axes perpendicular to $\hat{C}_4$, namely aa', bb', cc', and dd'.

(c) The plane of the paper is a mirror plane; since it is perpendicular to $\hat{C}_4$ (the principal axis), this element is called $\hat{\sigma}_h$.

(d) There are two types of vertical mirror planes—the planes $aa'z$ and $cc'z$, which pass through the corners of the square, and $bb'z$ and $dd'z$, which pass through the sides. Two of these are called $\hat{\sigma}_v$, two are called $\hat{\sigma}_d$; the choice as to which pair is called "vertical" and which "diagonal" is a matter of convention.

(e) There is a center of symmetry $\hat{\imath}$.

(f) There is another element of a type we have not yet discussed—the improper rotation, $\hat{S}_4$.   ∎

7. $\hat{S}_n$: The *improper rotation* operator. The requirements of group theory, to be discussed shortly, make it necessary to define one other type of symmetry operation. Several choices are possible; crystallographers prefer to define a rotation-inversion operation, but we shall follow the spectroscopists and define a rotation-reflection operation. The operation $\hat{S}_n$ is defined as a rotation about an axis by an angle $2\pi/n$ followed by a reflection in a mirror plane perpendicular to this axis:

$$\hat{S}_n = \hat{\sigma}_h \hat{C}_n$$

If $\hat{C}_n$ and $\hat{\sigma}_h$ are symmetry elements already, then their product, $\hat{S}_n$, must also be such (such was the case in the preceding example); but it is possible for $\hat{S}_n$ to be a symmetry element without necessarily requiring that $\hat{C}_n$ or $\hat{\sigma}_h$ be such.

Example: Consider two identical sticks which are placed one upon the other, perpendicular to each other.

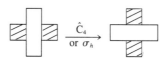

[Think of one stick as being above the plane of the paper while the other (shaded) is below.] The perpendicular axis is a $\hat{C}_2$ symmetry axis but not $\hat{C}_4$. There are two vertical mirror planes (one through each stick and perpendicular to the paper) but the plane of the paper ($\hat{\sigma}_h$) is not a symmetry element. However, the combination:

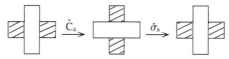

constitutes an $\hat{S}_4$ operation that *is* a symmetry element of this figure.   ∎

Exercise: Try to see two other symmetry elements for this figure—two $\hat{C}_2$ axes perpendicular to the one noted above. It may help to construct a cube and then to draw lines between opposite corners on two opposing faces:

Through the center of each face of the cube there is a $\hat{C}_2$ axis.   ∎

14.2 Groups

If two symmetry operations ($\hat{A}_1$ and $\hat{A}_2$) are symmetry elements of an object, then any product, $\hat{A}_3 = \hat{A}_2\hat{A}_1$, is also a symmetry element. Any set of operations for which any product of members of that set is a member of the set is called a *group*. All the symmetry operations of a molecule form such a group. The study of the mathematical properties of such a collection of symmetry elements is called *group theory*. In the time and space available, we shall see only a few simple elements of group theory and a few of its many applications in chemistry and physics; for those who wish to know more about this elegant and vital area, the short book by F. A. Cotton (ref. 1) is recommended.

Groups containing only symmetry elements that leave one point in the molecule invariant — that is, those described above — are called *point groups*. Group theory is also useful, indeed essential, in the study of crystals. A crystal, considered as an infinite lattice of atoms or molecules, has, in addition to point symmetry, symmetry elements (such as translations) that move a molecule into an identical adjacent lattice position. Groups that contain such operations are called *space groups*. Only point groups will be discussed here.

Each distinct group of symmetry elements has a name. There are two systems for naming point groups, the Hermann-Mauguin system favored by crystallographers and the Schoenflies system favored by molecular spectroscopists; we shall discuss only the latter. In the Schoenflies system, the names of the groups and the names of the operations are very similar; for example, $\hat{C}_2$ is an operation while C_2 is a group name; be careful not to confuse these two concepts.

Example: The symmetry elements of H_2O are:

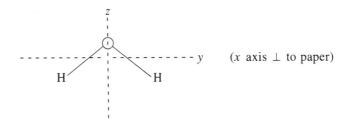

(*x* axis $\perp$ to paper)

(a) $\hat{C}_2(z)$, rotation about the *z* axis by 180°.
(b) $\hat{\sigma}_v(xz)$, reflection in the *xz* plane (perpendicular to paper).
(c) $\hat{\sigma}'_v(yz)$, reflection in the plane of the paper, *yz*.

That these elements form a group can be proven, starting with the effect of each operation on a point (x, y, z):

$$\hat{C}_2 \begin{bmatrix} x \\ y \\ z \end{bmatrix} = \begin{bmatrix} -x \\ -y \\ z \end{bmatrix}, \quad \hat{\sigma}_v \begin{bmatrix} x \\ y \\ z \end{bmatrix} = \begin{bmatrix} x \\ -y \\ z \end{bmatrix}, \quad \hat{\sigma}'_v \begin{bmatrix} x \\ y \\ z \end{bmatrix} = \begin{bmatrix} -x \\ y \\ z \end{bmatrix} \tag{14.3}$$

The product $\hat{\sigma}_v\hat{C}_2$ has the effect:

$$\hat{\sigma}_v\hat{C}_2\begin{bmatrix}x\\y\\z\end{bmatrix} = \hat{\sigma}_v\begin{bmatrix}-x\\-y\\z\end{bmatrix} = \begin{bmatrix}-x\\y\\z\end{bmatrix}$$

This is the same result as $\hat{\sigma}_v'$ would give; therefore we get the operator equality:

$$\hat{\sigma}_v\hat{C}_2 = \hat{\sigma}_v'$$

Proceeding in this manner, we can construct a multiplication table:

	$\hat{E}$	$\hat{C}_2$	$\hat{\sigma}_v(xz)$	$\hat{\sigma}_v'(yz)$
$\hat{E}$	$\hat{E}$	$\hat{C}_2$	$\hat{\sigma}_v$	$\hat{\sigma}_v'$
$\hat{C}_2$	$\hat{C}_2$	$\hat{E}$	$\hat{\sigma}_v'$	$\hat{\sigma}_v$
$\hat{\sigma}_v(xz)$	$\hat{\sigma}_v$	$\hat{\sigma}_v'$	$\hat{E}$	$\hat{C}_2$
$\hat{\sigma}_v'(yz)$	$\hat{\sigma}_v'$	$\hat{\sigma}_v$	$\hat{C}_2$	$\hat{E}$

This table contains no new operations, so these four operations form a group. The name of this group is C_{2v}. ∎

Naming Point Groups

It is not generally necessary to recognize all symmetry elements of a group in order to find the name of the group. Some elements are implied; for example: $\hat{C}_4$ implies $\hat{C}_4^2$ and $\hat{C}_4^3$; $\hat{C}_n$ together with $\hat{\sigma}_h$ implies $\hat{S}_n$. Even so, we need only identify certain essential elements in order to identify the group. The common groups are listed below. (Figure 14.1 shows some examples.)

Groups with no proper rotation axis.

C_1: contains only $\hat{E}$—that is, no symmetry at all. Example: CHFBrCl.

C_s: contains $\hat{E}$, $\hat{\sigma}$.

C_i: contains $\hat{E}$, $\hat{\imath}$.

S_n: contains $\hat{E}$, $\hat{S}_n$. Note that $\hat{S}_1 \equiv \hat{\sigma}$ and $\hat{S}_2 \equiv \hat{\imath}$, so these groups would be called C_s and C_i, respectively.

Groups with one proper rotation axis.

C_n: contains $\hat{E}$, $\hat{C}_n$ only (no mirror planes or other symmetry elements).

C_{nv}: contains $\hat{E}$, $\hat{C}_n$, and n vertical mirror planes called $\hat{\sigma}_v$. Examples: H_2O (above) was C_{2v}. NH_3 (not planar) has a $\hat{C}_3$ axis and three $\hat{\sigma}_v$ mirror planes and is C_{3v}.

C_{nh}: contains $\hat{E}$, $\hat{C}_n$, and $\hat{\sigma}_h$ (perpendicular to $\hat{C}_n$).

The dihedral groups.

D_n: contains $\hat{E}$, $\hat{C}_n$, and n $\hat{C}_2$ axes perpendicular to $\hat{C}_n$. The case $n = 2$ is rather

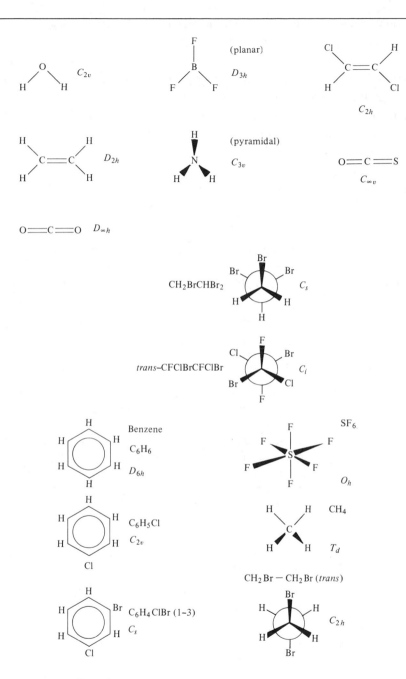

Figure 14.1   Examples of point groups

special, since it has three mutually perpendicular $\hat{C}_2$ axes; this group is some-times called V (German: Vierergruppe).

D_{nh}:   contains $\hat{E}$, $\hat{C}_n$, n $\hat{C}_2$ axes and a mirror plane perpendicular to the principal axis — $\hat{\sigma}_h$. It also has n vertical mirror planes and $\hat{\imath}$ if n is an even number. (If $n = 2$, this group is also called $V_h \equiv D_{2h}$.)

D_{nd}:   contains $\hat{E}$, $\hat{C}_n$, n $\hat{C}_2$ and n vertical mirror planes called $\hat{\sigma}_d$. It also contains an improper axis of double order ($\hat{S}_{2n}$) parallel to $\hat{C}_n$ and $\hat{\imath}$ if n is odd. (If $n = 2$, this group is also called $V_d \equiv D_{2d}$.)

Linear groups.   These are special cases of the above for $n = \infty$; that is, rotation about the axis by *any* angle is a symmetry element. There are two cases:

$\quad\quad C_{\infty v}$:   linear unsymmetrical. Examples: HCl, HCN .

$\quad\quad D_{\infty h}$:   linear symmetrical.   Examples: H_2, CO_2 .

Cubic groups.   These are groups with more than one axis $\hat{C}_n$ with $n \geq 3$. Those of chemical interest include:

$\quad\quad T_d$:   the symmetry of a regular tetrahedron.   Example: CH_4 .

$\quad\quad O_h$:   the symmetry of a regular octahedron.   Example: SF_6 .

$\quad\quad I_h$:   the symmetry of a regular icosahedron.   Example: $B_{12}H_{12}^{2-}$ ion .

The trick to identifying groups is to know which symmetry elements to look for, in what order, and when to stop. The flow chart (Figure 14.2) will aid you in this process. This chart is unique in that it allows you to make mistakes and get back on track; in particular, the perpendicular $\hat{C}_2$ axes in the D_{nd} groups are often difficult to see (a model may help), and this group is often misidentified as C_{nv}. You can avoid this error by looking for the element $\hat{S}_{2n}$ on the same axis as the principal axis $\hat{C}_n$; if you find $\hat{C}_n$, $\hat{S}_{2n}$, and n vertical mirror planes, the group *is* D_{nd} and the perpendicular C_2 axes *are* there whether you can see them or not. [*Hint:* They will bisect the vertical mirror planes.]

Example: Cut a square piece of cardboard. The symmetry elements of this figure were listed earlier, and you should be able to identify the group as D_{4h}.

Now place pins through each corner, all pointing up.

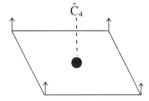

This destroys the horizontal plane of symmetry and the perpendicular $\hat{C}_2$ symmetry, and the group is now C_{4v} ($\hat{C}_4$ and four $\hat{\sigma}_v$ planes).

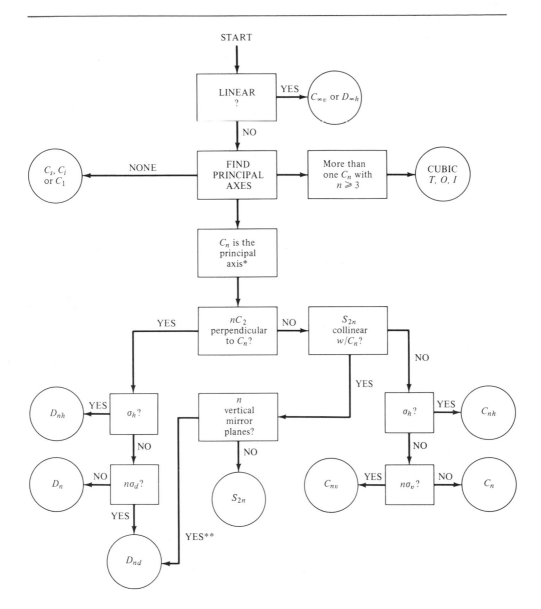

Figure 14.2 Flow diagram for determining point groups. The inversion operation is found in the groups $C_i = S_2$, C_{nh} (if n is even), D_{nh} (if n is even), D_{nd} (if n is odd), $D_{\infty h}$ (linear), S_6, and some of the cubic groups. The only cubic groups of much importance in chemistry are T_d, O_h, and I_h. Notes: *If there are three mutually perpendicular C_2 axes, choose the principal axis to the one that passes through the most (or the heaviest) atoms. **If such is the case, there *are* n perpendicular C_2 axes — try again to see them; building a model will certainly help. [J. Noggle, *J. Chem. Ed.*, **53**, 178 (1976).]

Now place the pins alternately up and down.

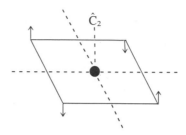

We no longer have a $\hat{C}_4$ axis, but there is clearly a $\hat{C}_2$ axis and two vertical mirror planes, but the group is *not* C_{2v}! Note that the $\hat{C}_2$ axis is also an $\hat{S}_4$ axis, so the group is D_{2d}. Look carefully: there *are* three $\hat{C}_2$ axes.   ∎

Exercise: What is the group designation of the two perpendicular sticks of the earlier example (p. 782)? (*Answer:* D_{2d})   ∎

Once mastered, the identification of point groups is a simple and entertaining pastime. You should practice, using the examples of Figure 14.1 and the problems at the end of this chapter, until perfect. A molecular model kit will be helpful—sometimes essential.

14.3   Symmetry of Functions

In the absence of degeneracies (to be discussed in Section 14.4), a wave function of a molecule must be either symmetric or antisymmetric with respect to all symmetry operations of the molecule. For example, the wave functions and MO's of H_2O (C_{2v} symmetry) must be such that:

$$\hat{C}_2\psi = \pm\psi, \qquad \hat{\sigma}_v\psi = \pm\psi, \qquad \hat{\sigma}'_v = \pm\psi$$

There are only four possible combinations, which are named as follows:

$$\hat{C}_2\psi = +\psi, \; \hat{\sigma}_v\psi = +\psi, \; \hat{\sigma}'_v\psi = +\psi: \quad A_1 \text{ type function}$$

$$\hat{C}_2\psi = +\psi, \; \hat{\sigma}_v\psi = -\psi, \; \hat{\sigma}'_v\psi = -\psi: \quad A_2 \text{ type function}$$

$$\hat{C}_2\psi = -\psi, \; \hat{\sigma}_v\psi = +\psi, \; \hat{\sigma}'_v\psi = -\psi: \quad B_1 \text{ type function}$$

$$\hat{C}_2\psi = -\psi, \; \hat{\sigma}_v \;\;\; = -\psi, \; \hat{\sigma}'_v\psi = +\psi: \quad B_2 \text{ type function} \qquad \textbf{(14.4)}$$

That these are the only permissible functions can be deduced from the C_{2v} multiplication table derived earlier.

Example: Is it possible to have a function that is antisymmetric with respect to all the operations of C_{2v}? If, for some function ψ:

$$\hat{C}_2\psi = -\psi \quad \text{and} \quad \hat{\sigma}_v\psi = -\psi$$

then both operations together give:

$$\hat{C}_2\hat{\sigma}_v\psi = \hat{C}_2(-\psi) = \psi$$

But we saw earlier that $\hat{C}_2\hat{\sigma}_v = \sigma_v'$, therefore this function *must* have:

$$\hat{\sigma}_v'\psi = +\psi$$

and must be a B_2-type function.

By this procedure, one can show that the combinations of symmetries given by Eq. (14.4) are the only ones permitted in C_{2v}. ∎

If we include the identity operation $\hat{E}$, for which all functions give $\hat{E}\psi = +\psi$, of course, these symmetries can be written as four-dimensional vectors — $(\hat{E}, \hat{C}_2, \hat{\sigma}_v, \hat{\sigma}_v')$:

$$\mathbf{A}_1 = (1, \quad 1, \quad 1, \quad 1)$$
$$\mathbf{A}_2 = (1, \quad 1, -1, -1)$$
$$\mathbf{B}_1 = (1, -1, \quad 1, -1)$$
$$\mathbf{B}_2 = (1, -1, -1, \quad 1) \tag{14.5}$$

These are listed together as a *character table* — see Table 14.1. The vectors of Eq. (14.5) are called *representations* and have the general symbol $\mathbf{\Gamma}$. (Table 14.1 also gives some examples of functions that have the symmetries indicated; the utility of these will be discussed later.)

Example: Show that the vectors of Eq. (14.5) are *orthogonal* to each other, and that, for any one of them, $\mathbf{\Gamma}_i \cdot \mathbf{\Gamma}_i = 4$.

Two vectors are orthogonal if their scalar product is zero (cf. Appendix IV). For $\mathbf{A}_2$ and $\mathbf{B}_2$ this product is:

$$\mathbf{A}_2 \cdot \mathbf{B}_2 = (1, 1, -1, -1) \cdot (1, -1, 1, -1) = 1 - 1 - 1 + (-1)^2 = 0$$

For $\mathbf{B}_2$, the scalar product with itself is:

$$\mathbf{B}_2 \cdot \mathbf{B}_2 = (1, -1, 1, -1) \cdot (1, -1, 1, -1) = (1)^2 + (-1)^2 + (1)^2 + (-1)^2 = 4$$

The reader should do the others as an exercise. ∎

Product Functions

If we have two functions — for example, f_{B_1} of B_1 symmetry and f_{B_2} of B_2 symmetry in C_{2v}, what will be the symmetry of the product $f_{B_1}f_{B_2}$? From the preceding definitions of the function types [Eq. (14.5)]:

$$\hat{C}_2(f_{B_1}f_{B_2}) = (-f_{B_1})(-f_{B_2}) = f_{B_1}f_{B_2}$$
$$\hat{\sigma}_v(f_{B_1}f_{B_2}) = (+f_{B_1})(-f_{B_2}) = -f_{B_1}f_{B_2}$$
$$\sigma_v'(f_{B_1}f_{B_2}) = (-f_{B_1})(+f_{B_2}) = -f_{B_1}f_{B_2}$$

The product therefore has A_2 symmetry.

Table 14.1   Character tables: C_{2v}, C_{2h}, D_{2h}

C_{2v}	$\hat{E}$	$\hat{C}_2(z)$	$\hat{\sigma}_v(xz)$	$\hat{\sigma}_v'(yz)$	Functions
$A_1(a_1)$	1	1	1	1	z, x^2, y^2, z^2
$A_2(a_2)$	1	1	-1	-1	xy
$B_1(b_1)$	1	-1	1	-1	x, xz
$B_2(b_2)$	1	-1	-1	1	y, yz

C_{2h}	$\hat{E}$	$\hat{C}_2$	$\hat{\imath}$	$\hat{\sigma}_h$	Functions
A_g	1	1	1	1	x^2, y^2, z^2, xy
B_g	1	-1	1	-1	xz, yz
A_u	1	1	-1	-1	z
B_u	1	-1	-1	1	x, y

D_{2h}	$\hat{E}$	$\hat{C}_2(z)$	$\hat{C}_2(y)$	$\hat{C}_2(x)$	$\hat{\imath}$	$\hat{\sigma}(xy)$	$\hat{\sigma}(xz)$	$\hat{\sigma}(yz)$	Functions
A_g	1	1	1	1	1	1	1	1	x^2, y^2, z^2
B_{1g}	1	1	-1	-1	1	1	-1	-1	xy
B_{2g}	1	-1	1	-1	1	-1	1	-1	xz
B_{3g}	1	-1	-1	1	1	-1	-1	1	yz
A_u	1	1	1	1	-1	-1	-1	-1	
B_{1u}	1	1	-1	-1	-1	-1	1	1	z
B_{2u}	1	-1	1	-1	-1	1	-1	1	y
B_{3u}	1	-1	-1	1	-1	1	1	-1	x

There is a quicker way to discover the symmetry of product functions. For the representation vectors of Eq. (14.5) we define the *direct product* operation as an element-by-element multiplication that forms a new vector:

$$(a_1, b_1, c_1, \ldots) \otimes (a_2, b_2, c_2, \ldots) = (a_1 a_2, b_1 b_2, c_1 c_2, \ldots) \qquad \textbf{(14.6)}$$

From this we see immediately that

$$\mathbf{B}_1 \otimes \mathbf{B}_2 = (1, -1, 1, -1) \otimes (1, -1, -1, 1) = (1, 1, -1, -1) = \mathbf{A}_2$$

This means that for any functions with C_{2v} symmetry, the product of any B_1 function and any B_2 function has A_2 symmetry:

$$f_{B_1} f_{B_2} = f_{A_2}$$

Exercise: Derive the following multiplication table for C_{2v}.

	A_1	A_2	B_1	B_2
A_1	A_1	A_2	B_1	B_2
A_2	A_2	A_1	B_2	B_1
B_1	B_1	B_2	A_1	A_2
B_2	B_2	B_1	A_2	A_1

■

The Totally Symmetric Representation

In any symmetry, one type of function is totally symmetric; that is, it is symmetric with respect to all the symmetry elements of the group. Depending on the group, totally symmetric functions may be called A_1, A_g, A_{1g}, and so on; in general we shall denote the totally symmetric representation as Γ_s. Note that the representation of Γ_s is all ones:

$$\Gamma_s = (1, 1, 1, \ldots)$$

For that reason (since $-1^2 = +1$) it will always be true that the product of any representation with itself is totally symmetric:

$$\Gamma \otimes \Gamma = \Gamma_s \tag{14.7}$$

Try this with C_{2v}; for example, prove that $\mathbf{B}_1 \otimes \mathbf{B}_1 = \mathbf{A}_1$, and so on. [Equation (14.7) must be modified for the degenerate case.]

Also, Γ_s times any of the vectors gives the same vector:

$$\Gamma_s \otimes \Gamma_i = \Gamma_i \tag{14.8}$$

In other words, Γ_s is effectively a unit operator. Again, in C_{2v}, prove $\mathbf{A}_1 \otimes \mathbf{B}_2 = \mathbf{B}_2$, and so on.

Molecular State Functions and Orbitals

A molecule of symmetry C_{2v}, for example, must have state functions whose symmetry is either A_1, A_2, B_1, or B_2 as described above. In molecular orbital theory, these functions are presumed to be products of one-electron functions, the molecular orbitals. If their products are required to have these symmetries, it will be very convenient if the MO's also have such symmetry; the symmetry of orbital functions is denoted with lower-case letters—for example, a_1, a_2, b_1, or b_2 in C_{2v} symmetry. An orbital configuration, such as:

$$(a_1)^2(b_1)^2$$

means the state function is a product of orbital functions (ϕ):

$$\psi = \phi_{a_1}(1)\phi_{a_1}(2)\phi_{b_1}(3)\phi_{b_2}(4)$$

The symmetry of the state function is, therefore:

$$\mathbf{A}_1 \otimes \mathbf{A}_1 \otimes \mathbf{B}_1 \otimes \mathbf{B}_1$$

(As an exercise, prove that this product is A_1; this means that the state function (ψ) has A_1 symmetry.) We shall discuss this point further in Section 14.6; before that, however, it will be necessary to say a few things about degenerate functions.

14.4   Degenerate Representations

Any group with at least one rotation axis $\hat{C}_n$ with $n \geq 3$ will have degenerate functions. We have already encountered degeneracies in linear molecules ($C_{\infty v}$ or

$D_{\infty h}$)—namely, the π and δ functions of diatomics. The occurrence of degeneracies gives rise to complications that we have avoided to this point by choosing our examples carefully; it also makes the theory a great deal more complicated. What follows is no more than a quick outline of some of the ways in which degenerate representations differ from nondegenerate ones. It is not intended to treat the subject thoroughly (for this, see ref. 1, for example) but only to provide a practical basis of facts so that, in our applications, such molecules can be included.

If two functions (ϕ_a and ϕ_b) are degenerate with respect to their eigenvalues for some operator (such as the molecular Hamiltonian), the effect of a symmetry operation ($\hat{A}$) of the group when operated on one of these eigenfunctions will be, in general, to turn it into a linear combination of the pair:

$$\hat{A}\phi_a = c_1\phi_a + c_2\phi_b$$
$$\hat{A}\phi_b = c_3\phi_a + c_4\phi_b \tag{14.9}$$

The reader may recall (Chapter 11, p. 622) that such a linear combination is still an eigenfunction with the same eigenvalue. Such a transformation cannot be represented by a simple multiplication by ± 1 [as in Eq. (14.5)] but must be represented by a matrix:

$$\begin{pmatrix} c_1 & c_2 \\ c_3 & c_4 \end{pmatrix}$$

Table 14.2 Character tables: C_{3v}, D_{3h}, T_d

C_{3v}	$\hat{E}$	$2\hat{C}_3$	$3\hat{\sigma}_v$	Functions[a]
A_1	1	1	1	$z, x^2 + y^2, z^2$
A_2	1	1	-1	
E	2	-1	0	$(x, y)(x^2 - y^2, xy)(xz, yz)$

D_{3h}	$\hat{E}$	$2\hat{C}_3$	$3\hat{C}_2$	$\hat{\sigma}_h$	$2\hat{S}_3$	$3\hat{\sigma}_v$	Functions[a]
A_1'	1	1	1	1	1	1	$z^2, x^2 + y^2$
A_2'	1	1	-1	1	1	-1	
E'	2	-1	0	2	-1	0	$(x, y)(x^2 - y^2, xy)$
A_1''	1	1	1	-1	-1	-1	
A_2''	1	1	-1	-1	-1	1	z
E''	2	-1	0	-2	1	0	(xz, yz)

T_d	$\hat{E}$	$8\hat{C}_3$	$3\hat{C}_2$	$6\hat{S}_4$	$6\hat{\sigma}_d$	Functions[a]
A_1	1	1	1	1	1	$x^2 + y^2 + z^2$
A_2	1	1	1	-1	-1	
E	2	-1	2	0	0	$(2z^2 - x^2 - y^2, x^2 - y^2)$
T_1	3	0	-1	1	-1	
T_2	3	0	-1	-1	1	$(x, y, z)(xy, xz, yz)$

[a]Groups in parenthesis are degenerate.

Recall from the last chapter that for diatomics (as for any linear molecule) the degenerate orbitals π_+ and π_- were exchanged by the $\hat{\sigma}_v$ operation; since $\hat{\sigma}_v\pi_+ = \pi_-$ and $\hat{\sigma}_v\pi_- = \pi_+$, the matrix operation for that transformation is:

$$\begin{pmatrix} 0 & 1 \\ 1 & 0 \end{pmatrix}\begin{pmatrix} \pi_+ \\ \pi_- \end{pmatrix} = \begin{pmatrix} \pi_- \\ \pi_+ \end{pmatrix}$$

Tables 14.2 and 14.3 give character tables for several groups that have degeneracies. For these groups the transformation can be represented by a number that is the *trace* (the sum of diagonal elements) of the transformation matrix. On these tables, symmetry elements of the same type (*class*) are listed together; for example in C_{3v} the three mirror planes $\hat{\sigma}_v$, $\hat{\sigma}_v'$, $\hat{\sigma}_v''$ are listed together as $3\hat{\sigma}_v$ while $\hat{C}_3$ and $\hat{C}_3^2$ are abbreviated as $2\hat{C}_3$. This is just a shorthand; if listed separately, the sets $(\hat{\sigma}_v, \hat{\sigma}_v', \hat{\sigma}_v'')$ and $(\hat{C}_3, \hat{C}_3^2)$ would have identical columns under them.

In nonlinear groups, doubly degenerate functions are called E; such an orbital is called e and can (like the π orbitals of Chapter 13) hold up to four electrons. Triply degenerate functions are called T (F in some references); a t-type orbital can (like the p-type atomic orbitals) hold up to six electrons.

The use of these character tables is pretty straightforward and not unlike our earlier examples. The symmetry of a product function can be deduced by taking the direct

Table 14.3   Character tables: D_{6h}, O_h

D_{6h}	$\hat{E}$	$2\hat{C}_6$	$2\hat{C}_3$	$\hat{C}_2$	$3\hat{C}_2'$	$3\hat{C}_2''$	$\hat{\imath}$	$2\hat{S}_3$	$2\hat{S}_6$	$\hat{\sigma}_d$	$3\hat{\sigma}_d$	$3\hat{\sigma}_v$	Functions
A_{1g}	1	1	1	1	1	1	1	1	1	1	1	1	$x^2 + y^2, z^2$
A_{2g}	1	1	1	1	-1	-1	1	1	1	1	-1	-1	
B_{1g}	1	-1	1	-1	1	-1	1	-1	1	-1	1	-1	
B_{2g}	1	-1	1	-1	-1	1	1	-1	1	-1	-1	1	
E_{1g}	2	1	-1	-2	0	0	2	1	-1	-2	0	0	(xz, yz)
E_{2g}	2	-1	-1	2	0	0	2	-1	-1	2	0	0	$(x^2 - y^2, xy)$
A_{1u}	1	1	1	1	1	1	-1	-1	-1	-1	-1	-1	
A_{2u}	1	1	1	1	-1	-1	-1	-1	-1	-1	1	1	z
B_{1u}	1	-1	1	-1	1	-1	-1	1	-1	1	-1	1	
B_{2u}	1	-1	1	-1	-1	1	-1	1	-1	1	1	-1	
E_{1u}	2	1	-1	-2	0	0	-2	-1	1	2	0	0	(x, y)
E_{2u}	2	-1	-1	2	0	0	-2	1	1	-2	0	0	

O_h	$\hat{E}$	$8\hat{C}_3$	$6\hat{C}_2$	$6\hat{C}_4$	$3\hat{C}_2$	$\hat{\imath}$	$6\hat{S}_4$	$8\hat{S}_6$	$3\hat{\sigma}_h$	$3\hat{\sigma}_d$	Functions
A_{1g}	1	1	1	1	1	1	1	1	1	1	$x^2 + y^2 + z^2$
A_{2g}	1	1	-1	-1	1	1	-1	1	1	-1	
E_g	2	-1	0	0	2	2	0	-1	2	0	$(3z^2 - r^2, x^2 - y^2)$
T_{1g}	3	0	-1	1	-1	3	1	0	-1	-1	
T_{2g}	3	0	1	-1	-1	3	-1	0	-1	1	(xz, yz, xy)
A_{1u}	1	1	1	1	1	-1	-1	-1	-1	-1	
A_{2u}	1	1	-1	-1	1	-1	1	-1	-1	1	
E_u	2	-1	0	0	2	-2	0	1	-2	0	
T_{1u}	3	0	-1	1	-1	-3	-1	0	1	1	(x, y, z)
T_{2u}	3	0	1	-1	-1	-3	1	0	1	-1	

product of the row vectors as before; however, now the result may be not one of the other representations, but a sum of those representations:

$$\Gamma_1 \otimes \Gamma_2 = \sum_i \Gamma_i \qquad (14.10)$$

If $\Gamma_1 = \Gamma_2$, the sum will *contain* the totally symmetric representation (Γ_s) as one of its terms. This is best illustrated with an example.

Example: In C_{3v} symmetry (Table 14.2) the product of E with itself is easily shown to be:

$$\mathbf{E} \otimes \mathbf{E} = (4, 1, 0)$$

This vector, which is not in the table, is an example of a *reducible representation* (the types we have been discussing so far are called *irreducible* representations) and must be some sum of the row vectors. In this case it may be apparent that the vector sum of A_1, A_2, and E will be:

$$\mathbf{A}_1 + \mathbf{A}_2 + \mathbf{E} = (4, 1, 0)$$

therefore:

$$\mathbf{E} \otimes \mathbf{E} = \mathbf{A}_1 + \mathbf{A}_2 + \mathbf{E}$$

What this result means is that a product of two E-type functions can be written as a sum of functions of type A_1, A_2, and E. ∎

Exercise: When nondegenerate representations (A or B types) are involved, direct products come out just as they did earlier (for example, in C_{2v}). Prove that, in T_d symmetry:

$$\mathbf{A}_2 \otimes \mathbf{E} = \mathbf{E}$$

$$\mathbf{A}_2 \otimes \mathbf{T}_1 = \mathbf{T}_2 \qquad \qquad ∎$$

14.5   Bonding Theory

A polyatomic molecule, having numerous electrons and nuclei, is clearly a very complex system, so the solution of the Schrödinger equation for the energy levels is no longer a topic that can be discussed readily at an elementary level. Nonetheless, the ideas that were developed in earlier chapters for solving simpler systems, especially H, He, H_2^+ and H_2, can be utilized to develop a useful picture of bonding and electronic structure in polyatomic molecules. There are basically two approaches to the discussion of the electronic structure of polyatomic molecules — namely, the localized two-center bond and symmetry orbitals.

The localized bond approach is useful because it simplifies the discussion of complex molecules by treating them as a sum of two-center bonds, which can be discussed much as we did for diatomics. Even the terminology is the same; bonds are called "σ" and "π" — although this notation is, strictly speaking, limited to linear molecules. It can provide useful answers to questions such as why CH_4 is tetrahedral rather than square-planar.

On the other hand, the symmetry orbital approach treats the electrons as belonging to the entire molecule. It is useful, indeed necessary, for discussing excited states and interpreting the electronic spectra of polyatomic molecules. It is also needed in cases such as conjugated organic molecules, where the localized bond picture is totally inadequate. (Benzene, a notorious example of the failure of the localized bond approach, will be discussed in Section 14.6.)

In this section we shall primarily discuss the localized bond approach, beginning with the experimental evidence for its validity.

Bond-Additive Properties of Molecules

The fundamental soundness of the localized bond picture of molecules is demonstrated by the fact that a number of molecular properties can be interpreted with good accuracy as a sum of bond contributions.

The energy of molecules, as measured by the energy required to break bonds in chemical reactions, photodissociation, and so on, can be interpreted in terms of average bond energies; Table 14.4 gives some values for this property. Such data are deduced from a variety of thermodynamic and spectroscopic results relating to the energy required to break a particular type of bond. The examples that follow will illustrate this concept.

Example: A reaction such as:

$$H_2 + H_2 + O_2 \longrightarrow H_2O + H_2O$$

can be viewed as a process in which two H—H and one O=O bond are broken, and four O—H bonds are made. The energy required will be estimated with the bond energies of Table 14.4.

$$\text{Bonds broken:} \quad 2(435) + 492 = 1362 \text{ kJ/mol}$$

$$\text{Bonds made:} \quad 4(464) = 1856 \text{ kJ/mol}$$

The reaction should therefore be exothermic by 494 kJ/mol. This energy corresponds roughly to ΔH_0^θ of the reaction as described in Chapters 5 and 6. The calculation above, therefore, predicts:

$$\Delta H_0^\theta = -494 \text{ kJ/mol}$$

for the formation of two moles of H_2O. The thermodynamic data of Table 6.4 give the accurate value for this quantity:

$$\Delta H_0^\theta = 2(-238.94) = -477.88 \text{ kJ/mol}$$

If the comparison is made to enthalpies at 298 K, then (from Table 6.1):

$$\Delta H = 2(\Delta H_{f, 298}^\theta) = -483.6 \text{ kJ}$$

[Which comparison is the more apt (298 K or 0 K) depends on the source of the "bond energies"; if the bond energy is derived from spectroscopic data (such as D_0 of Table 13.1), then enthalpies at 0 K are better; if the bond energy is derived from thermochemical data, then enthalpies at 298 K are better for purposes of comparison. In fact, the accuracy is not good enough to make the distinction significant.] ∎

Table 14.4   Average bond energies and dipoles

Bond	Energy (kJ/mol)	Dipole (debyes)	Bond	Energy (kJ/mol)	Dipole (debyes)	
C—C	344	0	H—H	435	0	(in H_2)
C—O	350	0.8	H—N	390	1.3	
C—N	293	0.5	H—O	464	1.5	
C=C	615	0	H—C	415	0.4	
C=O	724	2.5	C≡N	891	3.5	
C=N	615	—	N≡N	942	0	(in N_2)
C≡C	812	0	O=O	492	0	(in O_2)
C—Cl	326	1.5	Cl—Cl	239	0	(in Cl_2)

Example: Estimate the energy of the reaction:

$$CH_3—CH_3 \longrightarrow CH_2=CH_2 + H_2$$

using the average bond energies of Table 14.4.

Bonds broken:	1 C—C and 6 C—H.
Energy:	$344 + 6(415) = 2834$ kJ/mol
Bonds made:	1 C=C, 1 H—H, 4 C—H
Energy:	$615 + 435 + 4(415) = 2710$ kJ/mol

Net:

$$\Delta H_0^\theta = 124 \text{ kJ/mol}$$

The data of Table 6.4 give the accurate value:

$$\Delta H_0^\theta = 60.75 - (-69.12) = 129.17 \text{ kJ/mol} \qquad \blacksquare$$

From these examples we see that average bond energies can be used to estimate enthalpies of reactions. This method, however, is inferior to the methods of Chapter 6 and should be used only if no better data are available. However, the examples also illustrate that bond additivity of energies is a sound, if approximate, concept.

Another area in which bond additivity has been applied is the interpretation of the electrical properties of molecules. For example, the net dipole moment of a molecule can be considered as a vector sum of the bond dipoles of its constituent bonds; Table 14.4 gives some of these values. The same idea has been applied to molecular polarizabilities.

To the extent that properties such as bond energy and dipole moments are bond-additive, they form a tool for determining molecular structure; for example, one might be able to deduce what types of bonds are present, and in what direction they must point. However, this method is greatly inferior to spectroscopic methods for determining molecular structure; we shall discuss some of these spectroscopic methods later in this chapter.

Localized Molecular Orbitals

Localized molecular orbitals are linear combinations of atomic orbitals, one AO from each of the bonded atoms. For example, in water, the bonding MO's could be written as linear combinations of the orbitals of the two protons (which will be denoted H_A and H_B) and the orbitals from the oxygen. The oxygen $1s$ orbital can accurately be assumed to be localized on the oxygen atom and nonbonding. The most likely orbitals for bond formation will be the oxygen $2s$ and $2p$ orbitals and the $1s$ orbitals of the hydrogens. (Higher-energy orbitals can be neglected in a first approximation.) One possibility would be to assume that the oxygen orbitals $(O2s)$ and $(O2p_z)$ are non-bonding, and to construct MO's from LCAO using the p_x and p_y orbitals:

$$\sigma_1 = (H_A 1s) + (O2p_x)$$

$$\sigma_2 = (H_B 1s) + (O2p_y) \tag{14.11}$$

Then the configuration for the 10 electrons would be written:

$$H_2O(O1s)^2(O2s)^2(O2p_z)^2(\sigma_1)^2(\sigma_2)^2 \tag{14.12}$$

This configuration describes a product function that could reasonably serve as a starting point for calculating the energy of the ground state of this molecule. However, it is not the best starting point. One of the lessons of H_2^+ (Section 13.4) was that a strong bond results from the constructive overlap of the atomic orbitals, which will concentrate the electron density between the nuclei. The overlap of the orbitals in Eq. (14.11) will be optimum (for a given internuclear distance) if the H—O direction is in the direction of the p-orbital:

This would suggest that the configuration of Eq. (14.12) would have a minimum energy when the H—O—H bond angle is 90°; the bond angle is actually 105°. Of course, there are more interactions than just the overlap and the O—H nuclear repulsion—for example, the mutual repulsion of the H nuclei and of the electrons in the bond orbitals. But the point is that, even for a first approximation, the product function implied by Eq. (14.12) is a poor starting point.

Of course, as was mentioned in the discussion of HF in Chapter 13, there is no need to use "pure" atomic orbitals, so some combination of the available oxygen orbitals, preferably one that has a maximum amplitude in the actual bond direction, can be used; these are the *hybrid* orbitals.

The desirability of using hybrid atomic orbitals is illustrated more vividly by the common hydride of carbon—methane (CH_4). There is an abundance of experimental evidence that the four CH bonds of methane are exactly equivalent and point in the tetrahedral directions. Molecular orbitals that are linear combinations of the (H1s) with the carbon 2s or 2p orbitals are not at all realistic, since, for example, (H1s) + (C2s) is not equivalent to (H1s) + (C2p_z).

Hybrid Atomic Orbitals

LCAO-MO theory approximates molecular orbitals as linear combinations of a set of mathematical functions, the atomic orbitals, which are, in fact, eigenfunctions of the one-electron atom that were derived in Section 12.1. This choice is by no means unique: any complete set of orthogonal functions would have the same property. [You may recall the discussion in Chapter 11 (p. 590) regarding the expansion of functions in complete orthogonal sets.] However, the use of atomic orbitals to construct molecular wave functions is very popular and useful, because it reflects the chemist's intuition that atoms retain at least some of their identity when they are part of a compound.

But, even if the one-electron eigenfunctions are used, it was never required that we use the particular ones listed in Section 12.1. On several occasions it has been pointed out that any linear combination of degenerate functions was an equally good representation of those functions. (Recall the p_1 and p_{-1} orbitals vs. the p_x or p_y representations, or the π_+, π_- vs. π_x, π_y representations.) In the one-electron atom, the 2s and 2p functions, the 3s, 3p, and 3d functions, and so on were all degenerate. [Recall that the energy difference of the 2s and 2p orbitals arose (Section 12.3) as a result of ee repulsion in the *two*-electron atom.] Several linear combinations of these functions are very useful in bonding theory, and we shall discuss these next.

The *sp* Hybrids

The atomic orbitals were shown, in Section 12.1, to be products of a radial function (Table 12.1) and an angular function (Table 11.8); Eq. (12.6):

$$\psi_{nlm} = R_{nl}(r)Y_{lm}(\theta, \phi)$$

For s orbitals ($l = 0$), the angular function was just a constant: $Y_{0,0} = 1/\sqrt{4\pi}$. Thus the 2s function, for example, is:

$$(2s) = R_{2s}\frac{1}{\sqrt{4\pi}} \tag{14.13a}$$

The 2p functions are (cf. Table 11.8):

$$(2p_z) = R_{2p}\frac{1}{\sqrt{4\pi}}\sqrt{3}\cos\theta \tag{14.13b}$$

$$(2p_x) = R_{2p}\frac{1}{\sqrt{4\pi}}\sqrt{\frac{3}{2}}\sin\theta\cos\phi \tag{14.13c}$$

$$(2p_y) = R_{2p}\frac{1}{\sqrt{4\pi}}\sqrt{\frac{3}{2}}\sin\theta\sin\phi \tag{14.13d}$$

These functions are normalized and orthogonal to each other; any linear combination of them that is to be useful in bonding theory should maintain these properties. For example, the $2s$ and $2p_z$ functions could be combined as:

$$(spa) = a(2s) + b(2p_z)$$

$$(spb) = a(2s) - b(2p_z)$$

(The $3s$, $3p$ or $4s$, $4p$, ... sets could be used as well.) If the first function is to be normalized, we require:

$$\int |(spa)|^2 \, d\tau = 1$$

$$1 = a^2 \int (2s)^2 \, d\tau + b^2 \int (2p_z)^2 \, d\tau + 2ab \int (2s)(2p_z) \, d\tau$$

Since the atomic orbital functions are normalized, the first two integrals are equal to 1; because they are orthogonal, the third integral is zero. Therefore we require that the sum of squares of the coefficients be:

$$a^2 + b^2 = 1$$

If the functions (spa) and (spb) are to be orthogonal, we require:

$$\int (spa)(spb) \, d\tau = 0$$

$$= a^2 \int (2s)^2 \, d\tau - b^2 \int (2p)^2 \, d\tau = 0$$

Therefore $a^2 = b^2$. We conclude that:

$$a = b = \frac{1}{\sqrt{2}}$$

and the sp hybrids are:

$$(spa) = \frac{1}{\sqrt{2}}[(2s) + (2p_z)]$$

$$(spb) = \frac{1}{\sqrt{2}}[(2s) - (2p_z)] \qquad \textbf{(14.14)}$$

Example: Determine the shape of the functions of Eq. (14.14). Using Eq. (14.13), we find that the sp hybrids are:

$$(\text{const.})\,[R_{2s}(r) \pm R_{2p}(r) \sqrt{3} \cos \theta]$$

The positive combination will have a maximum in the $+z$ direction ($\theta = 0$, $\cos \theta = 1$) and a smaller value in the $-z$ direction ($\theta = 180°$, $\cos \theta = -1$); the negative combination will have exactly the opposite orientation.

The shape of a function $F(r, \theta)$ can be represented as a polar graph in which $|F|$ is plotted as the length of a radius vector vs. θ. This is conveniently done on polar-lined graph paper; if

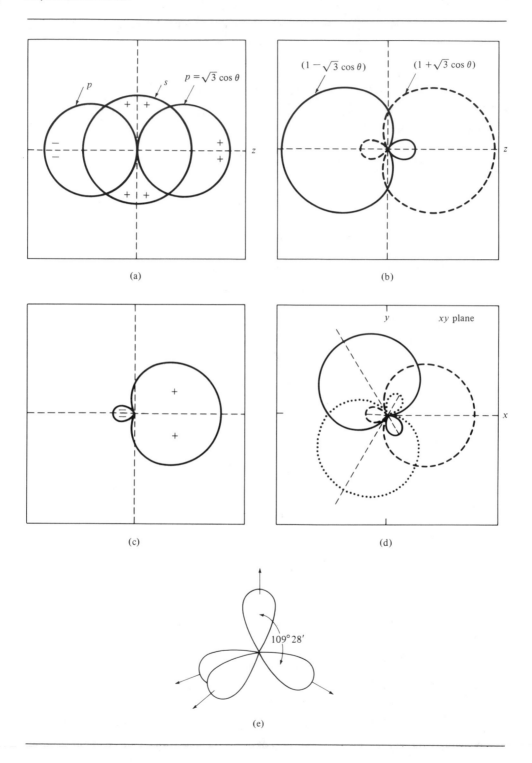

(a)

(b)

(c)

(d)

(e)

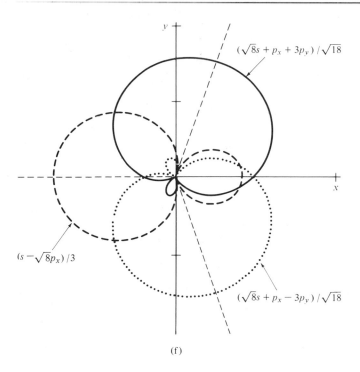

(f)

Figure 14.3 Hybrid orbitals. (a) The s and p orbitals shown separately. The signs (+ or −) indicate the sign of the orbital function in that area. (b) The sp hybrids. There are two such functions, shown by solid and dashed lines. (c) One of the sp^2 hybrids. (d) The three sp^2 hybrids shown superimposed. (e) The sp^3 hybrids; for simplicity, the back lobes [which look much like the negative lobe of panel (c)] are not shown. These orbitals point toward the corners of a tetrahedron and have a dihedral angle of 109°28′. (f) Another set of orbitals, which can be made from an s orbital and two p orbitals. The straight dashed lines running radially from the center indicate the direction in which the orbitals "point"—that is, the direction in which their amplitude is a maximum.

only Cartesian paper is available, a plot of the locus of:

$$X = |F| \cos \theta, \qquad Y = |F| \sin \theta$$

will give the same effect.

To make a polar plot of the hybrid orbitals, some value of r must be chosen, since R_{2s} and R_{2p} do not have any constant proportion. A convenient choice is the distance where $R_{2s} = R_{2p}$, where the shapes will be:

$$1 \pm \sqrt{3} \cos \theta \qquad\qquad \textbf{(14.15)}$$

The reader should make these plots for the sp hybrids. [See Figure 14.3(b).] ∎

Beryllium hydride (BeH_2) is known to be linear. The bonding in this compound can be described using localized LCAO-MO's that are linear combinations of the hydrogen $1s$ orbitals and sp hybrid orbitals of Be:

$$\sigma_1 = [c_1(H_A 1s) + c_2(Be spa)]$$
$$\sigma_2 = [c_1(H_B 1s) + c_2(Be spb)] \qquad \textbf{(14.16)}$$

The electronic configuration (six electrons) would then be written as:

$$BeH_2(Be1s)^2\sigma_1^2\sigma_2^2$$

The CH_2 radical, although it is not the most stable hydride of carbon, can be observed spectroscopically at high temperature; it was thought to be linear. [There is evidence that the ground state is bent; cf. R. A. Bernheim, H. W. Bernard, P. S. Wang, L. S. Wood, and P. S. Shell, *J. Chem. Phys.*, 53, 1280 (1970).] Linear bonding can be described using MO's like those of Eq. (14.16) and the p_x and p_y orbitals, which remain localized on carbon and are nonbonding; the two p orbitals are degenerate and are more properly called "π" in a linear molecule.

A reasonable configuration for the ground state is:

$$CH_2(C1s)^2(\sigma_1)^2(\sigma_2)^2(\pi)^2$$

with σ_1 and σ_2 defined as in Eq. (14.16). The nonbonded electrons in the π orbitals (that is, the degenerate p_x, p_y pair) are in an open shell; the discussion of oxygen in Chapter 13 (which also had a π^2 configuration) would lead us to expect that this molecule will be in a triplet state. However, such discussions are best pursued using the symmetry orbitals, so we shall return to CH_2 in the next section.

The sp^2 Hybrids

There are a number of combinations of two p functions (for example p_x and p_y) and an s function that are orthogonal to each other. One such combination, which produces three equivalent orbitals at an angle of 120° (that is, the functions are identical except for a rotation by 120°), are the sp^2 hybrids:

$$(sp^2 a) = \frac{(s) + \sqrt{2}(p_x)}{\sqrt{3}}$$

$$(sp^2 b) = \frac{\sqrt{2}(s) - (p_x) + \sqrt{3}(p_y)}{\sqrt{6}}$$

$$(sp^2 c) = \frac{\sqrt{2}(s) - (p_x) - \sqrt{3}(p_y)}{\sqrt{6}} \qquad \textbf{(14.17)}$$

These functions, pictured in Figure 14.3, lie in the xy plane and have maximum values in the directions $\phi = 0°$, 120°, 240°. (Of course, similar sets in the xz or yz plane could be made using the p_z function.)

Exercise: Combining Eq. (14.17) with Eqs. (14.13), the shapes of the $sp^2 a$ hybrid can be seen to be:

$$(sp^2 a) \propto 1 + \sqrt{3} \sin\theta \cos\phi$$

In the xy plane, $\theta = 90°$ and $\sin \theta = 1$; thus the shape is just like the sp hybrid:

$$1 + \sqrt{3} \cos \phi$$

except that the maximum is along the $+x$ axis. Make a polar plot to demonstrate that this is true. Also demonstrate that the other sp^2 hybrids point (have a maximum) in the directions $\phi = 120°$ and $\phi = 240°$. ∎

Boron hydride (BH_3) is known to be planar with three equivalent BH bonds and an HBH bond angle of 120°. The bonding in this molecule is well described using the LCAO-MO's:

$$\sigma_1 = [c_1(H_A 1s) + c_2(Bsp^2 a)]$$
$$\sigma_2 = [c_1(H_B 1s) + c_2(Bsp^2 b)]$$
$$\sigma_3 = [c_1(H_C 1s) + c_2(Bsp^2 c)] \qquad \textbf{(14.18)}$$

The electronic configuration in the ground state would then be:

$$BH_3(B1s)^2(\sigma_1)^2(\sigma_2)^2(\sigma_3)^2$$

The sp^3 Hybrids

All three p functions can be combined into four hybrids that are identical except for their orientation in space. These are:

$$(sp^3 1) = \tfrac{1}{2}[(s) + (p_x) + (p_y) + (p_z)]$$
$$(sp^3 2) = \tfrac{1}{2}[(s) + (p_x) - (p_y) - (p_z)]$$
$$(sp^3 3) = \tfrac{1}{2}[(s) - (p_x) + (p_y) - (p_z)]$$
$$(sp^3 4) = \tfrac{1}{2}[(s) - (p_x) - (p_y) + (p_z)] \qquad \textbf{(14.19)}$$

These functions will point to (have a maximum amplitude in the direction of) the vertices of a tetrahedron. The angle between the principal direction of each orbital is 109°28′.

Exercise: Demonstrate that the functions of Eq. (14.19) are normalized (presuming that the original atomic orbitals are normalized) and that they are mutually orthogonal. ∎

The four functions of Eq. (14.19) can be used to represent the four equivalent CH bonds of methane with four molecular orbitals:

$$\sigma_1 = [c_1(H_A 1s) + c_2(Csp^3 1)], \quad \ldots, \quad \sigma_4 = [c_1(H_D 1s) + c_2(Csp^3 4)]$$

Then the electronic configuration of methane would be

$$CH_4(C1s)^2(\sigma_1)^2(\sigma_2)^2(\sigma_3)^2(\sigma_4)^2$$

Other Molecules, Other Hybrids

The examples used so far have covered the first-row hydrides from beryllium to carbon, so the next logical example would be ammonia (NH_3). Ammonia has 10 electrons; two of these will be in the nonbonding ($N1s$) orbital, but since only six electrons are needed to bond with the three hydrogens, two other nonbonding electrons, a lone pair, will be present. Ammonia is known to be pyramidal, so the use of the sp^3 hybrid would be favored over, for example, the sp^2 hybrids plus a nonbonding p orbital. The localized MO's would be:

$$\sigma_1 = [c_1(H_A 1s) + c_2(Nsp^3 1)]$$
$$\sigma_2 = [c_1(H_B 1s) + c_2(Nsp^3 2)]$$
$$\sigma_3 = [c_1(H_C 1s) + c_2(Nsp^3 3)] \tag{14.20}$$

and the ($Nsp^3 4$) orbital would be nonbonding, as would ($N1s$). The ground-state configuration in this approximation would then be

$$NH_3(N1s)^2(Nsp^3 4)^2(\sigma_1)^2(\sigma_2)^2(\sigma_3)^2$$

This would predict an HNH bond angle of $109°28'$ for optimum overlap; the observed angle is $108°$.

The next first-row hydride, of course, is water, with two hydrogens and 10 electrons. Again LCAO molecular orbitals using sp^3 hybrids, such as those of Eq. (14.20), could be used to give a configuration with two lone pairs:

$$H_2O(O1s)^2(Osp^3 4)^2(Osp^3 3)^2(\sigma_1)^2(\sigma_2)^2$$

In this case the observed bond angle is $105°$ — rather different from the tetrahedral angle.

The bonding descriptions above for ammonia and water are, in any case, approximate, so the fact that the bonding overlap is not optimum at the observed bond angles is not of major consequence. However, the hybrid orbitals of Eqs. (14.14), (14.17) and (14.19) are not the only possibilities. In general an s orbital and three p orbitals can be combined to form four linear combinations:

$$\chi_1 = c_{1,1}(s) + c_{1,2}(p_z) + c_{1,3}(p_x) + c_{1,4}(p_y)$$
$$\chi_2 = c_{2,1}(s) + c_{2,2}(p_z) + c_{2,3}(p_x) + c_{2,4}(p_y)$$
$$\chi_3 = c_{3,1}(s) + c_{3,2}(p_z) + c_{3,3}(p_x) + c_{3,4}(p_y)$$
$$\chi_4 = c_{4,1}(s) + c_{4,2}(p_z) + c_{4,3}(p_x) + c_{4,4}(p_y) \tag{14.21}$$

There are 16 constants in these equations; however, four will be fixed by the requirement that the hybrid orbitals (χ) must be normalized and six will be fixed by the requirement that the functions be mutually orthogonal. There remain six constants that can be varied to give a variety of bond angles. One such possibility is pictured in Figure 14.3; the functions are:

$$\chi_1 = \frac{\sqrt{8}(s) + (p_x) + 3(p_y)}{\sqrt{18}}$$

$$\chi_2 = \frac{\sqrt{8}\,(s) + (p_x) - 3(p_y)}{\sqrt{18}}$$

$$\chi_3 = \frac{(s) - \sqrt{8}\,p_x}{3}$$

$$\chi_4 = (p_z) \tag{14.22}$$

If χ_1 and χ_2 were used to form LCAO molecular orbitals, optimum overlap would occur at a bond angle of 143°.

Elements in the third and higher rows of the periodic table have low-lying d orbitals that can be used for a localized bond picture. These are particularly useful for the transition metals. Table 14.5 lists several commonly used hybrids.

Other Types of Localized Orbitals

The bonding in ethylene

can be well described in terms of sp^2 carbon hybrids with bonds to two protons and to the other carbon. The perpendicular p orbitals (we shall assume that this direction is the z axis) can also be combined to form an LCAO-MO:

$$a(C_1 p_z) + b(C_2 p_z)$$

This orbital is generally called a "π" orbital by analogy to diatomic molecules, but this notation is unfortunate, since the symbol "π" rigorously denotes a degenerate *pair* of orbitals in a linear molecule. Whatever the notation, the fact that there are two bonds between the carbons neatly explains the fact that the C—C bond is rigid; that is, the CH_2 groups cannot rotate freely with respect to each other as can the —CH_3 groups of ethane.

For certain molecules—diborane, for example—it is useful to use LCAO-MO's that are combinations of three atomic orbitals on different centers; this is called a *three-center* bond.

Table 14.5 Some common hybrid orbitals

Coordination number	Hybrid description for central atom	Geometry	Bond angles
2	sp	Linear	180°
3	sp^2	Trigonal planar	120°
4	sp^3	Tetrahedral	109°28'
4	dsp^2	Square planar	90°
5	dsp^3	Trigonal bipyramid	120°, 90°
6	d^2sp^3	Octahedral	90°

14.6   Symmetry Orbitals

The wave function of a molecule must, as mentioned earlier, have the same symmetry as the potential field of the nuclei and thus be an eigenfunction of all of the symmetry operators of the molecule's point group. In molecular orbital theory, wave functions are approximated as products of the molecular orbital functions, and the discussion of Section 14.3 suggests that it would be advantageous if the orbital functions also had the symmetry of the molecule's point group.

As an example of this approach, we shall discuss the dihydrides of a second-row element (denoted X). Such a molecule, XH_2, may be either linear with point group $D_{\infty h}$ or nonlinear with point group C_{2v}:

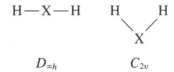

These cases must be discussed separately.

Linear XH₂

Any linear molecule with a center of symmetry will have the same symmetry elements as a homonuclear diatomic molecule; these were discussed in Chapter 13. Furthermore, the orbital names are as before, σ_g, σ_u, π_g, π_u, and so on, and the state functions are named in the same manner (Σ, Π, and so on).

Assuming that the $(X1s)$ orbital will be nonbonding, and neglecting high-energy atomic orbitals, we will construct the molecular orbitals as linear combinations of the two $(H1s)$, the $(X2s)$, and the three p orbitals of X. We therefore shall begin by classifying these atomic orbitals according to their symmetry. The conventional choice of a coordinate system with the z axis along the molecular axis is used.

The $1s$ and $2s$ orbitals of X clearly have σ_g symmetry; that is, they are even with respect to inversion through X, and cylindrically symmetric about the z axis.

The p_z orbital of X, the one that lies along the bonding axis, is also a σ-type function (having cylindrical symmetry about z) but is odd with respect to inversion through X; it is therefore of symmetry σ_u.

Since a linear molecule has cylindrical symmetry, the other two p orbitals must be degenerate. They are usually represented as p_x and p_y, a degenerate pair, but could be equally well represented as p_1 and p_{-1} (identical toroids with opposite sense of rotation). In either case, this degenerate pair of orbitals is of π type and is odd with respect to inversion; they are therefore properly classified as π_u.

The $1s$ orbitals of the protons are also σ types, but they do not lie at the center of symmetry (as do the X orbitals); in fact, these identical atoms are exchanged by symmetry operations such as $\hat{\imath}$ or $\hat{\sigma}_h$:

$$\hat{\imath}(H_A1s) = (H_B1s), \qquad \hat{\imath}(H_B1s) = (H_A1s)$$

Table 14.6   Atomic orbitals of XH_2 (linear) by symmetry

AO	Symmetry type
$(X1s)$	σ_g
$(X2s)$	σ_g
$(X2p_z)$	σ_u
$(X2p_x), (X2p_y)$	π_u (degenerate pair)
$\chi_g = (H_A1s) + (H_B1s)$	σ_g
$\chi_u = (H_A1s) - (H_B1s)$	σ_u

However, functions of the proper symmetry are easily obtained by a linear combination:

$$\chi_g = (H_A1s) + (H_B1s) \qquad (\text{type } \sigma_g)$$

$$\chi_u = (H_A1s) - (H_B1s) \qquad (\text{type } \sigma_u) \tag{14.23}$$

This is an example of a symmetry adapted linear combination (SALC). Table 14.6 summarizes this classification of the atomic orbitals.

Molecular Orbitals for Linear XH_2

If the LCAO-MO's are to have symmetry σ_g, σ_u, and so on, there is no surer way of achieving this than to take only linear combinations of atomic orbitals of the proper symmetry. This statement applies not only to approximate MO's but as well to the accurate ones; the correct MO's of such a molecule of σ_g symmetry, for example, will be linear combinations only of σ_g-type AO's—but a number of such orbitals would be needed.

The approximate MO's are now made as linear combinations of the atomic orbitals. The only linear combinations that are useful are those among AO's having the same symmetry, and the most important of these will be between AO's having similar energies. In the present case, the most important combinations are:

$$(1\sigma_g) = (X1s) \qquad\qquad (\text{nbo})$$

$$(2\sigma_g) = (X2s) + \chi_g \qquad\qquad (\text{bo})$$

$$(3\sigma_g) = (X2s) - \chi_g \qquad\qquad (\text{abo})$$

$$(1\sigma_u) = (X2p_z) + \chi_u \qquad\qquad (\text{bo})$$

$$(2\sigma_u) = (X2p_z) - \chi_u \qquad\qquad (\text{abo})$$

$$(1\pi_u) = (X2p_+) \quad \text{and} \quad (X2p_-) \qquad (\text{nbo, doubly degenerate}) \tag{14.24}$$

where bo, nbo, and abo denote bonding, nonbonding, and antibonding orbitals, respectively. These orbitals are diagrammed in Figure 14.4. Within a given symmetry type, the orbitals are numbered in order of increasing energy (for example,

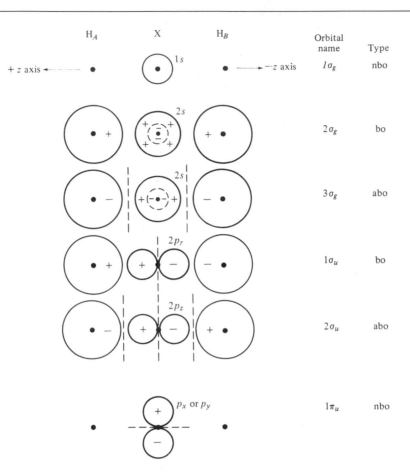

Figure 14.4 Molecular orbitals of linear XH_2 (schematic). X is a first-row element. The signs (+ or −) indicate the sign of the orbital function in that area. Dashed lines are nodes.

$1\sigma_g, 2\sigma_g, 3\sigma_g$). Here, as elsewhere, the functions with the larger number of nodes can be expected to be higher in energy. The filling order of the orbitals for linear H_2X is:

$$(1\sigma_g)\,(2\sigma_g)\,(1\sigma_u)\,(1\pi_u)\,(3\sigma_g)\,(2\sigma_u) \qquad (14.25)$$

The number of electrons will depend, of course, on the identity of the central atom X. If X = Be, there will be six electrons, and the configuration (and term symbol, determined as in Chapter 13) is:

$$\text{BeH}_2(1\sigma_g)^2(2\sigma_g)^2(1\sigma_u)^2\,^1\Sigma_g^+$$

The symmetry orbital description has several major advantages. First, it positions one very well for doing more accurate calculations, since the accurate MO's will have the same symmetry characteristics as the approximate ones. Also, this description is very well adapted for the discussion of excited electronic states and, hence, the electronic spectroscopy of these molecules. (However, note that excited states of linear molecules may not be linear; this is better discussed after we discuss the nonlinear MO's.) For example, some probable excited states of BeH_2 would be:

$$BeH_2(1\sigma_g)^2(2\sigma_g)^2(1\sigma_u)^1(1\pi_u)^1 \ {}^1\Pi_g \ \text{or} \ {}^3\Pi_g$$

$$BeH_2(1\sigma_g)^2(2\sigma_g)^2(1\sigma_u)^1(3\sigma_g)^1 \ {}^1\Sigma_u^+ \ \text{or} \ {}^3\Sigma_u^+$$

The selection rules are exactly as for homonuclear diatomics (Section 13.6), so the reader should see that only one of these excited states ($^1\Sigma_u^+$) represents a dipole-allowed transition from the ground state. If X is carbon, then the methylene radical configuration would be written as:

$$CH_2(1\sigma_g)^2(2\sigma_g)^2(1\sigma_u)^2(1\pi_u)^2$$

Note that this molecule, like O_2, is open-shell, since there are only two electrons in the degenerate pair of π orbitals. The same arguments as in Section 13.5 would demonstrate that the terms of this configuration will be $^3\Sigma_g^-$, $^1\Delta_g$, $^1\Sigma_g^+$, with the first being the ground state. (See the discussion of CH_2 on page 802.)

Nonlinear XH₂

As for the linear case, we begin by seeking the symmetry of the atomic orbitals, in this case in C_{2v} symmetry. For example, the symmetry operations of C_{2v} upon a p_x orbital on the central atom give:

rotation about z axis:	$\hat{C}_2(p_x) = -(p_x)$
reflection in the xz plane:	$\hat{\sigma}_v(p_x) = (p_x)$
reflection in the yz plane:	$\hat{\sigma}_v'(p_x) = -(p_x)$

Examination of Table 14.1 demonstrates that this function (p_x) is a function with b_1 symmetry. The reader should work out the symmetry of the other X atomic orbitals as an exercise; the answers are given in Table 14.7.

The hydrogen $1s$ orbitals are, again, exchanged by several symmetry operations, so SALC's must be used; these are:

$$\chi_1 = (H_A 1s) + (H_B 1s) \qquad \text{(symmetry } a_1)$$

$$\chi_2 = (H_A 1s) - (H_B 1s) \qquad \text{(symmetry } b_2) \qquad \textbf{(14.26)}$$

Table 14.7 Atomic orbitals of XH₂ (nonlinear) by symmetry

Operation	Orbitals						
	(X1s)	(X2s)	(X2p_z)	(X2p_x)	(X2p_y)	χ_1	χ_2
$\hat{C}_2(z)$	1	1	1	-1	-1	1	-1
$\hat{\sigma}_v(xz)$	1	1	1	1	-1	1	-1
$\hat{\sigma}_v(yz)$	1	1	1	-1	1	1	1
Symmetry type	a_1	a_1	a_1	b_1	b_2	a_1	b_2

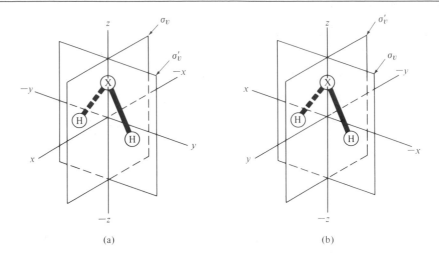

Figure 14.5 Conventions for C_{2v} symmetry. (a) The convention used in the bonding theory of planar C_{2v} molecules (such as XH_2, discussed in the text) places the atoms in the yz plane. (b) The convention used for vibrational spectroscopy of XH_2 molecules of C_{2v} symmetry places the atoms in the xz plane.

Molecular Orbitals for Nonlinear XH_2

The convention usually used in bonding theory is that a C_{2v} molecule such as XH_2 lies in the yz plane with the z axis passing through the X atom (Figure 14.5). This is a fairly arbitrary choice; molecular spectroscopists generally prefer the convention that the molecule lies in the xz plane; in fact, we shall use this convention later in this chapter. The principal effect of exchanging these two conventions is to exchange the names of orbitals b_1 and b_2 and states B_1 and B_2. This convention has already been used in naming the χ_2 orbital of Eq. (14.26) as "b_2"; if the other convention were used, the function would be named "b_1".

The LCAO-MO's can now be constructed by combining AO's having the same symmetry. With the help of Table 14.7, it may be apparent that the approximate molecular orbitals will be as below (Figure 14.6 shows these combinations schematically):

$$
\begin{aligned}
(1a_1) &= (X1s) & \text{(nbo)} \\
(2a_1) &= (X2s) + \chi_1 & \text{(bo)} \\
(3a_1) &= (X2p_z) + \chi_1 & \text{(bo)} \\
(4a_1) &= (X2p_z) - \chi_1 & \text{(abo)} \\
(1b_1) &= (X2p_x) & \text{(nbo)} \\
(1b_2) &= (X2p_y) + \chi_2 & \text{(bo)} \\
(2b_2) &= (X2p_y) - \chi_2 & \text{(abo)}
\end{aligned}
$$

(14.27)

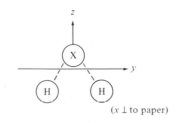

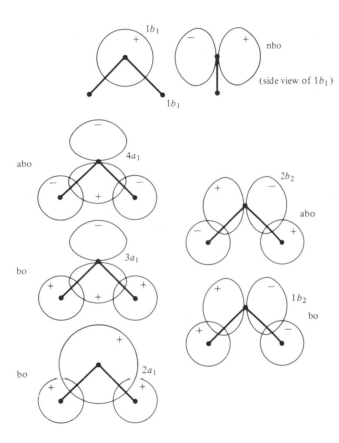

Figure 14.6 Molecular orbitals of nonlinear XH_2 (schematic). X is a first-row element, and the plane of the paper is *XY* (except for "side view"). For the $1b_1$ orbital (nbo), two views are given; this is just the p_x orbital of the X atom. (From G. Herzberg, *Molecular Spectra and Molecular Structure: III. Electronic Spectra of Polyatomic Molecules*, 1966: New York, Van Nostrand Reinhold Company, Fig. 124, p. 318)

The filling order for these orbitals is:

$$(1a_1)\,(2a_1)\,(1b_2)\,(3a_1)\,(1b_1)\,(4a_1) \tag{14.28}$$

Correlation of Linear and Nonlinear Orbitals

It may strike the reader as somewhat arbitrary that, in the symmetry orbital method, one decides the structure first (linear vs. nonlinear) and then writes an electronic structure. There seems to be no reason not to use the orbitals of Eq. (14.25) for H_2O, or those of Eq. (13.28) for BeH_2! In fact, the linear molecule is just a special case for an HXH bond angle of 180°, so these two descriptions must flow smoothly into each other. Figure 14.7 shows a *correlation* diagram for the orbitals of XH_2, plotting the orbital energy (schematically) as a function of the bond angle; this demonstrates how the C_{2v} orbitals of Eq. (14.27) correlate to the $D_{\infty h}$ orbitals of Eq. (14.24). (Such

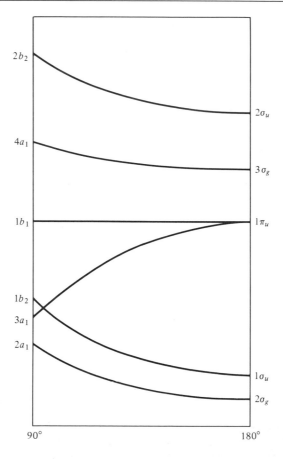

Figure 14.7   Walsh diagram for XH_2. This diagram shows the orbital energies as a function of the angle. This shows how the linear orbitals of Figure 14.4 and the nonlinear orbitals of Figure 14.6 correlated with each other. (From Herzberg, *Electronic Spectra of Polyatomic Molecules,* Fig. 125, p. 319)

diagrams are also called *Walsh diagrams*.) Note that the $C_{2v}(3a_1)$ and $(1b_1)$ orbitals become degenerate as the HXH angle approaches 180° and correlate to the $1\pi_u$ orbitals. As mentioned earlier, excited-state and ground-state geometries may differ, so charts such as Figure 14.7 are useful for interpreting molecular spectra.

Molecular State Symbols

The orbitals of Eq. (14.28) can be used to build up wave functions for the first-row dihydrides; for example:

$$H_2O(1a_1)^2(2a_1)^2(1b_2)^2(3a_1)^2(1b_1)^2\,^1A_1$$

Table 14.8   Electronic states of XH_2 [a]

Molecule	Lowest nonlinear state	First excited nonlinear state
BH_2	$(1a_1)^2(2a_1)^2(1b_2)^2(3a_1)^1\,^2A_1$	$\ldots(1b_2)^2(1b_1)\,^2B_1$
CH_2	$(1a_1)^2(2a_1)^2(1b_2)^2(3a_1)^2\,^1A_1$	$\ldots(3a_1)^1(1b_1)^1\,^3B_1,\,^1B_1$
NH_2	$(1a_1)^2(2a_1)^2(1b_2)^2(3a_1)^2(1b_1)^1\,^2B_1$	$\ldots(3a_1)^1(1b_1)^2\,^2A_1$
H_2O	$(1a_1)^2(2a_1)^2(1b_2)^2(3a_1)^2(1b_1)^2\,^1A_1$	$\ldots(1b_1)^1(4a_1)^1\,^3B_1,\,^1B_1$

Molecule	Lowest linear state	First excited linear state
BH_2	$(1\sigma_g)^2(2\sigma_g)^2(1\sigma_u)^2(1\pi_u)^1\,^2\Pi_u$	$\begin{cases}\ldots(1\sigma_u)^2(3\sigma_g)\,^2\Sigma_g^+\\ \ldots(1\sigma_u)^1(1\pi_u)^2\,^4\Sigma_u^-,\,^2\Sigma_u^-,\,^2\Sigma_u^+,\,^2\Delta_u\end{cases}$
CH_2	$(1\sigma_g)^2(2\sigma_g)^2(1\sigma_u)^2(1\pi_u)^2\,^3\Sigma_g^-$	$^1\Delta_g,\,^1\Sigma_g^+$ (same configuration)
NH_2	$(1\sigma_g)^2(2\sigma_g)^2(1\sigma_u)^2(1\pi_u)^3\,^2\Pi_u$	
H_2O	$(1\sigma_g)^2(2\sigma_g)^2(1\sigma_u)^2(1\pi_u)^4\,^1\Sigma_g^+$	

[a] Reference 3, pp. 343 and 341.

What this configuration means is that the ground-state wave function of H_2O is a product of six a_1-type functions, two b_1- and two b_2-type functions; the rules of Section 14.3 show that this product must have A_1 symmetry. Since electrons in the same spatial orbital — for example, $(1b_1)^2$ — must be paired in order to obey the Pauli principle, the state must be singlet. As for atomic and diatomic term symbols, the electron spin multiplicity, $g_S = 2S + 1$, is given as a left superscript. Accordingly the term symbol is $^1A_1$. Some other examples are shown in Table 14.8.

The procedure used for C_{2v} symmetry can be used for any type of molecule; complications will arise, however, when there are degenerate orbitals. There are some useful rules, which parallel directly those which we used for atoms (Section 12.7) and diatomic molecules (Section 13.5):

1. All closed shells are singlet (all electron spins paired) and totally symmetric ($\mathbf{\Gamma}_s$). The latter characteristic follows directly from the fact (mentioned earlier) that the product of any representation times itself is totally symmetric. In atoms, closed shells were $^1S_0$; in homonuclear diatomic molecules, closed shells were $^1\Sigma_g^+$; in polyatomics the symbol will vary according to the symmetry group: in C_{2v} (Table 14.2) closed shells are $^1A_1$; in C_{2h} (Table 14.2) closed shells are ^1A_g; in O_h symmetry (Table 14.3) closed shells are $^1A_{1g}$; and so on. As for atoms and diatomics, this rule simplifies the task of determining the term symbols of a molecule's electronic states, since filled inner shells can be ignored (remember that the product of any symmetry representation with $\mathbf{\Gamma}_s$ is the same representation — for example, $\mathbf{A}_1 \otimes \mathbf{B}_1 = \mathbf{B}_1$, and so on).

2. The state of the holes in a closed shell will be the same as the state of the electrons. For atoms this means that the states possible for p^4 are the same as for p^2, the states of d^6 are the same as for d^4, and so on (cf. Table 12.6). For diatomics, and any other linear molecule, the state symbol for π^3 is the same as for π^1. For polyatomic

Table 14.9   States of equivalent electrons for open-shell degenerate configurations

	Configuration	States
$C_{\infty v} (D_{\infty h})^a$	π^2	$^1\Sigma^+, \, ^1\Delta, \, ^3\Sigma^-$
C_{3v}	e^2	$^1A_1, \, ^1E, \, ^3A_2$
D_{6h}	$(e_{1g})^2$ or $(e_{1u})^2$	$^1A_{1g}, \, ^1E_{2g}, \, ^3A_{2g}$
	$(e_{2g})^2$ or $(e_{2u})^2$	$^1A_{1g}, \, ^1E_{2g}, \, ^3A_{2g}$
$T_d (O_h)^a$	e^2	$^1A_1, \, ^1E, \, ^3A_2$
	t_1^2, t_1^4	$^1A_1, \, ^1E, \, ^1T_2, \, ^3T_1$
	t_1^3	$^2E, \, ^2T_1, \, ^2T_2, \, ^4A_1$
	t_2^4, t_2^2	$^1A_1, \, ^1E, \, ^1T_2, \, ^3T_1$
	t_2^3	$^2E, \, ^2T_1, \, ^2T_2, \, ^4A_2$

^{a}For $D_{\infty h}$ and O_h, add g/u to symbols as appropriate; for example O_h, $e_u^2 \longrightarrow \, ^1A_{1g}, \, ^1E_g, \, ^3A_{2g}$.

E-type orbitals, a filled shell is e^4, so the state symbol for e^3 is the same as for e^1; for T-type orbitals, t^6 is a filled shell, so the states of t^5 are the same as t^1, the states of t^4 are the same as t^2.

Another phenomenon we encountered previously is that open shells involving degenerate orbitals will give rise to multiple states. For atoms, we demonstrated that the p^2 configuration could have states ^1D, ^1S, and ^3P (cf. Table 12.5 and Figure 12.10). For diatomics (and other linear molecules) the configuration π^2 has states $^3\Sigma^-$, $^1\Sigma^+$, and $^1\Delta$ (Section 13.5). The example done earlier, which showed that, in C_{3v} symmetry:

$$\mathbf{E} \otimes \mathbf{E} = \mathbf{A}_1 + \mathbf{A}_2 + \mathbf{E}$$

means that, in this symmetry, an e^2 configuration will correspond to three states having symmetry A_1, A_2, and E. Applying the Pauli principal to determine whether these states will be singlets or triplets is rather complicated; Table 14.9 gives results for some important cases.

Example: Benzene

One important area for which the symmetry orbital approach has been valuable is the interpretation of the electronic structure of conjugated organic molecules, of which benzene (C_6H_6) is the prototype.

The electronic structure of benzene (which has D_{6h} symmetry) can be described with a useful combination of the localized bond approach and symmetry orbitals. Of the 42 electrons in this molecule, 12 can be considered to be localized in the six non-bonding (C1s) orbitals. The $2s$, $2p_x$, and $2p_y$ orbitals of each carbon can be combined into sp^2 hybrids and used to form "σ-type" localized MO's for bonding to two adjacent carbons and the directly attached hydrogens. This "sigma framework" will utilize 24 electrons, which can be treated as "inner shell" and, for purposes of discussing the electronic spectra of benzene, can be ignored.

The remaining six electrons have the six carbon p_z orbitals available, and it is here that the symmetry orbital approach becomes valuable. The six p_z atomic orbitals can

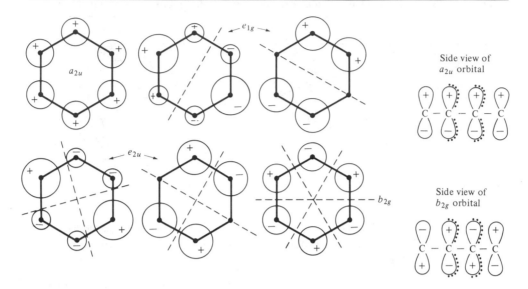

Figure 14.8 Molecular orbitals of benzene. The signs (+ or −) indicate the sign of the function above the plane. These are all combinations of p orbitals perpendicular to the plane, so the signs are reversed below the plane of the molecule (see side views provided). The size of each circle reflects the relative contribution of that AO to the MO.

be combined to form six molecular orbitals, which must have the symmetry of one or the other of the D_{6h} representation (Table 14.2); these are shown schematically in Figure 14.8.

The symmetry orbitals of benzene are, in order of energy, a_{2u}, e_{1g}, e_{2u}, and b_{2g}; these will be occupied by six electrons with the configuration:

$$C_6H_6[12 \text{ inner electrons}][24 \text{ “}\sigma\text{” electrons}](a_{2u})^2(e_{1g})^4 \, {}^1A_{1g}$$

[Figure 14.9 shows the orbital diagram for the outer six electrons.]

Even without providing a complete orbital description for the "σ" electrons, the configuration given above can be used to describe the excited electronic states that could be observed by absorption (uv) spectroscopy. This is often done using what could be called the "jumping-electron" model: the molecule absorbs a photon, and one of the e_{1g} electrons "jumps" to the e_{2u} orbital [the wavy line of Figure 14.9]. This is a useful picture, but it has some problems; the electronic configuration of the final state is:

$$C_6H_6[36 \text{ inner electrons}](a_{2u})^2(e_{1g})^3(e_{2u})^1$$

This is an open shell and will represent many electronic states. These states can be derived easily if we look at the hole in the inner shell rather than at the electrons; that is, the configuration $(e_{1g})^3$ can be treated as if it were $(e_{1g})^1$, and the configuration:

$$(e_{1g})^3(e_{2u})^1$$

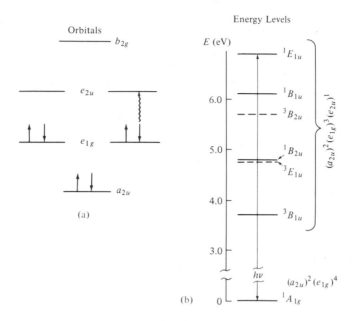

Figure 14.9 Orbitals and energy levels of benzene. (a) The orbital-energy diagram for the orbitals pictured in Figure 14.8. A transition to an excited state involves the promotion of one of the electrons to an unoccupied orbital (wavy line). (b) The energy-level diagram of benzene. The only dipole-allowed transition is shown by the vertical line. The location of states denoted with dashed lines is uncertain.

will have the same states as the configuration:

$$(e_{1g})^1 (e_{2u})^1$$

The states of this configuration can be derived in the same manner as we used for C_{3v} in an earlier example. It can be shown that, in C_{6h} symmetry, the direct product of E_{1g} and E_{2u} is:

$$\mathbf{E}_{1g} \otimes \mathbf{E}_{2u} = \mathbf{B}_{1u} + \mathbf{B}_{2u} + \mathbf{E}_{1u} \qquad \textbf{(14.29)}$$

Exercise: The reader with no more background in group theory than is provided by this chapter would probably find it difficult to derive Eq. (14.29); however, it is easy to prove that the equation is correct. Use the character table for D_{6h} to derive the vector that results from the direct product on the left-hand side; then add the vectors on the right-hand side to see that the result is the same. ∎

The significance of Eq. (14.29) is that the configuration $e_{1g}e_{2u}$, and also $(e_{1g})^3 e_{2u}$, will give rise to states of symmetry B_{1u}, B_{2u}, and E_{1u}. The electrons could be paired or unpaired, so both singlets and triplets are possible. Therefore the excited states that result when an electron "jumps" from e_{1g} to e_{2u} are:

$$^1B_{1u}, \ ^1B_{2u}, \ ^1E_{1u}, \ ^3B_{1u}, \ ^3B_{2u}, \ ^3E_{1u}$$

The energy levels of benzene are shown in Figure 14.9. (The position of those shown by dashed lines is uncertain.)

It should be obvious that the "jumping-electron" model for the excited states of benzene is a gross oversimplification. But there is a saving grace: the selection rules for dipole transitions (which will be explained in Section 14.8) show that, of the six excited states shown in Figure 14.9, only one has an allowed transition to the ground state, namely:

$$^1A_{1g} \longleftrightarrow {}^1E_{1u}$$

All other transitions to the ground state are forbidden. Thus, the jumping-electron model is not too far wrong, so long as you limit the discussion to absorption spectroscopy (as opposed, for example, to phosphorescence or emission spectroscopy). This is a not uncommon phenomenon, and the reason is simple: if a simplified model did not work for something, it would not be used.

14.7   The Born-Oppenheimer Approximation

As for diatomics, the Born-Oppenheimer approximation allows one to describe the molecule's energy in terms of the separate contributions for electronic, vibrational, and rotational energy. Furthermore, the electronic energy of the molecule must be expressed as a parametric function of the nuclear coordinates, $E_e(\mathbf{R})$.

The principal difference between the electronic energy of polyatomic molecules and that of diatomic molecules as discussed in Chapter 13 is in the number of nuclear coordinates required. The specification of the positions of N bodies in space — for example, the N nuclei of a molecule — requires $3N$ coordinates. Since the internal molecular energy is independent of the translational energy, it is sufficient to use a center-of-mass (COM) coordinate system, so that only $3N - 3$ coordinates are required. However, the electronic energy of a molecule will not (in field-free space) depend on the orientation of the molecule. A diatomic molecule has three internal (COM) coordinates, but two simply specify the direction of the internuclear vector in space; therefore the electronic energy is a function of only *one* coordinate, the internuclear distance (R); this is the case discussed in Chapter 13.

In the general case, an arrangement of N masses ($N > 2$) requires three coordinates to describe its orientation relative to a space-fixed COM coordinate system, so the electronic energy will be a function of $3N - 6$ nuclear coordinates. For example, a triatomic molecule (ABC) will have an electronic energy that is a function of three internal coordinates. These coordinates may be chosen in several ways — for example R_{AB}, R_{AC}, R_{BC}, or R_{AB}, R_{BC}, and the angle $\angle ABC$. In any case, the electronic energy $E_e(\mathbf{R})$ is a surface in four dimensions (one axis for E, three for the three coordinates) and, as such, is difficult to visualize. (Benzene, whose electronic energy is a function of 66 nuclear coordinates, is even more difficult to picture.) This problem is generally handled by looking at only a portion of the energy surface; in Chapter 10, for example, the three-atom interaction surface was discussed as a function of R_{AB} and R_{BC} with the angle fixed.

In any case, there will be some set of nuclear coordinates at which the electronic energy (for a given configuration) is a minimum; this is called the equilibrium structure (cf. R_e of Chapter 13). When, as above, we discuss "molecular structure," we mean this position.

Now, with these ideas, we can be more explicit about the division of the molecular energy permitted by the Born-Oppenheimer approximation; all of these have analogies in diatomic molecules, and most were discussed in Chapter 13.

1. *Electronic energy:* the energy of the electrons when the nuclei are at their equilibrium positions — that is, at the minimum in the $E_e(\mathbf{R})$ surface.
2. *Vibrational energy:* the energy due to the oscillations of the nuclei about their equilibrium positions.
3. *Rotational energy:* the energy due to the rigid rotation of the molecule about its COM.

The inadequacies of the Born-Oppenheimer approximation are compensated for by adding correction terms for vibration-rotation interactions (the α_e of Chapter 13) and vibration-electronic (vibronic) interactions.

Correspondingly, the spectroscopy of molecules can be divided into three subjects: (1) electronic spectroscopy (uv and visible), (2) vibrational spectroscopy (infrared), and (3) rotational spectroscopy (microwave).

Electronic spectroscopy of polyatomic molecules is a rather complex subject (cf. ref. 2), so we shall do little more than what has been done already. Before proceeding to the discussion of vibrations, we shall discuss selection rules, since this discussion applies equally to vibrational and electronic spectroscopy.

14.8 Selection Rules

As discussed earlier [cf. Eq. (11.100) and Section 13.9], a transition between two states ($\psi_1 \rightarrow \psi_2$) will be allowed by the electric dipole mechanism if one of the integrals:

$$I_x = \int \psi_1^* x \psi_2 \, d\tau, \quad I_y = \int \psi_1^* y \psi_2 \, d\tau, \quad I_z = \int \psi_1^* z \psi_2 \, d\tau \qquad \textbf{(14.30)}$$

is nonzero. To evaluate these integrals one must know the functional form of the wave function; this information is rarely available. Often, however, one can resolve the simple choice of zero/nonzero on the basis of symmetry, without knowing anything else about the states involved.

THEOREM. In any symmetry group, the integral over all space of a function F, which has one of the symmetries of that group:

$$I = \iiint\limits_{-\infty}^{\infty} F(x, y, z) \, dx \, dy \, dz \qquad \textbf{(14.31)}$$

will be zero unless the function is totally symmetric.

We shall prove this statement only for nondegenerate functions; the conclusion is generally true.

Let us assume that F is antisymmetrical for at least one operator ($\hat{S}$) of the group: $\hat{S}F = -F$. We now operate on Eq. (14.31) to get:

$$\hat{S}I = \int \hat{S}F \, d\tau = -\int F \, d\tau$$

But the integral (I) is just a number, so we therefore conclude:

$$I = \int F \, d\tau = -\int F \, d\tau$$

which can be true if and only if $I = 0$. Conversely, if F is symmetric with respect to *all* operators in the group, the integral is not necessarily zero.

For an integral such as I_x [Eq. (14.30)] to be nonzero, the integrand must be totally symmetric (Γ_s); if ψ_1 is Γ_1 and ψ_2 is Γ_2 and x is Γ_x, the triple direct product must be:

$$\Gamma_1 \otimes \Gamma_x \otimes \Gamma_2 = \Gamma_s \tag{14.32}$$

This will be true [see Eq. (14.7)] if:

$$\Gamma_1 \otimes \Gamma_2 = \Gamma_x \tag{14.33}$$

The general rule is that a transition $\psi_1 \leftrightarrow \psi_2$ will be electric-dipole allowed if the direct product of the representations of the two states has the same symmetry as x, y, or z.

Example: What transitions are permitted from a B_1 state in C_{2v} symmetry? From Table 14.1 we see that the function x is B_1, y is B_2, and z is A_1. The selection rules are: $\mathbf{B}_1 \otimes \mathbf{B}_1 = \mathbf{A}_1$ (z allowed), $\mathbf{B}_1 \otimes \mathbf{B}_2 = \mathbf{A}_2$ (forbidden), $\mathbf{B}_1 \otimes \mathbf{A}_2 = \mathbf{B}_2$ (y allowed), $\mathbf{B}_1 \otimes \mathbf{A}_1 = \mathbf{B}_1$ (x allowed). ∎

Exercise: Derive the selection rules for C_{2v} given below.

	A_1	A_2	B_1	B_2	
A_1	a	x	a	a	(a = allowed,
A_2	x	a	a	a	x = forbidden)
B_1	a	a	a	x	
B_2	a	a	x	a	

∎

14.9 Molecular Vibrations

To specify the positions of N nuclei requires $3N$ coordinates (called $\mathbf{R}$ collectively). The electronic wave function will be a parametric function of these coordinates, $\psi_e(\mathbf{R})$, as will be the energy $E_e(\mathbf{R})$; this energy forms the potential in which the nuclei move. There is some set of nuclear coordinates ($\mathbf{R}_e$) for which E_e will be a minimum—the equilibrium structure. Imagine a molecule in its equilibrium configuration as a potential surface with each nucleus in a potential well, vibrating about its

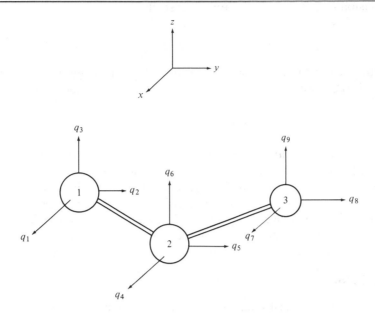

Figure 14.10 Displacement coordinates for a triatomic molecule. The motion of the three atoms from their equilibrium positions is measured by nine displacement coordinates, which define the changes of the Cartesian coordinates of each atom.

equilibrium position. This picture seems to suggest $3N$ vibrations, N nuclei moving in three orthogonal directions, but this is incorrect. If, for example, all the nuclei tried to move in the same direction at the same time, this would be a translation of the molecule as a whole, not a vibration; the electrons would move with the molecule, and no potential would oppose this motion. Obviously we need to look at the molecule as a whole, since the motions and potential of each nucleus depend on what the other nuclei are doing.

We analyze the vibrational motions of a molecule by defining a set of *displacement coordinates* q_i that measure the movement of each nucleus from its minimum-energy position when all other nuclei are in their equilibrium positions. These are (Figure 14.10):

Nucleus 1: $\quad q_1 = x_1 - x_{e1}, \quad q_2 = y_1 - y_{e1}, \quad q_3 = z_1 - z_{e1}$

Nucleus 2: $\quad q_4 = x_2 - x_{e2}, \quad q_5 = y_2 - y_{e2}, \quad q_6 = z_2 - z_{e2}$

Nucleus 3: $\quad q_7 = x_3 - x_{e3}, \quad q_8 = y_3 - y_{e3}, \quad q_9 = z_3 - z_{e3}$

and so on.

The electronic potential $E_e(q_1, \ldots, q_{3N})$ can be expanded in a Taylor series about the minimum (where all $\partial E_e / \partial q_i = 0$):

$$E_e(\mathbf{R}) = E_0 + \frac{1}{2} \sum_i \sum_j \left(\frac{\partial^2 E_e}{\partial q_i \, \partial q_j} \right)_{\text{eq.}} q_i q_j + \cdots \qquad \textbf{(14.34)}$$

This expansion defines a series of force constants:

$$k_{ij} = \left(\frac{\partial^2 E_e}{\partial q_i\, \partial q_j}\right)_{\text{eq.}} \tag{14.35}$$

Neglecting the anharmonic terms and subtracting off the electronic energy at the minimum, $E_e(\mathbf{R}_e)$, we get, from Eq. (14.34), the harmonic potential:

$$V = \frac{1}{2} \sum_{i,j} k_{ij}\, q_i\, q_j \tag{14.36}$$

When $i = j$, this potential is an obvious generalization of the diatomic case [Eq. (13.22)] with q_i replacing $x = R - R_e$; but what does k_{ij} mean when $i \neq j$? Because of the repulsive Coulombic potential between the like-charged nuclei, the force required to move a nucleus in a given direction will depend on the direction in which its neighbors are moving — greater if they are going toward each other and vice versa. This interaction is summarized in the cross constants, k_{ij}.

Normal Modes

A *normal coordinate* transformation seeks linear combinations of the displacement coordinates:

$$\xi_i = \sum_j a_{ij}\, q_j \tag{14.37}$$

such that the potential energy will have no cross terms and will be of the form:

$$V = \frac{1}{2} \sum_i c_{ii}\, \xi_i^2 \tag{14.38}$$

Then the Hamiltonian will be:

$$\hat{H} = \frac{1}{2} \sum_i \left(\frac{-\hbar^2}{b_{ii}} \frac{\partial^2}{\partial \xi_i^2} + c_{ii}\xi_i^2\right) \tag{14.39}$$

Note that the Hamiltonian also has no cross terms between the coordinates. In these equations the constants b_{ii} are functions of the masses of nuclei, in effect generalized reduced masses, and the constants c_{ii} are generalized force constants. The new variables ξ_i are the normal coordinates. For nonlinear molecules, six of the $3N$ terms of Eq. (14.39) will have zero force constants — that is, $c_{ii} = 0$; these are the three translations and three rotations. The remaining $3N - 6$ coordinates are called *genuine* vibrations. (The linear case will be discussed later.)

Because Eq. (14.39) is a sum of individual terms, one for each of the $3N - 6$ vibrations, it should be clear (following our earlier examples) that the vibrational energy will be a *sum* of terms:

$$E_{\text{vib}} = \sum_{i=1}^{3N-6} E_i \tag{14.40}$$

while the vibrational wave function is a *product:*

$$\Psi_{\text{vib}} = \prod_{i=1}^{3N-6} \psi_i \tag{14.41}$$

There will be $3N - 6$ separate equations, one for each normal mode of vibration:

$$\hat{H}_i \Psi_i = \left(\frac{-\hbar^2}{b_{ii}} \frac{\partial^2}{\partial \xi_i^2} + c_{ii}\xi_i^2 \right) \psi_i = E_i \psi_i \tag{14.42}$$

The Harmonic Oscillator Approximation

The mathematics of the normal coordinate transformation (ref. 4) may be unfamiliar to the reader, but the solution to Eq. (14.42) has been seen twice before (in Chapters 11 and 13), so the result, summarized next, should be no surprise.

We define a unitless variable:

$$Q_i = \frac{\xi_i}{\alpha_i} \tag{14.43a}$$

by analogy to $y = x/\alpha$, used in Chapter 11, with:

$$\alpha_i = \left(\frac{\hbar^2}{c_{ii} b_{ii}} \right)^{1/4} \tag{14.43b}$$

These substitutions give a Hamiltonian [from Eq. (14.42)] of the form:

$$\hat{H}_i = \left(Q_i^2 - \frac{\partial^2}{\partial Q_i^2} \right) \tag{14.44}$$

Comparison to Eq. (11.56) (with y instead of Q) or Eq. (13.24) (with $R - R_e$ in place of Q, and neglecting the anharmonic potential) will remind the reader that Eq. (14.44) is a harmonic-oscillator Hamiltonian, whose eigenvalues have been shown to be:

$$E_{vi} = (v_i + \tfrac{1}{2}) hc\omega_{ei} \tag{14.45a}$$

with the vibrational constant (one for each normal mode):

$$\omega_{ei} = \frac{1}{2\pi c} \sqrt{\frac{c_{ii}}{b_{ii}}} \tag{14.45b}$$

Conclusion: The total vibrational energy is a sum of the above:

$$E_{\text{vib}} = hc \sum_{i=1}^{3N-6} (v_i + \tfrac{1}{2}) \omega_{ei} \tag{14.46}$$

The wave functions are (Table 11.2):

$$(v = 0) \qquad \psi_{0i} = e^{-Q_i^2/2} \tag{14.47a}$$

$$(v = 1) \qquad \psi_{1i} = Q_i e^{-Q_i^2/2} \tag{14.47b}$$

$$(v = 2) \qquad \psi_{2i} = (4Q_i^2 - 2)e^{-Q_i^2/2} \qquad \textbf{(14.47c)}$$

$$(v = 3) \qquad \psi_{3i} = Q_i(8Q_i^2 - 12)e^{-Q_i^2/2} \qquad \textbf{(14.47d)}$$

and so on. With:

$$\Psi_{\text{vib}} = \prod_i \psi_{vi}$$

Table 14.10 lists experimental vibrational constants for some molecules; in this chapter we shall neglect the anharmonicities and other small corrections.

Table 14.10   Vibrational constants of some molecules

(Units: cm^{-1})

	ω_{e1}	ω_{e2}	ω_{e3}	ω_{e4}
Triatomic C_{2v}	a_1	a_1	b_1	
H_2O	3652	1595	3756	
D_2O	2666	1179	2784	
H_2S	2611	1290	2684	
CH_2	2968	1444	3000	
SO_2	1151	524	1361	
F_2O	830	490	1110	
Linear $C_{\infty v}$	σ^+	π	σ^+	
HCN	2089	712	3312	
ClCN	729	397	2201	
OCS	859	527	2079	
NNO	1285	589	2224	
Linear $D_{\infty h}$	σ_g^+	π_u	σ_u^+	
CO_2	1388	667	2349	
CS_2	657	397	1523	
Pyramidal C_{3v}	a_1	a_1	e	e
NH_3	3337	950	3414	1628
NO_3	2419	749	2555	1191
PH_3	2327	991	2421	1121
Tetrahedral T_d	a_1	e	t_2	t_2
CH_4	2914	1526	3020	1306
CD_4	2085	1054	2258	996
SiH_4	2187	978	2183	910
SiF_4	800	260	1022	420

Symmetry of Normal Modes

Discovering the exact forms of the normal coordinates is a rather involved procedure. In the next several pages we shall examine several examples, all of which are derived in one or more of the references at the end of the chapter. This will permit us to discuss a very important characteristic of those coordinates — their symmetry.

Each normal mode of a molecule can be classified according to its symmetry with respect to the effect of the operations of the point group upon it. We shall illustrate this using the normal modes of H_2O [Figure 14.11(a)]. It is important to note that the convention in vibrational spectroscopy is that planar C_{2v} molecules lie in the xz plane (σ_v as opposed to σ_v'); this is *opposite* to the convention usually used in MO theory! (See Figure 14.5.)

The arrows, for example, as shown by Figure 14.11, indicate the direction of the nuclear motion when the normal coordinate Q is positive; $-Q$ would have arrows in the opposite direction. The effect of $\hat{C}_2$ on Q_3 is, thus, to change it to $-Q_3$.

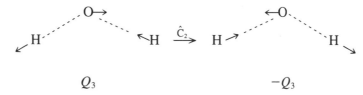

$$Q_3 \qquad\qquad -Q_3$$

The reader should examine the effect of the mirror planes in a similar manner; the results should be:

$$\left.\begin{aligned}
\hat{C}_2 Q_3 &= -Q_3 \\
\hat{\sigma}_v(xz) Q_3 &= Q_3 \\
\hat{\sigma}_v'(yz) Q_3 &= -Q_3
\end{aligned}\right\} \qquad \text{Therefore, } Q_3 \text{ has } b_1 \text{ symmetry.}$$

By this procedure, you should be able to show that Q_1 and Q_2 both have a_1 symmetry. (Note that lower-case letters are used for the normal coordinate symmetry.)

Linear Molecules

The $3N$ coordinates required to describe the positions of N atoms include three for the position of the COM and $3N - 3$ internal coordinates that describe the orientation (rotational position) of the molecule when the nuclei are in their equilibrium position and the vibrations of the nuclei about those equilibrium positions. The specification of the orientation of a rigid body in space requires, in general, three coordinates; these could be, for example, the rotation of the body about three orthogonal axes. If the body is linear, however, rotation about the z axis does not produce a distinguishable position, so only two coordinates are needed to describe its orientation in space. This leaves $3N - 5$ coordinates that will be genuine vibrations.

A triatomic molecule of the type AXA will, if it is nonlinear (for example, H_2O), have three translations, three rotations and three vibrations. If it is linear (for example, CO_2), it will have three translations, two rotations, and four vibrations. In effect, the

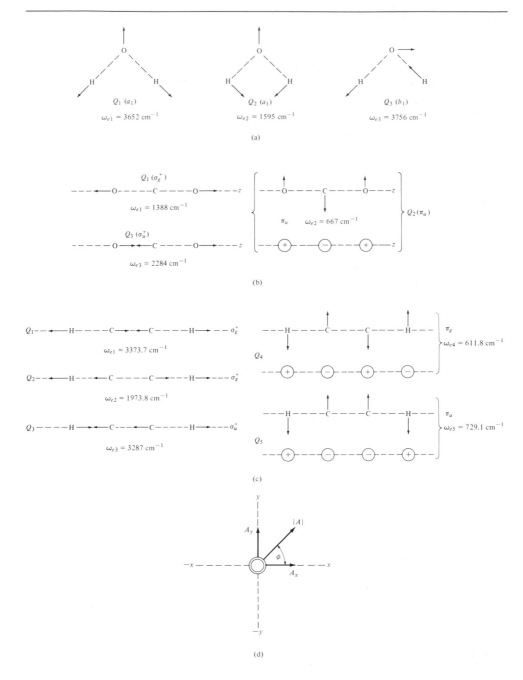

Figure 14.11 Normal modes of vibrations. The number given with each figure is the vibrational constant for that vibration. (a) Water. (b) Carbon dioxide. The signs (+ or −) denote motion into or out of the plane. (c) Acetylene. (d) An end view of the bending

normal coordinate transformation will give four coordinates with nonzero force constants. However, two of these coordinates will be *degenerate;* that is, their vibrational constants [Eq. (14.45)] will be the same. (The normal coordinates of CO_2 are shown in Figure 14.11.) An AXA molecule has only three vibrational *frequencies* in either case, linear or not; but if it is linear, one vibration is degenerate.

The degenerate bending vibration of CO_2 is represented usually [for example, in Figure 14.11(b)] as two bends in two orthogonal directions, such as x and y. Perhaps a better picture is that there is only one bending motion but *two coordinates* are required to describe it. If the amplitude of the vibration is A, these coordinates could be the Cartesian projections, A_x and A_y, or the magnitude $|A|$ together with the orientation ϕ of the plane of vibration. [See the diagram in Figure 14.11(d).]

The vibrations of linear molecules are named like the MO's; in $C_{\infty v}$ as σ^+, σ^-, π, and so on and in $D_{\infty h}$ as σ_g^+, σ_u^+, π_g, ...

Vibrational State Symmetry and Spectroscopy

A vibrational state will be indicated as an array of vibrational quantum numbers, one for each distinct mode, $(v_1, v_2, v_3, v_4, \ldots)$. Thus the ground state (all $v_i = 0$) is denoted $(0, 0, 0, \ldots)$ while a state with $v_1 = 0$, $v_2 = 2$, $v_3 = 1$ is denoted $(0, 2, 1, \ldots)$.

If the symmetry of a mode Q_i is Γ_i, the symmetry of Q_i^2 must be $\Gamma_i \otimes \Gamma_i = \Gamma_s$, totally symmetric. The ground state:

$$(0, 0, 0, \ldots) = e^{-Q_1^2/2} e^{-Q_2^2/2} e^{-Q_3^2/2} \ldots$$

is therefore always totally symmetric (A_1 in C_{2v}; states are represented by capital letters.)

Singly excited states such as

$$(1, 0, 0, \ldots) = Q_1 e^{-Q_1^2/2} e^{-Q_2^2/2} e^{-Q_3^2/2} \ldots$$

have the symmetry of the excited coordinate, Γ_1 in this case.

Doubly excited states such as:

$$(1, 1, 0, \ldots) = Q_1 Q_2 e^{-Q_1^2/2} e^{-Q_2^2/2} e^{-Q_3^2/2} \ldots$$

must have the symmetry of $Q_1 Q_2$ — that is, $\Gamma_1 \otimes \Gamma_2$. A $v = 2$ state [see Eq. (14.47c)] such as $(2, 0, 0, \ldots)$ is again totally symmetric.

---

vibrations. The motion has only one frequency but requires two coordinates to specify what is happening. The usual description is in terms of the Cartesian components of the displacement vector; this is the meaning of the degenerate vibrations of panels (b) and (c), which show pairs of vibrations for "in-plane" and "perpendicular-to-plane" vibrations. A somewhat better description is to define the amplitude of the vibration and the angle (ϕ) of the plane in which it is vibrating. In either case, the bending vibration requires two coordinates and is referred to as being "doubly degenerate" ($g = 2$).

IR Spectroscopy

Spectroscopic transitions are observed among these states, including:

fundamentals: $\qquad (0,0,0) \longrightarrow (1,0,0)\,, \qquad \tilde{\nu} \cong \omega_{e1}$

$\qquad\qquad\qquad\qquad (0,0,0) \longrightarrow (0,1,0)\,, \qquad \tilde{\nu} \cong \omega_{e2}$

and so forth

overtones: $\qquad\qquad (0,0,0) \longrightarrow (2,0,0)\,, \qquad \tilde{\nu} \cong 2\omega_{e1}$

combination bands: $\qquad (0,0,0) \longrightarrow (1,1,0)\,, \qquad \tilde{\nu} \cong \omega_{e1} + \omega_{e2}$

hot bands: $\qquad\qquad (1,0,0) \longrightarrow (2,0,0)\,, \qquad \tilde{\nu} \cong \omega_{e1}$

difference bands: $\qquad (1,0,0) \longrightarrow (0,1,0)\,, \qquad \tilde{\nu} \cong \omega_{e2} - \omega_{e1}$

Other things being equal (that is, all are dipole allowed transitions), the fundamentals will be the strongest absorption bands. Exceptions can occur due to *Fermi resonances;* if the frequency of a fundamental is very near to that of an overtone or combination band having the same symmetry, the weaker transition can "borrow" intensity from the fundamental and will be much stronger than expected.

The IR selection rules follow directly from what was said earlier. Since the ground state is totally symmetric, the selection rule for fundamentals is very simple — the vibrational coordinates must have the same symmetry as either x, y, or z in that group.

Another version of the rule for fundamentals is usually easier to visualize for linear molecules: the dipole moment of the molecule must change during the vibration. The reason for this requirement was discussed in Section 13.9. The centro-symmetric molecule CO_2 has no dipole moment in its equilibrium position. During the asymmetric stretch vibration ($O\rightarrow\leftarrow C\cdots O\rightarrow$), the molecule will have a dipole moment; therefore this mode is IR allowed. The symmetric stretch mode ($\leftarrow O \cdots C \cdots O \rightarrow$) has no dipole at any extension and is therefore IR forbidden. (The selection rules for IR of linear molecules in terms of symmetries, σ_g, etc., are the same as those given in Chapter 13 for electronic spectra.)

A normal mode whose fundamental is an allowed transition for the electric-dipole selection rule is called *IR active*.

The triatomic AXA molecule when nonlinear (C_{2v}, for example H_2O) has three frequencies and all are IR active (prove this using the symmetries and Table 14.1). The linear case ($D_{\infty h}$, for example CO_2) has also three frequencies but only two are IR active. This is a simple illustration of how IR can be used in structural studies.

The symmetry of the normal modes is clearly a very important property. It is not, as we may have implied, really necessary to actually solve the normal coordinate problem and to discover the actual forms of the normal modes in order to find the symmetries; group theory alone can provide this information directly. (See refs. 1, 2 and 4 for details.) Group theory can also predict the selection rules for the rotational structure of the IR bands; these will be either $\Delta J = \pm 1$ (*PR* branches) or $\Delta J = 0, \pm 1$ (*PQR* branches). As usual, our discussion refers primarily to gas-phase molecules; IR spectra of liquids, solutions, and solids are often very similar to those of the gas, but the rotational structure will usually be seen only for gases.

Table 14.11   IR bands of formaldehyde

Assignment	Description	Intensity	Frequency (cm^{-1})	
			H_2CO	D_2CO
$\omega_{e6}(b_2)$	Umbrella	Strong (PR)	1167	938
$\omega_{e5}(b_1)$	CHO bend	Strong (PR)	1280	990
$\omega_{e3}(a_1)$	HCH bend	Strong (PQR)	1503	1105
$\omega_{e2}(a_1)$	~CO str.	Very strong (PQR)	1743	1700
$2\omega_{e6}(A_1?)$	Overtone	Weak (PQR)	2081	
$\omega_{e1}(a_1)$	CH sym. str.	Strong (PQR)	2780	2056
$\omega_{e4}(b_1)$	CH asym. str.	Very strong (PR)	2870	2160
$2\omega_{e3}(A_1)$	Overtone	Strong (PQR)	2973	2208

Source: Reference 2.

Figure 14.12 shows some additional examples of normal coordinates. The IR bands of formaldehyde are listed in Table 14.11.

14.10   Raman Spectroscopy

Raman scattering spectroscopy, another method for studying the vibrations of molecules, can provide important information that is often not available from IR spectroscopy.

If a beam of light is incident upon a material, some of the light may be absorbed, but much of it will be scattered by the material. The most intense light scattering is caused by particulate matter when the particle size is of the order of the wavelength. However, even pure homogeneous materials will scatter light; this is called the Tyndall effect or *Rayleigh scattering*. This type of scattering is more effective for the shorter wavelengths of the visible spectrum; the Rayleigh scattering of sunlight in the upper atmosphere is responsible for the blue color of the sky.

Most of the light scattered by a homogeneous material has the same frequency as the incident beam. However, a small part of the scattered radiation will have a different frequency; this effect, observed first in 1928 by C. V. Raman and K. S. Krishnan, is called *Raman scattering*. This change in frequency is due to an exchange of energy between the photon and a molecule of the material. If the incident frequency is $\tilde{\nu}_i$ and the scattered frequency is $\tilde{\nu}_s$, then the process is:

$$\text{photon } (h\nu_i) + \text{molecule } (E_1) \longrightarrow \text{photon } (h\nu_s) + \text{molecule } (E_2)$$

The change in the molecule's energy is:

$$\frac{\Delta E}{hc} = \tilde{\nu}_i - \tilde{\nu}_s \tag{14.48}$$

If the molecule gains energy, $\tilde{\nu}_s < \tilde{\nu}_i$, the scattered light is called *Stokes* radiation; if the molecule loses energy, $\tilde{\nu}_s > \tilde{\nu}_i$, the radiation is called *anti Stokes*

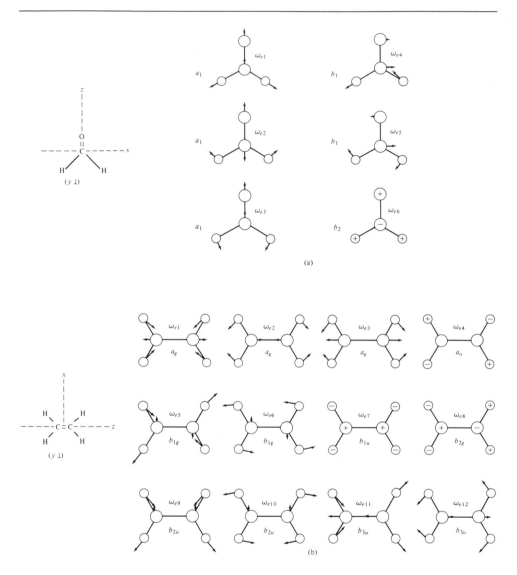

Figure 14.12 Normal vibrations. (a) Formaldehyde is assumed to lie in the *xz* plane; the *y* axis is perpendicular to the paper. (b) Ethylene lies in the *xz* plane, with the principal axis assumed to pass through the carbons; the *y* axis is perpendicular to the plane. (c) Normal vibrations of *cis*-CHCl═CHCl. (d) Normal vibrations of *trans*-CHCl═CHCl. The coordinate system for panels (c) and (d) is the same as for panel (b).

(c)

(d)

(Figure 14.13). (The same terminology is used in fluorescence, and in fact the techniques are quite similar. The distinction is that, in fluorescence, the photon is absorbed, then reemitted, while in the Raman effect it is only scattered. It is in fact preferred that $\bar{\nu}_i$ be a frequency that is not absorbed, otherwise fluorescence can obscure the Raman effect.) If the molecule loses or gains only rotational energy, the effect is called the rotational Raman effect; the change in frequency will be of the order of $\bar{B}_e$. If the molecule loses or gains vibrational energy, it is called the vibrational Raman effect (Figure 14.13).

Uses of Raman Spectroscopy

At one time Raman spectroscopy was a cumbersome and difficult technique; however, the advent of lasers, which provide an intense monochromatic light source for excitation, has greatly increased its use. Even so, the Raman technique is somewhat more difficult than IR; in particular the Raman scattering is very weak, so that gases are hard to study. In specific applications Raman may have advantages over IR; for example: (1) The selection rules are different, so vibrations that are IR inactive may be seen. (2) Liquid water is opaque to IR, so molecules in aqueous solutions cannot be studied by this technique; water is obviously transparent to visible light, so Raman scattering in that region is quite useful. (3) Some low-frequency vibrations fall into the far-IR region, which is difficult to study; these are more readily seen in the Raman spectrum. (4) The rotational Raman, in contrast to microwave spectroscopy, does not require the molecule to have a dipole moment; the rotational constants of homonuclear diatomics, for example, can be measured in this manner. Similarly, the vibrations of homonuclear diatomics can be studied by Raman but not by IR.

In this section we shall focus on the vibrational Raman. Some aspects of the rotational Raman are discussed in Chapter 15.

The Vibrational Raman Effect

Suppose, for purposes of illustration, a molecule has three vibrations; its vibrational state can then be characterized by the three vibrational quantum numbers, (v_1, v_2, v_3). If a molecule in its ground state $(0, 0, 0)$, interacts with a photon with frequency $\bar{\nu}_i$ and obtains enough energy to excite the first mode of vibration, then the energy balance is:

$$E(0,0,0) + hc\bar{\nu}_i = E(1,0,0) + hc\bar{\nu}_s$$

The change in the molecule's energy is:

$$E(1,0,0) - E(0,0,0) \cong hc\omega_{e1}$$

where ω_{e1} is the vibrational constant for the first normal mode; therefore:

$$\bar{\nu}_i - \bar{\nu}_s \cong \omega_{e1}$$

That is, the frequency *difference* is equal to the vibrational constant (but note that we have neglected the anharmonicity and rotations, so this is only an approximation). This is, of course, one of the Stokes lines.

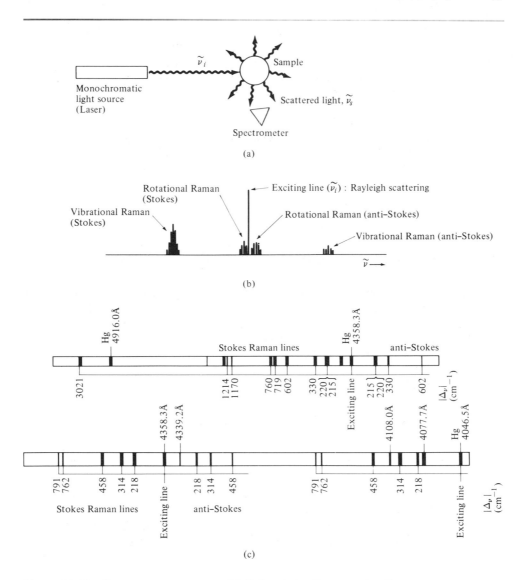

Figure 14.13 Raman spectroscopy. (a) Schematic diagram of a Raman spectrometer. The spectrometer is placed perpendicular to the excitation source so as to detect only scattered light. (b) Raman spectrum (schematic). (c) Raman spectra of CCl_4 and $CHCl_2Br$. These spectra were made using a mercury lamp as an excitation source. The numbers below the spectrum show the *shift* of the Raman line from the exciting line in cm^{-1}. (From G. Herzberg, *Molecular Spectra and Molecular Structure: II. Infrared and Raman Spectra of Polyatomic Molecules*, 1945: New York, Van Nostrand Reinhold Company, p. 250, Fig. 77)

The anti-Stokes vibrational Raman arises from the interaction of a photon with a molecule that is already vibrationally excited. For example:

$$hv_i + E(1, 0, 0) = hv_s + E(0, 0, 0)$$

In this case it is easily shown that:

$$\tilde{\nu}_i - \tilde{\nu}_s = -\omega_{e1}$$

That is, the anti-Stokes line is exactly opposite to the Stokes line with respect to $\tilde{\nu}_i$ [cf. Figure 14.13(b)]. Of course, at thermal equilibrium there will be fewer molecules in the $(1, 0, 0)$ state than in the $(0, 0, 0)$ state, so the anti-Stokes scattering will be somewhat weaker than the Stokes scattering—a great deal weaker if, as often happens, $hc\omega_e \gg kT$.

Selection Rules

For the example used above, the Raman scattered light (Stokes) could have as many as three frequencies, shifted from the incident frequency by ω_{e1}, ω_{e_2}, and ω_{e_3}). However, there are selection rules for Raman scattering as there were for absorption, so it is possible that not all lines may occur.

Vibrational modes that can exchange energy with a photon during scattering are called *Raman active*. In general a molecular transition from a state ψ_1 to a state ψ_2 will be Raman active if any of the integrals:

$$I = \int \psi_1^* \alpha \psi_2 \, d\tau$$

$$\alpha = x^2, y^2, z^2, xy, xz, yz \qquad \text{(or any additive combination of these)} \qquad \textbf{(14.49)}$$

is nonzero. [Except for some constants, the quantities α of Eq. (14.49) are elements of the *polarizability tensor;* we shall regard this as a technicality and pursue it no further.]

If the discussion is limited to the strongest lines of the Raman spectrum, the Stokes fundamentals, Eq. (14.49) gives rise to a very simple rule. If the integral of Eq. (14.49) is to be nonzero, the integrand $(\psi_1^* \alpha \psi_2)$ must be totally symmetric; that is, the direct product of the representation of $\psi_1(\Gamma_1)$, $\alpha(\Gamma_\alpha)$, and $\psi_2(\Gamma_2)$ must contain Γ_s, the totally symmetric representation:

$$\Gamma_1 \otimes \Gamma_\alpha \otimes \Gamma_2 = \Gamma_s \qquad \textbf{(14.50)}$$

Now, if ψ_1 is the ground vibrational state, $(0, 0, 0, \ldots)$ and ψ_2 is a singly excited state, then ψ_1 has symmetry Γ_s and ψ_2 has the symmetry of the excited normal mode (call it Q_i). If $\Gamma_1 = \Gamma_s$, then Eq. (14.50) will be true only if $\Gamma_\alpha = \Gamma_2$; in other words, a normal coordinate Q_i will be Raman active if and only if it has the same symmetry as one of the quadratic functions: $\alpha = x^2, y^2, z^2, xy, xz, yz$ (or any combination of these).

Raman and IR

A molecule with N atoms has $3N - 6$ (linear) or $3N - 5$ (nonlinear) vibrational coordinates. (However, because of degeneracies the number of distinct vibrational frequencies may be less than this number.) A vibration will be IR active if the normal coordinate has the same symmetry as x, y, or z; it will be Raman active if normal coordinate has the same symmetry as $\alpha = x^2$, y^2, z^2, xy, and so on. It is possible for a vibration to be both IR and Raman active or to be *inactive* (that is, neither IR nor Raman active). Once the symmetry of the vibrations is known, the selection rules are easily determined by inspection of the character tables.

For linear molecules, it was pointed out earlier that a vibration that changed the dipole moment of a molecule would be IR active. Correspondingly, a vibration will be Raman active if it changes the polarizability of a molecule. Generally the polarizability of a molecule is proportional to its size. For the case of CO_2 (Figure 14.11) the symmetric stretch:

$$\leftarrow O \quad C \quad O \rightarrow$$

changes the length of the molecule and, hence, its polarizability; this mode is Raman active. The asymmetric stretch:

$$O \rightarrow \leftarrow C \quad O \rightarrow$$

does not change the molecule's overall size and is Raman inactive.

A helpful rule for cases such as CO_2 is the *rule of mutual exclusion*. For any molecule that has a center of symmetry, all vibrational states are classified as either gerade (g) or ungerade (u). The ground state $(0, 0, 0, \ldots)$ is always gerade. The functions x, y, and z are all ungerade. Therefore only vibrations that are ungerade may be IR active. But the quadratic functions are all gerade; therefore [Eq. (14.50)] only gerade vibrations may be Raman active. Therefore, the rule is: for any molecule with a center of symmetry, no vibrational frequency will appear in both the IR and Raman spectrum.

Example: In C_{2v} symmetry (Table 14.1) there is a quadratic function of all symmetries (x^2, y^2, z^2 are A_1, xy is A_2, yz is B_1, yz is B_2), so any vibration of a C_{2v} molecule will be Raman active. In linear CO_2 (which has a center of symmetry) Q_2 and Q_3 are only IR active but the symmetric stretch (Q_1) is only Raman active. (Because of a Fermi resonance between ω_{e1} and the overtone $2\omega_{e2}$, there are actually *two* strong Raman bands for CO_2.) ∎

Example: With $3N - 6$ vibrations, one would expect vibrational spectra of large molecules to be very complex. Such is often the case, but high symmetry can reduce this number drastically. For example SF_6 has $3 \times 7 - 6 = 15$ normal coordinates. But there are only six distinct frequencies because of degeneracies: a_{1g} (775 cm^{-1}), e_g (664 cm^{-1}), t_{1u} (965 cm^{-1}), t_{1u} (617 cm^{-1}), t_{2g} (524 cm^{-1}), and t_{2u} (frequency uncertain). Of these, *two* (both t_{1u}) are IR active and *three* (a_{1g}, e_g, t_{2g}) are Raman active. One (t_{2u}) is inactive. Use the O_h character table (Table 14.3) to derive these activities for SF_6. ∎

IR and Raman spectroscopy together make it possible to determine the symmetry of molecules, since group theory can predict the symmetry and activity of bands for any possible structure. For example, an AX_4 molecule will have two IR bands (both t_2) and four Raman bands (a, e, t_2, t_2) if it is tetrahedral (T_d); but if it is square planar (D_{4h}), it will have three IR (a_{2u}, e_u, e_u) and three Raman (a_{1g}, b_{1g}, b_{2g}) bands. Furthermore, the D_{4h} molecule with a center of symmetry will have exclusive bands;

none will be both IR and Raman active. However, such band-counting structure determinations are often ambiguous because of missing bands (obscured by water vapor, out of range, and so on) or strong overtones (Fermi resonances).

14.11   Molecular Rotations

The moment of inertia (I) of a set of masses about any axis is a sum of all masses times the square of their perpendicular distances to the axis. If r_i is the perpendicular distance of a mass m_i from an axis, then:

$$I = \sum_i m_i r_i^2 \tag{14.51a}$$

For example, the moment of inertia about the z Cartesian axis is:

$$I_{zz} = \sum_i m_i(x_i^2 + y_i^2) \tag{14.51b}$$

A moment of inertia can be measured about any axis — that is, any direction in space. But one coordinate system is special — the principal axis system. Suppose I is calculated along all possible axes; then, if $1/\sqrt{I}$ is plotted along the various axes in any coordinate system, an ellipsoid will be obtained. The axes pointing toward the extrema of the ellipsoid are the principal axes of the molecule's rotational inertia. We shall denote these axes as X, Y, and Z. If a molecule has any symmetry, the directions of the principal axes are easily discovered from the following requirements: (a) the principal symmetry axis is always a principal inertia axis; (b) a plane of symmetry is always perpendicular to a principal axis; (c) the three principal axes are perpendicular to each other; (d) the origin will be at the center of mass.

Example: There are only a few groups for which symmetry does not uniquely determine the direction of all axes. One such is C_{2h}; for example:

One axis is (by rule b) perpendicular to the plane of the molecule so the other two are in that plane. They can be discovered by calculating I along various directions and plotting $1/\sqrt{I}$ vs. angle to get an ellipse. The heaviness of the Cl atoms would lead you to expect a near minimum I about an axis that passes through them. This simplifies the task; the maximum $1/\sqrt{I}$ will be near the Cl-Cl vector, and the other principal axis will be perpendicular to it.   ∎

In the principal-axis coordinate system, the rotational Hamiltonian is:

$$\hat{H} = \frac{1}{2}\left(\frac{\hat{J}_X^2}{I_{XX}} + \frac{\hat{J}_Y^2}{I_{YY}} + \frac{\hat{J}_Z^2}{I_{ZZ}}\right) \tag{14.52}$$

where the angular-momentum operators are as defined in Eqs. (11.71) (but with the symbol J instead of L).

The rotational energy levels can, of course, be measured experimentally by spectroscopy. If the molecule has a permanent dipole moment, then direct microwave spectroscopy can be used; molecules without a dipole moment can be studied by rotational Raman, rotational structure of its IR bands, or rotational structure of its electronic (uv) bands.

Next, we shall discuss some particular cases of rotational spectroscopy (limiting the discussion to molecules having no electronic angular momentum).

Linear Molecules

In this case $I_{XX} = I_{YY}$, and I_{ZZ} is effectively zero. The angular momentum about the Z axis, J_Z^2, is zero, so:

$$\hat{J}^2 = \hat{J}_X^2 + \hat{J}_Y^2$$

and the Hamiltonian is then [compare Eq. (11.96)]:

$$\hat{H} = \frac{1}{2I}\hat{J}^2$$

and the energy eigenvalue is:

$$E_J = \frac{J(J+1)\hbar^2}{2I} = hB_e J(J+1)$$

The rotational constant is defined as before [Eq. (13.42)]

$$B_e = \frac{h}{8\pi^2 I}$$

Some values for polyatomic molecules are given in Table 14.12.

In short, except for the definition of I, all linear molecules are exactly like the diatomic case discussed earlier. The moment of inertia of a linear polyatomic molecule is:

$$I = \sum_i m_i(z_i - z_0)^2 \tag{14.53}$$

where z_0 is the position of the COM. The COM is located by the requirement:

$$\sum_i m_i(z_i - z_0) = 0 \tag{14.54}$$

There is only one measurable quantity for the rotation of linear molecules, the rotational constant B_e, which gives one moment of inertia. The structure of polyatomic molecules is, therefore, underdetermined. Observation of isotopic species can, if the bond lengths can be assumed the same (a reasonable and frequently used approximation), give sufficient data to determine all bond lengths. For example, $^{16}O\,^{12}C\,^{32}S$ has $B_e = 6.081494$ GHz, while $^{16}O\,^{12}C\,^{34}S$ has $B_e = 5.93284$ GHz; this is sufficient

Table 14.12   Rotational constants[a]

Linear	B_e (GHz)	Symmetric tops	B (GHz)
HCN	44.31597	$^{14}NF_3$	10.68107
DCN	36.20740	$^{15}NF_3$	10.62935
$^{79}BrCN$	4.120198	PF_3	7.82001
$^{81}BrCN$	4.096788	$P^{35}Cl_3$	2.6171
OCS	6.081494	CH_3F	25.53591
$OC^{34}S$	5.93284	$CH_3{}^{35}Cl$	13.29295
$OC^{36}S$	5.79967	$CH_3{}^{37}Cl$	13.08824
^{18}OCS	5.70483		
$O^{13}CS$	5.69095		

Asymmetric tops	A (GHz)	B (GHz)	C (GHz)
HDS	290.300	145.200	94.130
O_3	106.530	13.349	11.843
CH_2O	282.106	38.834	34.004
CH_2F_2	49.138	10.604	9.249

[a]If no mass number is given, the most common isotope is intended; for example: ^{16}O, ^{32}S, ^{1}H, ^{14}N, ^{12}C.

information to determine (albeit with a lot of work) the OC and CS bond lengths. (It is best not to make these assumptions where the isotopes of hydrogen are involved. The bond lengths for CH and CD, for example, may differ by as much as 0.005 Å — a small difference, to be sure, but within the experimental error of this method.)

Symmetric Tops

When we stated that there is a principal axis perpendicular to every plane of symmetry and that these axes must be perpendicular to each other, it may have occurred to you that this could cause a problem, for example, in C_{3v} where there are three mirror planes at angles of 120°. Can the principal axes be perpendicular to all these planes? For such cases, it will always be found that the two moments of inertia in question are equal. In fact, the moment of inertia about any axis perpendicular to the $\hat{C}_3$ axis is the same — cf. the exercise below. Any molecule with an axis of rotation $\hat{C}_n$ with $n \geq 3$ will have two equal principal moments of inertia; such molecules are called symmetric tops.

Exercise: Draw an equilateral triangle and assign the same mass (m) to each corner. Measure graphically the moment of inertia about *any* axis in the plane of the triangle and through the center of the triangle (not just those passing through the corners). The answer should be $ma^2/2$ (a = length of side) regardless of the axis chosen.   ■

In a symmetric top, the unique axis (called Z) will always be the principal symmetry axis. We define two rotational constants:

$$A = \frac{h}{8\pi^2 I_{ZZ}}$$

$$B = \frac{h}{8\pi^2 I_{XX}} \tag{14.55}$$

($I_{YY} = I_{XX}$.) (The rotational constant A may be called C if it is smaller than B.) With these definitions, the Hamiltonian becomes:

$$\hat{H} = hB(\hat{J}_X^2 + \hat{J}_Y^2) + hA\hat{J}_Z^2$$

Using the definition $\hat{J}^2 = \hat{J}_X^2 + \hat{J}_Y^2 + \hat{J}_Z^2$, this becomes:

$$\hat{H} = hB\hat{J}^2 + h(A - B)\hat{J}_Z^2$$

from which the energy formula follows immediately:

$$E_{J,K} = hBJ(J + 1) + h(A - B)K^2 \tag{14.56}$$

In Eq. (14.56), K is the quantum number for the projection of the angular momentum on the molecule's principal axis (Z); K is an integer (like J) and it is required that $|K| \leq J$.

Measurement of A and B would be sufficient in some cases to determine the structure; for example, in NH_3 there are only two structural parameters, the NH distance and the HNH angle. However, the microwave selection rule is $\Delta J = 1$, $\Delta K = 0$, so the frequency for a $J \rightarrow J + 1$ transition is:

$$\nu = 2B(J + 1) \tag{14.57}$$

(exactly the same formula as for linear molecules). Therefore, the rotational constant A cannot be measured in this manner. Measurements on isotopically substituted molecules are usually necessary to determine all structural parameters of symmetric tops.

Spherical Tops

Any molecule with more than one axis $\hat{C}_n$ with $n \geq 3$ — that is, all those of cubic symmetry — will have three equal moments of inertia. In this case $I_{XX} = I_{YY} = I_{ZZ}$, and Eq. (14.52) becomes:

$$\hat{H} = \frac{1}{2I}(\hat{J}_x^2 + \hat{J}_y^2 + \hat{J}_z^2) = \frac{1}{2I}\hat{J}^2 \tag{14.58}$$

Therefore, the energy is:

$$E_J = \frac{\hbar^2}{2I}J(J + 1) \tag{14.59}$$

(Note that this is the same formula as for linear molecules.) Such molecules never have dipole moments, so direct rotational spectra are not observed. The rotational constant can be measured from the IR or Raman spectrum. Figure 14.14 shows the IR spectrum

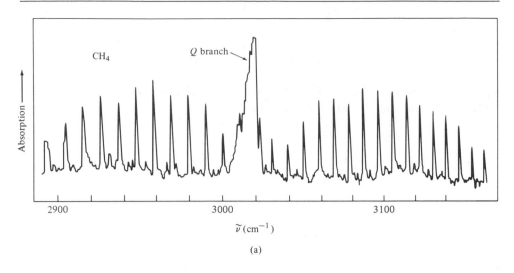

(a)

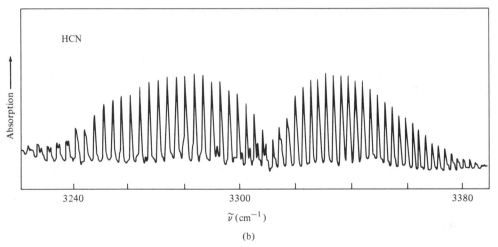

(b)

Figure 14.14 Infrared spectra of methane and HCN. Note the prominent Q branch in the spectrum of methane; these are the transitions for which the rotational quantum number (*J*) does not change. (From IUPAC, *Tables of Wavenumbers for the Calibration of IR Spectrometers*, 1961: Washington, D.C., Butterworth's)

of CH_4; the rotational structure is very similar to that observed for diatomic molecules (Figure 13.6), but note the presence of a Q branch ($\Delta J = 0$).

Asymmetric Tops

Molecules with three unequal moments of inertia are called asymmetric tops. These will include not just molecules with no symmetry but, for example, those of C_{2v}, C_{2h}, D_{2h}. The spectra and the calculations required to obtain moments of inertia from their spectra are usually complicated, and specialized texts should be consulted if details are required.

Postscript

We have now reached the end of the four chapters dealing with applications of quantum theory to chemical problems, and we have barely scratched the surface. We have attempted to establish a groundwork of concepts and vocabulary, so that the terminology that pervades chemistry can be understood and placed into perspective. We have not so much covered the subject as established a basis for further learning. It is hoped that the reader will be motivated to explore the subject more deeply through another course or by reading. Of the books mentioned, those by Karplus and Porter (Chapter 13, ref. 1) and Cotton (Chapter 14, ref. 1) are particularly recommended. Further applications, which illustrate how quantization effects macroscopic properties of matter, will be covered in Chapter 15.

References

1. F. Albert Cotton, *Chemical Applications of Group Theory*, 1971: New York, Wiley-Interscience.
2. G. Herzberg, *Molecular Spectra and Molecular Structure: II. Infrared and Raman Spectra of Polyatomic Molecules*, 1945: New York, Van Nostrand Reinhold Company.
3. G. Herzberg, *Molecular Spectra and Molecular Structure: III. Electronic Spectra and Electronic Structure of Polyatomic Molecules*, 1966: New York, Van Nostrand Reinhold Company.
4. E. B. Wilson, J. C. Decius, and P. C. Cross, *Molecular Vibrations*, 1955: New York, McGraw-Hill Book Co.
5. W. Gordy and R. L. Cook, *Microwave Molecular Spectra*, 1970: New York, Wiley-Interscience.

Problems

You may wish to build models for some of the symmetry problems.

14.1 Give the point-group symmetries of:

$$H-C\equiv C-H, \quad H-C\equiv C-Cl, \quad H-C\equiv N, \quad CCl_4$$

14.2 What are the point groups of CH_4, CH_3Cl, CH_2Cl_2, CCl_3H, PCl_3 (pyramidal), BF_3 (planar), CH_2FCl?

14.3 Give the point groups of (a) the isomers of dichloroethylene:

and (b) the isomers of dichlorobenzene:

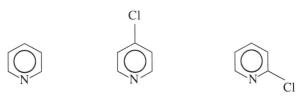

and (c) the pyradines:

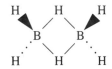

14.4   What is the point group of (a) diborane:

(b) diborane substituted in a terminal position, (c) diborane substituted in a bridge position, (d) diborane substituted in both bridge positions? [In parts (b), (c), and (d) assume that the substituent has the symmetry of a point.]

14.5   Sulfur often will occur as cyclic S_8 molecules. The structure is that of a puckered octagon:

$$
\begin{array}{ccc}
 & \text{S}\ \text{S}\ \text{S} & \\
\text{S} & & \text{S} \\
 & \text{S}\ \text{S}\ \text{S} &
\end{array}
$$

(alternate atoms up and down). What is the point group?

14.6   The dihedral angle in hydrogen peroxide:

$$
\begin{array}{c}
\text{O}-\text{O} \\
\diagup \qquad \diagdown \\
\text{H} \qquad\quad \text{H}
\end{array}
$$

is defined as the angle between the HO bonds, looking down the OO axis:

$$
\begin{array}{c}
\text{O} \\
\diagup\ \phi\ \diagdown \\
\text{H} \quad \text{H}
\end{array}
$$

What is the point group if $\phi = 0$, $\phi = 180°$, or $\phi =$ any other angle?

14.7   What is the point group of ethane (CH_3—CH_3) in the (a) eclipsed, (b) staggered conformation? What is the point group of 1,2-dichloroethane in the (c) *gauche:*

or (d) *trans:*

conformation?

14.8   What is the point group of boric acid:

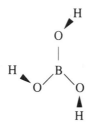

if: (a) the H's lie in the BO_3 plane, (b) the H's are above the plane in a propeller fashion?

14.9   Pi cyclopentadiene metal complexes have two cyclopentadiene rings (regular pentagons) with a metal atom "sandwiched" between. If the metal is Ru, the rings are eclipsed, but if the metal is Fe, the rings are staggered. What are the point groups?

14.10   In D_{2h} symmetry, what are the products $B_{2g} \otimes B_{3g}$, $B_{1u} \otimes B_{1g}$, $A_u \otimes B_{2u}$?

14.11   Find the following direct products:
(a) $A_2 \otimes E$ in C_{3v}.
(b) $A_1'' \otimes A_2''$ in D_{3h}.
(c) $A_2 \otimes T_1$ in T_d.

14.12   Enthalpies of formation as used in Chapter 6 refer to carbon (graphite) as the standard state, whereas the bond energies of Table 14.4 refer to C(gas). The heat of vaporization of graphite is needed; at 298 K this is:

$$C(\text{graphite}) \longrightarrow C(\text{gas}), \qquad \Delta H = 717 \text{ kJ}$$

Use this together with the average bond energies of Table 14.4 to estimate the ΔH of formation of the following molecules (compare Table 6.1): (a) CH_3OH, (b) $COCl_2$, (c) C_2H_4.

14.13   (a) Show that the sp^2 hybrids:

$$\chi_1 = \frac{\sqrt{5}(s) + 2(p_x) + 3(p_y)}{\sqrt{18}}$$

$$\chi_2 = \frac{\sqrt{5}(s) + 2(p_x) - 3(p_y)}{\sqrt{18}}$$

are normalized and orthogonal to each other. (b) What is the angle between the maxima of these hybrids? [*Hint:* The orbitals p_x and p_y have the properties of unit vectors along the x and y axes, respectively.] (c) There is, of course, a third orbital (χ_3), which must be normalized and orthogonal to these two. Derive the form and direction of this orbital.

14.14   Consider a hypothetical H_2X_2 molecule with C_{2h} symmetry:

$$\begin{matrix} H & & \\ & \searrow & \\ & X-X & \quad\quad (z \text{ axis perpendicular to paper}) \\ & & \searrow \\ & & H \end{matrix}$$

What are the proper symmetry designations of the following SALC's?

$$\chi_1 = (H_A 1s) + (H_B 1s), \quad\quad \chi_2 = (H_A 1s) - (H_B 1s)$$

$$\chi_3 = (X_A p_z) + (X_B p_z), \quad\quad \chi_4 = (X_A p_z) - (X_B p_z)$$

14.15   The symmetry orbitals for XH_3 with C_{3v} symmetry are (in order of energy):

$$(1a_1)(2a_1)(1e)(3a_1)(4a_1)(2e)$$

[Note that $(1a_1) \equiv (X1s)$, a nonbonding orbital.] What are the configurations and ground-state term symbols for (a) NH_3, (b) NH_3^+, (c) NH_3^-?

14.16   The symmetry orbitals for XH_3 with D_{3h} symmetry are (in order of energy):

$$(1a_1')(2a_1')(1e')(1a_2'')(3a_1')(2e')$$

What are the configurations and ground-state term symbols for (a) BH_3, (b) BH_3^+, (c) BH_3^-?

14.17   Water has a $^1A_1$ ground state (Table 14.8) and an excited state with a configuration . . . $(3a_1)^2(1b_1)^1(2b_2)^1$. What state(s) result from this configuration? Is it an allowed transition from the ground state to any of these states?

14.18   What are the state symbols for the ground state of NH_2^+, NH_2^-, BH_2^-, CH_2^-? (Assume these are nonlinear.)

14.19   The MO's of naphthalene (ref. 1, p. 164) give the following ground-state configuration: $(1b_{1u})^2(1b_{2g})^2(1b_{3g})^2(2b_{1u})^2(1a_u)^2$; the next (empty) orbital is $(2b_{2g})$.
(a) What are the symmetries of the ground and first excited states?
(b) Is this an allowed transition for absorption of a photon by electric dipole rules?

14.20   In benzene, what state(s) would result if an e_{1g} electron were excited to the b_{2g} orbital?

14.21   The C_s point-group character table is:

C_s	$\hat{E}$	$\hat{\sigma}$
A'	1	1
A''	1	-1

Which types of transitions are electric-dipole allowed?

$$A' \longleftrightarrow A'', \quad\quad A' \longleftrightarrow A', \quad\quad A'' \longleftrightarrow A''$$

14.22   Determine the symmetry of the following normal modes of *trans*-dichloroethylene (C_{2h}):

14.23 Calculate the energies of the lowest nine vibrational states of F_2O. [*Hint:* All will have energies less than 3000 cm^{-1}.]

14.24 In D_{6h} symmetry, what types of vibrations are IR active? Raman?

14.25 For formaldehyde (Figure 14.12) what are the symmetries of the vibrational states:

$$(0, 0, 0, 1, 0, 0), \quad (1, 0, 0, 0, 0, 1), \quad (0, 0, 0, 0, 1, 1)$$

Which of these will be dipole allowed for absorption from the ground state?

14.26 Estimate the frequency of the overtone band $(0, 0, 0) \rightarrow (2, 1, 1)$ in the IR spectrum of H_2O.

14.27 Which normal modes of acetylene (Figure 14.11) will be IR active?

14.28 In the IR spectrum of SO_2 the fundamental $(0, 0, 0) \rightarrow (0, 1, 0)$ is at 1290 cm^{-1} and the overtone $(0, 0, 0) \rightarrow (0, 2, 0)$ is at 2422 cm^{-1}. Calculate the vibrational constant and anharmonicity of this normal mode.

14.29 Ethylene (C_2H_4) has the following normal vibrations (Figure 14.12):

$$3 a_g, a_u, 2 b_{1g}, b_{1u}, b_{2g}, 2 b_{2u}, 2 b_{3u}$$

(a) Which are IR active?
(b) Which are Raman active?
(c) Are any inactive?
(d) Is the first overtone of a b_{1u} mode IR active?
(e) Is a combination band $a_u + b_{1g}$ IR active?

14.30 In a Raman experiment, SO_2 is irradiated with 600-nm light. What will be the wavelengths of the strongest lines (fundamentals) in the scattered light?

14.31 What are the directions of the principal inertial axes in dichloroethylene?

14.32 (a) Classify the following molecules as (1) asymmetric top, (2) symmetric top, (3) spherical top: CH_4, CH_3Cl, CH_2Cl_2, $CH_2 = CH_2$, benzene, SF_6.
(b) Which of these molecules will show a pure rotational (microwave) spectrum?

14.33 Calculate the principal moments of inertia of three masses (m) on the corners of an equilateral triangle (side length a).

14.34 The rotational constant (B) of BF_3 (a D_{3h} molecule) is 0.35 cm^{-1}. Calculate the BF bond length. (Use the results of Problem 14.33.)

14.35 The bond lengths for $Cl - C \equiv N$ (linear) are:

$$R_{C-Cl} = 1.629 \text{ Å}, \qquad R_{C-N} = 1.163 \text{ Å}$$

Calculate the moment of inertia of this molecule.

14.36   The rotation constant for $^{16}O^{12}C^{32}S$ is $\tilde{B} = 0.202864$ cm^{-1} and for $^{16}O^{12}C^{34}S$, $\tilde{B} = 0.197910$ cm^{-1}. Calculate the OC and CS bond lengths.

14.37   Ethylene has rotational constants:

$$\tilde{A} = 4.828 \text{ cm}^{-1}, \qquad \tilde{B} = 1.0012 \text{ cm}^{-1}, \qquad \tilde{C} = 0.8282 \text{ cm}^{-1}$$

What are the principal axes? Which is the "A" axis?

14.38   The rotational constant of CO_2 was measured from the rotational Raman spectrum and found to be $\tilde{B}_e = 0.3937$ cm^{-1}. Calculate the CO bond length.

14.39   Prove that any molecule that has a center of symmetry must have a zero dipole moment.

14.40   The moment of inertia of an A_4X tetrahedral molecule is:

$$I = \tfrac{8}{3} m_A r^2$$

where r is the AX bond length. Estimate the CH bond length in methane from its IR spectrum (Figure 14.14).

unless statistics lie he was
more brave than me: more blonde than you.

–e. e. cummings

15
Statistical Mechanics

Statistical mechanics, the formalism that makes the connection between particle mechanics and macroscopic behavior, was introduced in Chapter 5. In this chapter we shall take a new and somewhat more rigorous look at its foundations and then develop several new applications.

The presumption is that the reader has completed at least Chapters 11 through 13 on quantum theory and has covered Chapter 5 somewhat earlier. The reader who has not done so should cover Sections 5.1 through 5.4 before proceeding; in any event, a review of these sections is recommended.

15.1 The Canonical Ensemble

Consider a system of N particles in a rigid-walled container of volume V, in thermal contact with a thermostat at temperature T. This system must have an energy that is one of the quantized energies, E_i, allowed by quantum theory. Discussions in earlier chapters have shown that one particle has a large number of quantum states available to it, so the number of states available to a group of N particles, being all combinations of those of its constituents, is very large indeed.

Since our system is capable of exchanging energy (as heat) with the thermostat, it will fluctuate among these possible energy states. However, over a sufficiently long period of time, the N-particle system will have an average energy, $\langle E \rangle$, that is entirely characteristic of its macroscopic state as specified by the state variables N, T, V. Furthermore, if N is large, say 10^{20} or greater, the fluctuations about this average energy will be insignificant. To calculate the time-average properties of such a system would entail the specification of the dynamical variables of N particles, a task that rapidly becomes impossible as N becomes very large. This task can be avoided by a device invented by Gibbs — the *ensemble average*.

Suppose that our thermostat contains a very large number ($\tilde{N}$) of identical systems with the same temperature (T), volume (V) and number of particles (N). Each of these systems has a specific energy E_i, but all will not have the same energy. Such a collection of systems is called an ensemble; this particular collection, with N, T, V constant, is called a *canonical ensemble*. This is only one of several useful types of ensembles. Another is a collection of rigid-walled adiabatic systems that all have the same energy — constant N, E, V; this is called a *microcanonical ensemble,* and the temperature may fluctuate. A collection of systems at constant T, V, μ in which the number of particles (N) may fluctuate is called a *grand canonical ensemble*.

Of the $\tilde{N}$ systems in a canonical ensemble, the number that are in a particular energy state E_i will be denoted $\tilde{n}_i$. This will define the probability (p) of finding a system in a particular state as:

$$p_i = \frac{\tilde{n}_i}{\tilde{N}} \tag{15.1}$$

The average energy can then be calculated with an ensemble average:

$$\langle E \rangle = \sum_i p_i E_i \tag{15.2}$$

The replacement of a time average with an ensemble average of stationary systems is called the *ergodic hypothesis*. It is assumed that, if a system spends 1% of its time in a particular state, then a "snapshot" picture of a large number of identical systems will find 1% of them in that state. For equilibrium properties—the only type we shall discuss—the ergodic hypothesis is never in doubt; however, for irreversible processes over short times its use is arguable and has been the source of much learned writing.

The derivation of the probabilities (p_i) is nearly identical to that used in Section 5.3. It starts with the number of ways of assigning $\tilde{N}$ systems to the states with populations $\{\tilde{n}_i\}$:

$$W = \frac{\tilde{N}!}{\prod_i \tilde{n}_i} \tag{15.3}$$

Then one seeks a maximum in this probability subject to the constraints that the total number, $\tilde{N} = \sum_i \tilde{n}_i$, and average energy, $\langle E \rangle = \sum_i p_i E_i$, are constant. The use of the Lagrange method of undetermined multipliers (Section 5.3) gives the result:

$$p_i = \frac{e^{-\beta E_i}}{Z} \tag{15.4}$$

where the N-particle partition function (Z) is defined as:

$$Z = \sum_i e^{-\beta E_i} \tag{15.5}$$

The connection between the energies of individual particles (ε_i) as used in Chapter 5 to the energy of N independent particles (E_i) can be made in a straightforward manner. The connection between the one-particle partition function (z) of Chapter 5 and the N-particle function of Eq. (15.5) is somewhat more involved—we shall discuss this in the next section.

In many cases the number of energy states contributing to Z is so large, and the spacing between them so small, that it is possible to replace the sum of Eq. (15.5) by an integral; then:

$$Z = \int_0^\infty g(E) e^{-\beta E} \, dE \tag{15.6}$$

where $g(E)$ is the state density—the number of energy states in the interval E to $E + dE$.

Thermodynamic Properties

Thermodynamics deals with a set of state functions and the relationships among them; for example, with T and V as independent variables, some dependent state variables are $P(T, V)$, $U(T, V)$, $S(T, V)$, and $A(T, V)$ (together with auxiliary variables such as C_p, C_v, H, and G). Thermodynamics relates these variables to each other, but this process always involves empirical equations for some—for example, the equation of state $P(T, V)$ or the heat capacity $C_v^\theta(T)$. Now we shall develop the idea of the partition function $Z(N, T, V)$ as a sort of super state function from which all the others may be derived.

The partition function depends on the number of particles in the system (N); effectively this occurs because the state density is larger for a larger number of particles. The partition function will also depend on the size of the container (V); this occurs primarily because of the translational energies. Section 11.4 showed that the particle-in-a-box energy levels depend on the size of the box, the levels being more closely spaced in a larger box; this, of course, implies a volume dependence for E_i and Z. If you remember Chapter 5, you will not be surprised to learn that the temperature dependence of Z is via β; but that was never proven in general, so let's just say that Z is a function of β and discover later that this involves the temperature.

First we take the derivative of Z [Eq. (15.5)] with respect to β:

$$\left(\frac{\partial Z}{\partial \beta}\right)_{N,V} = \sum_i \frac{\partial}{\partial \beta} e^{-\beta E_i} = -\sum_i E_i e^{-\beta E_i}$$

If we divide this by Z we get, using $p_i = e^{-\beta E_i}/Z$:

$$\frac{1}{Z}\left(\frac{\partial Z}{\partial \beta}\right)_{N,V} = -\sum_i p_i E_i = -\langle E \rangle$$

The average energy of the system can be identified with the internal energy U as defined by the first law of thermodynamics; therefore:

$$U = -\frac{1}{Z}\left(\frac{\partial Z}{\partial \beta}\right)_{N,V} = -\left(\frac{\partial \ln Z}{\partial \beta}\right)_{N,V} \tag{15.7}$$

(Strictly, this should be $U - U_0$ as discussed in Section 5.4; here we shall simply choose the zero of energy such that $U_0 = 0$.)

The relationship of the partition function to the entropy is more subtle. Suppose that we may alter the energy of our system (reversibly) by either adding heat (q) or by doing mechanical work — that is, by altering the volume of the system. The first law of thermodynamics gives:

$$dU = dq + dw$$

Now, from the statistical definition of the energy [Eq. (15.2) with $U = \langle E \rangle$]:

$$U = \sum_i p_i E_i$$

we get:

$$dU = \sum_i E_i \, dp_i + \sum_i p_i \, dE_i \tag{15.8}$$

That is, the energy of the system can be changed either by changing the energy-state populations (p_i) or by moving the energy levels (E_i). The atomic and molecular energy levels discussed in Chapters 12, 13, and 14 were entirely characteristic of the molecules involved, and did not depend on T or V. (Actually they could be altered by extreme compression, but this factor is not significant except for extremely high pressures.) However, the translational states of Section 11.4 (the particle in a box) did depend on the size of the container and could be altered by changing the box size — that is, by doing mechanical work. Conversely we could argue that at constant volume the energy levels are constant and the only way to alter the average energy at constant volume is by adding heat, which changes the distribution among a fixed set of levels. These arguments lead us to conclude that the first term of Eq. (15.8), the change of U at constant volume, is the heat, while the second is the mechanical work. Therefore:

$$dq = \sum_i E_i \, dp_i$$

$$dw = \sum_i p_i \, dE_i \tag{15.9}$$

Remembering that the second law of thermodynamics defined a state function called the entropy by (for a reversible process):

$$dS = \frac{dq}{T}$$

where T was the thermodynamic temperature, we are ready to obtain a relationship between entropy and the partition function. We start by taking a general derivative of Z:

$$Z = \sum_i e^{-\beta E_i}$$

$$dZ = -\sum_i e^{-\beta E_i} \beta \, dE_i - \sum_i e^{-\beta E_i} E_i \, d\beta$$

Now we divide this by Z and use $p_i = e^{-\beta E_i}/Z$:

$$d(\ln Z) = -\beta \sum_i p_i \, dE_i - \left(\sum_i E_i p_i\right) d\beta$$

We can now identify the sum in the first term as the work and the sum in the second term as U:

$$d(\ln Z) = -\beta \, dw - U \, d\beta$$

The first law of thermodynamics gives:

$$dw = dU - dq$$

Therefore:

$$d(\ln Z) = \beta \, dq - \beta \, dU - U \, d\beta$$

$$\beta \, dq = d(\ln Z) + d(\beta U) \tag{15.10}$$

Now, the right-hand side of Eq. (15.10) involves only state functions; however, the heat is not a state function. We therefore conclude that β is an integrating factor for the reversible heat, and $\beta \, dq$ is a state function. The second law of thermodynamics shows that $1/T$ is the integrating factor for heat, so β must be proportional to $1/T$:

$$\beta = \frac{1}{kT} \tag{15.11}$$

where the proportionality constant (k) is called Boltzmann's constant. This is the same result as that of Section 5.5 but more general, since the derivation did not assume an ideal gas.

Now, inserting the definition of entropy into Eq. (15.10) with $\beta = 1/kT$, we get:

$$dS = k\,d(\ln Z) + d\!\left(\frac{U}{T}\right)$$

$$S = k\,\ln Z + \frac{U}{T} + \text{constant} \tag{15.12}$$

With the help of the third law of thermodynamics, the constant of integration of Eq. (15.12) can be set equal to zero, provided our system has a nondegenerate ground state (so $Z = 1$ at $T = 0$); remember that we assumed earlier that, at $T = 0$, the energy (U_0) was zero.

The remainder of the state functions can be derived easily. For example, $A = U - TS$ can be seen [with Eq. (15.12)] to be:

$$A = -kT\,\ln Z \tag{15.13}$$

Then, the relationship from Chapter 3 (Table 3.1):

$$P = \left(\frac{\partial A}{\partial V}\right)_T$$

gives:

$$P = kT\left(\frac{\partial \ln Z}{\partial V}\right)_{T,N} \tag{15.14}$$

Equation (15.7) may be written, with $T = 1/k\beta$, as:

$$U = kT^2\left(\frac{\partial \ln Z}{\partial T}\right)_{V,N} \tag{15.15}$$

Equations (15.12) to (15.15) demonstrate that the partition function is the source of all thermodynamic information for a macroscopic system; in that sense it plays a role similar to that of the wave function for particle mechanics. Thus, many physical problems can be reduced to that of finding the partition function; in general, this will be no easier than that of finding an accurate wave function.

15.2 Partition Function for Independent Particles

The problem of finding the energy levels of N interacting particles would begin with the Hamiltonian for an N-particle system; then one must find the eigenfunctions and eigenvalues of this Hamiltonian; this is clearly no easy task when $N = 1$ and close to impossible for $N = 10^{20}$. However, simplifications are possible — for example, using only pairwise interactions or a generalized average potential function. The problem is easiest if the particles are independent (no cross terms in the Hamiltonian), for then the energy of the N-particle system is just a sum of the individual energies (ε):

$$E = \sum_{n=1}^{N} \varepsilon(\text{particle } n)$$

This is the only case we shall discuss.

Each particle has a definite energy, which is one of the allowed energies (ε_i) that are the eigenvalues of its Hamiltonian. For a diatomic molecule, for example, the states are all the allowed combinations of its quantum numbers: n_x, n_y, n_z for translation, v for vibration, J for rotation, plus the quantum numbers needed to specify its electronic state. Each of these combinations (indexed as $i = 1, 2, 3, \ldots$) has a particular energy, which, in the Born-Oppenheimer approximation (Chapter 13) will be:

$$\varepsilon_n = \varepsilon(\text{elec}) + \varepsilon(\text{vib}) + \varepsilon(\text{rot}) + \varepsilon(\text{trans})$$

In a large collection of such particles, the number (n_i) that will be found in a state ε_i is given by Boltzmann's law (Chapter 5) as:

$$\frac{n_i}{N} = \frac{g_i e^{-\varepsilon_i/kT}}{z} \qquad (15.16)$$

$$z = \sum_j g_j e^{-\varepsilon_j/kT} \qquad (15.17)$$

The average energy is:

$$\langle E \rangle = \sum_i n_i \varepsilon_i = N \langle \varepsilon \rangle$$

The derivation that led to Eq. (15.7) will give the average energy of a single particle as:

$$\langle \varepsilon \rangle = kT^2 \frac{\partial \ln z}{\partial T}$$

Comparison to Eq. (15.15) shows:

$$\frac{\partial \ln Z}{\partial T} = N \frac{\partial \ln z}{\partial T}$$

This implies in turn that:

$$Z = az^N$$

where a is a constant, independent of temperature.

Consider, for simplicity, two particles (a and b); these have partition functions:

$$z_a = \sum_i e^{-\beta \varepsilon_{ai}}, \qquad z_b = \sum_i e^{-\beta \varepsilon_{bi}}$$

The product of these will be:

$$z_a z_b = \sum_i \sum_j e^{-\beta(\varepsilon_{ai} + \varepsilon_{bj})}$$

If a and b are distinguishable particles, this product is the correct two-particle partition function:

$$Z^{(2)} = z_a z_b$$

since the energy of the two particles is just:

$$E = \varepsilon_a + \varepsilon_b$$

For N different *distinguishable* particles we conclude that the N-particle partition function is:

$$Z = \prod_p z \, (\text{particle}) \tag{15.18}$$

If they are identical but distinguishable (for example, the atoms in a crystal lattice, which are distinguished by their position in the lattice), then z is the same for all, and:

$$Z = z^N \tag{15.19}$$

However, if a and b are indistinguishable, then the state (a in level i, b in level j) should not be counted in addition to the state (a in j, b in i). How many distinct two-particle states are there? Figure 15.1 shows the situation for the case that each particle has 10 quantum states. There are $10^2 = 100$ possible states, but the ones above the diagonal duplicate the ones below the diagonal, and we should not count both. What about the ones on the diagonal? There are two cases:

1. *Fermi-Dirac statistics.* No more than one particle of an identical set may occupy a given quantum state, so the diagonal states of Figure 15.1 should not be counted. This will be the case if the particles have $\frac{1}{2}$-integer spins — for example, electrons, protons, neutrons (all spin $\frac{1}{2}$) and various nuclei such as ^{7}Li (spin $\frac{3}{2}$). The Pauli exclusion principle (Chapter 12) is a manifestation of the same phenomenon for electrons. Such particles are called *fermions*.
2. *Bose-Einstein statistics.* Any number of identical particles may occupy a single quantum state, so the diagonal states of Figure 15.1 should be counted. This class includes all particles with a spin angular momentum having an integer quantum number — for example, deuterons (^{2}D, spin 1) the ^{6}Li nucleus (spin 3), the ^{4}He nucleus (spin zero), and photons. Such particles are called *bosons*.

Does it really matter? Of the 100 two-particle states in Figure 15.1, we should count 45 states for fermions and 55 states for bosons. If there were 100 states for each particle, there would be $100^2 = 10,000$ possible two-particle states, of which we should count 4950 for fermions or 5050 for bosons. If there were 10^6 one-particle states available, the score would be fermions 499,999,500,000, bosons 500,000,500,000. This is getting very close to half of the total possible states in both cases. We can now state a third alternative:

States of B

	1	2	3	4	5	6	7	8	9	10
1	*a*1 *b*1	*a*1 *b*2	*a*1 *b*3	*a*1 *b*4	*a*1 *b*5	*a*1 *b*6	*a*1 *b*7	*a*1 *b*8	*a*1 *b*9	*a*1 *b*10
2	*a*2 *b*1	*a*2 *b*2	*a*2 *b*3	*a*2 *b*4	*a*2 *b*5	*a*2 *b*6	*a*2 *b*7	*a*2 *b*8	*a*2 *b*9	*a*2 *b*10
3	*a*3 *b*1	*a*3 *b*2	*a*3 *b*3	*a*3 *b*4	*a*3 *b*5	*a*3 *b*6	*a*3 *b*7	*a*3 *b*8	*a*3 *b*9	*a*3 *b*10
4	*a*4 *b*1	*a*4 *b*2	*a*4 *b*3	*a*4 *b*4						
5	*a*5 *b*1	*a*5 *b*2	*a*5 *b*3		*a*5 *b*5					
6	*a*6 *b*1	*a*6 *b*2	*a*6 *b*3			*a*6 *b*6				
7	*a*7 *b*1	*a*7 *b*2	*a*7 *b*3				*a*7 *b*7			
8	*a*8 *b*1	*a*8 *b*2	*a*8 *b*3					*a*8 *b*8		
9	*a*9 *b*1	*a*9 *b*2	*a*9 *b*3						*a*9 *b*9	*a*9 *b*10
10	*a*10 *b*1	*a*10 *b*2	*a*10 *b*3						*a*10 *b*9	*a*10 *b*10

States of A

Figure 15.1   States of two particles with ten energy levels. The particles are denoted *a* and *b*; the energy levels are numbered. Thus, the state (*a*1, *b*2) means particle *a* is in energy level 1 while particle *b* is in level 2. If the particles are identical, the states (*a*1, *b*2) and (*a*2, *b*1) are not distinguishable.

3. *Boltzmann statistics*. The number of quantum states available is so large that multiple occupancy will occur rarely in proportion to the whole, so it doesn't matter whether or not we count doubly occupied states.

How many quantum states are there? A fair reading of the earlier chapters would suggest that there are an infinity; that cannot be quite true (eventually even the particle in a box would have enough energy to break out—the wall potential cannot be infinite), but the number *is* very, very large. But the key question is, how many are *available* — that is, how many have energies:

$$\varepsilon \lesssim kT$$

at the temperature of the system?

Actually, the requirements for Boltzmann statistics will be applicable for all but the lightest particles and/or the lowest temperatures. Exceptions include photons (which have no mass at all), electrons (such as the conduction electrons of a metal or in a plasma), and helium and hydrogen at low temperatures. It is possible that the major differences between the low-temperature behavior of the common isotope of helium (^{4}He, spin zero, a boson) and ^{3}He (spin $\frac{1}{2}$, a fermion) may be caused by statistical factors; for example, ^{4}He has a λ-type phase transition (Chapter 4, Figure 4.10) to a "superfluid" state at 2.19 K and boils at 4.2 K; ^{3}He boils at 3.2 K and shows no phase transition or tendency to become superfluid down to 1 K.

Only Boltzmann statistics will be considered here in detail. In this case, the two-particle partition function will be:

$$Z^{(2)} = \frac{z^2}{2}$$

If there are three identical particles, states such as (*a* in level 1, *b* in level 5, *c* in level 20) are not distinct from (*a* in 5, *b* in 20, *c* in 1); there are 3! such permutations, so the three-particle partition function (neglecting multiple occupancy) is:

$$Z^{(3)} = \frac{z^3}{3!}$$

The partition function for N identical particles that have many more quantum states available than there are particles is:

$$Z = \frac{z^N}{N!} \qquad (15.20)$$

This is commonly used with Stirling's approximation:

$$\ln N! = N \ln N - N$$

to give:

$$\ln Z = N \ln z - \ln N!$$
$$= N \ln z - N \ln N + N \ln e$$
$$\ln Z = N \ln \frac{ze}{N} \qquad (15.21)$$

Insertion of Eq. (15.21) into Eqs. (15.12) through (15.15) will give the equations relating z to the thermodynamic functions that are used in Chapter 5 (Table 5.4); practical applications regarding the thermodynamic properties of ideal gases will be found there.

15.3 The Translational Partition Function

The partition function for the translation of a particle is derived in Section 5.5 using classical mechanics; this derivation required the definition of a factor "h" as the size of a cell in phase space. The same result can be obtained using quantum mechanics, and the derivation is actually easier; furthermore, the factor "h" is clearly identified as Planck's constant.

The energy levels for a particle of mass m in a cubical box of length a were derived in Section 11.4; the result was:

$$\varepsilon_{n_x n_y n_z} = (n_x^2 + n_y^2 + n_z^2)\left(\frac{h^2}{8ma^2}\right)$$

The translational partition function is, therefore:

$$z_{tr} = \sum_{n_x=1}^{\infty} \sum_{n_y=1}^{\infty} \sum_{n_z=1}^{\infty} e^{-n_x^2 A} e^{-n_y^2 A} e^{-n_z^2 A} \tag{15.22}$$

with $A \equiv h^2/8ma^2kT$. For heavy particles in large boxes, the state density will be high enough that the sums of Eq. (15.22) can be replaced by integrals without error; thus:

$$z_{tr} = \int_0^{\infty} e^{-n_x^2 A}\, dn_x \int_0^{\infty} e^{-n_y^2 A}\, dn_y \int_0^{\infty} e^{-n_z^2 A}\, dn_z$$

This is a common integral found in most tables, so the result is easily shown to be:

$$z_{tr} = \left[\frac{1}{2}\left(\frac{\pi}{A}\right)^{1/2}\right]^3$$

$$= \left(\frac{\pi}{4A}\right)^{3/2}$$

$$= \left(\frac{2\pi ma^2kT}{h^2}\right)^{3/2}$$

Using the volume $V = a^3 = (a^2)^{3/2}$, this becomes:

$$z_{tr} = (2\pi mkT)^{3/2}\frac{V}{h^3} \tag{15.23}$$

This is the same result as derived in Section 5.5.

Exercise: Use Eq. (15.23) together with Eqs. (15.14) and (15.15) to derive the ideal gas law, $PV_m - RT$, and the equipartition value $U_m(\text{tr}) = \frac{3}{2}RT$. ∎

Density of States

We shall now develop the translational partition function in yet a third way, for this method will produce results of very general interest, which have a significance that goes beyond the particle in a box. Since we effectively assumed a continuum of states, it should be possible to use Eq. (15.6) to evaluate the partition function; this requires the evaluation of the state density function, $g(E)$.

Imagine (Figure 15.2) a space with axes labeled n_x, n_y, n_z, with marks at integer intervals. Each state (n_x, n_y, n_z) is represented by a point in this space; Figure 15.2 illustrates this for the state $(9, 6, 7)$. The radius vector to a point has a length:

$$n = \sqrt{n_x^2 + n_y^2 + n_z^2} \tag{15.24}$$

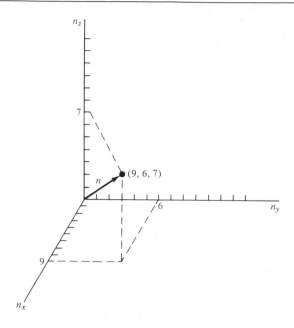

Figure 15.2   Point space. The state of a three-dimensional particle in a box is specified by three quantum numbers, n_x, n_y, and n_z. Each of these states is represented as a point in the positive octant of the coordinate system shown; the state $(9, 6, 7)$, meaning $n_x = 9$, $n_y = 6$, and $n_z = 7$, is shown as an example.

The energy of a state (n_x, n_y, n_z) is, therefore:

$$\varepsilon = n^2 \frac{h^2}{8ma^2}$$

However, the constant n is not necessarily an integer.

Now, imagine this figure from a long distance away. The points are very closely spaced and nearly form a continuum. The number of points with radius less than n will be the volume of an octant of a sphere:

$$\# \text{ points} = \frac{1}{8}\left(\frac{4\pi}{3}\right)n^3$$

The density of points between n and $n + dn$ will then be:

$$g(n)\,dn = \frac{\pi}{2}n^2\,dn \tag{15.25}$$

This is a very useful result, which we shall use twice again in this chapter. It can be used to calculate the translational partition function with [Eq. (15.6) for one particle]:

$$z = \int_0^\infty e^{-\varepsilon/kT}g(\varepsilon)\,d\varepsilon$$

$$\varepsilon = n^2 \frac{h^2}{8ma^2}, \qquad n = \varepsilon^{1/2} \left(\frac{h}{8ma^2} \right)^{-1/2}$$

$$d\varepsilon = 2n \frac{h^2}{8ma^2} dn = 2 \left(\frac{h}{8ma^2} \right)^{1/2} \varepsilon^{1/2} dn$$

From this it can be shown, after a bit of algebra, that:

$$g(\varepsilon)\, d\varepsilon = \frac{\pi}{2} n^2 \, dn = \frac{\pi}{4} \left(\frac{8ma^2}{h^2} \right)^{3/2} \varepsilon^{1/2} \, d\varepsilon \qquad \textbf{(15.26a)}$$

This represents the number of translational states between ε and $\varepsilon + d\varepsilon$. The partition function is then:

$$z_{\text{tr}} = \frac{\pi}{4} \left(\frac{8ma^2}{h^2} \right)^{3/2} \int_0^\infty e^{-\varepsilon/kT} \varepsilon^{1/2} \, d\varepsilon \qquad \textbf{(15.26b)}$$

A standard integral table gives the result

$$\int_0^\infty x^n e^{-ax} \, dx = \frac{\Gamma(n + 1)}{a^{n+1}}$$

The *gamma function,* $\Gamma(n + 1)$, is $n!$ if n is an integer (as such, this result was used frequently in Chapter 12); when $n = \frac{1}{2}$, the value of the gamma function is:

$$\Gamma(\tfrac{3}{2}) = \sqrt{\frac{\pi}{2}}$$

Therefore the partition function is:

$$z_{\text{tr}} = \frac{\pi^{3/2}}{8} \left(\frac{8ma^2 kT}{h^2} \right)^{3/2} \qquad \textbf{(15.26c)}$$

This can be shown to be identical to Eq. (15.23). The new results achieved by this derivation are the point density function, Eq. (15.25), and the fact that $g(\varepsilon)$ is proportional to $\sqrt{\varepsilon}$ [Eq. (15.26)]. [For a picture of $g(\varepsilon)$, look at Figure 11.9 from a distance.]

15.4 Nuclear Spins

Many atomic nuclei have an intrinsic angular momentum called spin — quite analogous to the electron spin discussed in Chapter 12. The spin quantum number is called I, analogous to the quantum number l of Section 11.6; the magnitude of the angular momentum will be:

$$\sqrt{I(I + 1)}\, \hbar$$

and the degeneracy (in the absence of a magnetic field) will be:

$$g_N = 2I + 1$$

Table 15.1   Nuclear spin quantum numbers

Nucleus	I	$g_I = 2I + 1$
^{12}C, ^{16}O, ^{32}S	0	1
^{1}H, ^{13}C, ^{19}F, ^{31}P, ^{195}Pt, ^{207}Pb	$\frac{1}{2}$	2
^{2}D, ^{14}N, ^{6}Li	1	3
^{7}Li, ^{11}B, ^{7}Li, ^{35}Cl, ^{37}Cl	$\frac{3}{2}$	4
^{27}Al, ^{66}Mn	$\frac{5}{2}$	6
^{10}B	3	7
^{115}In, ^{45}Sc	$\frac{7}{2}$	8

for the quantum numbers $M_I = I, \ldots, -I$. The quantum number I may be an integer or half-integer (Table 15.1) and is a characteristic of the nucleus. All nuclei with even mass number and even atomic number — for example, $^{12}_{6}C$ or $^{16}_{8}O$ — have $I = 0$; that is, they have no spin angular momentum.

The most obvious effect of the nuclear spin is to give the molecular states an additional degeneracy of $2I + 1$. There is a more subtle effect on the populations of the rotational levels, which we shall discuss shortly; if the rotational states of the molecule are in the high-temperature limit:

$$T \gg \theta_r$$

with the characteristic temperature for rotation defined as:

$$\theta_r = \frac{hc\tilde{B}}{k}$$

($\tilde{B}$ is the rotational constant as defined in Chapter 12), this effect will be negligible, and the rotational partition function will be as given in Section 5.6. In such a case, the primary effect of nuclear spin is on the entropy:

$$S_m(\text{nuclear spin}) = R \ln g_N \qquad \textbf{(15.27)}$$

For a single spin the nuclear degeneracy factor is $g_N = 2I + 1$; if there are several spins in the molecule, it is a product of such factors for each spin. For example, a spin $\frac{1}{2}$ has two states, $M_I = \frac{1}{2}$ (called α) and $-\frac{1}{2}$ (called β); it would add $R \ln 2$ to the entropy. Two nuclear spins $\frac{1}{2}$ (as in F_2, for example) have four possible states ($\alpha\alpha, \alpha\beta, \beta\alpha$, and $\beta\beta$), and will add $R \ln 4$ to the molar entropy.

Nuclear spin entropies are by no means negligible, yet they are uniformly ignored in compilations of thermodynamic entropies and free-energy functions (for example, Tables 3.2, 6.4, and 6.5). This ordinarily causes no problems, because the spins in the products and reactants will be the same, and the nuclear spin terms will cancel when ΔS_{rxn} is calculated.

Nuclear spins do have energy interactions with the molecule, as well as a Zeeman effect from the earth's magnetic field. These terms are so small, however, that they make no significant contribution to the energy or heat capacity, and the presumption that the $2I + 1$ states are degenerate causes no error. (Compare the discussion of electronic state degeneracy in Section 5.6.)

Ilya Prigogine (born 1917)

Is entropy "time's arrow"? Statistical mechanics makes the connection between mechanics and macroscopic processes; yet mechanical laws, both classical and quantum, are all invariant to time reversal, while, in the macroscopic world, there is a clear direction in time which gives meaning to the term "irreversible" as applied to many chemical and physical processes. Boltzmann explained this irreversibility as a progress from ordered to disordered states, with the disordered states being more probable simply because there were more of them; equilibrium thermodynamics is a simple case of seeking the most probable state, that of minimum energy and maximum entropy. Such ideas, the basis of reversible thermodynamics, fall far short of explaining many natural phenomena—the growth of a seed into a flower, the development of an embryo into a living being, and, over a longer time scale, the evolution of the species from single-celled organisms to dinosaurs or man; indeed, most biological processes at one stage or another show dramatic increases in order. Such processes are far from equilibrium and fundamentally dynamical, more akin to an oscillating steady-state than to an equilibrium phenomenon.

The development of order from disorder and the asymmetry of time are the subject of irreversible or nonequilibrium thermodynamics. Ilya Prigogine, who shares his time between the University of Texas and the Free University of Brussels in his native Belgium, received the 1977 Nobel Prize for his contributions to this field, especially the theory of dissipative structures. His short book, From Being to Becoming, *published in paperback by W. H. Freeman and Company (San Francisco), provides a useful, if somewhat compact, introduction to this subject.*

15.5 *Ortho* and *Para* Hydrogen

The high-temperature approximation for the rotational partition function is valid for all diatomic molecules at room temperature except hydrogen. The thermodynamic properties of hydrogen furnish a most dramatic example of the effect of nuclear spins on molecular statistics.

The wave function of a diatomic molecule in the Born-Oppenheimer approximation may be written as a product of factors for translation, electron motion, nuclear vibration, rotation, and (this is new) nuclear spin (abbreviated "ns"):

$$\Psi = \psi_{tr}\psi_{elec}\psi_{vib}\psi_{rot}\psi_{ns} \qquad (15.28)$$

A homonuclear diatomic has two identical nuclei, and its squared wave function, $|\Psi|^2$, must be invariant to the exchange of the labels on the nuclei (we shall use A and B as labels). This means that the wave function may be either *odd* or *even* with respect to this exchange (cf. the discussion of the Pauli exclusion principle, Section 12.4). Empirically it is found that all nuclei with integer spin (quantum number $I = 0, 1, 2, 3, \ldots$ — that is, bosons) must have wave functions that are *even* with respect to nuclear exchange, while nuclei with half-integer spins ($I = \frac{1}{2}, \frac{3}{2}, \frac{5}{2}, \ldots$ — that is, fermions) must have wave functions that are *odd* with respect to spin exchange. This property is called *parity*.

Referring to Eq. (15.28), we shall now discuss how the wave function of a homonuclear diatomic molecule will be affected by the exchange of the nuclear spin coordinates. (a) The translational part depends only on the COM coordinates and will not be affected. (b) The electronic part will not be affected *if* it is totally symmetric in the electron coordinates (that is, Σ_g^+); otherwise, Σ_g^- and Σ_u^+ functions are odd and Σ_u^- functions are even (not affected). (The situation with Π and Δ states is more complicated.) (c) The vibrational wave function depends only on the magnitude of the internuclear distance and is unaffected by nuclear exchange.

(d) The rotational wave functions are affected by nuclear exchange, but this requires a more detailed explanation. The rotational quantum numbers J, M define the wave function:

$$\psi_{\text{rot}}(J, M) = P_J^{|M|}(\theta) \, e^{iM\phi} \qquad (15.29)$$

(The associated Legendre polynomials were introduced in Section 11.7, and are tabulated in Table 11.8.) If a homonuclear diatomic molecule rotates by 180°, it will reach a position that is indistinguishable from its initial position; however, the nuclear coordinates are exchanged. Such a rotation affects the polar coordinates as:

$$\theta \to \pi - \theta, \qquad \phi \to \phi + \pi$$

The ϕ part of the rotational wave function becomes, upon exchange:

$$e^{iM\phi} \to e^{iM(\phi + \pi)} = e^{iM\phi} e^{iM\pi}$$

$$e^{iM\pi} = \cos M\pi + i \sin M\pi$$

Since M must be an integer, the sine term is zero; the cosine term will be $+1$ if M is even and -1 if M is odd.

Therefore the ϕ portion of the rotational wave function will, upon 180° rotation, become:

$$e^{iM\phi} \to \pm e^{iM\phi}$$

depending on the parity of M. The θ terms can be worked out with the following relationships from elementary trigonometry:

$$\cos \theta \to \cos (\pi - \theta) = -\cos \theta$$

$$\sin \theta \to \sin (\pi - \theta) = +\sin \theta$$

Exercise: The rotational wave function ψ_{JM} will, upon 180° rotation, either change sign (odd) or not (even). Use the relationships given above to demonstrate the following:

$$J = 0, \quad M = 0, \quad \psi = 1/\sqrt{4\pi}: \qquad \qquad \text{even}$$

$$J = 1, \quad M = 0 \quad \psi = (\text{const.}) \cos \theta: \qquad \text{odd}$$

$$J = 1, \quad M = 1 \qquad \psi = (\text{const.}) \sin \theta \, e^{i\phi}: \qquad\qquad \text{odd}$$

$$J = 2, \quad M = 0 \qquad \psi = (\text{const.}) (3 \cos^2 \theta - 1): \qquad \text{even}$$

$$J = 2, \quad M = 1 \qquad \psi = (\text{const.}) \sin \theta \cos \theta \, e^{i\phi}: \qquad \text{even}$$

$$J = 2, \quad M = 2 \qquad \psi = (\text{const.}) \sin^2 \theta \, e^{2i\phi}: \qquad \text{even}$$

The same will be true for the negative M values — for example:

$$J = 2, \quad M = -2 \qquad \psi = (\text{const.}) \sin^2 \theta \, e^{-2i\phi}: \qquad \text{even} \qquad \blacksquare$$

The general rule, implied by the exercise above, is that the rotational wave function will be even with respect to nuclear exchange (that is, 180° rotation) if J is even; it will be odd if J is odd.

(e) The nuclear spin wave function will obviously be affected by exchange — but how? Two nuclei with spins I will have a total nuclear spin angular momentum (symbol: T) that is the vector sum:

$$\mathbf{T} = \mathbf{I}_A + \mathbf{I}_B$$

Since $I_A = I_B = I$, the usual rules for the addition of angular momentum (Section 12.6) tell us that the allowed values for the total nuclear spin quantum number are:

$$T = 2I \; \ldots \; 0 \text{ (in integer steps)}$$

Each of these types of function has a statistical weight (degeneracy):

$$g = 2T + 1 \qquad\qquad\qquad \textbf{(15.30)}$$

The parity of the state of maximum spin ($T = 2I$) will always be even, and the parity will alternate as T drops in integer steps.

Example: If $I = \frac{1}{2}$, then $T = 1$ or 0. The $T = 1$ states are even and the $T = 0$ states are odd. If $I = 1$, then $T = 2$, 1, or 0. The $T = 2$ state is even, the $T = 1$ state is odd, and the $T = 0$ state is, again, even. If there is no nuclear spin ($I = 0$), the nuclear exchange parity is even.

The nuclear spin exchange is most easily illustrated for two spins $\frac{1}{2}$ (for example, H_2), since this is very similar to our discussion of two equivalent electrons (spin $\frac{1}{2}$, Section 12.6). Each spin $\frac{1}{2}$ has two functions for $M = \frac{1}{2}$ (called α) and $M = -\frac{1}{2}$ (called β). There are four wave functions for the two spins together; the symmetrized products are:

$$T = 0, \qquad \psi_{ns} = \alpha\beta - \beta\alpha, \qquad g = 1 \text{ (``singlet'')}$$

$$T = 1, \qquad \psi_{ns} = \begin{Bmatrix} \alpha\alpha \\ \alpha\beta + \beta\alpha \\ \beta\beta \end{Bmatrix}, \qquad g = 3 \text{ (``triplet'')}$$

It can be seen that the $T = 0$ (singlet) function is *odd* with respect to nuclear exchange, while the $T = 1$ (triplet) functions are *even* with respect to exchange of the identical nuclei. $\blacksquare$

Thus, the diatomic wave function can have its sign changed upon nuclear exchange in three places — electronic, rotational, or nuclear spin. The combinations will be more easily visualized if we define a *parity factor* (P) that is $+1$ for an even function

Table 15.2   Rules for nuclear spin exchange parity

$$P_{\text{elec}} = \begin{cases} +1 \text{ for } \Sigma_g^+ \text{ or } \Sigma_u^- \\ -1 \text{ for } \Sigma_u^+ \text{ or } \Sigma_g^- \end{cases}$$

$$P_{\text{rot}} = \begin{cases} +1 \text{ for } J \text{ even} \\ -1 \text{ for } J \text{ odd} \end{cases}$$

$$P_{\text{ns}} = \begin{cases} +1 \text{ for states with } T = 2I, 2I - 2, \text{ etc.} \\ -1 \text{ for states with } T = 2I - 1, 2I - 3, \text{ etc.} \end{cases}$$

The product $P_{\text{elec}} P_{\text{rot}} P_{\text{ns}}$ must be $+1$ for bosons, -1 for fermions.

and -1 for an odd function. The parity of the whole wave function [Eq. (15.28)] will then be:

$$P = P_{\text{elec}} P_{\text{rot}} P_{\text{ns}} \tag{15.31}$$

It is the factor P that must be equal to $+1$ for boson nuclei or -1 for fermion nuclei. The above rules are easily summarized in this notation; Table 15.2 does this.

Molecular Hydrogen

Hydrogen in its ground electronic state is $^1\Sigma_g^+$, so the electronic parity is $+1$. Therefore the product of the rotation and nuclear factors must be -1. Because of the close connection between rotations and nuclear spins it will be necessary to treat them together and define a partition function, z_{nsr}, for the nuclear spin-rotational energy levels.

The rotational energy of a diatomic molecule is given in Chapter 13 as:

$$\varepsilon_J = J(J + 1)hc\tilde{B}$$

This level has a degeneracy of $2J + 1$. The nuclear spin-rotation partition function will then be:

$$z_{\text{nsr}} = \sum_T \sum_{J=0} (2T + 1)(2J + 1)e^{-J(J+1)\theta_r/T} \tag{15.32}$$

$$\theta_r = \frac{hc\tilde{B}}{k}$$

Because of the required parity for fermions, the $T = 0$ state may go only with the even J values, while $T = 1$ states go with the odd values; therefore the partition function is:

$$z_{\text{nsr}} = \sum_{\substack{\text{even} \\ J}} (2J + 1)e^{-J(J+1)\theta_r/T} + 3\sum_{\substack{\text{odd} \\ J}} (2J + 1)e^{-J(J+1)\theta_r/T} \tag{15.33}$$

For hydrogen, $\theta_r = 85.35$ K, so only a few rotational states are populated at room temperature, and the high-temperature approximation used in Chapter 5 (replacing the sums by integrals) cannot be used here. However, this same fact makes it easy to evaluate Eq. (15.33) by simple summation.

The effect of nuclear spin on rotational populations can be experimentally verified in two ways. First, spectroscopic intensities for the rotational Raman (discussed in the next section) or the fine structure of the electronic band spectra (ultraviolet) will reflect the rotational-state populations. The population of the rotational states will be given by:

$$\frac{N_J}{N} = \frac{g_N(2J + 1)e^{-J(J+1)\theta_r/T}}{z_{nsr}} \tag{15.34}$$

For hydrogen, $g_N = 1$ for even J and $g_N = 3$ for odd J. A sample calculation is shown in Figure 15.3 — this should be contrasted to an earlier example (Figure 13.5) for the heteronuclear diatomic molecule CO.

Equation (15.33) can also be used to calculate the heat capacity of hydrogen. When

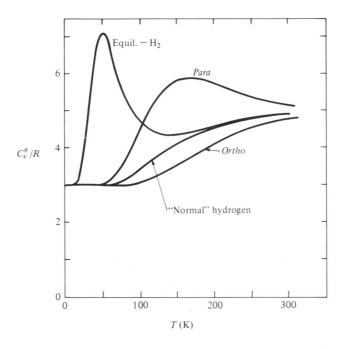

Figure 15.3 Heat capacity of hydrogen gas vs. temperature. "Normal" hydrogen is a mixture of $\frac{1}{4}$ *para* and $\frac{3}{4}$ *ortho,* and the proportion does not change with temperature. "Equil.-H$_2$" denotes an equilibrium mixture of *ortho* and *para* hydrogen — that is, the exchange is catalyzed so the proportions of *ortho* and *para* will change with temperature. The experimental points on the *"para"* curve are for a gas that was 95% *para* hydrogen. (From G. N. Lewis, M. Randall, K. S. Pitzer, and L. Brewer, *Thermodynamics,* 1961: New York, McGraw-Hill Book Co., Fig. 35-1, p. 597)

this was first done, the results disagreed with the observed values — especially at low temperature. The reason soon became clear. Hydrogen molecules in even or odd J states cannot readily exchange identities, because the nuclear spin state must also change — the only way they can change between the even and odd J manifolds is to dissociate into atoms and to come together again. In the absence of a catalyst, this exchange is very slow, and hydrogen is effectively a mixture of two types of molecules with quite different thermodynamic properties. Those molecules that are in the odd-J states ($T = 1$) are called *ortho*-hydrogen. Those molecules that are in the even-J states ($T = 0$) are called *para*-hydrogen. The partition functions are:

$$\text{\textit{ortho} hydrogen } (J \text{ odd}): \qquad z_{nsr} = 3 \sum_{odd} (2J + 1)e^{-J(J+1)\theta_r/T} \qquad \textbf{(15.35a)}$$

$$\text{\textit{para} hydrogen } (J \text{ even}): \qquad z_{nsr} = \sum_{even} (2J + 1)e^{-J(J+1)\theta_r/T} \qquad \textbf{(15.35b)}$$

Treated as a mixture, the heat capacity will be:

$$C_{vm}(H_2) = X_{ortho} C_{vm}(ortho) + X_{para} C_{vm}(para) \qquad \textbf{(15.36)}$$

The equilibrium molar ratio of the two species will be:

$$\frac{ortho}{para} = \frac{3 \sum_{odd} (2J + 1)e^{-J(J+1)\theta_r/T}}{\sum_{even} (2J + 1)e^{-J(J+1)\theta_r/T}} \qquad \textbf{(15.37)}$$

At room temperature or above, this ratio will be 3:1 (cf. Table 15.3), so "normal hydrogen" (as it is found in a tank) is 75% *ortho*. If normal hydrogen is cooled in the absence of a catalyst, it will remain a 3:1 mixture and its heat capacity can be calculated with Eqs. (15.35) and (15.36) (with $X_{ortho} = 0.75$, $X_{para} = 0.25$). If a catalyst is provided, "equilibrium hydrogen" will result, and the heat capacity calculated with Eq. (15.33) is correct. Pure *para* hydrogen can be made by exposing hydrogen to charcoal (a catalyst) at 20 K. If the catalyst is removed, the resulting gas (which will remain nearly all *para* as T is increased) has thermodynamic properties as calculated with Eq. (15.35b). Figure 15.4 contrasts the heat capacities of these various "types" of hydrogen.

Table 15.3   *Ortho*/*para* equilibrium in H_2

T (K)	Equilibrium mole fraction *Ortho*
20	0.002
50	0.229
100	0.613
300	0.749
500	0.750

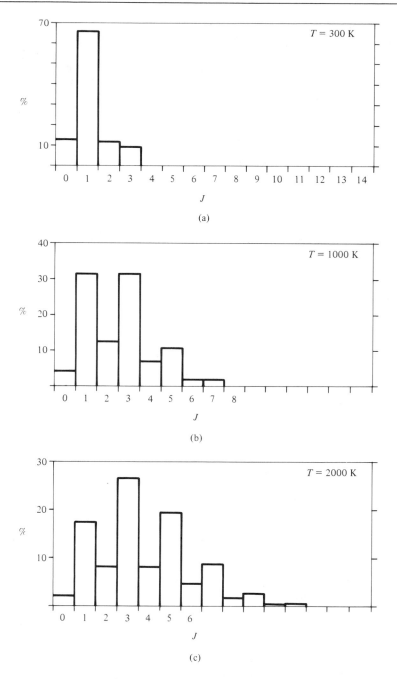

Figure 15.4 Populations of rotational states of hydrogen. *Ortho* hydrogen has only odd values for *J*, and *para* hydrogen has only even values for *J*. The alternation of intensities (most easily seen in the example for *T* = 2000 K) is due to the statistical weights of the nuclear spin states.

15.6 Other Homonuclear Diatomics

The effect of nuclear spin on rotational-level populations shows up for homonuclear diatomic molecules other than hydrogen, of course, but its effect on thermodynamics is observable only for the lightest molecules (for example, D_2) at low temperature. As an example of why this happens, let us look at the case of nitrogen (N_2, $\theta_r = 2.89$ K).

The common isotope of nitrogen is ^{14}N with $I = 1$. In $^{14}N_2$, each spin may have a projection quantum number $M = 1, 0, -1$, so there are nine combinations possible. These may be classified as to the total spin quantum number with values $T = 0$ ($g = 1$), $T = 1$ ($g = 3$) and $T = 2$ ($g = 5$). These nuclei are bosons, and the ground electronic state of N_2 is symmetric, so the $T = 0$ and $T = 2$ states ($5 + 1 = 6$ states in all) must go with the even J values, and the three $T = 1$ states with the odd J's. The partition function is therefore:

$$z_{nsr} = 6 \sum_{even} (2J + 1)e^{-J(J+1)\theta_r/T} + 3\sum_{odd} (2J + 1)e^{-J(J+1)\theta_r/T} \quad \textbf{(15.38)}$$

High Temperature Limit

In Section 5.6 it was demonstrated that in the high-temperature approximation ($T >> \theta_r$) the sum of rotational states could be replaced by an integral, and:

$$\sum_{J=0}^{\infty} (2J + 1)e^{-J(J+1)\theta_r/T} \cong \frac{T}{\theta_r} \quad \textbf{(15.39)}$$

For N_2, this limit can be expected to be valid at any temperature above the boiling point (77 K). Furthermore, the odd and even sums of Eq. (15.38) could be expected to be just half of the whole sum—Eq. (15.39). Therefore:

$$z_{nsr} = 6\left(\frac{T}{2\theta_r}\right) + 3\left(\frac{T}{2\theta_r}\right) = 9\left(\frac{T}{2\theta_r}\right)$$

But the 9 in this result is just the total nuclear spin degeneracy (g_N) and 2 is the symmetry number (σ) for a homonuclear diatomic (Section 5.6). In general, if $T >> \theta_r$:

$$z_{nsr} = g_N\left(\frac{T}{\sigma\theta_r}\right) \quad \textbf{(15.40)}$$

As mentioned earlier (Section 15.4), the nuclear factor (g_N) is usually omitted, and in this case the results given in Chapter 5 are correct.

The Rotational Raman Effect

The effect of nuclear spin does show up rather dramatically on the rotational-state populations, however; Figure 15.5 shows this calculation for nitrogen at 300 K. If this is compared to Figure 13.5 for the heteronuclear molecule CO (whose θ_r is nearly the same as N_2), the effect of the nuclear spin weights should be rather apparent. This type of intensity alternation is very obvious in spectroscopy—not IR, of course, because nitrogen has no dipole moment and is IR inactive.

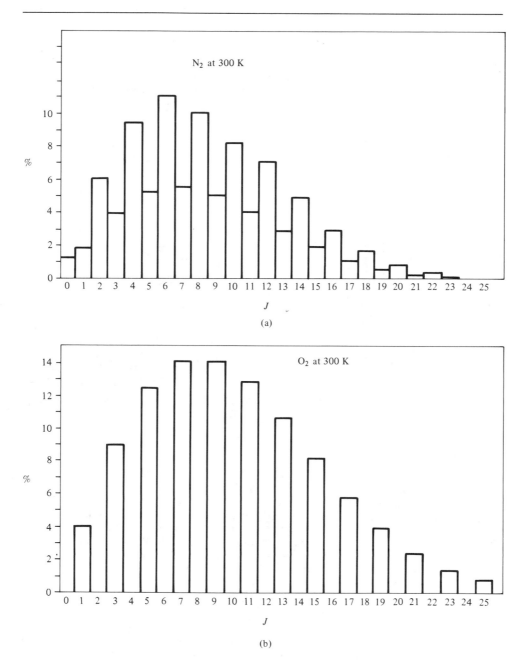

Figure 15.5 Populations of rotational states for nitrogen and oxygen at 300 K. The alternation of populations is due to nuclear spin statistics. The N-14 nucleus has a nuclear spin $I = 1$; the O-16 nucleus has no spin—that is, $I = 0$. (Actually, the rotational quantum number for oxygen should be called N; because oxygen has electron spin with $S = 1$, the symbol J is used to denote the total angular momentum, $\mathbf{J} = \mathbf{N} + \mathbf{S}$.)

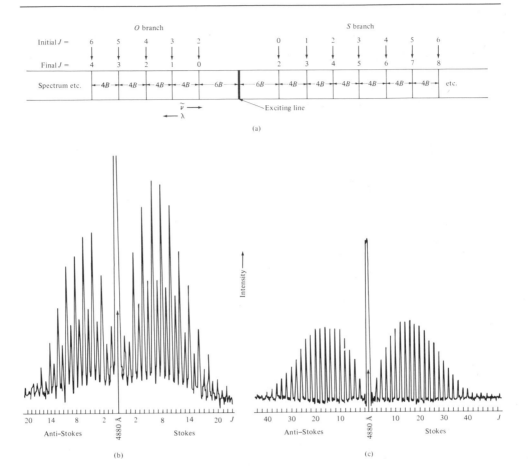

Figure 15.6 Rotational Raman spectrum of a diatomic molecule. (a) Schematic spectrum. The symbols in the spaces between the lines show the spacing (that is, the frequency difference of the lines to the right and to the left) as a multiple of the rotational constant B. For the "S" branch the J quantum number increases by two; for the "O" branch the J quantum number decreases by two. (b) Raman spectrum of N_2. (c) Raman spectrum of CO_2. [(b) and (c) are from H. M. G. Edwards, "High Resolution Raman Spectroscopy of Gases" in *Essays in Structural Chemistry* (A. J. Downs, D. A. Long and L. A. K. Snaveley, eds.) 1971: New York, Plenum Press, Fig. 6.1, p. 151 and Fig. 6.2, p. 152]

The Raman effect was discussed in Section 14.10; at that point, the discussion focused on the vibrational Raman, although the rotational Raman was briefly mentioned. Now we consider the case of a photon (frequency $\tilde{\nu}_i$) interacting with a homonuclear diatomic molecule and exchanging energy to alter its rotational state (leaving it in the same vibrational state). The scattered photon will have a frequency:

$$\tilde{\nu}_s = \tilde{\nu}_i - \frac{\Delta E_{rot}}{hc}$$

Using the rigid-rotor approximation, this is:

$$\tilde{\nu}_s = \tilde{\nu}_i - \tilde{B}[J'(J' + 1) - J''(J'' + 1)] \qquad (15.41)$$

In this equation, J'' is the initial quantum number (before the molecule interacts with the photon) and J' is the final state. The selection rules for this case require:

$$J' = J'' + 2 \quad \text{or} \quad J'' - 2$$

[The first case is called the S branch, and the second is called the O branch; remembering what the PQR branches were, this should make sense.] The resulting spectrum is shown schematically in Figure 15.6. The intensity of scattering for molecules originating in a particular J state will, of course, be proportional to N_J and thus reflect the pattern of alternating intensities shown in Figure 15.5. Figure 15.6(b) shows the Raman spectrum of nitrogen—note alternating intensities [cf. Figure 15.5 and Eq. (15.38)].

The case of $I = 0$ is most dramatic. If there is no nuclear spin, the nuclei are bosons and the product of the rotational and electronic parities must be $+1$. If the molecule's electronic state is Σ_g^+ or Σ_u^-, only even J levels will be populated; if the electronic state is Σ_u^+ or Σ_g^-, only odd J levels will be populated. In either case, alternate lines in the rotational Raman spectrum will be missing. Molecular oxygen ($^{16}O_2$, $^3\Sigma_g^-$, cf. Chapter 13) is such a case—see Figure 15.5. The same theory applies to all centrosymmetric linear molecules; Figure 15.6(c) shows the Raman spectrum of CO_2—note that all odd-J lines are missing.

How can one tell if peaks are missing in the rotational Raman? Referring to Figure 15.6, we see that, if all peaks are present, the ratio of the first spacing (between $\tilde{\nu}_i$ and the first scattered peak) to the second spacing (first to second scattered peak) is $6:4 = 1.50$. If the even peaks are missing, the ratio is $10:8 = 1.25$. If the odd peaks are missing, the ratio is $6:8 = 0.75$. Thus, without knowing the value of $\tilde{B}$, we can determine unambiguously which peaks are missing. Likewise, the alternating intensity pattern can be used to determine the statistical factors (g_{even} and g_{odd}) that will be characteristic of the spin quantum number I.

15.7 Heat Capacity of Solids

The applications we have seen so far, here and in Chapter 5, have concerned properties of ideal gases. The theory is equally straight-forward and successful in dealing with crystalline solids, and we shall now discuss that topic. The theories we shall discuss deal primarily with atomic or ionic solids (such as metals or NaCl), which consist of atoms or ions held in their lattice positions by fairly nondirectional forces due to their neighbors. Molecular solids, in which the lattice positions are occupied by covalently bonded aggregates of atoms, will have intramolecular motions (vibrations, rotations, and internal rotations) that will contribute to the thermodynamic functions in addition to the lattice motions. The contrast is much as we found between atomic and molecular gases (Chapter 5), except that the lattice vibrations replace the translational degrees of freedom.

Experimentally, atomic solids are found to have heat capacities (C_{vm}) that vary from zero at $T = 0$ to $3R$ per mole at high temperatures (the law of Dulong and Petit).

At any given temperature (below that where the law of Dulong and Petit applies) atomic solids may vary greatly in their heat capacity; for example, at 298 K, $C_{pm}(\text{Pb}) = 26.8$ J/K, (Au) 25.2 J/K, (Al) 24.4 J/K, (C, diamond) 8.62 J/K. (Note that C_p will be slightly greater than C_v—cf. Chapter 2. Also, metals may have some contributions due to the conduction electrons.)

Statistics of a Harmonic Oscillator

A body of mass μ bound in a harmonic potential:

$$V = \tfrac{1}{2}k(x - x_e)^2$$

where $x - x_e$ is the displacement from its minimum potential position, will, according to the theory of Section 11.5, have quantized energy levels given by the formulas:

$$\varepsilon_v = (v + \tfrac{1}{2})h\nu$$

$$v = 0, 1, 2, 3, 4, \ldots, \infty$$

$$\nu = \frac{1}{2\pi}\sqrt{\frac{k}{\mu}} \qquad (15.42)$$

The partition function is:

$$z = \sum_{v=0}^{\infty} e^{-(v+1/2)h\nu/kT}$$

$$= e^{-h\nu/2kT}\sum_{v=0}^{\infty}(e^{-u})^v$$

where $u = h\nu/kT$.

This sum can be evaluated using the Taylor series (Appendix I):

$$\frac{1}{1 - X} = \sum_{n=0}^{\infty} X^n$$

with $X = e^{-u}$. Therefore, the partition function of a harmonic oscillator is:

$$z = \frac{e^{-\varepsilon_0/kT}}{1 - e^{-u}} \qquad \left(\varepsilon_0 = \frac{h\nu}{2}\right) \qquad (15.43)$$

This partition function is referred to a zero of energy at the bottom of the potential well. It will be somewhat more convenient to choose the energy zero to the lowest vibrational energy level ($v = 0$), so that, in effect, the harmonic-oscillator energy is $vh\nu$, and ε_0 of Eq. (15.43) is equal to zero.

The average energy of the harmonic oscillator can be calculated using:

$$\bar{\varepsilon} = \frac{kT^2}{z}\frac{\partial z}{\partial T}$$

The result (for $\varepsilon_0 = 0$) is easily shown to be:

$$\bar{\varepsilon} = \frac{h\nu}{e^u - 1} = \frac{kTu}{e^u - 1} \tag{15.44}$$

where $u = h\nu/kT$.

Exercise: Show that the high-temperature limit of Eq. (15.44) is:

$$\bar{\varepsilon} = kT$$

This is the classical equipartition value. ∎

Einstein's Theory

The first major step in understanding the thermal properties of crystals was made by Einstein. He assumed that the atoms of the crystal were bound to their lattice position by a harmonic potential and could vibrate in three dimensions about that position; in other words, N atoms in a crystal acted like $3N$ harmonic oscillators with an average energy given by Eq. (15.44). For a mole ($N = L$, Avogadro's number):

$$U_m = 3L\bar{\varepsilon} = \frac{3LkTu}{e^u - 1} \tag{15.45}$$

where $u = h\nu_E/kT$; ν_E is the Einstein frequency and is a characteristic of the crystal lattice. The heat capacity is readily shown to be (with $R = Lk$):

$$C_{vm} = \left(\frac{\partial U_m}{\partial T}\right)_v$$

$$C_{vm} = \frac{3Ru^2 e^u}{(e^u - 1)^2} \tag{15.46}$$

The characteristic Einstein frequency is more often expressed as a characteristic temperature, the Einstein temperature:

$$\theta_E = \frac{h\nu_E}{k}, \qquad u = \frac{\theta_E}{T} \tag{15.47}$$

The prediction of the theory that the heat capacity of an atomic solid will be a universal function of θ/T, where θ is an empirical constant chosen for each case, is very close to correct. However Eq. (15.46) is not exact; it works well at high temperatures but rather poorly at low temperatures (Figure 15.7).

Exercise: Find the limit of Eq. (15.47) as T becomes large—that is, as $u \to 0$. (*Answer:* $C_{vm} = 3R$—that is, the Law of Dulong and Petit) ∎

Debye's Theory

Einstein's theory errs in assuming that the vibrations of the atoms in a crystal lattice are independent. This is not so; all the atoms are vibrating at the same time, but

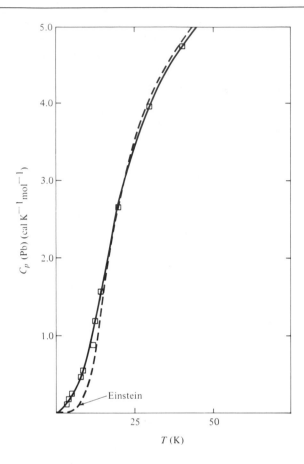

Figure 15.7   The molar heat capacity of lead vs. temperature. The dashed curve shows the theoretical calculation using Einstein's theory. (From Dole, *Introduction to Statistical Thermodynamics,* 1954: New York, Prentice-Hall, Inc., Fig. 10.6, p. 142)

the potential opposing the vibration of two neighbors will be greater if they move toward each other than if they moved away from each other. Debye's theory looks at the vibrations of the lattice as a whole, which result from the motions of the individual atoms.

In preparation, let us look at the standing waves on a beaded string (Figure 15.8). The amplitude oscillations of waves in one dimension are given by the formula:

$$\sin 2\pi(kx - \nu t)$$

where $k = 1/\lambda$ (λ is the wavelength). The wavelength and frequency are related via the wave velocity (c):

$$\nu = \frac{c}{\lambda} = kc$$

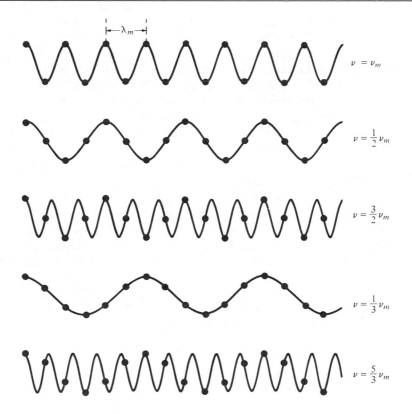

Figure 15.8 Waves on a beaded string. If only the motion of the beads is significant, the top drawing shows the maximum frequency (ν_m) that is physically significant. Subsequent drawings demonstrate that (looking at the beads only) frequencies higher than ν_m are identical to some frequency below ν_m.

The condition for standing waves on a string of length a is:

$$\lambda = \frac{2a}{n}, \qquad k = \frac{n}{2a}$$

where n is an integer.

If the string is continuous with the mass distributed in infinitesimal elements, there is no limit on the value of n or on the number of possible standing waves the string can support. But suppose that all the mass is concentrated in beads, which are separated by a distance d. The maximum meaningful oscillation frequency is that for which the beads move alternately up and down [Figure 15.8(a)]; we shall denote the maximum frequency as ν_m, with the minimum wavelength being:

$$\lambda_m = 2d$$

What about higher frequencies? Figure 15.8 shows a standing wave with $\nu = 1.5\ \nu_m$;

if we ignore the string (which is massless in any event) and look only at the bead displacements, this wave is exactly like $\nu = 0.5\nu_m$. Similarly, a frequency of $\nu = \frac{5}{3}\nu_m$ is indistinguishable from $\nu = \frac{1}{3}\nu_m$. All possible motions of the beads, ignoring the string, can be described in terms of waves with $\nu \le \nu_m$, $\lambda \ge \lambda_m$.

[The phenomenon just discussed is closely related to a problem in communications theory that occurs when a continuous waveform is to be represented by discrete points using an analog-to-digital converter (ADC). The Nyquist theorem states that the sampling rate of the digitizer must be twice the maximum frequency to be represented (corresponding to ν_m — see Figure 15.8). If you wished to make a digital high-fidelity recording with frequencies as high as 25 kHz, it would be necessary to use a 50-kHz digitizer. This fact was first pointed out by Harry Nyquist, a scientist working for AT&T, in his 1923 paper "Certain Factors Affecting Telegraph Speed." The theorem, as originally stated, was the obverse of the statement above. Nyquist showed that the number of discrete signals (bits, for example, the dots and dashes of a telegraphic code) that could be sent through a communications channel per second was equal to twice the bandwidth of the channel. Thus, a 3 kHz telephone line can transmit data at 6000 bits per second (Baud), while a 1 MHz broadcast channel can transmit data at 2,000,000 bits per second.]

If the preceding discussion is extended to a three-dimensional collection of beads, we have a model for the vibrations of an atomic lattice. The direction of a plane wave in three dimensions is specified by a propagation vector:

$$\mathbf{k} = (k_x, k_y, k_z)$$

and the amplitude oscillations are:

$$\sin 2\pi(k_x x + k_y y + k_z z - \nu t)$$

Let us assume a cubic crystal of length a; the condition for standing waves in three dimensions is:

$$k_x = \frac{n_x}{2a}, \qquad k_y = \frac{n_y}{2a}, \qquad k_z = \frac{n_z}{2a}$$

The constants n_x, n_y, and n_z must be integers. The wavelength of this wave is:

$$\frac{1}{\lambda} = \sqrt{k_x^2 + k_y^2 + k_z^2} = \frac{\sqrt{n_x^2 + n_y^2 + n_z^2}}{2a} \qquad \textbf{(15.48)}$$

With the definition:

$$n = \sqrt{n_x^2 + n_y^2 + n_z^2} \qquad \textbf{(15.49)}$$

(n is not necessarily an integer) the condition for standing waves is:

$$n = \frac{2a}{\lambda} \qquad \textbf{(15.50)}$$

If the crystal is large, n can be treated as a continuous variable, and the derivation used earlier to derive Eq. (15.25) will show that the number of waves between n and $n + dn$ is:

$$g(n)\,dn = \frac{\pi}{2}n^2\,dn$$

This can be used to derive the distribution function for wavelength; with Eq. (15.50):

$$g(n)\,dn = \frac{4\pi a^3}{\lambda^2}d\left(\frac{1}{\lambda}\right) \tag{15.51}$$

We prefer to deal with frequency. The propagation of waves in a solid material can be described in terms of three types of waves, one *longitudinal* (velocity c_l) and two *transverse* (velocity c_t). (For a string, Figure 15.8 illustrates one of the transverse waves; the other is for oscillations of the beads perpendicular to the paper. The longitudinal wave would keep the beads in line, but their separation would oscillate periodically; this is also called a *compressional* wave.) The wavelengths of these waves are:

$$\lambda_t = \frac{c_t}{\nu}, \qquad \lambda_l = \frac{c_l}{\nu}$$

Use of these relations in Eq. (15.51) gives the distribution function for frequencies between ν and $\nu + d\nu$ as:

$$g(\nu)\,d\nu = 4\pi a^3\left(\frac{2}{c_t^2} + \frac{1}{c_l^2}\right)\nu^2\,d\nu \tag{15.52}$$

The total number of vibrational modes of motion:

$$\int_0^{\nu_m} g(\nu)\,d\nu$$

must be the same as the total number of individual atomic vibrations — $3N$ for N atoms. Integration of Eq. (15.52) up to the maximum frequency (ν_m) gives:

$$3N = 4\pi a^3\left(\frac{2}{c_t^2} + \frac{1}{c_l^2}\right)\frac{\nu_m^3}{3}$$

This relationship is used to eliminate the wave velocities and crystal size (a^3) from Eq. (15.52) with the result:

$$g(\nu)\,d\nu = \frac{9N}{\nu_m^3}\nu^2\,d\nu \tag{15.53}$$

This can be converted into a distribution function for energy by multiplying by the average energy of a wave of frequency ν; assuming that each wave is a harmonic oscillator, this will be $\bar{\varepsilon}$ of Eq. (15.44):

$$dE = \bar{\varepsilon}g(\nu)\,d\nu = \frac{9Nh}{\nu_m^3}\frac{\nu^3\,d\nu}{e^{h\nu/kT} - 1}$$

With the usual substitution of $u = h\nu/kT$, this becomes:

$$dE = \frac{9Nk^4T^4}{h^3\nu_m^3}\frac{u^3\,du}{e^u - 1} \tag{15.54}$$

This is usually written in terms of the *Debye temperature:*

$$\theta_D = \frac{h\nu_m}{k}$$

$$dE = 9NkT\left(\frac{T}{\theta_D}\right)^3 \frac{u^3\, du}{e^u - 1} \tag{15.55}$$

(Note the contrast with Einstein's theory; Einstein assumes a *single* lattice vibration constant ν_E, while Debye assumes a *distribution* of frequencies with a maximum ν_m.)

The total energy summed over all frequencies, $0 < \nu < \nu_m$, will be the internal energy; it can be calculated by integrating Eq. (15.55) between:

$$0 < u < \frac{\theta_D}{T}$$

For one mole ($N = L$, $R = Lk$):

$$U_m = 9RT\left(\frac{T^3}{\theta_D^3}\right) \int_0^{\theta_D/T} \frac{u^3\, du}{e^u - 1} \tag{15.56}$$

The integral of Eq. (15.56) must be done numerically; values are tabulated in some advanced texts.

The derivation of the heat capacity in this theory starts with the average energy per vibration, Eq. (15.44). The heat capacity per vibration is $d\bar{\varepsilon}/dT$; this can be shown to be:

$$\frac{d\bar{\varepsilon}}{dT} = \frac{ku^2 e^u}{(e^u - 1)^2}$$

The Debye theory then gives:

$$C_v = \int_0^{\nu_m} \frac{d\bar{\varepsilon}}{dT} g(\nu)\, d\nu$$

The result for one mole ($N = L$) is:

$$C_{vm} = 3RD\,(\theta_D/T) \tag{15.57}$$

where D is the Debye heat-capacity integral:

$$D(\theta_D/T) = 3\left(\frac{T}{\theta_D}\right)^3 \int_0^{\theta_D/T} \frac{u^4 e^u\, du}{(e^u - 1)^2} \tag{15.58}$$

The integral of Eq. (15.58) must be done numerically. Values are tabulated in many advanced texts (see Figure 15.10). The Debye heat capacity fits experimental data very well with the appropriate choice for the constant θ_D; cf. Figure 15.9 and Table 15.4.

The low-temperature limit of the heat capacity is very interesting. This is most easily calculated starting with Eq. (15.56) for the internal energy. If $T << \theta_D$, the limit of the integral will be very large and can be extended to infinity without error:

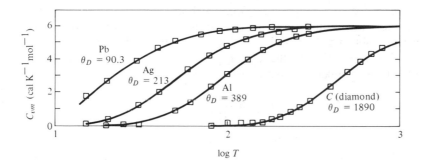

Figure 15.9 Heat capacities and the Debye theory. The heat capacities of four atomic solids are shown vs. log (T), together with theoretical values (solid lines) calculate using Debye's heat-capacity theory. (From Lewis et al. *Thermodynamics*, p. 56)

Table 15.4   Heat-capacity constants for solids

	Metals		Nonmetals	
	θ_D (K)	$(10^{+3}\gamma/R)(K^{-1})$		θ_D (K)
Ag	229	7.30–8.05	Diamond	1860
Al	375	16.5–17.5	KCl	227
Be	1160	2.7	NaCl	281
Na	160	21.6	AgCl	183
Ni	413	87.6	AgBr	144
Pt	233	81–83	CaF_2	474
Tl	430	40–43	FeS_2	645
Zn	235	6.3–7.6		

$$U_m(\text{low } T) = \frac{9RT^4}{\theta_D^3} \int_0^\infty \frac{u^3 \, du}{e^u - 1}$$

The required integral can be found on some of the more extensive integral tables:

$$\int_0^\infty \frac{u^3 \, du}{e^u - 1} = \frac{\pi^4}{15}$$

Using this, the heat capacity can be calculated as the derivative of U_m with the result

$$(\text{low } T) \qquad C_{vm} = \left(\frac{36\pi^4 R}{15\theta_D^3}\right) T^3 \qquad \textbf{(15.59)}$$

This is the famous Debye T^3 law (cf. Chapter 3) and is reasonably accurate for temperatures less than $\theta_D/10$.

The T^3 law is not accurate for all cases. For example, crystals such as graphite or boron nitride (BN) have a two-dimensional planar structure: a similar derivation to that above for standing waves in two dimensions shows that this structure should obey a T-squared law at low temperature (this has been verified experimentally). (This, of course, refers to BN in its graphite-like form; BN also has a diamond-like form that is used as an abrasive.) Also, the conduction electrons of metals give a heat-capacity contribution in addition to the Debye heat capacity that is linear in T; their heat capacity at low temperature is:

$$C_{vm} = \frac{36\pi^4 R}{15\theta_D^3} T^3 + \gamma T \tag{15.60}$$

Some values of the constant γ are given in Table 15.4.

As mentioned earlier, molecular solids may have heat-capacity contributions due to internal molecular motions; however, these will normally "freeze out" (become inactive) at low temperature, so the T^3 law may still be obeyed.

The limiting form of Eq. (15.57) at high temperature can be shown to give:

$$(T > \theta_D) \qquad C_{vm} = 3R\left(1 - \frac{\theta_D^2}{20T^2}\right) \tag{15.61}$$

The law of Dulong and Petit is obeyed when $T \gg \theta_D$ (cf. Problem 15.14).

For practical purposes, the Debye heat capacity is readily estimated when the characteristic temperature (θ_D) is known. (Similarly, measured heat capacities can be used to calculate θ_D.) At low temperature ($T \ll \theta_D$) Eq. (15.59) can be used. At high temperatures ($T > \theta_D$) Eq. (15.61) is accurate. For intermediate cases, Figure 15.10 can be used to calculate the heat-capacity integral needed in Eq. (15.57).

Debye's theory is clearly more realistic than Einstein's. However, the Einstein formulas (for other thermodynamic functions as well as heat capacity) are much simpler and easier to use and are reasonably accurate except at very low temperatures. Unfortunately, Einstein constants are not tabulated as widely as the Debye constants, but the estimate:

$$\theta_E \cong \left(\tfrac{3}{4}\right)\theta_D$$

(derived in Problem 15.17) works reasonably well.

15.8   Black-Body Radiation

A hot body may cool by emitting heat in the form of electromagnetic radiation. At moderate temperatures this radiation is primarily at wavelengths longer than visible light — that is, infrared. At higher temperatures the radiant energy shifts to shorter wavelengths and the body will glow red. At still higher temperatures the emitted radiation appears white, then blue.

In addition to the shift to shorter wavelengths, the total energy emitted increases with temperature. The total energy (at all wavelengths) emitted per unit area per unit time (e) is related to temperature by the Stefan-Boltzmann law:

$$e(T) = \sigma T^4 \tag{15.62}$$

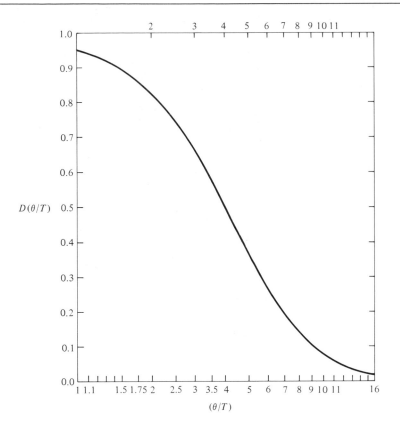

Figure 15.10 The Debye heat-capacity function. This graph can be used to calculate heat capacities of atomic solids for which the Debye characteristic temperature (θ) is known. Similarly, a single measurement of such a heat capacity (provided it is made in the temperature range where the heat capacity is changing significantly) can be used to calculate the Debye function (D, vertical scale) using Eq. (15.57); then this graph will give the value of the characteristic temperature and, hence, the heat capacity at other temperatures.

Experimentally, the Stefan-Boltzmann constant has a value:

$$\sigma = 5.67 \times 10^{-8} \quad \text{watts m}^{-2} \text{ K}^{-4}$$

This relationship was discovered empirically by J. Stefan (1879) and derived from thermodynamic considerations by L. Boltzmann (1884).

The radiation emitted by solid bodies is discussed in terms of an idealized object called a *black body,* which will absorb all radiation at any wavelength that strikes it. Experimentally and theoretically, a perfect black body can be closely approximated by a *hohlraum,* a hole in a solid material with black walls at a uniform temperature (Figure 15.11). The radiant energy in a hohlraum is studied via a small hole in the side.

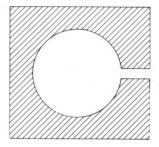

Figure 15.11   A hohlraum. A cavity with black walls inside a solid material behaves like a black body; the hole in the side is for observing the emitted radiation.

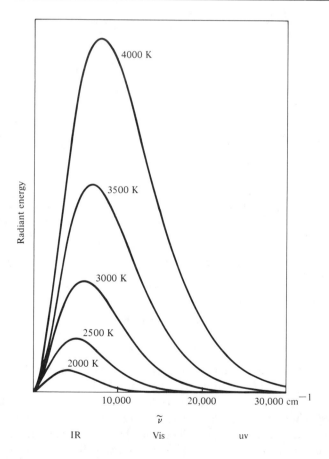

Figure 15.12   Radiant energy from a black body at various temperatures vs. wavenumber.

Some results for the frequency distribution of this energy are shown in Figure 15.12; note the peak, which shifts to shorter wavelength as the temperature is increased.

The explanation of the black-body radiation pattern was a major problem for physicists toward the end of the nineteenth century, and this was one of the problems that gave birth to the quantum theory. The rigorous explanation of this phenomenon treats the radiation in the hohlraum as a gas of massless particles called *photons*. Each photon is characterized by a frequency (ν) and has an energy

$$\varepsilon(\text{photon}) = h\nu$$

This derivation is rather involved and well beyond the scope of this chapter. In the first place, the number of photons in the hohlraum does not remain constant when the temperature is changed, so a grand canonical ensemble must be used. Also, photons are bosons, and the Boltzmann statistics we have used so far must be replaced by Bose-Einstein statistics. However, these theoretical developments were also beyond Max Planck, who first derived the correct black-body radiation law, and it is his derivation that we shall follow.

The Rayleigh-Jeans Law

Suppose we have a cubical cavity (hohlraum) of length a (volume, a^3). The derivation of the standing waves in this cavity is exactly that which we used earlier for the Debye theory of crystals (but the Rayleigh-Jeans derivation preceded Debye's by several years). A major difference is that with electromagnetic waves in a vacuum, there is no maximum frequency. Also electromagnetic waves must be transverse waves; there is no longitudinally polarized wave but only two transverse waves, polarized perpendicular to each other.

The condition for standing waves is:

$$\lambda = \frac{2a}{n}$$

where:

$$n = \sqrt{n_x^2 + n_y^2 + n_z^2}$$

and n_x, n_y, and n_z are integers. Treating n as a continuous variable, the number of such waves between n and $n + dn$ is given by Eq. (15.25):

$$g(n)\,dn = \frac{\pi}{2}n^2\,dn$$

Introducing the frequency, $\nu = c/\lambda$ for both the electric and magnetic waves, we get:

$$g(\nu)\,d\nu = 2\left(\frac{\pi}{2}\right)\left(\frac{2a\nu}{c}\right)^2\left(\frac{2a}{c}\right)dv$$

$$= \frac{8\pi a^3}{c^3}\nu^2\,dv$$

The total radiant energy density per unit volume will be:

$$\rho(\nu)\, d\nu = \frac{\bar{\varepsilon} g(\nu)\, d\nu}{V}$$

with the volume $V = a^3$ and the average energy per standing wave, $\bar{\varepsilon}$.

$$\rho(\nu)\, d\nu = \frac{8\pi}{c^3} \bar{\varepsilon} \nu^2\, d\nu \qquad\qquad (15.63)$$

So far, so good—the problem lies in the value of the average energy per standing wave, $\bar{\varepsilon}$. Rayleigh and Jeans assumed that this quantity would have the classical equipartition value:

$$\bar{\varepsilon} = kT$$

If this is substituted into Eq. (15.63), the Rayleigh-Jeans law results; this formula predicts that the radiation density at high frequency (short wavelengths) becomes *infinite*. This prediction, having firm theoretical grounds but contradicting both the experimental results and common sense, was called the "ultraviolet catastrophe."

Planck's Radiation Law

Planck avoided the ultraviolet catastrophe by assuming that each standing wave in the cavity was associated with some sort of oscillator in the wall that had a quantized energy:

$$\varepsilon_n = nh\nu$$

Then the derivation that led to Eq. (15.44) gives:

$$\bar{\varepsilon} = \frac{h\nu}{e^{h\nu/kT} - 1}$$

It is difficult to assign any physical reality to Planck's oscillators (the rigorous derivation does not need them), and it was this difficulty, probably, that limited the impact of Planck's theory on his contemporaries; the significance of this development was not realized until later. (Perhaps even Planck was not convinced; he never, until his dying day, fully accepted the quantum theory.)

If Eq. (15.44) is used in Eq. (15.63), the result is:

$$\rho(\nu)\, d\nu = \frac{8\pi k^4 T^4}{h^3 c^3} \frac{u^3\, du}{e^u - 1} \qquad\qquad (15.64)$$

where

$$u = \frac{h\nu}{kT}$$

This distribution function is in exact accord with the experimentally measured frequency distribution of the radiant energy.

The total radiant energy per unit volume (E/V) can be calculated from Eq. (15.64) by integrating over all frequencies:

$$0 < \nu < \infty$$

using the result:

$$\int_0^\infty \frac{x^3\,dx}{e^x - 1} = \frac{\pi^4}{15}$$

This is easily shown to be:

$$\frac{E}{V} = \frac{8\pi^5 k^4 T^4}{15 h^3 c^3} \qquad \textbf{(15.65)}$$

The radiation emitted by (or absorbed by) a black body per unit surface area (e) can be identified as the number of photons escaping through an aperture in the side of the hohlraum. This derivation is exactly analogous to that of Section 1.6 for the number of collisions of a gas with a wall (and Knudsen's formula). The number of collisions per unit area per unit time of a gas with a wall (Z_{wall}) is related to the particle density ($n^* = N/V$) and the average velocity by:

$$Z_{wall} = \frac{1}{4}\left(\frac{N}{V}\right)\bar{v}$$

[cf. Eq. (9.49)]. Now, the particles are photons and N/V is replaced by $\rho(\nu)$; also we replace the average velocity with the velocity of light (c). The radiant energy per unit area per unit time is then:

$$e(T) = \int_0^\infty \left(\frac{1}{4}\right)\rho(\nu)c\,d\nu = \left(\frac{c}{4}\right)\frac{E}{V}$$

With Eq. (15.65), this gives:

$$e(T) = \left(\frac{2\pi^5 k^4}{15 h^3 c^2}\right)T^4 \qquad \textbf{(15.66)}$$

This is the **Stefan-Boltzmann law** (cf. Problem 15.18).

Postscript

Aside from the more or less obvious lessons of this chapter, that statistical mechanics is a useful formalism for describing macroscopic properties in terms of microscopic properties (particle mechanics), several illustrations of the unity of scientific theory have been seen. The harmonic oscillator model, used in Chapter 5 to calculate thermodynamic properties of gases, and in Chapters 13 and 14 to explain molecular spectra, was used in this chapter to describe the heat capacity of solids and black-body radiation. Likewise, the formula for the density of integer points in space, Eq. (15.25), found application for evaluating the translational partition function, in the theory of the heat capacity of crystals, and in explaining the black-body radiation. Even more impressive is the generality of the idea, developed by Debye, that there had to be a maximum frequency for lattice vibrations in crystals; the same idea, developed independently a decade later by Nyquist, became the cornerstone of communications theory.

References

1. N. Davidson, *Statistical Mechanics,* 1962: New York, McGraw-Hill Book Co.
2. T. L. Hill, *Statistical Mechanics,* 1956: New York, McGraw-Hill Book Co.
Also see references at the end of Chapter 5.

Problems

15.1   The average translational energy of a particle can be defined with the state density function, $g(\varepsilon)$:

$$\langle \varepsilon \rangle = \frac{\displaystyle\int_0^\infty \varepsilon g(\varepsilon) e^{-\varepsilon/kT} \, d\varepsilon}{\displaystyle\int_0^\infty g(\varepsilon) e^{-\varepsilon/kT} \, d\varepsilon}$$

Use this to show that $\langle \varepsilon \rangle = \frac{3}{2}kT$, the equipartition value.

15.2   Use Eq. (15.26) to derive the Maxwell-Boltzmann law [Eq. (1.36)] for the distribution of molecular velocities of particles with energy:

$$\varepsilon = \tfrac{1}{2}mv^2$$

15.3   Calculate the contribution of nuclear spins to the molar entropy of $^{11}BF_3$.

15.4   Calculate the contribution of nuclear spins to the molar entropy of $^{31}PH_3$.

15.5   Calculate the value of the partition function, z_{nsr}, for H_2 at 100 K; compare your answer to the approximate value

$$z_{nsr} = \frac{g_N T}{\sigma \theta_r}$$

(Use $\theta_r = 85.35$ K.) Repeat the calculation at 500 K.

15.6   Calculate the percent *ortho* hydrogen in an equilibrium mixture at 200 K. (Use $\theta_r = 85.35$ K.)

15.7   Calculate the rotational internal energy per mole of *para* hydrogen at $T = 200$ K. Use $\theta_r = 85.35$ K and compare your answer to the equipartition value.

15.8   The deuterium nucleus has a spin $I = 1$. What is the nuclear spin-rotation partition function for D_2? What is the high-temperature limit?

15.9   Derive a formula for the S- and O-branch frequencies of a rotational Raman spectrum with exciting line $\bar{\nu}_i$.

15.10 What would be the ratio of spin statistical weights (observable from alternation of Raman intensities) for a homonuclear diatomic molecule with $I = \frac{5}{2}$? (Assume $^1\Sigma_g^+$ electronic state.)

15.11 What will be the ratio of the statistical-nuclear spin weights for the odd J/even J levels of a homonuclear diatomic molecule with $I = \frac{3}{2}$? (Assume the electronic state is $^1\Sigma_g^+$.)

15.12 For a rotational Raman spectrum of a homonuclear diatomic molecule, what are the ratios of the differences ($|\bar{\nu}_s - \bar{\nu}_i|$) of the second side band to the first if (a) all J's are present, (b) only even J's are present, (c) only odd J's are present?

15.13 The Raman spectrum of a homonuclear diatomic molecule is analyzed to give the rotational constant, $\theta_r = 3.00$ K. Analyze the S-branch relative intensity pattern below (for $T = 300$ K) to determine the nuclear spin.

J''	Rel. int.	J''	Rel. int.
0	1.00	4	7.37
1	4.90	5	13.6
2	4.71	6	8.54
3	10.4	7	14.2

15.14 Show that the Debye formula for the internal energy (per mole) of a crystal approaches $U_m = 3RT$ and that the heat capacity approaches $C_{vm} = 3R$ (the law of Dulong and Petit) as $T \rightarrow \infty$. [*Hint:* Use $e^u \cong 1 + u$ as $u \rightarrow 0$.]

15.15 The heat capacity of aluminum oxide at 5 K is $C_{pm} = 0.0012$ J/K. Use this value to estimate the Debye temperature for this material. Use this value to estimate the heat capacity at 10 K (obs. 0.0094 J/K).

15.16 Calculate the molar heat capacity (constant volume) of NaCl at 10 K, 100 K, and 300 K, using the Debye theory.

15.17 The Einstein formula is considerably more convenient to use than Debye's, but its constants (θ_E) are not usually tabulated. Show that if the Einstein frequency (ν_E) is interpreted as the average of the Debye frequencies, then:

$$\theta_E = \tfrac{3}{4}\theta_D$$

Compare these theories by calculating the heat capacity for $\theta_D/T = 3$.

15.18 Calculate the Stefan-Boltzmann constant (σ) from Eq. (15.66).

15.19 Show that the frequency at the maximum of the radiation density from a black body is:

$$\nu_{max} = \frac{2.821439kT}{h}$$

(This is one form of the Wien displacement law.)

15.20 Derive the black-body radiation law for wavelength, $\rho(\lambda)$.

APPENDIX I
Numerical Methods

This appendix discusses briefly a number of mathematical methods that can be used to solve various types of problems that occur in physical chemistry, particularly those involved in data treatment. For the most part, methods are limited to ones that can be used handily with a simple (nonprogrammable) calculator. If a programmable device such as a computer is to be used, there often are better methods available, and a text on numerical analysis should be consulted.

AI.1 Functions

When two quantities are related in such a way that a change in the value of one necessitates a change in the value of another, they are said to be functionally related. For example, the length of a metal bar is a function of its temperature, the pressure of a gas is a function of its volume, and so on.

A functional relationship can be expressed as a data table, a graph, or an equation. A data table may be generated from a series of discrete measurements — for example, if you measure the length (L) of a metal bar at a series of temperatures and get the results:

t (°C)	L (mm)
0	800.00
16	802.09
20	802.60
25	803.25
30	803.90

This table defines, for five points, a functional relationship, $L(t)$.

A graph can be obtained from a table by graphing the points and connecting them with a smooth curve (which, if some experimental error is possible, need not pass through all points). The data of the preceding example would fit very closely on a straight line. Graphical functions may also be generated directly if the experimental apparatus is equipped with an electrical recorder and an appropriate transducer.

The most common representation of a function, in mathematics texts at least, is an equation — for example,

$$f(x) = (3x^2 - 1)e^{-x^2}$$

Going from an equation to a graph or table is usually straightforward; the reverse process is not. Finding an equation to represent a data set is called curve fitting. This process usually starts with a proposed form for the equation containing one or more undetermined constants; this equation may have a theoretical basis or be totally empirical. Values are then chosen for the constants to make the calculated function pass as closely as possible through the observed points. This procedure will be discussed in more detail in Section AI.6.

An example of a semiempirical equation — one with some theoretical basis but with constants chosen to "fit" the experimental data — is:

$$\ln P = A + \frac{B}{T} + C \ln T \tag{AI.1a}$$

This equation could be used to represent the vapor pressure (P) of a liquid as a function of temperature (T). The same data set could also be fitted using a totally empirical equation, such as:

$$\log P = A + \frac{B}{T + C} \tag{AI.1b}$$

where again (A, B, C) are constants [but would have different values than they would in Eq. (AI.1a)].

Similarly, the heat capacity of a diatomic molecule can be represented by the theoretical equation:

$$C_{pm} = \tfrac{7}{2}R + \frac{R(\theta/T)^2 e^{\theta/T}}{(e^{\theta/T} - 1)^2} \tag{AI.2a}$$

where R is the gas constant and θ is a constant characteristic of the molecule (measurable by other methods). Such heat capacities may also be represented with a totally empirical equation:

$$C_{pm} = A + BT + CT^2 + DT^3 + ET^4 \tag{AI.2b}$$

Both of these equations are approximations that fit the observed data under limited circumstances; in the case of Eq. (AI.2b) this is typically $298 < T < 2000$ K.

Figure AI.1 illustrates some important properties of functions, including roots, area and slope, which will be discussed in subsequent sections.

AI.2 Power Series

Any analytic function can be fitted by an infinite power series:

$$f(x) = \sum_{n=0}^{\infty} a_n x^n = a_0 + a_1 x + a_2 x^2 + \cdots \tag{AI.3}$$

As a method of representing empirical data, this series is not useful unless it can be truncated after only a few terms (giving a *polynomial*). Equation (AI.2b) is an example of such a series. In order to be useful, a series representation must converge rapidly, either because the coefficients become small, $a_{n+1} \ll a_n$, or because $x \ll 1$ so succeeding powers become smaller and smaller.

It is often useful to represent equations as power series. This can be done using the Taylor series expansion about some point x_0, if the derivatives at that point are known

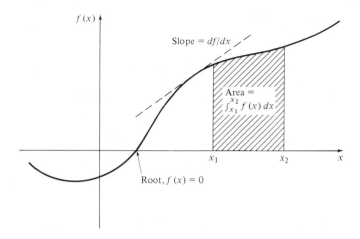

Figure AI.1 A function of one variable. This diagram illustrates three mathematical terms: (1) The *root* of a function $f(x)$ is the value of x for which $f(x) = 0$. (2) The derivative of the function, df/dx, at some value of x is the slope of the tangent line at that point. (3) The integral of the function from x_1 to x_2 is the area under the function between those coordinates.

or can be calculated:

$$f(x) = f(x_0) + \left(\frac{df}{dx}\right)_{x_0} (x - x_0) + \frac{1}{2}\left(\frac{d^2f}{dx^2}\right)_{x_0} (x - x_0)^2 + \cdots$$

$$+ \frac{1}{n!}\left(\frac{d^n f}{dx^n}\right)_{x_0} (x - x_0)^n \qquad \textbf{(AI.4)}$$

Several useful expansions follow:

$$e^x = 1 + x + \frac{x^2}{2!} + \frac{x^3}{3!} + \frac{x^4}{4!} + \cdots \qquad \textbf{(AI.5)}$$

$$\ln(1 + x) = x - \frac{x^2}{2} + \frac{x^3}{3} - \frac{x^4}{4} + \cdots \qquad (|x| \le 1) \quad \textbf{(AI.6)}$$

$$\frac{1}{1 + x} = 1 - x + x^2 - x^3 + x^4 - \cdots \qquad (|x| < 1) \quad \textbf{(AI.7)}$$

$$(1 + x)^{1/2} = 1 + \frac{x}{2} - \frac{x^2}{8} + \frac{x^3}{10} - \frac{5x^4}{128} + \frac{7x^5}{256} - \frac{21x^6}{1024} + \cdots \quad (|x| < 1) \quad \textbf{(AI.8)}$$

These expansions are most useful for $|x| \ll 1$ when they can be truncated after only a few terms.

Example: Series expansions are not usually used for computation, but a numerical example may help to show the limits of their utility. Let us estimate a reciprocal with three terms of Eq. (AI.7):

$$\frac{1}{1-x} \cong 1 + x + x^2$$

For several values of x, to five significant figures, we can calculate:

x	$1+x$	$1+x+x^2$	Exact
0.01	1.0100	1.0101	1.0101 ...
0.05	1.0500	1.0525	1.0526 ...
0.10	1.1000	1.1100	1.1111 ...
0.20	1.2000	1.2400	1.2500 ...
0.30	1.3000	1.3900	1.4286 ...

∎

Exercise: Calculate the value of the base of natural logarithms, e, using Eq. (AI.5). The exact value is $e = 2.718281828459045$. . . . ∎

Exercise: Find the limit of Eq. (AI.2a) as T becomes large. This could be done using L'Hopital's rule, but an expansion $e^x \sim 1 + x$ as x ($= \theta/T$) becomes small will work as well. (*Answer:* $9R/2$.) ∎

AI.3   Roots

Finding numerical roots is a technique of frequent value, but it is not always obvious when it should be used. Suppose we have a variable y that is related to x as:

$$y = \frac{3}{x} + \ln x + 17x^2$$

There is no problem to find the value of y for any given x, but what if we wish to find the value of x for which $y = 42$? We could do this by guessing — that is, keep trying values of x until one works. But there is an easier and faster way: state the problem as a function whose roots must be found — for example:

$$y = \frac{3}{x} + \ln x + 17x^2 = 42; \qquad x = ?$$

$$f(x) = \frac{3}{x} + \ln x + 17x^2 - 42 = 0$$

This now becomes a problem in finding a *root* of the function $f(x)$. (*Answer:* $x = 1.5264669$ gives $y = 42$.) No one who has passed high school algebra should have difficulty in finding the roots of:

$$f(x) = x^2 - 2x + 1 = 0$$

(*Answer:* $x = 1$); but finding the roots of the earlier equation is hardly more difficult than this if you have a calculator. Next we shall discuss two methods for finding numerical roots: Newton-Raphson and *regula falsi*.

Newton-Raphson Method

The Newton-Raphson method starts with an approximate value of the root (x_n) from which you calculate the value of the function $f(x_n)$ and its derivative $f'(x_n) = (df/dx)$. The next estimate of the root is then:

$$x_{n+1} = x_n - \frac{f(x_n)}{f'(x_n)} \qquad \textbf{(AI.9)}$$

This method will not converge unless your estimate is on the near side of any maximum or minimum of the function (Figure AI.2). For this reason it is important to "explore" the function by calculating $f(x)$ for a range of x to find the approximate value of the root. Often the physical situation will dictate the likely range of x; for example, in an equilibrium problem, if x represents a mole fraction, then clearly $0 < x < 1$. Convergence means that the successive values of x are not changing to the number of significant figures required; it does not require $f(x) = 0$ in any literal sense.

Example: Find a positive root (three significant figures) for

$$f(x) = x^3 + 302x - 6.52 \times 10^6 = 0$$

$$f' = 3x^2 + 302$$

It is useful to make a table of the results. First we "explore" the function.

x	$f(x)$	$f'(x)$	Comments
10	-5.52×10^6		—arbitrary first guess; since $f(0)$ is also negative, a larger value of x seems appropriate.
20	-6.51×10^6		—progress slow, try a much larger value.
100	-5.49×10^6		—still slow, try a bigger jump.
500	1.19×10^8		—at last $f(x)$ changes sign; there is a root between 100 and 500.
300	2.05×10^7		—guess again.
200	1.54×10^5	1.20×10^5	—now iterate with Eq. (AI.9); next $x =$ $200 - (1.54 \times 10^5)/(1.20 \times 10^5) = 187.$
187	7.57×10^4	1.05×10^5	
186	2.99×10^2	1.04×10^5	

The answer has obviously converged, even though $f(x) = 299$, not exactly zero; it is rarely necessary to get a precise result. [In this case, on a 13-digit calculator, $x = 186.27792$, $f(x) = -3 \times 10^{-6}$.] ∎

Exercise: Find a root $0 < x < 1$ for:

$$f(x) = x^3 + 3x^2 + 2x - 4$$

(*Answer:* $x = 0.796322$.) ∎

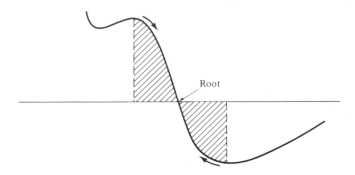

Figure AI.2 Region of convergence of the Newton-Raphson iteration. The Newton-Raphson method for finding roots will work only if the initial guess is nearer to the root than the nearest extremum (maximum, minimum). Inflection points may also cause convergence problems.

Exercise: Find the value of x for which $x = \cos x$. (Note that x must be in *radians*.) (*Answer:* $x = 0.7390851332$.)   ∎

Regula Falsi Method

The *regula falsi* method is a bit more cumbersome but may be preferred for functions that are difficult to differentiate or for programmed computations. It is also more likely to converge from "way out" than the Newton-Raphson method. First, explore the function until you find two values (x_0 and x_n) for which $f(x)$ has opposite signs. One of these (x_0) is called the pivot; initially the pivot may be either one. The next approximation of the root is:

$$x_{n+1} = \frac{x_n f_0 - x_0 f_n}{f_0 - f_n} \qquad \textbf{(AI.10)}$$

where $f_0 = f(x_0)$ and $f_n = f(x_n)$. This is best explained with an example.

Example: Find a root of $f(x) = x^3 + 3x^2 + 2x - 4$ for $0 < x < 1$. We readily calculate $f(0) = -4$, $f(1) = 2$. This fulfills our requirement that f_n and f_0 have opposite signs. Using either point as the pivot, we calculate the next approximation:

$$x_2 = \frac{1(-4) - 0(2)}{-4 - 2} = \frac{2}{3}$$

Then we calculate the value of the function at this point:

$$f\left(\frac{2}{3}\right) = -1.037037$$

We now choose the opposite point (the one for which f has the opposite sign) as pivot; that is, let $x_0 = 1$, $f_0 = 2$. Our iterative formula is then, from Eq. (AI.10):

$$x_{n+1} = \frac{2x_n - f_n}{2 - f_n}$$

The next several rounds of approximations are:

$$x_3 = 0.780488, \quad f_3 = -0.136098$$

$$x_4 = 0.794474, \quad f_4 = -0.016025$$

This is clearly converging, so let's write out the next step to see why:

$$x_5 = \frac{2(0.794474) - (-0.016025)}{2 - (-0.016025)} = 0.796107$$

As f_n gets smaller, the iterative formula is reducing to $x_{n+1} = x_n$. The reader should check the final iterations: $x_6 = 0.796297$, $x_7 = 0.796319$, $x_8 = 0.796322$, $x_9 = 0.796322$. ∎

Exercise: Find a root $0 < x < 1$ of $f(x) = x^3 + 2.7x^2 + 3.1x - 4.75$ by the *regula falsi* method. (*Answer:* 0.803206.) ∎

It is permissible to switch pivots during an iteration, and sometimes this may speed convergence. However, if the pivot is too close to the root, the denominator $(f_0 - f_n)$ may become too small for accurate computation. Usually the *regula falsi* will converge more surely, but more slowly, than the Newton-Raphson method. The *regula falsi* method is generally the best for use with computers; with a calculator, the Newton-Raphson method is usually easier.

AI.4 Integration

Numerical integration is required in two types of situations: (1) integration of a data table, or (2) integration of certain formula functions that cannot be done in closed form. A very simple function that cannot be integrated (between finite limits) without a numerical method is:

$$f(x) = e^{-x^2}$$

Numerical integration begins with a set of base points x_0, x_1, x_2, ..., for which the corresponding function values $f_0, f_1, f_2, \ldots$ are known. These may be calculated from a formula or be a set of experimental data. Some methods require that the base points be equally spaced—that is,

$$h = x_1 - x_0 = x_2 - x_1 = x_3 - x_2 = \ldots$$

and some do not. If you are integrating a formula, there is no reason not to choose equally spaced base points, but for data tables this may not be convenient.

Trapezoidal Rule

The simplest method is the trapezoidal rule, which uses a linear interpolation between base points:

$$\int_{x_0}^{x_1} f(x)\, dx \cong \tfrac{1}{2}(f_1 + f_0)(x_1 - x_0) \tag{AI.11a}$$

Note that $\tfrac{1}{2}(f_1 + f_0)$ is simply the average of that pair of points. This method is readily extended to multiple intervals, whether or not the points are equally spaced, by simply calculating the area of each segment and adding the results (Figure AI.3).

Example: From the table of specific heats below (for ice), calculate the heat:

$$q = \int_{T_1}^{T_2} c\, dT$$

that must be removed to cool ice from 0°C to −200°C.

t (°C)	c (cal/g deg)	Ave.	Δt	Product
−200	0.156			
		0.201	50	10.05
−150	0.246			
		0.289	50	14.45
−100	0.332			
		0.3835	60	23.01
− 40	0.435			
		0.4635	40	18.54
0	0.492			
			sum =	66.05 cal/g

Thus, 66 cal of heat are required per gram. The more accurate three-point rule [below, Eq. (AI.14)] gives 66.22 for this answer. ∎

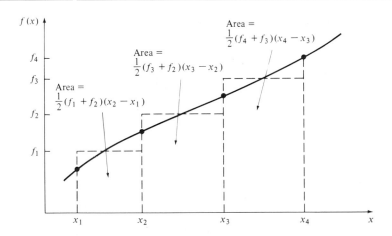

Figure AI.3   The trapezoidal rule. The area between x_1 and x_2 is approximately the area of a rectangle whose height is the average of the values of the function (f_1 and f_2) at those points. The total area (x_1 to x_4 in this case) is approximated by adding the estimates made by this method for each of the intervals.

For a series of equal-sized intervals (denoted h), it may be apparent that repeated use of the trapezoidal rule gives:

$$\int_{x_0}^{x_n} f(x)\,dz = \left(\frac{f_0}{2} + f_1 + f_2 + \cdots + f_{n-1} + \frac{f_n}{2}\right) h \qquad \textbf{(AI.11b)}$$

Simpson's Rule

If the data are available at constant increments of the independent variable, Simpson's rule may be used for greater accuracy; however, this rule requires an even number of intervals (an odd number of base points); it is:

(3 points):

$$\int_{x_0}^{x_2} f(x)\,dx = \frac{h}{3}[f_0 + 4f_1 + f_2]$$

(5 points):

$$\int_{x_0}^{x_4} f(x)\,dx = \frac{h}{3}[f_0 + 4f_1 + 2f_2 + 4f_3 + f_4] \qquad \textbf{(AI.12)}$$

($n + 1$ points, n even):

$$\int_{x_0}^{x_n} f(x)\,dx = \frac{h}{3}[f_0 + 4f_1 + 2f_2 + 4f_3 + 2f_4 + \cdots + 4f_{n-1} + f_n]$$

If, when integrating a tabular function, it happens that you have an even number of base points, the trapezoidal rule could be used for either the first or last interval, and Simpson's rule for the rest. [Which should you choose for trapezoidal? Choose either the one that contributes least to the integral or the one in which $f(x)$ changes the least; this will minimize the error.]

Example: Calculate $\int_0^1 e^{-x^2}\,dx$ using Simpson's rule. We choose $h = 0.1$ and calculate 11 base points.

x	$f(x)$	x	$f(x)$
0	1.000000	0.6	0.697676
0.1	0.990050	0.7	0.612626
0.2	0.960789	0.8	0.527292
0.3	0.913931	0.9	0.444858
0.4	0.852144	1.0	0.367879
0.5	0.778801		

The reader should use Simpson's rule to calculate:

$$\text{integral} = 0.746825$$

As with all such methods, the accuracy improves if h is decreased. For $h = 0.05$ (21 base points) Simpson's rule gives 0.746824, thus confirming the six-significant-figure accuracy of the calculation. ∎

Exercise: Calculate $\int_1^2 e^{-x^2}\,dx$ using $h = 0.25$ (five base points). [*Answer:* 0.135210. Accurate: 0.135256 ($h = 0.1$), 0.135257 ($h = 0.05$).] ∎

The Three-Point Rule

Doing three-point integrals when the base points are not equally spaced is more difficult. We use Lagrange's interpolation formula for three points (for $x_0 < x < x_2$):

$$f(x) = \frac{(x - x_1)(x - x_2)f_0}{(x_0 - x_1)(x_0 - x_2)} + \frac{(x - x_0)(x - x_2)f_1}{(x_1 - x_0)(x_1 - x_2)} + \frac{(x - x_0)(x - x_1)f_2}{(x_2 - x_0)(x_2 - x_1)} \quad \textbf{(AI.13)}$$

and the definitions (Figure AI.4):

$$a = x_1 - x_0, \qquad b = x_2 - x_1$$

The integral will be given by:

$$\text{area} = c_0 f_0 + c_1 f_1 + c_2 f_2 \quad \textbf{(AI.14)}$$

For the total area, integrating over $x_0 < x < x_2$:

$$\text{(total)} \qquad c_0 = \frac{(a + b)(2a - b)}{6a}$$

$$c_1 = \frac{(a + b)^3}{6ab}$$

$$c_2 = \frac{(a + b)(2b - a)}{6b} \quad \textbf{(AI.15)}$$

If there is an even number of base points, you may need a formula for part of the area. For integrating $x_0 < x < x_1$:

$$\text{(Area 1)} \qquad c_0 = \frac{a(2a + 3b)}{6(a + b)}, \qquad c_1 = \frac{a(3b + a)}{6b}, \qquad c_2 = \frac{-a^3}{6b(a + b)} \quad \textbf{(AI.16)}$$

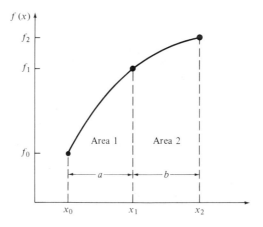

Figure AI.4   The three-point rule for integrals. This figure illustrates the definitions used in Eqs. (AI.15), (AI.16), and (AI.17).

and for $x_1 < x < x_2$:

(Area 2) $\quad c_0 = \dfrac{-b^3}{6a(a + b)}, \quad c_1 = \dfrac{b(3a + b)}{6a}, \quad c_2 = \dfrac{b(3a + 2b)}{6(a + b)}$ **(AI.17)**

These formulas will have the same degree of accuracy as Simpson's rule.

Evidently we have reached the limit of hand calculation, and a programmable device would be very useful. Computers usually have package programs for numerical integration, but they are often for integrating equations rather than data tables. The ultimate procedure for integration or differentiation (see next section) is to fit N points to an $(N - 1)$-degree polynomial and proceed from there. This is, however, a somewhat cumbersome procedure, not without its problems.

Mean Values

Another useful idea that we shall encounter in situations in which a mathematical relationship is defined by an integral is the *mean-value theorem*. For an interval $x_1 < x < x_2$, the equation:

$$\int_{x_1}^{x_2} y \, dx = \bar{y}[x_2 - x_1] \tag{AI.18}$$

defines a quantity ($\bar{y}$) called the *mean value* of y in the interval. This is merely a definition, one of several possible, for the term "mean value," but we shall find it useful in many situations. In particular, the trapezoidal rule [Eq. (AI.11)] may be used to get an approximation to $\bar{y}$:

$$\bar{y} \cong \tfrac{1}{2}(y_1 + y_2)$$

In some cases, the value of y at the middle of the interval may be used:

$$\bar{x} = \tfrac{1}{2}(x_1 + x_2)$$

$$\bar{y} \cong y(\bar{x})$$

Figure AI.5 contrasts these two approximations. In the approximate forms we have stated it, the integral mean-value theorem will work well for smoothly varying functions only. (It is exact for a linear function.)

AI.5 Differentiation

The simplest method for obtaining a numerical derivative is to reverse the definition:

$$\frac{df}{dx} = \lim_{\Delta x \to 0} \frac{\Delta f}{\Delta x}$$

for a finite, but small difference:

$$\frac{df}{dx} \cong \frac{f_2 - f_1}{x_2 - x_1} \tag{AI.19a}$$

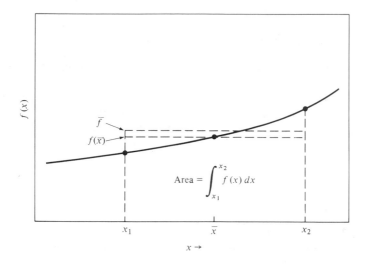

Figure AI.5   The integral mean value. The mean value of the function in the interval x_1 to x_2 is defined as the integral between these limits divided by the interval $\Delta x = x_2 - x_1$. For purposes of estimating the area, the mean value can be approximated as the average of the values of the function at the two limits (f_1 and f_2, cf. Figure A.3) or the value of the function at the middle of the interval—$f(\bar{x})$—that is, at the average value of the independent variable ($\bar{x}$).

The *differential mean-value theorem* (Figure AI.6) states that this equation is exact for *df/dx somewhere* in the interval $x_1 < x < x_2$. Typically, this theorem is correct but of little practical use. It becomes useful if, as will be true for smoothly varying functions, we realize that it will be reasonably accurate at the *center* of the interval.

Equation (AI.19a) can be used in two ways: to estimate a slope from two measured pairs of points (f_1, x_1) and (f_2, x_2), or to calculate an approximate derivative of an equation for $f(x)$. Differentiation of equation functions is rarely a problem unless you want to do it on a computer. For machine computation the formula:

$$\frac{df}{dx} = \frac{f(x + h) - f(x - h)}{2h} \qquad \textbf{(AI.19b)}$$

will do for some suitably small value of h. This formula is exact for linear or quadratic functions.

If it is desired to differentiate experimental data, another problem surfaces. Let us say we can measure the concentration (C) of some reacting chemical species at any time and wish to calculate the rate of reaction (dC/dt) at $t = 1$ sec. We could measure $C(0.9) = 0.875$ and $C(1.1) = 0.851$ and use Eq. (I.19) with $h = 0.1$:

$$\frac{dC}{dt} = \frac{0.851 - 0.875}{0.2} = -0.12$$

If we try to improve accuracy by making $h = 0.01$ and measure $C(0.99) = 0.864$,

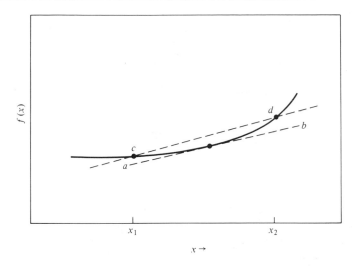

Figure AI.6 The differential mean-value theorem. The slope of the chord (*c-d*) will be the slope of the function at some point in the interval x_1 to x_2. For purposes of estimating numerical derivatives, it is assumed that the slope of the curve at the *middle* of the interval is equal to the slope of the chord. This estimate is exact for polynomial functions of order 2 or less (that is, for linear and quadratic functions).

$C(1.01) = 0.862$, we see that, when we subtract these quantities, there is only one significant figure! $dC/dt = -0.2$. Instead of improving accuracy, we have lost it. It is generally true that, for a set of data of a given accuracy, the integral of that set will be more accurate and the differential less accurate. As a method of data analysis, numerical differentiation should be avoided whenever possible.

Lagrange's three-point interpolation formula can be used to give the differential in the interval $x_0 < x < x_2$:

$$\frac{df}{dx} = \frac{(2x - x_1 - x_2)f_0}{a(a + b)} - \frac{(2x - x_0 - x_2)f_1}{ab} + \frac{(2x - x_0 - x_1)f_2}{b(a + b)} \qquad \textbf{(AI.20a)}$$

where a and b are defined as before, Figure AI.4. This formula is most accurate near the center of the interval. At $x = x_1$, Eq. (AI.20a) gives:

$$\left(\frac{df}{dx}\right)_{x_1} = \frac{a^2 f_2 - (a^2 - b^2)f_1 - b^2 f_0}{ab(a + b)} \qquad \textbf{(AI.20b)}$$

This simplifies to Eq. (AI.19) for equal intervals, $h = a = b$.

A more accurate set of formulas is available for five equally spaced points:

$$f'(x_0) = \frac{-25f_0 + 48f_1 - 36f_2 + 16f_3 - 3f_4}{12h} \qquad \textbf{(AI.21a)}$$

$$f'(x_1) = \frac{-3f_0 - 10f_1 + 18f_2 - 6f_3 + f_4}{12h} \qquad \textbf{(AI.21b)}$$

$$f'(x_2) = \frac{f_0 - 8f_1 + 8f_3 - f_4}{12h} \qquad \textbf{(AI.21c)}$$

$$f'(x_3) = \frac{-f_0 + 6f_1 - 18f_2 + 10f_3 + 3f_4}{12h} \qquad \textbf{(AI.21d)}$$

$$f'(x_4) = \frac{3f_0 - 16f_1 + 36f_2 - 48f_3 + 25f_4}{12h} \qquad \textbf{(AI.21e)}$$

These formulas are most accurate for the center point (x_2) and least accurate for the ends (x_0, x_4).

Example: The following data were simulated by direct integration of:

$$\frac{dC}{dt} = -0.159C^2$$

hence the rate can be calculated exactly. We shall test the formulas (AI.19) and (AI.21) by using them on the calculated points:

t	$C(t)$	
0.5	0.9264	
1.0	0.8628	
1.5	0.8074	(exact $dC/dt = -0.1037$)
2.0	0.7587	
2.5	0.7156	

The estimated rate at $t = 1.5$ can be calculated from the two points (0.5, 2.5) using Eq. (I.18); this gives rate $= -0.1054$. If we use the two points (1.0, 2.0), we get rate $= -0.1041$. The five-point formula gives:

$$\frac{dC}{dt} \cong \frac{0.9264 - 8(0.8628) + 8(0.7587) - 0.7156}{12(0.5)} = -0.1037 \qquad \blacksquare$$

Exercise: Calculate the rate for the preceding example at $t = 1$. [*Answer:* (points 0.5, 1.5) -0.1190, (all points) -0.1184, (exact) -0.1184.] $\qquad \blacksquare$

Exercise: In kinetics it is frequently useful to determine the *initial rate* of decomposition — that is, at $t = 0$. The "data" above were simulated assuming $C(0) = 1.0000$, so its initial rate is obviously -0.159 (from the generating equation). Calculate the initial rate using the five-point equation. (*Answer:* -0.15855.) $\qquad \blacksquare$

AI.6   Least Squares

An empirical or semiempirical formula (such as those discussed earlier) with m undetermined constants can be made to pass through m experimental points by the choice of appropriate values for the constants [presuming that we have a reasonably "correct" formula; a polynomial of degree ($m - 1$) will always work]. Given the inevitability of experimental errors, it is preferred to have $N > m$ points and to choose the constants so the curve passes "close" to all points; in the end, this means the constants are chosen to minimize the sum of the squares of the deviations of the observed points from the calculated line. Such a problem is overdetermined with *degrees of freedom* (DF):

$$\text{DF} = N - m \qquad\qquad \textbf{(AI.22)}$$

where N is the number of data points, and m is the number of constants to be determined.

Computer programs are available for this procedure for any reasonable number of constants and any type of function (not necessarily just polynomial functions). The procedure for linear functions — linear regression — is relatively simple and quite illustrative of the general procedure. It is also useful, since many physical problems can be put into linear form. This may require a change of variable; for example, rate constants (k) are distinctly nonlinear with temperature (T), but their logarithm ($\ln k$) vs. $1/T$ is reasonably linear, so these variables may be used for linear regression.

Linear Regression

Suppose we have N data pairs $\{y_i, x_i\}$ that we wish to fit to an equation of the form:

$$y = a + mx \qquad\qquad \textbf{(AI.23)}$$

with the slope (m) and intercept (a) to be chosen for "best fit" (Figure AI.7).

Given values for (a, b), we can calculate a value of y for each x_i; the deviation is:

$$\Delta_i = y_i - y(\text{calc.}) = y_i - (a + mx_i) \qquad\qquad \textbf{(AI.24a)}$$

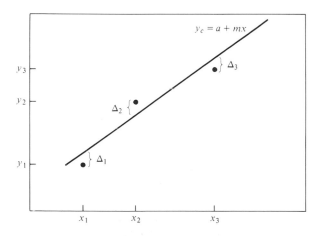

Figure AI.7 The least-squares line. This figure illustrates terms used in the discussion of least-squares fitting of a straight line to a set of points.

A "best fit" is obtained for the values (a, b) that minimize the sum:

$$R = \sum \Delta_i^2 = \sum (y_i - a - mx_i)^2 \tag{AI.24b}$$

(Here and hereafter a sum is presumed to be over the entire data set, $i = 1, \ldots, N$.)
A minimum will be found when:

$$\frac{dR}{da} = 0 \quad \text{and} \quad \frac{dR}{dm} = 0$$

$$\frac{dR}{da} = \sum 2(y_i - a - mx_i)(-1) = 0$$

$$\frac{dR}{dm} = \sum 2(y_i - a - mx_i)(-x_i) = 0$$

Noting that for N terms:

$$\sum_{i=1}^{N} a = Na$$

we can rearrange these equations to give:

$$\sum y_i = Na + m \sum x_i \tag{AI.25a}$$

$$\sum x_i y_i = a \sum x_i + m \sum x_i^2 \tag{AI.25b}$$

All of the sums required can be calculated from the data table $\{y_i, x_i\}$, so these are two equations in two unknowns; they are readily solved to give:

$$m = \frac{n \sum (x_i y_i) - \left(\sum x_i\right)\left(\sum y_i\right)}{D} \tag{AI.26a}$$

$$a = \frac{\left(\sum y_i\right)\left(\sum x_i^2\right) - \sum (x_i y_i)\left(\sum x_i\right)}{D} \tag{AI.26b}$$

$$D = N\left(\sum x_i^2\right) - \left(\sum x_i\right)^2 \tag{AI.26c}$$

Example: Fit the following data set to a linear equation.

i	x_i	y_i	x_i^2	$x_i y_i$	y_i^2
1	1	10	1	10	100
2	2	22	4	44	484
3	3	32	9	96	1024
4	4	40	16	160	1600
Sums	10	104	30	310	3208

(The need for the last sum will become apparent later.)

From these sums we calculate:

$$D = 4(30) - (10)^2 = 20$$

$$m = \frac{(4)(310) - (10)(104)}{20} = 10$$

$$a = \frac{(104)(30) - (310)(10)}{20} = 1$$

The best-fit equation is therefore:

$$y = 1 + 10x$$

Back calculation gives y(calc.) $= 11, 21, 31, 41$ for $i = 1$ to 4, respectively; from Eq. (AI.24b) we can calculate the residual, $R = 4$. This is the minimum value possible for the straight line passing through this data set. ■

With a real data set, doing these six sums can be a time-consuming process. In particular, if there are, say, four significant figures in the data, it is necessary to keep *at least* eight figures in the squares and sums. Fortunately, there are several inexpensive calculators that will calculate the six required sums (and the slope and intercept besides) with a single data entry.

Error Estimates

It is essential to obtain error estimates for the parameters (a, m), since they usually represent physical quantities. In our error analysis we shall assume that the experimental error is in y and that x can be measured to any precision required. This is often, but not invariably, the case.

First, we calculate the variance of the x points:

$$\sigma_x^2 = \frac{1}{N-1} \sum (x_i - \bar{x})^2$$

We can easily calculate $\bar{x} = (\sum x)/N$, but calculation with this equation requires a second entry of the data. Since we already know $\sum x$ and $\sum x^2$ from our calculation of the slope and intercept, the following equation, which can be derived in a straightforward manner, is more convenient:

$$\sigma_x^2 = \frac{1}{N-1} \left[\sum x^2 - \frac{\left(\sum x \right)^2}{N} \right] \tag{AI.27}$$

Similarly the variance of the y points is given by:

$$\sigma_y^2 = \frac{1}{N-1} \left[\sum y^2 - \frac{\left(\sum y \right)^2}{N} \right] \tag{AI.28}$$

Exercise: For the preceding example, calculate $\sigma_x = 1.2910$, $\sigma_y = 12.9615$. ■

Note that the "standard deviations" σ_x and σ_y do not tell us anything about "errors"; they refer only to the distribution (somewhat arbitrary) of the data points. The data could be perfectly linear and σ_x and σ_y would not be zero. (If you do linear regression on an automatic calculator, read the manual carefully; a common mistake is to assume that the "STD DEV" button tells you the error—it usually doesn't. But read on!)

From these standard deviations, we can calculate the *correlation factor r:*

$$r = \frac{m\sigma_x}{\sigma_y} \tag{AI.29}$$

This does tell you something about the errors, since $r = \pm 1$ means a perfect fit. But it is deceptive—$r = 0.999$ seems close enough to 1, but such a value could mean poor data in some cases. A better quantity is:

$$\varepsilon = \frac{1}{r}\left(\frac{1 - r^2}{N - 2}\right)^{1/2} \tag{AI.30}$$

The quantity 100ε corresponds roughly to the "% error" in the sense that a data set with 1% accuracy (loose definition) has $\varepsilon \cong 0.01$.

The residual (R) also gives an idea of the "goodness" of the fit. However, calculation using Eq. (AI.24) requires a second data entry on a calculator; an easier method to calculate R that uses only quantities we've calculated already is:

$$R = (N - 1)\sigma_y^2(1 - r^2) \tag{AI.31}$$

This method for calculating residuals is recommended for calculators which, typically, have a lot of digits and little memory. Computers typically have a lot of memory and few digits, so Eq. (AI.24) may be better. The two methods give the same answer only with infinite precision computation.

More to the point, the standard deviation of the slope is:

$$\sigma(m) = \varepsilon m \tag{AI.32}$$

The standard deviation of the intercept is:

$$\sigma(a) = \sigma(m)\left(\frac{\sum x_i^2}{N}\right)^{1/2} \tag{AI.33a}$$

Some calculators may tell you $\bar{x}$ and σ_x, but not $\sum x^2$, so an alternative form of Eq. (AI.33) is sometimes useful:

$$\sigma(a) = \sigma(m)\left[\bar{x}^2 + \frac{\sigma_x^2(N - 1)}{N}\right]^{1/2} \tag{AI.33b}$$

Exercise: Calculate the error parameters for the earlier example. [*Answer:* $r = 0.9960238$, $\varepsilon = 0.063$, $\sigma(m) = 0.63$, $\sigma(a) = 1.73$, $R = 4.00$. (Calculated with Eq. (I.31); earlier we got the same result using Eq. (AI.24).)] ∎

In reporting errors in the parameters (a, m) the standard deviation is a very conservative estimate. It is better to report a *confidence interval:*

$$\lambda = t_c\sigma \tag{AI.34}$$

Values of t_c (the "critical t factor") are listed in Table AI.1. (A commonly used rule-of-thumb is to estimate the "error" as twice the standard deviation; a glance at Table AI.1 will show that this is reasonable for 90% confidence and 5 or more data points.)

For our previous example with $N = 4$ (DF = 2) the 90% confidence interval is:

$$\lambda = 2.92\sigma = 2.92(0.63) = 1.8$$

The value of the slope would be reported as:

$$m = 10 \pm 1.8 \ (90\% \text{ confidence})$$

This has the meaning that it is probable (to 90% confidence) that the correct value of the slope lies between 8.2 and 11.8.

Exercise: Fit the following data to a straight line.

x:	2,	3,	4,	5,	6,	7
y:	13,	18,	20,	24,	30,	33

[*Answer:* $m = 4.00$, $\sigma(m) = 0.24$, $a = 5.00$, $\sigma(a) = 1.15$, $r = 0.992933$, $\varepsilon = 0.059$, $R = 4.00$.] ∎

Exercise: First-order kinetic data can be fitted to an equation of the form:

$$\ln C = \ln C_0 - kt$$

where C_0 is the initial concentration and k is the rate constant. Find the rate constant and 95% confidence interval of the data below.

t	C	$\ln C$
5	37.1	3.6136
10	29.8	3.3945
20	19.6	2.9755
30	12.3	2.5096
50	5.0	1.6094

Table AI.1   Critical t factors

Degrees of freedom	t_c (90%)	t_c (95%)	t_c (99%)
1	6.31	12.7	63.7
2	2.92	4.30	9.92
3	2.35	3.18	5.84
4	2.13	2.78	4.60
5	2.01	2.57	4.03
6	1.94	2.45	3.71
8	1.86	2.31	3.36
10	1.81	2.23	3.17
15	1.75	2.13	2.95
20	1.72	2.09	2.85
30	1.70	2.04	2.75
∞	1.64	1.96	2.58

[*Answer:* $N = 5$, DF $= 3$, $a = 3.84616$, $C_0 = e^a = 46.8$, $m = -0.0445924$, $r = -0.9998769$, $\varepsilon = 0.00906$, $\sigma(m) = 0.000404$, $\lambda = (3.18)(0.0000404) = 0.00128$. The final answer is:

$$k = (4.46 \pm 0.13) \times 10^{-2} \text{ (95\% confidence)]} \qquad \blacksquare$$

Exercise: Fit the vapor-pressure data for CCl_4 (below) to an equation of the form: $\ln P = A + B/T$.

t (°C)	P (torr)	$1/T$	$\ln P$
30	142.3	0.0032987	4.957937
50	314.4	etc.	etc.
70	621.1		
100	1463		

[*Answer:* $r = -0.999896$, $\varepsilon = 0.01$, $A = 17.389$, $B = -3765 \pm 112$ (90\% confidence).]

$\blacksquare$

In this treatment we have not distinguished between random errors and "systematic" or model-dependent errors. In the last example we might have gotten a more accurate slope if more points were available, but ultimately an accurate fit can be obtained only by a nonlinear model with more parameters [for example, Eq. (IA.1)]. The best way to detect such model deficiencies is to graph the function; systematic errors should be suspected if the deviations of the points are regular, as opposed to random. Least squares gives unbiased estimates of the parameters, but it is mindless—it will fit a hyperbola to a straight line if you ask it to.

References

1. B. Carnahan, H. A. Luther, and J. O. Wilkes, *Applied Numerical Analysis,* 1969: New York, John Wiley & Sons, Inc.
2. C. Ray Wylie, *Advanced Engineering Mathematics,* 4th ed., 1961: New York, McGraw-Hill Book Co.
3. I. Guttman and S. S. Wilkes, *Introductory Engineering Statistics,* 1965: New York, John Wiley & Sons, Inc.

APPENDIX II
Partial Derivatives

A function of one variable, $f(x)$, has a derivative that is defined as:

$$S(x) = \frac{df}{dx} = \lim_{\Delta x \to 0} \frac{f(x + \Delta x) - f(x)}{\Delta x} \quad \textbf{(AII.1)}$$

This defines a new function, $S(x) = df/dx$, whose significance is that its value at a particular point is the slope of $f(x)$ at that point. The change in f for changing x from x_1 to x_2 is:

$$f(x_2) - f(x_1) = \int_{x_1}^{x_2} S(x)\, dx \quad \textbf{(AII.2)}$$

For a function of two variables, $h(x, y)$, the partial derivatives are defined as:

$$S_x(x, y) = \left(\frac{\partial h}{\partial x}\right)_y = \lim_{\Delta x \to 0} \frac{h(x + \Delta x, y) - h(x, y)}{\Delta x}$$

$$S_y(x, y) = \left(\frac{\partial h}{\partial y}\right)_x = \lim_{\Delta y \to 0} \frac{h(x, y + \Delta y) - h(x, y)}{\Delta y} \quad \textbf{(AII.3)}$$

A function of two variables is a surface. A good physical example is the height of a mountain as a function of latitude (x) and longitude (y) as represented as contours of constant height on a topographical map (Figure AII.1). If you were standing on the side of a mountain, its slope would depend on the direction in which you looked. It is possible to define two orthogonal slopes: S_x, for the slope in the north/south direction (constant longitude, y); and S_y, the slope in the east/west direction (constant latitude, x). An infinitesimal movement in any direction can be expressed in terms of the x, y changes, dx and dy. The change of height for this movement (dh) is then expressed in terms of the slopes as:

$$dh = S_x\, dx + S_y\, dy \quad \textbf{(AII.4)}$$

The orthogonal slopes on this surface (S_x and S_y) are simply the partial derivatives of the function as defined by Eq. (AII.3), so:

$$\textbf{SLOPE FORMULA} \quad dh = \left(\frac{\partial h}{\partial x}\right)_y dx + \left(\frac{\partial h}{\partial y}\right)_x dy \quad \textbf{(AII.5a)}$$

[Here, and for the remainder of this appendix, important results such as Eq. (AII.5a) will be labeled by names that are used throughout the text.] This is readily generalized for more than two variables; if we have a function of n variables, $f(x_1, x_2, x_3, \ldots, x_n)$:

$$df = \frac{\partial f}{\partial x_1}\, dx_1 + \frac{\partial f}{\partial x_2}\, dx_2 + \frac{\partial f}{\partial x_3}\, dx_3 + \ldots + \frac{\partial f}{\partial x_n}\, dx_n \quad \textbf{(AII.5b)}$$

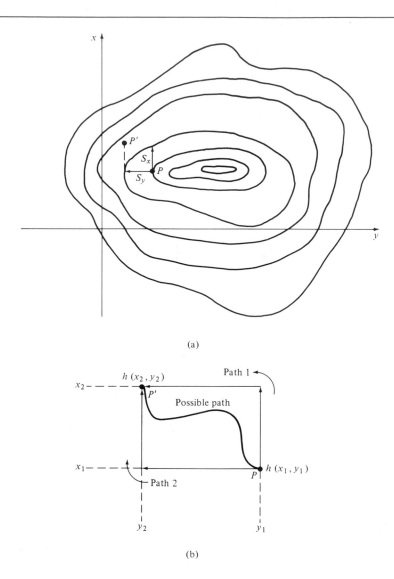

(a)

(b)

Figure AII.1 A function of two variables. (a) The function is shown as a contour plot, with the lines connecting points for which the function, $h(x, y)$, has a constant value. Shown for the point P are the two orthogonal slopes S_x for the slope parallel to the x axis and S_y for the slope parallel to the y axis. (b) Paths of integration. The integral of a state function, $h(x, y)$, is independent of the path and can, therefore, be calculated along any convenient path.

where each partial derivative is evaluated holding all other variables constant; for example, the first slope is:

$$\left(\frac{\partial f}{\partial x_1}\right)_{x_2, x_3, x_4, \ldots, x_n}$$

Path Integrals

For a finite change from point $P(x_1, y_1, h_1)$ of Figure AII.1 to point $P'(x_2, y_2, h_2)$, integration of Eq. (AII.4) requires the definition of a path; that is, since both x and y are changing, the function $y(x)$ defines exactly how this occurs. However, in walking down a mountain, the change in our altitude in moving from point P to point P' clearly does not depend on path. In such a case, dh of Eq. (II.4) is said to be a *perfect differential* and h is a *state function,* and the integral:

$$\Delta h = h_2 - h_1 = \int_{x_1, y_1}^{x_2, y_2} dh$$

is independent of the path. For purposes of computation, two convenient paths are (Figure AII.1):

1. Change x_1 to x_2 at constant y_1; change y_1 to y_2 at constant x_2;

$$\Delta h = \int_{x_1}^{x_2} S_x(x, y_1)\, dx + \int_{y_1}^{y_2} S_y(y, x_1)\, dy \qquad \textbf{(AII.6a)}$$

2. Change y_1 to y_2 at constant x_1; change x_1 to x_2 at constant y_2;

$$\Delta h = \int_{x_1}^{x_2} S_x(x, y_2)\, dx + \int_{y_1}^{y_2} S_y(y, x_1)\, dy \qquad \textbf{(AII.6b)}$$

Both of these paths must give the same answer for the change in h.

Reciprocal Variables

In the preceding example, $h(x, y)$ (Figure AII.1) the height was clearly the dependent variable while x and y were the independent variables. These variables are not reciprocal; that is, while we can express h as a function of x and y, we cannot express x or y as functions of h. If we specify a location—for example, latitude 38° north on the 122nd meridian—the altitude (h) at that location is uniquely defined (at least until the next earthquake). However, if we specified $x = 45°$ north and $h = 2000$ meters, that description could fit any number of locations in North America, Asia, or Europe, so there is no possibility of a functional relationship $y(x, h)$.

In thermodynamics we encounter a situation where we have a set of variables that *are* reciprocally related; that is, any may be considered the dependent variable and expressed as a function of any of the others. In such a case there is no distinction between *functions* and *variables,* since any may be either. For such a situation, there are a number of relationships that are very useful in thermodynamics; next we shall derive some of them.

Suppose we have four variables, w, x, y, z. Any two may be independent; then the other two will be functions of these two. For example, we could write $w(x, y)$, $z(x, y)$, but we could equally well write $x(y, z)$, $w(y, z)$, and so on. Among these four variables there are 24 possible slopes; for example:

$$\left(\frac{\partial x}{\partial y}\right)_w, \quad \left(\frac{\partial x}{\partial y}\right)_z, \quad \left(\frac{\partial z}{\partial x}\right)_w, \quad \left(\frac{\partial z}{\partial y}\right)_w, \quad \left(\frac{\partial z}{\partial w}\right)_x, \quad \left(\frac{\partial w}{\partial z}\right)_y$$

These slopes are not (and cannot be) all independent.

First, we start with the slope formula for $z(x, y)$:

$$dz = \left(\frac{\partial z}{\partial x}\right)_y dx + \left(\frac{\partial z}{\partial y}\right)_x dy$$

This formula is divided by the change in w, dw:

$$\frac{dz}{dw} = \left(\frac{\partial z}{\partial x}\right)_y \frac{dx}{dw} + \left(\frac{\partial z}{\partial y}\right)_x \frac{dy}{dw}$$

This formula is true for a change of variables in any direction, including a direction in which y does not change; in this case $dy = 0$ and the derivatives (dz/dw) and (dx/dw) become partial derivatives with y constant. Therefore:

CHAIN RULE $\qquad \left(\frac{\partial z}{\partial w}\right)_y = \left(\frac{\partial z}{\partial x}\right)_y \left(\frac{\partial x}{\partial w}\right)_y$ $\qquad$ **(AII.7)**

Note the similarity to the chain rule for total derivatives.

Again we start with the slope formula for $z(x, y)$, this time dividing by dz:

$$1 = \left(\frac{\partial z}{\partial x}\right)_y \frac{dx}{dz} + \left(\frac{\partial z}{\partial y}\right)_x \frac{dy}{dz} \qquad \textbf{(AII.8)}$$

For a particular path for which y is constant, $dy = 0$ and:

RECIPROCAL RULE $\qquad \left(\frac{\partial z}{\partial x}\right)_y \left(\frac{\partial x}{\partial z}\right)_y = 1$ $\qquad$ **(AII.9)**

Now we use two slope formulas; one for $z(x, y)$:

$$dz = \left(\frac{\partial z}{\partial x}\right)_y dx + \left(\frac{\partial z}{\partial y}\right)_x dy$$

and one that gives $y(x, z)$:

$$dy = \left(\frac{\partial y}{\partial x}\right)_z dx + \left(\frac{\partial y}{\partial z}\right)_x dz$$

The second of these can be used to eliminate dy in the first:

$$dz = \left(\frac{\partial z}{\partial x}\right)_y dx + \left(\frac{\partial z}{\partial y}\right)_x \left[\left(\frac{\partial y}{\partial x}\right)_z dx + \left(\frac{\partial y}{\partial z}\right)_x dz\right]$$

Now, collect the terms for dz and dx:

$$dz \left[1 - \left(\frac{\partial z}{\partial y} \right)_x \left(\frac{\partial y}{\partial z} \right)_x \right] = \left[\left(\frac{\partial z}{\partial x} \right)_y + \left(\frac{\partial z}{\partial y} \right)_x \left(\frac{\partial y}{\partial x} \right)_z \right] dx$$

The bracket on the left-hand side is zero by the reciprocal rule [Eq. (AII.9)]; on the right-hand side, the infinitesimal dx is small but not zero, so it must be true that:

$$\left(\frac{\partial z}{\partial x} \right)_y + \left(\frac{\partial z}{\partial y} \right)_x \left(\frac{\partial y}{\partial x} \right)_z = 0$$

Using the reciprocal rule, we can rearrange this as:

$$\textbf{CYCLIC} \atop \textbf{RULE} \qquad \left(\frac{\partial x}{\partial y} \right)_z = - \left(\frac{\partial x}{\partial z} \right)_y \left(\frac{\partial z}{\partial y} \right)_x \qquad \textbf{(AII.10)}$$

Partial derivatives such as $(\partial z/\partial x)_y$ and $(\partial z/\partial x)_w$ are not the same; they represent slopes in different directions. It is frequently necessary to relate these to other slopes. We start with $z(x, y)$:

$$dz = \left(\frac{\partial z}{\partial x} \right)_y dx + \left(\frac{\partial z}{\partial y} \right)_x dy$$

Then we divide by dx:

$$\frac{dz}{dx} = \left(\frac{\partial z}{\partial x} \right)_y + \left(\frac{\partial z}{\partial y} \right)_x \frac{dy}{dx}$$

This equation is true in any direction, including that for which w is constant; therefore:

$$\left(\frac{\partial z}{\partial x} \right)_w = \left(\frac{\partial z}{\partial x} \right)_y + \left(\frac{\partial z}{\partial y} \right)_x \left(\frac{\partial y}{\partial x} \right)_w \qquad \textbf{(AII.11)}$$

Example: While it is very helpful to understand these formulas and their derivations, their use can be relatively straightforward. Suppose we have four variables, $S, P, V,$ and T; the meaning of these variables (which will be given in the text) is irrelevant to this example. Now, suppose we need a relationship between the slopes $(\partial S/\partial V)_P$ and $(\partial S/\partial V)_T$; we can obtain this from Eq. (AII.11) by simply making the substitutions $S = z$, $V = x$, $P = w$, $T = y$:

$$\left(\frac{\partial S}{\partial V} \right)_P = \left(\frac{\partial S}{\partial V} \right)_T + \left(\frac{\partial S}{\partial T} \right)_V \left(\frac{\partial T}{\partial V} \right)_P \qquad \blacksquare$$

APPENDIX III
Coordinate Systems

AIII.1   Cartesian Coordinates

The Cartesian coordinate system, named for its inventor, René Descartes, specifies the position of a point in space by giving three numbers that are the projections of the point on the orthogonal axes x, y, z [Figure AIII.1(a)]. A right-handed system is always preferred; if the fingers of your right hand point from x to y, the thumb will point in the positive z direction.

An infinitesimal element of volume will be a cube having sides with length dx, dy, dz; it will be called (in this book) either $d\mathbf{r}$ or $d\tau$.

$$d\tau = dx\,dy\,dz \qquad \textbf{(AIII.1)}$$

This volume element will be useful for calculating integrals of functions of the coordinates, $F(x, y, z)$. In particular, the *integral over all space* of the function will be:

$$\int_{-\infty}^{\infty} dx \int_{-\infty}^{\infty} dy \int_{-\infty}^{\infty} F(x, y, z)\,dz \qquad \textbf{(AIII.2)}$$

The velocity of a particle can be defined as a vector with components $v_x = dx/dt$, $v_y = dy/dt$, $v_z = dz/dt$. Similarly the momentum is $p_x = mv_x$, and so on. The Cartesian vectors are:

$$\mathbf{r} = (x, y, z)$$
$$\mathbf{v} = (v_x, v_y, v_z)$$
$$\mathbf{p} = (p_x, p_y, p_z)$$

with volume elements:

$$d\mathbf{r} = dx\,dy\,dz = d\tau$$
$$d\mathbf{v} = dv_x\,dv_y\,dv_z$$
$$d\mathbf{p} = dp_x\,dp_y\,dp_z$$

AIII.2   Polar Coordinates

The position of a point in space can also be specified [Figure AIII.1(b)] by giving its distance (r) from the origin and its direction in space from the origin. According to the theorem of Pythagoras, generalized to three dimensions, the distance (r) from the origin to the point is:

$$r = \sqrt{x^2 + y^2 + z^2} \qquad \textbf{(AIII.3)}$$

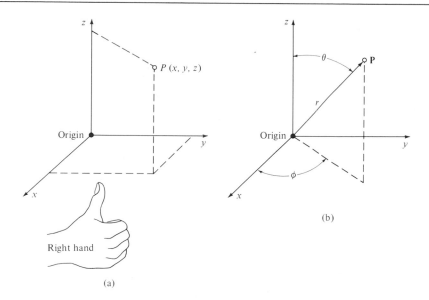

Figure AIII.1 Cartesian and polar coordinates. The right-hand rule illustrated: if the thumb of the right hand points in the positive z direction, the fingers should point from the x axis toward the y axis; such a coordinate system is called "right-handed." The angle θ is called the polar angle, and the angle ϕ is called the azimuthal angle.

The direction is given by two angles:

$$\text{(polar angle)} \qquad \theta = \cos^{-1} \frac{z}{r}$$

$$\text{(azimuthal angle)} \qquad \phi = \tan^{-1} \frac{y}{x} \qquad \textbf{(AIII.4)}$$

The inverse relationships are:

$$x = r \sin \theta \cos \phi$$
$$y = r \sin \theta \sin \phi$$
$$z = r \cos \theta \qquad \textbf{(AIII.5)}$$

The volume element is derived as follows: An element of area ($d\sigma$) on the surface of the sphere sweeps out a volume when moved radially by dr:

$$d\tau = d\sigma \, dr$$

The element of area is created by sweeping out a length of arc by changing θ; $dl_\theta = r \, d\theta$. This length is swept through $d\phi$ to make an area; but the radius for the arc along ϕ is not r but $(r \sin \theta)$ (Figure AIII.2); therefore $dl_\phi = r \sin \theta \, d\phi$, and:

$$d\sigma = r^2 \sin \theta \, d\theta \, d\phi \qquad \textbf{(AIII.6)}$$

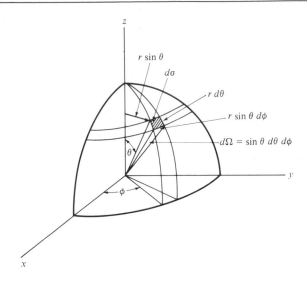

Figure AIII.2   Elements of surface area and solid angle in polar coordinates.

$$d\tau = r^2 \, dr \, \sin\theta \, d\theta \, d\phi \tag{AIII.7}$$

In this case, the range of variables is:

$$0 < r < \infty, \qquad 0 < \theta < \pi, \qquad 0 < \phi < 2\pi$$

and an integral of $F(r, \theta, \phi)$ *over all space* is:

$$\int_0^{2\pi} d\phi \int_0^{\pi} \sin\theta \, d\theta \int_0^{\infty} F(r, \theta, \phi) r^2 \, dr \tag{AIII.8}$$

It is common to define an *element of solid angle:*

$$d\Omega = \sin\theta \, d\theta \, d\phi \tag{AIII.9}$$

in which case the volume element is:

$$d\tau = r^2 \, dr \, d\Omega \tag{AIII.10}$$

Example: Calculate the volume of a sphere whose radius is R using Eq. (AIII.7).

$$V = \int_0^R r^2 \, dr \int_0^{\pi} \sin\theta \, d\theta \int_0^{2\pi} d\phi = \frac{R^3}{3}(2)(2\pi) = \frac{4\pi R^3}{3} \qquad \blacksquare$$

Exercise: Calculate the surface area of a sphere (radius R) using Eq. (AIII.6). (*Answer:* $4\pi R^2$.)   ∎

Velocities can be treated in a similar manner. In particular, the magnitude of the velocity (speed) is:

$$v = \sqrt{v_x^2 + v_y^2 + v_z^2}$$

and the volume element in velocity space is:

$$d\mathbf{v} = dv_x\,dv_y\,dv_z = v^2\,dv\,d\Omega \qquad \textbf{(AIII.11)}$$

AIII.3 Center-of-Mass Coordinates

We shall discuss only the case of two particles with mass m_1 and m_2. Specifying the position of two particles requires six coordinates:

$$\mathbf{r}_1 = (x_1, y_1, z_1)\,, \qquad \mathbf{r}_2 = (x_2, y_2, z_2) \qquad \textbf{(AIII.12)}$$

The center of mass (COM) of the particles is defined with respect to a line between them (Figure AIII.3) such that:

$$m_1\mathbf{r}_{c1} + m_2\mathbf{r}_{c2} = 0 \qquad \textbf{(AIII.13)}$$

In many cases it will be convenient to specify, in lieu of Eqs. (AIII.12), the position of the center of mass:

$$\mathbf{R} = (X_{\text{COM}}, Y_{\text{COM}}, Z_{\text{COM}}) \qquad \textbf{(AIII.14)}$$

and the *relative* positions of the particles $\mathbf{r}$. In effect $\mathbf{r}$ is a vector in a coordinate

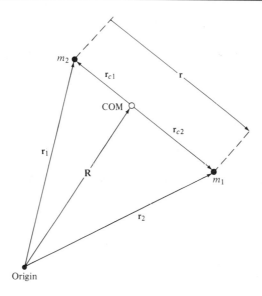

Figure AIII.3 Center-of-mass coordinates. This figure illustrates variables used in the text to relate the center-of-mass (COM) coordinate system to a space-fixed coordinate system.

system that is centered on the COM and is usually represented as:

$$\mathbf{r} = (r, \theta, \phi) \qquad \text{(AIII.15)}$$

in which r is the distance between the masses and (θ, ϕ) is the orientation of the vector between them.

We shall be particularly interested in two applications of COM coordinates: (1) The two masses are *attached* in a rigid or semirigid manner—for example, two atoms attached by a chemical bond to form a diatomic molecule (Chapters 2, 11, and 13). (2) We have two independent particles where only their *relative* motions are of concern—for example, bimolecular collisions (Chapter 9) and calculation of second virial coefficients (Chapter 1).

From Figure AIII.3 we can find the following vector relations:

$$\mathbf{r}_1 = \mathbf{R} + \mathbf{r}_{c1}, \qquad \mathbf{r}_2 = \mathbf{R} + \mathbf{r}_{c2}, \qquad \mathbf{r} = \mathbf{r}_{c2} - \mathbf{r}_{c1}$$

The last of these, together with the definition of the COM [Eq. (AIII.13)], can be used to show:

$$\mathbf{r}_{c2} = \frac{\mathbf{r}\mu}{m_2}, \qquad \mathbf{r}_{c1} = -\frac{\mathbf{r}\mu}{m_1} \qquad \text{(AIII.16)}$$

in which we have defined the *reduced mass:*

$$\frac{1}{\mu} = \frac{1}{m_1} + \frac{1}{m_2}, \qquad \mu = \frac{m_1 m_2}{m_1 + m_2} \qquad \text{(AIII.17)}$$

We shall later need the *total mass:*

$$M = m_1 + m_2 \qquad \text{(AIII.18)}$$

With the result Eq. (AIII.16) and the other vector sums we can get the relationship between the vectors (AIII.12) and the COM coordinates:

$$\mathbf{r}_1 = \mathbf{R} - \frac{\mathbf{r}\mu}{m_1}$$

$$\mathbf{r}_2 = \mathbf{R} + \frac{\mathbf{r}\mu}{m_2} \qquad \text{(AIII.19)}$$

The volume elements for two particles can be defined two ways:

$$d\mathbf{r}_1 \, d\mathbf{r}_2 = dx_1 \, dy_1 \, dz_1 \, dx_2 \, dy_2 \, dz_2$$

$$d\mathbf{R} \, d\mathbf{r} = dX \, dY \, dZ \, r^2 \, dr \, d\Omega$$

From calculus we learn:

$$d\mathbf{r}_1 \, d\mathbf{r}_2 = \left(\frac{\partial r_1}{\partial R} \frac{\partial r_2}{\partial r} - \frac{\partial r_2}{\partial R} \frac{\partial r_1}{\partial r} \right) d\mathbf{R} \, d\mathbf{r}$$

which, together with Eq. (AIII.19), gives:

$$d\mathbf{r}_1 \, d\mathbf{r}_2 = d\mathbf{R} \, d\mathbf{r} \qquad \text{(AIII.20)}$$

Velocities are handled in the same manner, with the definition of the velocity of the COM (**V**) and the relative velocity (**v**); in particular, Eqs. (AIII.19) apply for the corresponding velocities:

$$\mathbf{v}_1 = \mathbf{V} - \frac{\mathbf{v}\mu}{m_1}$$

$$\mathbf{v}_2 = \mathbf{V} + \frac{\mathbf{v}\mu}{m_2} \tag{AIII.21}$$

The momenta are particularly interesting, since:

$$\mathbf{p}_1 = m_1\mathbf{v}_1 = m_1\mathbf{V} - \mu\mathbf{v}$$

$$\mathbf{p}_2 = m_2\mathbf{v}_2 = m_2\mathbf{V} + \mu\mathbf{v}$$

$$\mathbf{p}_1 + \mathbf{p}_2 = (m_1 + m_2)\mathbf{V} = M\mathbf{V} \tag{AIII.22}$$

This shows that the total momentum is just $M\mathbf{V}$—the same as that of a total mass $M = m_1 + m_2$ centered at the COM. If one mass is an H atom ($m_1 = 1$) and the other is a fluorine ($m_2 = 19$), the translational motions of the diatomic molecule HF can be treated as a mass $M = 20$ placed exactly at the COM.

Exercise: Show that the kinetic energy:

$$T = \tfrac{1}{2}m_1v_1^2 + \tfrac{1}{2}m_2v_2^2$$

can be written as:

$$T = \tfrac{1}{2}MV^2 + \tfrac{1}{2}\mu v^2 \tag{AIII.23}$$

In terms of the momenta $P = MV$, $p = \mu v$, this is:

$$T = \frac{P^2}{2M} + \frac{p^2}{2\mu} \tag{AIII.24}$$

∎

The last exercise demonstrates that the motions of a pair of particles can always be broken up into two components: (1) the motion of the COM, (2) the *relative* motion in the COM system. In a coordinate system moving with the COM, the two particles can be treated as a single particle of mass μ.

Note that, by analogy to Eq. (AIII.20), the volume element in velocity space is:

$$d\mathbf{v}_1\,d\mathbf{v}_2 = d\mathbf{V}\,d\mathbf{v} \tag{AIII.25}$$

where v is the relative velocity and V is the velocity of the COM, and [from Eq. (AIII.11)]:

$$d\mathbf{v} = 4\pi v^2\,dv \tag{AIII.26}$$

These equations are used for the discussion of molecular collisions—Chapter 9.

APPENDIX IV
Vectors and Complex Numbers

AIV.1   Vectors

A vector is an ordered array of numbers

$$\mathbf{A} = (a_1, a_2, a_3, \ldots) \tag{AIV.1}$$

A three-dimensional vector is most common, in particular, the Cartesian vectors as in Appendix III, but there is no restriction to such a case. The algebra of vectors will be explained in terms of three-dimensional vectors; in most cases the generalization is straightforward.

Algebra

1. *Equality:* $\mathbf{A} = \mathbf{B}$ means $a_1 = b_1$, $a_2 = b_2$, $a_3 = b_3$; that is, each corresponding component must be equal.
2. *Addition:* $\mathbf{A} + \mathbf{B}$ produces a new vector ($\mathbf{C}$) whose components are

$$c_1 = a_1 + b_1, \qquad c_2 = a_2 + b_2,$$

   and so on.
3. *Multiplication:* There are three useful definitions of vector multiplication.
   (a) *Scalar:* $\mathbf{A} \cdot \mathbf{B} = a_1b_1 + a_2b_2 + a_3b_3$. This produces a scalar — that is, a simple number — as the result.
   (b) *Vector product or cross product:* This operation is generally useful only for Cartesian vectors. Its result is a new vector:

$$\mathbf{C} = \mathbf{A} \times \mathbf{B}$$

   with components:

$$c_1 = a_2b_3 - a_3b_2$$
$$c_2 = a_3b_1 - a_1b_3 \tag{AIV.2}$$
$$c_3 = a_1b_2 - a_2b_1$$

   (c) *Direct product:* $\mathbf{C} = \mathbf{A} \otimes \mathbf{B}$. This is an element-by-element multiplication to give a new vector; $c_1 = a_1b_1$, $c_2 = a_2b_2$, and so on.
4. *Magnitude.* The magnitude or length of a vector is:

$$|\mathbf{A}| = \sqrt{a_1^2 + a_2^2 + a_3^2 + \cdots} \tag{AIV.3}$$

   Note that this is related to the scalar product of a vector with itself:

$$|\mathbf{A}|^2 = \mathbf{A} \cdot \mathbf{A} = a_1^2 + a_2^2 + a_3^2 + \cdots \tag{AIV.4}$$

5. *Division* is not defined for vectors.

Example: If $\mathbf{A} - (3, 2, 1)$ and $\mathbf{B} = (1, 2, 3)$:

$$\mathbf{A} + \mathbf{B} = (4, 4, 4), \qquad \mathbf{A} \cdot \mathbf{B} = 3(1) + 2(2) + 1(3) = 10$$ ∎

Example: Find $(3, 2, 1) \times (1, 2, 3)$:

$$\left.\begin{array}{ll} \text{1st component:} & 2(3) - 1(2) = 4 \\ \text{2nd component:} & 1(1) - 3(3) = -8 \\ \text{3rd component:} & 3(2) - 2(1) = 4 \end{array}\right\} \begin{array}{l} \text{The answer is} \\ (4, -8, 4) \end{array}$$ ∎

Example: Find the length of $\mathbf{A} = (3, 2, 1)$.

$$|\mathbf{A}| = \sqrt{9 + 4 + 1} = \sqrt{14}$$ ∎

The following relationships are useful, but will not be proven:

$$\mathbf{A} \cdot \mathbf{B} = |\mathbf{A}||\mathbf{B}| \cos \theta; \qquad \textbf{(AIV.5)}$$

where θ is the smaller angle between the vectors. If the scalar product is zero, the vectors are perpendicular.

Example: Find the angle between the vectors of the earlier example. We saw that $\mathbf{A} \cdot \mathbf{B} = 10$ and $|\mathbf{A}| = \sqrt{14}$. Also $|\mathbf{B}| = \sqrt{14}$, so $\cos \theta = 10/14$ and $\theta = 44.4°$. ∎

The cross product gives a vector with magnitude:

$$|\mathbf{A} \times \mathbf{B}| = |\mathbf{A}||\mathbf{B}| \sin \theta \qquad \textbf{(AIV.6)}$$

and a direction that is perpendicular to both of the original vectors:

Example: We earlier found $(3, 2, 1) \times (1, 2, 3) = (4, -8, 4)$. We can prove this vector is perpendicular (and check our work) by the scalar product; for example:

$$(1, 2, 3) \cdot (4, -8, 4) = 1(4) - 2(8) + 3(4) = 0$$

In effect:

$$\mathbf{A} \cdot (\mathbf{A} \times \mathbf{B}) = 0$$

$$\mathbf{B} \cdot (\mathbf{A} \times \mathbf{B}) = 0$$ ∎

Example: The most important example of a vector product that we shall encounter is the definition of the angular momentum:

$$\mathbf{L} = \mathbf{r} \times \mathbf{p} \qquad \textbf{(AIV.7)}$$

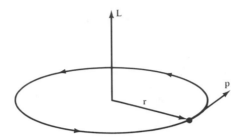

Figure AIV.1 Angular momentum. For the motion of a mass on a circular orbit, the angular momentum, $\mathbf{L} = \mathbf{r} \times \mathbf{p}$, is a vector perpendicular to the plane of the orbit. The linear momentum ($\mathbf{p}$) is tangent to the circle, and the position vector ($\mathbf{r}$) is a vector from the center of the circle to the position of the mass.

in terms of the Cartesian vectors:

$$\mathbf{r} = (x, y, z), \qquad \mathbf{p} = (p_x, p_y, p_z)$$

From our definition of the cross product [Eq. (AIV.2)] it can be shown, for $\mathbf{L} = (L_x, L_y, L_z)$:

$$L_x = yp_z - zp_y$$
$$L_y = zp_x - xp_z \qquad \qquad \textbf{(AIV.8)}$$
$$L_z = xp_y - yp_x$$

From Eq. (IV.6):

$$|\mathbf{L}| = |\mathbf{r} \times \mathbf{p}| = rp \sin \theta$$

For circular motion (Figure AIV.1), $\mathbf{r}$ and $\mathbf{p}$ are perpendicular, so $\sin \theta = 1$ and:

$$|\mathbf{L}| = mvr \qquad \qquad \textbf{(AIV.9)}$$

∎

AIV.2 Complex Numbers

A complex number can be looked upon as an ordered pair of real numbers, (a, b). Real numbers correspond to the points on a line, so a complex number corresponds to a point on a plane (Figure AIV.2) with a real axis and an imaginary axis.

If $z = (a, b)$, then $a = \text{Re}\,(z)$, the real part, and $b = \text{Im}\,(z)$, the imaginary part. The complex number can also be written as a multiple of unit vectors on the real axis (1 — the number one) and the imaginary axis (i):

$$z = a + ib \qquad \qquad \textbf{(AIV.10)}$$

The quantity i has the following properties:

$$i^2 = -1, \qquad i^3 = -i, \qquad i^4 = 1 \qquad \qquad \textbf{(AIV.11)}$$

—that is, it acts like $\sqrt{-1}$.

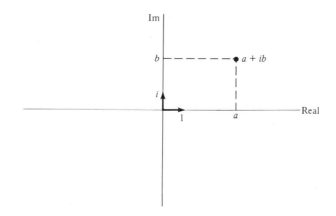

Figure AIV.2 Complex numbers. A complex number is an ordered pair of real numbers, $z = (a, b)$, and corresponds to one of the points on a plane. The number a is called the real part of the complex number, $a = \text{Re}(z)$, and the number b is called the imaginary part of the complex number, $b = \text{Im}(z)$.

Algebra

The algebra will be illustrated with two complex numbers:

$$z_1 = a_1 + ib_1, \qquad z_2 = a_2 + ib_2$$

1. *Equality:* $z_1 = z_2$ means $a_1 = a_2$ *and* $b_1 = b_2$. Both real and imaginary parts must be equal.
2. *Addition:* $z_1 + z_2 = (a_1 + a_2) + i(b_1 + b_2)$. Add real and imaginary parts separately.
3. *Multiplication:* This goes as in ordinary algebra except that $i^2 = -1$ and the real and imaginary parts must be collected together.

Example:

$$(3 + 2i)(2 + i) = 6 + 3i + 4i + 2i^2$$
$$= (6 - 2) + (3 + 4)i$$
$$= 4 + 7i \qquad \blacksquare$$

4. *Complex conjugate:* If $z = a + ib$, then $z^* = a - ib$. Change all i to $-i$.
5. *Magnitude:*

$$|z|^2 = zz^* \qquad \text{(AIV.12)}$$

If $z = a + ib$, $z^* = a - ib$, then:

$$zz^* = a^2 + b^2 \quad \text{and} \quad |z| = \sqrt{a^2 + b^2}$$

6. *Division:* The quotient $z_3 = z_1/z_2$ can be rationalized by multiplying the top and bottom by z_2^*:

$$z_3 = \frac{z_1}{z_2} = \frac{z_1 z_2^*}{|z_2|^2} \qquad \text{(AIV.13)}$$

Example: Divide $z_1 = 3 + 2i$ by $z_2 = 2 + i$. $|z_2|^2 = 4 + 1 = 5$:

$$\frac{(3 + 2i)(2 - i)}{(2 + i)(2 - i)} = \frac{8 + i}{5} = \frac{8}{5} + \frac{1}{5}i \qquad \blacksquare$$

Example: An interesting result of Eq. (AIV.13) is that:

$$\frac{1}{i} = \frac{1}{i}\frac{(-i)}{(-i)} = \frac{-i}{-i^2} = -i$$

The reciprocal of i is $-i$! $\blacksquare$

Polar Representation

A complex number can be written as:

$$z = a + ib = Ae^{i\phi} \qquad \text{(AIV.14)}$$

where the *amplitude A* is:

$$A = \sqrt{a^2 + b^2} \qquad \text{(AIV.15)}$$

and the *phase angle* is:

$$\phi = \tan^{-1}\frac{b}{a} \qquad \text{(AIV.16)}$$

Note that A is the same as the magnitude.

$$zz^* = (Ae^{i\phi})(Ae^{-i\phi}) = A^2$$

$$|z| = A$$

(See Figure AIV.3.)

A little geometry on Figure AIV.3 can derive a set of very important formulas called **Euler's relationships***:

$$e^{i\phi} = \cos\phi + i\sin\phi \qquad \text{(AIV.17)}$$

$$e^{-i\phi} = \cos\phi - i\sin\phi \qquad \text{(AIV.18)}$$

$$\cos\phi = \frac{1}{2}(e^{i\phi} + e^{-i\phi}) \qquad \text{(AIV.19)}$$

$$\sin\phi = \frac{1}{2i}(e^{i\phi} - e^{-i\phi}) \qquad \text{(AIV.20)}$$

Exercise: Prove:

$$e^{i\pi} = -1$$

*One of numerous things named after Leonhard Euler (1707–1783).

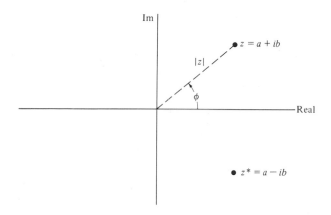

Figure AIV.3 Argand diagram for complex numbers. A complex number can be represented in terms of its polar coordinates in the complex plane (illustrated). The point corresponding to the complex conjugate of $z(z^*)$ is also shown.

This formula is notable for containing nearly all the major concepts of modern algebra — e, i, π, equality, and negative numbers. ∎

Exercise: If you took trigonometry, you probably derived the formula:

$$\sin 2x = 2 \sin x \cos x$$

Derive it now using Eqs. (AIV.19) and (AIV.20); it's a lot easier this way. ∎

Note the analogy between the complex equations (AIV.19) and (AIV.20) and the definitions of the hyperbolic functions for real numbers:

$$\cosh x = \frac{1}{2}(e^x + e^{-x})$$

$$\sinh x = \frac{1}{2}(e^x - e^{-x})$$

It is well known that quantities such as $x^2 - y^2$ can be factored as $(x + y)(x - y)$. The imaginary permits us to factor $x^2 + y^2 = (x + iy)(x - iy)$.

The real and imaginary parts may themselves be functions. For example, a function $\psi(x)$ may have a real part, $f(x)$, and an imaginary part, $g(x)$:

$$\psi(x) = f(x) + ig(x)$$
$$\psi^*(x) = f(x) - ig(x)$$
$$\psi\psi^* = f^2 + g^2$$

In this text, the principal applications of complex functions are found in Chapters 11 through 14. Euler's equations, (AIV.17) through (AIV.20) are especially important.

Answers to Problems

Chapter 1

1.1 28.85 g,
0.001287 g/cm^3

1.2 53.9 atm

1.3 3.23×10^{-8} cm

1.4 (a) 10.94 atm, (b) 10.70 atm, (c) 49.23 atm
(ideal gas), 44.53 atm (van der Waals)

1.5 10.61 (b) 42.84 atm

1.6 (a) 2.445 (b) 2.399 dm^3

1.7 16.897 dm^3

1.9 22.244 dm^3 atm,
$B = -.165$ dm^3

1.10 $B = -.100$ dm^3
$C = 0.004$ dm^6

1.11 $z = 1.0343$

1.12 (a) 3.3 cm^3 (b) 15.4 cm^3 (c) 12.5 cm^3

1.13 (a) 509 K (b) 370 K (c) 409 K (expt'l 410)

1.14 17 K, 13 K, 24 K (obs. 35 K)

1.15 $B = b - a/RT$

1.16 $t_c = 144°C$
$V_c = 124$ cm^3/mol

1.17 $P_c\sqrt{aR/216b^3}, V_c = 3b, T_c = \sqrt{8a/27bR}$

1.18 0.58 dm^3

1.19 $a = 5.196$ dm^3 atm,
$b = 0.0724$ dm^3
density 4.677 g/dm^3

1.20 $a = 2.891$ dm^6 atm,
$b = 0.04622$ dm$^3, P = 111.7$ atm
(ideal gas gives $P = 222$ atm)
(c) 2.130 dm^3

1.21 2.463×10^{19} cm^{-3}

1.22 $2.03 \times 10^{21}, 1.54 \times 10^{24}$ cm^{-2} s^{-1}

1.23 5.80×10^{-6} atm

1.24 $f(v)\,dv = (m/kT)$
$\exp(-mv^2/2kT)v\,dv$

1.25 765, 830 m/s

1.26 380, 412, 337 m/s

1.27 440 per million

1.28 1.7×10^{-4}

1.29 $r = 1.12246\sigma$

1.30 1.6×10^{-3} dm^6

1.31 $B = b_0[1 - (\varepsilon/kT) - \frac{1}{6}(\varepsilon/kT)^2 - \frac{1}{30}(\varepsilon/kT)^3 - \cdots$

1.34 (b) 404 K (square-well), 410 K (LJ),
410 K (observed)

1.35 $-22.96, 15.15$ cm^3 (b) 323.6 K

1.36 0.356

1.37 6.7 atm

1.38 345 atm

1.39 0.852 dm^3

1.40 $B = -568$ cm^3
$z = 0.9747$

1.41 -7.48 cm^3

1.42 84.65 atm
$P_{H_2} = 33.86, P_{CO} = 50.79$ atm

1.43 (a) 26.195 dm^3
(b) 3.750 atm

1.44 $\alpha = 1.012566/T - 280.663/T^2$

1.45 $\alpha = R(V_m - b)/$
$\{RTV_m - 2a[(V_m - b)/V_m]^2\}$

1.47 5.12×10^{-4} K^{-1}

1.48 1.15×10^{-5} atm^{-1}

1.49 13.6 atm

1.51 83 cm^3

1.52 2.9×10^{-4} K^{-1}

Chapter 2

2.1 (a) $w = 0, q = \Delta U = 312$ J
(b) $w = -208, q = 520, \Delta U = 312$ J

2.2 34,773 J

2.3 41,047 J

2.4 355 J

2.5 (a) 403 MPa (b) 852 Pa

2.6 (b) 1634 Pa (van der Waals, 732 Pa)

2.7 1200, 2000 J, 3.82 K, 7.8 dm^3, -800 J

2.8 14,617 J (b) 206 J/g

2.9 23.62 kJ

2.10 678 (trap.), 677 (Simp.)

2.11 8.25 kJ

2.12 57 J

2.13 -33 J

2.14 (c) -38 J (compare 2.13)

2.15 $+32.3$ kJ (with Berthelot B)

2.16 (a) 37.33 J/K (b) 38.26 J/K

2.17 (b) dm^3/K (c) 38.13 J/K

2.18 Zero; i.e., C_v is independent of V for van der Waals gas

2.19 33.13 J/K (33.42 if ideal)

2.20 8.51 J/K (obs. 9.15)

2.21 (1 atm) $1.004R$, (10 atm) $1.051R = 8.738$ J/K

2.22 (a) -1.8 kJ (b) -3.9 kJ (c) -4.7 kJ

2.23 -17.0 kJ

2.24 Ideal -4.09×10^4 J, van der Waals -3.91×10^4 J

2.25 (a) 75.4 K (b) 288 K

2.26 29.2 J/K

2.27 0.394 atm

2.28 (a) 267 (b) 320 (c) 400 K

2.29 108 K (ideal 109)

2.30 (a) -5.4 kJ (b) -3.0 kJ

2.31 $\sim 200°C$

2.32 303 K

2.33 $-2°$

2.34 0.54 K/atm (obs. 0.6475), $\Delta T = -26$ K

2.36 -63 K

2.37 (b) 0.30 K/atm (obs. 0.27)
(c) 866 K (obs. 621)

2.38 (a) 3.7 cm^3/K, 1.7 dm^3
(b) 3.22 cm^3/K, 1.53 dm^3

2.39 1.053 K/atm

2.40 (a) 0.271 K/atm (obs. 0.272)
(b) 658 K (obs. 621)

2.41 (b) 28 J/mole (c) $bRT/(V_m - b)$

2.42 27.16 kJ/mole

Chapter 3

3.1 (1 atm) 1.96×10^5
(15.3)3.66×10^5 J

3.2 (a) 19% (b) 52% (c) 68%

3.3 (a) 42 watts (b) 171 watts

3.4 171 watts, 683 watts

3.5 Area (PV) = work, Area (TS) = heat

3.6 680 J/K

3.7 55.05 J/K

3.8 $\Delta H = 41.7$ kJ, $\Delta S = 377$ J/K

3.9 (a) $q = 3.3$, $w = -0.9$, $\Delta U = 2.4$,
$\Delta H = 3.3$ (kJ), $\Delta S = 9.1$ (J/K)
(b) $q = 2.4$, $w = 0$, $\Delta U = 2.4$,
$\Delta H = 3.3$ (kJ), $\Delta S = 6.6$ (J/K)

3.10 10°C, $\Delta S = 21$ J/K

3.11 -7.62 J/K

3.12 148 J/K

3.13 7.12 J K^{-1} mol^{-1}

3.14 $\Delta S = 0$, 9.855, 17.289 J/K
$\Delta U = -5634$ J, -3000, 0 J

3.15 (kJ or

J/K)	w	q	ΔU	ΔH	ΔS
(a)	0	-1.25	-1.25	-2.08	-5.69
(b)	10.46	-10.46	0	0	-38.3
(c)	-0.83	2.08	1.25	2.08	6.5
(d)	-2.05	0	-2.05	-3.42	0

3.16 (a) 14.0 J/K (b) -13.7 J/K
(c) 15.4 and 1.7 J/K

3.17 (ideal gas) 19.14, (van der Waals) 19.28 J/K

3.18 Both 2287 J

3.19 $\Delta A = 2285$ J, $\Delta G = 2264$ J

3.22 $\kappa_S = 69.4 \times 10^{-6}$ atm^{-1}
$\kappa_T = 101 \times 10^{-6}$ atm^{-1},
$\gamma = 1.45$, $C_{vm} \approx 92$ J/K

3.24 $C_{vm} = 28.83$ J/K

3.25 0.25°C

3.26 0.42°C

3.27 145.52 J/K

3.28 192.3 J/K

3.29 -7.87 J/K

3.30 Real 171.8, ideal 173.4 J/K

3.31 162 J/K

3.32 (a) -374 J/K
(b) 0.7 MJ
(c) 2 MJ

3.33 (b) 2.47 J

3.34 1.568 J K^{-1} g^{-1}

3.35 1.4%

3.36 0.83°C

3.37 $\Delta l = 0.3$ cm

3.38 -1.5 MJ/m^3

3.41 1.1 g/cm^3

3.42 The temperature drops by roughly 0.3° per pull.

Chapter 4

4.1 205 torr; the value listed by Lange's *Handbook* (9th ed., 1956) is 210 torr.

4.2 750 torr

4.3 -213 J/mol

4.4 89.4 torr

4.5 (a) 27.09 kJ
(b) 6.73°C (actual 5.96°C)

4.6 42.54 kJ

4.7 $\Delta H = 30.57$ kJ $\pm$ 0.11 (90% confidence)

4.8 22.9 kJ

4.9 20.31 kJ/mol

4.10 52.0 kJ/mol

4.11 (a) 1.42 torr, 19.69 kJ/mol (b) 214 K

4.12 62.68 K

4.13 (a) $\ln (P/\text{torr}) = 35.956 - 970.62\,T$
$- 3.857 \ln T$

(b) 63.20 K

4.14 (c) 226 K (actual 232)

4.15 35.3 J K^{-1} mol^{-1}

4.16 4.7738 atm

4.17 10.09 torr

4.18 $A_s = 2.28 \times 10^{-2}$ J/m^2,
$U_s = 4.77 \times 10^{-2}$ J/m^2,
$S_s = 8.5 \times 10^{-5}$ (J/K)/m^2

4.19 $U_s = 0.125$ J/m^2,
$S_s = 1.75 \times 10^{-4}$ J K^{-1} m^{-2}

4.20 750 m^2, 54 J

4.21 (a) $U_s = 60.3 \times 10^{-3}$ J/m^2
(b) 0.18 J

4.22 762 torr

4.23 342.4°C

4.24 3.7 kJ/g

4.25 218.7 K

4.26 1.5×10^4 atm

4.27 $\Delta H(\text{sub}) = \Delta H(\text{vap}) + \Delta H(\text{fus})$

4.28 (vap) 105.2 kJ, (sub) 111.3 kJ

4.29 (sub) 51.02 kJ/mol, (vap) 44.93 kJ/mol,
(fus) 6.09 kJ/mol

4.30 (a) 13.04 kJ (b) 16.42 kJ (c) 3.38 kJ
(d) 109.9 K

4.31 (vap) 16.6 kJ, (sub) 26.5 kJ, (fus) 9.9 kJ

Chapter 5

5.1 24, 12, 34,650

5.2 (a) 2.356×10^{-8}
(b) 2.0074×10^{-9}
(c) 2.827×10^{-6}: ORGANIC is inherently
less probable than PCHEM!

5.3 (b) -0.161 J K^{-1} mol^{-1}

5.4 $R \ln 4 = 11.52$ J K^{-1}; $R \ln 6$

5.5 $T = 18$ K

5.6 1960.750/1960.735. The more accurate
formula gives 1953.902/1953.880.

5.7 $\exp(-1.8 \times 10^{13})$

5.8 (300 K)$N_1 = 268.941$, $N_0 = 731{,}059$

5.11 2.16×10^{21}

5.12 154.73 J K^{-1}

5.14 Exact 5.35, calc. 5

5.15 Exact 5.17, calc. 5

5.16 (3) 0.1921

5.17 (a) RT (b) $1.5RT$

5.19 (a) 4.628×10^4
(b) 101.8 J K^{-1}

5.20 1, 2, 2, 1, 3, 2

5.21 2, 4, 2, 6

5.23 $\theta_r = 7.58$ K. The high-T limit is valid
at its bp.

5.24 4.723

5.25 190.997 J K^{-1}

5.27 13.54, 14.47, 13.92 J K^{-1}

5.28 10,443 J

5.29 9924 J

5.30 191.407, 227.886 J K^{-1}

5.31 204.94, 242.279 J K^{-1}

5.32 Only HCl

5.33 (a) 33.731, 35.802, 36.971 J K^{-1}
(b) 37.41 J K^{-1}

5.34 27.645 J K^{-1}

5.35 73.158 J K^{-1}

5.36 211.95, 220.64 J K^{-1}

5.37 173.48 J K^{-1}, 8676 J,
176.77 J K^{-1}

5.38 71.75 J K^{-1}

5.39 221.66 J/K

5.41 $\phi° = 188.98$, $\phi' = 197.76$ J K^{-1} at 1000 K

5.42 3.61 cal K^{-1}, 15.11 J K^{-1}

Chapter 6

6.1 (a) $12\,C(s) + 11\,H_2(g) + \frac{11}{2}\,O_2(g) \longrightarrow$
(b) $Pb(s) + S(s, rh) + 2\,O_2(g) \longrightarrow$
(c) $Na(s) + \frac{1}{2}\,I_2(s) + \frac{3}{2}\,O_2(g) \longrightarrow$
(d) $2\,C(s) + 3\,H_2(g) + \frac{1}{2}\,O_2(g) \longrightarrow$

6.2 -486.99 kJ

6.3 362 kJ

6.4 -333.0 kJ

6.5 (1) -518.6 (2) -146.4 kJ

6.6 78.4 MJ/kg

6.7 122.47 kJ

6.8 -39.76 kJ

6.9 -229032 J

6.10 $\Delta H = -65877 - 30.09T$

6.11 2320 K

6.12 892 K

6.13 5.67

6.14 1.267

6.15 (a) 7.43×10^{-2}
(b) 1.14 atm

6.16 -448.35 kJ

6.17 -175 kJ

6.18 (a) 2.51 (b) 78%

6.19 7.2×10^6; yes

6.20 108 molecules/m^3

6.21 3.98×10^{-3}, 3.80×10^{-3}

6.22 688 K (658 actual)

6.23 (a)

$$\ln K_p = I - \frac{J}{T} + A \ln T$$

$$+ BT + \frac{C}{T^2}$$

(b)

$$\ln K_a = 7.7929 + \frac{4823}{T} - 3.062 \ln T$$

$$+ 0.01102\,T - \frac{12{,}328}{T^2}$$

6.24 (a)

$$\ln K_a = 14.535 + \frac{7923}{T} - 3.619 \ln T;$$

$$K(1000) = 0.079 \text{ (b) } 0.098$$

6.25 -88.07 kJ

6.26 3.5×10^5

6.27 (a) $\Delta H = 24 \pm 2$ kJ (b) 29.2 kJ
(c) 1.8%

6.28 (163.9 ± 5.8) kJ

6.29 $K(600) = 7.79$

6.30 $f(1000) = 1896$ atm

6.32 $\gamma(100) = 0.47$
$\gamma(500) = 0.31$
$\gamma(1000) = 0.54$

6.33 $\gamma(100) = 0.82$
$\gamma(500) = 0.42$
$\gamma(1000) = 0.34$

6.34 (b) 696 atm

6.35 (a) 1.34
(b)

$$\Gamma = \frac{PV_m}{RT}\left(1 - \frac{\kappa_T P}{2}\right)$$

value is the same at $P = 1000$.

6.36 $f(CH_4) = 252$ atm,
$f(CO_2) = 440$ atm

6.37 $K_p = 0.64$, 96% methane

6.38 $K_p = 0.277$

6.39 16.6%

6.40 95.8%

6.41 (1) 64% (2) 76% (3) 74%
(4) 74% (5) 75%

6.42 99.2%

6.43 $P = 39.3$ torr

6.44 25.8%

6.45 29% HI

6.46 89, 96%

6.47 (1000 K) 4.3%,
(1500 K) $1 \times 10^{-6}\%$

6.48 43%

6.49 29%

6.50 0.47 atm, total $P = 3.58$ atm

6.51 1.75 atm

6.52 Yes

6.53 89%

6.54 (a) $K_a = 1 \times 10^{-31}$
(b) 329 electrons/cm^3

6.55 The tetrahydrate

6.56 $P = 2 \times 10^{-23}$ atm

6.57 0.55 atm

6.59 -133.305 kJ

6.60 1.83×10^4 atm

Chapter 7

7.1 8.44

7.2 $X_2 = 0.0714$, $m = 0.500$

7.3 $\Delta V_{mix} = 2.8$ cm^3

7.4 -4.0 cm^3

7.5 NaCl (1.5) 20.2422, (0) 16.6253; H_2O
(1.5) 18.0068, (0) 18.0409 cm^3

7.6 $c = 4, p = 4, F = 2$

7.7 $c = 3, p = 4, F = 1$

7.8 (a) $c = 1, p = 1$
(b) $c = 1, p = 1$
(c) $c = 2, p = 1$
(d) $c = 2, p = 2$

7.9 (a) $P_B = 152, P_r = 280, P = 432$
(b) $X_T = 0.634$

7.10 $\Delta S_{mix} = 3.97$ J/K

7.11 (a) γ (EtOH) = 1.055, (iOct) 14.20
(b) 122.0

7.12 γ (iPr) = 2.18, (H_2O) 1.27
(b) 539 torr

7.13 At 868.4 torr, $\gamma_1 = 1.251$, $\gamma_2 = 1.937$

7.14 Point at $X_2 = 0.508$ gives $w/RT = -0.82$
and -0.45; not reasonable.

7.15 $w = 1149$ J

7.16 $\Delta G = -15905$ J,
$\Delta S = 56.84$ ($\Delta S^{ex} = 0.88$) J/K,
$\Delta H = 1040$ J

7.18 7.05 atm

7.19 0.514 atm

7.20 $\Delta G^{\theta} = 25.2$ kJ,
$\Delta H^{\theta} = 9.58$ kJ

7.21 (a) 23.38 g (b) $P^{\bullet}$ (H_2O) decreased 0.09 torr by solubility, increased 1.36 torr by pressure.

7.23 22.635 torr

7.24 0.990, 0.907, 0.647

7.25 2.144 K (kg/mol)

7.26 $\gamma = 0.60$

7.28 4.8 atm

7.29 1.24×10^5 g/mol

7.30 $M = 2.66 \times 10^5$ g/mol (ave.),
B (cyclohexane) $= 6.55 \times 10^{-4}$ m^3/kg^2,
B (benzene) $= 1.47 \times 10^{-5}$ m^3/kg^2

7.31 -6.19 kJ, -94 J/K,
-34.2 kJ

7.32 -27.39 kJ

7.33 -3.31 kJ

7.34 -373.3 kJ

7.35 -511.0 kJ

7.36 3.16

7.37 $K_a = 1.37 \times 10^{-5}$,
m (dimer) $= 1.37 \times 10^{-7}$ mol/kg

Chapter 8

8.1 12%

8.2 0.970, 0.899, 0.761

8.3 (a) 0.325
(b) 0.379
(c) 0.034
(d) 0.332

8.4 0.895 mol/kg

8.5 4.0 Å

8.7 (a) 0.29

8.8 0.76(0.74), 0.16(0.16),
0.043(0.047).
Note: In general, this equation cannot be relied upon for such concentrated solutions.

8.9 (b) 271.80 K

8.10 (a) 1.57×10^{-4}
(b) 1.69×10^{-2}
(c) 2.70×10^{-3}

8.11 2.0×10^{-5} (b) 1.1×10^{-4}

8.12 (a) 1.2×10^{-3} mol/kg
(b) 4.3×10^{-9}

8.13 9.9×10^{-8}

8.14 0.0112 mol/kg

8.15 (a) $\alpha = 0.805, p\mathrm{H} = 2.09$
(b) $\alpha = 0.830, p\mathrm{H} = 2.13$

8.16 (a) 2.6% (b) 3.8%

8.18 10^{57} atm required, not easily achieved.

8.19 15.6 Ω

8.20 0.049 mols/dm^3

8.21 (a) 118.7 cm^2 S, 0.66
(b) 153.4 cm^2 S, 0.48
(c) 172.6 cm^2 S, 0.43

8.22 (a) 7.618×10^{-4} cm^2 V^{-1} s^{-1}
(b) 6.546×10^{-4} cm^2 V^{-1} s^{-1}
(c) 6.368×10^{-4} cm^2 V^{-1} s^{-1}

8.23 46.7 cm^2 S

8.24 101.77 cm^2 S

8.25 126.97 ± 0.02 cm^2 S

8.26 1.84×10^{-4}

8.27 1.33×10^{-3}

8.28 (a) 0.910 (b) -0.145

8.29 1.0894 V

8.30 2.2337 V

8.31 -0.594 V

8.32 0.851

8.33 0.2223 ± 0.0002 V

8.34 0.5395 V

8.35 0.4121 V

8.36 $\Delta S^{\theta} = -48.15$ J/K,
$\Delta G^{\theta} = -6880$ J,
$\Delta H^{\theta} = -21.24$ kJ

8.37 S^{θ} (Cl$^-$, aq) $= 45.4$ J/K,
ΔG_f^{θ} (Cl$^-$, aq) $= -129.3$ kJ

8.38 (b) 0.5355 V
(c) 0.5345 V
(d) -8.3 kJ
(e) -95 kJ

8.39 (a) 0.615 (b) 7.26×10^{-6}

8.40 4.94×10^{-13}

8.41 0.0930 V

8.42 -3.202 V; no, Ba will react directly with acid.

8.43 1.006×10^{-14}

8.44 $[I_3^-] = 6.5 \times 10^{-4}$ mol/dm^3,
$[I^-] = 7.0 \times 10^{-4}$ mol/dm^3

8.45 (b) 7.8×10^{-4}

8.46 1.08, 1.09, 1.49, 2.23

8.47 -0.0155 V

8.49 (a) emf 1.312 V (b) 9.9 g

8.51 0.347

8.52 0.0297 V

8.53 0.0107 V

8.54 0.0076 V

Chapter 9

9.1 $3.66 \times 10^9 \text{ s}^{-1}$,
$3.08 \times 10^9 \text{ s}^{-1}$

9.2 6.1×10^{-6} cm

9.3 $\lambda = 1.3 \times 10^{-5}$ cm (300 K),
4.4×10^{-5} cm (1000 K)
$z = 9.5 \times 10^9 \text{ s}^{-1}$ (300 K),
$5.2 \times 10^9 \text{ s}^{-1}$ (1000 K)

9.4 6.8×10^{-6} cm,
5.2×10^{-3} cm, 5.16 cm

9.5 CO: $\bar{v} = 4.54 \times 10^4$ cm/s,
(rel) 6.42×10^4 cm/s
H_2: $\bar{v} = 1.69 \times 10^5$ cm/s,
(rel) 2.39×10^5 cm/s
CO-H_2 rel. vel. $= 1.75 \times 10^5$ cm/s

9.6 CO-CO, $Z = 2.58 \times 10^{28} \text{ cm}^{-3}$,
H_2-H_2, $Z = 5.58 \times 10^{28} \text{ cm}^{-3}$
H_2-CO, $Z = 10.9 \times 10^{28} \text{ cm}^{-3}$

9.7 $2.39 \times 10^{34} \text{ m}^{-3} \text{ s}^{-1}$

9.8 4 cm

9.9 2.71×10^{-10}

9.10 58 Å

9.11 6 hours

9.12 26 hours

9.13 1 atm, 300 K, $D = 0.192 \text{ cm}^2 \text{ s}^{-1}$
1 torr, 300 K, $D = 146 \text{ cm}^2 \text{ s}^{-1}$
1 atm, 1000 K, $D = 1.17 \text{ cm}^2 \text{ s}^{-1}$
1 torr, 1000 K, $D = 889 \text{ cm}^2 \text{ s}^{-1}$

9.14 36 minutes

9.15 solution $c = 0.82$,
solvent $c = 0.18$

9.16 24.3 days

9.17 $1.9 \text{ cm}^3/\text{min}$

9.18 $46.7 \text{ g cm}^{-1} \text{ s}^{-1}$

9.19 20×10^{-8} cm

9.20 $0.9248 \times 10^{-5} \text{ cm}^2 \text{ s}^{-1}$

9.21 vis.(Cl_2) $= 1.60$(vis. of C_2H_4)

9.22 $\sigma = 3.72 \times 10^{-8}$ cm

9.23 (a) 3.24 (b) 4.11
(c) 4.00 (all $\times 10^{-8}$ cm)

9.24 2.353×10^{-8} cm

9.25 $k_s = 19.2 \times 10^{-6} \text{ (g cm}^{-1} \text{ s}^{-1} \text{ K}^{-1/2})$,
$S = 139$ K,
$\sigma = 3.0 \times 10^{-8}$ cm,
$\varepsilon/k \cong 700$ K

9.26 20 mp

9.27 $M = 53 \times 10^3$ g/mol

9.28 0.002 cm/day

9.29 $x = 2.0 \times 10^{-13}$ sec

9.30 $M = 1.7 \times 10^5$ g/mol

9.31 (a) 2.390×10^{-14} sec
(b) 1.456×10^5 rpm

9.32 moves 3 mm, half-width 1 mm

9.33 7783 g/mol

9.34 5% greater at the bottom

Chapter 10

10.1 $\text{dm}^{3/2} \text{ mol}^{-1/2} \text{ sec}^{-1}$

10.2 1.5-order

10.3 $n = 2, k = (9.3 \pm 0.3)$
$\times 10^7 \text{ cm}^3 \text{ mol}^{-1} \text{ s}^{-1}$

10.4 $n = 1, k = (2.99 \pm 0.06) \times 10^{-2} \text{ min}^{-1}$

10.5 $n(OCl^-) = 1, n(I^-) = 1$,
$n(OH^-) = -1, k = 60.5 \text{ s}^{-1}$

10.6 $k = (1.93 \pm 0.01) \times 10^{-2} \text{ min}^{-1}$

10.7 32.7% decomposed

10.8 87.0 min

10.9 206 min

10.10 $k = (1.45 \pm 0.03) \times 10^{-3} \text{ min}^{-1}$

10.11 $k = (1.45 \pm 0.02) \times 10^{-2} \text{ min}^{-1}$

10.12 $k = (1.37 \pm 0.02) \times 10^{-2} \text{ min}^{-1}$

10.13 $k = (2.07 \pm 0.01) \times 10^{-2} \text{ min}^{-1}$

10.14 $A = 0.042, B = 0.142$

10.15 20.0 sec

10.16 $(6.47 \pm 0.08) \text{ dm}^3 \text{ mol}^{-1} \text{ min}^{-1}$

10.17 $k = (2.32 \pm 0.04) \times 10^{-5} \text{ torr}^{-1} \text{ min}^{-1}$

10.18 $23.7 \text{ dm}^3 \text{ mol}^{-1} \text{ min}^{-1}$

10.19 $(2.7 \pm 0.4) \times 10^{-3} \text{ dm}^3 \text{ mol}^{-1} \text{ min}^{-1}$

10.20 $kt = \dfrac{1}{a + p} \ln \dfrac{a(p + x)}{p(a - x)}$

10.21 For $n \neq 1$,
$C_0^{n-1}(n - 1)kt_f = f^{1-n} - 1$

10.22 $E_a \sim 50$ kJ/mol

10.23 $5.5 \times 10^{-12} \text{ s}^{-1}$

10.24 $r = -0.999484$,
$A = 2.2 \times 10^{13} \text{ s}^{-1}$,
$E_a = 101 \pm 2$ kJ/mol

10.25 $A = 1.0 \times 10^9 \text{ cm}^6 \text{ mol}^{-2} \text{ s}^{-1}$,
$E_a = -4.95 \pm 0.4$ kJ/mol

10.26 $r = -0.999681$,
$A = 8.8 \times 10^{10} \text{ dm}^3 \text{ mol}^{-1} \text{ s}^{-1}$,
$E_a = 186 \pm 3$ kJ/mol

10.27 $3.59 \times 10^{14} \text{ cm}^3 \text{ mol}^{-1} \text{ s}^{-1}$
(2×10^{13} obs., Table 10.2)

10.28 $\Delta S^\ddagger = 7.1 \text{ J/K}, \Delta H^\ddagger = 158 \text{ kJ/mol}$,

10.29 $\Delta S^{\ddagger} = 4.6$ J/K, $\Delta H^{\ddagger} = 37$ kJ/mol

10.30 $r = -0.999451$,
$\Delta S^{\ddagger} = 2$ J K^{-1} mol^{-1},
$\Delta H^{\ddagger} = 99 \pm 2$ kJ/mol

10.31 $r = -0.999702$,
$\Delta S^{\ddagger} = -183$ J K^{-1} mol^{-1},
$\Delta H^{\ddagger} = 180 \pm 3$ kJ/mol

10.32 24 cm^3/mol

10.33 Answer in text

10.34 5.108 s, 46.5%

10.35 Plot $1/k$ vs. $1/P$

10.36 $v = \dfrac{k_3 \phi^{1/2} I_a^{1/2} [Cl_2]}{\sqrt{k_4}}$

10.37 $v = \dfrac{k_2 k_3 [Br_2][H_2] \phi^{1/2} I_a^{1/2}}{\sqrt{k_5} \{k_3 [Br_2] + k_4 [HBr]\}}$

10.38 $v = k_e [N_2O_5]^{2/3}[O_3]^{2/3}$,
$k_e = (k_4 k_1^2 k_3^2 / 4 k_2^2)^{1/3}$

10.39 $v = k_e [ROOH][RH]$,
$k_e = k_2 / \sqrt{k_1 / 2 k_4}$

10.40 $v = k_e [SO_2Cl_2]^{3/2}[CuCl]^{1/2}$,
$k_e = k_3 (k_1 / 2 k_4)^{1/2}$

10.41 $v_p = k_p [\phi I_a / k_t]^{1/2}[M]$,
$\nu = k_p [M] / k_t \phi I_a$

10.42 $v_p = (k_p f k_i / k_{tr})[M][In]/[S]$,
$\nu = k_p [M] / k_{tr}[S]$

10.43 $\nu_p = 0.014$ mol dm^{-3} s^{-1},
$E_a = 84.5$ kJ/mol;
rate increases, ν decreases

10.44 $V_{max} = 1.56$ cm^3/g,
$b = 217$ torr^{-1}

10.45 $V_{max} = 34.2$ cm^3/g,
$b = 0.069$ torr^{-1}

10.46 $V_{max} = 12.2$ cm^3,
$b = 0.011$ torr^{-1},
$r = 0.997676$

10.47 $k = 0.107 \pm 0.001$.
Use only points in linear range.

10.48 $k = k_a S_0 b_{NO} / b_{O_2}$

10.50 $v_{max} = 2.5 \times 10^{-4}$,
$k_2 = 8.9 \times 10^4$ s^{-1},
$K_m = 9.9 \times 10^{-3}$ mol/dm^3

10.51 $v_{max} = 1.3 \times 10^{-5}$,
$K_m = 1.2 \times 10^{-3}$

10.52 $K_m \ln(S/S_0) + S = S_0 - k_2 E_0 t$

10.53 Intercept $= K_m / v_{max}$,
slope $= 1 / v_{max}$

10.54 Answer in text

10.55 This is uncompetitive inhibition; answer in text.

10.57 $k_2 = 1.4 \times 10^{11}$ dm^3 mol^{-1} s^{-1},
$k_1 = 2.5 \times 10^{-5}$ s^{-1}

10.59 $\tau = 1/(k_1 + k_{-1})$

10.60 $\tau = 1/[k_2(A_{eq} + B_{eq}) + k_{-2}(C_{eq} + D_{eq})]$

Chapter 11

11.3 $N = 4$, 486.27 nm;
$N = \infty$, 364.7 nm

11.4 2×10^8 cm/s, 1.74×10^{10} cm/s
(compare to c)

11.5 (a) $2y e^{-y^2/2}$ (b) zero

11.6 $\cos x$

11.7 $2x$, $2\hat{d}$

11.8 eigenvalue $= -1$

11.9 $a = \pm 1$

11.10 $A =$ any value, $b = 1$, eigenvalue $= 1$

11.11 eigenvalue $= -2, -6$

11.12 6, not an eigenfunction

11.13 eigenvalue $= a^2 + b^2 + c^2$

11.15 $\langle x^2 \rangle = a^2/3 - a^2/2n^2\pi^2$,
$\langle x \rangle = a/2$

11.16 $0.25 - \sin(n\pi/2)/2n\pi$

11.18 $\sqrt{2/a}$

11.19 $\delta x = 0.28a$, $\delta p = h/2a$,
$\delta p\, \delta x = 0.14h$

11.20 Total 17 states, highest 26

11.21 (b) 1.8×10^{-10} erg

11.22 $\bar{n} \sim 2 \times 10^6$

11.23 11.0 nm

11.24 $E = (n_x^2/a^2 + n_y^2/b^2 + n_z^2/c^2)(h^2/8m)$

11.25 $A_2^2 = 1/(8\sqrt{\pi}\,\alpha)$

11.26 (a) g s^{-2} (b) g (c) cm (d) cm$^{-1/2}$

11.28 15.7% outside

11.29 $k\alpha^2/4$

11.33 $-\hbar \sin\theta \exp(i\phi)$

11.34 $L^2 \psi = 0$

11.35 (b) 2.64×10^{-40} g cm^2

11.36 (a) $l = 1.5 \times 10^{29}$

11.37 $2\hat{L}_z \hbar$

11.38 $l = 2, m = 0$, eigenvalue $= 6\hbar^2$

11.40 $\Delta m = \pm 1$, particle must be charged

11.41 (a) 1100 nm, ultraviolet
(b) 412 nm

11.42 33.6 GHz, microwave

Chapter 12

12.1 eigenvalue $= \frac{1}{4}$

12.2 $r/a_0 = 7.1, 1.9$

12.3 $(32\pi)^{-1/2} a_0^{-3/2}$

12.4 0.6767

12.5 $3a_0^2$

12.6 $5.2336a_0/Z$

12.7 Z/a_0

12.9 $cm^{-3/2}, cm^{-3}, cm^{-1}$

12.12 (a) 121.5 nm (b) 656.5 nm

12.14 $\alpha - \beta$

12.15 $\alpha \pm i\beta$

12.16 74.06, 122.4 eV

12.17 0.744 eV

12.19 3.4 eV

12.20 20.61 eV

12.21 $J = 4, 3, 2, 1$

12.22 4, 3/2, 36

12.23 $^2S\,^2P\,^1S\,^2S\,^1S\,^4S\,^2D\,^2D$

12.25 $A = -68.73$ or -79.13

12.26 52.5

12.27 $g = 16, {}^6H_{15/2}$

12.28 (d) $\frac{5}{2}, \frac{3}{2}, \frac{1}{2}$ (e) 3, 2, 1

12.29 30,279, 33,210 cm^{-1}

12.30 149 cm^{-1}

12.31 ^2P ground $\Delta E = 112.0\ cm^{-1}$,
$^2P_{3/2} + {}^2P_{1/2}$ to $^2S_{1/2}$ is 1 and 2,
$^2D_{5/2}$ to $^2P_{3/2}$ is 4 or 5,
$^2D_{3/2}$ to $^2P_{3/2}$ is 5 or 4,
$^2P_{1/2}$ to $^2D_{3/2}$ is 3

12.32 Nine

12.33 Nine

12.34 Six, $\frac{4}{3}$

12.35 6.63 and 9.58 β

12.37 The line is split into two: $M = 0 \rightarrow M = 0$ and $M = \pm 1$

12.38 The ratio is $2j + 1$

12.39 A = $-61,500\ cm^{-1}$ (8 eV)

12.40 657

12.41 $T = 1, 2$

12.42 $\nu = A(I + \frac{1}{2})$,
2,298, 157,944 Hz

Chapter 13

13.1 $E = n^2h^2/2ma^2$

13.2 $\Delta E = \frac{3}{4}ba^4$; then
$\omega_e x_e \cong 3\hbar^2 b/k\mu$

13.3 $5.699 \times 10^5\ ergs/cm^2$

13.4 $-92.08\ kJ/mole$

13.5 Expand $e^x \sim 1 + x$

13.6 $\beta = 1.862\ \text{Å}^{-1}$

13.7 $\bar{\nu}_2 = 3\omega_e - 12\omega_e x_e$

13.8 2989.7, 51.9 cm^{-1}

13.9 2998.69, 47.29 cm^{-1}

13.10 4.552 eV

13.11 1.22×10^{12} cps

13.12 2.903 Å

13.13 0.921, 0.920 Å

13.14 $N_5/N_0 = 3.267$

13.15 30

13.16 2.99793×10^{10} cm/s

13.17 57,635.97 MHz, 0.18 MHz

13.18 $R = \bar{\nu}_1 + 2(J + 1)\bar{B}_e$
$- (J + 1)(2J + 5)\alpha_e$,
$P = \bar{\nu}_1 - 2J\bar{B}_e - J(2J - 3)\alpha_e$

13.19 1.913, 0.020 cm^{-1}

13.20 $T = 400 + 50$ K

13.21 HCl, $\bar{B}_0 = 10.441, \bar{D}_C \cong 0.0004\ cm^{-1}$

13.22 (1, 1, 4, 6, 3)

13.23 $^2\Pi_u(\frac{1}{2})\,^3\Sigma_g(1)\,^2\Pi_u(\frac{3}{2})\,^1\Sigma_g(2)$

13.24 BO $= 2(O_2)2.5(+)1.5(-)$

13.25 $^1\Sigma_g(1)\,^2\Pi_g(\frac{3}{2})\,^2\Sigma_u(\frac{1}{2})$

13.26 $^3\Pi_u(\pi_u^3\sigma_g)\,^3\Sigma_g^-\,(\pi_u^2\sigma_g^2)$

13.27 7.39×10^{-6}

13.28 $^2\Sigma^+$

13.29 $^2\Sigma^+$, BO = 2.5

13.30 $X(\sigma 1s)^2(\sigma^*1s)^2(\sigma^*2s)^2(\pi 2p)^4(\sigma 2p)^1$

13.31 $X \longrightarrow A, b \longrightarrow c$

13.32 $\lambda(0\text{-}0) = 110.87$ nm,
$\lambda(2\text{-}0) = 107.77$ nm

13.33 1102.1, 7.3 cm^{-1}

13.34 $\omega_e = 1793(ex), 1645(gr)$,
$\omega_e x_e = 22(ex), 13(gr)$

13.35 $D_0 = 5.127$ (5.080 accepted)

13.36 84%

13.37 1.525 eV

13.38 $B = 8.19, 300e/R_e = 6.106$ eV,
$A = 2.695 \times 10^3$ eV, $D_e = 3.83$ eV
(calculated)

13.39 4.22 eV (obs. 4.37)

13.40 5 lines

13.41 9.73 GHz

13.42 (b) $I = \mu_0/\sqrt{3}$

13.43 $\mu_0 = 1.08$ D

Chapter 14

14.1 $D_{\infty h}, C_{\infty v}, C_{\infty v}, T_d$

14.2 $T_d, C_{3v}, C_{2v}, C_{3v}, C_{3v}, D_{3h}, C_s$

14.3 (a) C_{2v}, C_{2h}, C_{2v} (b) C_{2v}, C_{2v}, D_{2h}
(c) C_{2v}, C_{2v}, C_s

14.4 $D_{2h}, C_s, C_{2v}, D_{2h}$

14.5 D_{4h}

14.6 C_{2v}, C_{2h}, C_2

14.7 $D_{3h}, D_{3d}, C_2, C_{2h}$

14.8 C_{3h}, C_3

14.9 D_{5h}, D_{5d}

14.10 B_{1g}, A_u, B_{2g}

14.11 (a) E (b) A_2' (c) T_2

14.12 (a) -226 kJ (Table 6.1: -201)
   (b) -174 kJ (-223)
   (c) $+29$ kJ (52.283)

14.13 (b) $112.6°$ (c) $\chi_3 = [2(s) - \sqrt{5}\,(p_x)]/3$

14.14 a_g, b_u, a_u, b_g

14.15 (a) $^1A_1$ (b) $^2A_1$ (c) $^2A_1$

14.16 (a) $^1A_1'$ (b) $^2E'$ (c) $^2A_2''$

14.17 $^3A_2, {}^1A_2$; both forbidden

14.18 $^1A_1, {}^1A_1, {}^1A_1, {}^2B_1$

14.19 $^1A_{1g}, {}^1B_{2u}, {}^3B_{2u}$; yes

14.20 $^1E_{2g}, {}^3E_{2g}$

14.21 All allowed

14.22 $a_u b_u$

14.23 (000) 1215 cm^{-1},
   (010) 1705,   (110) 2535,
   (100) 2045,   (030) 2685,
   (020) 2195,   (011) 2815,
   (001) 2325,   (200) 2875

14.24 R: a_{1g}, e_{1g}, e_{2g};
   IR: a_{2u}, e_{1u}

14.25 B_1, yes; B_2, yes; A_2, no.

14.26 Calc. 12,655 (obs. 12,151)

14.27 3 and 5

14.28 $\omega_e = 1448$ cm^{-1},
   $\omega_e x_e = 79$ cm^{-1}

14.29 (a) 5 IR (b) 6 R (c) one inactive
   (d) no (e) yes

14.30 645, 619, 653 nm

14.31 (a) In paper, perpendicular to C=C
   (b) perpendicular to paper (c) parallel to
   C=C. Smallest I (largest rotational
   constant) is about the axis passing through
   the carbons.

14.32 (a) 3, 2, 1, 1, 2, 3
   (b) CH_2Cl, CH_2Cl_2

14.33 $I_a = ma^2/2$ (axis in plane),
   $I_b = ma^2$ (axis perpendicular to plane)

14.34 1.30 Å

14.35 1.405×10^{-38} g cm^2

14.36 OC 1.16, CS 1.56 Å

14.37 A axis through C=C.

14.38 1.157 Å

14.40 1.1 Å

Chapter 15

15.3 28.815 J/K

15.4 23.05 J/K

15.5 100 K: (*para*) 1.030, (*ortho*) 1.634, (total)
   2.664, (approx) 2.344:
   500 K: 3.102, 9.306,
   (total) 12.408, (approx) 11.718

15.6 (*ortho*) 74.04%

15.7 $U = 1203$ J,
   equipart. $= 1663$ J

15.8

$$z_{nsr} = 6 \sum_{even} (2J + 1)e^{-x}$$
$$+ 3 \sum_{odd} (2J + 1)e^{-x},$$
$$x = J(J + 1)\theta_r/T;$$
$$\text{high } T: \tfrac{9}{2}(T/\theta_r)$$

15.9 $(S)\ \tilde{\nu}_s = \tilde{\nu}_i - (4J + 6)\tilde{B}$ for
   $J(\text{initial}) = 0, 1, 2, 3, \ldots$;
   $(O)\ \tilde{\nu}_s = \tilde{\nu}_i + (4J - 2)\tilde{B}$ for
   $J(\text{initial}) = 2, 3, 4, \ldots$

15.10 odd/even $= 7/5$

15.11 odd/even $= 10/6$

15.12 (a) 1.67 (b) 2.33 (c) 1.80

15.13 $I = \tfrac{3}{2}$

15.15 $\theta = 587$ K, $C_{vm} \cong C_{pm} = 0.0096$ J/K

15.16 0.0876 J/K, 17 J/K, 23.8 J/K

15.17 $C_{vm}/3R = 0.68$ (Debye, Figure 15.10),
   0.67 (Einstein)

15.18 5.6698×10^{-8} watt m^{-2} K^{-4}

15.20 $\rho(\lambda) = \dfrac{8\pi hc}{\lambda^5} \dfrac{1}{e^{hc/\lambda kT} - 1}$

Index

Atomic Numbers and Atomic Weights[a]

Element	Symbol	Z	Weight		Element	Symbol	Z	Weight
Actinium	Ac	89	227.0278		Mercury	Hg	80	200.59
Aluminum	Al	13	26.98154		Molybdenum	Mo	42	95.94
Americium	Am	95	(243)		Neodymium	Nd	60	144.24
Antimony	Sb	51	121.75		Neon	Ne	10	20.179
Argon	Ar	18	39.948		Neptunium	Np	93	237.0482
Arsenic	As	33	74.9216		Nickel	Ni	28	58.69
Astatine	At	85	(210)		Niobium	Nb	41	92.9064
Barium	Ba	56	137.33		Nitrogen	N	7	14.0067
Berkelium	Bk	97	(247)		Nobelium	No	102	(259)
Beryllium	Be	4	9.01218		Osinium	Os	76	190.2
Bismuth	Bi	83	208.9804		Oxygen	O	8	15.9994
Boron	B	5	10.81		Palladium	Pd	46	106.42
Bromine	Br	35	79.904		Phosphorus	P	15	30.97376
Cadmium	Cd	48	112.41		Platinum	Pt	78	195.08
Calcium	Ca	20	40.08		Plutonium	Pu	94	(244)
Californium	Cf	98	(251)		Polonium	Po	84	(209)
Carbon	C	6	12.011		Potassium	K	19	39.0983
Cerium	Ce	58	140.12		Praseodymium	Pr	59	140.9077
Cesium	Cs	55	132.9054		Promethium	Pm	61	(145)
Chlorine	Cl	17	35.453		Protactinium	Pa	91	231.0359
Chromium	Cr	24	51.996		Radium	Ra	88	226.0254
Cobalt	Co	27	58.9332		Radon	Rn	86	(222)
Copper	Cu	29	63.546		Rhenium	Re	75	186.207
Curium	Cm	96	(247)		Rhodium	Rh	45	102.9055
Dysprosium	Dy	66	162.50		Rubidium	Rb	37	85.4678
Einsteinium	Es	99	(252)		Ruthenium	Ru	44	101.07
Erbium	Er	68	167.26		Samarium	Sm	62	150.36
Europium	Eu	63	151.96		Scandium	Sc	21	44.9559
Fermium	Fm	100	(257)		Selenium	Se	34	78.96
Fluorine	F	9	18.998403		Silicon	Si	14	28.0855
Francium	Fr	87	(223)		Silver	Ag	47	107.868
Gadolinium	Gd	64	157.25		Sodium	Na	11	22.98977
Gallium	Ga	31	69.72		Strontium	Sr	38	87.62
Germanium	Ge	32	72.59		Sulfur	S	16	32.06
Gold	Au	79	196.9665		Tantalum	Ta	73	180.9479
Hafnium	Hf	72	178.49		Technetium	Tc	43	(98)
Helium	He	2	4.00260		Tellurium	Te	52	127.60
Holmium	Ho	67	164.9304		Terbium	Tb	65	158.9254
Hydrogen	H	1	1.0079		Thallium	Tl	81	204.383
Indium	In	49	114.82		Thorium	Th	90	232.0381
Iodine	I	53	126.9045		Thulium	Tm	69	168.9342
Iridium	Ir	77	192.22		Tin	Sn	50	118.69
Iron	Fe	26	55.847		Titanium	Ti	22	47.88
Krypton	Kr	36	83.80		Tungsten	W	74	183.85
Lanthanum	La	57	138.9055		Uranium	U	92	238.0289
Lawrencium	Lr	103	(260)		Vanadium	V	23	50.9415
Lead	Pb	82	207.2		Xenon	Xe	54	131.29
Lithium	Li	3	6.941		Ytterbium	Yb	70	173.04
Lutetium	Lu	71	174.967		Yttrium	Y	39	88.9059
Magnesium	Mg	12	24.305		Zinc	Zn	30	65.38
Manganese	Mn	25	54.9380		Zirconium	Zr	40	91.22
Mendelevium	Md	101	(258)					

[a]From *Pure Applied Chemistry,* 52, 2349 (1980). A value in parentheses is the mass number of the longest-lived isotope of the element.